Environmental CHEMISTRY

Environmental CHEMISTRY

S.C. BHATIA

B.E. (Chemical), M.B.A.

CBS Publishers & Distributors Pvt. Ltd.

New Delhi • Bengaluru • Chennai • Kochi • Kolkata • Mumbai
Hyderabad • Uttarakhand • Nagpur • Patna • Pune • Jharkhand

ISBN: 81-239-0826-1

First Edition: 2002
Reprint: 2003, 2006, 2007, 2008, 2010, 2011, 2013, 2018

Published by **Satish Kumar Jain** and produced by **Varun Jain** for
CBS Publishers & Distributors Pvt. Ltd.,
4819/XI Prahlad Street, 24 Ansari Road, Daryaganj, New Delhi - 110002
delhi@cbspd.com, cbspubs@airtelmail.in • www.cbspd.com
Ph.: 23289259, 23266861, 23266867 • Fax: 011-23243014

Corporate Office: 204 FIE, Industrial Area, Patparganj, Delhi - 110 092
Ph: 49344934 • Fax: 011-49344935
E-mail: publishing@cbspd.com • publicity@cbspd.com

Branches:
• *Bengaluru:* 2975, 17th Cross, K.R. Road, Bansankari 2nd Stage,
  Bengaluru - 70 • Ph: +91-80-26771678/79 • Fax: +91-80-26771680
  E-mail: cbsbng@gmail.com, bangalore@cbspd.com
• *Chennai:* No. 7, Subbaraya Street, Shenoy Nagar, Chennai - 600030
  Ph: +91-44-26681266, 26680620 • Fax: +91-44-42032115
  E-mail: chennai@cbspd.com
• *Kochi:* Ashana House, 39/1904, A.M. Thomas Road, Valanjambalam,
  Ernakulum, Kochi • Ph: +91-484-4059061-65
  Fax: +91-484-4059065 • E-mail: cochin@cbspd.com
• *Kolkata:* 6-B, Ground Floor, Rameshwar Shaw Road, Kolkata - 700014
  Ph: +91-33-22891126/7/8 • E-mail: kolkata@cbspd.com
• *Mumbai:* 83-C, Dr. E. Moses Road, Worli, Mumbai - 400018
  Ph: +91-9833017933, 022-24902340/41 • E-mail: mumbai@cbspd.com

Representatives:

• Hyderabad: 0-9885175004	• Nagpur: 0-9021734563
• Patna: 0-9334159340	• Pune: 0-9623451994
• Jharkhand: 0-9811541605	• Uttarakhand: 0-9716462459

Printed at:
Neekunj Print Process, Delhi (India)

Preface

Man's environment is under constant threat from his own activities. Man's expanding population, industrialisation, urbanisation and intensive agriculture have caused tremendous damage to our environment. Man's ignorance of laws of nature and his over-exploitation of natural resources have further aggravated the problem.

The changes in the environment have put the survival of man in danger. Preservation of the environment is, therefore, essential for the very existence of the human beings. Proper management of the environment is the only way to ensure sustained development of the society. It is therefore, essential to make the masses aware of the changes in the quality of our environment and strategies to prevent the situation from worsening further. Thus it is necessary to have some idea of the different methods used for pollution abatement. Information about the desired limit is also essential for calculating the extent to which pollution control is required. Proper monitoring would also help in focusing on the seriousness of the problem of environmental pollution and generate real concern among policy planners. An attempt, therefore, has been made in this book to introduce scientific monitoring, development of standards and their present position as well as different methods for pollution abatement.

Basic knowledge about environmental chemistry and pollution is absolutely essential for any scientist or engineer anywhere in the world. Any research and development project or any technological innovation have to be planned and executed by scientists and engineers. Any such enterprise will have to get prior clearance from ministry of environment or other statutory environment protection agency before it can take off. Environmental audit has been made compulsory for all industrial establishments in our country and many other countries in the world.

Thus it is obvious that basic principles regarding environmental chemistry and pollution control form an essential component of science and engineering curriculum at undergraduate and post-graduate levels. This book is expected to fill the gap of a much needed reference cum text book on environmental chemistry for students of B.Sc and M.Sc Environmental Chemistry and M Tech. (Environmental Engineering) and for the practising chemical engineer, mechanical engineer, civil and public health engineer, industrial chemist and research students involved in manufacturing or research and development with an interest in effluent treatment. Environmental pollution is a major hazard-facing the world today and there is an increasing awareness of the fact that a clean environment is necessary for smooth living and for the better health of human beings. Industrial projects have a profound influence on society and environment not only in terms of benefits but also in risks and hazards. The adverse impact on the environment is largely due to indiscriminate and unregulated exploitation of both renewable and unrenewable resources and the use and abuse of the environment as a sink for dumping the waste products of development activities.

Population explosion, industrialisation and agriculture have caused tremendous damage to our environment. Chapter 1 is devoted to man and environment. Chapter 2 deals with environment and sustainable development and it does not mean only control of pollution. It also means the way of living, sanitation our home and other activities including pollution control. Chapter 3 is devoted to environmental science, which is the application of knowledge from many disciplines to the study and management of the environment. It deals with the study of physical, chemical and biological conditions surrounding the living organisms which influence them internally or externally. Environmental chemistry is the science of chemical phenomenon occurring in the environment and involves the study of the sources, reactions, transport, effects and fates of chemical species in water, soil and air environment and the effects of the technology there on and chapter 4 is thus devoted to environmental chemistry and chemical cycles.

Chapter 5 explains the fundamentals of aquatic chemistry which involves processes like acid base, solubility oxidation reduction and complexation reactions. Chapters 7 and 8 are devoted to water pollution and water treatment. Various latest methods of treatment of industrial waste-water including removal of metals, toxic substances by absorption techniques and by membrane technology are also discussed. Atmospheric science deals with the movement of air masses in the atmosphere, atmospheric heat balance and atmospheric chemical reaction and thus chapter 8 concentrates on atmosphere and atmospheric chemistry. Chapter 9 is devoted to particles in the atmosphere – particulate is a term that has come to stand for particles in the atmosphere, although particulate matter or simply particles is preferred usage. Chapter 10 and 11 deals with gaseous inorganic air pollution and organic air pollutants. Various methods to control these pollutants are also discussed.

Chapter 12 explains photochemical smog which permeates atmosphere. Chapter 13 deals with green house effect which prevent the exit of reflected solar energy from the earth's atmosphere. Chapter 14 and 15 are devoted to hydrosphere and lithosphere. The geosphere, or solid earth is that part of the earth upon which humans live and from which they extract most of their food, minerals, fuels and thus, chapter 16 deals with geosphere and geochemistry. Soil and agricultural practices are strongly tied to the environment and cultivation of land and agricultural practices can influence both the atmosphere and hydrosphere. Keeping this in mind, chapter 17 is devoted to soil chemistry.

Chapter 18 explains the natural resources, energy and environment which are intimately related to the environment. Many specific chemicals in widespread use are hazardous because of their chemical reactivity, fire hazards, toxicities and other properties. Keeping this in mind, chapter 19 describes nature and sources of hazardous wastes, their chemical classes and generation, treatment and disposal. Chapter 20 is devoted to environmental chemistry of hazardous wastes – their origin to their point of entry into the environment, in geosphere, hydrosphere is discussed in detail.

Chapter 21 explains the reduction, treatment and disposal of hazardous wastes. Hazards waste management is a multidimensional issue, as water pollution, solid wastes and ground water hydraulics are all to be taken into account simultaneously. Various treatment methods and discharging of hazardous wastes are discussed in detail in this chapter. The increase in industrial chemical activities has resulted potential for major industrial disasters, prompting safety analysis and scientists to take a closer look at these aspects and thus chapter 22 is devoted to radioactivity in environment.

Biotechnology plays a significant role in waste treatment and recovery, restoration of environmental quality, reduction of energy consumption, pollution abatement and generation of useful organic chemicals from biconversion of biomass. Keeping this in mind chapter 23 describes the role of biotechnology in environmental protection.

Biochemical processes not only are profoundly influenced by chemical species in the environment, they largely determine the nature of these species, their degradation, and even their syntheses, particularly in the aquatic and soil environments – the study of such phenomena forms the basis of environmental chemistry. Keeping this in mind chapter 24 explains the various environmental aspects of biochemistry.

Toxicological chemistry is the science that deals with the chemical nature and reactions of toxic substances including their origins, uses and chemical aspects of exposure, fates and disposal. Toxicological chemistry centres on the relationship between the chemical nature of toxicants and their toxicological effects and thus chapter 25 and 26 are devoted to toxicological chemistry and toxicological chemistry of chemical substances. Chapter 27 concentrates on noise pollution a threat to the quality of our atmosphere and health hazard in cities, towns and factories as the ears are constantly assaulted by jarring high frequency noise range. Various methods of controlling and reducing industrial noise are also discussed. Chapter 28 explains the various environmental methods for chemical analysis. These methods help us for proper understanding of our environment as by using them, we can have clear idea of the identities and quantities of pollutants and other chemical species in air, water, soil and biological samples etc. Chapter 29 deals with environmental impact assessment, the objective of which is to identify the potential impacts (positive/adverse) of a development project on the surrounding environment.

Environmental auditing or the process of detecting waste of resources and environmental damage that can be avoided in any production activity is dealt in chapter 30. The need for the enactment of legislation on environmental protection has become urgent in view of increasing pollution and keeping this in mind chapter 31 introduces, explains and critically appraises the pollution control laws in existence.

The text is throughout supplemented with diagrams, figures and tables wherever needed. The treatment of all topics is in a cogent, lucid style aimed at enabling the reader to grasp the information quickly and easily. The glossary and index at the end of the book serve as quick references.

While painstaking care has gone into producing a useful and exhaustive reference textbook, the author would welcome any constructive criticism and a creative feedback from students, teachers and professionals in the industry and interaction that will certainly be mutually fruitful.

The author S C Bhatia, is a Chemical Engineer with management qualifications who has written several books on chemical and allied subjects, such as chemical process industries; perfumes, soaps, detergents and cosmetics; engineering chemistry; handbook of pollution and its control etc. At present, he is a renowned consultant in the field of environment, waste heat recovery (energy conservation) and petrochemicals.

Acknowledgements are due to Mr. Santosh Kumar Shrivastava and Mr. Harinder Singh Negi, the computer operators who worked long hours to complete the task of bringing out the book on time. The author is also indebited to Dr. M M Verma, Ms. Namita Das (M.Sc. Chemistry) for editing and proof-reading of the book. Acknowledgements are also due to Maj. Gen. H C Dua and Mr. Madhuresh Kumar for proof-reading.

S C Bhatia

Contents at a Glance

Contents

CHAPTER 1

Man and Environment

INTRODUCTION

Population explosion, industrialisation, urbanisation and intensive agriculture have caused tremendous damage to our environment. Man's ignorance of laws of nature and his over-exploitation of natural resources have further aggravated the problem. Fortunately, during the last few years, we have started realising our past mistakes and begun to make amends to prevent further degradation of our environment. A number of national and international conferences have been held during the last decade to debate the various issues involved. The most important and successful of such meetings was the 1992 Earth Summit, at Rio de Janeiro in Brazil where more than 100 Heads of Governments representing both developing and developed countries participated. The general consensus at this Summit was that environment and development should not be treated as contrary, but rather complementary to each other. This approach is essential for sustainable development. The spirit of Rio is reflected in Agenda 21, a non-binding action plan to meet the needs for global economic development with protection of the environment.

Environment means the surroundings in which we live. It is a life-sustaining system in which various living beings like animals, including man, birds, insects, micro-organisms like algae, fungi, protozoa, amoeba and non-living beings like air, water and soil are inter-related. Like man, his environment too is beautiful. The earth is a wonderful planet that has perennial sources of water to quench his thirst with their sweet water. Its atmosphere supplies pure air for him to breathe and has a natural ozone umbrella that protects him from sun's dangerous ultra-violet rays. It has a green carpet to utilise the carbon dioxide that we exhale to recycle into oxygen essential to sustain life on this planet. It has number of attractions like the rainbow to wonder at.

The atmosphere, the lithosphere and the hydrosphere form the biosphere in which life—be it man, animal or plant exists. The biosphere is not only a source of life sustaining elements but also a sink into which all waste products are dumped. From time immemorial, the biosphere is discharging faithfully its duty of recycling waste products to make good the loss so that every generation finds it the same as the one before it. But this self-cleaning and equilibrium maintenance of the biosphere is disastrously disturbed if waste products released into it exceed its capacity to purify herself. Of late, this is what is happening. We are loading it with enormous amounts of waste product that the biosphere is becoming more and more poisonous and soon a day will be reached when it becomes inhabitable.

1

Primitive man ate uncooked food available from plants, birds or animals within his reach. He ate the raw meat. He drank the water from the rivers. He lived in caves or huts made of mud, wood and leaves of some trees. This sort of living never polluted the environment. When Promethenes stole fire, man's travails began. He used it not only to cook food but also as a weapon to destroy the neighbour. With fire smoke issuing out was polluting the atmosphere; there was stink. It was in the beginning of the first century that the Roman philosopher Seneca complained about air pollution. This went on increasing until in the 20th century the Ganges became a death bed for all aquatic animals and the series of air pollution disasters affected millions all over the world.

ECOLOGY AND ENVIRONMENTAL SCIENCE

Man's environment is under constant threat from his own activities. Man's expanding population is the biggest challenge to the quality of the environment. The developments in industrial and agricultural sectors to provide food and other basic amenities to the increasing population have further deteriorated the environment. Uncontrolled mechanisation, over-exploitation of natural resources, deforestation and extensive use of chemical fertilisers and pesticides have brought about many changes in different components of the environment. On the other hand, nature has been striving hard to compromise with man and bear the on slaught of his activities. In fact, nature gave warning signals to man in the form of droughts and floods in many countries of the world. However, man did not care and continued to take the assistance of modern technology to fight with nature.

The changes in the environment have put the survival of man in danger. Preservation of the environment is, therefore, essential for the very existence of the human race. Proper management of the environment is the only way to ensure sustained development of the society. It is, therefore, essential to make the masses aware of the changes in the quality of our environment and strategies to prevent the situation from worsening further.

Although man has recently become conscious of the threat to environment by scientific and technological developments, our earliest ancestors were also aware of the concept of environment. They observed plants and animals, and learnt to protect themselves from harmful ones while making use of the others. They lived in a rich and competitive world closely tied to the environment. As hunters and gatherers, they knew a great deal about their surroundings. They knew the sources of water and use of plants and animals for food and medicinal purposes. The use of fire and various types of tools and weapons made their lives easier. Hunters and gatherers exploited the environment to fulfil their requirements. They cut down trees with axes and caused great changes in grass and shrubs with the use of fire. However, they could not cause excessive damage to the environment due to their low population, nomadic way of life and primitive technology.

The domestication of plants and animals gave impetus to agricultural technology. With the increase in productivity of land, crop production started increasing at a faster speed. Human population started increasing beyond the limits previously fixed by natural food supply. The developments in machinery, fertilisers, pesticides and high yielding varieties caused unprecedented increase in agricultural productivity. Overgrasing, widespread destruction of forests and intensive agricultural practices denuded the land and converted productive regions to barren areas. The discovery of new medicines and improved sanitation enhanced human survival and population began a rapid ascent.

It is, thus, quite amazing to recall the speed of development from simple tools made from stones to the development of wheel and cart, and ultimately from steam engine to solid fuel propelled rockets. As modern man progressed, he turned to resources such as wood, coal, minerals and fossil fuels. In the process, he learnt to master many forms of energy and matter, ultimately becoming manipulator of the environment with a power to change the destiny of the planet. The change from hunter-gatherer economy to that of agriculture and industry markedly changed the natural habitat. The shifting role of man in the environment placed heavy demands on air, water and natural resources. As man gained control over the environment, the link between him and nature weakened, and he began to consider himself separate from and superior to nature. It is in this context that we must understand the basic principles of ecology and environment so that we may find sustainable solutions to the problems faced by modern man.

ENVIRONMENT AND ECOLOGY

"Ecology" is the study of 'living organisms" in relation to the surroundings in which they live. The surrounds of the "organisms" are called "environment". This environment is made up of many components including purely physical features like climate, soil type (important for plants), as well as other living organisms and their effects. As such, environment may be defined as the sum total of biotic and abiotic factors influencing the response of organism.

The study of ecology helps us to understand how each and every human action affects our environment and thus, in the long run, affects ourselves. The word "ecosystem" applies to the whole community of organism and its environment as one unit. The ecologists normally consider the "environment" and the "community" it contains as a singly working system. Thus, the ecosystem consists of plants and animals and the physical environment in which they live.

The word 'environment", however, is defined in many other and different ways by the ecologists and environmentalists. By the term "environment", some of us imply a life support systems, and include those items like water, air, food, etc., which are absolutely essential for the sustenance of the aerobic life forms on this planet. This group of scientists believes that, all other items which may be conceived as constituting the environment are ancillary or complementary to the above life support system. As mentioned earlier, the ecologists are, however, a bit liberal to include the influence of other organisms as part of one's environment, and so classify the environment as "biotic" and "abiotic". The impact of major technological developments on environment are given in Table. 1.1.

BASIC CONCEPTS ON CHANGES IN THE ENVIRONMENT CAUSED BY MAN

The changes, brought in by the physical environmental processes on the earth's surface occur in such a way that equilibrium is maintained through negative feedback mechanism if man does not interfere in the natural state of physical environmental processes. Now man has emerged as a very important geomorphic agent and is capable of changing the earth's surface at a much faster rate than many of the natural processes. Thus it is important to study the role of man in changing the environmental processes because these processes affect the energy system, hydrological cycle, chemical element cycle, and sediment cycle which in turn maintain unity of biosphere ecosystem.

Man's Impact on Environmental Processes

The external environmental processes originate from the atmosphere and are basically related to solar energy which affects the basic elements of atmospheric processes. Man, by affecting solar radiation and thus the heat energy, may affect the processes of precipitation and air circulation which in turn would affect the environmental processes.

Table 1.1. Impact of major technological developments on environment.

Technological development	Environmental impact
Hunting and gathering	• People were knowledgeable about the environment and skilled in finding food and water. • They benefitted from their intelligence to manipulate tools and fire. • They lived healthy lives, were well fed and experienced low disease rates. • On the whole, they were exploitive of their resources. • Their widespread use of fire may have caused significant environmental damage. • The overall environmental impact was generally small because of low population density and lack of advanced technology.
Agriculture	• Farmers benefitted from new technologies to enhance crop yields. • They were knowledgeable about domestic animals and crops. • Diseases were more common among city dwellers due to increased population density. • They were highly exploitive of their resources. • The impact of subsistence level farming was significant. However, the impact of urban-based agriculture was much larger because of new technologies, trade in food products, increasing population, and lack of good land management practices.
Industry	• Industry relied on new technologies, energy, new forms of transportation and heavy input of materials. • Mass production and modern technology was transferred to the farm. • Industry is highly exploitive and devoted to maximum material output and consumption. • Impact is enormous and includes pollution, species extinction, waste production and dehumanisation. • Humans become subject to infectious diseases, and new industrial age diseases including ulcers, AIDS, heart diseases and mental illness. • Widespread environmental damage results from industry, agriculture and population growth.

Whether modifications and climatic change affected by man change and transform the very nature of environmental processes.

Man and Hydrological Processes

1. The basic input of the basin is rainfall which is intercepted first by vegetation and reaches the ground as "aerial streamlets" and through fall.

2. In the absence of vegetation, the rainfall reacts with the ground directly.

3. Some portion is lost to the atmosphere through evaporation.

4. The water available on the ground forms "surface storage" of which a sizeable portion moves down the slope as surface run-off, some portion is evaporated, some portion remains on the ground surface while some portion infiltrates downward into the soil to form soil moisture storage of which some reappears as seepage and springs via throughflow and interflow while some portion percolates downward to form "ground water storage" of which some portion reaches the channel though base flow, some portion moves upward as capillary rise to reach "soil moisture storage" and some portion is routed further downward through deep transfer.

5. "The channel storage" receives water from surface storage though surface run-off, from soil moisture storage, through interflow and throughflow and from ground water storage through base flow.

6. Thus the initial input of precipitation finds exist through two paths of output—(i) through evapo-transpiration from all types of storages as referred to above and (ii) through channel run-off from channel storage.

Man's Impact

Man affects and modifies the internal processes of hydrological regime of drainage basins in a variety of ways. These modifications have positive and negative effects.

The input of precipitation in the hydrological cycle of a drainage basin is modified through "cloud seeding" for induced precipitation (increase in input), atmospheric pollution (both increase and decrease in precipitation input), modified atmospheric circulation (e.g., urbanisation induces vertical convective currents and thus increases precipitation, forest clearance (decrease in precipitation), vegetation modification (change in precipitation) etc.

Surface storage is modified by land clearance, cultivation, urbanisation, land drainage, mining etc.. While surface run-off is increased due to deforestation and cultivation and is supplemented by additional input through channelled irrigation for crop-land and affluent disposal from urban areas. Infiltration is modified through devegetation (decrease in infiltration), urbanisation (decrease), afforestation and reforestation (increase) and irrigation (increase).

Soil moisture storage is positively affected by irrigation, planting of grasses and plants, artificial recharge, seepage from water supply systems, soakpits, cesspools etc. While it is negatively affected by land clearance through deforestation, burning of grasslands, urbanisation etc.

Ground water storage is modified through extraction of ground water for domestic use, and irrigational purposes while channel storage is modified through flood plain development. Channel modification (shortening and lengthening of channels), river regulation, construction of dams and reservoir etc.

The impacts of man's activities on different components of water basin may "include" increased flood hazard, and other changes in river regime, reduced availability of ground water, deterioration of water quality and widespread eutroplication of water bodies and river systems in response to increased nutrient loadings.

Man and Weathering and Mass-Movement Processes

Weathering is a natural process and is accomplished through various combinations of insolation, water frost, air pressure, oxygen, carbon dioxide, hydrogen, plants and animals.

Man being biological agent accelerates and decelerates the natural rate of weathering. Mining activities for extraction of minerals, blasting of hills and ridges for dam constructions and mineral extraction, quarrying for industrial (limestone for cement) and building materials etc. result in such a fast rate of disintegration of geometricals that this may be accomplished by natural weathering processes in thousands to millions of years. Man accelerates the rate of weathering on hill slopes by modifying the ground surface through deforestation. Deforestation slopes reduces the mechanical reinforcement and cohesion of unconsolidated geomaterials and thus increases slope instability which causes slope failure and mass movements of materials down the slope in the form of landslides, slumping and debris fall and slides. Man induced landslides due to deforestation have become common features in the foothills of the Himalayas.

Human activities causing mudflow and earthflow mainly fall into two categories e.g. (i) accumulation of waste soil and rock fragments arising out of mining activities into huge piles results in spontaneous lateral earth flow engulfing surrounding areas; and (ii) removal of support by undermining of natural masses of soil, regolith and rock. The constructional activities like dams, canals, etc. and construction at the base of hillslope leads to removal of support to slope and causes mass movement.

Man and Coastal Processes

The direct modifications of coastal processes by man include (i) disruption of wave motion and weakening of energy of coast-bound waves by injecting air bubble curtains; (ii) attempts to deflect or resist the effects of sea waves and currents by constructing sea walls, groynes (groins), break-waters (masonry walls to break sea-waves); (iii) trapping or import of sediments to replenish sea beaches; and (iv) plantation of trees to stabilise beaches and coastal dunes.

Man's attempt to reduce or stop coastal erosion and therefore to check retrogradation on the one hand and to promote deposition to encourage progradation on the other hand have not been successful because of complex nature of mechanisms of coastal processes, both erosional and depositional. These direct attempts of man to manipulate and modify coastal processes for specific purposes (to half erosion at harbours, to build beaches, to replenish already depleted beaches, to open inlets to encourage sea transport etc.) bring in changes in nearshore topography, mechanisms of wave and current action and coastal erosion, nature and pattern of sediment movement and deposition on the adjacent part of the coast where structural works have been initiated.

Man and River Process

Man, equipped with technological skill has attempted to metamorphose river channels in two ways (i) to train the rivers to get rid off their disastrous roles; and (ii) to develop the rivers as natural resources for developmental processes.

The modifications of channel processes by man may be grouped into two categories viz. (i) direct or intentional modifications for useful purposes; and (ii) indirect or inadvertent modifications which cause adverse effects for human society. Direct modifications of channel processes include flood control measures, channel improvements, dams and reservoir construction, stream channelisation, bank manipulation through channel control structures (revetment—blanket, revetment of rocks, concrete or other materials, previous revetment of open fence and baskets, solid fence and groynes of solid structures at right angles to channel flow, training structures—timber pile dikes, rock dikes, rock-filled pile dikes, artificial earthen levees, closure of secondary channels etc. and irrigation diversions (channels).

Human activities responsible for the aggradation and siltation of river valley due to increased sediment supply include accelerated soil erosion consequent upon extensive and intensive cultivation, deforestation, natural and deliberate forest fires, mining operation, urbanisation and highway construction etc., all of which not only increase sediment supply to cause channel aggradation but also increase suspended load of the channels which travels downstream and is deposited in lakes, and estuaries and thus adversely affects the aquatic life.

Man and Periglacial Processes

The periglacial areas are those which are in permanently frozen condition but their is no permanent ice cover on the ground surface. The most striking feature of periglacial areas is the "permafrost" (permanently frozen ground) and the "active layer" (i.e., diurnal freeze, during nights; and thaw, during day time) which is the uppermost layer of the periglacial area.

The degradation of permafrost through thermokarst or frost heaving largely depends on (i) the nature of geomaterials, whether unconsolidated or consolidated; (ii) content of ice present in the permafrost; (iii) nature and density of vegetation on the ground surface; (iv) insulating properties of surface vegetation; and (v) duration of sunlight and quantity of insolation during summer days.

The activities of man destabilise thermal condition of permafrosts and render them hazardous for human society in the following manners as such :

1. By removing of surface vegetation either through direct felling of trees or construction purposes.
2. By excavation activities for obtaining materials for construction purposes.
3. By regular driving of vehicles over the thawed permafrost surface during summer season.
4. By forest fire, either natural or deliberate.

It may be pointed out that terrain disturbance caused by man through burrow and excavation pits (for extraction of construction materials from the ground), movement of vehicles, road cuts etc. thins out the "active layer" whereas deforestation, cultivation, construction of roads, rails, buildings and drainage changes thicken the 'active layer.' In either case, the thermal equilibrium of permafrost is disturbed which induces thawing of frozen moistures of permafrost, release of excess ice and ultimately ground surface undergoes the process of subsidence.

Man and Sub-surface Processes

Man changes sub-surface condition by putting additional load on ground surface through construction of dams and reservoirs, highways and bridges, canals, injection of water undergound, building, irrigation etc. and by reducing load and pressure underground through water withdrawal, drilling mineral oil and natural gas, undergound mining, mining of solid material etc. It may be pointed out that any change in sub-surface environment occurs only when the impact of human activities exceeds the resulting force of the geomaterials.

Underground mining leads to diversion of undergound flow, disruption of water flow regimes, release of harmful gases, rockbursts, outbursts of geomaterials, subsidence cracks on the ground surface, ground surface disruptions etc.

Man and Pedological Processes

The apparent direct impact of man on total soil is total loss of all horizons of soil profile due to accelerated erosion consequent upon removal of surface vegetation mainly of slopy ground. The indirect human impacts include changes of soil properties of different soil horizons of soil profiles through various economic activities. It is therefore, necessary at the very outset to study the characteristics of differen horizons of soil profiles of different kinds of soils in various climatic and vegetation zones.

The introduction of modern technology involving heavy and huge farm machines results in th alteration of soil structure in certain conditions (wet silt soils) through compaction. Overgrazing c pasture with big animals (cow, herds, pigs etc.) deteriorates the structure of wet soils. Pesticides and herbicides, their persistence and concentration in the soil profiles change the soil properties by contaminating them.

The use of chemical fertilisers to enrich soils for increased food production adversely affect the productivity of soils but the leaching of nitrates and their movement to streams and lakes encourages unwanted plant growth which too adversely affects the aquatic organisms. Changes in plant cover in terms of changes of plant species also alter the chemical properties of soil profiles.

MAN AND HIS STYLE OF LIVING

Man is an important part of the biotic component of the environment and simultaneously he is also an important factor of the environment. Thus man plays important roles in the natural environmental system in different capacities such as biological or physical man", "social man", "economic man" and "technological man". All the natural functions of human beings such as birth, growth, health and deaths are affected and determined by the natural environment in the same manner as the cases of other organisms but man being most developed and advanced animal, both physically and mentally and hence technologically, is capable of making substantial changes in natural environment so as to make it suitable for his own living. As the skill and technologies of man developed with cultural development, his roles towards natural environment also changed progressively such as from user through modifier and changer to destroyer of the environment. So, it is the technology of man which has drastically changed the man-environment relationship from pre-historic period to the present most advanced industrial period. Modern technological man, intoxicated by highly advanced technology and materialistic viewpoints, has changed and is changing the environment for his vested interests to such an extent that even the very existence of human beings is threatened.

ENVIRONMENTAL SCIENCE

Environmental science is the application of knowledge from many disciplines to the study and management of the environment. It deals with the analysis of the conditions, circumstances and influences affecting life and how life in turn responds. 'Environ' means the surroundings and 'ment' means the actioning. Therefore, literally speaking, environmental science deals with the study of physical, chemical and biological conditions surrounding the living organisms which influence them internally or externally.

Ecology and environmental science are often viewed as synonymous, which means the environmentalists are frequently considered as ecologists. As a matter of fact, ecology is one of the disciplines of a much broader area of action covered by environmental science. Ecology deals with the processes dealing with what limits life, how living things interact with their surroundings and how living things make use of the resources. Environmental science encompasses many disciplines (Fig.1.1), each of which has its own concepts and principles. These concepts serve to unify scientific enquiry into a holistic understanding of the environment.

Environment denotes the sum total of physical and biological factors that directly influence the survival, growth, development and reproduction of organisms. Abiotic factors of environment include inorganic and organic compounds, physical factors and gradients such as temperature, moisture, wind, currents, tides, solar radiation, etc. Biotic factors include plants, animals and microbes.

The environment may be viewed as natural environment and man-made environment. The natural environment consists of air, water, soil, forests, wildlife, etc., whereas man-made environment consists of the work environment, housing, technology, transportation, utilities, settlements, etc. These environmental components can be considered as a resource to be exploited to fulfil the basic physical needs of man. Man's desire for ultimate joy and comfort has led him to exploit the natural environment to the extent of reducing its capacities for self-stabilisation. As a consequence of this outright disregard of the impact of these activities on the natural processes, numerous environmental problems have risen.

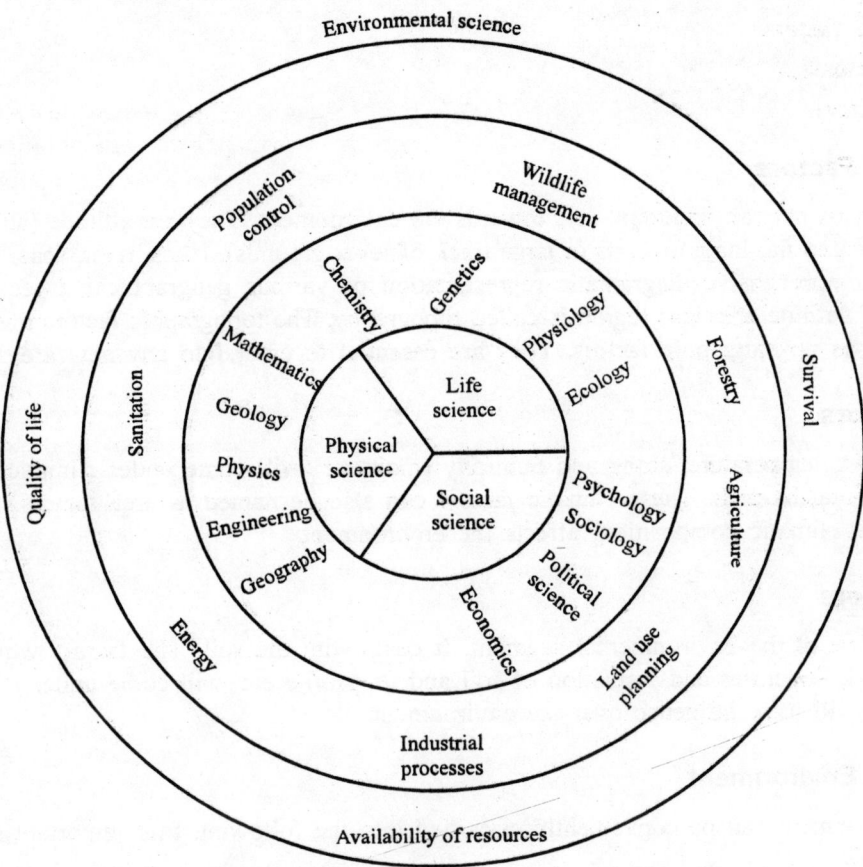

Fig. 1.1. Scope of environmental science.

Classically, the total environment can be divided into the atmosphere, hydrosphere, lithosphere (geosphere) and the biosphere. These divisions are somewhat arbitrary, because they are completely interrelated.

Significance of Environmental Science

The continuing increase of human population and destruction of natural environment with pollutants has awakened the public to the study of environment. Unfortunately, word ecology became identified in the public mind with the much broader problem of human environment, and ecology came to mean everything and anything about the environment. The science of ecology is concerned with the environments of all animals and plants, and is not solely concerned with humans. As such ecology has much to do to address some of the broader questions about humans and their environment. Ecology should be to environmental science, as physics to engineering.

FACTORS AFFECTING THE ENVIRONMENT

The following are the three important factors which influence the environment of living organisms.

1. Topographic factors.
2. Climatic factors.
3. Edaphic factors.

Topographic Factors

Topographic factors play an important role towards the environment. They are altitude (above sea level), plateaus (an elevated flat lands), plains (a large track of level grounds), lakes, river, seas, marine region, valleyes. The comprehensive diagramatic representation of various geographical aspects towards the identification of definite area (or) region is called topography. The topographic factors mentioned above are also named as physiographic factors. They are essential to enter into any new arena.

Climatic Factors

Atmosphere, light, temperature, smog and humidity (moisture) will come under climatic factors which cause environmental hazards. These climatic factors can also be named as area factors. Any deviation from the general climatic composition, affects the environment.

Edaphic Factors

Lithosphere is one of the environmental segment. It deals with the soil. The factors which deals with the nature of soil, structures and formation of soil and its profile etc. will come under edaphic factors. Edaphic factors will have influence over the environment.

Segments of Environment

The entire environment can be conveniently classified into the following four important segments viz.,

1. Lithosphere.
2. Biosphere.
3. Hydrosphere.
4. Atmosphere.

Lithosphere

The solid phase diaphram related to the environment is collectively called-as lithosphere. It is the mantle of rocks constituting the earth's crust. The earth is considered as a cold, spherical solid planet of the solar system, which spins on its own axis and revolves around the sun at a certain constant distance. Lithosphere covers the entire solid component of the earth. It is a complex mixture containing minerals, rocks, soil, organic compounds, air and H_2O. Lithosphere is one of the environmental segment having layer type of structure. It contains mainly three layers viz., (i) crust; (ii) mantle; and (iii) core.

Biosphere

The zone of living organisms which denotes mutual interaction of organisms related to lithosphere, hydrosphere and atmosphere with the environment is called biosphere. It represents the relationship between living organisms (plants and animal) in 10,000 metres below sea-level and 6000 metres above sea-level. Under natural circumstances plants and animals influence each other's life directly (or) indirectly. Green plants will release life supporting oxygen into the atmosphere through photosynthesis reaction in which glucose, starch type of food materials are prepared in the presence of light. Animals inhale O_2 during respiration and give out CO_2 which is utilised by plants during glucose and starch formation. In the atmosphere the equivalence ratio of O_2 and CO_2 will be protected by the plants and animals existence.

Hydrosphere

The region of the environment related to water is hydrosphere. 80% of the earth's surface is covered with water. Water being the universal solvent, is essential to all life.

Hydrosphere includes sea, rivers, oceans, lakes, ponds, streams, glaciers, polar ice caps, ground water etc. Among the available water source 97% is from oceans, 2% from glaciers and polar ice caps. The remaining 1% water is only useful to living organisms towards drinking and also towards agricultural activities etc. So only 1% of the water resource is found to be fresh in nature.

Atmosphere

Atmosphere is the region of air which covers the earth to a height of about 500 kms from earth's surface. It is the protective thick gaseous mantle, surrounding the earth, which sustains life on earth and saves it from unfriendly environment of outer space. So, atmosphere is considered as the cover of air that envelopes the earth.

The atmosphere contains mixture of N_2, O_2, CO_2, H_2O, SO_2, NH_3, NO, H_2S, CH_4, C_2H_6, C_2H_4, C_3H_6, O_3, O_2^+, NO^+, O^+ etc. Atmosphere will protect the living organisms on the earth surface from the harmful rays coming from the sun. The atmosphere will absorb harmful cosmic rays and electomagnetic radiations from the sun and emit only visible, infra-red and radio rays to the earth surface. Hence the atmosphere plays a vital role in maintaining the heat balance of the earth.

The atmosphere is sub-divided into four regions, of varying altitudes viz., troposphere, stratosphere mesosphere and thermosphere.

ISSUES TO BE OBSERVED

1. Environmental perception (power to understand) and people awareness towards the environment should be observed.
2. Control of environmental pollution.
3. Environmental education and training centres.
4. Population growth rate and its comparison.
5. Impact of environmental science due to the advanced technology and its effect on air, water and soil.
6. Control of environmental degradation.
7. Control towards over-consumption.
8. Environmental engineering sector.

STEPS TO BE TAKEN BY THE MANAGEMENT

1. Environmental awareness among the public should be created and make the people simultaneously to involve in the environmental programmes conducted by the management.

2. The management should be alert towards deforestation and if necessary the management should complaint against the people causing deforestation and see that the deforestation should be strictly prohibited.

3. Environmental education and training should be included in the syllabus at school, college and university level by giving appropriate weightage.

4. The information related to the tree plantation scheme should be given to the public and see that the scheme should be implemented effectively by the effective involvement of the people which are young and energetic.

5. Every person is made to recognise the environmental protection as their moral responsibility.

6. The environmental hindrance raised out of unequality towards the utility of natural resources, rapid urbanisation, installation of pollution oriented industries, population explosion, indiscriminate cutting of plants, trees, cleaning of the jungles, forests. etc. will be conveyed to the public through different communication skills. The remedies for such issues should be suggested and see that such suggestions will be implemented.

7. Complete understanding over the biological, physical, social and cultural environmental problems will be imparted among the minds of the public to meet social justice.

8. Contact programmes and guest lectures were conducted by inviting industrialists, eminent doctors, social scientists, technical engineers etc. to minimise the environmental pollution in total.

9. Central environmental education institute and its branches should be established in the country and suggestions to protect environment, were given to the people on all seasons during every calendar year.

10. An annual plan should be prepared in relation to the regional, state and national level environmental issues.

11. Towards providing jobs in various environmental education and training centres, preference will be given to those having the knowledge of environment.

12. Environmental management should be alert, active and non-selfish motive towards solving the entire environmental issues.

CHAPTER 2

Environment and Sustainable Development

INTRODUCTION

Development generally means economic growth and advancement related to mass production of various commodities and consumer goods. The emphasis is always on increasing the production of commercial produce even at the risk of polluting the environment. For instance, agriculturalists are always interested in bringing more and more of forest area under pasture or agriculture and for herding the increasing number of meat or milk yielding livestock. In this process of development oriented progress, there is environmental instability and degradation.

Human existence is directly linked to the fulfilment of basic needs—fresh air to breathe, clean water to drink and food for long term survival. The Constitution of India enjoins upon the State that at least the basic needs of the people must be fully met. This objective is not difficult to be achieved because India is richly endowed with natural resources including wide range of mineral resources distributed throughout the country. Nevertheless, it has been observed that with each passing Five Year Plan the number of people below the poverty line has gone up. It becomes imperative, therefore, that the limited natural resources must be converted into goods and services most efficiently so that the needs of a larger number of people can be met from the same resource base.

Man has mostly been engaged in faster economic exploitation of valuable resources without any serious consideration of the ecological aspects and its consequences. He has misused most of the precious fossil fuels. Increasing use of fertilisers and pesticides in agriculture has aggravated the problems of water pollution. Deforestation has led to soil erosion, loss of soil fertility and recurring floods. Industrialisation and urbanisation have accelerated the problems of waste land, sanitation and source of clear air. In addition to this there are growing problems of feeding and housing due to ever increasing human population.

Biodiversity means the variety of biological entities inhibiting the earth—wild plant and animals, micro-organisms, domesticated animals and plants, and even genetic material like seeds and germplasm. In fact, it is the very basis of our living. In the modern age, the life support system faces its greatest ever threat. Habitat destruction, hunting, pollution, displacement natives and a host of other human made forces have already pushed thousands of species and varieties into the twilight zone of extinction, with many more following day by day. In addition to this, it represents as unprecedented erosion in humanity's food, medicinal, economic, and cultural resource base.

13

The relation between humanity and bio-sphere is threatened today. The destructive impact of the poor majority the world over struggling to remain alive coupled with affluent minority consuming the world's resources are undermining the very means by which man can survive and flourish.

Increasing ecological imbalance is being created year after year, the beneficiary insects, plants and animals are getting eliminated during heavy application of pesticides. Excessive use of fertilisers with disregard to the other consequences leading to backlash to environmental and social pollution. Agro-industry is one of the main causes of pollution, for example, in Punjab agro based pollution is quite high. Over exploitation of life support system has shown the seeds of social revolution. Watershed forests are mercilessly cut to create land for agriculture, urbanisation and other purposes, causing silting of rivers and lakes. National and international capacities are ill organised and fragmented i.e., split up as agriculture, forestry, fishery, wild life and so on. The gain by one sector may be negated by the loss by the other.

Due to excessive deforestation, many wild animal species have already become extinct or are in process to face extinction. The basic understanding of rules governing feeding inter-relationship in a biotic community has highlighted the abuse of forest and wild life. The need for conservation of forest and wild life conservation is now fully recognised. Biodiversity conservation has already attained global importance.

The economic development plans and policies for conservation of nature aim at improving the quality of life. The former mainly considers the material aspects of the quality of life, whereas the latter covers the material, ecological and aesthetic aspects. The gains of economic development are realised quickly in comparison to conservation of plans. The conservation approach considers the developmental activities of man in relation to their impact on natural biota, climatic factors, water and soil resources; the economic approach takes into account methods which yield profits quickly. The natural conservation approach is preferred by the ecologists because it is scientifically sound and rational and has prospects for the long term future of man.

In the process of development, we may notice some harms or deteriorating environment conditions unless the precautionary measures are taken in advance. Does that mean we shall stop all developmental activities ? Every time we plan for a new project or take any developmental activity, we are being opposed by environmental lobby in spite of the fact that we also need the developmental processes for our survival and to compete with the process of development in other advanced countries.

As and when we construct dams, bridges and roads, we are facing agitation against these because it harms the environment. When we start thermal power station and set up industries like chemical industries, we face agitations since these pollute the environment. Similarly when we go for nuclear power plants we get opposition for radio-activity. Does that mean we shall have no power and take no developmental activity ? Does that mean such environmental authorities and agencies are against our development ? If it is so, then whose interest is protected ?

We certainly need such projects and activities for our development and survival. But at the same time we need pollution free environment for our safety and longivity. We need development but a sustainable development. All our development projects should be pollution free. There is need to develop eco-friendly projects. The effluents or wastes from the industries are to be treated properly before commencing of such projects. Wherever possible they are to be recirculated to produce by-products.

CONVERSION PROCESS

Our natural resources base comprises of the air, water, energy, space and land (flora and fauna, minerals, etc.) which are crucial for the very survival of mankind. It needs to be underlined that none of these natural resources can be manufactured or created in our laboratories. The harnessing and utilisation of these natural resources must, therefore, be done for sustainable development rather than for achieving short-term objectives.

The general trend over the last 50 years, however, has been on maximum exploitation of the natural resources with the attendant pollution of air, water and land being accepted as unavoidable. Since, our very survival depends upon the health of these natural resources, the corner stone in any environmental policy has to be optimisation of the finite natural resources so as to meet the present and future requirements of the society on a sustainable basis. Linkages among the natural resources are indicated in Fig. 2.1 along with criticality of maintaining the integrity for human welfare. These natural resources get utilised in one form or another in all our development activities. The extent of conversion of natural resources into useful goods and services is dictated by the efficiency of the conversion process. What is not converted into productive output, is thrown back into the eco-system in the form of gaseous, liquid or solid pollutants as indicated in Fig. 2.2.

The two most important factors necessary for improving the efficiency of the "conversion process" are: (i) state-of-the-art technology; and (ii) updated management packages.

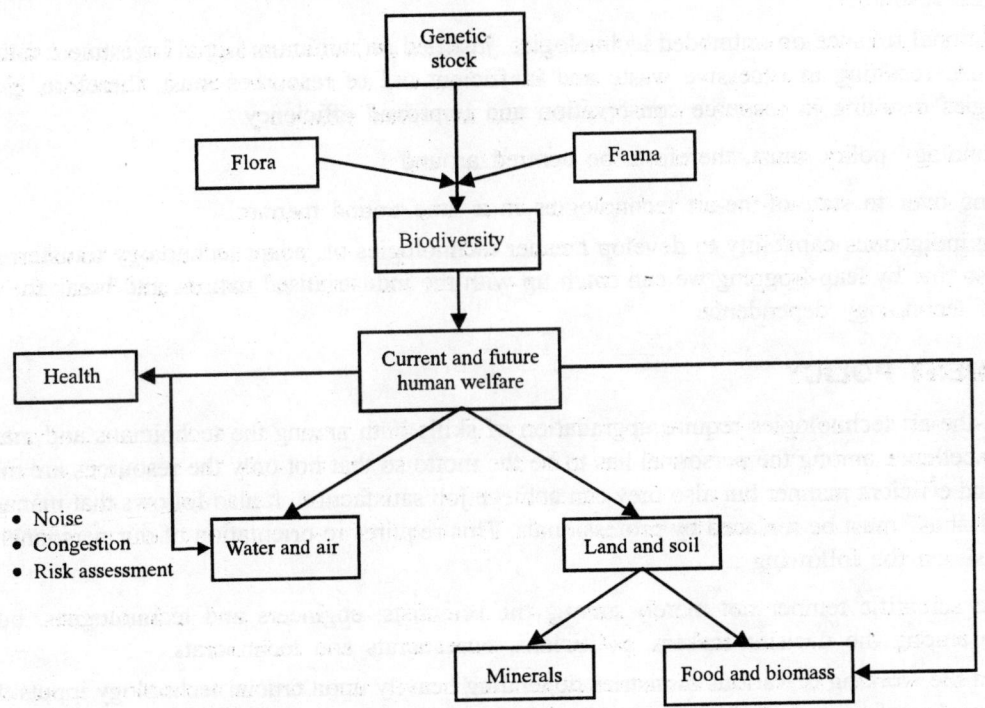

Fig. 2.1. Relationship between natural resources.

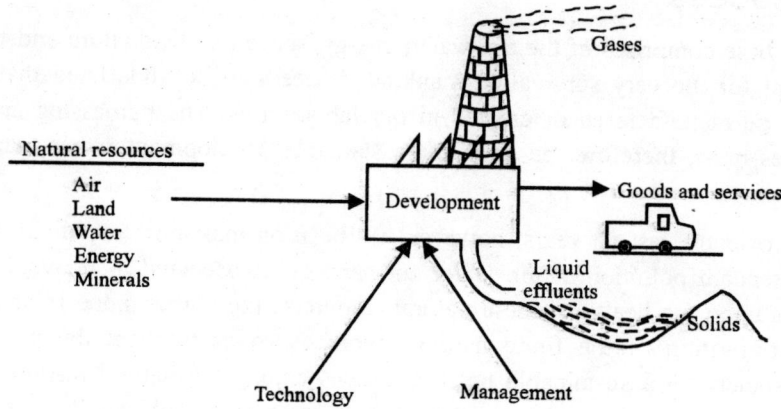

Fig. 2.2. Resource conversion for development.

TECHNOLOGY POLICY

The state-of-the-art technologies are a more efficient production process to achieve :

1. Waste prevention and reduction by lesser consumption of raw materials.
2. Modification and upgradation of the technological process for making possible the optimal utilisation of natural resources.

Our traditional reliance on outmoded technologies, justified on minimum initial investment rather than life cycle cost, resulting in excessive waste and inefficient use of resources must, therefore, give way to technologies resulting in resource conservation and improved efficiency.

The technology policy must, therefore, be centred around :

1. Switching over to state-of-the-art technologies in a time bound manner.
2. Creating indigenous capability to develop cleaner technologies or; adapt technology transferred from abroad so that by leap-frogging we can catch up with the industrialised nations and break the vicious cycle of technology dependence.

MANAGEMENT POLICY

The state-of-the-art technologies require upgradation of skills both among the technicians and managers. Pursuit of excellence among the personnel has to be the motto so that not only the resources are managed in a better and efficient manner but also they can achieve job satisfaction. It also follows that management largely by "Babus" must be replaced by professionals. This requires re-orientation of our personnel policy with emphasis on the following :

1. Inculcate scientific temper not merely among the scientists, engineers and technologists, but even more so among the decision-makers, politicians, bureaucrats and technocrats.
2. Re-orient the working of various ministries depending heavily upon crucial technology inputs through induction of professionals fully conversant with the subject at all levels of the decision-making.
3. Revise personnel policy to facilitate result-oriented recruitment and promotion by discarding the present time-scale promotion of personnel from various services.

4. All positions of the level of Director and above to be filled on the basis of competence and merit only through open competition.

5. Adoption of a system of contractual hiring and firing of personnel on the basis of performance by fixing the responsibility and accountability.

6. Facilitate easy mobility of professionals among the university, laboratory, industry and ministries on the basis of goal-oriented programmes and contractual appointments.

7. Re-orient the educational system to encourage creativity by imparting skills on the basis of aptitude and interest rather than mere acquisition of degrees.

8. Shift away from a regime of subsidies to meaningful productivity and quality improvement through pursuit of excellence at all levels.

9. Ensure transparency in decision-making with 'right to know' made operational so that the public can, if needed, know-how and why a decision has been arrived at.

10. Enshrine pursuit of excellence through suitable incentives and actively discourage sycophancy through disincentives.

OTHER POLICY INITIATIVES

Our development process in the last four decades has resulted in serious socio-economic as well as environmental implications which, amongst others, include :

1. Low economic growth rate with the population below poverty line on the increase.

2. Accelerated destruction of the biodiversity and the genetic stock.

3. Pollution of air and water and degradation of land.

4. Polarisation of the society into "haves" and have-nots" with a real possibility of law and order problem in not too distant a future.

Consequently, a major chunk of precious natural resources is going waste rather than contributing to the well being and prosperity of the people. Indeed, the environmental degradation and pollution is depriving the poor of even such gifts of nature as fresh air and clean water. The situation is bleak and is likely to worsen if the current resource use trend is not reversed soon through structural economic changes brought about by the use of "cleaner technologies" and "professional management". Policy initiatives are, therefore, required in the following areas :

1. Management of professionals.

2. Harnessing of sectoral policies.

3. Natural resource based planning.

Having covered the first two areas earlier, we may now examine the implications of sectoral policies and the need for their harmonisation with the example of forest and industrial policies.

FOREST POLICY AND INDUSTRIAL POLICY HARMONISATION

In order to achieve rapid economic growth we adopted in free India the strategy of industrialisation with special incentive schemes being introduced from time to time to direct industries into the industrially backward regions. The Planning Commission introduced an incentive programme for promoting of establishment of industries in the so-called "no industry districts". Introduction of omnibus incentive packages, however, leads to a rather anomalous situation as the objectives of forest policy come in direct conflict with the promotion of industrialisation in these backward districts. Fig. 2.3 indicates the areas eligible for incentives under the industrialisation policy. Indeed, while the forest policy calls for protection of a given area to conserve the remaining forest cover; the industrial policy is offering incentives to destroy that very forest cover. Thus, there is a clear case for harmonising the two policies to ensure that the scarce natural resources are not destroyed.

Harmonisation is required not merely between forest and industrial policies but between forest and many other sectoral policies because the intersectoral linkages of the forest policy impinge upon many sectoral policies as indicated in Fig. 2.3.

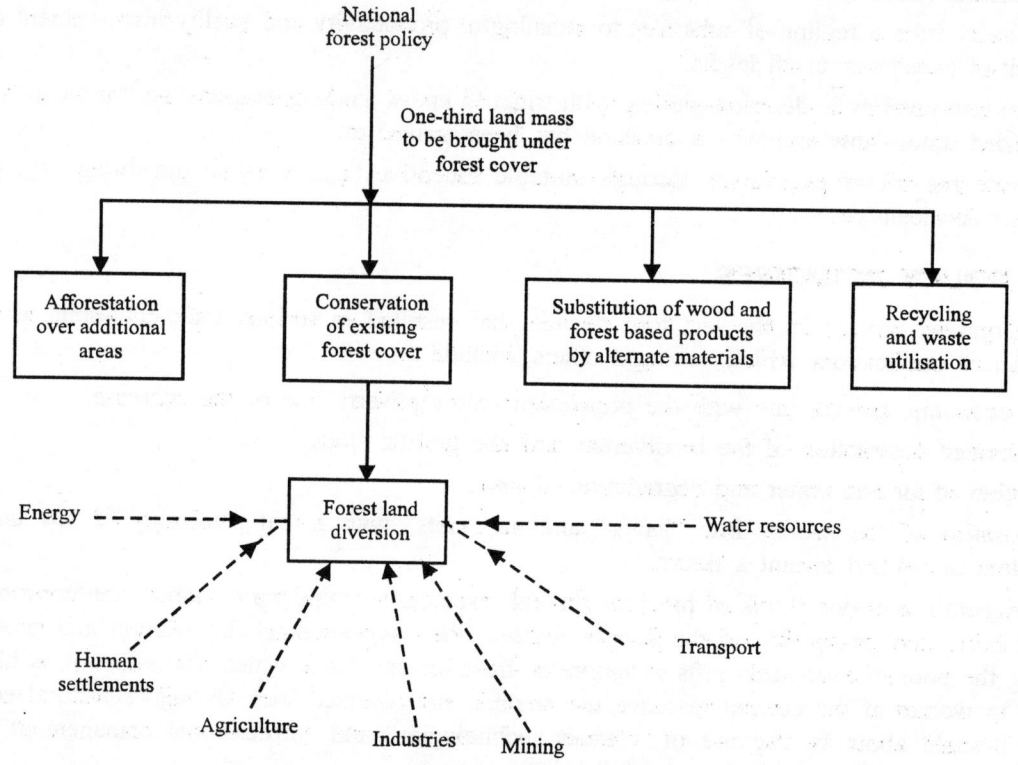

Fig. 2.3. Inter-sectoral linkages.

RESOURCE CONSERVATION AND MANAGEMENT

Contrary to general belief, the best use for some of the biological resources like biodiversity, genetic stock etc. is their conservation. Fragile ecosystems and those with high species richness and biodiversity need special protection.

The most serious threats to species is habitat destruction which includes replacement of the entire habitat by settlements, crop lands, pastures, plantations, mines, quarries, reservoirs etc. Two other serious threats are over-exploitation and the effect of introducing exotic species. It is imperative, therefore, to take steps for careful scrutiny of development activities so that the primary objective of human welfare through a balanced mix of conservation of biodiversity and development of natural resources can be achieved and sustained.

Diversity characterises most living organisms and biodiversity is commonly used to describe the number variety and variability of living organisms. The earth supports almost 5–100 million or more species though only an estimated 17 million species have been described so far. This legacy is the result of nearly 3 billion years of evolution. Biological resources are extremely labile entities and once lost cannot be replaced at any cost. Plant biodiversity as a national and global resource is extremely valuable but is poorly understood, inadequately documented and often wasted. The preservation of biodiversity is both a matter of investment and insurance to :

1. Sustain and improve agricultural, forestry and fisheries production.
2. Act as a buffer against harmful environmental changes.
3. Provide raw materials for scientific and industrial innovation.
4. Safeguard transferring the biological riches to posterity.

One of the most fundamental contribution of the plant diversity for human survival is in the supply of world's food. Out of the estimated 250,000 species of flowering plants, nearly 3000 are regarded as food source. Around 20 plant species have been domesticated for food and most of these have become crops of major economic importance. Even though agricultural practices have been evolving over the last 11–12,000 years, most of the world food crops were domesticated and widely dispersed only about 2000 years ago. Originally, the plants were consumed directly from the wild and gathering of wild produce continues throughout the world even today. Many of the species from which crop plants have been selected still survive in the wild. These surviving plants, together with closely related species, comprise the wild relatives of crops which continue to evolve under natural conditions and constitute a very valuable resource through which improvements in yield, nutrition characteristics, disease and past resistance and environmental adaptation in crops can be brought about.

Although relatively few plants are globally used for food production, plant resources provide a varied source of nutrition needs at a local level. For example, in one region of Peru alone fruits of 193 species are regularly consumed and of these 120 species are collected exclusively from the wild. Medicinal plant species also are still harvested, to a large extent from the wild. The World Health Organisation has listed over 21,000 plants reported to be of medicinal use throughout the world. It is estimated that nearly 80% of the people in developing countries rely on traditional medicines based on widely growing medicinal plants, roots and herbs. The international pharmaceutical companies depend upon more than 119 pure chemical substances extracted from some 90 species of higher plants for producing drugs and medicines throughout the world.

A significant proportion of the international demand for ornamental plants such as Orchids, Cacti and even Succulents is still satisfied from collections from the wild. Biodiversity the world over is in peril because the habitats are threatened due to the so-called development programmes. A recent study of threats facing the mammals of Australia and Americas indicates that out of 119 species of mammals, 76% are threatened by habitat loss and modification followed by over exploitation. It has been predicted that at the 1992 rate of deforestation, some 2–8% of the world's species will be threatened with extinction in the next 25 years.

It is important to understand that the distribution of biodiversity is not uniform across the globe. Some habitats, for example, tropical forests, possess a greater density of species than others. The richness of these habitats can be gauged by recent extensive studies and surveys carried out indicating that 18 places on earth—termed as Hot Spots—support nearly 50,000 endemic plant species, about 20% of the world's total flora, but comprise only 0.5% of the earth-surface.

STRATEGIES FOR SUSTAINABLE DEVELOPMENT

The appropriate strategies are to be formulated and adopted not only on national but at international level also for sustainable development in an eco-environment. The need of the hour is to evolve suitable methods of pollution control, conservation of energy, conservation of wastes, soil and metallic minerals and at the same time to increase the carrying capacity of the earth in relation to man.

In order to protect the environment among pre-requisites for prolonged better and sustainable relationship, is the conservation of living resources. The human use of biosphere for the benefit of the present generation should be properly managed, while maintaining its potential to meet the needs of future generations. There should be judicious use of resources.

In developed and ecologically conscious countries, it is the "total ecosystem conservation policy" which gains credence over the "special interest conservation policy". In case of developing and under developed countries, the situation is unfortunately the reverse. However, in India, there has been of late a shift from special interest to the total ecosystem conservation policy.

Since the total ecological balance is at a very critical stage, the effective corrective measures are required urgently in the foreseable future. Some of the measures include :

1. Reforestation of developed and geomorphically vulnerable hill terraces, low land flood plains as riverine tracts with indegenious tree species together with compilation of complete survival and growth data over the years.

2. Development of sustainable food and firewood resources on wastelands around inhabitation on appropriate patterns is required in order to stop forest grazing.

3. Evolving favourable public opinion for the implementation of various sustainable development programmes. The local population should be made aware of the direct and indirect impact of the ecological imbalances, emenating from fast evolving 'bad lands' so the local inhabitants who are part and parcel of this ecosystem, need to be educated and involved in the habitat resurrection programmes.

4. Ideally, an engineer, a forester, an agronomist, a range manager, a wild life specialist, a land scape architect, an economist and an ecologist should effectively participate in any development planning of a city of industrial growth. They should keep in view the basic ecological principles which govern functions and stability of ecosystems.

5. Augmenting water resources of the region by developing small check dams across gulleys. It will help in checking fast run-off and mass wasting of sub-stratum during monsoons, and also serve as reservoir for human consumption and for irrigation by the local inhabitants.

6. Every aspect of a development project must have environmental pros and cons before it is launched.

7. There is need for reservation of prime quality crop land for crops and adoption of management practices to maintain the productivity of crop land.

8. We must preserve as many varieties as possible of domesticated and other economic or useful plants, animals and micro-organisms and their wild relatives.

9. There should be proper control and regulation of living resources utilisation so that it is sustainable.

10. Careful allocation and management of timber concessions, protection of watersheds, maintenance of support system of fisheries, prevention of species extinction are other remedial measures for sustainable development in the environment.

11. The existing environmental Acts and Laws are to be updated so that no one who releases pollution should escape punishment. Pollution Control Boards are to be made more active and are to be given more powers. The chairman and other members of the Boards must possess the knowledge and experience of environmental education.

12. We must encourage environment studies as a separate subject in order to have a professional approach to the environmental problems.

The following priorities at national and international level are suggested to encourage sustainable development in the environment.

Priorities at National Level

The suggested measures in this regard are :

1. Preparation and implementation of national and sub-national conservation strategies to protect the environment.

2. Adoption of anticipatory environmental policies.

3. An advance assessment of the likely environmental effects of all major developmental activities.

4. Establishment of soil and water conservation body at the policy making level.

5. Environmental education programmes and campaigns are the need of the hour.

Priorities at International Level

The suggested measures and remedies at international level to sustain development are :

1. Implementation of international conservation conventions.

2. Multilateral and bilateral assistance for reforestation and restoration of degraded environment.

3. Cooperation programmes for conservation of tropical forests.

Environment and sustainable development does not mean only control of pollution. It also means the way of living, sanitation our home and rituals and all other activities including population control. Every citizen must live with sound health and get quality food for his survival. On the whole for sustainable development we need holistic approach to have an environment free of pollution for progressive future and national development.

If the suggested measures are implemented vigorously at national as well as at international level, the eco-friendly sustainable development will be self evident within short span of time.

Thus, human survival depends upon efficient management of natural resources which are limited and becoming scarce. Environmental management also emphasises optimal resource management. In order to make our development process environmentally compatible action must be taken to resolve complex issues relating to policies, project planning and organisational and administrative set up. The tasks involved are neither easy nor likely to be accepted by the vested interests. Pursuit excellence, however, is a pre-requisite to achieve the objective of sustainable management through most efficient resource management.

CHAPTER 3

Environmental Science and Technology

INTRODUCTION

Environmental science in its broadest sense is the science of the complex interactions that occur among the terrestrial, atmospheric, aquatic, living and anthropological environments. It includes all the disciplines, such as chemistry, biology, ecology, sociology, and government, that affect or describe these interactions. Thus, environmental science is defined as the study of the earth, air, water, and living environments and the effect of technology thereon. To a significant degree, environmental science has evolved from investigations of the ways by which, and places in which living organisms carry out their life cycles. This is the discipline of natural history, which in recent times has evolved into ecology, the study of environmental factors that affect organisms and how organisms interact with these factors and with each other.

For better or for worse, the environment in which all humans must live has been affected irreversibly by technology. Therefore, technology is considered strongly in terms of how it affects the environment and in the ways by which, applied intelligently by those knowledgeable in environmental science, it can serve, rather than damage, this Earth upon which all living beings depend for their welfare and existence. Air, water, earth, life and technology are strongly interconnected as shown in Fig. 3.1.

Traditionally, environmental science is divided among the study of the atmosphere, the hydrosphere, the geosphere, and the biosphere. The atmosphere is the thin layer of gases that cover earth's surface. In addition to its role as a reservoir of gases, the atmosphere moderates earth's temperature, absorbs energy and damaging ultraviolet radiation from the sun, transports energy away from equatorial regions, and serves as a pathway for vapour-phase movement of water in the hydrologic cycle. The hydrosphere contains earth's water. Over 97% of earth's water is in oceans, and most of the remaining freshwater is in the form of ice. Therefore, only a relatively small percentage of the total water on earth is actually involved with terrestrial, atmospheric, and biological processes. Exclusive of seawater, the water that circulates through environmental processes and cycles occurs in the atmosphere, underground as groundwater, and as surface water in streams, rivers, lakes, ponds, and reservoirs. The geosphere consists of the solid earth, including soil, which supports most plant life. The part of the geosphere that is directly involved with environmental processes through contact with the atmosphere, the hydrosphere, and living things is the solid lithosphere. The lithosphere varies from 50 to 100 km in thickness. The most important part of it insofar as interactions with the other spheres of the environment are concerned is its thin outer skin composed largely of lighter silicate-based minerals and called the crust. All living entities on earth compose the biosphere. Living organisms and the aspects of the environment pertaining directly to them are called biotic, and other portions of the environment are abiotic.

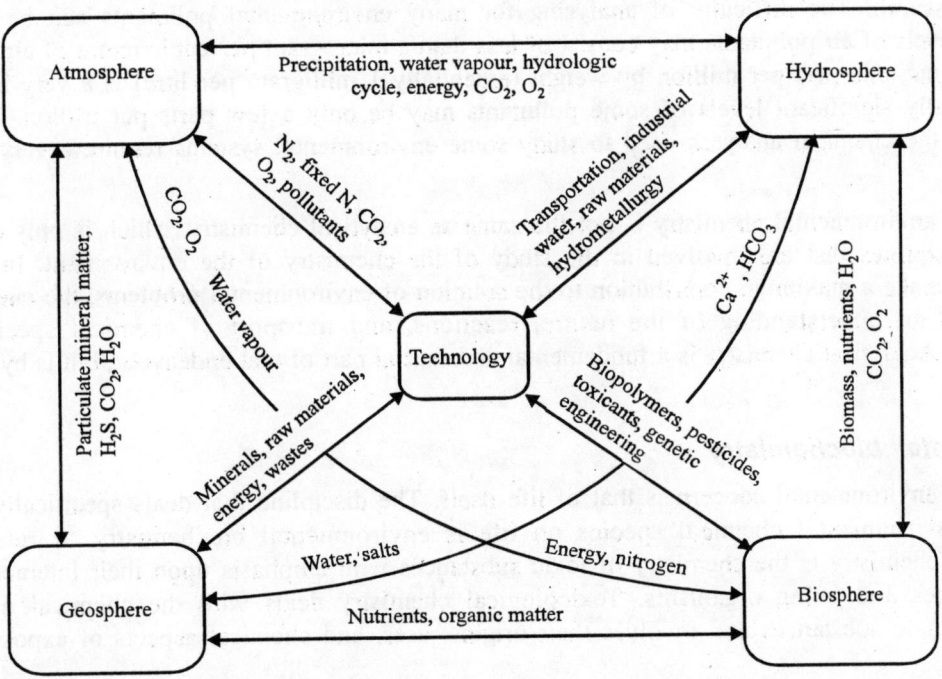

Fig. 3.1. Illustration of the close relationships among the air, water, and earth environments with each other and with living systems, as well as the tie-in with technology.

To a large extent, the strong interactions among living organisms and the various spheres of the abiotic environment are best described by cycles of matter that involve biological, chemical, and geological processes and phenomena. Such cycles are called biogeochemical cycles, and are discussed in more detail in this chapter.

Environmental Chemistry and Environmental Biochemistry

Environmental chemistry involves a study of Freon reactions in the stratosphere or an analysis of PCB deposits in ocean sediments. It also covers the chemistry and biochemistry of volatile and soluble organometallic compounds biosynthesised by anaerobic bacteria. Literally thousands of other examples of environmental chemical phenomena could be given. Environmental chemistry may be defined as the study of the sources, reactions, transport, effects, and fates of chemical species in water, soil, and air environments and the effects of technology thereon.

There are some things that environmental chemistry is not. It is not just the same old chemistry with a different cover and title. Because it deals with natural systems, it is more complicated and difficult than "pure" chemistry. By building on an ever-increasing body of knowledge, the environmental chemist can make educated guesses as to how environmental systems will behave.

Chemical analysis in environmental chemistry

One of environmental chemistry's major challenges is the determination of the nature and quantity of specific pollutants in the environment. Thus, chemical analysis is a vital first step in environmental chemistry research. The difficulty of analysing for many environmental pollutants can be awesome. Significant levels of air pollutants may consist of less than a microgram per cubic metre of air. For many water pollutants, one part per million by weight (essentially 1 milligram per litre) is a very high value. Environmentally significant levels of some pollutants may be only a few parts per trillion. Thus, it is obvious that the chemical analyses used to study some environmental systems require a very low limit of detection.

However, environmental chemistry is not the same as analytical chemistry, which is only one of the many subdisciplines that are involved in the study of the chemistry of the environment. In 'order for chemistry to make a maximum contribution to the solution of environmental problems, the chemist must work toward an understanding of the nature, reactions, and transport of chemical species in the environment. Analytical chemistry is a fundamental and crucial part of that endeavour, but is by no means all of it.

Environmental biochemistry

The ultimate environmental concern is that of life itself. The discipline that deals specifically with the effects of environmental chemical species on life is environmental biochemistry. A related area, toxicological chemistry is the chemistry of toxic substances with emphasis upon their interactions with biologic tissues and living organisms. Toxicological chemistry deals with the chemical nature and reactions of toxic substances and involves their origins, uses, and chemical aspects of exposure, fates, and disposal.

Water, Air, Earth, Life and Technology

In light of the definitions given above, it is now possible to consider environmental chemistry from the viewpoint of the interactions among water, air, earth, life, and technology already outlined in Fig. 3.1. These five environmental "spheres" and the interrelationships among them are summarised here.

Water and the hydrosphere

Water, with a deceptively simple chemical formula of H_2O, is a vitally important substance in all parts of the environment. Water covers about 70% of earth's surface. It occurs in all spheres of the environment—in the oceans as a vast reservoir of saltwater, on land as surface water in lakes and rivers, underground as groundwater, in the atmosphere as water vapour, and in the polar ice caps as solid ice. Water is an essential part of all living systems and is the medium from which life evolves and in which life exists.

Energy and matter are carried through various spheres of the environment by water. Water leaches soluble constituents from mineral matter and carries them to the ocean or leaves them as mineral deposits some distance from their sources. Water carries plant nutrients from soil into the bodies of plants by way of plant roots. Solar energy absorbed in the evaporation of ocean water is carried as latent heat and released inland. The accompanying release of latent heat provides the energy that largely carries heat from equatorial regions toward earth's poles, and powers massive storms.

Air and the atmosphere

The atmosphere is a protective blanket which nurtures life on the earth and protects it from the hostile environment of outer space. It is the source of carbon dioxide for plant photosynthesis and of oxygen for respiration. It provides the nitrogen that nitrogen-fixing bacteria and ammonia-manufacturing industrial plants use to produce chemically-bound nitrogen, an essential component of life molecules. As a basic part of the hydrologic cycle the atmosphere transports water from the oceans to land, thus acting as the condenser in vast solar-powered still. The atmosphere serves a vital protective function, absorbing harmful ultraviolet radiation from the sun and stabilising earth's temperature. Atmospheric science deals with the movement of air masses in the atmosphere, atmospheric heat balance, and atmospheric chemical composition and reactions.

Earth

The geosphere, or solid earth, is that part of the earth upon which humans live and from which they extract most of their food, minerals, and fuels. The earth is divided into layers, including the solid iron-rich inner core, molten outer core, mantle, and crust. Environmental science is most concerned with the lithosphere, which consists of the outer mantle and the crust. The latter is the earth's outer skin that is accessible to humans. It is extremely thin compared to the diameter of the earth, ranging from 5 to 40 km thick.

Geology

Geology is the science of the geosphere. As such, it pertains mostly to the solid mineral portions of earth's crust. But it must also consider water, which is involved in weathering rocks and in producing mineral formations; the atmosphere and climate, which have profound effects on the geosphere and interchange matter and energy with it; and living systems, which largely exist on the geosphere and in turn have significant effects on it. Geological science uses chemistry to explain the nature and behaviour of geological materials, physics to explain their mechanical behaviour, and biology to explain the mutual interactions between the geosphere and the biosphere. Modern technology; for example, the ability to move massive quantities of dirt and rock around, has a profound influence on the geosphere.

The most important part of the geosphere for life on earth is soil formed by the disintegrative weathering action of physical, geochemical, and biological processes on rock. It is the medium upon which plants grow, and virtually all terrestrial organisms depend upon it for their existence. The productivity of soil is strongly affected by environmental conditions and pollutants.

Life

Biology is the science of life. It is based on biologically synthesised chemical species, many of which exist as large molecules, called *macromolecules*. As living beings, the ultimate concern of humans with their environment is the interaction of the environment with life. Therefore, biological science is a key component of environmental science and environmental chemistry. As already discussed, ecology is the study of environmental factors that affect organisms and how organisms interact with these factors and with each other.

Technology

Technology refers to the ways in which humans do and make things with materials and energy. In the modern era, technology is, to a large extent, the product of engineering, based on scientific principles.

Science deals with the discovery, explanation, and development of theories pertaining to interrelated natural phenomena of energy, matter, time, and space. Based on the fundamental knowledge of science, engineering provides the plans and means to achieve specific practical objectives. Technology uses these plans to carry out the desired objectives.

It is essential to consider technology, engineering, and industrial activities in studying environmental science because of the enormous influence that they have on the environment. Humans will use technology to provide the food, shelter, and goods that they need for their well-being and survival. The challenge is to interweave technology with considerations of the environment and ecology so that the two are mutually advantageous, rather than in opposition to each other.

Technology, properly applied, is an enormously positive influence for environmental protection. The most obvious such application is in air and water pollution control. Necessary as "end-of-pipe" measures are for the control of air and water pollution, it is much better to use technology in manufacturing processes to prevent the formation of pollutants. Technology is being used increasingly to develop highly efficient processes of energy conversion, renewable energy resource utilisation, and conversion of raw materials to finished goods with minimum generation of hazardous waste by-products. In the transportation sector, properly applied technology in areas such as high speed train transport can enormously increase the speed, energy efficiency, and safety of means for moving people and goods.

Until very recently, technological advances were made largely without heed to environmental impacts. Now, however, the greatest technological challenge is to reconcile technology with environmental consequences. The survival of human-kind and of the planet that supports it now requires that the established two-way interaction between science and technology becomes a three-way relationship including environmental protection.

Ecology

Ecology is the science that deals with the relationships between living organisms with their physical environment and with each other. Ecology can be approached from the viewpoints of (i) the environment and the demands it places on the organisms in it or (ii) organisms and how they adapt to their environmental conditions. An ecosystem consists of an assembly of mutually interacting organisms and their environment in which materials are interchanged in a largely cyclical manner. An ecosystem has physical, chemical, and biological components along with energy sources and pathways of energy and materials interchange. The environment in which a particular organism lives is called its habitat. The role of an organism in a habitat is called its niche.

For the study of ecology, it is often convenient to divide the environment into four broad categories. The terrestrial environment is based on land and consists of *biomes*, such as grasslands, savannas, deserts, or one of several kinds of forests. The freshwater environment can be further subdivided between standing-water habitats (lakes, reservoirs) and running-water habitats (streams, rivers). The oceanic marine environment is characterised by saltwater and may be divided broadly into the shallow waters of the continental shelf composing the neritic zone and the deeper waters of the ocean that constitute the oceanic region. An environment in which two or more kinds of organisms exist together to their mutual benefit is termed a symbiotic environment.

A particularly important factor in describing ecosystems is that of populations consisting of numbers of a specific species occupying a specific habitat. Populations may be stable, or they may grow exponentially as a population explosion. A population explosion that is unchecked, results in resource depletion, waste accumulation, and predation culminating in an abrupt decline called a population crash. Behaviour in areas such as hierarchies, territoriality, social stress, and feeding patterns plays a strong role in determining the fates of populations.

Two major subdivisions of modern ecology are ecosystem ecology, which views ecosystems as large units, and population ecology, which attempts to explain ecosystem behaviour from the properties of individual units. In practice, the two approaches are usually merged. Descriptive ecology describes the types and nature of organisms and their environment, emphasising structures of ecosystems and communities and dispersions and structures of populations. Functional ecology explains how things work in an ecosystem, including how population responds to environmental alteration and how matter and energy move through ecosystems.

An understanding of ecology is essential in the management of modern industrialised societies in ways that are compatible with environmental preservation and enhancement. The branch of ecology that deals with predicting the impacts of technology and development and making recommendations so that these activities will have minimum adverse impact, or even positive impacts, on ecosystems may be termed applied ecology.

Energy and Cycles of Energy

Biogeochemical cycles and virtually all other processes on earth are driven by energy from the sun. The sun acts as a so-called blackbody radiator with an effective surface temperature of 5780 K (absolute temperature in which each unit is the same as a Celsius degree, but with zero taken at absolute zero). It transmits energy to earth as electromagnetic radiation with a maximum energy flux at about 500 nanometers, which is in the visible region of the spectrum. A one square metre area perpendicular to the line of solar flux at the top of the atmosphere receives energy at a rate of 1,340 watts; sufficient, for example, to power an electric iron. This is called the solar flux.

Energy in natural systems is transferred by heat, which is the form of energy that flows between two bodies as a result of their difference in temperature, or by work, which is a transfer of energy that does not depend upon a temperature difference. Such transfers are governed by the laws of thermodynamics. The first law of thermodynamics states that, although energy may be transferred or transformed, it is conserved and is not lost. Chemical energy in the food ingested by organisms is converted by metabolic processes to work or heat that can be utilised by the organisms, but there is no net gain or loss of energy overall. The second law of thermodynamics describes the tendency toward disorder in natural systems. It demonstrates that each time energy is transformed, some is wasted in the sense that it cannot be utilised for work. Thus, for example, only a fraction of the energy that organisms derive from metabolising food can be converted to work; the rest is dissipated as heat.

Light and electromagnetic radiation

Electromagnetic radiation, particularly light, is of utmost importance in considering energy in environmental systems. Therefore, the following important points related to electromagnetic radiation should be noted :

1. Energy can be carried through space at the speed of light, 3.00×10^8 metres per second (m/s) in a vacuum, by electromagnetic radiation, which includes visible light, ultraviolet radiation, infrared radiation, microwaves, radio waves, gamma rays, and x-rays.

2. Electromagnetic radiation has a wave character. The waves move at the speed of light, c, and have characteristics of wavelength (λ), amplitude, and frequency (v, Greek "*nu*") as illustrated in Fig. 3.2.

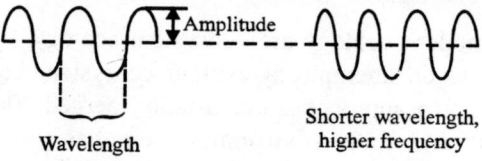

Fig. 3.2. Electromagnetic radiation.

3. The wavelength is the distance required for one complete cycle and the frequency is the number of cycles per unit time. They are related by the following equation:

$$v \, \lambda = c$$

where v is in units of cycles per second (s^{-1}, a unit called the hertz, Hz) and λ is in metres (m).

4. In addition to behaving as a wave, electromagnetic radiation also has characteristics of particles.

5. The dual wave/particle nature of electromagnetic radiation is the basis of the quantum theory of electromagnetic radiation, which states that radiant energy may be absorbed or emitted only in discrete packets called quanta or photons. The energy, 'E', of each photon is given by the equation

$$E = hv$$

where 'h' is Planck's constant, (i.e. 6.63×10^{-34} J-s (joule × second)).

6. From the preceding, it is seen that the energy of a photon is higher when the frequency of the associated wave is higher (and the wavelength shorter).

Energy flow and photosynthesis in living systems

Whereas materials are recycled through ecosystems, the flow of useful energy may be viewed as essentially a one-way process. Incoming solar energy can be regarded as high-grade energy, because it can cause useful reactions to occur, the most important of which in living systems is photosynthesis. As shown in Fig. 3.3, solar energy captured by green plants energises chlorophyll, which in turn, powers metabolic processes that produce carbohydrates from water and carbon dioxide. These carbohydrates are repositories of stored chemical energy that can be converted to heat and work by metabolic reactions with oxygen in organisms. Ultimately, most of the energy is converted to low-grade heat, which is eventually re-radiated away from earth by infrared radiation.

Energy utilisation

During the last two centuries, the human impact on energy utilisation and conversion has been enormous and has resulted in many of the environmental problems now facing humankind. This time period has seen a transition from the almost exclusive use of energy captured by photosynthesis and utilised as biomass (food to provide muscle power, wood for heat), to the use of fossil fuels for about 90 per cent, and nuclear energy for about 5 per cent of all energy employed commercially. Fossil fuel consumption is divided primarily among petroleum, natural gas, and coal. These sources of energy are limited and their pollution potential is high. The mining of coal and the extraction of petroleum is environmentally disruptive; the combustion of high-sulphur coal releases acidic sulphur dioxide to the atmosphere; and all fossil fuels produce carbon dioxide, a greenhouse gas. Therefore, it will be necessary to move toward the utilisation of alternate energy sources, particularly those that are renewable. Prominent among these is solar energy, and biomass will, to a degree, come back as an energy source. Nuclear energy, with modern safe and efficient reactors is gaining increasing attention as a reliable, environmentally friendly energy source. The study of energy utilisation is crucial in the environmental sciences.

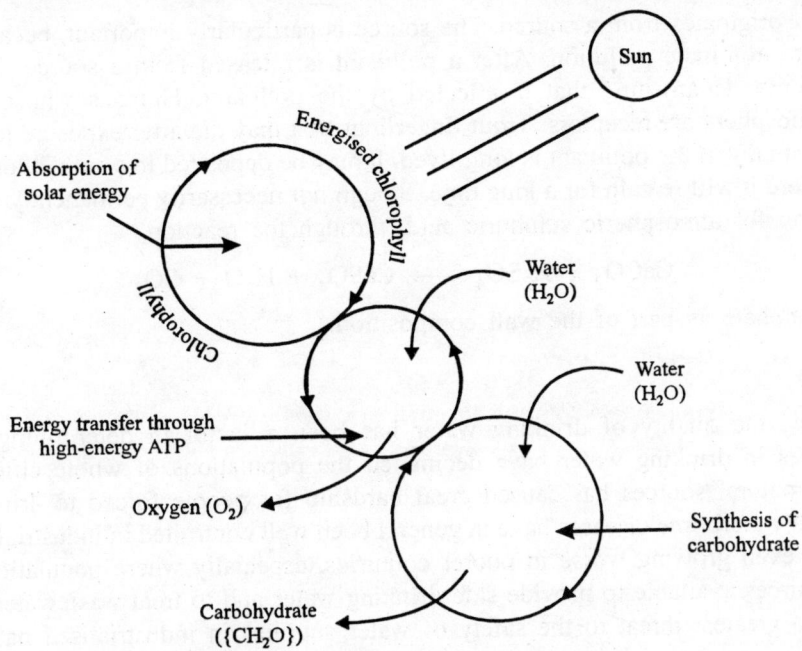

Fig. 3.3. Energy conversion and transfer by photosynthesis.

Human Impact and Pollution

The demands of increasing population coupled with the desire of most people for a higher material standard of living are resulting in worldwide pollution on a massive scale. Environmental pollution can be divided among the categories of water, air, and land pollution. All three of these areas are linked. For example, some gases emitted to the atmosphere can be converted to strong acids by atmospheric chemical processes, fall to the earth as acid rain, and pollute water with acidity. Improperly discarded hazardous wastes can leach into groundwater that is eventually released as polluted water into streams.

Some definitions pertaining to pollution

In some cases, pollution is a clear-cut phenomenon, whereas in others, it lies largely in the eyes of the beholder. Toxic organochlorine solvent residues leached into water supplies from a hazardous waste chemical dump are pollutants in anybody's view. However, loud rock music amplified to a high decibel level by the sometimes questionable miracle of modern electronics is pleasant to some people and a very definite form of noise pollution to others. Frequently, time and place determine what may be called a pollutant. The phosphate that the sewage treatment plant operator has to remove from waste-water is chemically the same as the phosphate that the farmer a few miles away has to buy at high prices for fertiliser. Most pollutants are, in fact, resources gone to waste; as resources become more scarce and expensive, economic pressure will almost automatically force solutions to many pollution problems.

A reasonable definition of a pollutant is a substance present in greater than natural concentration as a result of human activity that has a net detrimental effect upon its environment or upon something of value in that environment. Contaminants, which are not classified as pollutants unless they have some detrimental effect, cause deviations from the normal composition of an environment.

Every pollutant originates from a source. The source is particularly important, because it is generally the logical place to eliminate pollution. After a pollutant is released from a source, it may act upon a receptor. The receptor is anything that is affected by the pollutant. Humans whose eyes smart from oxidants in the atmosphere are receptors. Trout fingerlings that may die after exposure to dieldrin in water are receptors. Eventually, if the pollutant is long-lived, it may be deposited in a sink, a long-time repository of the pollutant. Here it will remain for a long time, though not necessarily permanently. Thus, a limestone wall may be a sink for atmospheric sulphuric acid, through the reaction,

$$CaCO_3 + H_2SO_4 \longrightarrow CaSO_4 + H_2O + CO_2$$

which fixes the sulphate as part of the wall composition.

Water pollution

Throughout history, the quality of drinking water has been a factor in determining human welfare. Waterborne diseases in drinking water have decimated the populations of whole cities. Unwholesome water polluted by natural sources has caused great hardship for people forced to drink it or use it for irrigation. Although waterborne diseases have in general been well controlled in industrialised nations, they are prevalent, and even growing worse in poorer countries, especially where population pressures have overtaxed the resources available to provide safe drinking water and to treat wastewater. Currently, toxic chemicals pose the greatest threat to the safety of water supplies in industrialised nations.

Many of the chemicals have contaminated water supplies. Two examples are insecticide and herbicide runoff from agricultural land, and industrial discharge into surface waters. Another serious problem is the threat to groundwater from waste chemical dumps and landfills, storage lagoons, treating ponds, and other facilities.

Water pollution should be a concern of every citizen. Understanding the sources, interactions, and effects of water pollutants is essential for controlling pollutants in an environmentally safe and economically acceptable manner. Above all, an understanding of water pollution and its control depends upon a basic knowledge of aquatic environmental science.

Air pollution

Inorganic air pollutants consist of many kinds of substances. Many solid and liquid substances may become particulate air contaminants. Another important class of inorganic air pollutants consists of oxides of carbon, sulphur, and nitrogen. Carbon monoxide is a directly toxic material that is fatal at relatively small doses. Carbon dioxide is a natural and essential constituent of the atmosphere that is required for plants for photosynthesis. However, CO_2 may turn out to be the most deadly air pollutant of all, because of its potential as a greenhouse gas that might cause devastating global warming. Oxides of sulphur and nitrogen are acid-forming gases that can cause acid precipitation. Ammonia, hydrogen chloride, and hydrogen sulphide, are also inorganic air pollutants.

A number of gaseous inorganic pollutants enter the atmosphere as the result of human activities. Those added in the greatest quantities are carbon monoxide (CO), sulphur dioxide (SO_2), nitric oxide (NO), and nitrogen dioxide (NO_2). Other inorganic pollutant gases include ammonia (NH_3), nitrous oxide (N_2O), hydrogen sulphide (H_2S), elemental chlorine (Cl_2), hydrogen chloride (HCl), and hydrogen fluoride (HF). Substantial quantities of some of these gases are added to the atmosphere each year by human activities. Globally, atmospheric emissions of carbon monoxide, sulphur oxides, and nitrogen oxides are of the order of one to several hundred million tonnes per year.

Organic pollutants are common atmospheric contaminants that may have a strong effect upon atmospheric quality. Such pollutants may come from both natural and artificial sources. In some cases, contaminants from both kinds of sources interact to produce a pollution effect. This occurs, for example, when terpene hydrocarbons evolved from citrus and conifer trees interact with nitrogen oxides from automobiles to produce photochemical smog.

The effects of organic pollutants in the atmosphere may be divided into two major categories. The first consists of direct effects, such as cancer caused by exposure to vinyl chloride. The second is the formation of secondary pollutants, especially photochemical smog. In the case of pollutant hydrocarbons in the atmosphere, the latter is the more important effect. In some localised situations, particularly the workplace, direct effects of organic air pollutants may be equally important.

Some air pollutants, particularly those that may result in irreversible global warming or destruction of the protective stratospheric ozone layer, are of such a magnitude that they have the potential to threaten life on earth.

Pollution of the geosphere and hazardous wastes

The most serious kind of pollutant that is likely to contaminate the geosphere, particularly soil, consists of hazardous wastes. A simple definition of a hazardous waste is that it is a potentially dangerous substance that has been discarded, abandoned, neglected, released or designated as a waste material, or one that may interact with other substances to pose a threat. In a simple sense a hazardous waste is a material that has been left where it may cause harm if encountered.

Although humans have been exposed to hazardous substances such as noxious volcanic gases even in prehistorical times, the modern industrial era has seen creation of problems with hazardous wastes that pose real threats to the environment and humankind.

Technology : The Problems it Poses and the Solutions it Offers

Modern technology has provided the means for massive alteration of the environment and pollution of the environment. However, technology, intelligently applied with a strong environmental awareness, also provides the means for dealing with problems of environmental pollution and degradation.

Some of the major ways in which modern technology has contributed to environmental alteration and pollution are the following :

1. Agricultural practices that have resulted in intensive cultivation of land, drainage of wetlands, irrigation of arid lands, and application of herbicides and insecticides.

2. Manufacturing of huge quantities of industrial products that consume vast amounts of raw materials and produce large quantities of air pollutants, water pollutants, and hazardous waste by-products.

3. Extraction and production of minerals and other raw materials with accompanying environmental disruption and pollution.

4. Energy production and utilisation with environmental effects that include disruption of soil by strip mining, pollution of water by release of saltwater from petroleum production, and emission of air pollutants, such as acid-rain-forming sulphur dioxide.

5. Modern transportation practices, particularly reliance on the automobile, that cause scarring of land surfaces from road construction, emission of air pollutants, and greatly increased demands for fossil fuel resources.

Despite all of the problems that it raises, technology based on a firm foundation of environmental science can be very effectively applied to the solution of environmental problems. One important example of this is the redesign of basic manufacturing processes to minimise raw material consumption, energy use, and waste production. Consider a generalised manufacturing process shown in Fig. 3.4 with proper design the environmental acceptability of such a process can be greatly enhanced. In some cases, raw materials and energy sources can be chosen in ways that minimise environmental impact. If the process involves manufacture of a chemical, it may be possible to completely alter the reactions used so that the entire operation is more environmentally friendly. Raw materials and water may be recycled to the maximum extent possible. Best available technologies may be employed to minimise air, water, and solid waste emissions.

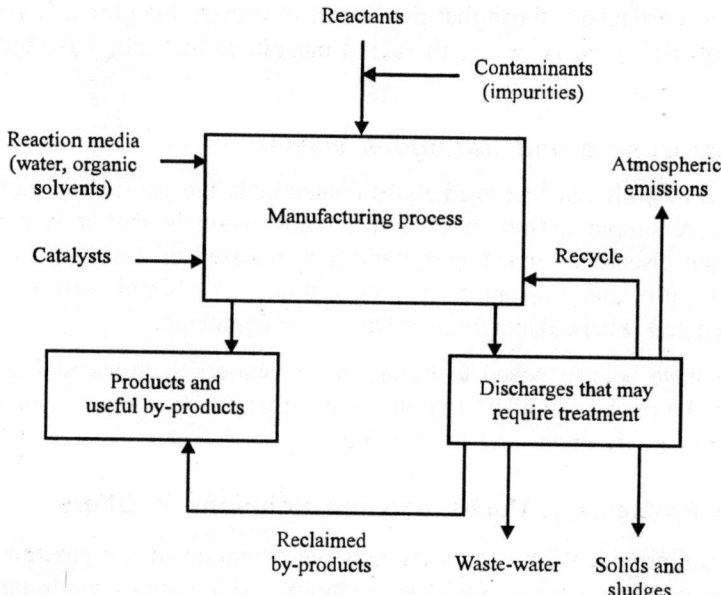

Fig. 3.4. A manufacturing process viewed from the standpoint of minimisation of environmental impact.

There are numerous ways in which technology can be applied to minimise environmental impact. Among these are the following :

1. Use of state-of-the-art computerised control to achieve optimum energy efficiency, maximum utilisation of raw materials, and minimum production of pollutant by-products.

2. Use of materials that minimise pollution problems; for example, heat-resistant materials that enable use of high temperatures for efficient thermal processes.

3. Application of processes and materials that enable maximum materials recycling and minimum waste product production; for example, advanced membrane processes for wastewater treatment to enable water recycling.

4. Application of advanced biotechnologies, such as in the biological treatment of wastes.

5. Use of best available catalysts for efficient synthesis.

6. Use of lasers for precision machining and processing to minimise waste production.

CHAPTER 4

Environmental Chemistry and Chemical Cycles

INTRODUCTION

Environmental chemistry is the science of chemical phenomenon occurring in the environment. It is multibranched science involving chemistry, physics, life sciences, agriculture, public health, botany and medical sciences etc. and may be defined as *the study of the sources, reactions, transport effects and fates of chemical species in the water, soil and air environment*. It is a major and important part of environmental studies and the effecs of human activities upon environmental segments such as atmosphere, hydrosphere, lithosphere and biosphere.

One of the main objectives of environmental chemistry is to determine the nature as well as quantity of specific pollutants in the environment. A basic understanding of the fundamental concepts of environmental chemistry is, therefore, must for all chemists and non-chemists who are engaged in the study of environmental and related science.

Since the amount of environmental pollutants in natural samples is appreciably small (e.g. the levels of air pollutants may be less than a microgram per cubic metre of air or level of many water pollutants may be a part per million by weight or even less than it), the environmental chemist requires high sensitivity and accuracy in determining such environmental pollutants.

Thus, environmental chemistry as already defined, is the study of the *sources, reactions, transport, effects, and fates of chemical species in water, soil and air environments and the effects of technology thereon.*

Aquatic chemistry is the branch of environmental chemistry that deals with chemical phenomena in water. To a large extent, aquatic chemistry addresses chemical phenomena in so-called "natural waters", consisting of water in streams, lakes, oceans, underground aquifers, and other places where the water is rather freely exposed to the atmosphere, soil, rock, and living systems. "Fundamentals of Aquatic Chemistry" discusses some of the fundamental phenomena that apply to chemical species dissolved in water, particularly acid-base reactions, and complexation. Oxidation-reduction phenomena in water and many important aquatic chemical interactions occur between species dissolved in water and those in gaseous, solid, and immiscible liquid phases. One thing that clearly distinguishes the chemistry of natural waters from water isolated, contained, and purified by humans is the strong influence of micro-organisms on aquatic chemistry.

Atmospheric chemistry

Atmospheric chemistry deals with chemical phenomena in the atmosphere. To understand these processes it is first necessary to have a basic knowledge of the structure and composition of the atmosphere. Photochemical reactions occur when electromagnetic radiation from the sun energises gas molecules forming reactive species that initiate chain reactions that largely determine key atmospheric chemical phenomena. Organic species in the atmosphere result in some important pollution phenomena that also influence inorganic species.

Geosphere

Geosphere is the part of atmospheric chemistry and Geochemistry which outlines the physical nature and chemical characteristics of the geosphere and introduces some basic geochemistry. Soil is a uniquely important part of the geosphere that is essential to life on earth. Human activities have such a profound effect on the environment that it is convenient to invoke a fifth sphere of the environment called the "anthrosphere".

Biosphere

The zone of living organisms which denotes mutual interactions of organisms related to lithosphere, hydrosphere and atmosphere with the environment is called biosphere. It represents the relationship between living organisms (plants and animal) in 10,000 metres below sea-level and 6000 metres above sea-level.

Analytical chemistry

Analytical chemistry is uniquely important in the study of environmental chemistry as it applies to environmental chemistry.

Water : Quality, Quantity, and Chemistry

Waterborne diseases such as cholera and typhoid killed millions of people in the past. Some of these diseases still cause great misery in less developed countries. Ambitious programmes of dam and dike construction have reduced flood damage, but they have had a number of undesirable side effects in some areas, such as inundation of farmland by reservoirs and unsafe dams prone to failure. Aquatic chemistry must consider groundwater and water in rivers, lakes, estuaries, and oceans, as well as the phenomena that determine the distribution and circulation of chemical species in natural waters. Study of aquatic chemistry requires some understanding of the sources, transport, characteristics, and composition of water. The chemical reactions that occur in water and the chemical species found in it are strongly influenced by the environment in which the water is found. The chemistry of water exposed to the atmosphere is quite different from that of water at the bottom of a lake. Micro-organisms play an essential role in determining the chemical composition of water. Thus, in discussing water chemistry, it is necessary to consider the many general factors that influence this chemistry.

The study of water is known as hydrology, and is divided into a number of subcategories. Limnology is the branch of the science dealing with the characteristics of freshwater, including biological properties as well as chemical and physical properties. Oceanography is the science of the ocean and its physical and chemical characteristics. The chemistry and biology of the earth's vast oceans are unique because of the ocean's high salt content, great depth, and other factors.

Sources and uses of water : The hydrologic cycle

The world's water supply is found in the five parts of the hydrologic cycle (Fig. 4.1). A large portion of the water is found in the oceans. Another fraction is present as water vapour in the atmosphere (clouds). Some water is contained in the solid state as ice and snow in snowpacks, glaciers, and the polar ice caps. Surface water is found in lakes, streams, and reservoirs. Groundwater is located in aquifers underground.

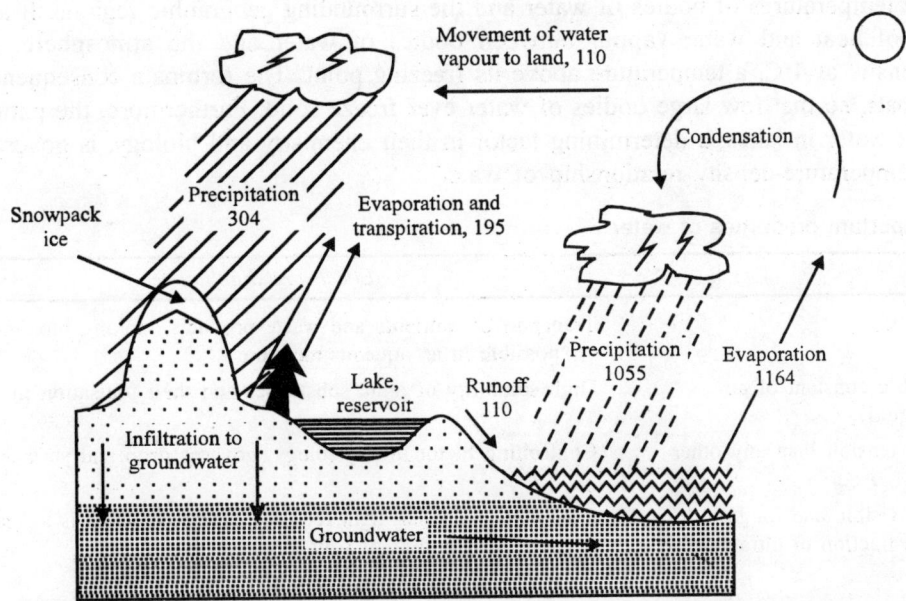

Fig. 4.1. The hydrologic cycle, quantities of water in trillions of litres per day.

There is a strong connection between the hydrosphere, where water is found, and the lithosphere, which is that part of the geosphere accessible to water. Human activities affect both. For example, disturbance of land by conversion of grasslands or forests to agricultural land or intensification of agricultural production may reduce vegetation cover, decreasing transpiration (loss of water vapour by plants) and affecting the micro-climate. The result is increased rain runoff, erosion, and accumulation of silt in bodies of water. The nutrient cycles may be accelerated, leading to nutrient enrichment of surface waters. This, in turn, can profoundly affect the chemical and biological characteristics of bodies of water.

The water that humans use is primarily fresh surface water and groundwater, both of which sources may differ from each other significantly. In arid regions, a small fraction of the water supply comes from the ocean, a source that is likely to become more important as the world's supply of freshwater dwindles relative to demand. Saline or brackish groundwaters may also be utilised in some areas.

The properties of water, a unique substance

Water has a number of unique properties that are essential to life, many of which are due to water's ability to form hydrogen bonds. These characteristics are summarised in Table 4.1. Water is an excellent solvent for many materials; thus, it is the basic transport medium for nutrients and waste products in life processes. The extremely high dielectric constant of water relative to other liquids has a profound effect upon its solvent properties, in that most ionic materials are dissociated in water. With the exception of liquid ammonia, water has the highest heat capacity of any liquid or solid, $1 \text{ cal} \times g^{-1} \times deg^{-1}$. Because of this high heat capacity, a relatively large amount of heat is required to change appreciably the temperature of a mass of water; hence, a body of water can have a stabilising effect upon the temperature of nearby geographic regions. In addition, this property prevents sudden large changes of temperature in large bodies of water and thereby protects aquatic organisms from the shock of abrupt temperature variations. The extremely high heat of vapourisation of water, 585 cal/g at 20°C, likewise stabilises the temperatures of bodies of water and the surrounding geographic regions. It also influences the transfer of heat and water vapour between bodies of water and the atmosphere. Water has its maximum density at 4°C, a temperature above its freezing point. The fortunate consequence of this fact is that ice floats, so that few large bodies of water ever freeze solid. Furthermore, the pattern of vertical circulation of water in lakes, a determining factor in their chemistry and biology, is governed largely by the unique temperature-density relationship of water.

Table 4.1. Important properties of water.

Property	Effects and Significance
Excellent solvent	Transport of nutrients and waste products, making biological processes possible in an aqueous medium
Highest dielectric constant of any common liquid	High solubility of ionic substances and their ionisation in solution
Higher surface tension than any other liquid	Controlling factor in physiology; governs drop and surface phenomena
Transparent to visible and longer-wavelength fraction of ultraviolet light	Colourless, allowing light required for photosynthesis to reach considerable depths in bodies of water
Maximum density as a liquid at 4°C	Ice floats; vertical circulation restricted in stratified bodies of water
Higher heat of evaporation than any other material	Determines transfer of heat and water molecules between the atmosphere and bodies of water
Higher latent heat of fusion than any other liquid except ammonia	Temperature stabilised at the freezing point of water
Higher heat capacity than any other liquid except ammonia	Stabilisation of temperatures of organisms and geographical regions

Characteristics of bodies of water

The physical condition of a body of water strongly influences the chemical and biological processes that occur in water. Surface water occurs primarily in streams, lakes, and reservoirs. Wetlands are flooded areas in which the water is shallow enough to enable growth of bottom-rooted plants. Estuaries constitute another type of body of water, consisting of arms of the ocean into which streams flow. The mixing of fresh and salt water gives estuaries unique chemical and biological properties. Estuaries are the breeding grounds of much marine life, which makes their preservation very important.

Water's unique temperature-density relationship results in the formation of distinct layers within non-flowing bodies of water, as shown in Fig. 4.2. During the summer a surface layer (epilimnion) is, heated by solar radiation and, because of its lower density, floats upon the bottom layer, or hypolimnion. This phenomenon is called thermal stratification. When an appreciable temperature difference exists between the two layers, they do not mix but behave independently and have very different chemical and biological properties. The epilimnion, which is exposed to light, may have a heavy growth of algae. As a result of exposure to the atmosphere and (during daylight hours) because of the photosynthetic activity of algae, the epilimnion contains relatively higher levels of dissolved oxygen and generally is aerobic. In the hypolimnion, bacterial action on biodegradable organic material may cause the water to become anaerobic (lacking dissolved oxygen). As a consequence, chemical species in a relatively reduced form tend to predominate in the hypolimnion.

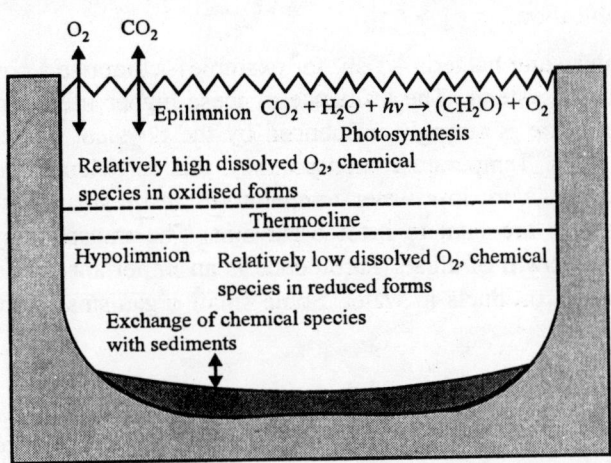

Fig. 4.2. Stratification of a lake.

The shear-plane, or layer between epilimnion and hypolimnion, is called the thermocline. During the autumn, when the epilimnion cools, a point is reached at which the temperatures of the epilimnion and hypolimnion are equal. This disappearance of thermal stratification causes the entire body of water to behave as a hydrological unit, and the resultant mixing is known as overturn. An overturn also generally occurs in the spring. During the overturn, the chemical and physical characteristics of the body of water become much more uniform, and a number of chemical, physical, and biological changes may result. Biological activity may increase from the mixing of nutrients. Changes in water composition during overturn may cause disruption in water-treatment processes.

Aquatic life

The living organisms (biota) in an aquatic ecosystem may be classified as either autotrophic or heterotrophic. Autotrophic organisms utilise solar or chemical energy to fix elements from simple, non-living inorganic material into complex life molecules that compose living organisms. Algae are typical autotrophic aquatic organisms. Generally, CO_2, NO_3^-, and $H_2PO_4^-/HPO_4^{2-}$ are sources of C, N, and P, respectively, for autotrophic organisms. Organisms that utilise solar energy to synthesise organic matter from inorganic materials are called producers.

Heterotrophis

Heterotrophic organisms utilise the organic substances produced by autotrophic organisms as energy sources and as the raw materials for the synthesis of their own biomass. Decomposers (or reducers) are a sub-class of the heterotrophic organisms and consist of chiefly bacteria and fungi, which ultimately break down material of biological origin to the simple compounds originally fixed by the autotrophic organisms.

The ability of a body of water to produce living material is known as its productivity. Productivity results from a combination of physical and chemical factors. Water of low productivity generally is desirable for water supply or for swimming. Relatively high productivity is required for the support of fish. Excessive productivity can result in choking by weeds and can cause odour problems. The growth of algae may become quite high in very productive waters, with the result that the concurrent decomposition of dead algae reduces oxygen levels in the water to very low values. This set of conditions is commonly called eutrophication.

Life forms higher than algae and bacteria – fish, for example—comprise a comparatively small fraction of the biomass in most aquatic systems. The influence of these higher life forms upon aquatic chemistry is minimal. However, aquatic life is strongly influenced by the physical and chemical properties of the body of water in which it lives. Temperature, transparency, and turbulence are the three main physical properties affecting aquatic life. Very low water temperatures result in very slow biological processes, whereas very high temperatures are fatal to most organisms. The transparency of water is particularly important in determining the growth of algae. Turbulence is an important factor in mixing processes and transport of nutrients and waste products in water. Some small organisms (plankton) depend upon water currents for their own mobility.

Dissolved oxygen

Dissolved oxygen (DO) frequently is the key substance in determining the extent and kinds of life in a body of water. Oxygen deficiency is fatal to many aquatic animals such as fish. The presence of oxygen can be equally fatal to many kinds of anaerobic bacteria.

Biochemical oxygen demand (BOD)

Biochemical oxygen demand is another important water-quality parameter. It refers to the amount of oxygen utilised when the organic matter in a given volume of water is degraded biologically. A body of water with a high biochemical oxygen demand, and no means of rapidly replenishing the oxygen, obviously cannot sustain organisms that require oxygen.

Carbon dioxide is produced by respiratory processes in waters and sediments and can also enter water from the atmosphere. Carbon dioxide is required for the photosynthetic production of biomass by algae and in some cases is a limiting factor. High levels of carbon dioxide produced by the degradation of organic matter in water can cause excessive algal growth and productivity.

The levels of nutrients in water frequently determine its productivity. Aquatic plant life requires an adequate supply of carbon (CO_2), nitrogen (nitrate), phosphorus (orthophosphate), and trace elements such as iron. In many cases, phosphorus is the limiting nutrient and is generally controlled in attempts to limit excess productivity. The salinity of water also determines the kinds of life forms present. Irrigation waters may pick up harmful levels of salt. Marine life obviously requires or tolerates salt water, whereas many freshwater organisms are intolerant of salt.

Aquatic chemistry

Aquatic environmental chemical phenomena involve processes familiar to chemists, including acid-base, solubility, oxidation-reduction, and complexation reactions. Although most aquatic chemical phenomena are discussed here from the thermodynamic (equilibrium) viewpoint, it is important to keep in mind that kinetics "rates of reactions" are very important in aquatic chemistry. Biological processes play a key role in aquatic chemistry. For example, algae undergoing photosynthesis can raise the pH of water by removing aqueous CO_2, thereby converting HCO_3^- ion to CO_3^{2-} ion, which reacts with Ca^{2+} in water to precipitate $CaCO_3$.

Acid base

Acid-base phenomena in water, involve loss and acceptance of H^+ ion. Many species act as acids in water by releasing H^+ ion, others act as bases by accepting H^+ ions, and the water molecule itself does both. An important species in the acid-base chemistry of water is bicarbonate ion, HCO_3^-, which may act as either an acid or a base:

$$HCO_3^- \rightleftharpoons CO_3^{2-} + H^+ \qquad \text{...(4.1)}$$

$$HCO_3^- + H^+ \rightleftharpoons CO_2(aq) + H_2O \qquad \text{...(4.2)}$$

Metal ions in water are bound with water molecules, so they are said to be hydrated. A typical example of a hydrated metal ion is dissolved calcium ion, $Ca(H_2O)_6^{2+}$. Some other species in water bind much more strongly than water to metal ions. For example, one or more cyanide ions can bond to dissolved iron (II):

$$Fe(H_2O)_6^{2+} + CN^- \rightleftharpoons FeCN(H_2O)_5^+ + H_2O \qquad \text{...(4.3)}$$

This phenomenon is called complexation, the species that binds with the metal ion, CN^- in the example above, is called a ligand, and the product in which the ligand is bound with the metal ion is a complex or complex ion. A special case of complexation in which a ligand bonds in two or more places to a metal ion is called chelation.

Oxidation-reduction reactions occur for a number of species in water (Fig. 4.3). These reactions are usually mediated by bacteria. In anaerobic water, reduced species tend to predominate. An example is illustrated in Fig. 4.3 for the reduction of sulphate ion to H_2S by microbial action on organic matter, designated $[CH_2O]$.

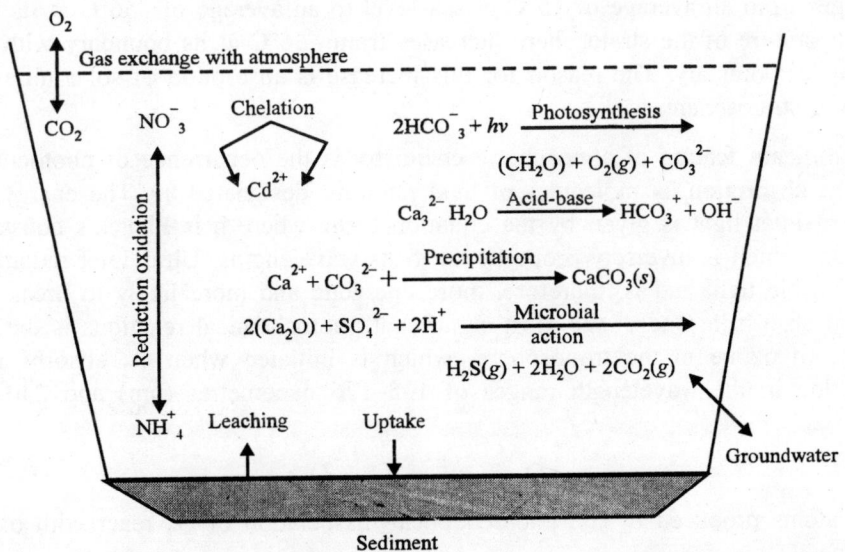

Fig. 4.3. Major aquatic chemical processes.

The interactions with other phases are particularly important in aquatic chemistry. These include both the kinetics and thermodynamics (equilibrium) of gas dissolution and evolution. Examples are oxygen dissolving into water from the atmosphere and carbon dioxide evolved from water after its production by bacterial metabolism. Precipitation and dissolution of solids are important; an example is dissolution of calcium carbonate by dissolved carbon dioxide:

$$CaCO_3(s) + CO_2(aq) + H_2O \rightleftharpoons Ca^{2+} + 2HCO_3^- \qquad ...(4.4)$$

Very small colloidal particles are important species in water. Of the order of a micrometre or less in size, colloids are stabilised as suspensions in water by their attraction for water molecules or by their electrical charges. One of their most important characteristics is their huge surface area relative to volume.

Exchange of species between water solution and solid or colloidal phases serves to remove solutes from, and add them to water. Poorly soluble organic substances, for example, may be held by organic rich sediments. Ion-exchange processes serve to exchange ionic solutes, particularly cations, between water and sediments.

The Atmosphere and Atmospheric Chemistry

The atmosphere consists of the thin layer of mixed gases covering the earth's surface. Exclusive of water, atmospheric air is 78.1% (by volume) nitrogen, 21.0% oxygen, 0.9% argon, and 0.03% carbon dioxide. Normally, air contains 1–3% water vapour by volume. In addition, air contains a large variety of trace level gases at levels below 0.002%, including neon, helium, methane, krypton, nitrous oxide, hydrogen, xenon, sulphur dioxide, ozone, nitrogen dioxide, ammonia, and carbon monoxide.

The atmosphere is divided into several layers on the basis of temperature. Of these, the most significant are the troposphere extending in altitude from the earth's surface to approximately 11 kilometres (km) and the stratosphere from about 11 km to approximately 50 km. The temperature of the troposphere ranges from an average of 15°C at sea level to an average of –56°C at its upper boundary. The average temperature of the stratosphere increases from –56°C at its boundary with the troposphere to –2°C at its upper boundary. The reason for this increase is absorption of solar ultraviolet energy by ozone (O_3) in the stratosphere.

The most significant feature of atmospheric chemistry is the occurrence of photochemical reactions resulting from the absorption by molecules of light photons, designated hv. The energy, E, of a photon of visible or ultraviolet light is given by the equation $E = hv$ where h is Planck's constant and v is the frequency of light, which is inversely proportional to its wavelengths. Ultraviolet radiation has a higher frequency than visible light and is, therefore, more energetic and more likely to break chemical bonds in molecules that absorb it. One of the most significant photochemical reactions is the one responsible for the presence of ozone in the troposphere, which is initiated when O_2 absorbs highly energetic ultraviolet radiation in the wavelength ranges of 135–176 nanometres (nm) and 240–260 nm in the stratosphere:

$$O_2 + hv \rightarrow O + O \qquad ...(4.5)$$

The oxygen atoms produced by the photochemical dissociation of O_2 react with oxygen molecules to produce ozone, O_3,

$$O + O_2 + M \rightarrow O_3 + M \qquad ...(4.6)$$

where, M is a third body, such as a molecule of N_2, which absorbs excess energy from the reaction. The ozone that is formed is very effective in absorbing ultraviolet radiation in the 220–330 nm wavelength range, which causes the temperature increase observed in the stratosphere. The ozone serves as a very valuable filter to remove ultraviolet radiation from the sun's rays. If this radiation reached the earth's surface, it would cause skin cancer and other damage to living organisms.

Gaseous oxides in the atmosphere

Oxides of carbon, sulphur, and nitrogen are important constituents of the atmosphere and are pollutants at higher levels. Of these, carbon dioxide, CO_2, is the most abundant. It is a natural atmospheric constituent, and it is required for plant growth. However, the level of carbon dioxide in the atmosphere, now at about 350 parts per million (ppm) by volume, is increasing by about 1 ppm per year. This increase in atmospheric CO_2 may well cause general atmospheric warming—the "greenhouse effect"—with potentially very serious consequences for the global atmosphere and for life on earth. Though not a global threat, carbon monoxide, CO, can be a serious health threat, because it prevents blood from transporting oxygen to body tissues.

The two most serious nitrogen oxide air pollutants are nitric oxide, NO, and nitrogen dioxide, NO_2, collectively denoted as "NO_x". These tend to enter the atmosphere as NO, and photochemical processes in the atmosphere convert NO to NO_2. Further reactions can result in the formation of corrosive nitrate salts or nitric acid, HNO_3. Nitrogen dioxide is particularly significant in atmospheric chemistry because of its photochemical dissociation by light with a wavelength less than 430 nm to produce highly reactive O atoms. This is the first step in the formation of photochemical smog. Sulphur dioxide, SO_2, is a reaction product of the combustion of sulphur-containing fuels, such as high-sulphur coal. Part of this sulphur dioxide is converted in the atmosphere to sulphuric acid, H_2SO_4, normally the predominant contributor to acid precipitation.

Hydrocarbons and photochemical smog

The most abundant hydrocarbon in the atmosphere is methane, CH_4, released from underground sources as natural gas and produced by the fermentation of organic matter. Methane is one of the least reactive atmospheric hydrocarbons and is produced by diffuse sources, so that its participation in the formation of pollutant photochemical reaction products is minimal. The most significant atmospheric pollutant hydrocarbons are the reactive ones produced as automobile exhaust emissions. In the presence of NO, under conditions of temperature inversion, low humidity, and sunlight, these hydrocabons produce undesirable photochemical smog manifested by the presence of visibility-obscuring particulate matter, oxidants such as ozone, and noxious organic species such as aldehydes.

Particulate matter

Particles ranging from aggregates of a few molecules to pieces of dust readily visible to the naked eye are commonly found in the atmosphere. Some atmospheric particles, such as sea salt formed by the evaporation of water from droplets of sea spray, are natural and even beneficial atmospheric constituents. Very small particles called condensation nuclei serve as bodies for atmospheric water vapour to condense upon and are essential for the formation of precipitation.

Colloidal-sized particles in the atmosphere are called aerosols. Those formed by grinding up bulk matter are known as dispersion aerosols, whereas particles formed from chemical reactions of gases are condensation aerosols; the latter tend to be smaller. Smaller particles are in general the most harmful because they have a greater tendency to scatter light and are the most respirable (tendency to be inhaled into the lungs). Much of the mineral particulate matter in a polluted atmosphere is in the form of oxides and other compounds produced during the combustion of high-ash fossil fuel. Smaller particles of fly ash enter furnace flues and are efficiently collected in a properly equipped stack system. However, some fly ash escapes through the stack and enters the atmosphere. Unfortunately, the fly ash thus released tends to consist of smaller particles that do the most damage to human health, plants, and visibility.

Geosphere and Soil

The geosphere, or solid earth, is that part of the earth upon which humans live and from which they extract most of their food, minerals, and fuels. Once thought to have an almost unlimited buffering capacity against the perturbations of humankind, the geosphere is now known to be rather fragile and subject to harm by human activities, such as mining, acid rain, erosion from poor cultivation practices, and disposal of hazardous wastes. It may be readily seen that the preservation of the geosphere in a form suitable for human habitation is one of the greatest challenges facing humankind.

Soil

Soil consists of a large variety of material composing the uppermost layer of the earth's crust upon which plants grow. In addition to solids, soil contains air and water. Typically, soil solids consist of about 95% mineral matter and 5% organic material, although the proportions vary widely. Soils are formed by the weathering (physical and chemical disintegration) of parent rocks as the result of interactive geological, hydrological, and biological processes. Soils are porous and are vertically stratified into horizons through the action of water, organisms, and weathering processes. Soils are open systems that undergo continual exchange of matter and energy with the atmosphere, hydrosphere, and biospehere. The most active and important part of soil is topsoil, the layer in which plants are rooted and in which most biological activity occurs.

Biosphere

Biosphere is the name given to that part of the environment consisting of organisms and living biological material. Virtually all of the biosphere is contained by the geosphere and hydrosphere in the very thin layer where these environmental spheres interface with the atmosphere. There are some specialised life forms at extreme depths in the ocean, but these are still relatively close to the atmospheric interface.

The biosphere strongly influences, and in turn is strongly influenced by, the other parts of the environment. It is believed that organisms were responsible for converting earth's original reducing atmosphere to an oxygen-rich one, a process that also resulted in formation of massive deposits of oxidised minerals, such as iron in deposits of Fe_2O_3. Photosynthetic organisms remove CO_2 from the atmosphere, thus preventing runaway greenhouse warming of earth's surface. Organisms strongly influence bodies of water, producing biomass required for life in the water and mediating oxidation-reduction reactions in the water. Organisms are strongly involved with weathering processes that break down rocks in the geosphere and convert rock matter to soil. Lichens, consisting of symbiotic (mutually advantageous) combinations of algae and fungi, attach strongly to rocks; they secrete chemical species that slowly dissolve the rock surface and retain surface moisture that promotes rock weathering.

The most important aspect of the biosphere is plant photosynthesis, which fixes solar energy ($h\nu$) and carbon from atmospheric CO_2 in the form of high-energy biomass,

$$CO_2 + H_2O \xrightarrow{h\nu} (CH_2O) + O_2(g) \qquad ...(4.7)$$

In so doing, plants and algae function as autotrophic organisms, those that utilise solar or chemical energy to fix elements from simple, non-living inorganic material into complex life molecules that compose living organisms. The opposite process, biodegradation, breaks down biomass either in the presence of oxygen (aerobic respiration),

$$\{CH_2O\} + O_2(g) \rightarrow CO_2 + H_2O \qquad ...(4.8)$$

or absence of oxygen (anaerobic respiration):

$$2\{CH_2O\} \rightarrow CO_2(g) + CH_4(g) \qquad ...(4.9)$$

Both aerobic and anaerobic biodegradation get rid of biomass and return carbon dioxide to the atmosphere. The latter reaction is the major source of atmospheric methane. Non-degraded remains of these processes constitute organic matter in aquatic sediments and in soils, which has an important influence on the characteristics of these solids. Carbon that was originally fixed photosynthetically forms the basis of all fossil fuels in the geosphere.

There is a strong interconnection between the biosphere and the anthrosphere. Humans depend upon the biosphere for food, fuel, and raw materials. Human influence on the biosphere continues to change it drastically. Fertilisers, pesticides, and cultivation practices have vastly increased yields of biomass, grains, and food. Destruction of habitat is resulting in the extinction of vast numbers of species, in some cases even before they are discovered. Bioengineering of organisms with recombinant DNA technology and older techniques of selection and hybridisation are causing great changes in the characteristics of organisms, and promise to result in even more striking alterations in the future. It is the responsibility of humankind to make such changes intelligently and to protect and nurture the biosphere.

Matter and Cycles of Matter

Cycles of matter (Fig. 4.4), often based on elemental cycles, are of utmost importance in the environment. Global geochemical cycles can be regarded from the viewpoint of various reservoirs, such as oceans, sediments, and the atmosphere, connected by conduits through which matter moves continuously. The movement of a specific kind of matter between two particular reservoirs may be reversible or irreversible. The fluxes of movement for specific kinds of matter vary greatly as do the contents of such matter in a specified reservoir.

Cycles of matter would occur even in the absence of life on earth, but are strongly influenced by life forms, particularly plants and micro-organisms. Organisms participate in biogeochemical cycles, which describe the circulation of matter, particularly plant and animal nutrients, through ecosystems. As part of the carbon cycle, atmospheric carbon in CO_2 is fixed as biomass, and as part of the nitrogen cycle, atmospheric N_2 is fixed in organic matter. The reverse of these kinds of processes is mineralisation in which biologically bound elements are returned to inorganic states. Biogeochemical cycles are ultimately powered by solar energy, fine-tuned and directed by energy expended by organisms. In a sense, the solar-energy-powered hydrologic cycle (Fig. 4.1) acts as an endless conveyer belt to move materials essential for life through ecosystems.

Fig. 4.4 shows a general cycle with all five spheres or reservoirs in which matter may be contained. As noted at the beginning of this chapter, human activities now have such a strong influence on materials cycles that it is useful to refer to the "anthrosphere" along with the other environmental "spheres" as a reservoir of materials. Using Fig. 4.4 as a model, it is possible to arrive at any of the known elemental cycles. Some of the numerous possibilities for materials exchange are summarised in Table 4.2.

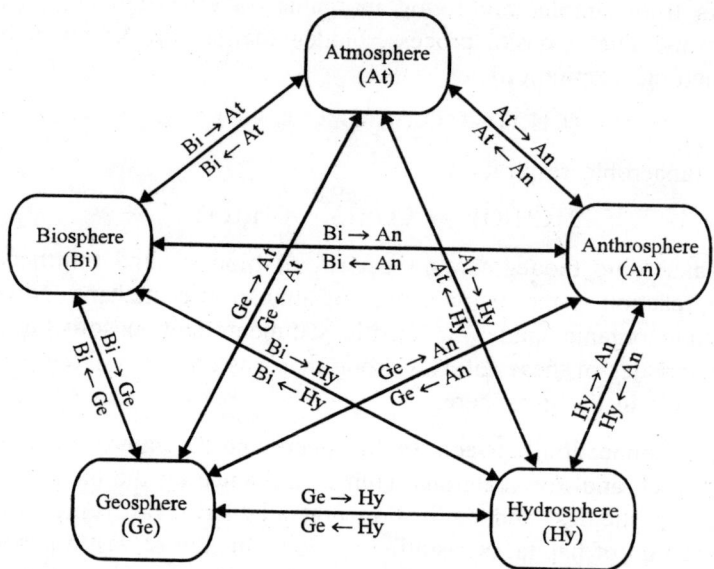

Fig. 4.4. General cycle showing interchange of matter among the atmosphere, biosphere, anthrosphere, geosphere, and hydrosphere.

Table 4.2. Interchange of materials among the possible spheres of the environment.

To/From →	Atmosphere	Hydrosphere	Biosphere	Geosphere	Anthrosphere
Atmosphere	—	H_2O	O_2	H_2S, particles	SO_2, CO_2
Hydrosphere	H_2O	—	{CH_2O}	Mineral solutes	Water pollutants
Biosphere	O_2, CO_2	H_2O	—	Mineral nutrients	Fertilisers
Geosphere	H_2O	H_2O	Organic matter	—	Hazardous wastes
Anthrosphere	O_2, N_2	H_2O	Food	Minerals	—

Endogenic and exogenic cycles

Materials cycles may be divided broadly between endogenic cycles, which predominantly involve subsurface rocks of various kinds, and exogenic cycles, which occur largely on earth's surface, and usually have an atmospheric component. These two kinds of cycles are broadly outlined in Fig. 4.5. In general, sediment and soil can be viewed as being shared between the two cycles and constitute the predominant interface between them.

Most biogeochemical cycles can be described as elemental cycles involving nutrient elements, such as carbon, nitrogen, oxygen, phosphorus, and sulphur. Many are exogenic cycles in which the element in question spends part of the cycle in the atmosphere – O_2 for oxygen, N_2 for nitrogen, CO_2 for carbon. Others, notably the phosphorus cycle, do not have a gaseous component and are endogenic cycles. All sedimentary cycles involve salt solutions or soil solutions that contain dissolved substances leached from weathered minerals; these substances may be deposited as mineral formations, or they may be taken up by organisms as nutrients.

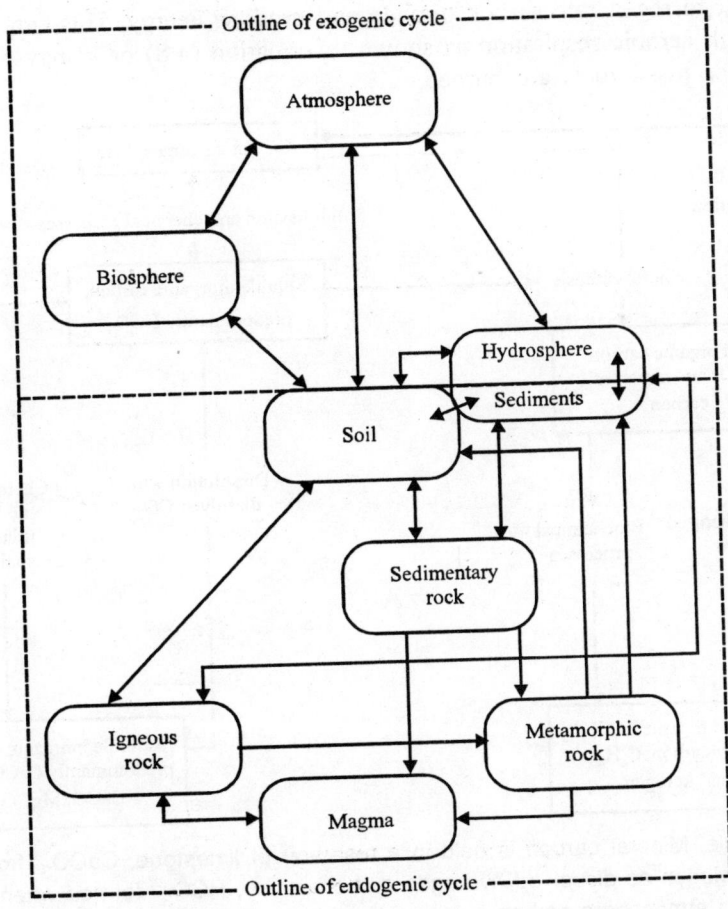

Fig. 4.5. General outline of exogenic and endogenic cycles.

Carbon cycle

Carbon is circulated through the carbon cycle, shown in Fig. 4.6. This cycle shows that carbon may be present as gaseous atmospheric CO_2, constituting a relatively small, but highly significant, portion of global carbon. Some of the carbon is dissolved in surface water and groundwater as HCO_3^- or molecular $CO_2(aq)$. A very large amount of carbon is present in minerals, particularly calcium and magnesium carbonates, such as $CaCO_3$. Photosynthesis fixes inorganic C as biological carbon, represented as (CH_2O), which is a constituent of all life molecules. Another fraction of carbon is fixed as petroleum and natural gas, with a much larger amount as hydrocarbonaceous kerogen (the organic matter in oil shale), coal, and lignite, represented as C_xH_{2x}. Manufacturing processes are used to convert hydrocarbons to xenobiotic compounds with functional groups containing halogens, oxygen, nitrogen, phosphorus, or sulphur. Though a very small amount of total environmental carbon, these compounds are particularly significant because of their toxicological chemical effects.

An important aspect of the carbon cycle is that it is the cycle by which solar energy is transferred to biological systems and ultimately to the geosphere and anthrosphere as fossil carbon and as fossil fuels. Organic, or biological carbon, (CH_2O), is contained in energy-rich molecules that can react with molecular oxygen, O_2, to regenerate carbon dioxide and produce energy. This can occur biochemically in an organism through aerobic respiration as shown in Equation (4.8) or it may occur as combustion, such as when wood or fossil fuels are burned.

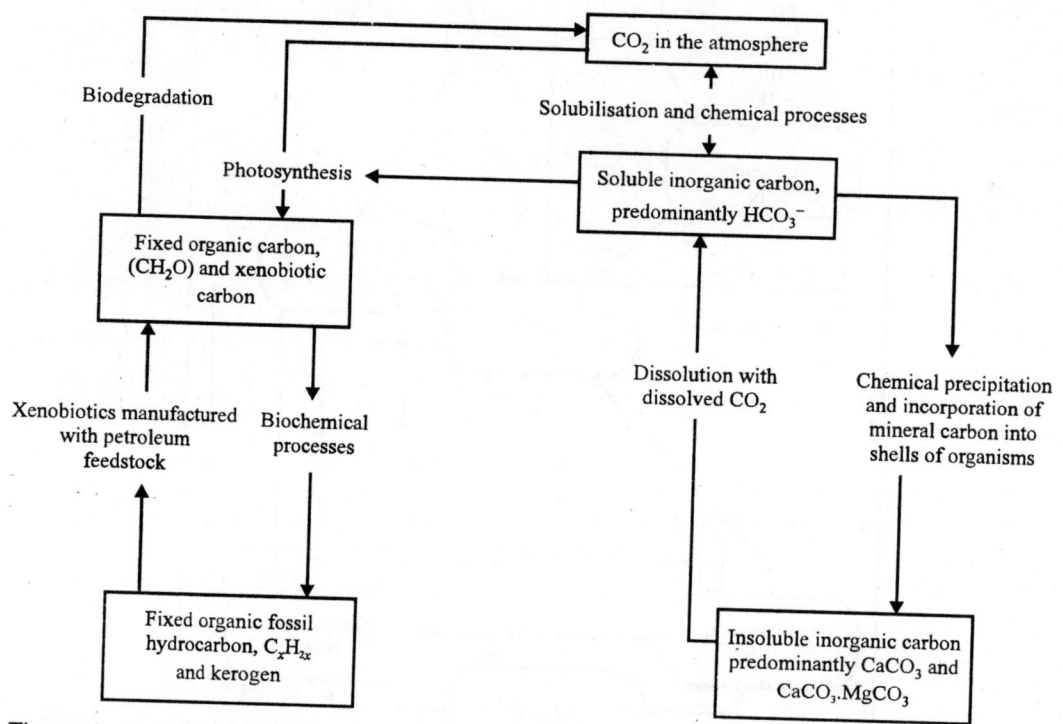

Fig. 4.6. The carbon cycle. Mineral carbon is held in a reservoir of limestone, $CaCO_3$, from which it may be leached into a mineral solution as dissolved hydrogen carbonate ion, HCO_3^- formed when dissolved $CO_2(aq)$ reacts with $CaCO_3$. In the atmosphere carbon is present as carbon dioxide, CO_2. Atmospheric carbon dioxide is fixed as organic matter by photosynthesis and organic carbon is released as CO_2 by microbial decay of organic matter.

Micro-organisms are strongly involved in the carbon cycle, mediating crucial biochemical reactions discussed later in this section. Photosynthetic algae are the predominant carbon-fixing compounds in water; as they consume CO_2, the pH of the water is raised, enabling precipitation of $CaCO_3$ and $CaCO_3.MgCO_3$. Organic carbon fixed by micro-organisms is transformed by biogeochemical processes to fossil petroleum, kerogen, coal, and lignite. Micro-organisms degrade organic carbon from biomass, petroleum, and xenobiotic sources, ultimately returning it to the atmosphere as CO_2.

Hydrocarbons, such as those in crude oil, and some synthetic hydrocarbons are degraded by micro-organisms. This is an important mechanism for eliminating pollutant hydrocarbons, such as those that are accidentally spilled on soil or water. Biodegradation can also be used to treat carbon-containing compounds in hazardous wastes.

Nitrogen cycle

As shown in Fig. 4.7, nitrogen occurs prominantly in all the spheres of the environment. The atmosphere is 78% by volume elemental nitrogen, N_2, and constitutes an inexhaustible reservoir of this essential element. Nitrogen, though constituting much less of biomass than carbon or oxygen, is an essential constituent of proteins. The N_2 molecule is very stable so that breaking it down to atoms that can be incorporated with inorganic and organic chemical forms of nitrogen is the limiting step in the nitrogen cycle. This does occur by highly energetic processes in lightning discharges that produce nitrogen oxides. Elemental nitrogen is also incorporated into chemically bound forms, or fixed, by biochemical processes by microorganisms. The biological nitrogen is mineralised to the inorganic form during the decay of biomass. Large quantities of nitrogen are fixed synthetically under high temperature and high pressure conditions according to the following overall reaction:

$$N_2 + 3H_2 \rightarrow 2NH_3 \qquad \qquad ...(4.10)$$

The production of gaseous N_2 and N_2O by micro-organisms and the evolution of these gases to the atmosphere completes the nitrogen cycle through a process called denitrification.

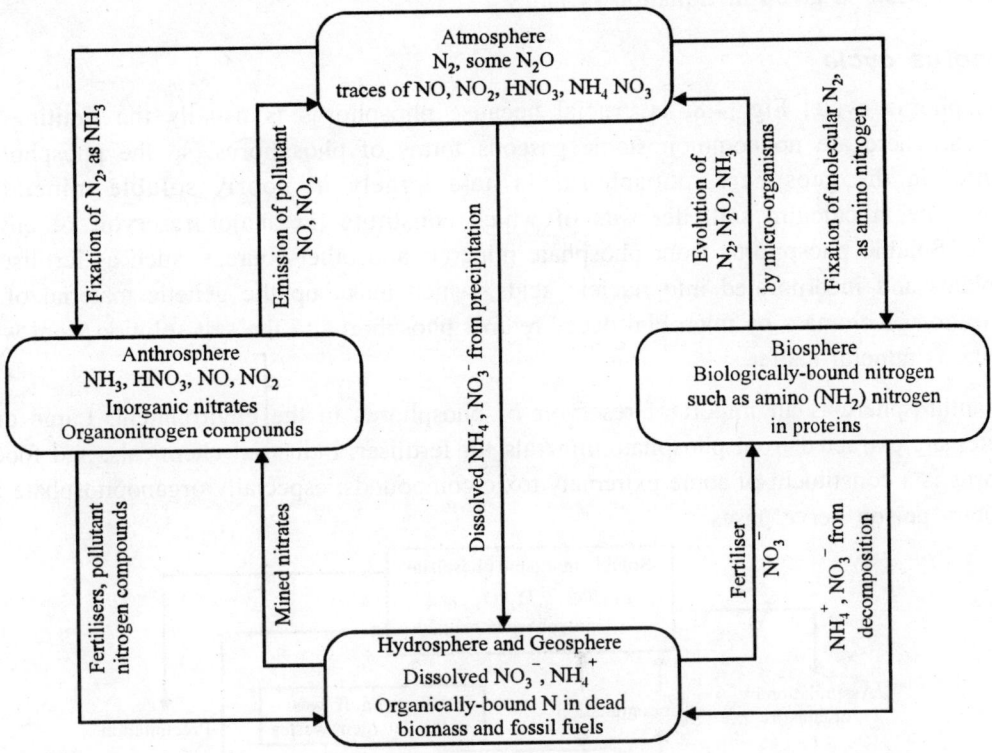

Fig. 4.7. The nitrogen cycle.

Oxygen cycle

The oxygen cycle involves the interchange of oxygen between the elemental form of gaseous O_2, contained in a huge reservoir in the atmosphere, and chemically bound O in CO_2, H_2O, and organic matter. It is strongly tied with other elemental cycles, particularly the carbon cycle. Elemental oxygen becomes chemically bound by various energy-yielding processes, particularly combustion and metabolic processes in organisms. It is released in photosynthesis. This element readily combines with, and oxidises other species, such as carbon in aerobic respiration (Equation 4.8) or carbon and hydrogen in the combustion of fossil fuels, such as methane:

$$CH_4 + 2O_2 \rightarrow CO_2 + 2H_2O \qquad \qquad ...(4.11)$$

Elemental oxygen also oxidises inorganic substances, such as iron(II) in minerals:

$$4FeO + O_2 \rightarrow 2Fe_2O_3 \qquad \qquad ...(4.12)$$

A particularly important aspect of the oxygen cycle is stratospheric ozone, O_3. It is relatively in small concentration of ozone in the stratosphere more than 10 kilometres high in the atmosphere filters out ultraviolet radiation in the wavelength range of 220–330 nm, thus protecting life on earth from the highly damaging effects of this radiation.

The oxygen cycle is completed when elemental oxygen is returned to the atmosphere. The only significant way in which this is done is through photosynthesis mediated by plants. The overall reaction for photosynthesis is given in Equation (4.7).

Phosphorus cycle

The phosphorus cycle, Fig. 4.8, is crucial because phosphorus is usually the limiting nutrient in ecosystems. There are no common stable gaseous forms of phosphorus, so the phosphorus cycle is endogenic. In the geosphere, phosphorus is held largely in poorly soluble minerals, such as hydroxyapatite, a calcium salt, deposits of which constitute the major reservoir of environmental phosphate. Soluble phosphorus from phosphate minerals and other sources, such as fertilisers, is taken up by plants and incorporated into nucleic acids, which make up the genetic material of organisms. Mineralisation of biomass by microbial decay returns phosphorus to the salt solution from which it may precipitate as mineral matter.

The anthrosphere is an important reservoir of phosphorus in the environment. Large quantities of phosphates are extracted from phosphate minerals for fertiliser, industrial chemicals, and food additives. Phosphorus is a constituent of some extremely toxic compounds, especially organophosphate insecticides and military poison nerve gases.

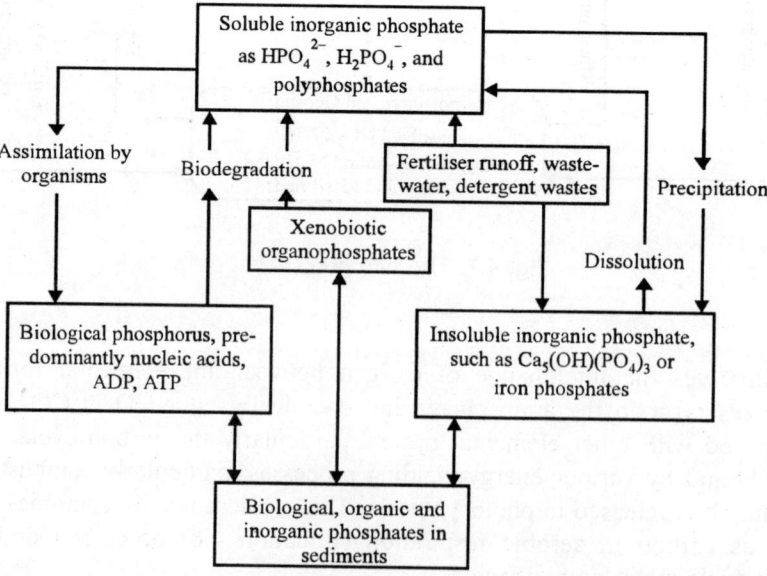

Fig. 4.8. The phosphorus cycle.

Sulphur cycle

The sulphur cycle shown in Fig. 4.9 illustrates the main features of a bio-geochemical cycle in a specific ecosystem. The cycling and reservoir pools, the chemical forms of the elements, and the organisms involved are shown in Fig. 4.9. Sulphate (SO_4) in the water is the principal available form that is reduced by autotrophic plants and incorporated into proteins, sulphur being an essential constituent of certain amino acids. When animals excrete, or the bodies of plants and animals are decomposed by heterotrophic microorganisms, sulphate may be returned to the water or hydrogen sulphide (H_2S) is released. Some of the H_2S is then reconverted to sulphate by specialised sulphur bacteria. Some of these bacteria are called chemosynthetic organisms, because they obtain their own energy from the chemical oxidation of inorganic compounds (in this case oxidation of sulphide to sulphur, and so on) instead of from light as do photosynthetic organisms or from organic matter as do heterotrophic organisms. The green sulphur bacteria are photosynthetic, but since H_2S is oxidised rather than H_2O, as in regualr photosynthesis, free oxygen is not released. Thus, it is relatively complex in that it involves several gaseous species, poorly soluble minerals, and several species in solution. It is tied with the oxygen cycle in that sulphur combines with oxygen to form gaseous sulphur dioxide, SO_2, an atmospheric pollutant, and soluble sulphate ion, SO_4^{2-}. Among the significant species involved in the sulphur cycle are gaseous hydrogen sulphide, H_2S; mineral sulphides, such as PbS; sulphuric acid, H_2SO_4, the main constituent of acid rain; and biologically bound sulphur in sulphur-containing proteins.

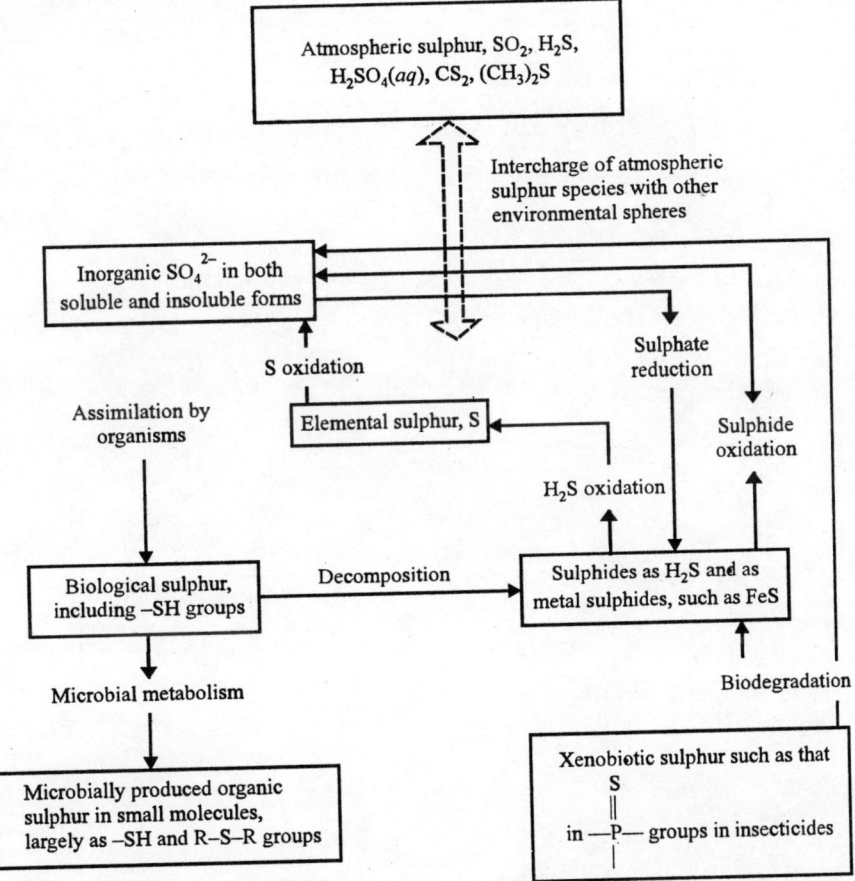

Fig. 4.9. The sulphur cycle.

Insofar as pollution is concerned, the most significant part of the sulphur cycle is the presence of pollutant SO_2 gas and H_2SO_4, in the atmosphere. The former is a somewhat toxic gaseous air pollutant evolved in the combustion of sulphur-containing fossil fuels. The major detrimental effect of sulphur dioxide in the atmosphere is its tendency to oxidise in the atmosphere to produce sulphuric acid.

Chapter 5

Fundamentals of Aquatic Chemistry (Aquatic Microbial Chemistry)

INTRODUCTION

To understand water pollution, it is first necessary to have an appreciation of chemical phenomena that occurs in water. The current chapter expands upon the fundamentals of aquatic chemistry with emphasis upon acid-base and complexation phenomena.

Aquatic environmental chemical phenomena involve processes like acid-base, solubility, oxidation-reduction, and complexation reactions. It is also important to keep in mind that kinetics–rates of reactions–are very important in aquatic chemistry. Biological processes play a key role in aquatic chemistry. For example, algae undergoing photosynthesis can raise the pH of water by removing aqueous CO_2, thereby converting HCO^{3-} ion to CO_3^{2-} ion; this ion in turn reacts with Ca^{2+} in water to precipitate $CaCO_3$.

The study of chemical processes in water is rarely easy. Even under carefully controlled condition in a laboratory, the investigation of chemical species in water can be very difficult. Generally, the media used in laboratory investigations are made up to constant ionic strength with a relatively inert electrolyte. This minimises effects arising from differences in factors such as activity coefficients or, when potentiometric measurements are made, liquid junction potentials at the solution-reference electrode interface. The temperature of the medium may be regulated to one hundredth of a degree, or even more accurately if necessary. If an equilibrium constant is to be measured, the system may be allowed to reach equilibrium over a very long period. However, the wide variation among values given for the same constants attests to the difficulty of describing even relatively simple chemical system under the most carefully controlled conditions.

Compared to the laboratory, it is much more difficult to describe chemical phenomena in natural water systems. Such systems are very complex and a description of their chemistry must take many variables into consideration. In addition to water, these systems contain mineral phases, gas phases, and organisms. As open, dynamic systems, they have variable inputs and outputs of energy and mass. Therefore, except under unusual circumstances, a true equilibrium condition is not obtained, although an approximately steady-state aquatic system frequently exists. Most metals found in natural waters do not exist as simple hydrated cations in the water, an oxyanions often are found as polynuclear species, rather than as simple monomers. The nature of chemical species in water containing bacteria or algae is strongly influenced by the action of these organism. Thus, an exact description of the chemistry of a natural water system based upon acid-base, solubility, and complexation equilibrium constants, redox potential, pH, and other chemical parameter is not possible. Therefore, the systems must be described by simplified models, often based around equilibrium chemical concepts. Though not exact, nor entirely realistic, such models can yield useful generalisation and insights pertaining to the nature of aquatic chemical processes and provide guidelines for the description and measurement of natural water systems. Though greatly simplified, such models are very helpful in visualising the conditions that determine chemical species and their reactions in natural waters and waste-waters.

Water Molecule

Some of the special characteristics of water include its polar character, tendency to form hydrogen bonds, and ability to hydrate metal ions. Water's properties can best be understood by considering the structure and bonding of the water molecule :

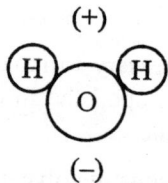

The water molecule is made up of two hydrogen atoms bonded to an oxygen atom. The three atoms are not in a straight line; instead, as shown above, they form an angle of 105°.

Because of water's bent structure and the fact that the oxygen atom attracts the negative electrons more strongly than do the hydrogen atoms, the water molecule behaves like a body having opposite electrical charges at either end or pole. Such a body is called dipole. Due to the fact that it has opposite charges at opposite ends, the water dipole may be attracted to either positively or negatively charged ions. For example, when NaCl dissolves in water to form positive Na^+ ions and negative Cl^- ions, the positive sodium ions are surrounded by water molecules with their negative ends pointed at the ions, and the chloride ions are surrounded by water molecules with their positive ends pointing at the negative ions, as shown in Fig. 5.1. This kind of attraction for ions is the reason why water dissolves many ionic compounds and salts that do not dissolve in other liquids.

In addition to being a polar molecule, the water molecule has another important property which gives it many of its special characteristics: the ability to form hydrogen bonds. Hydrogen bonds are a special type of bond that can form between the hydrogen in one water molecule and the oxygen in another water molecule. This bonding takes place because the oxygen has a partial negative charge and the hydrogen, a partial positive charge. Hydrogen bonds, shown in Fig. 5.2, as dashed lines, hold the water molecules together in large groups.

Hydrogen bonds also help to hold some solute molecules or ions in solution. This happens when hydrogen bonds form between the water molecules and hydrogen, nitrogen, or oxygen atoms on the solute molecule (Fig. 5.2). Hydrogen bonding also aids in retaining extremely small particles called colloidal particles in suspension in water.

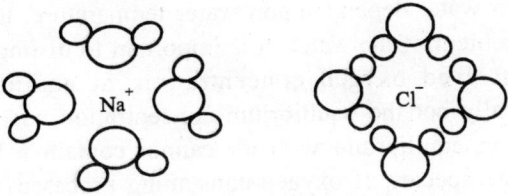

Fig. 5.1. Polar water molecules surrounding Na^+ ion (left) and Cl^- ion (right).

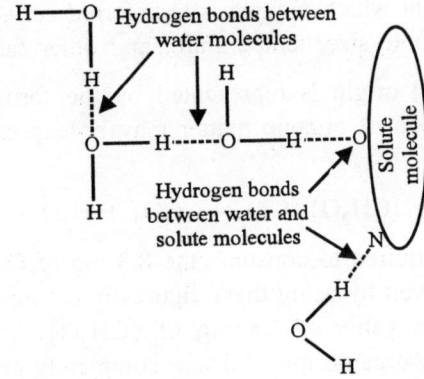

Fig. 5.2. Hydrogen bonding between water molecules and between water molecules and a solute molecule in solution.

GASES IN WATER

Dissolved gases—O_2 for fish and CO_2 for photosynthetic algae—are crucial to the welfare of living species in water. Some gases in water can also cause problems, such as the death of fish from bubbles of nitrogen formed in the blood of fish exposed to water supersaturated with N_2.

The solubilities of gases in water are calculated with *Henry's law*, which states that the solubility of a gas in a liquid is proportional to the partial pressure of that gas in contact with the liquid.

Oxygen in Water

Without an appreciable level of dissolved oxygen, many kinds of aquatic organisms cannot exist in water. Dissolved oxygen is consumed by the degradation of organic matter in water. Many fish kills are caused not from the direct toxicity of pollutant but from a deficiency of oxygen because of its consumption in the biodegradation of pollutants.

Most elemental oxygen comes from the atmosphere, which is 20.95% oxygen by volume of dry air. Therefore, the ability of a body of water to reoxygenate itself by contact with the atmosphere is an important characteristic. Oxygen is produced by the photosynthetic action of algae, but this process is really not an efficient means of oxygenating water because some of the oxygen formed by photosynthesis during the daylight hours is lost at night when the algae consume oxygen as part of their metabolic processes. When the algae die, the degradation of their biomass also consumes oxygen.

The solubility of oxygen in water depends upon water temperature, the partial pressure of oxygen in the atmosphere, and the salt content of the water. It is important to distinguish between oxygen solubility, which is the maximum dissolved oxygen concentration, at equilibrium, and dissolved oxygen concentration, which is generally not the equilibrium concentration and is limited by the rate at which oxygen dissolves. The water in equilibrium with air cannot contain a high level of dissolved oxygen compared to many other solute species. If oxygen-consuming processes are occurring in the water, the dissolved oxygen level may rapidly approach zero unless some efficient mechanism for the reaeration of water is operative, such as turbulent flow in a shallow stream or air pumped into the aeration tank of an activated sludge secondary waste treatment facility. The problem becomes largely one of kinetics, in which there is a limit to the rate at which oxygen is transferred across the air-water interface. This rate depends upon turbulence, air bubble size, temperature, and other factors.

If organic matter of biological origin is represented by the formula $\{CH_2O\}$, the consumption of oxygen in water by the degradation of organic matter may be expressed by the following biochemical reaction:

$$\{CH_2O\} + O_2 \rightarrow CO_2 + H_2O \qquad \qquad ...(5.1)$$

The weight of organic material required to consume the 8.3 mg of O_2 in a litre of water in equilibrium with the atmosphere at 25°C is given by using these figures in a simple stoichiometric calculation based on equation. 5.1, which yields a value of 7.8 mg of $\{CH_2O\}$. Thus, the microorganism-mediated degradation of only 7 or 8 mg of organic material can completely consume the oxygen in one litre of water initially saturated with air at 25°C. The depletion of oxygen to levels below those that will sustain aerobic organism requires the degradation of even less organic matter at higher temperatures (where the solubility of oxygen is less) or in water not initially saturated with atmospheric oxygen. Furthermore, there are no common aquatic chemical reactions that replenish dissolved oxygen; except for oxygen provided by photosynthesis, it must come from the atmosphere.

The temperature effect on the solubility of gases in water is especially important in the case of oxygen. The solubility of oxygen in water decreases from 14.74 mg/L at 0°C to 7.03 mg/L at 35°C. At higher temperatures, the decreased solubility of oxygen, combined with the increased respiration rate of aquatic organisms, frequently causes a condition in which a higher demand for oxygen accompanied by lower solubility of the gas in water results in severe oxygen depletion.

WATER ACIDITY AND CARBON DIOXIDE IN WATER

Acidity as applied to natural water and waste-water is the capacity of the water to neutralise OH^-; it is analogous to alkalinity, the capacity to neutralise H^+. Although virtually all water has some alkalinity, acidic water is not frequently encountered, except in cases of severe pollution. Acidity generally results from the presence of weak acids, particularly CO_2, but sometimes including others, such as $H_2PO_4^-$, H_2S, proteins, and fatty acids. Acidic metal ions, particularly Fe^{3+}, may also contribute to acidity.

From the pollution standpoint, strong acids are the most important contributors to acidity. The term free mineral acid is applied to strong acids such as H_2SO_4 and HCl in water. Acid mine water is a common water pollutant that contains an appreciable concentration of free mineral acid. Whereas total acidity is determined by titration with base to the phenolphthalein endpoint (pH 8.2), free mineral acid is determined by titration with base to the methyl orange endpoint (pH 4.3).

The acidic character of some hydrated metal ions may contribute to acidity:

$$Al(H_2O)_6^{3+} \rightleftharpoons Al(H_2O)_5OH^{2+} + H^+ \qquad \qquad ...(5.2)$$

Some industrial wastes, such as spent steel pickling liquor, contain acidic metal ions and often some excess strong acid. The acidity of such wastes must be measured in calculating the amount of lime, or other chemicals, that must be added to neutralise the acid.

Carbon Dioxide in Water

The most important weak acid in water is carbon dioxide, CO_2. Because of the presence of carbon dioxide in air and its production from microbial decay of organic matter, dissolved CO_2 is present in virtually all natural waters and waste-waters. Rainfall from even an absolutely unpolluted atmosphere is slightly acidic due to the presence of dissolved CO_2. Carbon dioxide, and its ionisation products, bicarbonate ion (HCO_3^-), and carbonate ion (CO_3^{2-}), have an extremely important influence upon the chemistry of water. Many minerals are deposited as salts of the carbonate ions. Algae in water utilise dissolved CO_2 in the synthesis of biomass. The equilibrium of dissolved CO_2 with gaseous carbon dioxide in the atmosphere,

$$CO_2 \text{ (water)} \rightleftharpoons CO_2 \text{ (atmosphere)} \qquad \qquad ...(5.3)$$

and equilibrium of CO_3^{2-} ion between aquatic solution and solid carbonate minerals,

$$MCO_3 \text{ (slightly soluble carbonate salt)} \rightleftharpoons M^{2+} + CO_3^{2-} \qquad ...(5.4)$$

have a strong buffering effect upon the pH of water.

Carbon dioxide is only about 0.035% by volume of normal dry air. As a consequence of the low level of atmospheric CO_2, water totally lacking in alkalinity (capacity to neutralise H^+) in equilibrium with the atmosphere contains only a very low level of carbon dioxide. However, the formation of HCO_3^- and CO_3^{2-} greatly increases the solubility of carbon dioxide. High concentrations of free carbon dioxide in water may adversely affect respiration and gas exchange of aquatic animals. It may even cause death and should not exceed level of 25 mg/L in water.

A large share of the carbon dioxide found in water is a product of the break-down of organic matter by bacteria. Even algae, which utilise CO_2 in photosynthesis, produce it through their metabolic processes in the absence of light. As water seeps through layers of decaying organic matter while infiltrating the ground, it may dissolve a great deal of CO_2 produced by the respiration of organisms in the soil. Later, as water goes through limestone formations, it dissolves calcium carbonate because of the presence of the dissolved CO_2;

$$CaCO_3(s) + CO_2(aq) + H_2O \rightleftharpoons Ca^{2+} + 2HCO_3^- \qquad \qquad ...(5.5)$$

This process is the one by which limestone caves are formed. The implications of the above reaction for aquatic chemistry are discussed in greater detail in further section of this chapter.

The concentration of gaseous CO_2 in the atmosphere varies with location and season; it is increasing by about 1 part per million (ppm) by volume per year. For purposes of calculation here, the concentration of atmospheric CO_2 will be taken as 350 ppm (0.0350%) in dry air. At 25°C, water in equilibrium with unpolluted air containing 350 ppm carbon dioxide has a $CO_2(aq)$ concentration of 1.146×10^{-5} M and this value will be used for subsequent calculations.

Although CO_2 in water is often represented as H_2CO_3, the equilibrium constant for the reaction

$$CO_2(aq) + H_2O \rightleftharpoons H_2CO_3 \qquad ...(5.6)$$

is only around 2×10^{-3} at 25°C, so just a small fraction of the dissolved carbon dioxide is actually present as H_2CO_3. In this text, non-ionised carbon dioxide in water will be designated simply as CO_2, which in subsequent discussions will stand for the total of dissolved molecular CO_2 and undissociated H_2CO_3.

The CO_2, HCO_3^-, CO_3^{2-} system in water may be described by the equations,

$$CO_2 + H_2O \rightleftharpoons HCO_3^- + H^+ \qquad ...(5.7)$$

$$K_{a1} = \frac{[H^+][HCO_3^-]}{[CO_2]} = 4.45 \times 10^{-7} \qquad pK_{a1} = 6.35 \qquad ...(5.8)$$

$$HCO_3^- \rightleftharpoons CO_3^{2-} + H^+ \qquad ...(5.9)$$

$$K_{a2} = \frac{[H^+][CO_3^{2-}]}{[HCO_3^-]} = 4.69 \times 10^{-11} \qquad pK_{a2} = 10.33 \qquad ...(5.10)$$

where $pK_a = -\log K_a$. The predominant species formed by CO_2 dissolved in water depends upon pH. This is best shown by a distribution of species diagram with pH as a master variable as illustrated in Fig. 5.3. Such a diagram shows the major species present in solution as a function of pH. For CO_2 in aqueous solution, the diagram is a series of plots of the fractions present as CO_2, HCO_3^-, and CO_3^{2-} as a function of pH. These fractions, designated as α_x, are given by the following expressions:

$$\alpha_{CO_2} = \frac{[CO_2]}{[CO_2] + [HCO_3^-] + [CO_3^{2-}]} \qquad ...(5.11)$$

$$\alpha_{HCO_3^-} = \frac{[HCO_3^-]}{[CO_2] + [HCO_3^-] + [CO_3^{2-}]} \qquad ...(5.12)$$

$$\alpha_{CO_3^{2-}} = \frac{[CO_3^{2-}]}{[CO_2] + [HCO_3^-] + [CO_3^{2-}]} \qquad ...(5.13)$$

Substitution of the expressions for K_{a1} and K_{a2} into the α expressions gives the fractions of species as a functions of acid dissociation constants and hydrogen ion concentration:

$$\alpha_{CO_2} = \frac{[H^+]^2}{[H^+]^2 + k_{a1}^2[H^+] + K_{a1}K_{a2}} \qquad ...(5.14)$$

$$\alpha_{HCO_3^-} = \frac{K_{a1}[H^+]}{[H^+]^2 + k_{a1}[H^+] + K_{a1}K_{a2}} \qquad ...(5.15)$$

$$\alpha_{CO_3^{2-}} = \frac{K_{a1}K_{a2}}{[H^+]^2 + K_{a1}[H^+] + K_{a1}K_{a2}} \qquad ...(5.16)$$

Calculations from these expressions show the following :

1. For pH significantly below pK_{a1}, α_{CO_2} is essentially 1.
2. When pH = pK_{a1}, α_{CO_2} = $\alpha_{HCO_3^-}$.
3. When pH = $^{1/2}(pK_{a1} + pK_{a2})$, $\alpha_{HCO_3^-}$ is at its maximum value of 0.98.
4. When p^H = pK_{a2}, $\alpha_{HCO_3^-}$ = $\alpha_{CO_3^{2-}}$.
5. For pH significantly above pK_{a2}, $\alpha_{CO_3^{2-}}$ is essentially 1.

The distribution of species diagram in Fig. 5.3, shows that hydrogen carbonate (bicarbonate) ion (HCO^{3-}) is the predominant species in the pH range found in most waters, with CO_2 predominating in more acidic waters.

As mentioned above, the value of $[CO_2(aq)]$ in water at 25° C in equilibrium with air that is 350 ppm CO_2 is 1.146×10^{-5} M. The carbon dioxide dissociates partially in water to produce equal concentrations of H^+ and HCO_3^-;

$$CO_2 + H_2O \rightleftharpoons HCO_3^- + H^+ \qquad \qquad ...(5.17)$$

The concentrations of H^+ and HCO_3^- are calculated from K_{a1} :

$$K_{a1} = \frac{[H^+][HCO_3^-]}{[CO_2]} = \frac{[H^+]^2}{1.146 \times 10^{-5}} = 4.45 \times 10^{-7} \qquad \qquad ...(5.18)$$

$$[H^+] = [HCO_3^-] = (1.146 \times 10^{-5} \times 4.45 \times 10^{-7})^{1/2} = 2.25 \times 10^{-6}$$

$$pH = 5.65$$

This calculation explains why pure water that has equilibrated with the unpolluted atmosphere is slightly acidic with a pH somewhat less than 7.

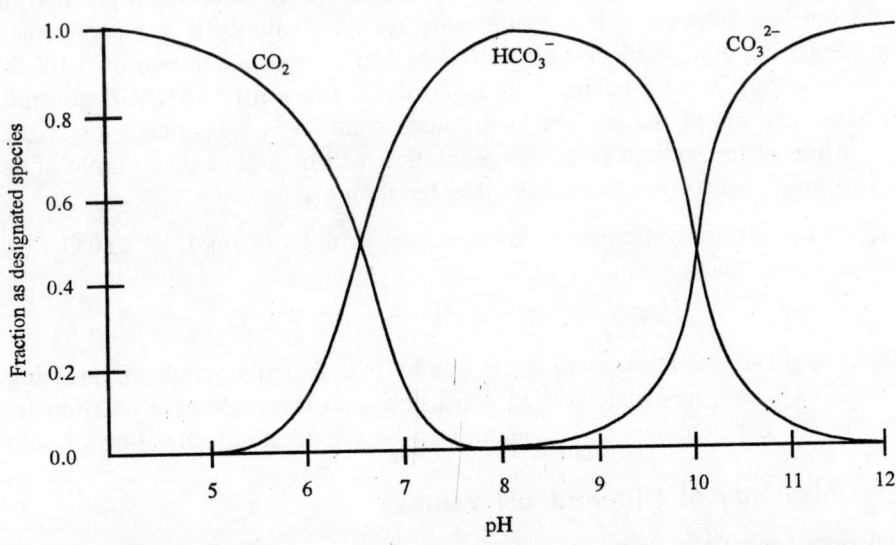

Fig. 5.3. Distribution of species diagram for the CO_2, HCO_3^-, CO_3^{2-} system in water.

ALKALINITY

The capacity of water to accept H^+ ions (protons) is called alkalinity. Alkalinity is important in water treatment and in the chemistry and biology of natural water. Frequently, the alkalinity of water must be known to calculate the quantities of the chemicals to be added in treating the water. Highly alkaline water often has a high pH and generally contains elevated levels of dissolved solids. These characteristics may be detrimental for water to be used in boilers, food processing, and municipal water systems. Alkalinity serves as a pH buffer and reservoir for inorganic carbon, thus helping to determine the ability of a water to support algal growth and other aquatic life, so it can be used as a measure of water fertility. Generally, the basic species responsible for alkalinity in water are bicarbonate ion, carbonate ion, and hydroxide ion:

$$HCO_3^- + H^+ \rightarrow CO_2 + H_2O \qquad \qquad ...(5.19)$$

$$CO_3^{2-} + H^+ \rightarrow HCO_3^- \qquad \qquad ...(5.20)$$

$$OH^- + H^+ \rightarrow H_2O \qquad \qquad ...(5.21)$$

Other, usually minor, contributors to alkalinity are ammonia and the conjugate bases of phosphoric, silicic, boric, and organic acids.

At pH values below 7, $[H^+]$ in water detracts significantly from alkalinity, and its concentration must be subtracted in computing the total alkalinity. Therefore, the following equation is the complete equation for alkalinity in a medium where the only contributors to it are HCO_3^-, CO_3^{2-}, and OH^-:

$$[alk] = [HCO_3^-] + 2\,[CO_3^{2-}] + [OH]^- - [H^+] \qquad \qquad ...(5.22)$$

Alkalinity generally is expressed as phenolphthalein alkalinity, corresponding to titration with acid to the pH at which HCO_3^- is the predominant carbonate species (pH 8.3), or total alkalinity, corresponding to titration with acid to the methyl orange endpoint (pH 4.3), where both bicarbonate and carbonate species have been converted to CO_2.

It is important to distinguish between high basicity, manifested by an elevated pH, and high alkalinity, the capacity to accept H^+. Whereas pH is an intensity factor, alkalinity is a capacity factor. This may be illustrated by comparing a solution of 1.00×10^{-3} M NaOH with a solution of 0.100 M HCO_3^-. The sodium hydroxide solution is quite basic, with a pH of 11, but a litre of it will neutralise only 1.00×10^{-3} mole of acid. The pH of the sodium bicarbonate solution is 8.34, much lower than that of the NaOH. However, a litre of the sodium bicarbonate solution will neutralise 0.100 mole of acid; therefore, its alkalinity is 100 times that of the more basic NaOH solution.

In engineering terms, alkalinity frequently is expressed in units of mg/L of $CaCO_3$, based upon the following acid-neutralising reaction:

$$CaCO_3 + 2H^+ \rightarrow Ca^{2+} + CO_2 + H_2O \qquad \qquad ...(5.23)$$

The equivalent weight of calcium carbonate is one-half its formula weight. Expressing alkalinity in terms of mg/L of $CaCO_3$ can, however, lead to confusion, and the preferable notation for the chemist is equivalents/L, the number of moles of H^+ neutralised by the alkalinity in a litre of solution.

Contributors to Alkalinity at Different pH Values

Natural water typically has an alkalinity, designated here as "[alk]," of 1.00×10^{-3} equivalents per litre (eq/L) meaning that the alkaline solutes in 1 litre of the water will neutralise 1.00×10^{-3} moles of acid.

The contributions made by different species to alkalinity depend upon pH. This is shown here by calculation of the relative contributions to alkalinity of HCO_3^-, CO_3^{2-}, and OH^- at pH 7.00 and pH 10.00. First, for water at pH 7.00, $[OH^-]$ is too low to make any significant contribution to the alkalinity. Furthermore, as shown by Fig. 5.3, at pH 7.00 $[HCO_3^-] >> [CO_3^{2-}]$. Therefore, the alkalinity is due to HCO_3^- and $[HCO_3^-] = 1.00 \times 10^{-3}$ M. Substitution into the expression for K_{a1} shows that at pH 7.00 and $[HCO_3^-] = 1.00 \times 10^{-3}$ M, the value of $[CO_2(aq)]$ is 2.25×10^{-4} M, somewhat higher than the value that arises from water in equilibrium with atmospheric air, but readily reached due to the presence of carbon dioxide from bacterial decay in water and sediments.

Consider next the case of water with the same alkalinity, 1.00×10^{-3} eq/L that has a pH of 10.00. At this higher pH both OH^- and CO_3^{2-} are present at significant concentrations compared to HCO_3^- and the following may be calculated:

$$[alk] = [HCO_3^-] + 2[CO_3^{2-}] + [OH^-] = 1.00 \times 10^{-3} \qquad ...(5.24)$$

The concentration of CO_3^{2-} is multiplied by 2 because each CO_3^{2-} ion can neutralise $2H^+$ ions. The other two equations that must be solved to get the concentrations of HCO_3^-, CO_3^{2-} and OH^- are

$$[OH^-] = \frac{K_w}{[H^+]} = \frac{1.00 \times 10^{-14}}{1.00 \times 10^{-10}} = 1.00 \times 10^{-4} \qquad ...(5.25)$$

and

$$[CO_3^{2-}] = \frac{K_{a2}[HCO_3^-]}{[H^+]} \qquad ...(5.26)$$

Solving these three equations gives $[HCO_3^-] = 4.64 \times 10^{-4}$ M and $[CO_3^{2-}] = 2.18 \times 10^{-4}$ M. Therefore, the contributions to the alkalinity of this solution are the following:

$$4.64 \times 10^{-4} \text{ eq/L from } HCO_3^-$$

$$2 \times 2.18 \times 10^{-4} = 4.36 \times 10^{-4} \text{ eq/L from } CO_3^{2-}$$

$$1.00 \times 10^{-4} \text{ eq/L from } OH^-$$

$$\overline{alk = 1.00 \times 10^{-3} \text{ eq/L}}$$

Dissolved Inorganic Carbon and Alkalinity

The values given above can be used to show that at the same alkalinity value the concentration of total dissolved inorganic carbon, $[C]$,

$$[C] = [CO_2] + [HCO_3^-] + [CO_3^{2-}] \qquad ...(5.27)$$

varies with pH. At pH 7.00,

$$[C]_{pH7} = 2.25 \times 10^{-4} + 1.00 \times 10^{-3} + 0 = 1.22 \times 10^{-3} \qquad ...(5.28)$$

whereas at pH 10.00,

$$[C]_{pH10} = 0 + 2.18 \times 10^{-4} + 4.64 \times 10^{-4} = 6.82 \times 10^{-4} \qquad ...(5.29)$$

The calculation above shows that the dissolved inorganic carbon concentration at pH 10.00 is only about half that at pH 7.00. This is because at pH 10 major contributions to alkalinity are made by CO_3^{2-} ion, each of which has twice the alkalinity of each HCO_3^- ion, and by OH^-, which does not contain any carbon. The lower inorganic carbon concentration at pH 10 shows that the aquatic system can donate dissolved inorganic carbon for use in photosynthesis with a change in pH but none in alkalinity. This pH dependent difference in dissolved inorganic carbon concentration represents a significant potential source of carbon for algae growing in water which fix carbon by the overall reactions

$$CO_2 + H_2O + h\nu \rightarrow \{CH_2O\} + O_2 \qquad \text{...(5.30)}$$

and

$$HCO_3^- + H_2O + h\nu \rightarrow \{CH_2O\} + OH^- + O_2 \qquad \text{..(5.31)}$$

As dissolved inorganic carbon is used up to synthesise biomass, $\{CH_2O\}$, the water becomes more basic. The amount of inorganic carbon that can be consumed before the water becomes too basic to allow algal reproduction is proportional to the alkalinity. In going from pH 7.00 to pH 10.00 the amount of inorganic carbon consumed from 1.00 L of water having an alkalinity of 1.00×10^{-3} eq/L is

$$[C]_{pH7} \times 1\ L - [C]_{pH10} \times 1L =$$
$$1.22 \times 10^{-3}\ \text{mol} - 6.82 \times 10^{-4}\ \text{mol} = 5.4 \times 10^{-4}\ \text{mol} \qquad \text{...(5.32)}$$

This translates to an increase of 5.4×10^{-4} mol/L of biomass. Since the formula mass of $\{CH_2O\}$ is 30, the weight of biomass produced amounts to 16 mg/L. Assuming no input of additional CO_2, at higher alkalinity more biomass is produced for the same change in pH, whereas at lower alkalinity less is produced. Because of this effect, biologists use alkalinity as a measure of water fertility.

Influence of Alkalinity on CO_2 Solubility

The increased solubility of carbon dioxide in water with an elevated alkalinity can be illustrated by comparing its solubility in pure water (alkalinity 0) to its solubility in water initially containing 1.00×10^{-3} M NaOH (alkalinity 1.00×10^{-3} eq/L). The number of moles of CO_2 that will dissolve in a litre of pure water from the atmosphere containing 350 ppm carbon dioxide is

$$\text{Solubility} = [CO_2(aq)] + [HCO_3^-] \qquad \text{...(5.33)}$$

Substituting values calculated above gives

$$\text{Solubility} = 1.146 \times 10^{-5} + 2.25 \times 10^{-6} = 1.371 \times 10^{-5}\ \text{M}$$

The solubility of CO_2 in water initially 1.00×10^{-3} M in NaOH is about 100-fold higher because of uptake of CO_2 by the reaction

$$CO_2(aq) + OH^- \rightleftharpoons HCO_3^- \qquad \text{...(5.34)}$$

so that

$$\text{Solubility} = [CO_2(aq)] + [HCO_3^-]$$
$$= 1.146 \times 10^{-5} + 1.00 \times 10^{-3} = 1.01 \times 10^{-3}\ \text{M} \qquad \text{...(5.35)}$$

CALCIUM AND OTHER METALS IN WATER

Metal ions in water, commonly denoted M^{n+}, exist in numerous forms. A bare metal ion, Ca^{2+} for example, cannot exist as a separate entity in water. In order to secure the highest stability of their outer electron shells, metal ions in water are bonded, or coordinated, to other species. These may be water molecules or other stronger bases (electron-donor partners) that might be present. Therefore, metal ions in water solution are present in forms such as the hydrated metal cation $M(H_2O)_x^{n+}$. Metal ions in aqueous solution seek to reach a state of maximum stability through chemical reactions including acid-base.

$$Fe(H_2O)_6^{3+} \rightleftharpoons FeOH(H_2O)_5^{2+} + H^+ \qquad \qquad ...(5.36)$$

precipitation,

$$Fe(H_2O)_6^{3+} \rightleftharpoons Fe(OH)_3(s) + 3H_2O + 3H^+ \qquad \qquad ...(5.37)$$

and oxidation-reduction reactions :

$$Fe(H_2O)_6^{2+} \rightleftharpoons Fe(OH)_3(s) + 3H_2O + e^- + 3H^+ \qquad \qquad ...(5.38)$$

These all provide means through which metal ions in water are transformed to more stable forms. Because of reactions such as these and the formation of dimeric species, such as $Fe_2(OH)_2^{4+}$, the concentration of simple hydrated $Fe(H_2O)_6^{3+}$ ion in water is vanishingly small; the same holds true for many other hydrated metal ions dissolved in water.

Behaviour of Metal Ions in Water

The properties of metal dissolved in water depend largely upon the nature of metal species dissolved in the water. Therefore, *speciation* of metal plays a crucial role in their environmental chemistry in natural waters and waste-waters. In addition to the hydrated metal ions, for example, $Fe(H_2O)_6^{3+}$ and hydroxy species such as $FeOH(H_2O)_5^{2+}$ discussed above, metals may exist in water reversibly bound to inorganic anions or to organic compounds as metal complexes, or they may be present as *organometallic* compounds containing carbon-to-metal bonds. The solubilities, transport properties, and biological effects of such species are often vastly different from those of the metal ions themselves. Subsequent sections of this chapter consider metal species with an emphasis upon metal complexes. Special attention is given to *chelation*, in which particularly strong metal complexes are formed.

Hydrated Metal Ions as Acids

Hydrated metal ions, particularly those with a charge of +3 or more, tend to lose H^+ ion from the water molecules bound to them in aqueous solution, and fit the definition of *Bronsted acids*. (According to the Bronsted definition, acids are H^+ donors and bases are H^+ acceptors). The acidity of a metal ion increases with charge and decreases with increasing radius. As shown by the reaction,

$$Fe(H_2O)_6^{3+} \rightleftharpoons Fe(H_2O)_5OH^{2+} + H^+ \qquad \qquad ...(5.39)$$

hydrated iron ion is a relatively strong acid, with K_{a1} of 8.9×10^{-4}. Hydrated trivalent metal ions, such as iron, generally are minus at least one hydrogen ion at neutral pH values or above. For tetravalent metal ions, the completely protonated forms, $M(H_2O)_x^{4+}$, are rare even at very low pH values. Commonly, O^{2-} is coordinated to tetravalent metal ions; an example is the vanadium species, VO^{2+}. Generally, divalent metal ions do not lose a hydrogen ion at pH values below 6, whereas monovalent metal ions such as Na^+ do not act as acids at all, and exist in water solution as simple hydrated ions.

The tendency of hydrated metal ions to behave as acids may have a profound effect upon the aquatic environment. A good example is acid mine water, which derives part of its acidic character from the character of hydrated iron :

$$Fe(H_2O)_6^{3+} \rightleftharpoons Fe(OH)_3(s) + 3H^+ + 3H_2O \qquad \qquad ...(5.40)$$

Hydroxide, OH^-, bonded to a metal ion, may function as a bridging group to join two or more metal together through the following dehydration-dimerisation:

$$2Fe(H_2O)_5OH^{2+} \rightarrow (H_2O)_4\ Fe \underset{\underset{H}{O}}{\overset{\overset{H}{O}}{\diamond}} Fe(H_2O)_4{}^{4+} + 2H_2O \qquad ...(5.41)$$

Among the metals other than iron(III) forming polymeric species with OH^- as a bridging group are Al, Be, Bi, Ce, Co, Cu, Ga, Mo, Pb, Sc, Sn, and U. Additional hydrogen ions may be lost from water molecules bonded to the dimers, furnishing OH^- groups for further bonding and leading to the formation of polymeric hydrolytic species. If the process continues, colloidal hydroxy polymers are formed, and finally precipitates are produced. This process is thought to be the general one by which hydrated iron (III) oxide, $Fe_2O_3.x(H_2O)$, [also called ferric hydroxide, $Fe(OH)_3$] is precipitated from solutions containing iron (III).

Calcium in Water

Of the cations found in most fresh-water systems, calcium generally has the highest concentration. The chemistry of calcium, although complicated enough, is simpler than that of the transition metal ions found in water. Calcium is a key element in many academical processes, and minerals constitute the primary sources of calcium ion in waters. Among the primary contribution minerals are gypsum, $CaSO_4.2H_2O$; anhydrite, $CaSO_4$; dolomite, $CaMg(CO_3)_2$; and calcite and aragonite, which are different mineral forms of $CaCO_3$.

Calcium ion, along with magnesium and sometimes iron ion, accounts for water hardness. The most common manifestation of water hardness is the curdy precipitate formed by soap in hard water. Temporary hardness is due to the presence of calcium and bicarbonate ion in water and may be eliminated by boiling the water :

$$Ca^{2+} + 2HCO_3^- \rightleftharpoons CaCO_3(s) + CO_2(g) + H_2O \qquad ...(5.42)$$

Increased temperature may force this reaction to the right by evolving CO_2 gas, and a white precipitate of calcium carbonate may form in boiling water having temporary hardness. Water containing a high level of carbon dioxide readily dissolves calcium from its carbonate minerals :

$$CaCO_3(s) + CO_2(aq) + H_2O \rightleftharpoons Ca^{2+} + 2HCO_3^- \qquad ...(5.43)$$

When this reaction is reversed and CO_2 is lost from the water, calcium carbonate deposits are formed. The concentration of CO_2 in water determines the extent of dissolution of calcium carbonate. The carbon dioxide that water may gain by equilibration with the atmosphere is not sufficient to account for the levels of calcium dissolved in natural waters, especially groundwaters. Rather, the respiration of micro-organisms degrading organic matter in water, sediments, and soil,

$$\{CH_2O\} + O_2 \rightarrow CO_2 + H_2O \qquad ...(5.44)$$

accounts for the very high levels of CO_2 and HCO_3^- observed in water. This is an extremely important factor in aquatic chemical processes and geochemical transformations.

Dissolved Carbon Dioxide and Calcium Carbonate Minerals

The equilibrium between dissolved carbon dioxide and calcium carbonate minerals is important in determining several natural water chemistry parameters such as alkalinity, pH, and dissolved calcium concentration (Fig. 5.4). For freshwater, the typical figures quoted for the concentrations of both HCO_3^- and Ca^{2+} are 1.00×10^{-3} M. It may be shown that these are reasonable values when the water is in equilibrium with limestone, $CaCO_3$, and with atmospheric CO_2. The concentration of CO_2 in water in equilibrium with air has already been calculated as 1.146×10^{-5} M. The other constants needed to calculate $[HCO_3^-]$ and $[Ca^{2+}]$ are the acid dissociation constants for CO_2:

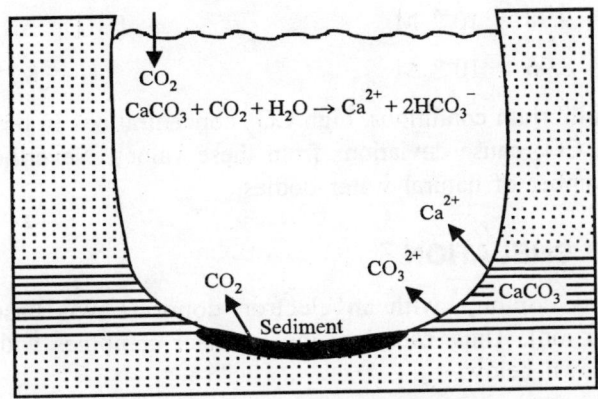

Fig. 5.4. Carbon dioxide-calcium carbonate equilibria.

$$K_{a1} = \frac{[H^+][HCO_3^-]}{[CO_2]} = 4.45 \times 10^{-7} \qquad \qquad ...(5.45)$$

the acid dissociation constant of HCO_3^-:

$$K_{a2} = \frac{[H^+][CO_3^{2-}]}{[HCO_3^-]} = 4.69 \times 10^{-11} \qquad \qquad ...(5.46)$$

and the solubility product of calcium carbonate (calcite):

$$K_{sp} = [Ca^{2+}][CO_3^{2-}] = 4.47 \times 10^{-9} \qquad \qquad ...(5.47)$$

The reaction between calcium carbonate and dissolved CO_2 is

$$CaCO_3(s) + CO_2(aq) + H_2O \rightleftharpoons Ca^{2+} + 2HCO_3^- \qquad \qquad ...(5.48)$$

for which the equilibrium expression is the following:

$$K' = \frac{[Ca^{2+}][CO_3^{2-}]^2}{[CO_2]} = \frac{K_{sp}K_{a1}}{K_{a2}} = 4.24 \times 10^{-5} \qquad \qquad ...(5.49)$$

The stoichiometry of Reaction 5.48 gives a bicarbonate ion concentration that is twice that of calcium. Substitution of 1.146×10^{-5} for $[CO_2]$ into the expression for K' yields values of 4.99×10^{-4} M for $[Ca^{2+}]$ and 9.98×10^{-4} M for $[HCO_3^-]$. Substitution into the expression for K_{sp} yields 8.96×10^{-6} M for $[CO_3^{2-}]$. When known concentrations are substituted into the product $K_{a1}K_{a2}$,

$$K_{a1}K_{a2} = \frac{[H^+]^2[CO_3^{2-}]}{[CO_2]} = 2.09 \times 10^{-17} \qquad \qquad ...(5.50)$$

a value of 5.17×10^{-9} M is obtained for $[H^+]$ (pH 8.29). The alkalinity is essentially equal to the bicarbonate ion concentration, which is much higher than that of CO_3^{2-} or OH^-.

To summarise, for water in equilibrium with solid calcium carbonate and atmospheric CO_2, the following concentrations are calculated:

$$[CO_2] \quad = 1.146 \times 10^{-5} \text{ M} \qquad\qquad [Ca^{2+}] = 4.99 \times 10^{-4} \text{ M}$$

$$[HCO_3^-] = 9.98 \times 10^{-4} \text{ M} \qquad\qquad [H^+] = 5.17 \times 10^{-9} \text{ M}$$

$$[CO_3^{2-}] = 8.96 \times 10^{-6} \text{ M} \qquad\qquad\qquad pH = 8.29$$

Factors such as non-equilibrium conditions, high CO_2 concentrations in bottom regions, and increased pH due to algal uptake of CO_2 cause deviations from these values. Nevertheless, they are close to the values found in a large number of natural water bodies.

COMPLEXATION AND CHELATION

A metal ion in water may combine with an electron donor (Lewis base) to form a complex or coordination compound (or ion). Thus, cadmium ion in water combines with a ligand, cyanide ion, to form a complex ion as shown below:

$$Cd^{2+} + CN^- \;\rightleftharpoons\; CdCN^+ \qquad\qquad ...(5.51)$$

Additional cyanide ligands may be added to form the progressively weaker (more easily dissociated) complexes $Cd(CN)_2$, $Cd(CN)_3^-$, and $Cd(CN)_4^{2-}$.

In this example, the cyanide ion is a unidentified ligand, which means that it possesses only one site that bonds to the cadmium metal ion. Complexes of unidentate ligands are of relatively little importance in solution in natural waters. Of considerably more importance are complexes with chelating agents. A chelating agent has more than one atom that may be bonded to a central metal ion at one time to form a ring structure. Thus, pyrophosphate ion, $P_2O_7^{4-}$, bonds to two sites on calcium ion to form a chelate:

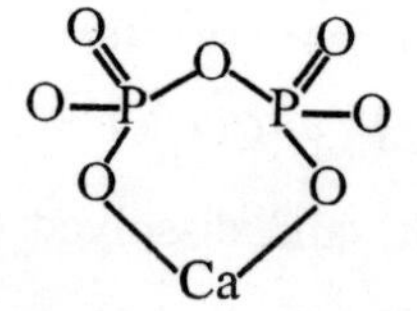

In general, since a chelating agent may bond to a metal ion in more than one place simultaneously, chelates are more stable than complexes involving unidentate ligands. Stability tends to increase with number of chelating sites available on the ligand.

Structures of metal chelates take a number of different forms, all characterised by rings in various configurations. The structure of a tetrahedrally coordinated chelate of nitrilotriacetate ion is shown in Fig. 5.5. The ligands found in natural waters and waste-waters contain a variety of functional groups which can donate the electrons required to bond the ligand to a metal ion. Among the most common of these groups are :

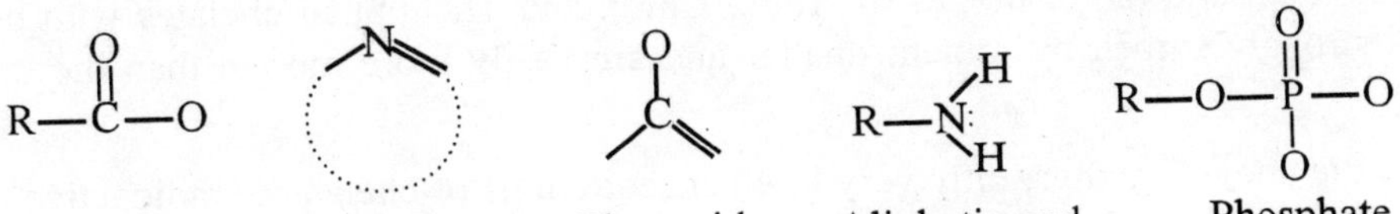

Carboxylate   Heterocyclic   Phenoxide   Aliphatic and   Phosphate
               nitrogen   aromatic amino

These ligands complex most metal ions found in unpolluted waters and biological systems (Mg^{2+}, Ca^{2+}, Mn^{2+}, Fe^{2+}, Fe^{3+}, Cu^{2+}, Zn^{2+}, VO^{2+}). They also bind to contaminant metal ions such as Co^{2+}, Ni^{2+}, Sr^{2+}, Cd^{2+}, and Ba^{2+}.

Complexation may have a number of effects, including reaction of both ligands and metals, Among the ligand reactions are oxidation-reduction, decarboxylation, and hydrolysis. Complexation may cause changes in the oxidation state of the metal and may result in a metal becoming solubilised from an insoluble compound. The formation of insoluble complex compounds removes metal ions from solution.

Complex compounds of metal such as iron (in haemoglobin) and magnesium (in chlorophyll) are vital to life process. Naturally occurring chelating agents, such as humic substances and amino acids, are found in water and soil.

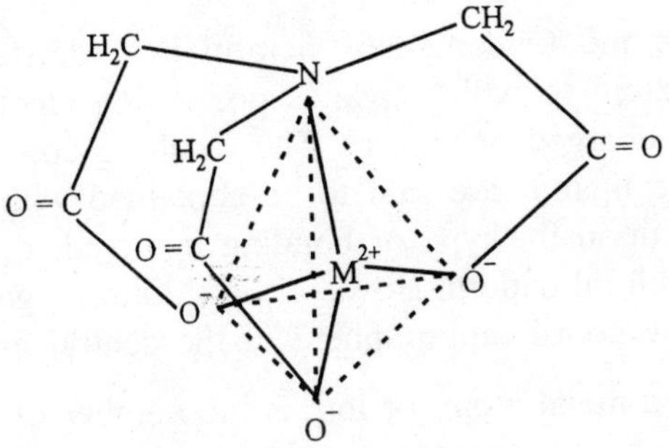

Fig. 5.5. Nitrilotriacetate chelate of a divalent metal ion in a tetrahedral configuration.

The high concentration of chloride ion in seawater results in the formation of some chloro complexes. Synthetic chelating agents such as sodium tripolyphosphate, sodium ethylenediaminetetraacetate (EDTA), sodium nitrilotriacetate (NTA), and sodium citrate are produced in large quantities for use in metal-plating baths, industrial water treatment, detergent formulations, and food preparation. Small quantities of these compounds enter aquatic systems through waste discharges.

Occurrence and Importance of Chelating Agents in Water

Chelating agents are common potential water pollutants. These substances can occur in sewage effluent and industrial waste-water such as metal plating waste-water. Chelates formed by the strong chelating agent ethylenediaminetetraacetate have been shown to greatly increase the migration rate of radioactive ^{60}Co from pits and trenches used by the Oak Ridge National Laboratory in Oak Ridge, Tennessee for disposal of intermediate-level radioactive waste. EDTA was used as a cleaning and solubilising agent for the decontamination of hot cells, equipment, and reactor components. Analysis of water from sample wells in the disposal pits showed EDTA concentration of 3.4×10^{-7} M. The presence of EDTA 12–15 years after its burial attests to its low rate of biodegradation. In addition to cobalt, EDTA strongly chelates radioactive plutonium and radioisotopes of Am^{3+}, Cm^{3+}, and Th^{4+}. Such chelates with negative charges are much less strongly sorbed by mineral matter and are vastly more mobile than the unchelated metal ions.

Contrary to the above findings, only very low concentrations of chelatable radioactive plutonium were observed in groundwater near the Idaho Chemical Processing Plant's low-level waste disposal well. No plutonium was observed in wells at any significant distance from the disposal well. The waste processing procedure used was designed to destroy any chelating agents in the waste prior to disposal, and no chelating agents were found in the water pumped from the test wells.

Complexing agents in waste-water are of concern primarily because of their ability to solubilise heavy metals from plumbing and from deposits containing heavy metals. Complexation may increase the leaching of heavy metals from waste disposal sites and reduce the efficiency with which heavy metals are removed with sludge in conventional biological waste treatment. Removal of chelated iron is difficult with conventional municipal water treatment processes. Iron and perhaps several other essential micronutrient metal ions are kept in solution by chelation in algal cultures, The availability of chelating agents may be a factor in determining algal growth. The yellow-brown colour of some natural waters is due to naturally-occurring chelates of iron.

BONDING AND STRUCTURE OF METAL COMPLEXES

This section discusses some of the fundamentals helpful in understanding complexation in water. A complex consists of a central atom to which ligands possessing electron-donor properties are bonded. The ligands may be negatively charged or neutral. The resulting complex may be neutral or may have positive or negative charge. The ligands are said to be contained within the coordination sphere of the central metal atom. Depending upon the type of bonding involved, the ligands within the coordination sphere are held in a definite structural pattern. However, in solution, ligands of many complexes exchange rapidly between solution and the coordination sphere of the central metal ion.

The coordination number of a metal atom, or ion, is the number of ligand electron-donor groups that are bonded to it. The most common coordination numbers are 2,4, and 6. Polynuclear complexes contain two or more metal atoms joined together through bridging ligands, frequently OH. An example of such a complex is the dinuclear complex of iron.

$$(H_2O)_4Fe \underset{O}{\overset{O}{\diamond}} Fe(H_2O)_4{}^{4+}$$

The nature and strength of bonds in metal complexes are crucial in determining their behaviour and stability. In forming complexes, electron pairs are donated to the central metal ion by the ligands. These electron pairs fill empty orbitals on the metal ion, thus allowing the electron distribution of the metal ion to approach that of a noble gas.

Selectivity and Specificity in Chelation

Although chelating agents are never entirely specific for a particular metal ion, some complicated chelating agents of biological origin approach almost complete specificity for certain metal ions. One example of such a chelating agent is ferrichrome, synthesised by, and extracted from fungi, which forms extremely stable chelates with iron. It has been observed that cyanobacteria of the *Anabaena* species secrete appreciable quantities of iron-selective hydroxamate chelating agents during periods of heavy algal bloom. These organisms readily take up iron chelated by hydroxamate-chelated iron, whereas some competing green algae, such as *Scenedesmus*, do not. Thus, the chelating agent serves a dual function of promoting the growth of certain cyanobacteria while suppressing the growth of competing algae, allowing the former to predominate.

CALCULATIONS OF SPECIES CONCENTRATIONS

The stability of complex ions in solution is expressed in terms of formation constants. These can stepwise formation constants (K expressions) representing the bonding of individual ligands to a metal ion or overall formation constants (β expressions) representing the binding of two or more ligands to a metal ion. These concepts are illustrated for complexes of zinc ion with ammonia by the following:

$$Zn^{2+} + NH_3 \longrightarrow ZnNH_3^{2+} \qquad \qquad ...(5.52)$$

$$K_1 = \frac{[ZnNH_3^{2+}]}{[Zn^{2+}][NH_3]} = 3.9 \times 10^2 \text{ (Stepwise formation constant)} \qquad ...(5.53)$$

$$ZnNH_3^{2+} + NH_3 \longrightarrow Zn(NH_3)_2^{2+} \qquad \qquad ...(5.54)$$

$$K_2 = \frac{[Zn(NH_3)_2^{2+}]}{[ZnNH_3^{2+}][NH_3]} = 2.1 \times 10^2 \qquad \qquad ...(5.55)$$

$$Zn^{2+} + 2NH_3 \longrightarrow Zn(NH_3)_2^{2+} \qquad \qquad ...(5.56)$$

$$\beta_2 = \frac{[Zn(NH_3)_2^{2+}]}{[Zn^{2+}][NH_3]^2} = K_1K_2 = 8.2 \times 10^4 \text{ (Overall formation constant)} \qquad ...(5.57)$$

(For $Zn(NH_3)_3^{2+}$, $\beta_3 = K_1K_2K_3$ and for $Zn(NH_3)_4^{2+}$, $\beta_4 = K_1K_2K_3K_4$).

The following sections show some calculations involving chelated metal ions in aquatic systems. In addition to the complexation itself, consideration must be given to competition of H^+ for ligands, competition among metal ions for ligands, competition among different ligands for metal ions, and precipitation of metal ions by various precipitants.. Not the least of the problems involved in such calculations is the lack of accurately known values of equilibrium constants to be used under the conditions being considered, a factor which can yield questionable results from even the most elegant computerised calculations. Furthermore, kinetic factors are often quite important. Such calculations, however, can be quite useful to provide an overall view of aquatic systems in which complexation is important and as general guidelines to determine areas in which more data should be obtained.

COMPLEXATION BY DEPROTONATED LIGANDS

In most circumstances metal ions and hydrogen ions compete for ligands, making the calculation of species concentrations more complicated. Before going into such calculations, however, it is instructive to look at an example in which the ligand has lost all ionisable hydrogen. At pH values of 11 or above, EDTA is essentially all in the completely ionised tetranegative form, Y^{4-}, illustrated below:

(Y^+)

Consider a waste-water with an alkaline pH of 11 containing copper at a total level of 5.0 mg/L and excess uncomplexed EDTA at a level of 200 mg/L (expressed as the disodium salt, $Na_2H_2C_{10}H_{12}O_8N_2 \cdot 2H_2O$, formula weight of 372). At this pH uncomplexed EDTA is present as ionised Y^{4-}. The questions to be asked are: Will most of the copper be present as the EDTA complex? If so, what will be the equilibrium concentration of the hydrated copper(II) ion, Cu^{2+}? To answer the former question it is first necessary to calculate the molar concentration of uncomplexed excess EDTA, Y^{4-}. Since disodium EDTA with a formula weight of 372 is present at 200 mg/L (ppm), the total molar concentration of EDTA as Y^{4-} is 5.4×10^{-4} M. The formation constant K_1 of the copper-EDTA complex CuY^{2-} is given by

$$K_1 = \frac{[CuY^{2-}]}{[Cu^{2+}][Y^{4-}]} = 6.3 \times 10^{18} \qquad \qquad ...(5.58)$$

The ratio of complexed copper to uncomplexed copper is

$$\frac{[CuY^{2-}]}{[Cu^{2+}]} = [Y^{4-}]K_1 = 5.4 \times 10^{-4} \times 6.3 \times 10^{18} = 3.3 \times 10^{15} \qquad ...(5.59)$$

and therefore, essentially all of the copper is present as the complex ion. The molar concentration of total copper in a solution containing 5.0 mg/L copper is 7.9×10^{-5} M, which in this case is essentially all in the form of the EDTA complex. The very low concentration of uncomplexed, hydrated copper ion is given by

$$[Cu^{2+}] = \frac{[CuY^{2-}]}{K_1[Y^{4-}]} = \frac{7.9 \times 10^{-5}}{6.3 \times 10^{18} \times 5.4 \times 10^{-4}} = 2.3 \times 10^{-20} \qquad ...(5.60)$$

It is seen that in the medium described, the concentration of hydrate copper ion is extremely low compared to total copper ion. Any phenomenon in solution that depends upon the concentration of the hydrated copper ion (such as a physiological effect or an electrode response) would be very different in the medium described, as compared to the effect observed if all of the copper at a level of 5.0 mg/L were present as Cu^{2+} in a more acidic solution and in the absence of complexing agent. The phenomenon of reducing the concentration of hydrated metal ion to very low value through the action of strong chelating agents is one of the most important effects of complexation in natural aquatic systems.

COMPLEXATION BY PROTONATED LIGANDS

Generally, complexing agent, particularly chelating compounds, are conjugate bases of Bronsted acids. As examples, NH_3 is the conjugate base of the NH_4^+ acid cation, and glycinate anion, $H_2NCH_2CO_2^-$, is the conjugate base of glycine, $^+H_3NCH_2CO_2^-$. Therefore, in many cases hydrogen ion competes with metal ions for a ligand, so that the strength of chelation depends upon pH. In the nearly neutral pH range usually encountered in natural waters, most organic ligands are present in a conjugated acid form.

In order to understand the competition between hydrogen ion and metal ion for a ligand it is useful to know the distribution of ligand species as a function of pH. Consider nitrilotriacetic acids, commonly designated H_3T, as an example. The trisodium salt of this compound, (NTA) is used as a detergent phosphate substitute and is a strong chelating agent. Biological processes are required' for NTA degradation and under some conditions it persists for long time in water. Given the ability of NTA to solubilise and transport heavy metal ions, this material is of considerable environmental concern.

Nitrilotriacetic acid, H_3T, loses hydrogen ion in three steps to form the nitrilotriacetate anion, T^{3-}, the structural formula of which is

The T^{3-} species may coordinate through three $-CO_2^-$ groups and through the nitrogen atom, as shown in Fig. 5.5. The stepwise ionisation of H_3T is given by the following equilibria :

$$H_3T \rightleftharpoons H^+ + H_2T^- \qquad \qquad ...(5.61)$$

$$K_{a1} = \frac{[H^+][H_2T^-]}{[H_3T]} = 2.18 \times 10^{-2} \qquad pK_{a1} = 1.66 \qquad ...(5.62)$$

$$H_2T^- \rightleftharpoons H^+ + HT^{2-} \qquad \qquad ...(5.63)$$

$$K_{a2} = \frac{[H^+][HT^{2-}]}{[H_2T^-]} = 1.12 \times 10^{-3} \qquad pK_{a2} = 2.95 \qquad ...(5.64)$$

$$HT^{2-} \rightleftharpoons H^+ + T^{3-} \qquad \qquad ...(5.65)$$

$$K_{a3} = \frac{[H^+][T^{3-}]}{[HT^{2-}]} = 5.25 \times 10^{-11} \qquad pK_{a3} = 10.28 \qquad ...(5.66)$$

These equilibrium expressions show that uncomplexed NTA may exist in solution as any one of the four species H_3T, H_2T^-, HT^{2-}, or T^{3-}, depending upon the pH of the solution. As was shown for the $CO_2/HCO_3^-/CO_3^{2-}$ system in Fig. 5.3, fractions of NTA species can be illustrated graphically by a diagram of the distribution-of-species with pH as a master (independent) variable. The key points used to plot such a diagram for NTA are given in Table 5.1. and the plot of fractions of species (α values) as a function of pH is shown in Fig. 5.6. Examination of the plot shows the complexing anion T^{3-} is the predominant species only at relatively high pH values, much higher than usually would be encountered in natural waters. The HT^{2-} species has an extremely wide range of predominance, however, spanning the entire normal pH range of ordinary fresh waters.

Table 5.1. Fractions of NTA species at selected pH values.

pH value	αH_3T	αH_2T	αHT^{2-}	αHT^{3-}
pH below 1.00	1.00	0.00	0.00	0.00
pH = pK_{a1}	0.49	0.49	0.02	0.00
pH = $1/2(pK_{a1} + pK_{a2})$	0.16	0.68	0.16	0.00
pH = pK_{a2}	0.02	0.49	0.49	0.00
pH = $1/2(pK_{a2} + pK_{a3})$	0.00	0.00	1.00	0.00
pH = pK_{a3}	0.00	0.00	0.50	0.50
pH above 12	0.00	0.00	0.00	1.00

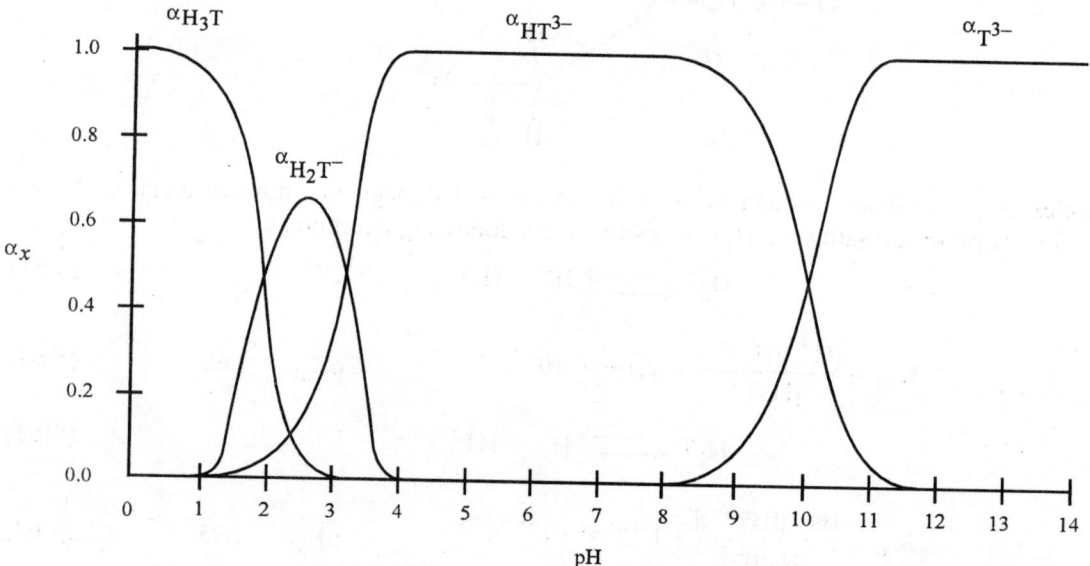

Fig. 5.6. Plot of fraction of species α_x as a function of pH for NTA species in water.

SOLUBILISATION OF LEAD ION FROM SOLIDS BY NTA

A major concern regarding the widespread introduction of strong chelating agents such as NTA into aquatic ecosystems from sources such as detergents or electroplating wastes is that of possible solubilisation of toxic heavy metals from solids through the action of chelating agents. Experimentation is required to determine whether this may be a problem, but calculations are helpful in predicting probable effects. The extent of solubilisation of heavy metals depends upon a number of factors, including the stability of the metal chelates, the concentration of the complexing agent in the water, pH, on the nature of the insoluble metal deposit. Several example calculations are given here.

Consider first the solubilisation of lead from solid $Pb(OH)_2$ by NTA at pH 8.00. As illustrated in Fig. 5.6. essentially all uncomplexed NTA is present as HT^{2-} ion at pH 8.00. Therefore, the solubilisation reaction is

$$Pb(OH)_2(s) + HT^{2-} \rightleftarrows PbT^- + OH^- + H_2O \qquad \qquad ...(5.67)$$

which may be obtained by adding the following reactions;

$$Pb(OH)_2(s) \rightleftarrows Pb^{2+} + 2OH^- \qquad \qquad ...(5.68)$$

$$K_{sp} = [Pb^{2+}][OH^-]^2 = 1.61 \times 10^{-20} \qquad \qquad ...(5.69)$$

$$HT^{2-} \rightleftarrows H^+ + T^{3-} \qquad \qquad ...(5.70)$$

$$K_{a3} = \frac{[H^+][T^{3-}]}{[HT^{2-}]} = 5.25 \times 10^{-11} \qquad \qquad ...(5.71)$$

$$Pb^{2+} + T^{3-} \rightleftarrows PbT^- \qquad \qquad ...(5.72)$$

$$K_f = \frac{[PbT^-]}{[Pb^{2+}][T^{3-}]} = 2.45 \times 10^{11} \qquad \qquad ...(5.73)$$

$$H^+ + OH^- \rightleftarrows H_2O \qquad \qquad ...(5.74)$$

$$\frac{1}{K_w} = \frac{1}{[H^+][OH^-]} = \frac{1}{1.00 \times 10^{-14}} \qquad \qquad ...(5.75)$$

$$Pb(OH)_2(s) + HT^{2-} \rightleftarrows PbT^- + OH^- + H_2O \qquad \qquad ...(5.76)$$

$$K = \frac{[PbT^-][OH^-]}{[HT^{2-}]} = \frac{K_{sp}K_{a3}K_f}{K_w} = 2.07 \times 10^{-5} \qquad \qquad ...(5.77)$$

Assume that a sample of water contains 25 mg/L of $N(CH_2CO_2Na)_3$, the trisodium NTA salt, formula weight 257. The total concentration of both complexed and uncomplexed NTA is 9.7×10^{-15} mmol/mL. Assuming a system in which NTA at pH 8.00 is in equilibrium with solid $Pb(OH)_2$, the NTA may be primarily in the uncomplexed form, HT^{2-}, or in the lead complex, PbT^-. The predominant species may be determined by calculating the ratio of $[PbT^-]$ to $[HT^{2-}]$ from the expression for K, noting that at pH 8.00, $[OH^-] = 1.00 \times 10^{-6}$ M :

$$\frac{[PbT^-]}{[HT^{2-}]} = \frac{K}{[OH^-]} = \frac{2.07 \times 10^{-5}}{1.00 \times 10^{-6}} = 20.7 \qquad \qquad ...(5.78)$$

Since $[PbT^-]/[HT^{2-}]$ is approximately 20 to 1, most of the NTA in solution is present as the lead chelate. The concentration of PbT^- on a molar basis is just slightly less than the 9.7×10^{-5} mmols/mL total NTA present. The atomic weight of lead is 207 so that the concentration of lead in solution is approximately 20 mg/L. This reaction is pH-dependent so that the fraction of NTA chelated decreases with increasing pH.

Reaction of NTA with Metal Carbonate

Carbonates are common forms of heavy metal ion solids. Solid lead carbonate, $PbCO_3$, is stable within the pH region and alkalinity conditions often found in natural waters and waste-waters. An example similar to one in the preceding section may be worked, assuming that equilibrium is established with $PbCO_3$ rather than with solid $Pb(OH)_2$. In this example it is assumed that 25 mg/L of trisodium NTA is in equilibrium with $PbCO_3$ at pH 7.00 and a calculation is made to determine whether the lead will be complexed appreciably by the NTA. The carbonate ion, CO_3^{2-}, reacts with H^+ to form HCO_3^-. As already discussed the acid-base equilibrium reactions for the $CO^2/HCO_3^-/CO_3^{2-}$ systems are :

$$CO_2 + H_2O \rightleftharpoons HCO_3^- + H^+ \qquad \qquad ...(5.79)$$

$$K'_{a1} = \frac{[H^+][HCO_3^-]}{[CO_2]} = 4.45 \times 10^{-7} \qquad \qquad pK'_{a1} = 6.35 \qquad ...(5.80)$$

$$HCO_3^- \rightleftharpoons CO_3^{2-} + H^+ \qquad \qquad ...(5.81)$$

$$K'_{a2} = \frac{[H^+][CO_3^{2-}]}{[CJCO_3^-]} = 4.65 \times 10^{-11} \qquad \qquad pK'_{a2} = 10.33 \qquad ...(5.82)$$

where the acid dissociation constants of the carbonate species are designated as K'_a to distinguish them from the acid dissociation constants of NTA. Fig. 5.3 shows that within a pH range of about 7 to 10 the predominant carbonic species is HCO_3^-; therefore, the CO_3^{2-} released by the reaction of NTA with $PbCO_3$ will go into solution as HCO_3^-:

$$PbCO_3(s) + HT^{2-} \rightleftharpoons PbT^- + HCO_3^- \qquad \qquad ...(5.83)$$

This reaction and its equilibrium constant are obtained as follows:

$$PbCO_3(s) \rightleftharpoons Pb^{2+} + CO_3^{2-} \qquad \qquad ...(5.84)$$

$$K_{sp} = [Pb^{2+}][CO_3^{2-}] = 1.48 \times 10^{-13} \qquad \qquad ...(5.85)$$

$$Pb^{2+} + T^{3-} \rightleftharpoons PbT^- \qquad \qquad ...(5.86)$$

$$K_f = \frac{[PbT^-]}{[Pb^{2+}][T^{3-}]} = 2.45 \times 10^{11} \qquad \qquad ...(5.87)$$

$$HT^{2-} \rightleftharpoons H^+ + T^{3-} \qquad \qquad ...(5.88)$$

$$K_{a3} = \frac{[H^+][T^{3-}]}{[HT^{2-}]} = 5.25 \times 10^{-11} \qquad \qquad ...(5.89)$$

$$CO_3^{2-} + H^+ \rightleftharpoons HCO_3^- \qquad \qquad ...(5.90)$$

$$\frac{1}{K'_{a2}} = \frac{[HCO_3^-]}{[CO_3^{2-}][H^+]} = \frac{1}{4.69 \times 10^{-11}} \qquad \qquad ...(5.91)$$

$$PbCO_3\ (s) + HT^{2-} \rightleftharpoons PbT^- + HCO_3^- \qquad \qquad ...(5.92)$$

$$K = \frac{[PbT^-][HCO_3^-]}{[HT^{2-}]} = \frac{K_{sp}K_{a3}K_f}{K'_{a2}} = 4.06 \times 10^{-2} \qquad \qquad ...(5.93)$$

From the expression for K, it may be seen that the degree to which $PbCO_3$ is solubilised as PbT^- depends upon the concentration of HCO_3^-. Although this concentration will vary appreciably, the figure commonly used to describe natural waters is a bicarbonate ion concentration of 1.00×10^{-3}. Using this value the following may be calculated:

$$\frac{[PbT^-]}{[HT^{2-}]} = \frac{K}{[HCO_3^-]} = \frac{4.06 \times 10^{-2}}{1.00 \times 10^{-3}} = 40.6 \qquad \qquad ...(5.94)$$

Thus, under the given conditions, most of the NTA in equilibrium with solid $PbCO_3$ would be present as the lead complex. As in the previous example, at a trisodium NTA level of 25 mg/L the concentration of soluble lead(II) would be approximately 20 mg/L. At relatively higher concentrations of HCO_3^-, the tendency to solubilise lead would be diminished, whereas at lower concentrations of HCO_3^-, NTA would be more effective in solubilising lead.

Effect of Calcium Ion upon the Reaction of Chelating Agents with Slightly Soluble Salts

Chelatable calcium ion, Ca^{2+}, which is generally present in natural waters and waste-waters, competes for the chelating agent with a metal in a slightly soluble salt, such as $PbCO_3$. At pH 7.00, the reaction between calcium ion and NTA is

$$Ca^{2+} + HT^{2-} \rightleftharpoons CaT^- + H^+ \qquad \qquad ...(5.95)$$

described by the following equilibrium expression :

$$K' = \frac{[CaT^-][H^+]}{[Ca^{2+}][HT^{2-}]} = 1.48 \times 10^8 \times 5.25 \times 10^{11} = 7.75 \times 10^{-3} \qquad \qquad ...(5.96)$$

The value of K' is the product of the formation constant of CaT^-, (1.48×10^8) and K_{a3} of NTA, 5.25×10^{-11}. The fraction of NTA bound as CaT^- depends upon the concentration of Ca^{2+} and the pH. Typically, $[Ca^{2+}]$ in water is 1.00×10^{-3} M. Assuming this value and pH 7.00, the ratio of NTA present in solution as the calcium complex to that present as HT^{2-} is:

$$\frac{[CaT^-]}{[HT^{2-}]} = \frac{[Ca^{2+}]}{[H^+]}K' = \frac{1.00 \times 10^{-3}}{1.00 \times 10^{-7}} \times 7.75 \times 10^{-3} \qquad \qquad ...(5.97)$$

$$\frac{[CaT^-]}{[HT^{2-}]} = 77.5$$

Therefore, most of the NTA in equilibrium with 1.00×10^{-3} M Ca^{2+} would be present as the calcium complex, CaT^-, which would react with lead carbonate as follows:

$$PbCO_3(s) + CaT^- + H^+ \rightleftharpoons Ca^{2+} + HCO_3^- + PbT^- \qquad \qquad ...(5.98)$$

$$K'' = \frac{[Ca^{2+}][HCO_3^-][PbT^-]}{[CaT^-][H^+]} \qquad \qquad ...(5.99)$$

Equation 5.98 may be obtained by subtracting equation 5.95 from equation 5.83, and its equilibrium constant may be obtained by dividing the equilibrium constant of equation 5.95 by that of equation 5.93:

$$PbCO_3(s) + HT^{2-} \rightleftharpoons PbT^- + HCO_3^- \qquad \qquad ...(5.100)$$

$$K = \frac{[PbT^-][HCO_3^-]}{[HT^{2-}]} = \frac{K_sK_{a3}K_f}{K_{a2}'} = 4.06 \times 10^{-2} \qquad \qquad ...(5.101)$$

$$-(Ca^{2+} + HT^{2-} \rightleftharpoons CaT^- + H^+ \qquad \qquad ...(5.102)$$

$$K = \frac{[CaT^-][H^+]}{[Ca^{2+}][HT^{2-}]} = 7.75 \times 10^{-3} \qquad \qquad ...(5.103)$$

$$PbCO_3(s) + CaT^- + H^+ \rightleftharpoons Ca^{2+} + HCO_3^- + PbT^- \qquad \qquad ...(5.104)$$

$$K'' = \frac{K}{K'} = \frac{4.06 \times 10^{-2}}{7.75 \times 10^{-3}} = 5.24 \qquad \qquad ...(5.105)$$

Having obtained the value of K'', it is now possible to determine the distribution of NTA between PbT$^-$ and CaT$^-$. Thus, for water containing NTA chelated to calcium at pH 7.00 a concentration of HCO$_3^-$ of 1.00×10^{-3}, a concentration of Ca$_2^+$ of 1.00×10^{-3}, and in equilibrium with solid PbCO$_3$, the distribution of NTA between the lead complex and the calcium complex is:

$$\frac{[PbT^-]}{[CaT^-]} - \frac{[H^+]K''}{[Ca^{2+}][HCO_3^-]} = \frac{1.00 \times 10^{-7} \times 5.24}{1.00 \times 10^{-3} \times 1.00 \times 10^{-3}} = 0.524$$

It may be seen that only about 1/3 of the NTA would be present as the lead chelate whereas under the identical conditions, but in the absence of Ca^{2+}, approximately all of the NTA in equilibrium with solid PbCO$_3$ was chelated to NTA. Since the fraction of NTA present as the lead chelate is directly proportional to the solubilisation of PbCO$_3$, differences in calcium concentration will affect the degree to which NTA solubilises lead from lead carbonate.

Competing Equilibria Between Chloride and NTA Ligands in Seawater

The formation of chloro complexes is an additional consideration when chelation of metal ions is considered in seawater as shown in Fig. 5.7 for the chelation of cadmium with NTA. This scheme illustrates the complicated interactions involved in determining the degree of chelation of a heavy metal ion (cadmium) in seawater. It can be seen that the presence of excess chloride results in the formation of chloro complexes; higher concentrations of H$^+$ break up the cadmium-NTA chelate (CdT$^-$) by protonation of the ligand; and both Ca^{2+} and Mg^{2+} compete for the NTA ligands.

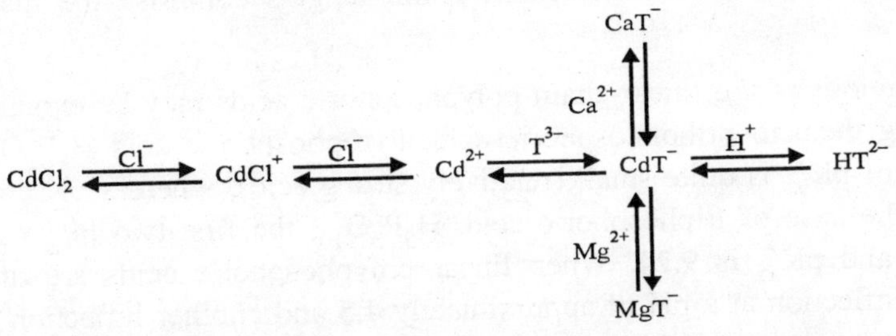

Fig. 5.7. Chelation scheme for cadmium in seawater.

POLYPHOSPHATES IN WATER

Phosphorus occurs as many oxoanions, anionic forms in combination with oxygen. Some of these are strong complexing agents. Since about 1930, salts of polymeric phosphorus oxoanions have been used increasingly for water treatment, for water softening, and as detergent builders. When used for water treatment, polyphosphates "sequester" calcium ion in a soluble or suspended form. The effect is to reduce the equilibrium concentration of calcium ion and prevent the precipitation of calcium carbonate in installations such as water pipes and boilers. Furthermore, when water is softened properly with polyphosphates, calcium does not form precipitates with soaps or interact detrimentally with detergents. The simplest form of phosphate is orthophosphate, PO_4^{3-} :

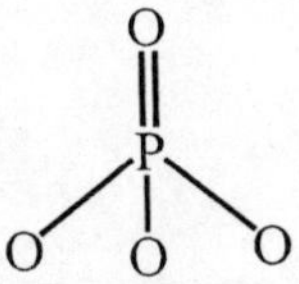

The orthophosphate ion possesses three sites for attachment of H^+. orthophosphate acid, H_3PO_4, has a pK_{a1} of 2.17, pK_{a2} of 7.31, and a pK_{a3} of 12.36. Much of the orthophosphate in natural waters originates from the hydrolysis of polymeric phosphate species.

Pyrophosphate ion, $P_2O_7^{4-}$, is the first of a series of unbranched chain polyphosphates produced by the condensation of orthophosphate :

$$2PO_4^{3-} + H_2O \rightleftharpoons P_2O_7^{4-} + 2OH^- \qquad \qquad ...(5.106)$$

A long series of linear polyphosphates may be formed, the second of which is triphosphate ion, $P_3O_{10}^{5-}$. These species consist of PO_4 tetrahedra with adjacent tetrahedra sharing a common oxygen atom at one corner. The structural formulae of the acidic forms, $H_4P_2O_7$ and $H_5P_3O_{10}$, are :

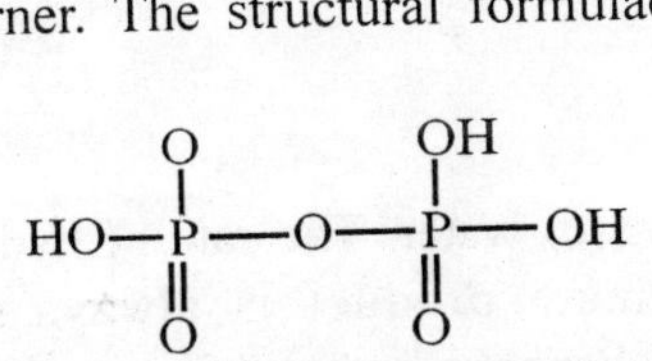

Pyrophosphoric
(diphosphoric) acid

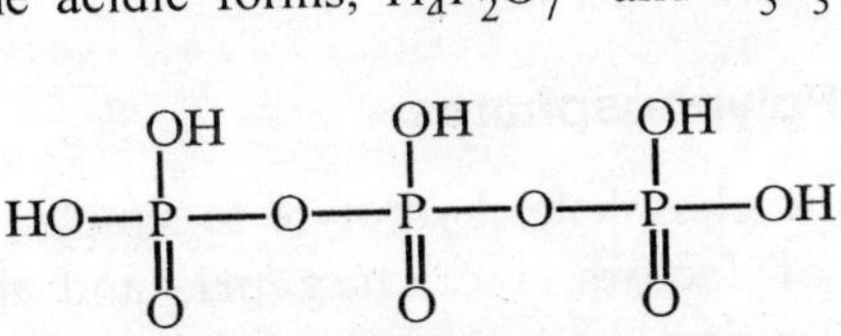

Triphosphoric acid

It is easy to visualise the longer chains composing the higher linear polyphosphates. Vitreous sodium phosphates are mixtures consisting of linear phosphate chains with from 4 to approximately 18 phosphorus atoms each. Those with intermediate chain lengths comprise the majority of the species present.

The acid-base behaviour of the linear-chain polyphosphoric acids may be explained in terms of their structure by comparing them to orthophosphoric acid. Pyrophosphoric acid $H_4P_2O_7$, has four ionisable hydrogens. The value of pK_{a1} is quite small (relatively strong acid), whereas pK_{a2} is 2.64, pK_{a3} is 6.76, and pK_{a2} is 9.42. In the case of triphosphoric acid, $H_5P_3O_{10}$, the first two pK_a values are small, pK_{a3} is 2.30, pK_{a4} is 6.50 and pK_{a5} is 9.24. When linear polyphosphoric acids are titrated with base, the titration curve has an inflection at a pH of approximately 4.5 and another inflection at a pH close to 9.5. To understand these phenomena consider the ionisation of triphosphoric acid below:

Stepwise ionisation of
triphosphoric acid

Each P atom in the polyphosphate chain is attached to an –OH group that has one readily ionisable hydrogen that is readily removed in titrating to the first equivalence point. The end phosphorus atoms have two OH groups each. One of the OH groups on an end phosphorus atom has a readily ionisable hydrogen, whereas the other loses its hydrogen much less readily. Therefore, one mole of triphosphoric acid, $H_5P_3O_{10}$ loses three moles of hydrogen ion at a relatively low pH (below 4.5), leaving the $H_2P_3O_{10}^{3-}$ species with two ionisable hydrogens. At intermediate pH values (below 9.5), an additional two moles of "end hydrogens" are lost to form the $P_3O_{10}^{5-}$ species. Titration of a linear-chain polyphosphoric acid up to pH 4.5 yields the number of moles of phosphorus atoms per mole of acid and titration from pH 4.5 to pH 9.5 yields the number of end phosphorus atoms. Orthophosphoric acid, H_3PO_4, differs from the linear chain polyphosphoric acids in that it has a third ionisable hydrogen, which is removed only in extremely basic media.

Hydrolysis of Polyphosphates

All of the polymeric phosphates hydrolyse to simpler products in water. The rate of hydrolysis depends upon a number of factors, including pH, and the ultimate product is always some form of orthophosphate. The simplest hydrolytic reaction of a polyphosphate is that of pyrophosphoric acid to orthophosphoric acid:

$$H_4P_2O_7 + H_2O \rightarrow 2H_3PO_4 \qquad \qquad ...(5.107)$$

Researchers have found evidence that algae and other micro-organisms catalyse the hydrolysis of polyphosphates. Even in the absence of biological activity, polyphosphates hydrolyse chemically at a reasonable rate in water so that there is much less concern about the possibility of their transporting heavy metal ions than is the case with organic chelating agents such as NTA or EDTA, which must depend upon microbial degradation for their decomposition.

Complexation by Polyphosphates

In general, chain phosphates are good complexing agents and even form complexes with alkali-metal ions. Ring phosphates, which have alternate P and O atoms in ring structure, form much weaker complexes than do chain species. The different chelating abilities of chain and ring phosphates are due to structural hindrance of bonding by the ring polyphosphates.

COMPLEXATION BY HUMIC SUBSTANCES

The most important class of complexing agents that occur naturally are the humic substances. These are degradation-resistance materials formed during the decomposition of vegetation that occur as deposits in solid, marsh sediments, peat, coal, lignite, or in almost any location where large quantities of vegetation have decayed. They are commonly classified on the basis of solubility. If a material containing humic substances is extracted with strong base, and the resulting solution is acidified, the products are: (i) a non-extractable plant residue called humin; (ii) a material that precipitates from the acidified extract, called humic acid; and (iii) an organic material that remains in the acidified solution, called fulvic acid. Because of their acid-base, sorptive, and complexing properties, both the soluble and insoluble humic substances have a strong effect upon the properties of water. In general, fulvic acid dissolves in water and exerts its effects as the soluble species. Humin and humic acid remain insoluble and affect water quality through exchange of species, such as cations or organic materials, with water.

Humic substances are high-molecular-weight, polyelectrolytic macromolecule. Molecular weights range from a few hundred for fulvic acid to tens of thousands for the humic acid and humin fractions. These substances contain a carbon skeleton with a high degree of aromatic character and with a large percentage of the molecular weight incorporated in functional groups, most of which contain oxygen. The elementary composition of most humic substances is within the following ranges: C, 45-55%, O, 30-45%, H, 3-6%; N, 1-5%; and S, 0-1%. The terms humin, humic acid, and fulvic acid do not refer to single compounds but to a wide range of compounds of generally similar origin, with many properties in common.

Some feeling for the nature of humic substances may be obtained by considering the structure of a hypothetical molecule of fulvic acid below:

This structure is typical of the type of compound composing fulvic acid. The compound has a formula weight of 666, and its chemical formula may be represented by $C_{20}H_{15}(CO_2H)_6(OH)_5(CO)_2$. As shown in the hypothetical compound, the functional groups that may be present in fulvic acid are carboxyl, phenolic hydroxyl, alcoholic hydroxyl, and carbonyl. The functional groups vary with the particular acid sample. Approximate ranges in units of milliequivalents per gram of acid are: total acidity, 12-14; carboxyl, 8-9; phenolic hydroxyl, 3-6; alcoholic hydroxyl, 3-5; and carbonyl, 1-3. In addition, some methoxyl groups, $-OCH_3$, may be encountered at low levels.

Their binding of metal ions is one of the most important environmental qualities of humic substances. This binding can occur as chelation between a carboxyl group and a phenolic hydroxyl group, as chelation between two carboxyl groups, or as complexation with a carboxyl group (Fig. 5.8).

Iron and aluminium are very strongly bound to humic substances, whereas magnesium is rather weakly bound. Other common ions, such as Ni^{2+}, Pb^{2+}, Ca^{2+} and Zn^{2+}, are intermediate in their binding to humic substances.

The role played by soluble fulvic-acid complexes of metal in natural waters is not well known. They probably keep some of the biological important transition-metal ions in solution and are particularly involved in iron solubilisation and transport. Fulvic acid-type compounds are associated with colour in water. These yellow materials, called *Gelbstoffe*, frequently are encountered along with soluble iron.

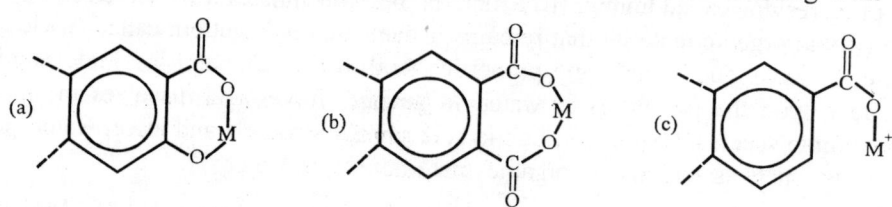

Fig. 5.8. Binding of a metal ion, M^{2+}, by humic substances (a) by chelation between carboxyl and phenolic hydroxyl; (b) by chelation between two carboxyl groups; and (c) by complexation with carboxyl group.

Insoluble humic substances, the humins and humic acids, effectively exchange cations with water and may accumulate large quantities of metals. Lignite coal, which is largely a humic-acid material, tends to remove some metal ions from water.

Special attention has been given to humic substances, following the discovery of trihalomethanes (THMs, such as chloroform and dibromochloromethane) in water supplies. It is now generally believed that these suspected carcinogens can be formed in the presence of humic substances during the disinfection of raw municipal drinking water by chlorination. The humic substances produce THMs by reaction with chlorine. The formation of THMs can be reduced by removing as much of the humic material as possible prior to chlorination.

COMPLEXATION AND REDOX PROCESSES

Complexation may have a strong effect upon equilibrium by shifting reactions, such as that for the oxidation of lead,

$$Pb \underset{\longleftarrow}{\overset{\longrightarrow}{\rule{2cm}{0pt}}} Pb^{2+} + 2e^- \qquad \qquad ...(5.108)$$

strongly to the right by binding to the product ion, thus cutting its concentration down to very low levels.

Of perhaps more importance is the fact that upon oxidation,

$$M + 1/2\ O_2 \rightleftarrows MO \qquad \qquad ...(5.109)$$

many metals form self-protective coatings of oxides, carbonates, or other insoluble species, which prevent further chemical reaction. Copper and aluminium roofing and structural iron are examples of materials which are thus self-protecting. A chelating agent in contact with such metals can result in continual dissolution of the protective coating so that the exposed metal corrodes readily. For example, chelating agents in waste-water may increase the corrosion of metal plumbing, thus adding heavy metals to effluents. Solutions of chelating agents employed to clean metal surfaces in metal plating operation have a similar effect.

CHAPTER 6

Water Pollution

INTRODUCTION

Water whether from underground or surface sources, found in nature is polluted. The pollution may be due to various reasons, namely, from sewage, industrial wastes or from natural contaminants. Such water, if supplied directly without treatment, may not be used by the consumers due to psychological, aesthetic or physiological reasons. Waterborne epidemic may spread by supply of such water. Various diseases that may be transmitted through water have been discussed earlier. Natural water may also contain high amount of turbidity, organic load from decaying leaves, vegetations or animals. Minerals in solution or suspension forms may contaminate water.

Water will require treatment for all the above reasons. Most important is the necessity of removing the germs of diseases. For palatability, it must be free from unpleasant tastes and odours; from the aesthetic point of view, it should have an inviting appearance. It should be useful for domestic needs as well as for a wide variety of industrial purposes.

Degree and methods of treatment depend on the nature of source, quality of water of the source, purpose for which it should be supplied. Quality of domestic water may need different standard than that of an industry. Again, quality requirement of one industry may differ from that of the other. All these factors determine which treatment process to be employed and which may not be necessary.

The various treatment methods and the nature of impurities removed by employing them are given below :

Process	Impurity removal
Aeration	Tastes and odour removal, oxygen deficiency.
Screening	Floating matter.
Plain sedimentation	Large suspended solids.
Coagulation	Fine particles.
Filtration	Colloidal particles, microorganisms.
Activated carbon	Elements causing tastes and odours.
Softening	Hardness.
Disinfection	Living organisms including pathogens.

It is not that all the treatment processes tabulated above will be required for a treatment plant. The treatment process selected will depend on the quality of water at the source and nature of water required.

In case of water taken from a surface source, generally the treatment units required are plain sedimentation, coagulation, filtration and disinfection to make the water fit for domestic use.

Ground water available in nature generally is free from suspended solids and micro-organisms. They contain more of dissolved salts and quite often cause the problem due to hardness, iron content, etc. Such waters may need treatment like only softening or iron removal etc. But apparently whether required or not, the municipal water supply meant for domestic purposes, irrespective of the type of source, should be properly disinfected. This is a preventive measure against bacterial contamination and it guards against 'after' growth.

Purification works in a public water supply system must be selected and designed to deliver water that is already not so, has been made (i) hygienically safe; (ii) aesthetically attractive and palatable; and (iii) economically satisfactory for the uses to which it is to be put.

NATURE AND TYPES OF WATER POLLUTANTS

Throughout history, the quality of drinking water has been a factor in determining human welfare. Fecal pollution of drinking water has frequently caused waterborne diseases that have decimated the populations of whole cities. Unwholesome water polluted by natural sources has caused great hardship for people forced to drink it or use it for irrigation.

Today there are still occasional epidemics of bacterial and viral diseases caused by infectious agents carried in drinking water—ominously, a major outbreak of deadly cholera. Currently, waterborne toxic chemicals pose the greatest threat to the safety of water supplies in industrialised nations. This is particularly true of groundwater, which exceeds in volume the flow of all rivers, lakes, and streams. In some areas, the quality of groundwater is subject to a number of chemical threats. There are many possible sources of chemical contamination. These include wastes from industrial chemical production, metal plating operations, and pesticide runoff from agricultural lands. Some specific pollutants include industrial chemicals such as chlorinated hydrocarbons; heavy metals, including cadmium, lead, and mercury; saline water; bacteria, particularly coliforms; and general municipal and industrial wastes.

Recently, there has been a tremendous growth in the manufacture and use of synthetic chemicals. Many of the chemicals have contaminated water supplies. Two examples are insecticide and herbicide runoff from agricultural land, and industrial discharge into surface waters. Most serious, though, is the threat to groundwater from waste chemical dumps and landfills, storage lagoons, treating ponds, and other facilities.

It is clear that water pollution should be a concern of every citizen. Understanding the sources, interactions, and effects of water pollutants is essential for controlling pollutants in an environmentally safe and economically acceptable manner. Above all, an understanding of water pollution and its control depends upon a basic knowledge of aquatic environmental chemistry. Water pollution may be studied much more effectively with a sound background in the fundamental properties of water, aquatic microbial reactions, sediment, water interactions, and other factors involved with the reactions, transport, and effects of these pollutants.

Water pollutants can be divided among some general classifications, as summarised in Table 6.1. Most of these categories of pollutants, and several subcategories, are discussed in this chapter.

Table 6.1. General types of water pollutants.

Class of pollutant	Significance
Trace elements	Health, aquatic biota
Heavy metals	Health, aquatic biota
Organically-bound metals	Metal transport
Radionuclides	Toxicity
Inorganic pollutants	Toxicity, aquatic biota
Asbestos	Human health
Algal nutrients	Eutrophication
Acidity, alkalinity, salinity (in excess)	Water quality, aquatic life
Trace organic pollutants	Toxicity
Polychlorinated biphenyls	Possible biological effects
Pesticides	Toxicity, aquatic biota, wildlife
Petroleum wastes	Effect on wildlife, esthetics
Sewage, human, animal wastes	Water quality, oxygen levels
Biochemical oxygen demand	Water quality, oxygen levels
Pathogens	Health effects
Detergents	Eutrophication, wildlife, esthetics
Chemical carcinogens	Incidence of cancer
Sediments	Water quality, aquatic biota, wildlife
Taste, odour, and colour	Esthetics

ELEMENTAL POLLUTANTS

Trace element is a term that refers to those elements that occur at very low levels of a few parts per million or less in a given system. The term trace substance is a more general one applied to both elements and chemical compounds. Table 6.2 summarises the more important trace elements encountered in natural waters. Some of these are recognised as nutrients required for animal and plant life. Of these, many are essential at low levels, but toxic at higher levels. This is typical behaviour for many substances in the aquatic environment, a point that must be kept in mind in judging whether a particular element is beneficial or detrimental. Some of these elements, such as lead or mercury, have such toxicological and environmental significance.

Some of the heavy metals are among the most harmful of the elemental pollutants. These elements are in general the transition metals and some of the representative elements, such as lead and tin, in the lower right-hand corner of the periodic table.

Heavy metals include essential elements like iron as well as toxic metals like cadmium and mercury. Most of them have a tremendous affinity for sulphur and disrupt enzyme function by forming bonds with sulphur groups in enzymes. Protein carboxylic acid ($-CO_2H$) and amino ($-NH_2$) groups are also chemically bound by heavy metals. Cadmium, copper, lead, and mercury ions bind to cell membranes, hindering transport processes through the cell wall. Heavy metals may also precipitate phosphate biocompounds or catalyse their decomposition.

Table 6.2. Important trace elements in natural waters.

Element	Sources	Effects and significance
Arsenic	Mining by-product, chemical waste	Toxic, possibly carcinogenic
Beryllium	Coal, industrial wastes	Toxic
Boron	Coal, detergents, wastes	Toxic
Chromium	Metal plating	Essential as Cr, toxic as Cr
Copper	Metal plating, mining, industrial waste	Essential trace element, toxic to plants and algae at higher levels
Fluorine	Natural geological sources, wastes, water additive	Prevents tooth decay at around 1 mg/l, toxic at higher levels
Iodine	Industrial wastes, natural brines, seawater intrusion	Prevents goiter
Iron	Industrial wastes, corrosion, acid mine water, microbial action	Essential nutrient, damages fixtures by staining
Lead	Industrial waste, mining, fuels,	Toxic, harmful to wildlife
Manganese	Industrial wastes, acid mine water, microbial action	Toxic to plants, damages fixtures by staining
Mercury	Industrial waste, mining, coal	Toxic, mobilised as methyl mercury compounds by anaerobic bacteria
Molybdenum	Industrial wastes, natural sources	Essential to plants, toxic to animals
Selenium	Natural sources, coal	Essential at lower levels, toxic at higher levels
Zinc	Industrial waste, metal plating, plumbing	Essential element, toxic to plants at higher levels

Some of the metalloids, elements on the borderline between metals and non-metals, are significant water pollutants. Arsenic, selenium, and antimony are of particular interest.

Inorganic chemical manufacture has the potential to contaminate water with trace elements. Among the industries regulated for potential trace element pollution of water are those producing chlor-alkali, hydrofluoric acid, sodium dichromate (sulphate process and chloride ilmenite process), aluminium fluoride, chrome pigments, copper sulphate, nickel sulphate, sodium bisulphate, sodium hydrosulphate, sodium bisulphite, titanium dioxide, and hydrogen cyanide.

HEAVY METALS

Cadmium

Pollutant cadmium in water may arise from industrial discharges and mining wastes. Cadmium is widely used in metal plating. Chemically, cadmium is very similar to zinc, and these two metals frequently undergo geochemical processes together. Both metals are found in water in the +2 oxidation state.

The effects of acute cadmium poisoning in humans are very serious. Among them are high blood pressure, kidney damage, destruction of testicular tissue, and destruction of red blood cells. It is believed that much of the physiological action of cadmium arises from its chemical similarity to zinc. Specifically, cadmium may replace zinc in some enzymes, thereby altering the stereo-structure of the enzyme and impairing its catalytic activity. Disease symptoms ultimately result.

Cadmium and zinc are common water and sediment pollutants in harbours surrounded with industrial installations. Concentrations of more than 100 ppm dry weight sediment have been found in harbour sediments. Typically, during periods of calm in the summer when the water stagnates, the anaerobic bottom layer of harbour water has a low soluble Cd concentration because microbial reduction of sulphate produces sulphide,

$$2\{CH_2O\} + SO_4^{2-} + H^+ \rightarrow 2CO_2 + HS^- + 2H_2O \qquad \ldots(6.1)$$

which precipitates cadmium as insoluble cadmium sulphide :

$$CdCl^+ \; (chloro \; complex \; in \; seawater) + HS^- \rightarrow CdS(s) + H^+ + Cl^- \qquad \ldots(6.2)$$

Mixing of bay water from outside the harbour and harbour water by high winds during the winter results in desortion of cadmium from harbour sediments by aerobic bay water. This dissolved cadmium is carried out in the bay, where it reacts with suspended solid materials, which then become incorporated with the bay sediments. This is an example of the sort of complicated interaction of hydraulic, chemical solution-solid, and microbiological factors involved in the transport and distribution of a pollutant in an aquatic system.

Lead

Inorganic lead arising from a number of industrial and mining sources occurs in water in the +2 oxidation state. Lead from leaded gasoline used to be a major source of atmospheric and terrestrial lead, much of which eventually enters natural water systems. In addition to pollutant sources, lead-bearing limestone and galena (PbS) contribute lead to natural waters in some locations.

Despite greatly increased total use of lead by industry, evidence from hair samples and other sources indicates that body burdens of this toxic metal have decreased during recent decades. This may be the result of less lead used in plumbing and other products that come in contact with food or drink.

Acute lead poisoning in humans causes severe dysfunction in the kidneys, reproductive system, liver, and the brain and central nervous system. Sickness or death results. Lead poisoning from environmental exposure is thought to have caused mental retardation in many children. Mild lead poisoning causes anaemia. The victim may have headaches and sore muscles and may feel generally fatigued and irritable.

Except in isolated cases, lead is probably not a major problem in drinking water, although the potential exists in cases where old lead pipe is still in use. Lead used to be a constituent of solder and some pipe-joint formulations, so that household water does have some contact with lead.

Mercury

Because of its toxicity, mobilisation as methylated forms by anaerobic bacteria, and other pollution factors, mercury generates a great deal of concern as a heavy-metal pollutant. Mercury is found as a trace component of many minerals, with continental rocks containing an average of around 80 parts per billion, or slightly less, of this element. Cinnabar, red mercuric sulphide, is the chief commercial mercury ore. Fossil fuel coal and lignite contain mercury, often at levels of 100 parts per billion or even higher, a matter of some concern with increased use of these fuels for energy resources.

Metallic mercury is used as an electrode in the electrolytic generation of chlorine gas, in laboratory vacuum apparatus, and in other applications. Large quantities of inorganic mercury(I) and mercury(II) compounds are used annually. Organic mercury compounds used to be widely applied as pesticides, particularly fungicides. These mercury compounds include aryl mercurials such as phenyl mercuric dimethyldithiocarbamate (used in paper mills as a slimicide and as a mould retardant for paper), and alkylmercurials such as ethylmercuric chloride, C_2H_5HgCl, used as a seed fungicide. The alkyl mercury compounds tend to resist degradation and are generally considered to be more of an environmental threat than either the aryl or inorganic compounds.

Mercury enters the environment from a large number of miscellaneous sources related to human use of the element. These include discarded laboratory chemicals, batteries, broken thermometers, lawn fungicides, amalgam tooth fillings, and pharmaceutical products. Taken individually, each of these sources may not contribute much of the toxic metal, but the total effect can be substantial. Sewage effluent sometimes contains up to 10 times the level of mercury found in typical natural waters.

Among the toxicological effects of mercury are neurological damage, including irritability, paralysis, blindness, or insanity; chromosome breakage; and birth defects. The milder symptoms of mercury poisoning, such as depression and irritability, have a psychopathological character. Therefore, mild mercury poisoning may escape detection. Some forms of mercury are relatively nontoxic and have been used as medicines, in the treatment of syphilis, for example, for centuries. Other forms of mercury, particularly organic compounds, are highly toxic.

The unexpectedly high concentrations of mercury found in water and in fish tissues result from the formation of soluble monomethylmercury ion, CH_3Hg^+, and volatile dimethylmercury, $(CH_3)_2Hg$, by anaerobic bacteria in sediments. Mercury from these compounds becomes concentrated in fish lipid (fat) tissue and the concentration factor from water to fish may exceed 10^3. The methylating agent by which inorganic mercury is converted to methylmercury compounds is methylcobalamin, a vitamin B_{12} analog:

$$HgCl_2 \xrightarrow{\text{Methylcobalamin}} CH_3HgCl + Cl^- \qquad\qquad ...(6.3)$$

It is believed that the bacteria that synthesise methane produce methylcobalamin as an intermediate in the synthesis. Thus, waters and sediments in which anaerobic decay is occurring, provide the conditions under which methylmercury production occurs. In neutral or alkaline waters, the formation of dimethylmercury, $(CH_3)_2Hg$, is favoured. This volatile compound can escape to the atmosphere.

METALLOIDS

The most significant water pollutant metalloid element is arsenic, a toxic element that has been the chemical villain of more than a few murder plots. Acute arsenic poisoning can result from the ingestion of more than about 100 mg of the element. Chronic poisoning occurs with the ingestion of small amounts of arsenic over a long period of time. There is some evidence that this element is also carcinogenic.

Arsenic occurs in the earth's crust at an average level of 2–5 ppm. The combustion of fossil fuels, particularly coal, introduces large quantities of arsenic into the environment, much of it reaching natural waters. Arsenic occurs with phosphate minerals and enters into the environment along with some phosphorus compounds. The most common of these are lead arsenate/sodium arsenite, Na_3AsO_3, $Pb_3(AsO_4)_2$; and Paris Green, $Cu_3(AsO_3)_2$. Another major source of arsenic is mine tailings. Arsenic produced as a by-product of copper, gold, and lead refining greatly exceeds the commercial demand for arsenic, and it accumulates as waste material.

Like mercury, arsenic may be converted to more mobile and toxic methyl derivatives by bacteria, according to the following reactions :

$$H_3AsO_4 + 2H^+ + 2e^- \rightarrow H_3AsO_3 + H_2O \qquad \text{...(6.4)}$$

$$H_3AsO_3 \xrightarrow{\text{Methylcobalamin}} CH_3AsO(OH)_2 \qquad \text{...(6.5)}$$

Methylarsinic
acid

$$CH_3AsO(OH)_2 \xrightarrow{\text{Methylcobalamin}} (CH_3)_2AsO(OH) \qquad \text{...(6.6)}$$

(Dimethylarsinic acid)

$$(CH_3)_2AsO(OH) + 4H^+ + 4e^- \rightarrow (CH_3)_2AsH + 2H_2O \qquad \text{...(6.7)}$$

ORGANICALLY BOUND METALS AND METALLOIDS

There are two major types of metal-organic interactions to be considered in an aquatic system. The first of these is complexation, usually chelation when organic ligands are involved. A reasonable definition of complexation by organics applicable to natural water and waste-water systems is a system in which a species is present that reversibly dissociates to a metal ion and an organic complexing species as a function of hydrogen ion concentration :

$$ML + 2H^+ \rightleftharpoons M^{2+} + H_2L \qquad \text{...(6.8)}$$

In this equation, M^{2+} is a metal ion and H_2L is the acidic form of a complexing — frequently chelating — ligand, L^{2-}, illustrated here as a compound that has 2 ionisable hydrogens.

Organometallic compounds, on the other hand, contain metals bound to organic entities by way of a carbon atom and do not dissociate reversibly at lower pH or greater dilution. Furthermore, the organic component, and sometimes the particular oxidation state of the metal involved, may not be stable apart from the organometallic compound.

The interaction of trace metals with organic compounds in natural waters is too vast an area to cover in detail. However, it may be noted that metal-organic interactions may involve organic species of both pollutant (such as EDTA) and natural (such as fulvic acids) origin. These interactions are influenced by, and sometimes play a role in, redox equilibria; formation and dissolution of precipitates; colloid formation and stability; acid-base reactions; and microorganism-mediated reactions in water. Metal-organic interactions may increase or decrease the toxicity of metals in aquatic ecosystems, and they have a strong influence on the growth of algae in water.

Organotin Compounds

Of all the metals, tin has the greatest number of organometallic compounds in commercial use. In addition to synthetic organotin compounds, methylated tin species can be produced biologically in the environment.

Major industrial uses of organotin compounds include applications of tin compounds in fungicides, acaricides, disinfectants, antifouling paints, stabilisers to lessen the effects of heat and light in PVC plastics, catalysts, and precursors for the formation of films of SnO_2 on glass. Tributyl tin chloride and related tributyl tin (TBT) compounds have bactericidal, fungicidal, and insecticidal properties and are of particular environmental significance because of growing use as industrial biocides. In addition to tributyl tin chloride, other tributyl tin compounds used as biocides include the hydroxide, the naphthenate, bis(tributyltin) oxide, and tris(tributylstannyl) phosphate. A major use of TBT is in boat and ship hull coating to prevent the growth of fouling organisms. Other applications include preservation of wood, leather, paper, and textiles. Antifungal TBT compounds are used as slimicides in cooling tower water. Because of their applications near or in contact with bodies of water, organotin compounds are potentially significant water pollutants.

INORGANIC SPECIES

Among the inorganic water pollutants, cyanide ion, CN^-, is probably the most important. Others include ammonia, carbon dioxide, hydrogen sulphide, nitrite, and sulphite.

Cyanide

Cyanide, a deadly poisonous substance, exists in water as HCN, a weak acid with a K_a of 6×10^{-10}. The cyanide ion has a strong affinity for many metal ions, forming relatively less-toxic ferrocyanide, $Fe(CN)_6^{4-}$, with iron, for example. Volatile HCN is very toxic and has been used in gas chamber executions in some countries.

Cyanide is widely used in industry, especially for metal cleaning and electroplating. It is also one of the main gas and coke scrubber effluent pollutants from gas works and coke ovens. Cyanide is widely used in certain mineral-processing operations.

Excessive and Other Inorganic Pollutants

Excessive levels of ammoniacal nitrogen cause water-quality problems. Ammonia is the initial product of the decay of nitrogenous organic wastes, and its presence frequently indicates the presence of such wastes. It is a normal constituent of low-pE groundwaters and is sometimes added to drinking water, where it reacts with chlorine to provide residual chlorine.

Hydrogen sulphide

Hydrogen sulphide, H_2S, is a product of the anaerobic decay of organic matter containing sulphur. It is also produced in the anaerobic reduction of sulphate by microorganisms and is evolved as a gaseous pollutant from geothermal waters. Wastes from chemical plants, paper mills, textile mills, and tanneries may also contain H_2S. Its presence is easily detected by its characteristic rotten-egg odour. The sulphide ion has tremendous affinity for many heavy metals.

Carbon dioxide

Free carbon dioxide, CO_2, is frequently present in water at high levels due to decay of organic matter. It is also added to softened water during water treatment as part of a recarbonation process. Excessive carbon dioxide levels may make water more corrosive and may be harmful to aquatic life.

Nitrite ion

Nitrite ion, NO_2^-, occurs in water as an intermediate oxidation state of nitrogen. Its pE range of stability is relatively narrow. Nitrite is added to some industrial process water to inhibit corrosion.

Sulphite ion

SO_3^{2-}, is found in some industrial waste-waters. Sodium sulphie is commonly added to boiler feedwaters as an oxygen scavenger :

$$2SO_3^{2-} + O_2 \rightarrow 2SO_4^{2-}$$ ...(6.9)

Asbestos in Water

The toxicity of inhaled asbestos is well established. The fibres scar lung tissue and cancer eventually develops.

ALGAL NUTRIENTS AND EUTROPHICATION

The term eutrophication, describes a condition of lakes or reservoirs involving excess algal growth, which may eventually lead to severe deterioration of the body of water. The first step in eutrophication of a body of water is an input of plant nutrients from watershed runoff or sewage. The nutrient-rich body of water then produces a great deal of plant biomass by photosynthesis, along with a smaller amount of animal biomass. Dead biomass accumulates in the bottom of the lake, where it partially decays, recycling nutrients carbon dioxide, phosphorus, nitrogen, and potassium. If the lake is not too deep, bottom-rooted plants begin to grow, accelerating the accumulation of solid material in the basin. Eventually a marsh is formed, which finally fills in to produce a meadow or forest.

Eutrophication is often a natural phenomenon; for instance, it is basically responsible for the formation of huge deposits of coal and peat. However, human activity can greatly accelerate the process. In most cases, the single plant nutrient most likely to be limiting is phosphorus, and it is generally named as the culprit in excessive eutrophication. Household detergents are a common source of phosphate in waste-water, and eutrophication control has concentrated upon eliminating phosphates from detergents, removing phosphate at the sewage-treatment plant, and preventing phosphate-laden sewage effluents (treated or untreated) from entering bodies of water.

ACIDITY, ALKALINITY AND SALINITY

Aquatic biota are sensitive to extremes of pH. Largely because of osmotic effects, they cannot live in a medium having a salinity to which they are not adapted. Thus, a fresh-water fish soon succumbs in the ocean, and sea fish normally cannot live in fresh water. Excess salinity soon kills plants not adapted to it. There are, of course, ranges in salinity and pH in which organisms live and these ranges frequently may be represented by a reasonably symmetrical curve, along the fringes of which an organism may live without really thriving. These curves do not generally exhibit a sharp cutoff at one end or the other, as does the high-temperature end of the curve representing the growth of bacteria as a function of temperature.

The most common source of pollutant acid in water is acid mine drainage. The sulphuric acid in such drainage arises from the microbial oxidation of pyrite or other sulphide minerals. The values of pH encountered in acid-polluted water may fall below 3, a condition deadly to most forms of aquatic life except the culprit bacteria mediating the pyrite and iron oxidation. Industrial wastes frequently contribute strong acid to water. Sulphuric acid produced by the air oxidation of pollutant sulphur dioxide enters natural waters as acidic rainfall. In cases where the water does not have contact with a basic mineral, such as limestone, the water pH may become dangerously low.

Excess alkalinity, and frequently accompanying high pH, generally are not introduced directly into water by human activity. However, in many geographic areas, the soil and mineral strata are alkaline and impart a high alkalinity to water. Human activity can aggravate the situation; for example, by exposure of alkaline overburden from strip mining to surface water or groundwater. Excess alkalinity in water is manifested by a characteristic fringe of white salts at the edges of a body of water or on the banks of a stream.

Water salinity may be increased by a number of human activities. Water passing through a municipal water system inevitably picks up salt from a number of processes; for example, recharging water softeners with sodium chloride. Salts can leach from spoil piles. One of the major environmental constraints on the production of shale oil, for example, is the high percentage of leachable sodium sulphate in piles of spent shale. Careful control of these wastes is necessary to prevent further saline pollution of water in areas where salinity is already a problem.

The water evaporates in the dry summer heat, leaving a salt-laden area behind which no longer supports much plant growth. With time, these areas spread, destroying the productivity of crop land.

OXYGEN, OXIDANTS, AND REDUCTANTS

Oxygen is a vitally important species in water. In water, oxygen is consumed rapidly by the oxidation of organic matter, $\{CH_2O\}$:

$$\{CH_2O\} + O_2 \xrightarrow{\text{Microorganisms}} CO_2 + H_2O \qquad ...(6.10)$$

Unless the water is reaerated efficiently, as by turbulent flow in a shallow stream, it rapidly becomes depleted in oxygen and will not support higher forms of aquatic life. In addition to the microorganism-mediated oxidation of organic matter, oxygen in water may be consumed by the biooxidation of nitrogenous material

$$NH_4^+ + 2O_2 \rightarrow 2H^+ + NO_3^- + H_2O \qquad ...(6.11)$$

and by the chemical or biochemical oxidation of chemical reducing agents :

$$4Fe^{2+} + O_2 + 10H_2O \rightarrow 4Fe(OH)_3(s) + 8H^+ \qquad ...(6.12)$$

$$2SO_3^{2-} + O_2 \rightarrow 2SO_4^{2-} \qquad ...(6.13)$$

All these processes contribute to the deoxygenation of water.

The degree of oxygen consumption by microbially-mediated oxidation of contaminants in water is called the biochemical oxygen demand (or biological oxygen demand), BOD. This parameter is commonly measured by determining the quantity of oxygen utilised by suitable aquatic microorganisms during a five-day period. Despite the somewhat arbitrary five-day period, this test remains a respectable measure of the short-term oxygen demand exerted by a pollutant.

The addition of oxidisable pollutants to streams produces a typical oxygen sag curve as shown in Fig. 6.1. Initially, a well-aerated, unpolluted stream is relatively free of oxidisable material; the oxygen level is high; and the bacterial population is relatively low. With the addition of oxidisable pollutant, the oxygen level drops because reaeration cannot keep up with oxygen consumption. In the decomposition zone, the bacterial population rises. The septic zone is characterised by a high bacterial population and very low oxygen levels. The septic zone terminates when the oxidisable pollutant is exhausted, and then the recovery zone begins. In the recovery zone, the bacterial population decreases and the dissolved oxygen level increases until the water regains its original condition.

Although BOD is a reasonably realistic measure of water quality insofar as oxygen is concerned, the test for determining it is time-consuming and cumbersome to perform. Total organic carbon (TOC), is frequently measured by catalytically oxidising carbon in the water and measuring CO_2 that is evolved. It has become popular because TOC determination is easily performed instrumentally.

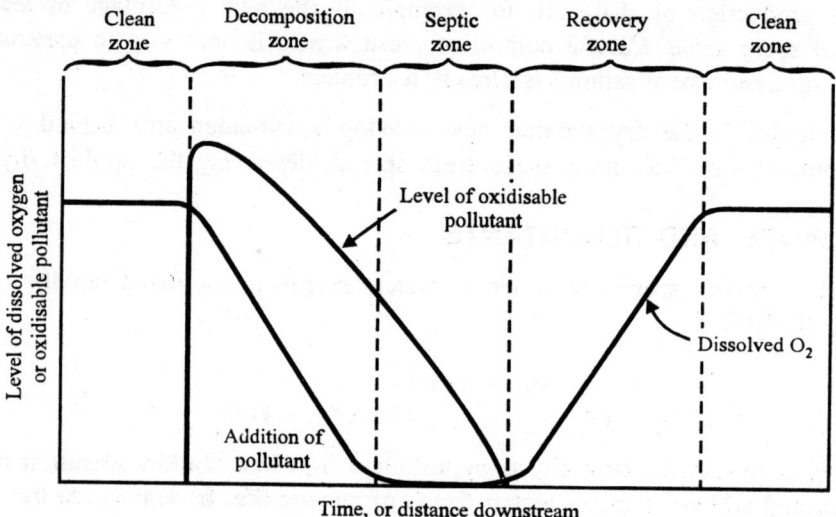

Fig. 6.1. Oxygen sag curve resulting from the addition of oxidisable pollutant material to a stream.

ORGANIC POLLUTANTS

Sewage

As shown in Table 6.3, sewage from domestic, commercial, food-processing, and industrial sources contains a wide variety of pollutants, including organic pollutants. Some of these pollutants, particularly oxygen-demanding substances, oil, grease, and solids, are removed by primary and secondary sewage-treatment processes. Others, such as salts, heavy metals, and refractory (degradation-resistant) organics, are not efficiently removed.

Disposal of inadequately treated sewage can cause severe problems. For example, offshore disposal of sewage results in the formation of beds of sewage residues. Municipal sewage typically contains about 0.1% solids, even after treatment, and these settle out in the ocean in a typical pattern. The warm sewage water rises in the cold hypolimnion and is carried in one direction or another by tides or currents. It does not rise above the thermocline; instead, it spreads out as a cloud from which the solids rain down on the ocean floor. Aggregation of sewage colloids is aided by dissolved salts in seawater, thus promoting the formation of sludge-containing sediment.

Table 6.3. Some of the primary constituents of sewage from a city sewage system.

Constituent	Potential sources	Effects in water
Oxygen-demanding substances	Mostly organic materials, particularly human feces	Consume dissolved oxygen
Refractory organics	Industrial wastes, household products	Toxic to aquatic life
Viruses	Human wastes	Cause disease (possibly cancer); major deterrent to sewage recycle through water systems
Detergents	Household detergents	Esthetics, prevent grease and oil removal, toxic to aquatic life
Phosphates	Detergents	Algal nutrients
Grease and oil	Cooking, food processing, industrial wastes	Esthetics, harmful to some aquatic life
Salts	Human wastes, water softeners, industrial wastes	Increase water salinity
Heavy metals	Industrial wastes, chemical laboratories	Toxicity
Chelating agents	Some detergents, industrial wastes	Heavy metal ion solubilisation and transport
Solids	All sources	Esthetics, harmful to aquatic life

Another major disposal problem with sewage is the sludge produced as a product of the sewage treatment process. This sludge contains organic material which continues to degrade slowly; refractory organics; and heavy metals. The amounts of sludge produced are truly staggering. A major consideration in the safe disposal of such amounts of sludge is the presence of potentially dangerous components such as heavy metals. Careful control of sewage sources is needed to minimise sewage pollution problems. Particularly, heavy metals and refractory organic compounds need to be controlled at the source to enable use of sewage, or treated sewage effluents, for irrigation, recycle to the water system, or groundwater recharge.

Soaps, Detergents, and Detergents Builders

Soaps, detergents, and associated chemicals are potential sources of organic pollutants. These pollutants are discussed briefly here.

Soaps

Soaps are salts of higher fatty acids, such as sodium stearate, $C_{17}H_{35}COO^-Na^+$. The cleaning action of soap results largely from its emulsifying power and its ability to lower the surface tension of water. This concept may be understood by considering the dual nature of the soap anion. An examination of its structure shows that the stearate ion consists of an ionic carboxyl "head" and a long hydrocarbon "tail":

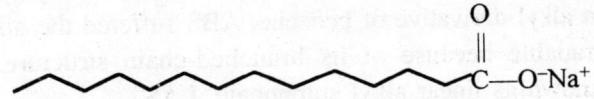

In the presence of oils, fats, and other water-insoluble organic materials, the tendency is for the "tail" of the anion to dissolve in the organic matter, whereas the "head" remains in aquatic solution. Thus, the soap emulsifies, or suspends, organic material in water. In the process, the anions form colloidal soap micelles. The primary disadvantage of soap as a cleaning agent comes from its reaction with divalent cations to form insoluble salts of fatty acids.

If sufficient soap is used, all of the divalent cations may be removed by their reaction with soap, and the water containing excess soap will have good cleaning qualities. This is the approach commonly used when soap is employed with unsoftened water in the bathtub or wash basin, where the insoluble calcium and magnesium salts can be tolerated. However, in applications such as washing clothings, the water must be softened by the removal of calcium and magnesium or their complexation by substances such as polyphosphates.

Although the formation of insoluble calcium and magnesium salts has resulted in the essential elimination of soap as a cleaning agent for clothings, dishes, and most other materials, it has distinct advantages from the environmental standpoint. As soon as soap gets into sewage or an aquatic system, it generally precipitates as calcium and magnesium salts. Hence, any effects that soap might have in solution are eliminated. With eventual biodegradation, the soap is completely eliminated from the environment. Therefore, aside from the occasional formation of unsightly scum, soap does not cause any substantial pollution problems.

Detergents

Synthetic detergents have good cleaning properties and do not form insoluble salts with "hardness ions" such as calcium and magnesium. Such synthetic detergents have the additional advantage of being the salts of relatively strong acids, and, therefore, they do not precipitate out of acidic waters as insoluble acids, an undesirable characteristic of soaps.

The key ingredient of detergents is the surfactant or surface-active agent, which acts in effect to make water "wetter" and a better cleaning agent. Surfactants concentrate at interfaces of water with gases (air), solids (dirt), and immiscible liquids (oil). They do so because of their amphiphilic structure, meaning that one part of the molecule is a polar or ionic group (head) with a strong affinity for water, and the other part is a hydrocarbon group (tail) with an aversion to water. This kind of structure is illustrated below for the structures of surfactant sodium dodecylsulphate:

$$Na^+O-\overset{\displaystyle O}{\underset{\displaystyle O}{\overset{\|}{\underset{\|}{S}}}}-O-\overset{H}{\underset{H}{C}}-\overset{H}{\underset{H}{C}}-\overset{H}{\underset{H}{C}}-\overset{H}{\underset{H}{C}}-\overset{H}{\underset{H}{C}}-\overset{H}{\underset{H}{C}}-\overset{H}{\underset{H}{C}}-\overset{H}{\underset{H}{C}}-\overset{H}{\underset{H}{C}}-\overset{H}{\underset{H}{C}}-\overset{H}{\underset{H}{C}}-\overset{H}{\underset{H}{C}}-H$$

This compound is an alkyl sulphate surfactant, a type widely used in a variety of shampoos, cosmetics, cleaners, and laundry detergent formulations.

Until the early 1960s, the most common surfactant used was an alkyl benzene sulphonate, ABS, a sulphonation product of an alkyl derivative of benzene. ABS suffered the distinct disadvantage of being only very slowly biodegradable because of its branched-chain structure. ABS was replaced by a biodegradable surfactant known as linear alkyl sulphonate, LAS.

LAS, α–benzenesulphonate, has the general structure :

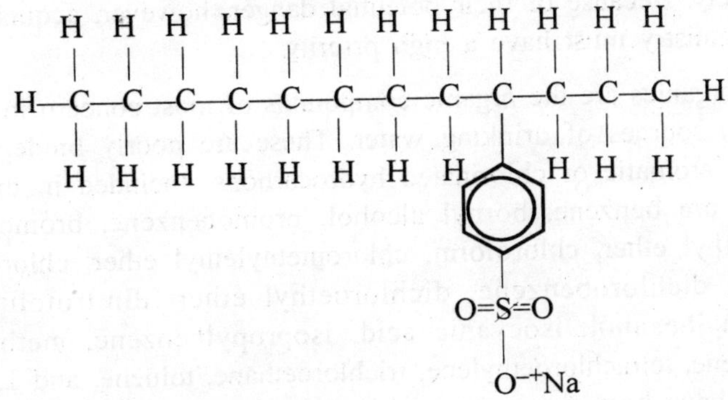

H—C—C—C—C—C—C—C—C—C—C—C—C—H (with H atoms above and below each carbon, and a benzene ring attached bearing $O=S=O$ and $O^-{}^+Na$)

where the benzene ring may be attached at any point on the alkyl chain except at the ends. LAS is more biodegradable than ABS because the alkyl portion of LAS is not branched and does not contain the tertiary carbon which is so detrimental to biodegradability. Since LAS has replaced ABS in detergents, problems arising from the surface-active agents in the detergents (such as toxicity to fish fingerlings) have greatly diminished and the levels of surface-active agents found in water have decreased markedly.

Most of the environmental problems currently attributed to detergents do not arise from the surface-active agents, which basically improve the wetting qualities of water. The builders added to detergents continued to cause environmental problems for a longer time. However, builders bind to hardness ions, making the detergent solution alkaline and greatly improving the action of the detergent surfactant. A commercial solid detergent contains only 10–30% surfactant. In addition, some detergents still contain polyphosphates added to complex calcium and to function as a builder. Other ingredients include ion exchangers, alkalies (sodium carbonate), anticorrosive sodium silicates, amide foam stabilisers, soil-suspending carboxymethylcellulose, bleaches, fabric softeners, enzymes, optical brighteners, fragrances, dyes, and diluent sodium sulphate. Of these materials, the polyphosphates have caused the most concern as environmental pollutants, although these problems have largely been resolved.

Increasing demands on the performance of detergents have led to a growing use of enzymes in detergent formulations destined for both domestic and commercial applications. To a degree, enzymes can take the place of chlorine and phosphates, both of which can have detrimental environmental consequences. Lipases and cellulases appear to be the most rapidly growing segments of the detergent enzyme market.

Biorefractory Organic Pollutants

Millions of tonnes of organic compounds are manufactured globally each year. Significantly quantities of several thousand such compounds appear as water pollutants. Most of these compounds, particularly the less biodegradable ones, are substances to which living organisms have not been exposed until recent years. Frequently, their effects upon organisms are not known, particularly for long-term exposures at very low levels. The potential of synthetic organics for causing genetic damage, cancer, or other ill effects is uncomfortably high. On the positive side, organic pesticides enable a level of agricultural productivity without which millions would starve. Synthetic organic chemicals are increasingly taking the place of natural products in short supply. Thus it is that organic chemicals are essential to the operation of a modern society. Because of their potential danger, however, acquisition of knowledge about their environmental chemistry must have a high priority.

Biorefractory organics are the organic compounds of most concern in waste-water, particularly when they are found in sources of drinking water. These are poorly biodegradable substances, prominent among which are aromatic or chlorinated hydrocarbons. Included in the list of biorefractory organic industrial wastes are benzene, bornyl alcohol, bromobenzene, bromochlorobenzene, butylbenzene, camphor, chloroethyl ether, chloroform, chloromethylethyl ether, chloronitrobenzene, chloropyridine, dibromobenzene, dichlorobenzene, dichloroethyl ether, dinitrotoluene, ethylbenzene, ethylene dichloride, 2-ethylhexanol, isocyanic acid, isopropylbenzene, methylbiphenyl, methyl chloride, nitrobenzene, styrene, tetrachloroethylene, trichloroethane, toluene, and 1,2-dimethoxybenzene. Many of these compounds have been found in drinking water, and some are known to cause taste and odour problems in water. Biorefractory compounds are not completely removed by biological treatment, and water contaminated with these compounds must be treated by physical and chemical means, including air stripping, solvent extraction, ozonation, and carbon adsorption.

PESTICIDES IN WATER

Chemicals used in the control of invertebrates include insecticides, molluscides for the control of snails and slugs, and nematicides for the control of microscopic roundworms. Vertebrates are controlled by rodenticides which kill rodents, avicides used to repel birds, and piscicides used in fish control. Herbicides are used to kill plants. Plant growth regulatores, defoliants, and plant desiccants are used for various purposes in the cultivation of plants. Fungicides are used against fungi, bactericides against bacteria, slimicides against slime-causing organisms in water, and algicides against algae. However, insecticides and fungicides are the most important pesticides with respect to human exposure in food because they are applied shortly before or even after harvesting. The potential exists for large quantities of pesticides to enter water either directly, in applications such as mosquito control, or indirectly, primarily from drainage of agricultural lands.

Natural Product Insecticides, Pyrethrins, and Pyrethroids

Several significant classes of insecticides are derived from plants. These include nicotine from tobacco, rotenone extracted from certain legume roots, and pyrethrins. Because of the ways that they are applied and their biodegradabilities, these substances are unlikely to be significant water pollutants.

Pyrethrins, and their synthetic analogs, represent both the oldest and newest of insecticides. Extracts of dried chrysanthemum or pyrethrum flowers, which contain pyrethrin I and related compounds, have been known for their insecticidal properties for a long time, and may have been used as botanical insecticides in China almost 2000 years ago. The most important commercial sources of insecticidal pyrethrins are chrysanthemum varieties grown in Kenya. Pyrethrins have several advantages as insecticides, including facile enzymatic degradation, which makes them relatively safe for mammals; ability to rapidly paralyse ("knock down") flying insects; and good biodegradability characteristics.

Synthetic analogs of the pyrethrins, pyrethroids, have been widely produced as insecticides during recent years. The first of these was allethrin, and another common example is fenvalerate.

DDT and Organochlorine Insecticides

Chlorinated hydrocarbon or organochlorine insecticides are hydrocarbon compounds in which various numbers of hydrogen atoms have been replaced by Cl atoms. Once the most commonly used insecticides, these have been largely phased out of general use because of their toxicities, and particularly their persistence and accumulation in food chains. They are discussed briefly here largely because of their historical interest but their residues in soils and sediments still contribute to water pollution.

Of the organochlorine insecticides, the most notable has been DDT [dichlorodiphenyltrichloroethane or 1,1,1-trichloro-2,2-bis(4-chlorophenyl)ethane]. It is very persistent insecticide and accumulates in food chains. For some time, methoxychlor was a popular DDT substitute, reasonably biodegradable, and with a low toxicity to mammals. Structurally similar chlordane, aldrin, dieldrin/endrin, and heptachlor, all now banned for application in the U.S., share common characteristics of high persistence and suspicions of potential carcinogenicity. Toxaphene is a mixture of up to 177 individual compounds produced by chlorination of camphene, a terpene isolated from pine trees to give a material that contains about 68% Cl and has an empirical formula of $C_{10}H_{10}Cl_8$. This compound had the widest use of any agricultural insecticide, particularly on cotton. A mixture of five isomers, 1,2,3,4,5,6-hexachlorocyclohexane has been widely produced for insecticidal use. Only the gamma isomer is effective as an insecticide, whereas the other isomers give the product a musty odour and tend to undergo bioaccumulation. A formulation of the essentially pure gamma isomer has been marketed as the insecticide called lindane.

Organophosphate insecticides

Organophosphate insecticides are insecticidal organic compounds that contain phosphorus, most of which are organic derivatives of orthophosphoric acid. Some of the organophosphate insecticides are esters of orthophosphoric acid, the most familiar example of which is paraoxon. More commonly, insecticidal phosphorus compounds are phosphorothionate compounds, such as parathion or chlorpyrifos.

The toxicities of organophosphate insecticides vary a great deal. For example, as little as 120 mg of parathion has been known to kill an adult human and a dose of 2 mg has killed a child. Most accidental poisonings have occurred by absorption through the skin. In contrast, malathion shows how differences in structural formula can cause pronounced differences in the properties of organophosphate pesticides. Malathion has two carboxyester linkages which are hydrolysable by carboxylase enzymes to relatively non-toxic products.

Carbamates

Carbamate pesticides have been widely used because some are more biodegradable than the formerly popular organochlorine insecticides and have lower dermal toxicities than most common organophosphate pesticides.

Carbaryl has been widely used as an insecticide on lawns or gardens. It has a low toxicity to mammals. Carbofuran has a high water solubility and acts as a plant systemic insecticide. As a plant systemic insecticide, it is taken up by the roots and leaves of plants so that insects are poisoned by the plant material on which they feed. Pirimicarb has been widely used in agriculture as a systemic aphicide. Unlike many carbamates, it is rather persistent, with a strong tendency to bind to soil.

The toxic effects of carbamates to animals are due to the fact that these compounds inhibit acetylcholinesterase. Unlike some of the organophosphate insecticides, they do so without the need for undergoing a prior biotransformation and are therefore classified as direct inhibitors. Their inhibition of acetylcolinesterase is relatively reversible. Loss of acetylcholinesterase inhibition activity may result from hydrolysis of the carbamate ester, which can occur metabolically.

Herbicides

Bipyridilium compounds

A bipyridilium compound contains two pyridine rings per molecule. The two important pesticidal compounds of this type are the herbicides diquat and paraquat; other members of this class of herbicides include chlormequat, morfamquat, and difenzoquat. Applied directly to plant tissue, these compounds rapidly destroy plant cells and give the plant a frost-bitten appearance.

Because of its widespread use as a herbicide, the possibility exists of substantial paraquat contamination of food. Drinking water contamination by paraquat has also been observed.

Herbicidal heterocyclic nitrogen compounds

A number of important herbicides contain three heterocyclic nitrogen atoms in ring structures and are therefore called triazines. Triazine herbicides inhibit photosynthesis. Selectivity is gained by the inability of target plants to metabolise and detoxify the herbicide. The most long established and common example of this class is atrazine, widely used on corn, and a widespread water pollutant in corn-growing regions. Another member of this class is metrabuzin, which is widely used on soyabeans, sugarcane, and wheat.

Chlorophenoxy herbicides

The chlorophenoxy herbicides, including 2,4-D and 2,4,5-trichlorophenoxyacetic acid (2,4,5-T) are manufactured on a large scale for weed and brush control and as military defoliants. At one time the latter was of particular concern because of contaminant TCDD present as a manufacturing by-product.

Substituted amide herbicides

A diverse group of herbicides consists of substituted amides. Prominent among these are propanil, applied to control weeds in rice fields, and alachor, widely applied to fields to kill germinating grass and broad-leaved weed seedlings.

Nitroaniline herbicides

Nitroaniline herbicides are characterised by the presence of NO_2 and a substituted NH_2 group on a benzene ring. This class of herbicides is widely represented in agricultural applications and includes benefin, oryzalin, pendimethalin, and fluchoralin.

Miscellaneous herbicides

A wide variety of chemicals have been used as herbicides, and have been potential water pollutants. In addition to the classes of compounds discussed above, other types of herbicides include substituted ureas, carbamates, and thiocarbamates.

By-Products of Pesticide Manufacture

A number of water pollution and health problems have been associated with the manufacture of organochlorine pesticides. For example, degradation-resistant hexachlorobenzene, is used as a raw material for the synthesis of other pesticides and has often been found in water.

The most notorious by-products of pesticide manufacture are polychlorinated dibenzodioxins. These species have a high environmental and toxicological significance. Of the dioxins, the most notable pollutant and hazardous waste compound is 2,3,7,8-tetrachlorodibenzo-p-dioxin (TCDD), often referred to simply as "dioxin". Because of its properties, TCDD is a stable, persistent environmental pollutant and hazardous waste constituent of considerable concern. It has been identified in some municipal incineration emissions, and has been a widespread environmental pollutant from improper waste disposal.

POLYCHLORINATED BIPHENYLS

Polychlorinated biphenyls (PCB compounds) have been found throughout the world in water, sediments, bird tissue, and fish tissue. These compounds constitute an important class of special wastes. Polychlorinated biphenyls have very high chemical, thermal, and biological stability; low vapour pressure; and high dielectric constants. These properties have led to the use of PCBs as coolant-insulation fluids in transformers and capacitors; for the impregnation of cotton and asbestos; as plasticisers; and as additives to some epoxy paints.

Several chemical formulations have been developed to substitute for PCBs in electrical applications. Disposal of PCBs from discarded electrical equipment and other sources remains a problem, particularly since PCBs can survive ordinary incineration by escaping as vapours through the smokestack. However, they can be destroyed by special incineration processes.

Askarel

Askarel is the generic name of PCB-containing dielectric fluids in transformers. These fluids are 50–70% PCBs and may contain 30–50% trichlorobenzenes (TCBs).

Biodegradation of PCBs

The biodegradation of PCBs in a river provides an interesting example of microbial degradation of environmental chemicals. As a result of the dumping of PCBs in certain river, these virtually insoluble, dense, hydrophobic materials accumulated in the river's sediments, causing serious concern about their effects on water quality as a result of their bioaccumulation in fish. Methods of removal, such as edging, were deemed prohibitively expensive and likely to cause severe contamination and disposal problems.

Although it is well known that aerobic bacteria could degrade PCBs with only one or two Cl atom constituents, most of the PCB congeners discharged to the sediments had multiple chlorine atom constituents, specifically an average of 3.5 Cl atoms per PCB molecule at the time the PCBs are discharged. Since the PCB products tend to stay in anaerobic surroundings, some assistance is required to provide oxygen to finish the biodegradation aerobically by introducing aerobic bacteria acclimated to PCB biodegradation, along with the oxygen and nutrients required for their growth.

RADIONUCLIDES IN THE AQUATIC ENVIRONMENT

The massive production of radionuclides (radioactive isotopes) by weapons and nuclear reactors has been accompanied by increasing concern about the effects of radioactivity upon health and the environment. Radionuclides are produced as fission products of heavy nuclei of such elements as uranium or plutonium. They are also produced by the reaction of neutrons with stable nuclei. These phenomena are illustrated in Fig. 6.2. Radionuclides are formed in large quantities as waste products in nuclear power generation. Their ultimate disposal is a problem that has caused much controversy regarding the widespread use of nuclear power. Artificially produced radionuclides are also widely used in industrial and medical applications, particularly as "tracers". With so many possible sources of radionuclides, it is impossible to entirely eliminate radioactive contamination of aquatic systems. Furthermore, radionuclides may enter aquatic systems from natural sources. Therefore, the transport, reactions, and biological concentration of radionuclides in aquatic ecosystems are of great importance to the environmental chemist.

Radionuclides differ from other nuclei in that they emit ionising radiation – alpha particles, beta particles, and gamma rays. The most massive of these emissions is the alpha particle, a helium nucleus of atomic mass 4, consisting of two neutrons and two protons. The symbol for an alpha particle is $_2^4\alpha$. An example of alpha production is found in the radioactive decay of uranium-238:

$$_{92}^{238}U \rightarrow {}_{90}^{234}Th + {}_2^4\alpha$$

...(6.14)

This transformation consists of a uranium nucleus, atomic number 92 and atomic mass 238, losing an alpha particle, atomic number 2 and atomic mass 4 to yield a thorium nucleus, atomic number 90 and atomic mass 234.

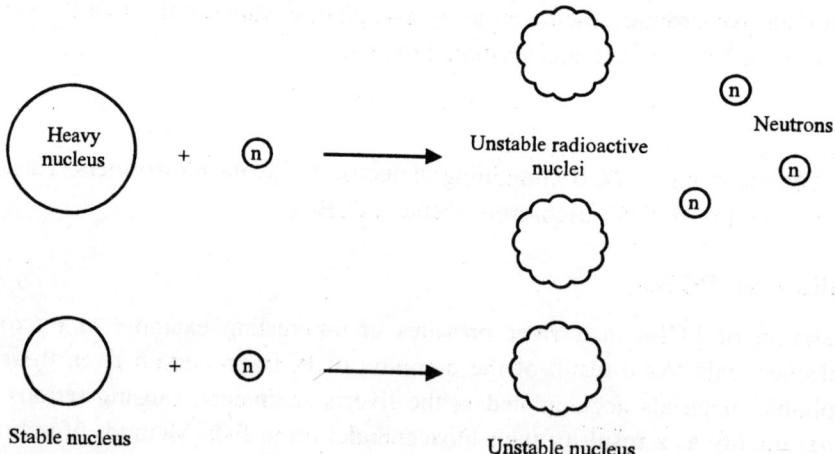

Fig. 6.2. A heavy nucleus, such as that of ^{235}U, may absorb a neutron and break up (undergo fission), yielding lighter radioactive nuclei. A stable nucleus may absorb a neutron to produce a radioactive nucleus.

Beta radiation consists of either highly energetic, negative electrons, which are designated $_{-1}^{0}\beta$, or positive electrons, called positrons and designated $_1^0\beta$. A typical beta emitter, chlorine-38, may be produced by irradiating chlorine with neutrons. The chlorine-37 nucleus, natural abundance 24.5% absorbs a neutron to produce chlorine-38 and gamma radiation :

$$_{17}^{37}Cl + {}_0^1n + {}_{17}^{38}Cl + \gamma$$

...(6.15)

The chlorine-38 nucleus is radioactive and loses a negative beta particle to become an argon-38 nucleus:

$$^{38}_{17}\text{Cl} \rightarrow\ ^{38}_{18}\text{Ar} +\ ^{0}_{-1}\beta \qquad ...(6.16)$$

Since the negative beta has essentially no mass and a -1 charge, the stable product isotope, argon-38, has the same mass and a charge 1 greater than chlorine-38.

Gamma rays are electromagnetic radiation similar to X-rays, though more energetic. Since the energy of gamma radiation is often a well-defined property of the emitting nucleus, it may be used in some cases for the qualitative and quantitiative analysis of radionuclides.

The primary effect of alpha particles, beta particles, and gamma rays upon materials is the production of ions; therefore, they are called ionising radiation. Due to their large size, alpha particles do not penetrate matter deeply, but cause an enormous amount of ionisation along their short path of penetration.

Therefore, alpha particles present little hazard outside the body, but are very dangerous when ingested. Although beta particles are more penetrating than alpha particles, they produce much less ionisation per unit path length. Gamma rays are much more penetrating than particulate radiation. Their degree of penetration is proportional to their energy.

Radiation damages living organisms by initiating harmful chemical reactions in tissues. For example, bonds are broken in the macromolecules that carry out life processes. In cases of acute radiation poisoning, bone marrow, which produces red blood cells is destroyed and the concentration of red blood cells is diminished. Radiation-induced genetic damage is of great concern. Such damage may not become apparent until many years after exposure. As humans have learned more about the effects of ionising radiation, the dosage level considered to be safe has steadily diminished.

Some radionuclides found in water, primarily radium and potassium-40, originate from natural sources, particularly leaching from minerals. Others come from pollutant sources, primarily nuclear power plants and testing of nuclear weapons.

The levels of radionuclides found in water typically are measured in units of picocuries/litre, where a curie is 3.7×10^{10} disintegrations per second, and a picocurie is 1×10^{-12} that amount, or 3.7×10^{-2} disintegrations per second (2.2 disintegrations per minute). The radionuclide of most concern in drinking water is radium.

As the use of nuclear power has increased, the possible contamination of water by fission-product radioisotopes has become more of a cause for concern. (If nations continue to refrain from testing nuclear weapons above ground, it is hoped that radioisotopes from this source will contribute only minor amounts of radioactivity to water.) Table 6.4 summarises the major natural and artificial radionuclides likely to be encountered in water.

Transuranic elements are of growing concern in the ocean environment. These alpha emitters are long-lived and highly toxic. As their production increases, so does the risk of environmental contamination. Included among these elements are various isotopes of neptunium, plutonium, americium, and curium. Specific isotopes, with half-lives in years given in parentheses, are: Np-237 (2.14×10^{6}); Pu-236 (2.85); Pu-238 (87.8); Pu-239 (2.44×10^{4}); Pu-240 (6.54×10^{3}); Pu-241 (15); Pu-242 (3.87×10^{5}); Am-241 (433); Am-243 (7.37×10^{6}); Cm-242 (0.22); and Cm-244 (17.9).

Table 6.4. Radionuclides in water.

Radionuclide	Half-life	Nuclear Reaction, Description, Source
Naturally occurring and from cosmic reactions		
Carbon-14	5730 years	$^{14}N(n,p)$ ^{14}C,[a] thermal neutrons from cosmic or nuclear-weapon sources reacting with N_2.
Silicon-32	~ 300 years	^{40}Ar (p,x) ^{32}Si, nuclear spallation (splitting of the nucleus) of atmospheric argon by cosmic-ray protons
Potassium-40	~ $1/4 \times 10^9$ years	0.0119% of natural potassium.
Naturally occurring from ^{238}U series		
Radium-226	1620 years	Diffusion from sediments, atmosphere
Lead-210	21 years	$^{226}Ra \rightarrow$ 6 steps $\rightarrow$ ^{210}Pb
Thorium-230	75,200 years	$^{238}U \rightarrow$ 3 steps $\rightarrow$ ^{230}Th produced *in situ*
Thorium-234	24 days	$^{238}U \rightarrow$ ^{234}Th produced *in situ*
From reactor and weapons fission		
Strontium-90	28 years	These are the fissions-product radioisotopes of greatest significance because of their high yields and biological activity.
Iodine-131	8 days	
Cesium-137	30 years	
Barium-140	13 days	The isotopes from barium-140 through krypton-85 are listed in generally decreasing order of fission yield.
Zirconium-95	65 days	
Cerium-141	33 days	
Strontium-89	51 days	
Ruthenium-103	40 days	
Krypton-85	10.3 years	
Cobalt-60	5.25 years	From non-fission neutron reactions in reactors
Manganese-54	310 years	From nonfission neutron reactions in reactors
Iron-55	2.7 years	$^{56}Fe(n, 2n)$ ^{55}Fe, from high-energy neutrons acting on iron in weapon hardware.
Plutonium-239	24,300 years	^{238}U (n, γ) ^{239}Pu, neutron capture by uranium.

[a] This notation denotes the isotope nitrogen-14 reacting with a neutron, n, giving off a proton, p, and forming the isotope carbon-14; other nuclear reactions may be similarly deduced from the notation shown. (Note that × represents nuclear fragments from the spallation reaction).

CHAPTER 7

Water Treatment

INTRODUCTION

The treatment of water may be divided into three major categories: (i) purification for domestic use; (ii) treatment for specialised industrial applications; and (iii) treatment of waste-water to make it acceptable for release or reuse.

The type and degree of treatment are strongly dependent upon the source and intended use of the water. Water for domestic use must be thoroughly disinfected to eliminate disease-causing microorganisms, but may contain appreciable levels of dissolved calcium and magnesium (hardness). Water to be used in boilers may contain bacteria but must be quite soft to prevent scale formation. Waste-water being discharged into a large river may require less rigorous treatment than water to be reused in an arid region. As world demand for limited water resources grows, more sophisticated and extensive means will have to be employed to treat water.

Most physical and chemical processes used to treat water involve similar phenomena, regardless of their application to the three main categories of water treatment listed above. Therefore, after introductions to water treatment for municipal use, industrial use, and disposal, each major kind of treatment process is discussed as it applies to all of these applications.

A schematic diagram of a typical municipal water treatment plant is shown in Fig. 7.1. This particular facility treats water containing excessive hardness and a high level of iron. The raw water taken from wells first goes to an aerator. Contact of the water with air removes volatile solutes such as hydrogen sulphide, carbon dioxide, methane, and volatile odorous substances such as methane thiol (CH_3SH) and bacterial metabolites. Contact with oxygen also aids iron removal by oxidising soluble iron to insoluble iron. The addition of lime as CaO or $Ca(OH)_2$ after aeration raises the pH and results in the formation of precipitates containing the hardness ions Ca^{2+} and Mg^{2+}. These precipitates settle from the water in a primary basin. Much of the solid material remains in suspension and requires the addition of coagulants (such as ferric and aluminium sulphates, which form gelatinous metal hydroxides) to settle the colloidal particles. Activated silica or synthetic polyeleectrolytes may also be added to stimulate coagulation or flocculation. The settling occurs in a secondary basin after the addition of carbon dioxide to lower the pH. Sludge from both the primary and secondary basins is pumped to a sludge lagoon. The water is finally chlorinated, filtered, and pumped to the city water mains.

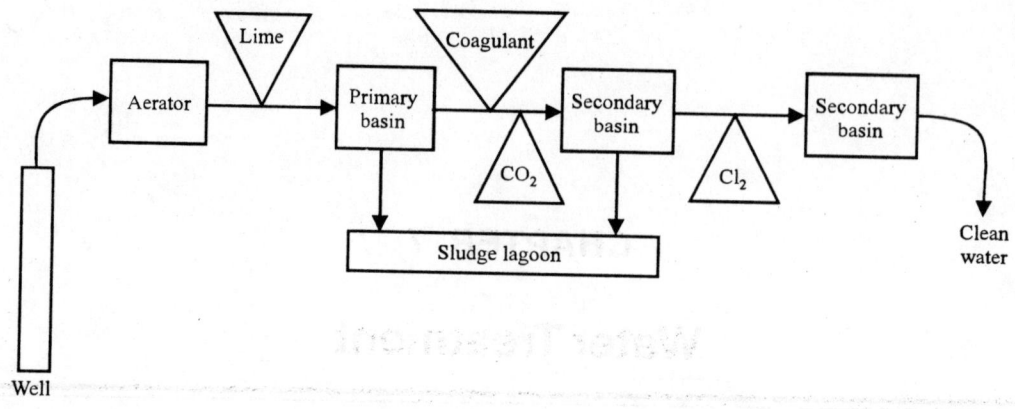

Fig. 7.1. Schematic diagram of a municipal water treatment plant.

TREATMENT OF WATER FOR INDUSTRIAL USE

Water is widely used in various process applications in industry. Other major industrial uses are boiler feedwater and cooling water. The kind and degree of treatment of water in these applications depends upon the end use. As examples: cooling water may require only minimal treatment; removal of corrosive substances and scale-forming solutes is essential for boiler feedwater; and water used in food processing must be free of pathogens and toxic substances. Improper treatment of water for industrial use can cause problems, such as corrosion, scale formation, reduced heat transfer in heat exchangers, reduced water flow, and product contamination. These effects may cause reduced equipment performance or equipment failure, increased energy costs due to inefficient heat utilisation or cooling, increased costs for pumping water, and product deterioration. Obviously, the effective treatment of water at minimum cost for industrial use is a very important area of water treatment.

Numerous factors must be taken into consideration in designing and operating an industrial water treatment facility. These include the following :

1. Water requirement.
2. Quantity and quality of available water sources.
3. Sequential use of water (successive uses for applications requiring progressively lower water quality).
4. Water recycle.
5. Discharge standards.

External treatment, usually applied to the plant's entire water supply, uses processes such as aeration, filtration, and clarification to remove material from water that may cause problems. Such substances include suspended or dissolved solids, hardness, and dissolved gases. Following this basic treatment, the water may be divided into different streams, some to be used without further treatment and the rest to be treated for specific applications. Internal treatment is designed to modify the properties of water for specific applications. Examples of internal treatment include the following :

1. Reaction of dissolved oxygen with hydrazine or sulphite.
2. Addition of chelating agents to react with dissolved Ca^{2+} and prevent formation of calcium deposits.
3. Addition of precipitants, such as phosphate used for calcium removal.

4. Treatment with dispersants to inhibit scale..

5. Addition of inhibitors to prevent corrosion.

6. Adjustment of pH.

7. Disinfection for food processing uses or to prevent bacterial growth in cooling water.

SEWAGE TREATMENT

Typical municipal sewage contains oxygen-demanding materials, sediments, grease, oil, scum, pathogenic bacteria, viruses, salts, algal nutrients, pesticides, refractory organic compounds, heavy metals and an astonishing variety of flotsam ranging from children's socks to sponges. It is the job of the waste treatment plant to remove as much of this material as possible.

Several characteristics are used to describe sewage. These include turbidity (international turbidity units); suspended solids (ppm); total dissolved solids (ppm); acidity (H^+ ion concentration or pH); and dissolved oxygen (in ppm O_2). Biochemical oxygen demand is used as a measure of oxygen-demanding substances.

Current processes for the treatment of waste-water may be divided into three main categories of primary treatment, secondary treatment, and tertiary treatment, each of which is discussed separately. Also discussed are total waste-water treatment systems, based largely upon physical and chemical processes.

Primary Waste Treatment

Primary treatment of waste-water consists of the removal of insoluble matter such as grit, grease, and scum from water. The first step in primary treatment normally is screening. Screening removes or reduces the size of trash and large solids that get into the sewage system. These solids are collected on screens and scraped off for subsequent disposal. Most screens are cleaned with power rakes. Comminuting devices shred and grind solids in the sewage. Particle size may be reduced to the extent that the particles can be returned to the sewage flow.

Grit in waste-water consists of such materials as sand and coffee grounds which do not biodegrade well and generally have a high settling velocity. Grit removal is practised to prevent its accumulation in other parts of the treatment system, to reduce clogging of pipes and other parts, and to protect moving parts from abrasion and wear. Grit normally is allowed to settle in a tank under conditions of low flow velocity, and it is then scraped mechanically from the bottom of the tank.

Primary sedimentation removes both settleable and floatable solids. During primary sedimentation, there is a tendency for flocculent particles to aggregate for better settling, a process that may be aided by the addition of chemicals. The material that floats in the primary settling basin is known collectively as grease. In addition to fatty substances, the grease consists of oils, waxes, free fatty acids, and insoluble soaps containing calcium and magnesium. Normally, some of the grease settles with the sludge and some floats to the surface, where it may be removed by a skimming device.

Secondary Waste Treatment by Biological Processes

The most obvious harmful effect of biodegradable organic matter in waste-water is BOD, consisting of a biochemical oxygen demand for dissolved oxygen by microorganism-mediated degradation of the organic matter. Secondary waste-water treatment is designed to remove BOD, usually by taking advantage of the same kind of biological processes that would otherwise consume oxygen in water receiving the waste-water. Secondary treatment by biological processes takes many forms, but consists basically of the following: Microorganisms provided with added oxygen are allowed to degrade organic material in solution or in suspension until the BOD of the waste has been reduced to acceptable levels. The waste is oxidised biologically under conditions controlled for optimum bacterial growth and at a site where this does not influence the environment.

One of the simplest biological waste treatment processes is the trickling filter (Fig. 7.2) in which waste-water is sprayed over rocks or other solid support material covered with microorganisms. The structure of the trickling filter is such that contact of the waste-water with air is allowed and degradation of reorganic matter occurs by the action of the microorganisms.

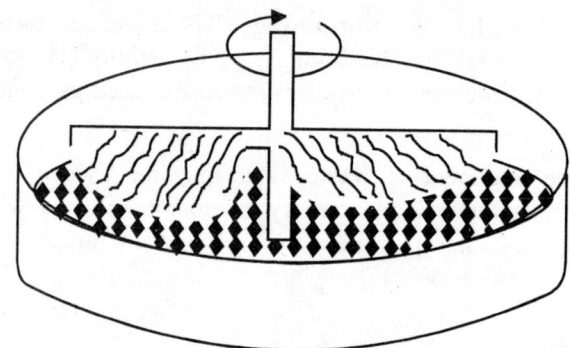

Fig. 7.2. Trickling filter for secondary waste treatment.

Rotating biological reactors, another type of treatment system, consist of groups of large plastic discs mounted close together on a rotating shaft. The device is positioned so that at any particular instant, half of each disc is immersed in waste-water and half exposed to air. The shaft rotates constantly, so that the submerged portion of the discs is always changing. The discs, usually made of high-density polyethylene or polystyrene, accumulate thin layers of attached biomass, which degrades organic matter in the sewage. Oxygen is absorbed by the biomass and by the layer of waste-water adhering to it during the time that the biomass is exposed to air.

Both trickling filters and rotating biological reactors are examples of fixed-film biological (FFB) or attached growth processes. The greatest advantage of these processes is their low energy consumption. The energy consumption is minimal because it is not necessary to pump air or oxygen into the water, as is the case with the popular activated sludge process described below. The trickling filter has long been a standard means of waste-water treatment, and a number of waste-water treatment plants use trickling filters at present.

The activated sludge process, Fig. 7.3, is probably the most versatile and effective of all waste treatment processes. Microorganisms in the aeration tank convert organic material in waste-water to microbial biomass and CO_2. Organic nitrogen is converted to ammonium ion or nitrate. Organic phosphorus is converted to orthophosphate. The microbial cell matter formed as part of the waste degradation processes is normally kept in the aeration tank until the microorganisms are past the log phase of growth, at which point, the cells flocculate relatively well to form settleable solids. These solids settle out in a settller and a fraction of them is discarded. Part of the solids, the return sludge, is recycled to the head of the aeration tank and comes into contact with fresh sewage. The combination of a high concentration of "hungry" cells in the return sludge and a rich food source in the influent sewage provides optimum conditions for the rapid degradation of organic matter.

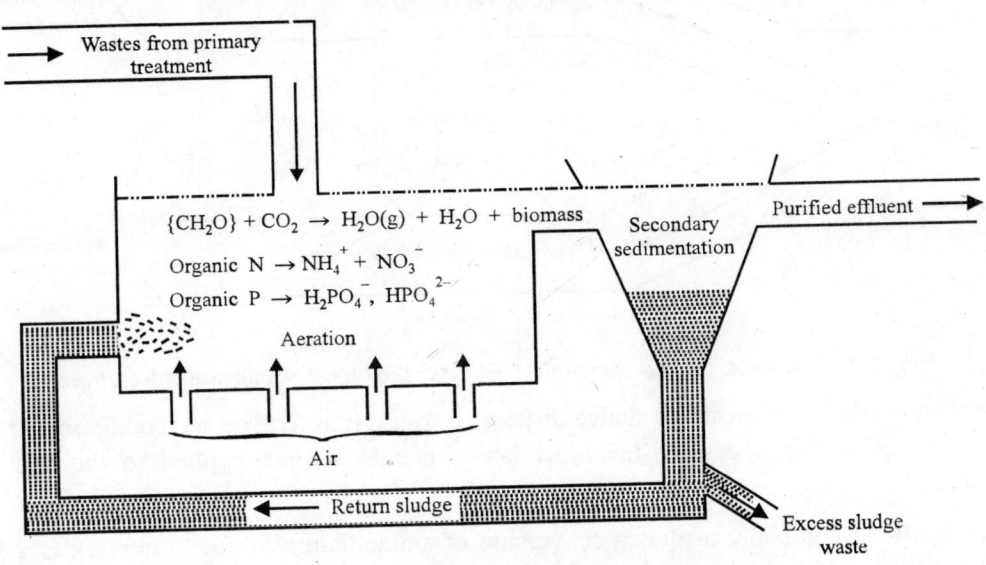

$${CH_2O} + CO_2 \rightarrow H_2O(g) + H_2O + biomass$$

Organic N $\rightarrow NH_4^+ + NO_3^-$

Organic P $\rightarrow H_2PO_4^-, HPO_4^{2-}$

Fig. 7.3. Activated sludge process.

The degradation of organic matter that occurs in an activated sludge facility also occurs in streams and other aquatic environments. However, in general, when a degradable waste is put into a stream, it encounters only a relatively small population of microorganisms capable of carrying out the degradation process. Thus, several days may be required for the buildup of a sufficient population of organisms to degrade the waste. In the activated sludge process, continual recycling of active organisms provides the optimum conditions for waste degradation, and a waste may be degraded within the very few hours that it is present in the aeration tank.

The activated sludge process provides two pathways for the removal of BOD, as illustrated schematically in Fig. 7.4 BOD may be removed by (1) oxidation of organic matter to provide energy for the metabolic processes of the microorganisms, and (2) synthesis, incorporation of the organic matter into cell mass. In the first pathway, carbon is removed in the gaseous form as CO_2. The second pathway provides for removal of carbon as a solid in biomass. That portion of the carbon converted to CO_2 is vented to the atmosphere and does not present a disposal problem. The disposal of waste sludge, however, is a problem, primarily because it is only about 1% solids and contains many undesirable components. Normally, partial water removal is accomplished by drying on sand filters, vacuum filtration, or centrifugation. The dewatered sludge may be incinerated or used as land fill. To a certain extent, sewage sludge may be digested in the absence of oxygen by methane-producing anaerobic bacteria to produce methane and carbon dioxide,

$$2{CH_2O} \rightarrow CH_4 + CO_2 \qquad \qquad ,...(7.1)$$

a process that reduces both the volatile-matter content and the volume of the sludge by about 60%. A carefully designed plant may produce enough methane to provide for all of its power needs.

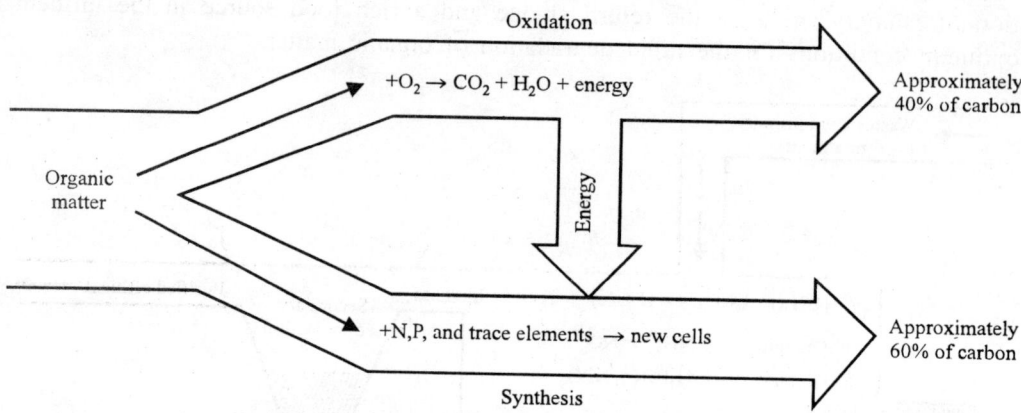

Fig. 7.4. Pathways for the removal of BOD in biological waste-water treatment.

One of the most desirable means of sludge disposal is to use it to fertilise and condition soil. However, care has to be taken that excessive levels of heavy metals are not applied to the soil as sludge contaminants.

Nitrification (the microbially mediated conversion of ammonium nitrogen to nitrate); is a significant process that occurs during biological waste treatment. Ammonium ion is normally the first inorganic nitrogen species produced in the biodegradation of nitrogenous organic compounds. It is oxidised, under the appropriate conditions, first to nitrite by Nitrosomonas bacteria,

$$2NH_4^+ + 3O_2 \rightarrow 4H^+ + 2NO_2^- + 2H_2O \qquad \text{...(7.2)}$$

then to nitrate by Nitrobacter:

$$2NO_2^- + O_2 \rightarrow 2NO_3^- \qquad \text{...(7.3)}$$

These reactions occur in the aeration tank of the activated sludge plant and are favoured in general by long retention times, low organic loadings, large amounts of suspended solids, and high temperatures. Nitrification can reduce sludge settling efficiency because the denitrification reaction :

$$4NO_3^- + 5\{CH_2O\} + 4H^+ \rightarrow 2N_2(g) + 5CO_2(g) + 7H_2O \qquad \text{...(7.4)}$$

occurring in the oxygen-deficient settler causes bubbles to form on the sludge floc (aggregated sludge particles), making it so buoyant that it floats to the top. This prevents settling of the sludge and increases the organic load in the receiving waters. Under the appropriate conditions, however, advantage can be taken of this phenomenon to remove nutrient nitrogen from water.

Tertiary Waste Treatment

Unpleasant as the thought may be, many people drink used water – water that has been discharged from a municipal sewage treatment plant or from some industrial process. This raises serious questions about the presence of pathogenic organisms or toxic substances in such water. Obviously, there is a great need to treat waste-water in a manner that makes it amenable to reuse. This requires treatment beyond the secondary processes.

Tertiary waste treatment (sometimes called advanced waste treatment) is a term used to describe a variety of processes performed on the effluent from secondary waste treatment. The contaminants removed by tertiary waste treatment fall into the general categories of : (i) suspended solids; (ii) dissolved organic compounds; and (iii) dissolved inorganic materials, including the important class of algal nutrients.

Each of these categories presents its own problems with regard to water quality. Suspended solids are primarily responsible for residual biological oxygen demand in secondary sewage effluent waters. The dissolved organics are the most hazardous from the standpoint of potential toxicity. The major problem with dissolved inorganic materials is that presented by algal nutrients, primarily nitrates and phosphates. In addition, potentially hazardous toxic metals may be found among the dissolved inorganics.

In addition to these chemical contaminants, secondary sewage effluents often contains a number of disease-causing microorganisms, requiring disinfection in cases where humans may later come into contact with the water. Among the bacteria that may be found in secondary sewage effluent are organisms causing tuberculosis, dysenteric bacteria (*Bacillus dysenteriae, shigella dysenteriae, shigella paradysenteriae, Proteus vulgaris*), cholera bacteria (*Vibrio cholerae*), bacteria causing mud fever (*Leptopsira icterohemorrhagiae*), and bacteria causing typhoid fever (*Salmonella typhosa, salmonella paratyphi*).

In addition, viruses causing diarrhoea, eye infections, infectious hepatitis, and polio may be encountered. Ingestion of sewage still causes disease, even in more developed nations.

Physical-Chemical Treatment of Municipal Waste-water

Complete physical-chemical waste-water treatment systems offer both advantages and disadvantages relative to biological treatment systems. The capital costs of these facilities can be less than those of biological treatment facilities, and they usually require less land. They are better able to cope with toxic materials and overloads.

However, they require careful operator control and consume relatively large amounts of energy. Basically, a physical-chemical treatment process involves :

1. Removal of scum and solid objects.
2. Clarification, generally with addition of a coagulant, and frequently with the addition of other chemicals (such as lime for phosphorus removal).
3. Filtration to remove filtrable solids;
4. Activated carbon adsorption.
5. Disinfection.

The basic steps of a complete physical-chemical waste-water treatment facility are shown in Fig. 7.5.

INDUSTRIAL WASTE-WATER TREATMENT

Waste-water to be treated must be characterised fully, particularly with a thorough chemical analysis of possible waste constituents and their chemical and metabolic products. The biodegradability of waste-water constituents should also be determined. The options available for the treatment of waste-water are summarised briefly in this section.

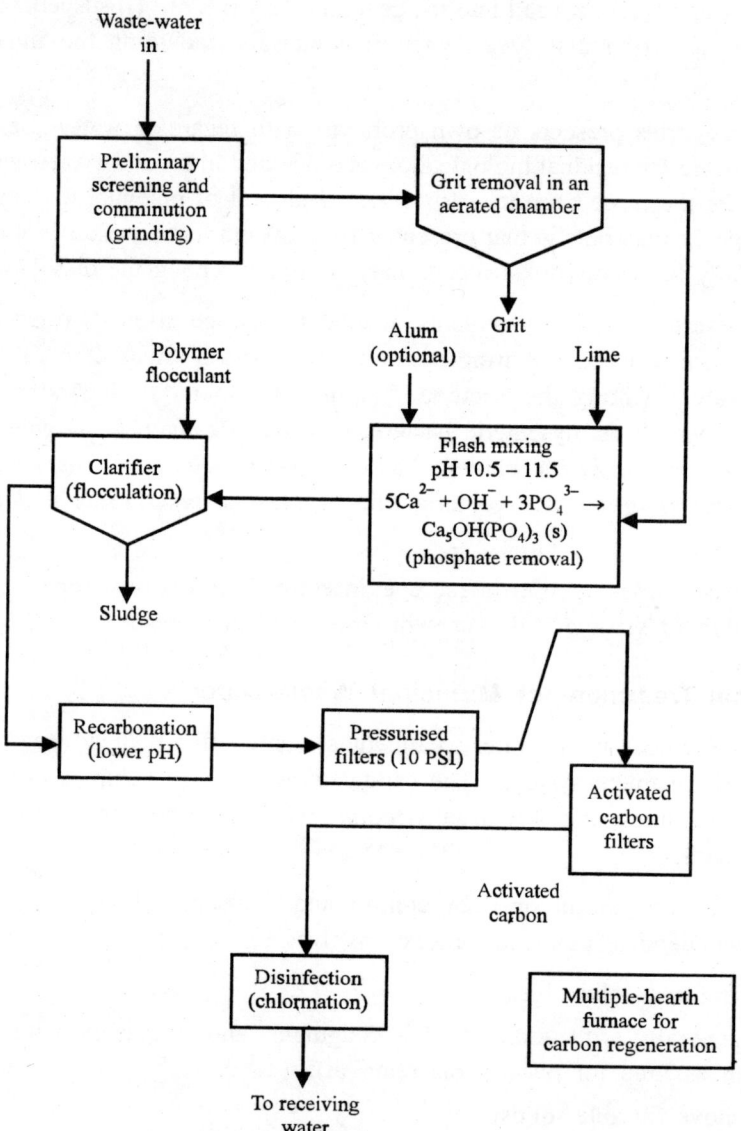

Fig. 7.5. Major components of a complete physical-chemical treatment facility for municipal waste-water.

One of two major ways of removing organic wastes is biological treatment by an activated sludge, or related process. It may be necessary to acclimate microorganisms to the degradation of constituents that are not normally biodegradable. Consideration needs to be given to possible hazards of biotreatment sludges, such as those containing excessive levels of heavy metal ions. The other major process for the removal of organics from waste-water is sorption by activated carbon, usually in columns of granular activated carbon. Activated carbon and biological treatment can be combined with the use of powdered activated carbon in the activated sludge process. The powdered activated carbon sorbs some constituents that may be toxic to microorganisms and is collected with the sludge. A major consideration with the use of activated carbon to treat waste-water is the hazard that spent activated carbon may present from the wastes it retains. These hazards may include those of toxicity or reactivity, such as those posed by explosives manufacture wastes sorbed to activated carbon. Regeneration of the carbon is expensive and can be hazardous in some cases.

Waste-water can be treated by a variety of chemical processes, including acid/base neutralisation, precipitation, and oxidation/reduction. In some cases, these treatment steps must precede biological treatment; for example, waste-water exhibiting extremes of pH must be neutralised in order for microorganisms to thrive in it. Cyanide in the waste-water may be oxidised with chlorine and organics with ozone, hydrogen peroxide promoted with ultraviolet radiation, or dissolved oxygen at high temperatures and pressures. Heavy metals may be precipitated with base, carbonate, or sulphide.

Waste-water can be treated by several physical processes. In some cases, simple density separation and sedimentation can be used to remove water-immiscible liquids and solids. Filtration is frequently required and floatation by gas bubbles generated on particle surfaces may be useful. Waste-water solutes can be concentrated by evaporation, distillation, and membrane processes, including reverse osmosis, hyperfiltration, and ultrafiltration. Organic constituents can be removed by solvent extraction, air stripping, or steam stripping.

Synthetic resins are useful for removing some pollutant solutes from waste-water. Organophilic resins have proved useful for the removal of alcohols; aldehydes; ketones; hydrocarbons; chlorinated alkanes, alkenes, and aryl compounds; esters, including phthalate esters; and pesticides. Cation exchange resins are effective for the removal of heavy metals.

REMOVAL OF SOLIDS

Relatively large solid particles are removed from water by simple settling and filtration. A special type of filtration procedure known as microstraining is especially effective in the removal of the very small particles. These filters are woven from stainless steel wire so fine that it is barely visible. This enables preparation of filters with openings only 60–70 μm across. These openings may be reduced to 5–15 μm by partial clogging with small particles, such as bacterial cells. The cost of this treatment is likely to be substantially lower than the costs of competing processes. High flow rates at low back pressures are normally achieved.

The removal of colloidal solids from water usually requires coagulation. Salts of aluminium and iron are the coagulants most often used in water treatment. Of these, alum or filter alum is most commonly used. This substance is a hydrated aluminium sulphate, $Al_2(SO_4)_3 \cdot 18H_2O$. When this salt is added to water, the aluminium ion hydrolyses by reactions that consume alkalinity in the water, such as :

$$Al(H_2O)_6^{3+} + 3HCO_3^- \rightarrow Al(OH)_3(s) + 3CO_2 + 6H_2O \qquad \qquad ...(7.5)$$

The gelatinous hydroxide thus formed carries suspended material with it as it settles. In addition, however, it is likely that positively charged hydroxyl-bridged dimers and higher polymers are formed which interact specifically with colloidal particles, bringing about coagulation. Metal ions in coagulants also react with virus proteins and destroy up to 99% of the virus in water.

Anhydrous iron sulphate added to water forms ferric hydroxide in a reaction analogous to Eq. 7.5. An advantage of iron sulphate is that it works over a wide pH range of approximately 4–11. Hydrated iron sulphate, $FeSO_4.7H_2O$, or copperas, is also commonly used as a coagulant. It forms a gelatinous precipitate of hydrated iron oxide; in order to function, it must be oxidised to iron by dissolved oxygen in the water at a pH higher than 8.5, or by chlorine, which can oxidise iron at lower pH values.

Sodium silicate partially neutralised by acid aids coagulation, particularly when used with alum. The chemical mechanism by which this activated silica operates is still not known with certainty.

Natural and synthetic polyelectrolytes are used in flocculating particles. Among the natural compounds so used are starch and cellulose derivatives, proteinaceous materials, and gúms composed of polysaccharides. More recently, selected synthetic polymers that are effective flocculants have come into use. Neutral polymers and both anionic and cationic polyelectrolytes have been used successfully as flocculants in various applications.

Coagulation-filtration is a much more effective procedure than filtration alone for the removal of suspended material from water. As the term implies, the process consists of the addition of coagulants that aggregate the particles into larger size particles, followed by filtration. Either alum or lime, often with added polyelectrolytes, is most commonly employed for coagulation.

The filtration step of coagulation-filtration is usually performed on a medium such as sand or anthracite coal. Often, to reduce clogging, several media with progressively smaller interstitial spaces are used. One example is the rapid sand filter, which consists of a layer of sand supported by layers of gravel particles, the particles becoming progressively larger with increasing depth. The substance that actually filters the water is coagulated material that collects in the sand. As more material is removed, accumulation of the coagulated material eventually clogs the filter and must be removed by back-flushing.

An important class of solids that must be removed from waste-water consists of suspended solids in secondary sewage effluent that arise primarily from sludge that was not removed in the settling process. These solids account for a large part of the BOD in the effluent and may interfere with other aspects of tertiary waste treatment. For example, these solids may clog membranes in reverse osmosis water treatment processes. The quantity of material involved may be rather high. Processes designed to remove suspended solids often will remove 10–20 mg/L of organic material from secondary sewage effluent. In addition, a small amount of the inorganic material is removed as well.

REMOVAL OF CALCIUM AND OTHER METALS

Calcium and magnesium salts, which generally are present in water as bicarbonates or sulphates, cause water hardness. One of the most common manifestations of water hardness is the insoluble "curd" formed by the reaction of soap with calcium or magnesium ions. Although ions that cause water hardness do not form insoluble products with detergents, they do adversely affect detergent performance. Therefore, calcium and magnesium must be complexed or removed from water for detergents to function properly.

Another problem caused by hard water is the formation of mineral deposits. For example, when water containing calcium and bicarbonate ions is heated, insoluble calcium carbonate is formed :

$$Ca^{2\oplus} \ 2HCO_3^- \rightarrow CaCO_3(s) + CO_2(g) + H_2O$$

...(7.6)

This product coats the surfaces of hot water systems, clogging pipes and reducing heating efficiency. Dissolved salts such as calcium and magnesium bicarbonates and sulphates can be especially damaging in boiler feedwater. Clearly, the removal of water hardness is essential for many uses of water.

Several processes are used for softening water. On a large scale, such as in community water-softening operations, the lime-soda process is used. This process involves the treatment of water with lime, $Ca(OH)_2$, and soda ash, Na_2CO_3. Calcium is precipitated as $CaCO_3$ and magnesium as $Mg(OH)_2$. When the calcium is present primarily as "bicarbonate hardness," it can be removed by the addition of $Ca(OH)_2$ alone :

$$Ca^{2+} + 2HCO_3^- + Ca(OH)_2 \rightarrow 2CaCO_3(s) + 2H_2O \qquad ...(7.7)$$

When bicarbonate ion is not present at substantial levels, a source of CO_3^{2-} must be provided at a high enough pH to prevent conversion of most of the carbonate to bicarbonate. These conditions are obtained by the addition of Na_2CO_3. For example, calcium present as the chloride can be removed from water by the addition of soda ash :

$$Ca^{2+} + 2Cl^- + 2Na^+ + CO_3^{2-} \rightarrow CaCO_3(s) + 2Cl^- + 2Na^+ \qquad ...(7.8)$$

Note that the removal of bicarbonate hardness results in a net removal of soluble salts from solution, whereas removal of non-bicarbonate hardness involves the addition of at least as many equivalents of ionic material as are removed. The precipitation of magnesium as the hydroxide requires a higher pH than the precipitation of calcium as the carbonate :

$$Mg^{2+} + 2OH^- \rightarrow Mg(OH)_2(s) \qquad ...(7.9)$$

The high pH required may be provided by the basic carbonate ion from soda ash :

$$CO_3^{2-} + H_2O \rightarrow HCO_3^- + OH^- \qquad ...(7.10)$$

Some large-scale, lime-soda softening plants make use of the precipitated calcium carbonate product as a source of additional lime. The calcium carbonate is first heated to at least 825°C to produce quicklime, CaO :

$$CaCO_3 + heat \rightarrow CaO + CO_2(g) \qquad ...(7.11)$$

The quicklime is then slaked with water to produce calcium hydroxide :

$$CaO + H_2O \rightarrow Ca(OH)_2 \qquad ...(7.12)$$

The water softened by lime-soda softening plants usually suffers from two defects. First, because of super-saturation effects, some $CaCO_3$ and $Mg(OH)_2$ usually remain in solution. If not removed, these compounds will precipitate at a later time and cause harmful deposits or undesirable cloudiness in water. The second problem results from the use of highly basic sodium carbonate, which gives the product water an excessively high pH, up to pH 11. To overcome these problems, the water is recarbonated by bubbling CO_2 into it. The carbon dioxide converts the slightly soluble calcium carbonate and magnesiumhydroxide to their soluble bicarbonate forms :

$$CaCO_3(s) + CO_2 + H_2O \rightarrow Ca^{2+} + 2HCO_3^- \qquad ...(7.13)$$

$$Mg(OH)_2(s) + 2CO_2 \rightarrow Mg^{2+} + 2HCO_3^- \qquad ...(7.14)$$

The CO_2 also neutralises excess hydroxide ion :

$$OH^- + CO_2 \rightarrow HCO_3^- \qquad ...(7.15)$$

The pH generally is brought within the range 7.5–8.5 by recarbonation. The source of CO_2 used in the recarbonation process may be from the combustion of carbonaceous fuel. Scrubbed stack gas from a power plant frequently is utilised. Water adjusted to a pH, alkalinity, and Ca^{2+} concentration very close to $CaCO_3$ saturation is labelled *chemically stabilised*. It neither precipitates $CaCO_3$ in water mains, which can clog the pipes, nor dissolves protective $CaCO_3$ coatings from the pipe surfaces. Water with Ca^{2+} concentration much below $CaCO_3$ saturation is called an *aggressive* water.

Calcium may be removed from water very efficiently by the addition of orthophosphate :

$$5Ca^{2+} + 3PO_4^{3-} + OH^- \rightarrow Ca_5OH(PO_4)_3(s) \qquad \qquad ...(7.16)$$

It should be pointed out that the chemical formation of a slightly soluble product for the removal of undesired solutes such as hardness ions, phosphate, iron, and manganese must be followed by sedimentation in a suitable apparatus. Frequently, coagulants must be added, and filtration employed for complete removal of these sediments.

Water may be purified by ion exchange, the reversible transfer of ions between aquatic solution and a solid material capable of bonding ions. The removal of NaCl from solution by two ion exchange reactions is a good illustration of this process. First the water is passed over a solid cation exchanger in the hydrogen form, represented by H^{+-} {Cat(s)}:

$$H^{+-} \{Cat(s)\} + Na^+ + Cl^- \rightarrow Na^{+-} \{Cat(s)\} + H^+ + Cl^- \qquad \qquad ...(7.17)$$

Next the water is passed over an anion exchanger in the hydroxide ion form, represented by OH^{-+} {An(s)}:

$$OH^{-+} \{An(s)\} + H^+ + Cl^- \rightarrow Cl^{-+} \{An(s)\} + H_2O \qquad \qquad ...(7.18)$$

Thus, the cations in solution are replaced by hydrogen ion and the anions by hydroxide ion, yielding water as the product.

The softening of water by ion exchange does not require the removal of all ionic solutes, just those cations responsible for water hardness. Generally, therefore, only a cation exchanger is necessary. Furthermore, the sodium rather than the hydrogen form of the cation exchanger is used, and the divalent cations are replaced by sodium ion. Sodium ion at low concentrations is harmless in water to be used for most purposes, and sodium chloride is a cheap and convenient substance with which to recharge the cation exchangers.

A number of materials have ion-exchanging properties. Among the minerals especially noted for their ion exchange properties are the aluminium silicate minerals, or zeolites. An example of a zeolite which has been used commercially in water softening is glauconite, $K_2(MgFe)_2Al_6(Si_4O_{10})_3(OH)_{12}$. Synthetic zeolites have been prepared by drying and crushing the white gel produced by mixing solutions of sodium silicate and sodium aluminate.

The discovery of synthetic ion exchange resins composed of organic polymers with attached functional groups marked the beginning of modern ion exchange technology. Structures of typical synthetic ion exchangers are shown in Figs. 7.6 and 7.7. The cation exchanger shown in Fig. 7.6 is called a strongly acidic cation exchanger because the parent $-SO_3^-H^+$ group is a strong acid. When the functional group binding the cation is the $-CO_2^-$ group, the exchange resin is called a weakly acidic cation exchanger, because the $-CO_2H$ group is a weak acid. Fig. 7.7 shows a strongly basic anion exchanger in which the functional group is a quaternary ammonium group, $-N^+(CH_3)_3$. In the hydroxide form, $-N+(CH_3)_3OH^-$, the hydroxide ion is readily released; hence the exchanger is classified as strongly basic.

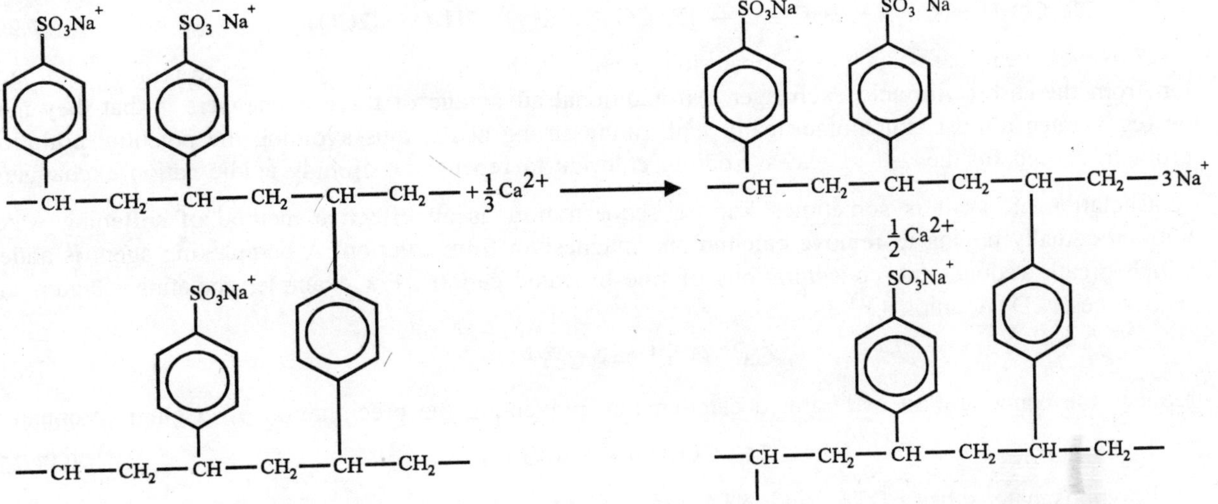

Fig. 7.6. Strongly acidic cation exchanger. Sodium exchange for calcium in water is shown.

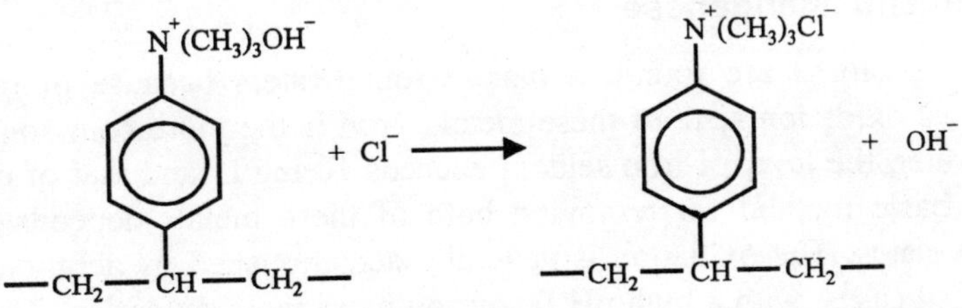

Fig. 7.7. Strongly basic anion exchanger. Chloride exchange for hydroxide ion is shown.

The water-softening capability of a cation exchanger is shown in Fig. 7.6, where sodium ion on the exchanger is exchanged for calcium ion in solution. The same reaction occurs with magnesium ion. Water softening by cation exchange is now a widely used, effective, and economical process.

In many areas having a low water flow, however, it is not likely that home water softening by ion exchange may be used universally without some deterioration of water quality arising from the contamination of waste-water by sodium chloride. such contamination results from the periodic need to regenerate a water softener with sodium chloride, in order to displace calcium and magnesium ions from the resin and replace these hardness ions with sodium ions :

$$Ca^{2+} \{Cat(s)\}_2 + 2Na^+ + 2Cl^- \rightarrow 2Na^+ \{Cat(s)\} + Ca^{2+} + 2Cl^- \qquad ...(7.19)$$

During the regeneration process, a large excess of sodium chloride must be used – several pounds for a home water softener. Appreciable amounts of dissolved sodium chloride can be introduced into sewage by this route.

Strongly acidic cation exchangers are used for the removal of water hardness. Weakly acidic cation exchangers having the $-CO_2H$ group as a functional group are useful for removing alkalinity. Alkalinity generally is manifested by bicarbonate ion. This species is a sufficiently strong base to neutralise the acid of a weak acid cation exchanger :

$$2R\text{–}CO_2H + Ca^{2+} + 2HCO_3^- \rightarrow [R\text{–}CO_2^-]_2Ca^{2+} + 2H_2O + 2CO_2 \qquad ...(7.20)$$

However, weak bases such as sulphate ion or chloride ion are not strong enough to remove hydrogen ion from the carboxylic acid exchanger. An additional advantage of these exchangers is that they may be regenerated almost stoichiometrically with dilute strong acids, thus avoiding the potential pollution problem caused by the use of excess sodium chloride to regenerate strongly acidic cation exchangers.

Chelation or, as it is sometimes known, sequestration, is an effective method of softening water without actually having to remove calcium and magnesium from solution. A complexing agent is added which greatly reduces the concentrations of free hydrated cations. For example, chelating calcium ion with excess EDTA anion (Y^{4-}),

$$Ca^{2+} + Y^{4-} \rightarrow CaY^{2-} \qquad ...(7.21)$$

reduces the concentration of hydrated calcium ion, preventing the precipitation of calcium carbonate :

$$Ca^{2+} + CO_3^{2-} \rightarrow CaCO_3(s) \qquad ...(7.22)$$

Polyphosphate salts, EDTA, and NTA are chelating agents commonly used for water softening. Polysilicates are used to complex iron.

Removal of Iron and Manganese

Soluble iron and manganese are found in many groundwaters because of reducing conditions which favour the soluble +2 oxidation state of these metals. Iron is the more commonly encountered of the two metals. In groundwater, the level of iron seldom exceeds 10 mg/L, and that of manganese is rarely higher than 2 mg/L. The basic method for removing both of these metals depends upon oxidation to higher insoluble oxidation states. The oxidation is generally accomplished by aeration. The rate of oxidation is pH-dependent in both cases, with a high pH favouring more rapid oxidation. The oxidation of soluble Mn to insoluble MnO_2 is a complicated process. It appears to be catalysed by solid MnO_2, which is known to adsorb Mn. This adsorbed Mn is slowly oxidised on the MnO_2 surface.

Chlorine and potassium permanganate are sometimes employed as oxidising agents for iron and manganese. There is some evidence that organic chelating agents with reducing properties hold iron in a soluble form in water. In such cases, chlorine is effective because it destroys the organic compounds and enables the oxidation of iron.

In water with a high level of carbonate, $FeCO_3$ and $MnCO_3$ may be precipitated directly by raising the pH above 8.5 by the addition of sodium carbonate or lime. This approach is less popular than oxidation, however.

Relatively high levels of insoluble iron and manganese frequently are found in water as colloidal material which is difficult to remove. These metals may be associated with humic colloids or "peptising" organic material that binds to colloidal metal oxides, stabilising the colloid.

Heavy metals such as copper, cadmium, mercury, and lead are found in waste-waters from a number of industrial processes. Because of the toxicity of many heavy metals, their concentrations must be reduced to very low levels prior to release of the waste-water. A number of approaches are used in heavy metals removal.

Lime treatment removes heavy metals as insoluble hydroxides, basic salts, or coprecipitated with calcium carbonate or ferric hydroxide. This process does not completely remove mercury, cadmium, or lead, so their removal is aided by addition of sulphide (most heavy metals are sulphide-seekers):

$$Cd^{2+} + S^{2-} \rightarrow CdS(s) \qquad \qquad ...(7.23)$$

Heavy chlorination is frequently necessary to break down metal-solubilising ligands. Lime precipitation does not normally permit recovery of metals and is sometimes undesirable from the economic viewpoint.

Electrodeposition (reduction of metal ions to metal by electrons at an electrode), reverse osmosis, and ion exchange are frequently employed for metal removal. Solvent extraction using organic-soluble chelating substances is also effective in removing many metals. Cementation, a process by which a metal deposits by reaction of its ion with a more readily oxidised metal, may be employed :

$$Cu^{2+} + Fe \text{ (iron scrap)} \rightarrow Fe^{2+} + Cu \qquad \qquad ...(7.24)$$

Activated carbon adsorption effectively removes some metals from water at the part per million level. Sometimes a chelating agent is sorbed to the charcoal to increase metal removal.

Even when not specifically designed for the removal of heavy metals, most waste treatment processes remove appreciable quantities of the more troublesome heavy metals encountered in waste-water. Biological waste treatment effectively removes metals from water. These metals accumulate in the sludge from biological treatment, so sludge disposal must be given careful consideration.

Various physical-chemical treatment processes effectively remove heavy metals from waste-waters. One such treatment is lime precipitation followed by activated-carbon filtration. Activated-carbon filtration may also be preceded by treatment with iron chloride to form an iron hydroxide floc, which is an effective heavy metals scavenger. Similarly, alum, which forms aluminium hydroxide, may be added prior to activated-carbon filtration.

The form of the heavy metal has a strong effect upon the efficiency of metal removal. For instance, chromium (VI) is normally more difficult to remove then chromium (III). Chelation may prevent metal removal by solubilising metals.

In the past, removal of heavy metals has been largely a fringe benefit of waste-water treatment processes. Currently, however, more consideration is being given to design and operating parameters that specifically enhance heavy-metals removal as part of waste-water treatment.

REMOVAL OF DISSOLVED ORGANICS

Very low levels of exotic organic compounds in drinking water are suspected of contributing to cancer and other maladies. Water disinfection processes, which by their nature involve chemically rather severe conditions, particularly of oxidation, have a tendency to produce disinfection by-products. Some of these are chlorinated organic compounds produced by chlorination of organics in water, especially humic substances. Removal of organics to very low levels prior to chlronation has been found to be effective in preventing trihalomethane formation. Another major class of disinfection by-products consists of organo-oxygen compounds, such as aldehydes, carboxylic acids, and oxoacids.

In addition to disinfection by-products, many organic compounds survive, or are produced by, secondary waste-water treatment. Almost half of these are humic substances with a molecular-weight range of 1000–5000. Among the remainder are found ether-extractable materials, carbohydrates, proteins, detergents, tannins, and lignins. The humic compounds, because of their high molecular weight and anionic character, influence some of the physical and chemical aspects of waste treatment. The ether-extractables contain many of the compounds that are resistant to biodegradation and are of particular concern regarding potential toxicity, carcinogenicity, and mutagenicity. In the ether extract are found many fatty acids, hydrocarbons of the *n*-alkane class, naphthalene, diphenylmethane, diphenyl, methylnaphthalene, isopropylbenzene, dodecylbenzene, phenol, dioctylphthalate, and triethylphosphate.

The standard method for the removal of dissolved organic material is adsorption on activated carbon, a product that is produced from a variety of carbonaceous materials, including wood, pulp-mill char, peat, and lignite. The carbon is produced by charring the raw material anaerobically below 600°C followed by an activation step consisting of partial oxidation. Carbon dioxide may be employed as an oxidising agent at 600–700°C.

$$CO_2 + C \rightarrow 2CO \qquad \qquad ...(7.25)$$

or the carbon may be oxidised by water at 800–900°C :

$$H_2O + C \rightarrow H_2 + CO \qquad \qquad ...(7.26)$$

These processes develop porosity, increase the surface area, and leave the carbon atoms in arrangements that have affinities for organic compounds.

Activated carbon comes in two general types : granulated activated carbon, consisting of particles 0.1–1 mm in diameter, and powdered activated carbon, in which most of the particles are 50–100 μm in diameter.

The exact mechanism by which activated carbon holds organic materials is not known. However, one reason for the effectiveness of this material as an adsorbent is its tremendous surface area. A solid cubic foot of carbon particles may have a combined pore and surface area of approximately 10 square miles!

Although interest is increasing in the use of powdered activated carbon for water treatment, currently granular carbon is more widely used. It may be employed in a fixed bed, through which water flows downward. Accumulation of particulate matter requires periodic backwashing. An expanded bed in which particles are kept slightly separated by water flowing upward may be used with less chance of clogging.

Economics requires regeneration of the carbon, which is accomplished by heating it to 950°C in a steam-air atmosphere. This process oxidises adsorbed organics and regenerates the carbon surface, with an approximately 10% loss of carbon.

Removal of organics may also be accomplished by adsorbent synthetic polymers. Such polymers have hydrophobic surfaces and strongly attract relatively insoluble organic compounds, such as chlorinated pesticides. The porosity of these polymers is up to 50% by volume, and the surface area may be as high as 850 m^2/g. They are readily regenerated by solvents such as isopropanol and acetone. Under appropriate operating conditions, these polymers remove virtually all nonionic organic solutes.

Oxidation of dissolved organics holds some promise for their removal. Ozone, hydrogen peroxide, molecular oxygen (with or without catalysts), chlorine and its derivatives, permanganate, or ferrate can be used. Electrochemical oxidation may be possible in some cases. A promising new development is the use of high-energy electron beams produced by high-voltage electron accelerators to destroy organic compounds in water.

REMOVAL OF DISSOLVED INORGANICS

In order for complete water recycling to be feasible, inorganic-solute removal is essential. The effluent from secondary waste treatment generally contains 300–400 mg/l more dissolved inorganic material than does the municipal water supply. It is obvious, therefore, that 100% water recycle without removal of inorganics would cause the accumulation of an intolerable level of dissolved material. Even when water is not destined for immediate reuse, the removal of the inorganic nutrients phosphorus and nitrogen is highly desirable to reduce eutrophication downstream. In some cases, the removal of toxic trace metals is needed.

One of the most obvious methods for removing inorganics from water is distillation. Unfortunately, the energy required for distillation is generally too high for the process to be economically feasible. Furthermore, volatile materials such as ammonia and odorous compounds are carried over to a large extent in the distillation process, unless special preventive measures are taken. Freezing produces a very pure water, but is considered uneconomical with present technology. Membrane processes considered most promising for bulk removal of inorganics from water are electrodialysis, ion exchange, and reverse osmosis. (Other membrane processes used in water purification are nanofiltration, ultra-filtration, microfiltration, and dialysis.)

Electrodialysis

Electrodialysis consists of applying a direct current across a body of water separated into vertical layers by membranes alternately permeable to cations and anions. Cations migrate toward the cathode and anions toward the anode. Cations and anions both enter one layer of water, and both leave the adjacent layer. Thus, layers of water enriched in salts alternate with those from which salts have been removed. The water in the brine-enriched layers is recirculated to a certain extent to prevent excessive accumulation of brine. The principles involved in electrodialysis treatment are shown in Fig. 7.8.

Although the relatively small ions constituting the salts dissolved in waste-water readily pass through the membranes, large organic ions (proteins, for example) and charged colloids migrate to the membrane surfaces, often fouling or plugging the membranes and reducing efficiency. In addition, growth of microorganisms on the membranes can cause fouling.

Experience with pilot plants indicates that electrodialysis has the potential to be a practical and economical method of removing up to 50% of the dissolved inorganics from secondary sewage effluent, once the effluent has been carefully pretreated to eliminate fouling substances. Such a level of efficiency would permit repeated recycle of water without dissolved inorganic materials reaching unacceptably high levels.

Ion Exchange

The ion exchange process used for removal of inorganics consists of passing the water successively over a solid cation exchanger and a solid anion exchanger, which replace cations and anions by hydrogen ion and hydroxide ion, respectively. The net result is that each equivalent of salt is replaced by a mole of water. For the hypothetical ionic salt MX, the reactions are :

$$H^{+-} \{Cat(s)\} + M^+ + X^- \rightarrow M^{+-} \{Cat(s)\} + H^+ + X^- \qquad ...(7.27)$$

$$OH^{-+} \{An(s)\} + H^+ + X^- \rightarrow X^{-+} \{An(s)\} + H_2O \qquad ...(7.28)$$

where $^-$ $\{Cat(s)\}$ represents the solid cation exchanger and $^+$ $\{An(s)\}$ represents the solid anion exchanger. The cation exchanger is regenerated with strong acid and the anion exchanger with strong base.

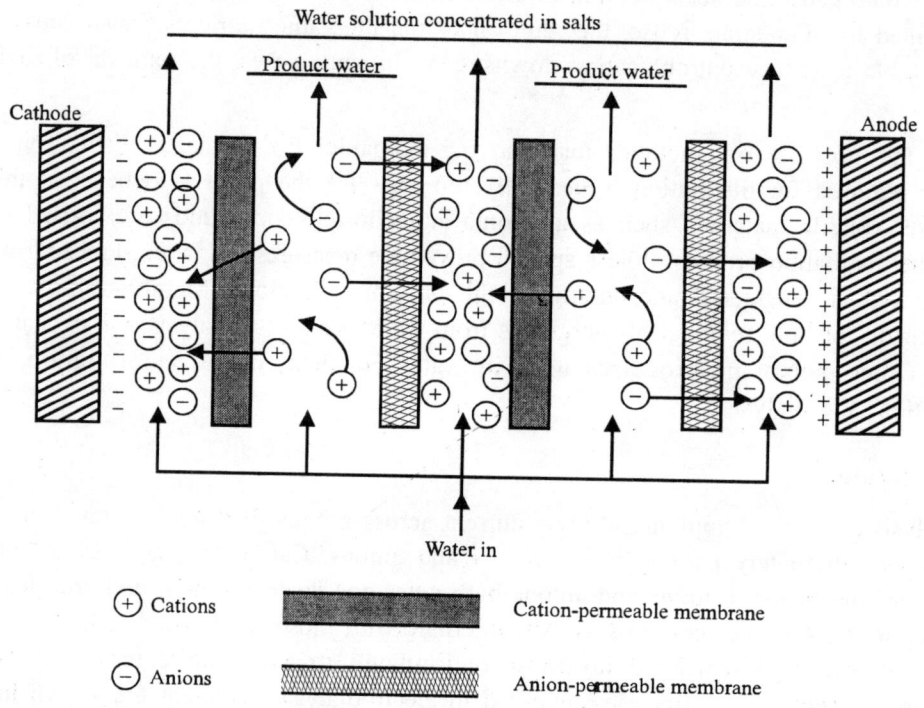

Fig. 7.8. Electrodialysis apparatus for the removal of ionic material from water.

Demineralisation by ion exchange generally produces water of a very high quality. Unfortunately, some organic compounds in waste-water foul ion exchangers, and microbial growth on the exchangers can diminish their efficiency. In addition, regeneration of the resins is expensive, and the concentrated wastes from regeneration require disposal in a manner that will not damage the environment.

Reverse Osmosis

Reverse osmosis is a very useful technique for the purification of water. Basically, reverse osmosis consists of forcing pure water through a semipermeable membrane that allows the passage of water but not of other material. This process depends on the preferential sorption of water on the surface of the membrane, which is composed of porous cellulose acetate or polyamide. Pure water from the sorbed layer is forced through pores in the membrane under pressure. If the thickness of the sorbed water layer is d, the pore diameter for optimum separation should be 2d. The optimum pore diameter depends upon the thickness of the sorbed pure water layer and may be several times the diameters of the solute and solvent molecules. Therefore, reverse osmosis is not a simple sieve separation or ultrafiltration process. The principle of reverse osmosis is illustrated in Fig. 7.9.

Phosphorus Removal

Advanced waste treatment normally requires removal of phosphorus to reduce algal growth. Algae may grow at PO_4^{3-} levels as low as 0.05 mg/l. Growth inhibition requires levels well below 0.5 mg/L. Since municipal wastes typically contain approximately 25 mg/l of phosphate (as orthophosphates, polyphosphates, and insoluble phosphates), the efficiency of phosphate removal must be quite high to prevent algal growth. This removal may occur in the sewage treatment process (i) in the primary settler; (ii) in the aeration chamber of the activated sludge unit; or (iii) after secondary waste treatment.

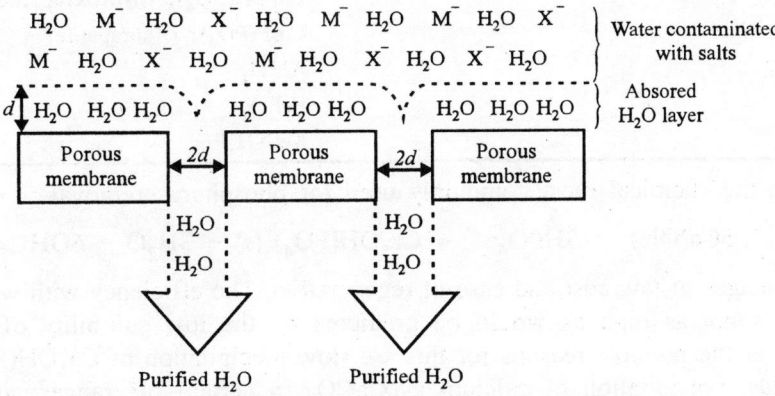

Fig. 7.9. Solute removal from water by reverse osmosis.

Normally, the activated sludge process removes about 20% of the phosphorus from sewage. Thus, an appreciable fraction of largely biological phosphorus is removed with the sludge. Detergents and other sources contribute significant amounts of phosphorus to domestic sewage and considerable phosphate ion remains in the effluent. However, some wastes, such as carbohydrate wastes from sugar refineries, are so deficient in phosphorus that supplementation of the waste with inorganic phosphorus is required for proper growth of the microorganisms degrading the wastes.

Under some sewage plant operating conditions, much greater than normal phosphorus removal has been observed. In such plants, characterised by high dissolved oxygen and high pH levels in the aeration tank, removal of 60–90% of the phosphorus has been attained, yielding two or three times the normal level of phosphorus in the sludge. In a conventionally operated aeration tank of an activated sludge plant, the CO_2 level is relatively high because of release of the gas by the degradation of organic material. A high CO_2 level results in a relatively low pH, due to the presence of carbonic acid. The aeration rate generally is not very high because oxygen is transferred relatively more efficiently from air when the dissolved oxygen levels in water are relatively low. Therefore, the aeration rate normally is not high enough to sweep out sufficient dissolved carbon dioxide to bring its concentration down to low levels. Thus, the pH generally is low enough that phosphate is maintained primarily in the form of the $H_2PO_4^-$ ion. However, at a higher rate of aeration in a relatively hard water, the CO_2 is swept out, the pH rises, and reactions such as the following occur :

$$5Ca^{2+} + 3HPO_4^{2-} + H_2O \rightarrow Ca_5OH(PO_4)_3(s) + 4H^+ \qquad ...(7.29)$$

The precipitated hydroxyapatite or other form of calcium phosphate is incorporated in the sludge floc. Reaction 7.29 is strongly hydrogen ion-dependent, and an increase in the hydrogen ion concentration drives the equilibrium back to the left. Thus, under anaerobic conditions when the sludge medium becomes more acidic due to higher CO_2 levels, the phosphate returns to solution.

Chemically, phosphate is most commonly removed by precipitation. Some common precipitants and their products are shown in Table 7.5. Precipitation processes are capable of at least 90–95% phosphorus removal at reasonable cost.

Table 7.5. Chemical precipitants for phosphate and their products.

Precipitant(s)	Products
$Ca(OH)_2$	$Ca_5OH(PO_4)_3$ (hydroxyapatite)
$Ca(OH)_2 + NaF$	$Ca_5F(PO_4)_3$ (fluorapatite)
$Al_2(SO_4)_3$	$AlPO_4$
$FeCl_3$	$FePO_4$
$MgSO_4$	$MgNH_4PO_4$

Lime, $Ca(OH)_2$, is the chemical most commonly used for phosphorus removal :

$$5Ca(OH)_2 + 3HPO_4^{2-} \rightarrow Ca_5OH(PO_4)_3(s) + 3H_2O + 6OH^- \quad ...(7.30)$$

Lime has the advantages of low cost and ease of regeneration. The efficiency with which phosphorus is removed by lime is not as high as would be predicted by the low solubility of hydroxyapatite, $Ca_5(OH(PO_4)_3$. Some of the possible reasons for this are slow precipitation of $Ca_5OH(PO_4)_3$; formation of non-settling colloids; precipitation of calcium as $CaCO_3$ in certain pH ranges; and the fact that phosphate may be present as condensed phosphates (polyphosphates) which form soluble complexes with calcium ion. Phosphate can be removed from solution by adsorption on some solids, particularly activated alumina, Al_2O_3. Removals of up to 99.9% of orthophosphate have been achieved with this method.

Nitrogen Removal

Next to phosphorus, nitrogen is the algal nutrient most commonly removed as part of advanced waste-water treatment. The techniques most often used for nitrogen removal are summarised in Table 4.6. Nitrogen is municipal waste-water generally is present as organic nitrogen or ammonia. Ammonia is the primary nitrogen product produced by most biological waste treatment processes. This is because it is expensive to aerate sewage sufficiently to oxidise the ammonia to nitrate through the action of nitrifying bacteria. If the activated sludge process is operated under conditions such that the nitrogen is maintained in the form of amonia, the latter may be stripped in the form of NH_3 gas from the water by air. For ammonia stripping to work, the ammoniacal nitrogen must be converted to volatile NH_3 gas, which requires a pH substantially higher than the pK_a of the NH_4^+ ion. In practice, the pH is raised to approximately 11.5 by the addition of lime (which, as noted above also serves to remove phosphate). The ammonia is stripped from the water by air.

Nitrification followed by denitrification is a promising technique for the removal of nitrogen from waste-water. The first step is an essentially complete conversion of ammonia and organic nitrogen to nitrate under strongly aerobic conditions, achieved by more extensive than normal aeration of the sewage:

$$NH_4^+ + 2O_2 \text{ (Nitrifying bacteria)} \rightarrow NO_3^- + 2H^+ + H_2O \quad ...(7.31)$$

The second step is the reduction of nitrate to nitrogen gas. This reaction is also bacterially catalysed and requires a carbon source and a reducing agent such as methanol, CH_3OH.

$$6NO_3^- + 5CH_3OH + 6H^+ \text{ (Denitrifying bacteria)} \rightarrow 3N_2(g) + 5CO_2 + 13H_2O \qquad ...(7.32)$$

The denitrification process may be carried out either in a tank or on a carbon column. In pilot plant operation, conversions of 95% of the ammonia to nitrate and 86% of the nitrate to nitrogen have been achieved.

Table 7.6. Common processes for the removal of nitrogen from waste-water.

Process	Principles and conditions
Air stripping ammonia	Ammonium ion is the initial product of biodegradation of nitrogenous waste. It is removed by raising the pH to approximately 11 with lime and stripping ammonia gas from the water by air in a stripping tower. Scaling, icing, and air pollution are major disadvantages.
Ammonium ion exchange	This is an attractive alternative to air stripping, made possible by the development of clinoptilolite, a natural zeolite selective for ammonia : Na^+ (clinoptilolite) + NH_4^+ $\rightarrow$ Na^+ + NH_4^+ (clinoptilolite). Regnerated with sodium or calcium salts.
Biosynthesis	The production of biomass in the sewage treatment system and its subsequent removal from the sewage effluent result in a net loss of nitrogen from the system.
Nitrification-denitrification	Several schemes are based on the conversion of ammonium nitrogen to nitrate under aerobic conditions, $2\ NH_4^+ + 3O_2 \xrightarrow{\text{Nitrosomonas}} 4\ H^+ + 2NO_2^- \oplus 2\ H_2O$ $2\ NO_2^- + O_2 \xrightarrow{\text{Nitrobacter}} 2\ NO_3^-$ followed by production of elemental nitrogen (denitrification): $4\ NO_3^- + 5\ \{CH_2O\} + 4H^+ \xrightarrow[\text{bacteria}]{\text{dinitrifiying}} 2\ N_2(g) + 5\ CO_2(g) + 7\ H_2O$ Denitrification may be accomplished in an anaerobic activated sludge system or in an anaerobic column. Some times additional organic matter (methanol) is added.
Chlorination	Reaction of ammonium ion and hypochlorite (from chlorine) results in denitrification by chemical reactions : $NH_4^+ + HOCl \rightarrow NH_2Cl + H_2O + H^+$ $2\ NH_2Cl + HOCl \rightarrow N_2(g) + 3\ H^+ + 3\ Cl^- + H_2O$

Sludge

Perhaps the most pressing water treatment problem at this time has to do with sludge collected or produced during water treatment. Finding a safe place to put the sludge or a use for it has proved troublesome, and the problem is aggravated by the growing numbers of water treatment systems.

Improper disposal of wastes continues to be a subject of public and governmental concern. Some sludge is present in waste-water and may be collected from it. Such sludge includes human wastes, garbage grindings, organic wastes and inorganic silt and grit from stormwater runoff, and organic and inorganic wastes from commercial and industrial sources. There are two major kinds of sludge generated in a waste treatment plant. The first of these is organic sludge from activated sludge, trickling filter, or rotating biological reactors. The second is inorganic sludge from the addition of chemicals, such as in phosphorus removal. Most commonly, sewage sludge is subjected to anaerobic digestion in a digester designed to allow bacterial action to occur in the absence of air. This reduces the mass and volume of sludge and ideally results in the formation of a stabilised humus. Disease agents are also destroyed in the process.

Following digestion, sludge is generally conditioned and thickened to concentrate and stabilise it and make it more dewaterable. Relatively inexpensive processes, such as gravity thickening, may be employed to get the moisture content down to about 95%. Sludge may be further conditioned chemically by the addition of iron or aluminium salts, lime, or polymers. Sludge dewatering is employed to convert the sludge from an essentially liquid material to a damp solid containing not more than about 85% water. This may be accomplished on sludge drying beds consisting of layers of sand and gravel. Mechanical devices may also be employed, including vacuum filtration, centrifugation, and filter presses. Heat may be used to aid the drying process.

Some of the alternatives for the ultimate disposal of sludge include land spreading, ocean dumping, and incineration. Each of these choices has disadvantages, such as the presence of toxic substances in sludge spread on land, or the high fuel cost of incineration. Some of the undesirable components found in sewage sludge are shown in Table 7.7. This table refers to sludge from a primary settler, although many of the same components are found in secondary settler sludges.

Table 7.7. Undesirable components typically found in sewage sludge.

Component	Level, ppm by dry weight Unless otherwise stated
Organics	
PCB	1–105
DDT	0–1 (found much less frequently now)
DDD	0–0.5 (found much less frequently now)
Dieldrin	0–2
Aldrin	0–16
Phenol	Sometimes encountered
Heavy metals	
Cadmium	0–100
Lead	up to 400
Mercury	3–15
Chromium	up to 700
Copper	80–1000
Nickel	25–400
Zinc	300–2000 (common deterrent to use of sludge as a soil conditioner due to its toxicity to plants).
Pathogenic microorganisms	
Human viruses	Generally present
Salmonella (in raw sludge)	500 viable cells/100 ml
Salmonella (in digested sludge)	30 viable cells/100 ml
Fecal coliforms (raw sludge)	1×10^7 viable cells/100 ml
Fecal coliforms (digested sludge)	4×10^5 viable cells/100 ml

Rich in nutrients, waste sewage sludge contains around 5% N, 3% P, and 0.5% K on a dry-weight basis and can be used to fertilise and condition soil. The humic material in the sludge improves the physical properties and cation-exchange capacity of the soil. Among the factors limiting this application of sludge are excess nitrogen pollution of runoff water and groundwater, survival of pathogens, and the presence of heavy metals in the sludge.

Possible accumulation of heavy metals is of the greatest concern insofar as the use of sludge on cropland is concerned. Sewage sludge is an efficient heavy metals scavenger. On a dry basis, sludge samples from industrial cities have shown levels of up to 9,000 ppm zinc, 6,000 ppm copper, 600 ppm nickel, and up to 800 ppm cadmium! These and other metals tend to remain immobilised in soil by chelation with organic matter, adsorption on clay minerals, and precipitation as insoluble compounds, such as oxides or carbonates. However, increased application of sludge on cropland has caused distinctly elevated levels of zinc and cadmium in both leaves and grain of corn. Therefore, caution has been advised in heavy or prolonged application of sewage sludge to soil. The problem of heavy metals in sewage sludge is one of the many reasons for not allowing mixture of wastes to occur prior to treatment. Sludge does, however, contain nutrients which should not be wasted, given the possibility of eventual fertiliser shortages. Prior control of heavy metal contamination from industrial sources should greatly reduce the heavy metal content of sludge and enable it to be used more extensively on soil.

An increasing problem in sewage treatment arises from sludge sidestreams. These consist of water removed from sludge by various treatment processes. Sewage treatment processes can be divided into mainstream treatment processes (primary clarification, trickling filter, activated sludge, and rotating biological reactor) and sidestream processes. During sidestream treatment, sludge is dewatered, degraded, and disinfected by a variety of processes, including gravity thickening, dissolved air floatation, anaerobic digestion, aerobic digestion, vacuum filtration, centrifugation, belt-filter press filtration, sand-drying-bed treatment, sludge-lagoon settling, wet air oxidation, pressure filtration, and Purifax treatment. Each of these produces a liquid by-product sidestream which is circulated back to the mainstream. These add to the biochemical oxygen demand and suspended solids of the mainstream.

A variety of chemical sludges are produced by various water treatment and industrial processes. Among the most abundant of such sludges is alum sludge produced by the hydrolysis of Al salts used in the treatment of water, which creates gelatinous aluminium hydroxide :

$$Al^{3+} + 3OH^-(aq) \rightarrow Al(OH)_3(s) \qquad \qquad ...(7.33)$$

Alum sludges normally are 98% or more water and are very difficult to dewater.

Both iron II and iron III compounds are used for the precipitation of impurities from waste-water via the precipitation of $Fe(OH)_3$. The sludge contains $Fe(OH)_3$ in the form of soft, fluffy precipitates that are difficult to dewater beyond 10 or 12% solids. The addition of either lime, $Ca(OH)_2$, or quicklime, CaO, to water is used to raise the pH to about 11.5 and cause the precipitation of $CaCO_3$, along with metal hydroxides and phosphates. Calcium carbonate is readily recovered from lime sludges and can be recalcined to produce CaO, which can be recycled through the system.

Metal hydroxide sludges are produced in the removal of metals such as lead, chromium, nickel, and zinc from waste-water by raising the pH to such a level that the corresponding hydroxides or hydrated metal oxides are precipitated. The disposal of these sludges is a substantial problem because of their toxic heavy metal content. Reclamation of the metals is an attractive alternative for these sludges.

Pathogenic (disease-causing) microorganisms may persist in the sludge left from the treatment of sewage. Many of these organisms present potential health hazards, and there is risk of public exposure when the sludge is applied to soil. Therefore, it is necessary both to be aware of pathogenic microorganisms in municipal waste-water treatment sludge and to find means of reducing the hazards caused by their presence.

The most significant organisms in municipal sewage sludge include the following : (i) indicators, including fecal and total coliform; (ii) pathogenic bacteria, including *Salmonellae* and *Shigellae*; (iii) enteric (intestinal) viruses, including enterovirus and poliovirus; and (iv) parasites, such as *Entamoeba histolytica* and *Ascaris lumbricoides*.

Several ways are recommended to significantly reduce levels of pathogens in sewage sludge. Aerobic digestion involves aerobic agitation of the sludge for periods of 40 to 60 days (longer times are employed with low sludge temperatures). Air drying involves draining and/or drying of the liquid sludge for at least three months in a layer 20–25 cm thick. This operation may be performed on under rained sand beds or in basins. Anaerobic digestion involves maintenance of the sludge in an anaerobic state for periods of time ranging from 60 days at 20°C to 15 days at temperatures exceeding 35°C. Composting involves mixing dewatered sludge cake with bulking agents subject to decay, such as wood chips or shredded municipal refuse, and allowing the action of bacteria to promote decay at temperatures ranging up to 45–65°C. The higher temperatures tend to kill pathogenic bacteria. Finally, pathogenic organisms may be destroyed by lime stabilisation in which sufficient lime is added to raise the pH of the sludge to 12 or higher.

WATER DISINFECTION

Chlorine is the most commonly used disinfectant employed for killing bacteria in water. When chlorine is added to water, it rapidly hydrolyses according to the reaction :

$$Cl_2 + H_2O \rightarrow H^+ + Cl^- + HOCl \qquad \qquad ...(7.34)$$

which has the following equilibrium constant :

$$K = \frac{[H^+][Cl^-][HOCl]}{[Cl_2]} = 4.5 \times 10^{-4} \qquad \qquad ...(7.35)$$

Hypochlorous acid, HOCl, is a weak acid that dissociates according to the reaction,

$$HOCl \rightleftarrows H^+ + OCl^- \qquad \qquad ...(7.36)$$

with an ionisation constant of 2.7×10^{-8}. From the above it can be calculated that the concentration of elemental Cl_2 is negligible at equilibrium above pH 3 when chlorine is added to water at levels below 1.0 g/l.

Sometimes, hypochlorite salts are substituted for chlorine gas as a disinfectant. Calcium hypochlorite, $Ca(OCl)_2$, is commonly used. The hopochlorites are safer to handle than gaseous chlorine.

The two chemical species formed by chlorine in water, HOCl and OCl^-, are known as free available chlorine. Free available chlorine is very effective in killing bacteria. In the presence of ammonia, monochloramine, dichloramine, and trichloramine are formed :

$$NH_4^+ + HOCl \rightarrow NH_2Cl \text{ (monochloramine)} + H_2O + H^+ \qquad \qquad ...(7.37)$$

$$NH_2Cl + HOCl \rightarrow NHCl_2 \text{ (dichloramine)} + H_2O \qquad ...(7.38)$$

$$NHCl_2 + HOCl \rightarrow NCl_3 \text{ (trichloramine)} + H_2O \qquad ...(7.39)$$

The chloramines are called combined available chlorine. Chlorination practice frequently provides for formation of combined available chlorine which, although a weaker disinfectant than free available chlorine, is more readily retained as a disinfectant throughout the water distribution system. Too much ammonia in water is considered undesirable because it exerts demand for chlorine.

At sufficiently high Cl:N molar ratios in water containing ammonia, some HOCl and OCl⁻ remains unreacted in solution, and a small quantity of NCl_3 is formed. The ratio at which this occurs is called the breakpoint. Chlorination beyond the breakpoint ensures disinfection. It has the additional advantage of destroying the more common materials that cause odour and taste in water.

At moderate levels of NH_3–N (approximately 20 mg/l), when the pH is between 5.0 and 8.0, chlorination with a minimum 8:1 weight ratio of Cl to NH_3–nitrogen produces efficient denitrification:

$$NH_4^+ + HOCl \rightarrow NH_2Cl + H_2O + H^+ \qquad ...(7.40)$$

$$2NH_2Cl + HOCl \rightarrow N_2(g) + 3H^+ + 3Cl^- + H_2O \qquad ...(7.41)$$

This reaction is used to remove pollutant ammonia from waste-water. However, problems can arise from chlorination of organic wastes. Typical of such by-products is chloroform, produced by the chlorination of humic substances in water.

Chlorine is used to treat water other than drinking water. It is employed to disinfect effluent from sewage treatment plants, as an additive to the water in electric power plant cooling towers, and to control microorganisms in food processing.

Chlorine Dioxide

Chlorine dioxide, ClO_2, is an effective water disinfectant that is of particular interest because, in the absence of impurity Cl_2, it does not produce impurity trihalomethanes in water treatment. In acidic and neutral water, respectively, the two half-reactions for ClO_2 acting as an oxidant are the following :

$$ClO_2 + 4H^+ + 5e^- \rightleftharpoons Cl^- + 2H_2O \qquad ...(7.42)$$

$$ClO_2 + e^- \rightleftharpoons ClO_2^- \qquad ...(7.43)$$

In the neutral pH range, chlorine dioxide in water remains largely as molecular ClO_2 until it contacts a reducing agent with which to react. Chlorine dioxide is a gas that is violently reactive with organic matter and explosive when exposed to light. For these reasons, ClO_2 is not shipped, but is generated onsite by processes such as the reaction of chlorine gas with solid sodium hypochlorite :

$$2NaClO_2(s) + Cl_2(g) \rightleftharpoons 2ClO_2(g) + 2NaCl(s) \qquad ...(7.44)$$

A high content of elemental chlorine in the product may require its purification to prevent unwanted side-reactions from Cl_2.

As a water disinfectant, chlorine dioxide does not chlorinate or oxidise ammonia or other nitrogen-containing compounds. Some concern has been raised over possible health effects of its main degradation by-products, ClO_2^- and ClO_3^-.

NATURAL WATER PURIFICATION PROCESSES

Virtually all of the materials that waste treatment processes are designed to eliminate may be absorbed by soil or degraded in soil. In fact, most of these materials are essential for soil fertility. Waste-water may provide the water that is essential to plant growth, in addition to the nutrients – phosphorus, nitrogen and potassium – usually provided by fertilisers. Waste-water also contains essential trace elements and vitamins. Stretching the point a bit, the degradation of organic wastes provides the CO_2 essential for photosynthetic production of plant biomass.

Soil may be viewed as a natural filter for wastes. Most organic matter is readily degraded in soil and, in principle, soil constitutes an excellent treatment system–primary, secondary and tertiary–for water. Soil has physical, chemical, and biological characteristics can it to bring about enable waste-water detoxification, biodegradation, chemical decomposition, and physical and chemical fixation. A number of soil characteristics are important in determining its use for land treatment of wastes. These characteristics include physical form, ability to retain water, aeration, organic content, acid-base characteristics, and oxidation-reduction behaviour. Soil is a natural medium for a number of living organisms that may have an effect upon biodegradation of waste-waters, including those that contain industrial wastes. Of these, the most important are bacteria, including those from the genera *Agrobacterium*, *Arthrobacteri*, *Bacillus*, *Flavobacterium*, and *Pseudomonas*. Actinomycetes and fungi are important in decay of vegetable matter and may be involved in biodegradation of wastes. Other unicellular organisms that may be present in or on soil are protozoa and algae. Soil animals, such as earthworms, affect soil parameters such as soil texture. The growth of plants in soil may have an influence on its waste treatment potential in such aspects as uptake of soluble wastes and erosion control. If soil treatment systems are not properly designed and operated, odour can become an overpowering problem.

Industrial Waste-water Treatment by Soil

Wastes that are amenable to land treatment are biodegradable organic substances, particularly those contained in municipal sewage and in waste-water from some industrial operations, such as food processing. However, through acclimation over a long period of time, soil bacterial cultures may develop that are effective in degrading normally recalcitrant compounds that occur in industrial waste-water. Acclimated microorganisms are found particularly at contaminated sites, such as those where soil has been exposed to crude oil for many years.

A variety of enzyme activities are exhibited by microorganisms in soil that enable them to degrade synthetic substances. Even sterile soil may show enzyme activity due to extracellular enzymes secreted by microorganisms in soil. Some of these enzymes are hydrolase enzymes such as those that catalyse the hydrolysis of organophosphate compounds.

Land treatment is mostly used for petroleum refining wastes and is applicable to the treatment of fuels and wastes from leaking underground storage tanks. It can also be applied to biodegradable organic chemical wastes, including some organohalide compounds. Land treatment is not suitable for the treatment of wastes containing acids, bases, toxic inorganic compounds, salts, heavy metals, and organic compounds that are excessively soluble, volatile, or flammable.

$$R-O-\overset{\overset{\displaystyle X}{\|}}{\underset{\underset{\displaystyle R}{|}}{\underset{\displaystyle O}{|}}{P}}-O-Ar \xrightarrow[\text{Phosphatase enzyme}]{H_2O}$$

...(7.45)

$$R-O-\overset{\overset{\displaystyle X}{\|}}{\underset{\underset{\displaystyle R}{|}}{\underset{\displaystyle O}{|}}{P}}-OH-HOAr$$

+ {O} $\xrightarrow[\text{oxidase}]{\text{Diphenol}}$ + H_2O

...(7.46)

WATER REUSE AND RECYCLE

Water reuse and recycling are becoming much more common as demands for water exceed supply. Unplanned reuse occurs as the result of waste effluents entering receiving waters or groundwater and subsequently being taken into a water distribution system. Planned reuse utilises waste-water treatment systems deliberately designed to bring water up to standards required for subsequent applications. The term direct reuse refers to water that has retained its identity from a previous application; reuse of water that has lost its identity is termed indirect reuse. The distinction also needs to be made between recycling and reuse. Recycling occurs internally before water is ever discharged. An example is condensation of steam in a steam power plant followed by return of the steam to boilers. Reuse occurs when water discharged by one user is taken up, for example, from a river, by another user.

Reuse of water continues to grow because of two major factors. The first of these is lack of supply of water. The second is that widespread deployment of modern water treatment processes significantly enhances the quality of water available for reuse. These two factors come into play in semi-arid regions in countries with advanced technological bases. For example, Israel, which is dependent upon irrigation for essentially all its agriculture, reuses about 2/3 of the country's sewage effluent for irrigation, whereas the U.S., where water is relatively more available, uses only about 2.4% of its water for this purpose.

Since drinking water and water used for food processing requires the highest quality of all large applications, intentional reuse for potable water is relatively less desirable, though widely practised unintentionally or out of necessity. This leaves three applications with the greatest potential for reuse :

1. *Irrigation for cropland, golf courses, and other applications requiring water for plant and grass growth:* This is the largest potential application for reused water and one that can take advantage of plant nutrients, particularly nitrogen and phosphorus, in water.

2. *Cooling and process water in industrial applications:* For some industrial applications, relatively low quality water can be used and secondary sewage effluent is a suitable source.

3. *Groundwater recharge:* Groundwater can be recharged with reused water either by direct injection into an aquifer or by applying the water to land, followed by percolation into the aquifer. The latter, especially, takes advantage of biodegradation and chemical sorption processes to further purify the water.

It is inevitable that water recycle and reuse will continue to grow. This trend will increase the demand for water treatment, both qualitatively and quantitatively. In addition, it will require more careful consideration of the original uses of water to minimise water deterioration and enhance its suitability for reuse.

BIOCHEMICAL PROCESSES FOR ORGANIC WASTES

Organic waste materials arise in immense quantities every year—as agricultural residues of crops, animal and human wastes, food processing by products and wastes, wastes in processing fish, tannery wastes and so on. Agricultural residues are burnt or composted which is a natural biochemical process to yield manure. Human wastes in big towns and cities are treated in sewage plants with production of biogas as a byproduct in large cities. More often raw sewage is dumped into rivers or the sea with deleterious consequences. Some of the industrial wastes are subject to biochemical processes to yield useful products—in other cases the treatment is a must for preventing pollution even if there are no positive returns. The garbage from cities is another major waste and except in a few cases, such municipal garbage is mostly dumped as landfill. It is estimated that agricultural residues can provide a significant part of farm's energy need and it is a recurring renewable source. But processing costs are too high and the energy values too costly at present. In this brief summary some of these wastes and the potential for processing them by biochemical processes is reviewed.

The organic wastes are of a wide variety in content, quantities availability over the year, collection costs, composition of convertible material and contaminants etc. Most agricultural residues are of low value and high in volume posing serious problems in handling for any continuous biochemical processing or even for pyrolysis. The same is the case in forestry wastes which however are a little more compact and drier. Although much effort is put in, it is doubtful if crop residues such as the fibrous stalks or sugarcane trash will find any useful or workable processing to yield fuel values during this century. To the extent that they are composted instead of burnt in the fields and thus undergo the natural carbon cycle through soil microorganisms, they provide useful nutrients for crops. More of forest wastes, particularly lumbering wastes may be burnt for generation of power for the timber-based industries and biochemical processing does not seem to be on the cards in the near future. Hay is of course a cattle feed and this is likely to continue and hay may not figure as a waste for biochemical processing. Corn cobs is a big item in countries like USA which cultivate enormous quantities of maize but except for a small quantity, used for producing furfural by chemical conversion, no biochemical processing seems likely.

There are some products or residues of particular interest to India residues from oilseeds such as coconut husk and shell, groundnut shells, rice bran and paddy husk, banana stems, coconut water and cashew apple juice, cashew-nut shell liquid, sugarcane bagasse, orange and other fruit process wastes, the list is endless. Some of these are being processed and utilised mainly by non-biochemical processes. Coconut husk is one item which is subjected to a natural process of retting under water to remove the material sticking to the fibres. There is need to study the biochemistry of this process and the microorganisms which are responsible or assist in the process so that a streamlined version of quick retting can be evolved. Jute fibres are also subject to a similar retting process but much simpler because the material extraneous to the fibre is much less. There is very little recovery of coconut water at the large processes of coconut to copra but perhaps the quantity is too small for concentration. Fermentation to a liquor is profitable but too much circumscribed by excise regulations.

Cashew-Nut Shell liquid (CNSL) has been the subject of much research work due to its phenolic components. CNSL is processed into grades suitable for resins for paint and other industries. Biochemical processing of this material is not feasible and the chemical processing is appropriate.

Paddy husk is said to be an ideal source of furfural by chemical conversion but otherwise it is only a fuel and presently direct biochemical processing is not feasible. If the husk is first chemically converted into soluble pentoses, a subsequent processing to ethanol or acids is possible but unlikely to become one of commercial value. Rice bran is also a valuable material for its oil and protein content and solvent extraction plants are already processing the same. There is no chemical process for recovery of the oil but the degradation of the oil in the bran is an enzymatic process and attempts to inhibit the activity of this enzyme in the bran can help in better recoveries and quality of the oil. The residual cake is mostly exported and here the problem is one of total elimination of fungus growth.

Banana Stems is a wasted resource but here again biochemical processing is not likely to be of any use. Attempts to make use of the fibre as a paper base is a possibility but here again quantities at one location are too small and the material contains too much water to be suitable for collection or storage.

Fruit processing wastes are a serious problem for disposal. Orange skins are a good source of pectin by chemical recovery system. The sugar content of wastes is too low for any useful fermentation and the more appropriate process will be one of anaerobic degradation to yield methane. The entire process for anaerobic digestion of semi solid wastes particularly those containing fibrous constituents needs very detailed study as no off-the-shelf system is available for such wastes.

Whey is a waste product of milk processing - for butter or cheese. It contains the sugar lactose and milk proteins and can be a serious pollutant unless properly treated. The low concentration makes it costly to evaporate for recovery of the constituents. The sugar lactose is also not fermented by ordinary yeast to yield alcohol but lactose can be converted to lactic acid in high yields. Here again the low concentration makes it costly to make pure lactic acid from whey. Unfortunately the processing of milk for cheese or butter is not organised in India on such a scale as to promote byproduct recovery. But it is of interest to mention an elegant biochemical process which uses a specially developed yeast which ferments lactose and is linked to ultrafiltration or membrane separation to separate the protein from lactose (due to large difference in molecular weights). The protein part is concentrated to give a cattle or poultry feed additive.

Starch factories using maize or tapioca have waste products which again create serious pollution problems. In the case of maize the corn steep liquor is concentrated and serves as a nutrient for fermentation in the production of antibiotics. The oil from maize germ is an important edible oil and widely used. The problem waste is the washings in the recovery of starch and this can be converted into a sugary syrup but at a cost which may not be recovered fully. In processing tapioca also the washings can be collected and processed but the skin and refuse are not useful. The various residues can be handled by anaerobic digestion. Biochemical processing does not figure in starch/glucose factories except in the use of enzymes for starch conversion.

Brewing industry is an important biochemical process industry in that the enzymes developed during germination of barley are themselves used in the conversion of starch to sugars and in the subsequent fermentation process. The spent grains after starch conversion and extraction is an important cattle feed and therefore does not call for any processing except drying in the event of requiring storage.

Molasses from sugar factories is a valuable waste or rather by-product for processing by fermentation or biochemical means into useful products. But these conversion processes also create residual wastes which need biochemical processes of disposal. The main constituent of molasses is soluble sugars which are easily converted into ethyl alcohol and most of the molasses in India are handled in this manner. Molasses can serve equally well for the production or single cell protein (SCP) in the form of dry yeast— the process being under high aeration with added nutrients and conditions which stimulate cell growth instead of the dismutation of glucose to ethyl alcohol and carbon dioxide which prevails under anaerobic conditions. Unfortunately India is yet to mobilise this technology, for supplementing the protein needs of the country. However compressed yeast is being produced for use in bakeries.

Another biochemical conversion is of the sugars in molasses to citric acid in one plant in India uses the latest technology of deep aerobic fermentation with a special strain of aspergillus niger after a sophisticated pretreatment of molasses to eliminate some undesirable metal ions. Molasses can also be processed by other organisms to yield glutamic acid which as the sodium salt is a food flavouring agent (seldom used in India) and also into the essential amino acid lysine. Neither of these processes is in use in India.

Molasses illustrates the case of a waste material which contains many desirable components together with undesirable ones in traces or larger proportions necessitating specialised treatment and use of supplemen's for some critical processes such as that of citric acid of SCP. It also illustrates the need for more processing of the waste after the primary production process.

The spent liquor after distillation of alcohol is much a secondary waste of large volume and high BOD. Most of it is lagooned for slow anaerobic removal of the BOD through microbes specially developed by additions of cow dung. This treatment is unsatisfactory and attempts to have regular anaerobic digestion in tanks with collection of methane gas for fuel use, have been bogged down by the poor activity of the microbes and consequent unduly long hold-up time. Many claims of improved versions of anaerobic digestion with holding time of only a week and hence high rate of gas production per unit tank volume have been made recently in other countries and could be adopted for India.

Pulp and paper manufacture also generates some wastes which have to be handled biochemically. The pulping process dissolves out the lignin but in a condition for use as fuel after evaporation. When alkali is used for pulping there is no possibility of microbial action and the alkali also has to be recovered. The sulphite pulp liquor has been a traditional source of fermentable sugars and alcohol production as well as of SCP. But sulphite process is out of vogue due to cost. The main non-cellulose fibres have yet to succeed. Recently efforts to isolate enzymes which attack the ligno cellulose bonds are being made. The pentosans can be removed by a mild pulping but using pentosans for useful products other than SCP has been difficult. Recently a special yeast—Pachysolan Tanophylus—has been-isolated and acclimatised to ferment pentoses (Xylose) to ethyl alcohol. The rates are slow and concentration of alcohol are also low so that conversion costs are high. Other organisms to convert pentoses to organic acids are available but again costs for recovery are high.

The pulping process for bagasse can be carried out by newer methods such as the use of alcoholic caustic when it is claimed that pentosans and lignins can be independently recovered leaving the cellulose intact. In such a case pentosans can be first concentrated, hydrolysed and subjected to fermentation by special yeasts or bacteria to useful products.

There is a lot of effort to convert cellulose into soluble carbohydrates for further processing into ethyl alcohol. The cellulose splitting enzymes of organisms like Trichoderma species have been isolated from special strains with high activity. But so far the efforts have succeeded only in purer forms of waste cellulose like newspapers and the efforts may not lead to viable operations. Biochemical processing of wood wastes to get pure cellulose with the removal of other components as such or in a partly degraded form for use in some conversion process is still a long way off. Probably gasification of wood wastes and conversion to methanol—non-biochemical processing—may be the easiest way out There are however organisms which have been isolated which can convert such wastes into ethyl alcohol at a fairly high temperature and their potential is under study. (Clostridium Thermocellum and Thermo saccharolyticus).

The more significant biochemical process for widespread use is the generation of methane from cowdung. The material is convenient to handle in a slurry form and there is not much of suspended material to clog up so that small domestic plants are easy to work if simple precautions are taken. Large plants for dairy farms are not difficult to design and there is scope for such units throughout the country to stop the practice of drying and burning of cowdung cakes. The problems arise in linking such units to the utilisation of the liquid manure directly. The modification of cowdung gas plants to handle other agricultural residues such as straw or trash has not been successful and must be an area of intense effort as it could be a good source of convenient fuel for villages with no loss of manure values. The subject of bacterial processes for organic wastes is an area of growing importance from the point of view of recovery of useful products or energy and of avoiding pollution.

CHAPTER 8

Atmosphere and Atmospheric Chemistry

INTRODUCTION

The atmosphere is a protective blanket which nurtures life on the earth and protects it from the hostile environment of outer space. The atmosphere is the source of carbon dioxide for plant photosynthesis and of oxygen for respiration. It provides the nitrogen that nitrogen-fixing bacteria and ammonia-manufacturing plants use to produce chemically-bound nitrogen, an essential component of life molecules. As a basic part of the hydrologic cycle (Fig. 8.1) the atmosphere transports water from the oceans to land, thus acting as the condenser in a vast solar-powered still.

Unfortunately, the atmosphere also has been used as a dumping ground for many pollutant materials – ranging from sulphur dioxide to refrigerant Freon – a practice which causes damage to vegetation and materials, shortens human life, and alters the characteristics of the atmosphere itself.

The atmosphere serves a vital protective function. It absorbs most of the cosmic rays from outer space and protects organisms from their effects. It also absorbs most of the electromagnetic radiation from the sun, allowing transmission of significant amounts of radiation only in the regions of 300–2500 nm (near-ultraviolet, visible, and near-infrared radiation) and 0.01–40 m (radio waves). By absorbing electromagnetic radiation below 300 nm the atmosphere filters out damaging ultraviolet radiation that would otherwise be very harmful to living organisms.

Furthermore, because it reabsorbs much of the infrared radiation by which absorbed solar energy is re-emitted to space, the atmosphere stabilises the Earth's temperature, preventing the tremendous temperature extremes that occur on planets and moons lacking substantial atmospheres.

PHYSICAL CHARACTERISTICS OF THE ATMOSPHERE

Atmospheric science deals with the movement of air masses in the atmosphere, atmospheric heat balance, and atmospheric chemical composition and reactions. In order to understand atmospheric chemistry and air pollution, it is important to have an overall appreciation of the atmosphere, its composition, and physical characteristics (Table 8.1).

In addition to components in Table 8.1 atmospheric air may contain 0.1 to 5% water by volume, with a normal range of 1 to 3%. Atmospheric trace gases in dry air near ground level are given in Table 8.2.

Table 8.1. Composition of unpolluted dry air and the approximate total masses of the different constituents of the atmosphere. Many trace gases are not listed.

Constituent	Molecular formula	Volume fraction	Total mass (millions of metric tonnes)
Nitrogen	N_2	78.09%	3,850,000,000
Oxygen	O_2	80.94%	1,180,000,000
Argon	Ar	0.93%	65,000,000
Carbon dioxide	CO_2	0.038%	8,500,000
Neon	Ne	18 ppm	64,000
Helium	He	5.8 ppm	3,700
Methane	CH_4	1.3 ppm	3.700
Krypton	Kr	1 ppm	15,000
Hydrogen	H_2	0.5 ppm	180
Nitrous oxide	N_2O	0.85 ppm	1,900
Carbon monoxide	CO	0.1 ppm	500
Ozone	O_3	0.08 ppm	800
Sulphur dioxide	SO_2	0.001 ppm	11
Nitrogen dioxide	NO_2	0.001 ppm	8

Table 8.2. Atmospheric trace gases in dry air near ground level.

Gas or species	Volume per cent	Major sources	Process for removal from the atmosphere
CH_4	1.6×10^{-4}	Biogenic	Photochemical
CO	$\sim 1.8 \times 10^{-5}$	Photochemical, anthropogenic	Photochemical
N_2O	3×10^{-5}	Biogenic	Photochemical
NO_x	10^{-10}–10^{-6}	Photochemical, lightning, anthropogenic	Photochemical
HNO_3	10^{-9}–10^{-7}	Photochemical	Washed out by precipitation
NH_3	10^{-8}–10^{-7}	Biogenic	Photochemical, washed out by precipitation
H_2	5×10^{-5}	Biogenic, photochemical	Photochemical
H_2O_2	10^{-8}–10^{-6}	Photochemical	Washed out by precipitation
HO·	10^{-13}–10^{-10}	Photochemical	Photochemical
HO_2·	10^{-11}–10^{-9}	Photochemical	Photochemical
H_2CO	10^{-8}–10^{-7}	Photochemical	Photochemical
CS_2	10^{-9}–10^{-8}	Anthropogenic, biogenic	Photochemical
OCS	10^{-8}	Anthropogenic, biogenic, photochemical	Photochemical
SO_2	$\sim 2 \times 10^{-8}$	Anthropogenic, photochemical, volcanic	Photochemical
I_2	0-trace	—	—
CCl_2F_2	2.8×10^{-5}	Anthropogenic	Photochemical
H_3CCCl_3	$\sim 1 \times 10^{-8}$	Anthropogenic	Photochemical

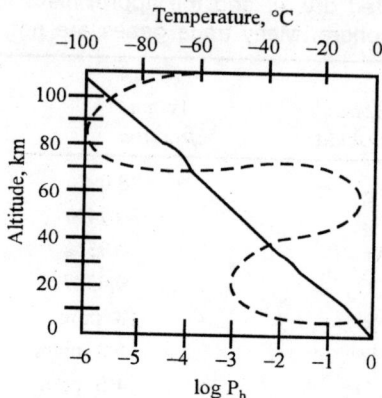

Fig. 8.1. Variation of pressure (solid line) and temperature (dashed line) with altitude.

Variation of Pressure and Density with Altitude

It is well known that, the density of the atmosphere decreases sharply with increasing altitude as a consequence of the gas laws and gravity. More than 99% of the total mass of the atmosphere is found within approximately 30 km (about 20 miles) of the earth's surface. Such an altitude is miniscule compared to the earth's diameter, so it is not an exaggeration to characterise the atmosphere as a "tissue-thin" protective layer. Although the total mass of the global atmosphere is huge, approximately 5.14×10^{15} metric tons, it is still only about one millionth of the earth's total mass.

The fact that atmospheric pressure decreases as an approximately exponential function of altitude largely determines the characteristics of the atmosphere. Ideally, in the absence of mixing and at a constant absolute temperature, T, the pressure at any given height, P_h, is given in the exponential form,

$$P_h = P_0 e^{-Mgh/RT} \qquad \qquad ...(8.1)$$

where P_0 is the pressure at zero altitude (sea level); M is the average molar mass of air (28.97 g/mole in the troposphere); g is the acceleration of gravity (981 cm $\times$ sec^{-2} at sea level); h is the altitude in cm; and R is the gas constant (8.314 $\times 10^7$ erg $\times$ deg^{-1} $\times$ mole^{-1}). These units are given in the CGS (centimetre-gram-sec) system for consistency; altitude can be converted to metres or kilometres as appropriate.

The factor RT/Mg is defined as the scale height, which represents the increase in altitude by which the pressure drops by e^{-1}. At an average sea level temperature of 288 K, the scale height is 8×10^5 cm or 8 km; at an altitude of 8 km, the pressure is only about 39% that at sea level.

Conversion of Equation 8.1 to the logarithmic (base 10) form and expression of h in km yields

$$\text{Log } P_h = \text{Log } P_0 - \frac{Mgh \times 10^5}{2.303RT} \qquad \qquad ...(8.2)$$

and taking the pressure at sea level to be exactly 1 atm gives the following expression :

$$\text{Log } P_h = \frac{Mgh \times 10^5}{2.303RT} \qquad \qquad ...(8.3)$$

Plots of P_h and temperature *versus* altitude are shown in Fig. 8.1. The plot of P_h is non-linear because of variations arising from non-linear variations in temperature with altitude that are discussed later in this section and from the mixing of air masses.

The characteristics of the atmosphere vary widely with altitude, time (season), location (latitude), and even solar activity. Extremes of pressure and temperature are illustrated in Fig. 8.1. At very high altitudes normally reactive species, such as atomic oxygen, O, persist for long periods of time. That occurs because the pressure is very low at these altitudes so that the distance travelled by a reactive species before it collides with a potential reactant–its mean free path–is quite high. A particle with a mean free path of 1×10^{-6} cm at sea level has a mean free path greater than 1×10^{6} cm at an altitude of 500 km, where the pressure is lower by many orders of magnitude.

Stratification of the Atmosphere

As shown in Fig. 8.2, the atmosphere is stratified on the basis of temperature/density relationships resulting from interrelationships between physical and photochemical (light-induced chemical phenomena) processes in air.

The lowest layer of the atmosphere extending from sea level to an altitude of 10–16 km is the troposphere, characterised by a generally homogeneous composition of major gases other than water and decreasing temperature with increasing altitude from the heat-radiating surface of the earth. The upper limit of the troposphere, which has a temperature minimum of about –56°C varies in altitude by a kilometre or more with atmospheric temperature, underlying terrestrial surface, and time. The homogeneous composition of the troposphere results from constant mixing by circulating air masses. However, the water vapour content of the troposphere is extremely variable because of cloud formation, precipitation, and evaporation of water from terrestrial water bodies.

The very cold temperature of the tropopause layer at the top of the troposphere serves as a barrier that causes water vapour to condense to ice so that it cannot reach altitudes at which it would photodissociate through the action of intense high-energy ultraviolet radiation. If this happened, the hydrogen produced would escape the earth's atmosphere and be lost. (Much of the hydrogen and helium originally present in the earth's atmosphere was lost by this process).

The atmospheric layer directly above the troposphere is the stratosphere, in which the temperature rises to a maximum of about –2°C with increasing altitude. This phenomenon is due to the presence of ozone, O_3, which may reach a level of around 10 ppm by volume in the mid-range of the stratosphere. The heating effect is caused by the absorption of ultraviolet radiation energy by ozone, a phenomenon discussed later in this chapter.

The absence of high levels of radiation-absorbing species in the mesosphere immediately above the stratosphere results in a further temperature decrease to about –92°C at an altitude around 85 km. The upper regions of the mesosphere and higher define a region called the exosphere from which molecules and ions can completely escape the atmosphere. Extending to the far outer reaches of the atmosphere is the thermosphere, in which the highly rarified gas reaches temperatures as high as 1200°C by the absorption of very energetic radiation of wavelengths less than approximately 200 nm by gas species in this region.

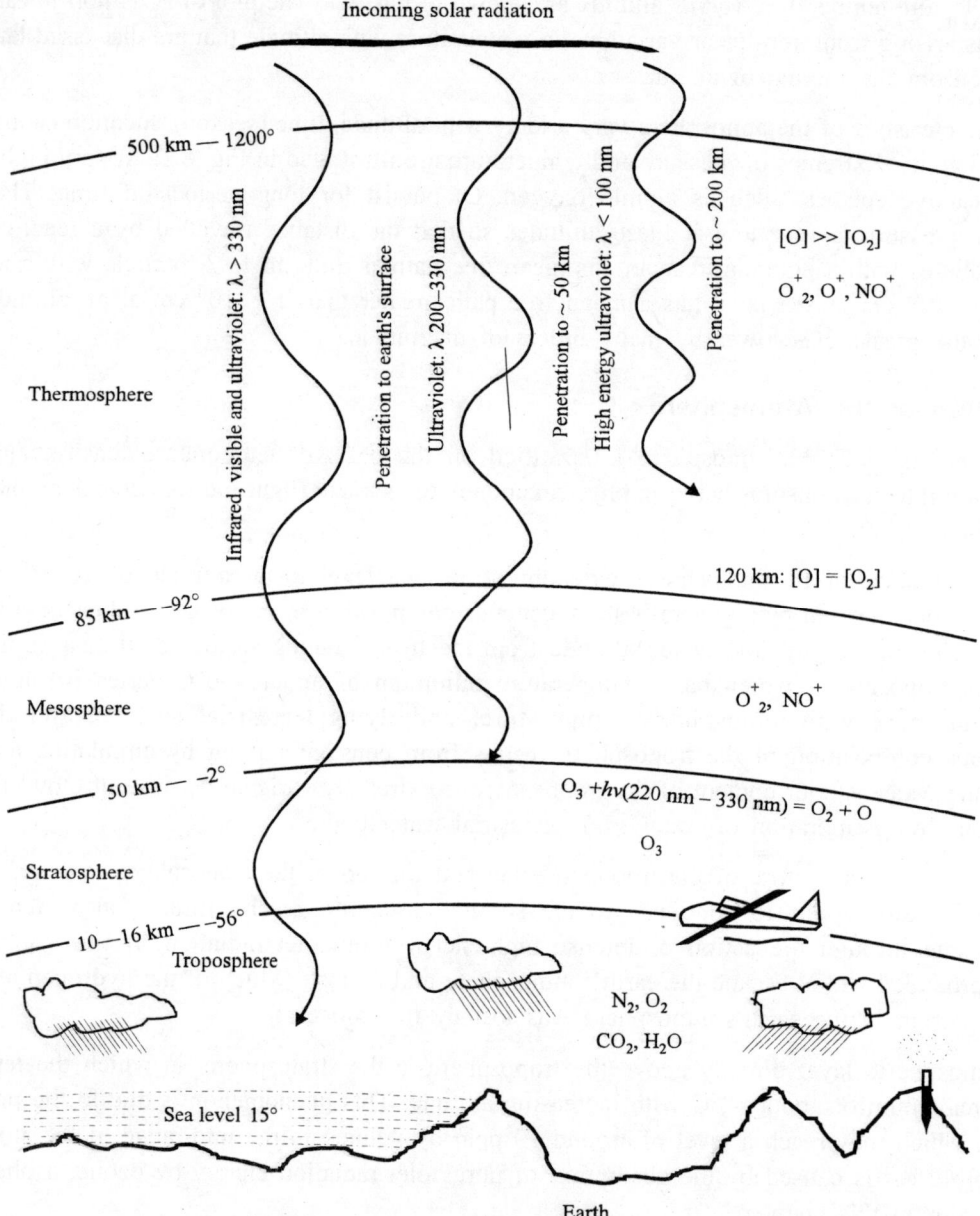

Fig. 8.2. Major regions of the atmosphere (not to scale).

ENERGY TRANSFER IN THE ATMOSPHERE

The physical and chemical characteristics of the atmosphere and the critical heat balance of the earth are determined by energy and mass transfer processes in the atmosphere.

Incoming solar energy is largely in the visible region of the spectrum. The shorter wavelength blue solar light is scattered relatively more strongly by molecules and particles in the upper atmosphere, which is why the sky is blue as it is viewed by scattered light. Similarly, light that has been transmitted through scattering atmospheres appears red, particularly around sunset and sunrise, and under circumstances in which the atmosphere contains a high level of particles. The solar energy flux reaching the atmosphere is huge, amounting to 1.34×10^3 watts per square metre (19.2 kcal per minute per square metre) perpendicular to the line of solar flux at the top of the atmosphere, as illustrated in Fig. 8.3. This value is the solar constant, and may be termed insolation, which stands for "incoming solar radiation". If all this energy reached the earth's surface and were retained, the planet would have vapourised long ago. As it is, the complex factors involved in maintaining the earth's heat balance within very narrow limits are crucial to retaining conditions of climate that will support present levels of life on earth. The great changes of climate that resulted in ice ages during some periods, or tropical conditions during others, were caused by variations of only a few degrees in average temperature. Marked climate changes within recorded history have been caused by much smaller average temperature changes. The mechanisms by which the Earth's average temperature is retained within its present narrow range are complex and not completely understood, but the main features are explained here.

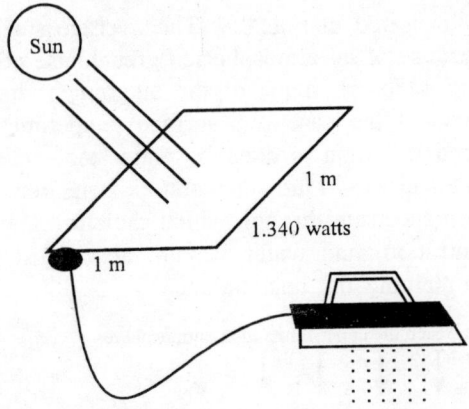

Fig. 8.3. Solar flux at the distance of the earth from the sun is 1.34×10^3 watts/m^2.

About half of the solar radiation entering the atmosphere reaches the Earth's surface either directly or after scattering by clouds, atmospheric gases, or particles. The remaining half of the radiation is either reflected directly back or absorbed in the atmosphere and its energy radiated back into space at a later time as infrared radiation. Most of the solar energy reaching the surface is absorbed and it must be returned to space in order to maintain heat balance. In addition, a very small amount of energy (less than 1% of that received from the sun) reaches the earth's surface by convection and conduction processes from the Earth's hot mantle, and this, too, must be lost.

Energy transport, which is crucial to eventual re-radiation of energy from the Earth is accomplished by three major mechanisms–conduction, convection, and radiation. Conduction of energy occurs through the interaction of adjacent atoms or molecules without the bulk movement of matter. Convection involves the movement of whole masses of air, which may be either relatively warm or cold. It is the mechanism by which abrupt temperature variations occur when large masses of air move across an area. As well as carrying sensible heat due to the kinetic energy of molecules, convection carries latent heat in the form of water vapour which releases heat as it condenses. An appreciable fraction of the earth's surface heat is transported to clouds in the atmosphere by conduction and convection before being lost ultimately by radiation.

Radiation of energy in earth's atmosphere occurs through electromagnetic radiation in the infrared region of the spectrum. As the only way in which energy is transmitted through a vacuum, radiation is the means by which all energy lost from the planet to maintain its heat balance is ultimately returned to space. The electromagnetic radiation that carries energy away from the earth is of a much longer wavelength than the sunlight that brings energy to the earth. This is a crucial factor in maintaining the earth's heat balance and one susceptible to upset by human activities. The maximum intensity of incoming radiation occurs at 0.5 micrometres (500 nanometres) in the visible region, with essentially none outside the range of 0.2 μm to 3 μm. This range encompasses the whole visible region and small parts of the ultraviolet and infrared adjacent to it. Outgoing radiation is in the infrared region, with maximum intensity at about 10 μm, primarily between 2 μm and 40 μm. Thus the earth loses energy by electromagnetic radiation of a much longer wavelength (lower energy per photon) than the radiation by which it receives energy.

Earth's Radiation Budget

The earth's radiation budget is illustrated in Fig. 8.4. The average surface temperature is maintained at a relatively comfortable 15°C because of an atmospheric "greenhouse effect" in which water vapour and, to a lesser extent carbon dioxide, reabsorb much of the outgoing radiation and reradiate about half of it back to the surface. Were this not the case, the surface temperature would average around −18°C. Most of the absorption of infrared radiation is done by water molecules in the atmosphere. Absorption is weak in the regions 7–8.5 μm and 11–14 μm and non-existent between 8.5 μm and 11 μm, leaving a "hole" in the infrared absorption spectrum through which radiation may escape. Carbon dioxide, though present at a much lower concentration than water vapour, absorbs strongly between 12 μm and 16.3 μm, and plays a key role in maintaining the heat balance.

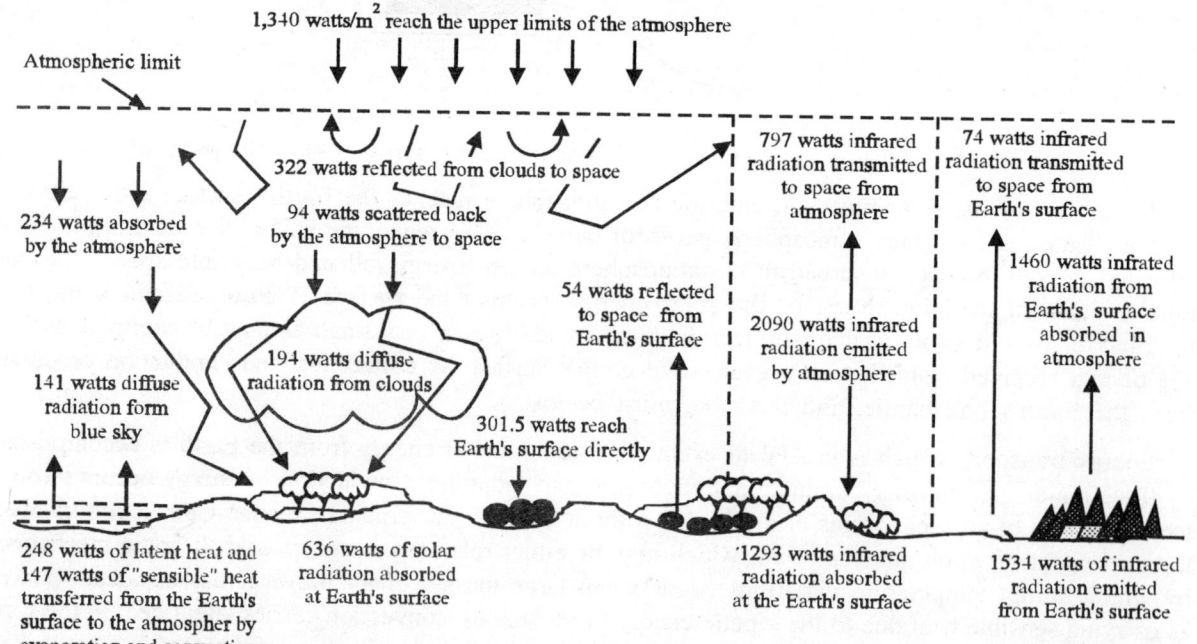

Fig. 8.4. Earth's radiation budget expressed on the basis of portions of the 1,340 watts/m² composing the solar flux.

There is concern that an increase in the carbon dioxide level in the atmosphere could prevent sufficient energy loss to cause a perceptible and damaging increase in the Earth's temperature. This phenomenon, is popularly known as the greenhouse effect and may occur from elevated CO_2 levels caused by increased use of fossil fuels and the destruction of massive quantities of forests.

An important aspect of solar radiation that reaches Earth's surface is the percentage reflected from the surface, described as *albedo*. Albedo is important in determining earth's heat balance in that absorbed radiation heats the surface, and reflected radiation does not. Albedo varies spectacularly with the surface. At the two extremes, freshly fallen snow has an albedo of 90% because it reflects 9/10 of incoming radiation, whereas freshly plowed black topsoil has an albedo of only about 2.5%.

ATMOSPHERIC MASS TRANSFER, METEOROLOGY, AND WEATHER

Meteorology is the science of atmospheric phenomena, encompassing the study of the movement of air masses as well as physical forces in the atmosphere—heat, wind, and transitions of water, primarily liquid to vapour, or vice versa. Meteorological phenomena affect, and in turn are affected by, the chemical properties of the atmosphere. For example, meteorological phenomena determine whether or not power plant stack gas heavily laced with sulphur dioxide is dispersed high in the atmosphere, with little direct effect upon human health, or settles as a choking chemical blanket in the vicinity of the power plant.

Short-term variations in the state of the atmosphere are described as weather. The weather is defined in terms of seven major factors: temperature, clouds, winds, humidity, horizontal visibility (as affected by fog, etc.), type and quantity of precipitation, and atmospheric pressure. All of these factors are closely interrelated. Longer term variations and trends within a particular geographical region in those factors that compose weather are described as climate.

Atmospheric Water in Energy and Mass Transfer

The deriving force behind weather and climate is the distribution and ultimate re-radiation to space of solar energy. A large fraction of solar energy is converted to latent heat by evaporation of water into the atmosphere. As water condenses from atmospheric air, large quantities of heat is released. This is a particularly significant means for transferring energy from the ocean to land. Solar energy falling on the ocean is converted to latent heat by the evaporation of water, then the water vapour moves inland where it condenses. The latent heat released when the water condenses warms the surrounding land mass.

Atmospheric water can be present as vapour, liquid, or ice. The water content of air can be expressed as humidity. Relative humidity, expressed as a percentage, describes the amount of water vapour in the air as a ratio of the maximum amount that the air can hold at that temperature. Air with a given relative humidity can undergo any of several processes to reach the saturation point at which water vapour condenses in the form of rain or snow. For this condensation to happen, air must be cooled below a temperature called the dew point, and condensation nuclei must be present. These nuclei are hygroscopic substances such as salts, sulphuric acid droplets, and some organic materials, including bacterial cells. Air pollution in some form is now an important source of condensation nuclei.

The liquid water in the atmosphere is present largely in clouds. Clouds normally form when rising, adiabatically cooling air can no longer hold water in the vapour form, and the water forms very small aerosol droplets. Clouds may be classified in three major forms. Cirrus clouds occur at great altitudes and have a thin feathery appearance. Cumulus clouds are detached masses with a flat base and frequently a "bumpy" upper structure. Stratus clouds occur in large sheets and may cover all of the sky visible from a given point as overcast. Clouds are important absorbers and reflectors of radiation (heat). Their formation is affected by human activities, especially particulate matter pollution and emission of deliquescent gases, such as SO_2 and HCl.

The formation of precipitation from the very small droplets of water that compose clouds is a complicated and important process. Cloud droplets normally take somewhat longer than a minute to form by condensation. They average about 0.04 mm across and do not exceed 0.2 mm in diameter. Raindrops range from 0.5 to 4 mm in diameter. Condensation processes do not form particles large enough to fall as precipitation (rain, snow, sleet, or hail). The small condensation droplets must collide and coalesce to form precipitation-size particles. When droplets reach a threshold diameter of about 0.04 mm, they grow more rapidly by coalescence with other particles than by condensation of water vapour.

Air Masses

Distinct air masses are a major feature of the troposphere. These air masses are uniform and are horizontally homogeneous. Their temperature and water-vapour content are particularly uniform. These characteristics are determined by the nature of the surface over which a large air mass forms. Polar continental air masses form over cold land regions; polar maritime air masses form over polar oceans. Air masses originating in the tropics may be similarly classified as tropical continental air masses or tropical maritime air masses. The movement of air masses and the conditions in them may have important effects upon pollutant reactions, effects, and dispersal.

Solar energy received by earth is largely redistributed by the movement of huge masses of air with different pressures, temperatures, and moisture contents separated by boundaries called fronts. Horizontally moving air is called wind, whereas vertically moving air is referred to as an air current. Atmospheric air moves constantly, with behaviour and effects that reflect the laws governing the behaviour of gases. First of all, gases will move horizontally and/or vertically from regions of high atmospheric pressure to those of low atmospheric pressure. Furthermore, expansion of gases causes cooling, whereas compression causes warming. A mass of warm air tends to move from earth's surface to higher altitudes, where the pressure is lower; in so doing, it expands adiabatically (that is, without exchanging energy with its surroundings) and becomes cooler. If there is no condensation of moisture from the air, the cooling effect is about 10°C per 1000 metres of altitudes, a figure known as the dry adiabatic lapse rate. A cold mass of air at a higher altitude does the opposite; it sinks and becomes warmer at about 10°C/1000 m. Often, however, when there is sufficient moisture in rising air, water condenses from it, releasing latent heat. This partially counteracts the cooling effect of the expanding air, giving a moist adiabatic lapse rate of about 6°C/1000 m. Parcels of air do not rise and fall, or even move horizontally, in a completely uniform way, but exhibit eddies, currents, and various degrees of turbulence.

As noted above, wind is air moving horizontally, whereas air currents are created by air moving up or down. Wind occurs because of differences in air pressure from high pressure regions to low pressure areas. Air currents are largely convection currents formed by differential heating of air masses. Air that is over a solar-heated land mass is warmed, becomes less dense, therefore rises, and is replaced by cooler and more dense air. Wind and air currents are strongly involved with air pollution phenomena. Wind carries and disperses air pollutants. Prevailing wind direction is an important factor in determining the areas most affected by an air pollution source. Wind is an important renewable energy resource.

Furthermore, wind plays an important role in the propagation of life by dispersing spores, seeds, and organisms, such as spiders.

Topographical Effects

Topography, the surface configuration and relief features of the earth's surface may strongly affect winds and air currents. Differential heating and cooling of land surfaces and bodies of water can result in local convective winds, including land breezes and sea breezes at different times of the day along the seashore, as well as breezes associated with large bodies of water inland. Mountain topography causes complex and variable localised winds. The masses of air in mountain valleys heat up during the day causing upslope winds and cool off at night causing downslope winds. Upslope winds flow over ridge tops in mountainous regions. The blocking of wind and of masses of air by mountain formations some distance inland from seashores can trap bodies of air, particularly when temperature inversion conditions occur.

Movement of Air Masses

Basically, weather is the result of the interactive effects of (i) redistribution of solar energy, (ii) horizontal and vertical movement of air masses with varying moisture contents; and (iii) evaporation and condensation of water, accompanied by uptake and release of heat. To see how these factors determine weather – and ultimately climate – on a global scale, first consider the cycle illustrated in Fig. 8.5. This Fig. shows solar energy being absorbed by a body of water, and causing some water to evaporate.

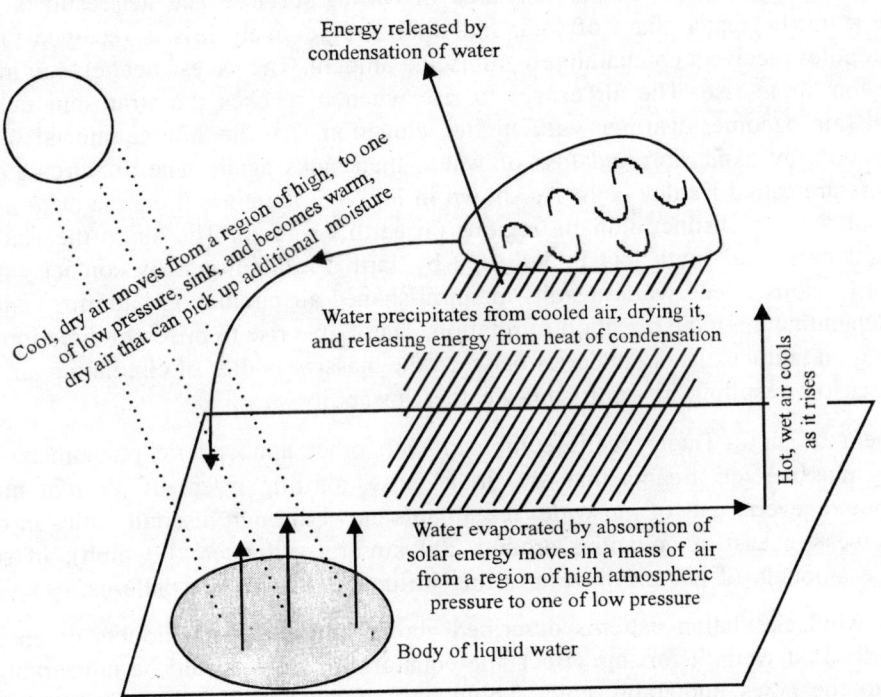

Fig. 8.5. Circulation patterns involved with movement of air masses and water; uptake and release of solar energy as latent heat in water vapour.

The warm, moist mass of air thus produced moves from a region of high pressure to one of low pressure, and cools by expansion as it rises in what is called a convection column. As the air cools, water condenses from it, and energy is released; this is a major pathway by which energy is transferred from the earth's surface to high in the atmosphere. As a result of condensation of water and loss of energy, the air is converted from warm, moist air to cool, dry air. Furthermore, the movement of the parcel of air to high altitudes results in a degree of "crowding" of air molecules and creates a zone of relatively high pressure high in the troposphere at the top of the convection column. This air mass in turn moves from the upper-level region of high pressure to one of low pressure; in so doing, it subsides, thus creating an upper-level low pressure zone, and becomes warm, dry air in the process. The pileup of this air at the surface creates a surface high pressure zone, where the cycle described above began. The warm dry air in this surface high pressure zone again picks up moisture, and the cycle begins again.

Global Weather

The factors discussed above that determine and describe the movement of air masses are involved in the massive movement of air, moisture, and energy that occurs globally. The central feature of global weather is the redistribution of solar energy that falls unequally on earth at different latitudes (relative distances from the equator and poles). Consider Fig. 8.6, sunlight, and the energy flux from it, is most intense at the equator because, averaged over the seasons, solar radiation comes in perpendicular to earth's surface at the equator. With increasing distance from the equator (higher latitudes) the angle is increasingly oblique and more of the energy-absorbing atmosphere must be traversed, so that progressively less energy is received per unit area of earth's surface. The net result is that equatorial regions receive a much greater share of solar radiation, progressively less is received farther from the equator, and the poles receive a comparatively miniscule amount. The excess heat energy in the equatorial regions causes the air to rise. The air ceases to rise when it reaches the stratosphere because in the stratosphere the air becomes warmer with higher elevation. As the hot equatorial air rises in the troposphere, it cools by expansion and loss of water, then sinks again. The air circulation patterns in which this occurs are called Hadley cells. As shown in Fig. 8.6, there are three major groupings of these cells, which result in very distinct climatic regions on earth's surface. The air in the Hadley cells does not move straight north and south, but is deflected by earth's rotation and by contact with the rotating earth; this is the Coriolis effect, which results in spiral-shaped air circulation patterns, called cyclonic or anti-cyclonic, depending upon the direction of rotation. These give rise to different directions of prevailing winds, depending on latitude. The boundaries between the massive bodies of circulating air shift markedly over time and season, resulting in significant weather instability.

The movement of air in Hadley cells combined with other atmospheric phenomena results in the development of massive jet streams that are, in a sense, shifting rivers of air that may be several kilometres deep and several tens of km wide. Jet streams move through discontinuities in the tropopause generally from west to east at velocities around 200 km/hr (well over 100 mph); in so doing, they redistribute huge amounts of air and have a strong influence on weather patterns.

The air and wind circulation patterns described above shift massive amounts of energy over long distances on earth. If it weren't for this effect, the equatorial regions would be unbearably hot, and the regions closer to the poles intolerably cold. About half of the heat that is redistributed is carried as sensible heat by air circulation, almost 1/3 is carried by water vapour as latent heat, and the remaining approximately 20% by ocean currents.

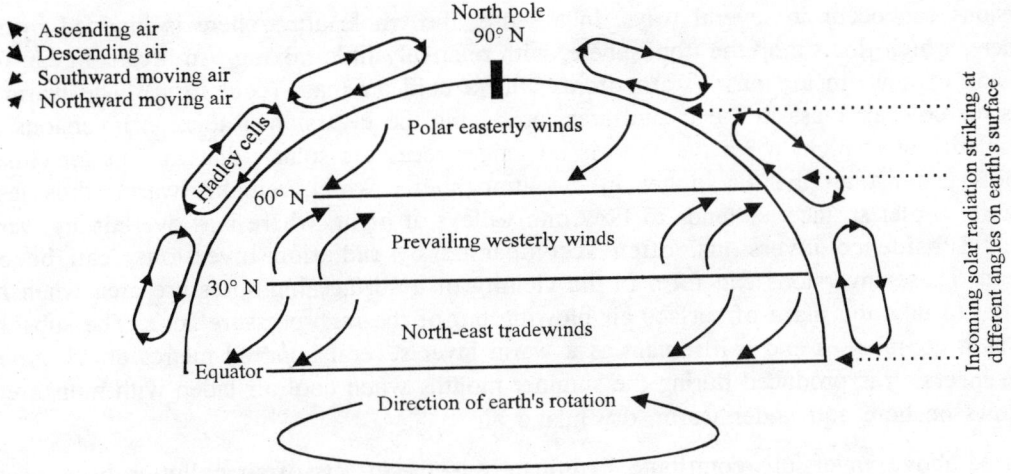

Fig. 8.6. Global circulation of air in the northern hemisphere.

Weather Fronts and Storms

As noted earlier, the interface between two masses of air that differ in temperature, density, and water content is called a front. A mass of cold air moving such that it displaces one of warm air is a cold front, and a mass of warm air displacing one of cold air is a warm front. Since cold air is more dense than warm air, the air in a cold mass of air along a cold front pushes under warmer air. This causes the warm, moist air to rise, so that water condenses from it. The condensation of water releases energy, so that the air rises further. The net effect can be formation of massive cloud formations (thunderheads) that may reach stratospheric levels. These spectacular thunderheads may produce heavy rainfall and even hail, and sometimes violent storms with strong winds, including tornadoes. Warm fronts cause somewhat similar effects as warm, moist air pushes over colder air. However, the front is usually much broader, and the weather effects milder, typically resulting in widespread drizzle, rather than intense rainstorms.

Swirling cyclonic storms, such as typhoons, hurricanes, and tornadoes, are created in low pressure areas by rising masses of warm, moist air. As such air cools, water vapour condenses, and the latent heat released warms the air more, sustaining and intensifying its movement upward in the atmosphere. Air rising from surface level creates a low pressure zone, into which surrounding air moves. The movement of the incoming air assumes a spiral pattern, thus causing a cyclonic storm.

INVERSIONS AND AIR POLLUTION

The complicated movement of air across the earth's surface is a crucial factor in the creation and dispersal of air pollution phenomena. When air movement ceases, stagnation can occur with a resultant build-up of atmospheric pollutants in localised regions. Although the temperature of air relatively near the earth's surface normally decreases with increasing altitude, certain atmospheric conditions can result in the opposite condition–increasing temperature with increasing altitude. Such conditions are characterised by high atmospheric stability and are known as temperature inversions. Because they limit the vertical circulation of air, temperature inversions result in air stagnation and the trapping of air pollutants in localised areas.

Inversions can occur in several ways. In a sense, the whole atmosphere is inverted by the warm stratosphere, which floats atop the troposphere, with relatively little mixing. An inversion can form from the collision of a warm air mass (warm front) with a cold air mass (cold front). The warm air mass overrides the cold air mass in the frontal area, producing the inversion. Radiation inversions are likely to form in still air at night when the earth is no longer receiving solar radiation. The air closest to the earth cools faster than the air higher in the atmosphere, which remains warm, thus less dense. Furthermore, cooler surface air tends to flow into valleys at night, where it is overlain by warmer, less dense air. Subsidence inversions, often accompanied by radiation inversions, can become very widespread. These inversions can form in the vicinity of a surface high-pressure area when high-level air subsides to take the place of surface air blowing out of the high-pressure zone. The subsiding air is warmed as it compresses and can remain as a warm layer several hundred metres above ground level. A marine inversion is produced during the summer months when cool air laden with moisture from the ocean blows onshore and under warm, dry inland air.

As noted above, inversions contribute significantly to the effects of air pollution because, as shown in Fig. 8.7, they prevent mixing of air pollutants, thus keeping the pollutants in one area. This not only prevents the pollutants from escaping, but also acts like a container in which additional pollutants accumulate. Furthermore, in the case of secondary pollutants formed by atmospheric chemical processes, such as photochemical smog, the pollutants may be kept together so that they react with each other and with sunlight to produce even more noxious products.

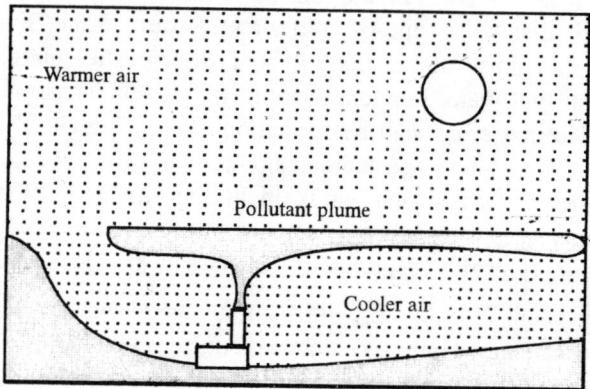

Fig. 8.7. Illustration of pollutants trapped in a temperature inversion.

GLOBAL CLIMATE AND MICROCLIMATE

Perhaps the single most important influence on earth's environment is climate, consisting of long-term weather patterns over large geographical areas. As a general rule, climatic conditions are characteristic of a particular region. This does not mean that climate remains the same throughout the year, of course, because it varies with season. One important example of such variation is the monsoon, seasonal variations in wind patterns between oceans and continents. The climates of Africa and the Indian subcontinent are particularly influenced by monsoons. In the latter, for example, summer heating of the Indian land mass causes air to rise, thereby creating a low pressure area that attracts warm, moist air from the ocean. This air rises on the slopes of the Himalayan mountains, which also block the flow of colder air from the north, moisture from the air condenses, and monsoon rains carrying enormous amounts of precipitation fall. Thus, from May until into August, summer monsoon rains fall in India,

Bangladesh, and Nepal. Reversal of the pattern of winds during the winter months causes these regions to have a dry season, but produces winter monsoon rains in the Philippine islands, Indonesia, New Guinea, and Australia.

Summer monsoon rains are responsible for tropical rain forests in Central Africa. The interface between this region and the Sahara Desert varies over time. When the boundary is relatively far north, rain falls on the Sahel desert region at the interface, crops grow, and the people do relatively well. When the boundary is more to the south, a condition which may last for several years, devastating droughts and even starvation may occur.

It is known that there are fluctuations, cycles, and cycles imposed on cycles in climate. The causes of these variations are not completely understood, but they are known to be substantial, and even devastating to civilisation. The last ice age, which ended only about 10,000 years ago and which was preceded by several similar ice ages, produced conditions under which much of the present land mass of the Northern Hemisphere was buried under thick layers of ice and uninhabitable. A "mini-ice age" occurred during the 1300s, causing crop failures and severe hardship in northern Europe. In modern times the El-Niño-Southern Oscillation occurs with a period of several years when a large, semi-permanent tropical low pressure area shifts into the Central Pacific region from its more common location in the vicinity of Indonesia. This shift modifies prevailing winds, changes the pattern of ocean currents, and affects upwelling of ocean nutrients with profound effects on weather, rainfall, and fish and bird life over a vast area of the Pacific from Australia to the west coasts of South and North America.

Human Modifications of Climate

Although earth's atmosphere is huge and has an enormous ability to resist and correct for detrimental change, it is possible that human activities are reaching a point at which they may be adversely affecting climate. One such way is by emission of large quantities of carbon dioxide and other greenhouse gases into the atmosphere, such that global warming may occur and cause substantial climatic change. Another way is through the release of gases, particularly chlorofluorocarbons (Freons) that may cause destruction of vital stratospheric ozone. Human effects on climate are "The Endangered Global Atmosphere".

Microclimate

The preceding section described climate on a large scale, ranging up to global dimensions. The climate that organisms and objects on the surface are exposed to close to the ground, under rocks, and surrounded by vegetation, is often quite different from the surrounding macroclimate. Such highly localised climatic conditions are termed the microclimate. Microclimate effects are largely determined by the uptake and loss of solar energy very close to earth's surface and by the fact that air circulation due to wind is much lower at the surface. During the day, solar energy absorbed by relatively bare soil heats the surface, but is lost only slowly because of very limited air circulation at the surface. This provides a warm blanket of surface air several cm thick, and an even thinner layer of warm soil. At night, radiative loss of heat from the surface of soil and vegetation can result in surface temperatures several degrees colder than the air about 2 metres above ground level. These lower temperatures result in condensation of dew on vegetation and the soil surface, thus providing a relatively more moist microclimate near ground level. Heat absorbed during early morning evaporation of the dew tends to prolong the period of cold experienced right at the surface.

Vegetation substantially affects microclimate. In relatively dense growths, circulation may be virtually zero at the surface because vegetation severely limits convection and diffusion. The crown surface of the vegetation intercepts most of the solar energy, so that maximum solar heating may be a significant distance up from earth's surface. The region below the crown surface of vegetation thus becomes one of relatively stable temperature. In addition, in a dense growth of vegetation, most of the moisture loss is not from evaporation from the soil surface, but rather from transpiration from plant leaves. The net result is the creation of temperature and humidity conditions that provide a favorable living environment for a number of organisms, such as insects and rodents.

Another factor influencing microclimate is the degree to which the slope of land faces north or south. South-facing slopes of land in the northern hemisphere receive greater solar energy. Advantage has been taken of this phenomenon in restoring land strip-mined for brown coal in Germany by terracing the land so that the terraces have broad south slopes, and very narrow north slopes. On the south-sloping portions of the terrace, the net effect has been to extend the short summer growing season by several days, thereby significantly increasing crop productivity. In areas where the growing season is longer, better growing conditions may exist on a north slope because it is less subject to temperature extremes and to loss of water by evaporation and transpiration.

Effects of Urbanisation on Microclimate

A particularly marked effect on microclimate is that induced by urbanisation. In a rural setting, vegetation and bodies of water have a moderating effect, absorbing modest amounts of solar energy and releasing it slowly. The stone, concrete, and asphalt pavement of cities have an opposite effect, strongly absorbing solar energy, and re-radiating heat back to the urban microclimate. Rainfall is not allowed to accumulate in ponds, but is drained away as rapidly and efficiently as possible. Human activities generate significant amounts of heat, and produce large quantities of CO_2 and other greenhouse gases that retain heat. The net result of these effects is that a city is capped by a heat dome in which the temperature is as much as $5°$ warmer than in the surrounding rural areas, so that large cities have been described as "heat islands". The rising warmer air over a city brings in a breeze from the surrounding area and causes a local greenhouse effect that probably is largely counterbalanced by reflection of incoming solar energy by particulate matter above cities. Overall, compared to climatic conditions in nearby rural surroundings, the city microclimate is warmer, foggier and overlain with more cloud cover a greater percentage of the time, subject to more precipitation, though generally less humid.

CHEMICAL AND PHOTOCHEMICAL REACTIONS IN THE ATMOSPHERE

Fig. 8.8 represents some of the major atmospheric chemical processes, which are discussed under the topic of atmospheric chemistry. The study of atmospheric chemical reactions is difficult. One of the primary obstacles encountered in studying atmospheric chemistry is that the chemist generally must deal with incredibly low concentrations, so that the detection and analysis of reaction products is quite difficult. Simulating high-altitude conditions in the laboratory can be extremely hard because of interferences, such as those from species given off from container walls under conditions of very low pressure. Many chemical reactions that require a third body to absorb excess energy occur very slowly in the upper atmosphere, where there is a sparse concentration of third bodies, but occur readily in a container whose walls effectively absorb energy. Container walls may serve as catalysts for some important reactions, or they may absorb important species and react chemically with the more reactive ones.

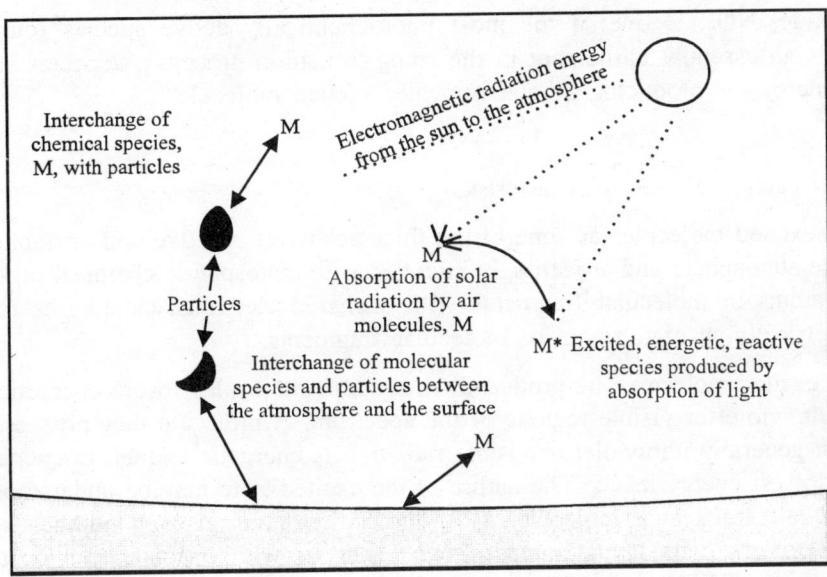

Fig. 8.8. Representation of major atmospheric chemical processes.

Atmospheric chemistry involves the unpolluted atmosphere, highly polluted atmospheres, and a wide range of gradations in between. The same general phenomena govern all, and produce one huge atmospheric cycle, in which there are numerous sub-cycles. Gaseous atmospheric chemical species fall into the following somewhat arbitrary and overlapping classifications: inorganic oxides (CO, CO_2, NO_2, SO_2), oxidants (O_3), reductants (CO, SO_2, H_2S), organics (in the unpolluted atmosphere, CH_4 is the predominant organic species, whereas alkanes, alkenes, and aryl compounds are common around sources of organic pollution), photochemically active species (NO_2, formaldehyde), acids (H_2SO_4), bases (NH_3), salts (NH_4HSO_4), and unstable reactive species (electronically excited NO_2, $HO\cdot$. radical). In addition, both solid and liquid particles play a strong role in atmospheric chemistry as sources and sinks for gas-phase species, as sites for surface reactions (solid particles), and as bodies for aqueous-phase reactions (liquid droplets). Two constituents of utmost importance in atmospheric chemistry are radiant energy from the sun, predominantly in the ultraviolet region of the spectrum, and the hydroxyl radical, $HO\cdot$. The former provides a way to pump a high level of energy into a single gas molecule to start a series of atmospheric chemical reactions, and the latter is the most important reactive intermediate and "currency" of daytime atmospheric chemical phenomena; NO_3 radicals are important intermediates in nighttime atmospheric chemistry.

Photochemical Processes

The absorption of light by chemical species can bring about reactions, called photochemical reactions, which do not otherwise occur under the conditions (particularly the temperature) of the medium in the absence of light. Thus, photochemical reactions, even in the absence of a chemical catalyst, occur at temperatures much lower than those which otherwise would be required. Photochemical reactions, which are induced by intense solar radiation, play a very important role in determining the nature and ultimate fate of a chemical species in the atmosphere.

Nitrogen dioxide, NO_2, is one of the most photochemically active species found in a polluted atmosphere and is an essential participant in the smog-formation process. A species such as NO_2 may absorb light of energy $h\nu$, producing an electronically excited molecule,

$$NO_2 + h\nu \rightarrow NO_2^* \qquad \qquad ...(8.4)$$

designated in the reaction above by an asterisk.

Electronically excited molecules are one of the three relatively reactive and unstable species that are encountered in the atmosphere and are strongly involved with atmospheric chemical processes. The other two species are atoms or molecular fragments with unshared electrons, called free radicals, and ions consisting of electrically-charged atoms or molecular fragments.

Electronically excited molecules are produced when stable molecules absorb energetic electromagnetic radiation in the ultraviolet or visible regions of the spectrum. A molecule may possess several possible excited states, but generally ultraviolet or visible radiation is energetic enough to excite molecules only to several of the lowest energy levels. The nature of the excited state may be understood by considering the disposition of electrons in a molecule. Most molecules have an even number of electrons. The electrons occupy orbitals, with a maximum of two electrons with opposite spin occupying the same orbital. The absorption of light may promote one of these electrons to a vacant orbital of higher energy. In some cases the electron thus promoted retains a spin opposite to that of its former partner, giving rise to an excited singlet state. In other cases the spin of the promoted electron is reversed, so that it has the same spin as its former partner; this gives rise to an excited triplet state.

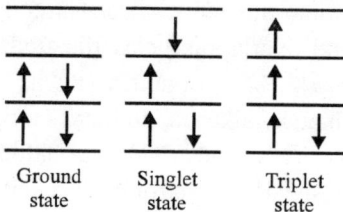

Ground   Singlet   Triplet
state   state   state

These excited states are relatively energised compared to the ground state and are chemically reactive species. Their participation in atmospheric chemical reactions, such as those involved in smog formation, will be discussed later in detail.

In order for a photochemical reaction to occur, light must be absorbed by the reacting species. If the absorbed light is in the visible region of the sun's spectrum, the absorbing species is coloured. Coloured NO_2 is a common example of such a species in the atmosphere. Normally, the first step in a photochemical process is the activation of the molecule by the absorption of a single unit of photochemical energy characteristic of the frequency of the light called a quantum of light. The energy of one quantum is equal to the product $h\nu$, where h is Planck's constant, 6.63×10^{-27} erg.s (6.63×10 J·s), and ν is the frequency of the absorbed light in s^{-1} (inversely proportional to its wavelength, λ).

The reactions that occur following absorption of a photon of light to produce an electronically excited species are largely determined by the way in which the excited species loses its excess energy. This may occur by one of the following processes :

1. Loss of energy to another molecule or atom (M) by physical quenching, followed by dissipation of the energy as heat

$$O_2^* + M \rightarrow O_2 + M \text{ (higher translational energy)} \qquad \qquad ...(8.5)$$

2. Dissociation of the excited molecule (the process responsible for the predominance of atomic oxygen in the upper atmosphere)

$$O^*_2 \rightarrow O + O \qquad \qquad ...(8.6)$$

3. Direct reaction with another species

$$O^*_2 + O_3 \rightarrow 2O_2 + O \qquad \qquad ...(8.7)$$

4. Luminescence consisting of loss of energy by the emission of electromagnetic radiation

$$NO_2^* \rightarrow NO_2 + h\nu \qquad \qquad ...(8.8)$$

If the re-emission of light is almost instantaneous, luminescence is called fluorescence and if it is significantly delayed, the phenomenon is phosphorescence. Chemiluminescence is said to occur when the excited species (such as NO_2 below) is formed by a chemical process :

$$O_3 + NO \rightarrow NO_2^* + O_2 \text{ (higher energy)} \qquad \qquad ...(8.9)$$

5. Intermolecular energy transfer in which an excited species transfers energy to another species, which becomes excited by the transfer

$$O_2^* + Na \rightarrow O_2 + Na^* \qquad \qquad ...(8.10)$$

A subsequent reaction by the second species is called a photosensitised reaction.

6. Intramolecular transfer in which energy is transferred within a molecule

$XY \rightarrow XY^\dagger$ (where † denotes another excited state of the same molecule) ...(8.11)

7. Spontaneous isomerisation as in the conversion of o-nitrobenzaldehyde to o-nitrosobenzoic acid, a reaction used in chemical actinometers to measure exposure to electromagnetic radiation :

$$...(8.12)$$

8. Photoionisation through loss of an electron

$$N_2^* \rightarrow N_2^+ + e^- \qquad \qquad ...(8.13)$$

Electromagnetic radiation absorbed in the infrared region is not sufficiently energetic to break chemical bonds, but does cause the receptor molecules to gain vibrational and rotational energy. The energy absorbed as infrared radiation ultimately is dissipated as heat and raises the temperature of the whole atmosphere. As already discussed, the absorption of infrared radiation is very important in the earth's acquiring heat from the sun and in the retention of energy radiated from the earth's surface.

Ions and Radicals in the Atmosphere

One of the characteristics of the upper atmosphere which is difficult to duplicate under laboratory conditions is the presence of significant levels of electrons and positive ions. Because of the rarefied conditions, these ions may exist in the upper atmosphere for long periods before recombining to form neutral species.

At altitudes of approximately 50 km and up, ions are so prevalent that the region is called the ionosphere. The presence of the ionosphere has been known since about 1901, when it was discovered that radio waves could be transmitted over long distances, where the curvature of the earth makes line-of-sight transmission impossible. These radio waves bounce off the ionosphere.

Ultraviolet light is the primary producer of ions in the ionosphere. In darkness, the positive ions slowly recombine with free electrons. The process is especially rapid in the lower regions of the ionosphere, where the concentration of species is relatively high. Thus, the lower limit of the ionosphere lifts at night and makes possible the transmission of radio waves over much greater distances.

The Earth's magnetic field has a strong influence upon the ions in the upper atmosphere. Probably the best-known manifestation of this phenomenon is found in the Van Allen belts, which consist of two belts of ionised particles encircling the earth. If they are visualised as two doughnuts, then the axis of the earth's magnetic field extends through the holes in the doughnuts. In the inner belt, the highly energetic ionising radiation consists of protons. In the outer belt, it consists of electrons. A schematic diagram of the Van Allen belts is shown in Fig. 8.9.

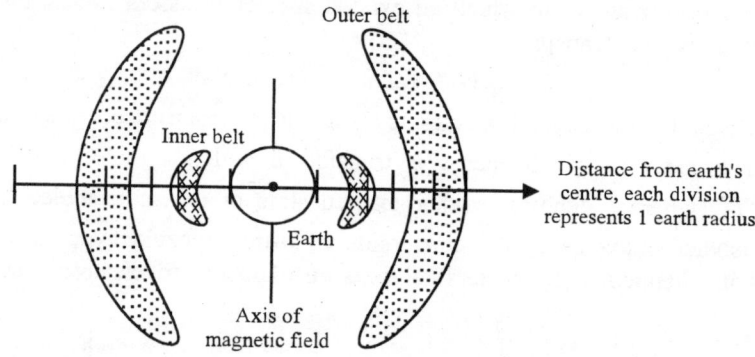

Fig. 8.9. Cross-section of the Van Allen belts encircling the earth.

Although ions are produced in the upper atmosphere primarily by the action of energetic electromagnetic radiation, they may also be produced in the troposphere by the shearing of water droplets during precipitation. The shearing may be caused by the compression of descending masses of cold air or by strong winds over hot, dry land masses. The last phenomenon is known as the foehn, sharav. These hot, dry winds cause severe discomfort. The ions produced by them consist of electrons and positively charged molecular species.

Free Radicals

In addition to forming ions by photoionisation, energetic electromagnetic radiation in the atmosphere may produce atoms or groups of atoms with unpaired electrons called free radicals :

$$H_3C-\overset{\displaystyle O}{\overset{\displaystyle \|}{C}}-H + h\nu \longrightarrow H_3C\cdot + \cdot\overset{\displaystyle O}{\overset{\displaystyle \|}{C}}-H \qquad ...(8.14)$$

Free radicals are involved with most significant atmospheric chemical phenomena and are of the utmost importance in the atmosphere. Because of their unpaired electrons and the strong pairing tendencies of electrons under most circumstances, free radicals are highly reactive. The upper atmosphere is so rarefied, however, that at very high altitudes radicals may have half-lives of several minutes, or even longer. Radicals can take part in chain reactions in which one of the products of each reaction is a radical.

Eventually, through processes such as reaction with another radical, one of the radicals in a chain is destroyed and the chain ends :

$$H_3C\cdot + H_3C\cdot \rightarrow C_2H_6 \qquad\qquad ...(8.15)$$

This process is a chain-terminating reaction. Reactions involving free radicals are responsible for smog formation.

Free radicals are quite reactive; therefore, they generally have short lifetimes. It is important to distinguish between high reactivity and instability. A totally isolated free radical or atom would be quite stable. Therefore, free radicals and single atoms from diatomic gases tend to persist under the rarefied conditions of very high altitudes because they can travel long distances before colliding with another reactive species. However, electronically excited species have a finite, generally very short, lifetime because they can lose energy through radiation without having to react with another species.

Hydroxyl and Hydroperoxyl Radicals in the Atmosphere

As illustrated in Fig. 8.10, the hydroxyl radical, $HO\cdot$, is the single most important reactive intermediate species in atmospheric chemical processes. It is formed by several mechanisms. At higher altitudes it is produced by photolysis of water :

$$H_2O + h\nu \rightarrow HO\cdot + H \qquad\qquad ...(8.16)$$

In the presence of organic matter, hydroxyl radical is produced in abundant quantities as an intermediate in the formation of photochemical smog. To a certain extent in the atmosphere, and for laboratory experimentation, $HO\cdot$ is made by the photolysis of nitrous acid vapour :

$$HONO + h\nu \rightarrow HO\cdot + NO \qquad\qquad ...(8.17)$$

In the relatively unpolluted troposphere, hydroxyl radical is produced as the result of the photolysis of ozone,

$$O_3 + h\nu \ (\lambda < 315 \ nm) \rightarrow O^* + O_2 \qquad\qquad ...(8.18)$$

followed by the reaction of a fraction of the excited oxygen atoms with water molecules :

$$O^* + H_2O \rightarrow 2HO\cdot \qquad\qquad ...(8.19)$$

Involvement of the hydroxyl radical in chemical transformations of a number of trace species in the atmosphere is summarised in Fig. 8.10. Among the important atmospheric trace species that react with hydroxyl radical are carbon monoxide, sulphur dioxide, hydrogen sulphide, methane, and nitric oxide.

Hydroxyl radical is most frequently removed from the troposphere by reaction with methane or carbon monoxide :

$$CH_4 + HO\cdot \rightarrow H_3C\cdot + H_2O \qquad\qquad ...(8.20)$$

$$CO + HO\cdot \rightarrow CO_2 + H \qquad\qquad ...(8.21)$$

The highly reactive methyl radical, $H_3C\cdot$, reacts with O_2,

$$H_3C\cdot + O_2 \rightarrow H_3COO\cdot \qquad\qquad ...(8.22)$$

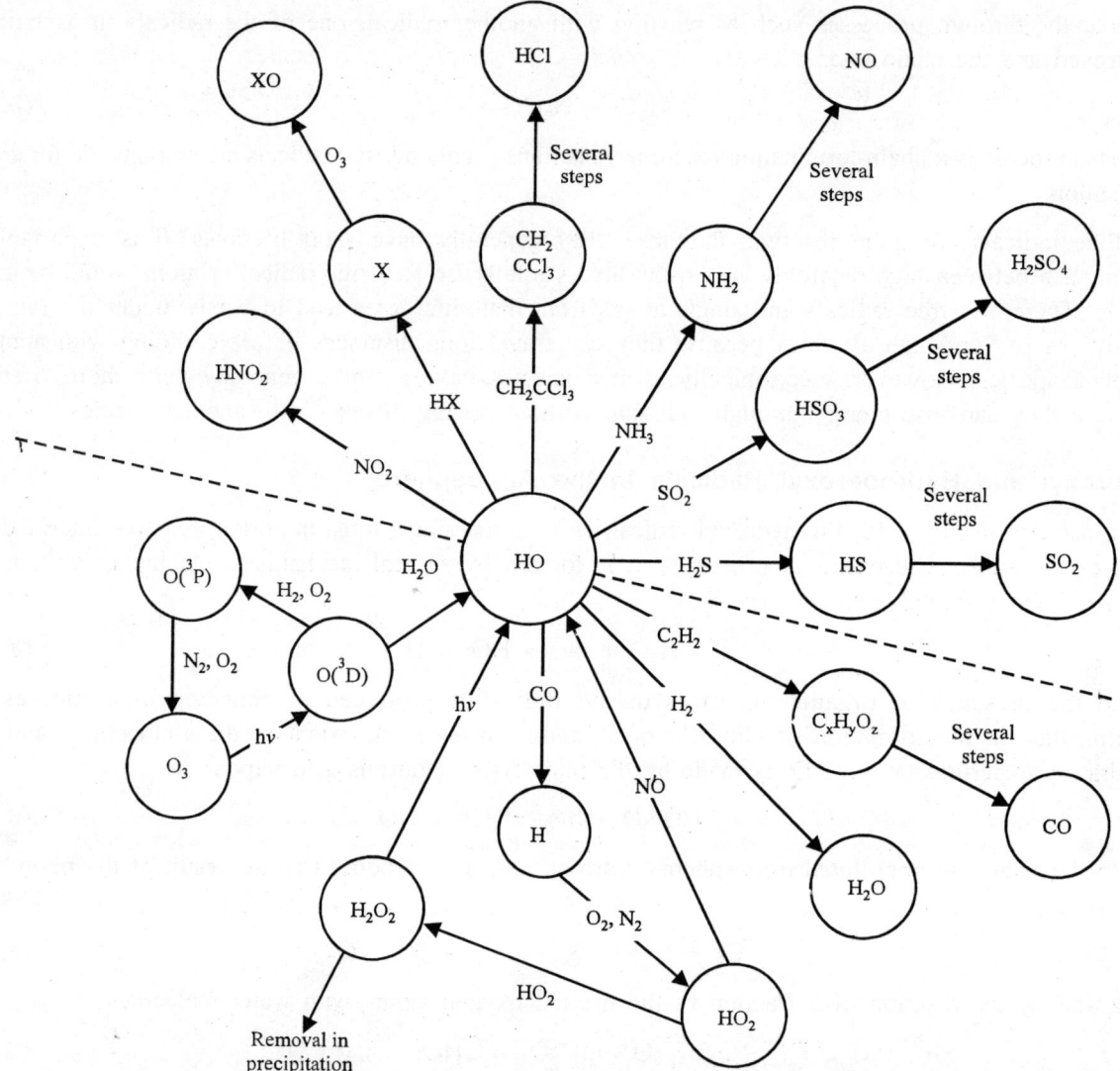

Fig. 8.10. Control of trace gas concentrations by HO· radical in the troposphere. Processes below the dashed line are those largely involved in controlling the concentrations of HO· in the troposphere; those above the line control the concentrations of the associated reactants and products. Reservoirs of atmospheric species are shown in circles, reactions denoting conversion of one species to another are shown by arrows, and the reactants or photons needed to bring about a particular conversion are shown along the arrows. Hydrogen halides are denoted by HX and hydrocarbons by H_xY_y.

to form methylperoxyl radical, $H_3COO·$. The hydrogen atom produced in Equation 8.21 reacts with O_2 to produce hydroperoxyl radical :

$$H + O_2 \rightarrow HOO· \qquad \qquad ...(8.23)$$

The hydroperoxyl radical can undergo chain termination reactions, such as

$$HOO· + HO· \rightarrow H_2O + O_2 \qquad \qquad ...(8.24)$$

$$HOO\cdot + HOO\cdot \rightarrow H_2O_2 + O_2 \qquad \qquad ...(8.25)$$

or reactions that regenerate hydroxyl radical :

$$HOO\cdot + NO \rightarrow NO_2 + HO\cdot \qquad \qquad ...(8.26)$$

$$HOO\cdot + O_3 \rightarrow 2O_2 + HO\cdot \qquad \qquad ...(8.27)$$

The global concentration of hydroxyl radical, averaged diurnally and seasonally, is estimated to range from 2×10^5 to 1×10^6 radicals per cm^3 in the troposphere. Because of the higher humidity and higher incident sunlight, which result in elevated O* levels, the concentration of HO· is higher in tropical regions. The southern hemisphere probably has about a 20% higher level of HO· than does the northern hemisphere because of greater production of anthropogenic HO· — consuming CO in the northern hemisphere.

The hydroperoxyl radical, HOO·, is an intermediate in some important chemical reactions. In addition to its production by the reactions discussed above, in polluted atmospheres, hydroperoxyl radical is made by the following two reactions, starting with photolytic dissociation of formaldehyde to produce a reactive formyl radical :

$$HCHO + h\nu \rightarrow H + H\dot{C}O \qquad \qquad ...(8.28)$$

$$H\dot{C}O + O_2 \rightarrow HOO\cdot + CO \qquad \qquad ...(8.29)$$

The hydroperoxyl radical reacts more slowly with other species than does the hydroxyl radical. The kinetics and mechanisms of hydroperoxyl radical reactions are difficult to study because it is hard to retain these radicals free of hydroxyl radicals.

Chemical and Biochemical Processes in Evolution of the Atmosphere

It is now widely believed that the earth's atmosphere originally was very different from its present state and that the changes were brought about by biological activity and accompanying chemical changes. Approximately 3.5 billion years ago, when the first primitive life molecules were formed, the atmosphere was probably free of oxygen and consisted of a variety of gases, such as carbon dioxide, water vapour, and perhaps even methane, ammonia, and hydrogen. The atmosphere was bombarded by intense, bond-breaking ultraviolet light, which, along with lightning and radiation from radionuclides, provided the energy to bring about chemical reactions that resulted in the production of relatively complicated molecules, including even amino acids and sugars. From the rich chemical mixture in the sea, life molecules evolved. Initially, these very primitive life forms derived their energy from fermentation of organic matter formed by chemical and photochemical processes, but eventually they gained the capability to produce organic matter, "$\{CH_2O\}$," by photosynthesis :

$$CO_2 + H_2O + h\nu \rightarrow \{CH_2O\} + O_2(g) \qquad \qquad ...(8.30)$$

Photosynthesis released oxygen, thereby setting the stage for the massive biochemical transformation that resulted in the production of almost all the atmosphere's oxygen.

The oxygen initially produced by photosynthesis was probably quite toxic to primitive life forms. However, much of this oxygen was converted to iron oxides by reaction with soluble iron(II) :

$$4Fe^{2-} + O_2 + 4H_2O \rightarrow 2Fe_2O_3 + 8H^- \qquad \qquad ...(8.31)$$

This resulted in the formation of enormous deposits of iron oxides, the existence of which provides major evidence for the liberation of free oxygen in the primitive atmosphere.

Eventually, enzyme systems developed that enabled organisms to mediate the reaction of waste-product oxygen with oxidisable organic matter in the sea. Later, this mode of waste-product disposal was utilised by organisms to produce energy by respiration, which is now the mechanism by which non-photosynthetic organisms obtain energy.

In time, O_2 accumulated in the atmosphere, providing an abundant source of oxygen for respiration. It had an additional benefit in that it enabled the formation of an ozone shield. The ozone shield absorbs bond-rupturing ultraviolet light. With the ozone shield protecting tissue from destruction by high-energy ultraviolet radiation, the earth became a much more hospitable environment for life, and life forms were enabled to move from the sea to land.

ACID-BASE REACTIONS IN THE ATMOSPHERE

Acid-base reactions occur between acidic and basic species in the atmosphere. The atmosphere is normally at least slightly acidic because of the presence of a low level of carbon dioxide, which dissolves in atmospheric water droplets and dissociates slightly :

$$CO_2(g) \xrightarrow{\text{water}} CO_2(aq) \qquad \qquad ...(8.32)$$
$$CO_2(aq) + H_2O \rightarrow H^+ + HCO_3^- \qquad \qquad ...(8.33)$$

Atmospheric sulphur dioxide forms a somewhat stronger acid when it dissolves in water :

$$SO_2(g) + H_2O \rightarrow H^+ + HSO_3^- \qquad \qquad ...(8.34)$$

In terms of pollution, however, strongly acidic HNO_3 and H_2SO_4 formed by the atmospheric oxidation of N oxides, SO_2, and H_2S are much more important because they lead to the formation of damaging acid rain.

As reflected by the generally acidic pH of rainwater, basic species are relatively less common in the atmosphere. Particulate calcium oxide, hydroxide, and carbonate can get into the atmosphere from ash and ground rock, and can react with acids such as in the following reaction :

$$Ca(OH)_2(s) + H_2SO_4(aq) \rightarrow CaSO_4(s) + 2H_2O \qquad \qquad ...(8.35)$$

The most important basic species in the atmosphere is gas-phase ammonia, NH_3. The major source of atmospheric ammonia is from biodegradation of nitrogen-containing biological matter and from bacterial reduction of nitrate :

$$NO_3^-(aq) + 2\{CH_2O\}(biomass) + H^+ \rightarrow NH_3(g) + 2CO_2 + H_2O \qquad \qquad ...(8.36)$$

Ammonia is particularly important as a base in the atmosphere because it is the only water-soluble base present at significant levels in the atmosphere. Dissolved in atmospheric water droplets, it plays a strong role in neutralising atmospheric acids :

$$NH_3(aq) + HNO_3(aq) \rightarrow NH_4NO_3(aq) \qquad \qquad ...(8.37)$$
$$NH_3(aq) + H_2SO_4(aq) \rightarrow NH_4HSO_4(aq) \qquad \qquad ...(8.38)$$

These reactions have three effects: (i) They result in the presence of NH_4^+ ion in the atmosphere as dissolved or solid salts; (ii) they serve in part to neutralise acidic constituents of the atmosphere; and (iii) they produce relatively corrosive ammonium salts.

REACTIONS OF ATMOSPHERIC OXYGEN

Some of the primary features of the exchange of oxygen among the atmosphere, lithosphere, hydrosphere, and biosphere are summarised in Fig. 8.11. The oxygen cycle is critically important in atmospheric chemistry, geochemical transformations, and life processes.

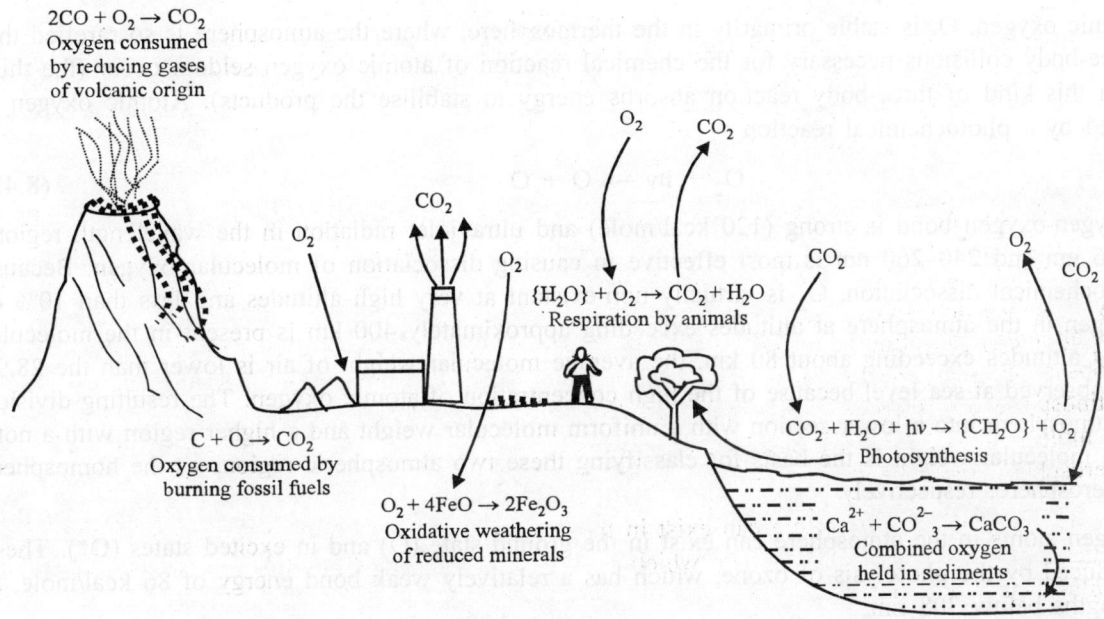

$$h\nu$$

$$O_3 + h\nu \rightarrow O + O_2$$
$$O + O_2 + M \rightarrow O_3 + M$$
$$O_2 + h\nu \rightarrow O + O$$

ozone shield: absorption of ultraviolet radiation from 220 nm to 330 nm

$$2CO + O_2 \rightarrow CO_2$$
Oxygen consumed by reducing gases of volcanic origin

O_2 CO_2

CO_2

O_2 O_2 CO_2

O_2
CO_2

$\{H_2O\} + O_2 \rightarrow CO_2 + H_2O$
Respiration by animals

$C + O \rightarrow CO_2$
Oxygen consumed by burning fossil fuels

$CO_2 + H_2O + h\nu \rightarrow \{CH_2O\} + O_2$
Photosynthesis

$O_2 + 4FeO \rightarrow 2Fe_2O_3$
Oxidative weathering of reduced minerals

$Ca^{2+} + CO_3^{2-} \rightarrow CaCO_3$
Combined oxygen held in sediments

Fig. 8.11. Oxygen exchange among the atmosphere, lithosphere, hydrosphere, and biosphere.

Oxygen in the troposphere plays a strong role in processes that occur on the earth's surface. Atmospheric oxygen takes part in energy-producing reactions, such as the burning of fossil fuels :

$$CH_4 \text{ (in natural gas)} + 2O_2 \rightarrow CO_2 + 2H_2O \qquad \qquad \text{...(8.39)}$$

Atmospheric oxygen is utilised by aerobic organisms in the degradation of organic material. Some oxidative weathering processes consume oxygen, such as

$$4FeO + O_2 \rightarrow 2Fe_2O_3 \qquad \qquad \text{...(8.40)}$$

Oxygen is returned to the atmosphere through plant photosynthesis :

$$CO_2 + H_2O + h\nu \rightarrow \{CH_2O\} + O_2 \qquad \qquad ...(8.41)$$

All molecular oxygen now in the atmosphere is thought to have originated through the action of photosynthetic organisms, which shows the importance of photosynthesis in the oxygen balance of the atmosphere. It can be shown that most of the carbon fixed by these photosynthetic processes is dispersed in mineral formations as humic material; only a very small fraction is deposited in fossil fuel beds. Therefore, although combustion of fossil fuels consumes large amounts of O_2, there is no danger of running out of atmospheric oxygen.

Molecular oxygen is somewhat unusual in that its ground state is a triplet state with two unpaired electrons, designated here as $^3O_2$, which can be excited to singlet molecular oxygen, designated here as $^1O_2$. The latter can be produced by several processes, including direct photochemical excitation, transfer of energy from other electronically excited molecules, ozone photolysis, and high-energy oxygen-producing reactions.

Because of the extremely rarefied atmosphere and the effects of ionising radiation, elemental oxygen in the upper atmosphere exists to a large extent in forms other than diatomic O_2. In addition to O_2, the upper atmosphere contains oxygen atoms, O; excited oxygen molecules, O_2^*, and ozone, O_3.

Atomic oxygen, O, is stable primarily in the thermosphere, where the atmosphere is so rarefied that the three-body collisions necessary for the chemical reaction of atomic oxygen seldom occur (the third body in this kind of three-body reaction absorbs energy to stabilise the products). Atomic oxygen is produced by a photochemical reaction :

$$O_2 + h\nu \rightarrow O + O \qquad \qquad ...(8.42)$$

The oxygen-oxygen bond is strong (120 kcal/mole) and ultraviolet radiation in the wavelength regions 135–176 nm and 240–260 nm is most effective in causing dissociation of molecular oxygen. Because of photochemical dissociation, O_2 is virtually non-existent at very high altitudes and less than 10% of the oxygen in the atmosphere at altitudes exceeding approximately 400 km is present in the molecular form. At altitudes exceeding about 80 km, the average molecular weight of air is lower than the 28.97 g/mole observed at sea level because of the high concentration of atomic oxygen. The resulting division of the atmosphere into a lower section with a uniform molecular weight and a higher region with a non-uniform molecular weight is the basis for classifying these two atmospheric regions as the homosphere and heterosphere, respectively.

Oxygen atoms in the atmosphere can exist in the ground state (O) and in excited states (O*). These are produced by the photolysis of ozone, which has a relatively weak bond energy of 86 kcal/mole, at wavelengths below 308 nm,

$$O_3 + h\nu \ (\lambda < 308 \text{ nm}) \rightarrow O^* + O_2 \qquad \qquad ...(8.43)$$

or by highly energetic chemical reactions such as

$$O + O + O \rightarrow O_2 + O^* \qquad \qquad ...(8.44)$$

Excited atomic oxygen emits visible light at wavelengths of 636 nm, 630 nm, and 558 nm. This emitted light is partially responsible for airglow, a very faint electromagnetic radiation continuously emitted by the earth's atmosphere. Although its visible component is extremely weak, airglow is quite intense in the infrared region of the spectrum.

Oxygen ion, O^+, which may be produced by ultraviolet radiation acting upon oxygen atoms,

$$O + h\nu \rightarrow O^+ + e^- \qquad \qquad ...(8.45)$$

is the predominant positive ion in some regions of the ionosphere. It may react with molecular oxygen or nitrogen,

$$O^+ + O_2 \rightarrow O_2^+ + O \qquad \qquad ...(8.46)$$

$$O^+ + N_2 \rightarrow NO^+ + N \qquad \qquad ...(8.47)$$

to form other positive ions.

In intermediate regions of the ionosphere, O_2^+ is produced by absorption of ultraviolet radiation at wavelengths of 17–103 nm. This diatomic oxygen ion can also be produced by the photochemical reaction of low-energy X-rays,

$$O_2 + h\nu \rightarrow O_2^+ + e^- \qquad \qquad ...(8.48)$$

and by the following reaction :

$$N_2^+ + O_2 \rightarrow N_2 + O_2^+ \qquad \qquad ...(8.49)$$

Ozone, O_3, has an essential protective function because it absorbs harmful ultraviolet radiation in the stratosphere and serves as a radiation shield, protecting living beings on the earth from the effects of excessive amounts of such radiation. It is produced by a photochemical reaction,

$$O_2 + h\nu \rightarrow O + O \qquad \qquad ...(8.50)$$

(where the wavelength of the exciting radiation must be less than 242.4 nm), followed by a three-body reaction,

$$O + O_2 + M \rightarrow O_3 + M(\text{increased energy}) \qquad \qquad ...(8.51)$$

in which M is another species, such as a molecule of N_2 or O_2, which absorbs the excess energy given off by the reaction and enables the ozone molecule to stay together. The region of maximum ozone concentration is found within the range of 25–30 km high in the stratosphere where it may reach 10 ppm.

Ozone absorbs ultraviolet light very strongly in the region 220–330 nm. If this light were not absorbed by ozone, severe damage would result to exposed forms of life on the Earth. Absorption of electromagnetic radiation by ozone converts the radiation's energy to heat and is responsible for the temperature maximum encountered at the boundary between the stratosphere and the mesosphere at an altitude of approximately 50 km. The reason that the temperature maximum occurs at a higher altitude than that of the maximum ozone concentration arises from the fact that ozone is such an effective absorber of ultraviolet light, so that most of this radiation is absorbed in the upper stratosphere, where it generates heat, and only a small fraction reaches the lower altitudes, which remain relatively cool.

The overall reaction,

$$2O_3 \rightarrow 3O_2 \qquad \qquad ...(8.52)$$

is favoured thermodynamically so that ozone is inherently unstable. Its decomposition in the stratosphere is catalysed by a number of natural and pollutant trace constituents, including NO, NO_2, H, HO·, HOO·, ClO, Cl, Br, and BrO. Ozone decomposition also occurs on solid surfaces, such as metal oxides and salts produced by rocket exhausts.

Although the mechanisms and rates for the photochemical production of ozone in the stratosphere are reasonably well known, the natural pathways for ozone removal are less well understood. In addition to undergoing decomposition by the action of ultraviolet radiation, stratospheric ozone reacts with atomic oxygen, hydroxyl radical and NO :

$$O_3 + h\nu \rightarrow O_2 + O \qquad \text{...(8.53)}$$

$$O_3 + O \rightarrow O_2 + O_2 \qquad \text{...(8.54)}$$

$$O_3 + HO\cdot \rightarrow O_2 + HOO\cdot \qquad \text{...(8.55)}$$

The HO· radical is regenerated from HOO· by the reaction,

$$HOO\cdot + O \rightarrow HO\cdot + O_2 \qquad \text{...(8.56)}$$

$$O_3 + NO \rightarrow NO_2 + O_2 \qquad \text{...(8.57)}$$

The NO consumed in this reaction is regenerated from NO_2,

$$NO_2 + O \rightarrow NO + O_2 \qquad \text{...(8.58)}$$

and some NO is produced from N_2O :

$$N_2O + O \rightarrow 2NO \qquad \text{...(8.59)}$$

Recall that N_2O is a natural component of the atmosphere and is a major product of the denitrification process by which fixed nitrogen is returned to the atmosphere in gaseous form.

Ozone is an undesirable pollutant in the troposphere. It is toxic to animals and plants, and it also damages materials, particularly rubber.

REACTIONS OF ATMOSPHERIC NITROGEN

The 78% by volume of nitrogen contained in the atmosphere constitutes an inexhaustible reservoir of that essential element. A small amount of nitrogen is fixed in the atmosphere by lightning, and some is also fixed by combustion processes, particularly in internal combustion and turbine engines.

Before the use of synthetic fertilisers reached its current high levels, chemists were concerned that denitrification processes in the soil would lead to nitrogen depletion on the earth. Now, with millions of tonnes of synthetically fixed nitrogen being added to the soil each year, major concern has shifted to possible excess accumulation of nitrogen in soil, fresh water, and the oceans.

Unlike oxygen, which is almost completely dissociated to the monatomic form in higher regions of the thermosphere, molecular nitrogen is not readily dissociated by ultraviolet radiation. However, at altitudes exceeding approximately 100 km, atomic nitrogen is produced by photochemical reactions :

$$N_2 + h\nu \rightarrow N + N \qquad \text{...(8.60)}$$

Other reactions which may produce monatomic nitrogen are :

$$N_2^+ + O \rightarrow NO^+ + N \qquad \text{...(8.61)}$$

$$NO^+ + e^- \rightarrow N + O \qquad \text{...(8.62)}$$

$$O^+ + N_2 \rightarrow NO^+ + N \qquad \text{...(8.63)}$$

As shown in Reactions, 8.57 and 8.58, NO is involved in the removal of stratospheric ozone and is regenerated by the reaction of NO_2 with atomic O, itself a precursor to the formation of ozone. An ion formed from NO, the NO^+ ion, is one of the predominant ionic species in the so-called E region of the ionosphere. A plausible sequence of reactions by which NO^+ is formed is the following :

$$N_2 + h\nu \rightarrow N_2^+ + e^- \qquad \qquad ...(8.64)$$

$$N_2^+ + O \rightarrow NO^+ + N \qquad \qquad ...(8.65)$$

In the lowest (D) region of the ionosphere, which extends from approximately 50 km in altitude to approximately 85 km, NO^+ is produced directly by ionising radiation :

$$NO + h\nu \rightarrow NO^+ + e^- \qquad \qquad ...(8.66)$$

In the lower part of this region, the ionic species N_2^+ is formed through the action of galactic cosmic rays :

$$N_2 + h\nu \rightarrow N_2^+ + e^- \qquad \qquad ...(8.67)$$

Pollutant oxides of nitrogen, particularly NO_2, are key species involved in air pollution and the formation of photochemical smog. For example, NO_2 is readily dissociated photochemically to NO and reactive atomic oxygen :

$$NO_2 + h\nu \rightarrow NO + O \qquad \qquad ...(8.68)$$

This reaction is the most important primary photochemical process involved in smog formation.

ATMOSPHERIC CARBON DIOXIDE

Although only about 0.035% (350 ppm) of air consists of carbon dioxide, it is the atmospheric "non-pollutant" species of most concern. As already mentioned, carbon dioxide, along with water vapour, is primarily responsible for the absorption of infrared energy re-emitted by the earth so that some of this energy is reradiated back to the earth's surface. Current evidence suggests that changes in the atmospheric carbon dioxide level will substantially alter the earth's climate through the greenhouse effect.

Valid measurements of overall atmospheric CO_2 can only be taken in areas remote from industrial activity. Such areas include Antarctica and the top of Mauna Loa Mountain in Hawaii. Measurements of carbon dioxide levels in these locations over the last 30 years suggest an annual increase in CO_2 of about 1 ppm per year.

The most obvious factor contributing to increased atmospheric carbon dioxide is consumption of carbon-containing fossil fuels. In addition, release of CO_2 from the biodegradation of biomass and uptake by photosynthesis are important factors determining overall CO_2 levels in the atmosphere. The role of photosynthesis is illustrated in Fig. 8.12, which shows a seasonal cycle in carbon dioxide levels in the northern hemisphere. Maximum values occur in April and minimum values in late September or October. These oscillations are due to the "photosynthetic pulse", influenced most strongly by forests in middle latitudes. Forests have a much greater influence than other vegetation because trees carry out more photosynthesis. Furthermore, forests store enough fixed, but readily oxidisable carbon in the form of wood and humus to have a marked influence on atmospheric CO_2 content. Thus, during the summer months, forest trees carry out enough photosynthesis to reduce the atmospheric carbon dioxide content markedly. During the winter, metabolism of biota, such as bacterial decay of humus, releases a significant amount of CO_2. Therefore, the current worldwide destruction of forests and conversion of forest lands to agricultural uses contributes substantially to a greater overall increase in atmospheric CO_2 levels.

With current trends, it is likely that global CO_2 levels will double by the middle of the next century, which may well raise the earth's mean surface temperature by 1.5 to 4.5°C. Such a change might have more potential to cause massive irreversible environmental changes than any other disaster short of global nuclear war.

Chemically and photochemically, carbon dioxide is a comparatively insignificant species because of its relatively low concentrations and low photochemical reactivity. The one significant photochemical reaction that carbon dioxide undergoes, and a major source of CO at higher altitudes, is the photodissociation of CO_2 by energetic solar ultraviolet radiation in the stratosphere :

$$CO_2 + h\nu \rightarrow CO + O \qquad \qquad ...(8.69)$$

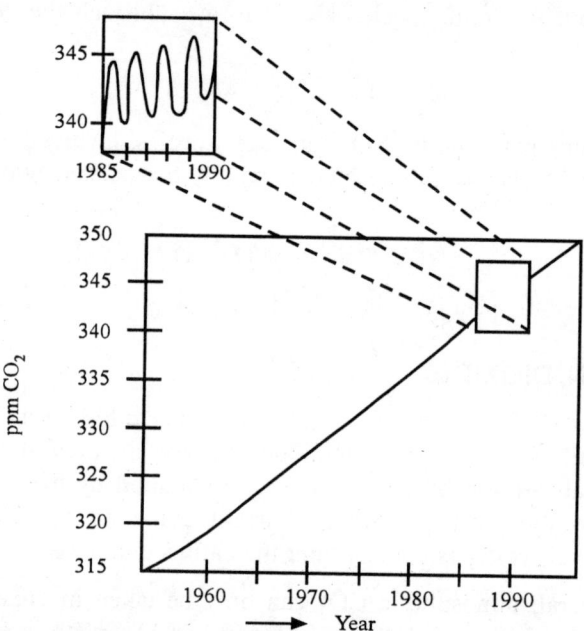

Fig. 8.12. Increases in atmospheric CO_2 levels in recent years. The inset illustrates seasonal variations in the northern hemisphere.

ATMOSPHERIC WATER

The water vapour content of the troposphere is normally within a range of 1–3% by volume with a global average of about 1%. However, air can contain as little as 0.1% or as much as 5% water. The percentage of water in the atmosphere decreases rapidly with increasing altitude. Water circulates through the atmosphere in the hydrologic cycle.

Water vapour absorbs infrared radiation even more strongly than does carbon dioxide, thus greatly influencing the earth's heat balance. Clouds formed from water vapour reflect light from the sun and have a temperature-lowering effect. On the other hand, water vapour in the atmosphere acts as a kind of "blanket" at night, retaining heat from the earth's surface by absorption of infrared radiation. As already discussed, water vapour and the heat released and absorbed by transitions of water between the vapour state and liquid or solid are strongly involved in atmospheric energy transfer. Condensed water vapour in the form of very small droplets is of considerable concern in atmospheric chemistry. The harmful effects of some air pollutants – for instance, the corrosion of metals by acid-forming gases – requires the presence of water which may come from the atmosphere. Atmospheric water vapour has an important influence upon pollution-induced fog formation under some circumstances. Water vapour interacting with pollutant particulate matter in the atmosphere may reduce visibility to undesirable levels through the formation of very small atmospheric aerosol particles.

The cold tropopause serves as a barrier to the movement of water into the stratosphere. Thus little water is transferred from the troposphere to the stratosphere, and the main source of water in the stratosphere is the photochemical oxidation of methane :

$$CH_4 + 2O_2 + h\nu \xrightarrow[\text{(several steps)}]{} CO_2 + 2H_2O \qquad ...(8.70)$$

The water thus produced serves as a source of stratospheric hydroxyl radical as shown by the following reaction :

$$H_2O + h\nu \rightarrow HO\cdot + H \qquad ...(8.71)$$

ATMOSPHERIC PARTICLES

Particles are common significant components of the atmosphere, particularly the troposphere. Colloidal-sized particles in the atmosphere are called aerosols. These particles originate in nature from sea sprays, smokes, dusts, and the evaporation of organic materials from vegetation. Other typical particles of natural origin in the atmosphere are bacteria, fog, pollen grains, and volcanic ash.

Many important atmospheric phenomena involve aerosol particles, including electrification phenomena, cloud formation, and fog formation. Particles help determine the heat balance of the earth's atmosphere by reflecting light. Probably the most important function of particles in the atmosphere is their action as nuclei for the formation of ice crystals and water droplets. Current efforts at rain-making are centred around the addition of condensing particles to atmospheres supersaturated with water vapour. Dry ice was used in early attempts; now silver iodide, which forms huge numbers of very small particles, is used.

As illustrated in Fig. 8.13, particles are involved in many chemical reactions in the atmosphere. Neutralisation reactions, which occur most readily in solution, may take place in water droplets suspended in the atmosphere. Small particles of metal oxides and carbon have a catalytic effect on oxidation reactions. Particles may also participate in oxidation reactions induced by light.

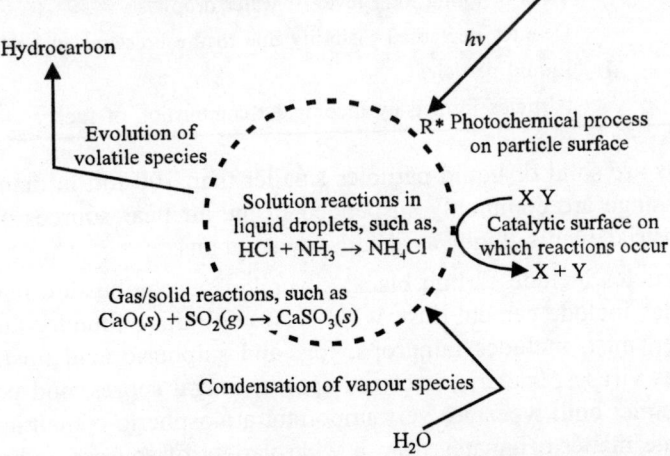

Fig. 8.13. Atmospheric chemical processes involving particles.

CHAPTER 9

Particles in the Atmosphere

INTRODUCTION

Particles in the atmosphere, which range in size from about one-half millimetre (the size of sand or drizzle) down to molecular dimensions, are made up of an amazing variety of materials and discrete objects that may consist of either solids or liquid droplets. A number of terms are commonly used to describe atmospheric particles; the more important of these are summarised in Table 9.1. Particles abound in the atmosphere. Even the Arctic, remote from sources of industrial pollution, is afflicted with an "Arctic Haze" of airborne particles from October to May each year. Particulates is a term that has come to stand for particles in the atmosphere, although *particulate matter* or simply *particles*, is preferred usage. Particulate matter makes up the most visible and obvious form of air pollution.

Table 9.1. Important terms describing atmospheric particles.

Term	Meaning
Aerosol	Colloidal-sized atmospheric particle
Condensation aerosol	Formed by condensation of vapours or reactions of gases
Dispersion aerosol	Formed by grinding of solids, atomisation of liquids, or dispersion of dusts.
Fog	Term denoting high level of water droplets
Haze	Denotes decreased visibility due to the presence of particles
Mists	Liquid particles
Smoke	Particles formed by incomplete combustion of fuel

Atmospheric aerosols are solid or liquid particles smaller than 100 μm in diameter. Pollutant particles in the 0.001 to 10 μm range are commonly suspended in the air near sources of pollution, such as the urban atmosphere, industrial plants, highways, and power plants.

Very small, solid particles include carbon black, silver iodide, combustion nuclei, and sea-salt nuclei (Fig. 9.1). Larger particles include cement dust, wind-blown soil dust, foundry dust, and pulverised coal. Liquid particulate matter, mist, includes raindrops, fog, and sulphuric acid mist. Some particles are of biological origin, such as viruses, bacteria, bacterial spores, fungal spores, and pollen. Particulate matter may be organic or inorganic; both types are very important atmospheric contaminants. As discussed later in this chapter, particulate matter originates from a wide variety of sources and processes, ranging from simple grinding of bulk matter to complicated chemical or biochemical syntheses. The effects of particulate matter are also widely varied. Either by itself, or in combination with gaseous pollutants, particulate matter may be detrimental to human health. Atmospheric particles may damage materials, reduce visibility, and cause undesirable aesthetic effects.

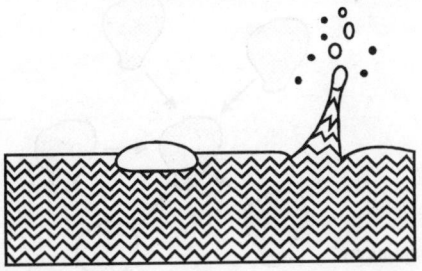

Fig. 9.1. Bursting bubbles in seawater form small liquid aerosol particles. Evaporation of water from aerosol particles results in the formation of small solid particles of sea-salt nuclei.

For the most part, aerosols consist of carbonaceous material, metal oxides and glasses, dissolved ionic species (electrolytes), and ionic solids. The predominant constituents are carbonaceous material, water, sulphate, nitrate, ammonium nitrogen, and silicon. The composition of aerosol particles varies significantly with size. The very small particles tend to be acidic and often originate from gases, such as from the conversion of SO_2 to H_2SO_4. Larger particles tend to consist of materials generated mechanically, such as by the grinding of limestone, and have a greater tendency to be basic.

PHYSICAL BEHAVIOUR OF PARTICLES IN THE ATMOSPHERE

As shown in Fig. 9.2, atmospheric particles undergo a number of processes in the atmosphere. Small colloidal particles are subject to *diffusion processes*. Smaller particles *coagulate* together to form larger particles. *Sedimentation* or *dry deposition* of particles, which have often reached sufficient size to settle by coagulation, is one of two major mechanisms for particle removal from the atmosphere. The other is *scavenging* by raindrops and other forms of precipitation. Particles also react with atmospheric gases.

Physical Processes for Particle Formation

Dispersion aerosols, such as dusts, formed from the disintegration of larger particles are usually above 1 μm in size. Typical processes for forming dispersion aerosols include evolution of dust from coal grinding, formation of spray in cooling towers, and blowing of dirt from dry soil.

Many dispersion aerosols originate from natural sources, such as sea spray, windblown dust, and volcanic dust. However a vast variety of human activities break up material and disperse it to the atmosphere. "All terrain" vehicles churn across desert lands, coating fragile desert plants with layers of dispersed dust. Quarries and rock crushers spew out plumes of ground rock. Cultivation of land has made it much more susceptible to dust-producing wind erosion.

However, since much more energy is required to break material down into small particles than is required for or released by the synthesis of particles through chemical synthesis or the adhesion of smaller particles, most dispersion aerosols are relatively large. Larger particles tend to have fewer harmful effects than smaller ones. As examples, larger particles are less *respirable* in that they do not penetrate so far into the lungs as smaller ones and larger particles are relatively more easy to remove from air pollution effluent sources.

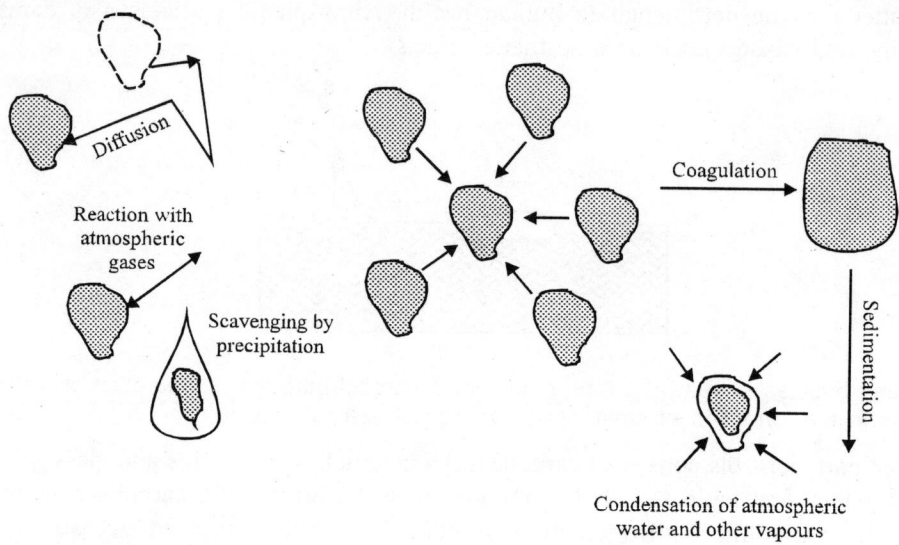

Fig. 9.2. Processes that particles undergo in the atmosphere.

Chemical Processes for Particle Formation

Most chemical processes that produce particles are combustion processes, including fossil-fuel-fired power plants; incinerators; home furnaces, fireplaces, and stoves; cement kilns; internal combustion engines; forest, brush, and grass fires; and active volcanoes. Particles from combustion sources tend to occur in a size range below 1 μm. Such very small particles are particularly important because they are most readily carried into the alveoli of lungs and they are likely to be enriched in more hazardous constituents, such as toxic heavy metals and arsenic. The latter characteristic can enable use of small particle analysis for tracking sources of particulate pollutants.

Inorganic particles

Metal oxides constitute a major class of inorganic particles in the atmosphere. These are formed whenever fuels containing metals are burned. For example, particulate iron oxide is formed during the combustion of pyrite-containing coal :

$$3FeS_2 + 8O_2 \rightarrow Fe_3O_4 + 6SO_2 \qquad \ldots(9.1)$$

Organic vanadium in residual fuel oil is converted to particulate vanadium oxide. Part of the calcium carbonate in the ash fraction of coal is converted to calcium oxide and is emitted to the atmosphere through the stack :

$$CaCO_3 + heat \rightarrow CaO + CO_2 \qquad \ldots(9.2)$$

A common process for the formation of aerosol mists involves the oxidation of atmospheric sulphur dioxide to sulphuric acid, a hygroscopic substance that accumulates atmospheric water to form small liquid droplets :

$$2SO_2 + O_2 + 2H_2O \rightarrow 2H_2SO_4 \qquad \ldots(9.3)$$

In the presence of basic air pollutants, such as ammonia or calcium oxide, the sulphuric acid reacts to form salts :

$$H_2SO_4(\text{droplet}) + 2\ NH_3(g) \rightarrow (NH_4)_2SO_4\ (\text{droplet}) \qquad ...(9.4)$$

$$H_2SO_4(\text{droplet}) + CaO(s) \rightarrow CaSO_4\ (\text{droplet}) + H_2O \qquad ...(9.5)$$

Under low-humidity conditions, water is lost from these droplets and a solid aerosol is formed.

The preceding examples show several ways in which solid or liquid inorganic aerosols are formed by chemical reactions. Such reactions constitute an important general process for the formation of aerosols, particularly the smaller particles.

Organic particles

A significant portion of organic particulate matter is produced by internal combustion engines in complicated processes that involve pyrosynthesis and nitrogenous compounds. These products may include nitrogen-containing compounds and oxidised hydrocarbon polymers. Lubricating oil and its additives may also contribute to organic particulate matter. A study of particulate matter emitted by gasoline auto engines (with and without catalysts) and diesel truck engines measured more than 100 compounds quantitatively. Among the prominent classes of compounds found were n-alkanes, n-alkanoic acids, benzaldehydes, benzoic acids, azanaphthalenes, polycyclic aromatic hydrocarbons, oxygenated PAHs, pentacyclic triterpanes, and steranes (the last two classes of hydrocarbons are multi-ringed compounds characteristic of petroleum that enter exhaust gases from lubricating oil).

PAH synthesis

The organic particles of greatest concern are PAH hydrocarbons, which consist of condensed ring aromatic molecules. The most often cited example of a PAH compound is benzo(a)pyrene, a compound that the body can metabolise to a carcinogenic form.

PAHs may be synthesised from saturated hydrocarbons under oxygen-deficient conditions. Hydrocarbons with very low molecular masses, including even methane, may act as precursors for the polycyclic aromatic compounds. Low-molar-mass hydrocarbons form PAHs by pyrosynthesis. This happens at temperatures exceeding approximately 500°C at which carbon-hydrogen and carbon-carbon bonds are broken to form free radicals. These radicals undergo dehydrogenation and combine chemically to form aromatic ring structures, which are resistant to thermal degradation.

Polycyclic aromatic compounds may be formed from higher alkanes present in fuels and plant materials by the process of pyrolysis, the "cracking" of organic compounds to form smaller and less stable molecules and radicals.

The Composition of Inorganic Particles

Fig. 9.3 illustrates the basic factors responsible for the composition of inorganic particulate matter. In general, the proportions of elements in atmospheric particulate matter reflect relative abundances of elements in the parent material. The source of particulate matter is reflected in its elemental composition, taking into consideration chemical reactions that may change the composition. For example, particulate matter largely from ocean spray origin in a coastal area receiving sulphur dioxide pollution may show anomalously high sulphate and corresponding low chloride content. The sulphate comes from atmospheric oxidation of sulphur dioxide to form non-volatile ionic sulphate, whereas some chloride originally from the NaCl in the sea-water may be lost from the solid aerosol as volatile HCl.

$$2SO_2 + O_2 + 2H_2O \rightarrow 2H_2SO_4 \qquad \qquad ...(9.6)$$

$$H_2SO_4 + 2NaCl(particulate) \rightarrow Na_2SO_4(particulate) + 2HCl \qquad ...(9.7)$$

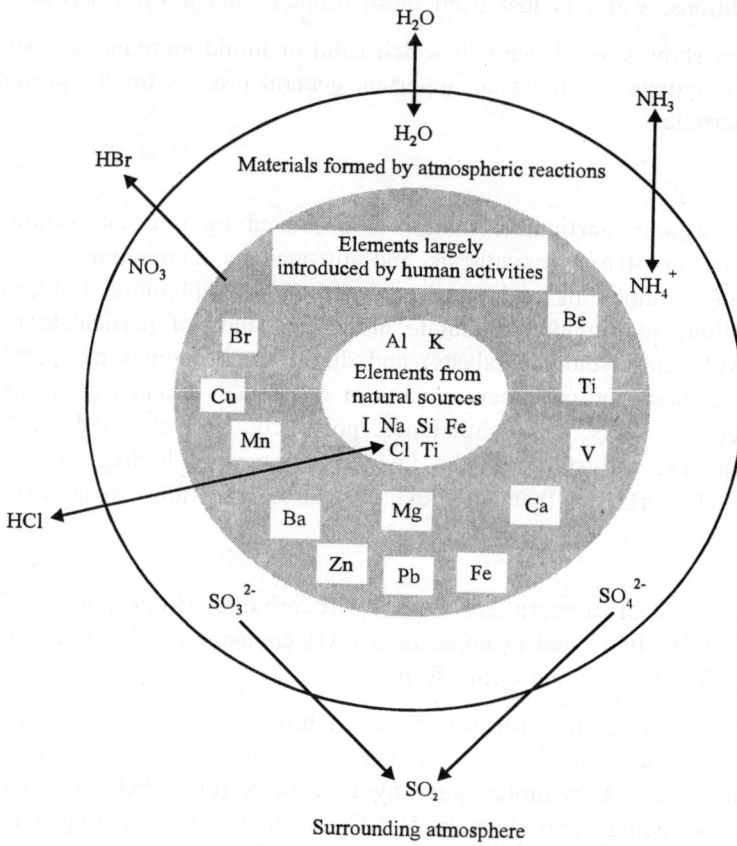

Fig. 9.3. Some of the components of inorganic particulate matter and their origins.

The chemical composition of atmospheric particulate matter is quite diverse. Among the constituents of inorganic particulate matter found in polluted atmospheres are salts, oxides, nitrogen compounds, sulphur compounds, various metals, and radionuclides. In coastal areas, sodium and chlorine get into atmospheric particles as sodium chloride from sea spray. The major trace elements that typically occur at levels above 1 µg/m³ in particulate matter are aluminium, calcium, carbon, iron, potassium, sodium, and silicon; note that most of these tend to originate from terrestrial sources. Lesser quantities of copper, lead, titanium, and zinc and even lower levels of antimony, beryllium, bismuth, cadmium, cobalt, chromium, cesium, lithium, manganese, nickel, selenium, strontium, and vanadium are commonly observed. The likely sources of some of these elements are given below :

1. *Al, Fe, Ca, Si:* Soil erosion, rock dust, coal combustion.
2. *C:* Incomplete combustion of carbonaceous fuels.

3. *Na, Cl:* Marine aerosols, chloride from incineration of organohalide polymer wastes.

4. *Sb, Se:* Very volatile elements, possibly from the combustion of oil, coal, or refuse.

5. *V:* Combustion of residual petroleum (present at very high levels in residues from crude oil).

6. *Zn:* Tends to occur in small particles, probably from combustion.

7. *Pb:* Combustion of leaded fuels and wastes containing lead.

Particulate carbon as soot, carbon black, coke, and graphite originates from auto and truck exhausts, heating furnaces, incinerators, power plants, and steel and foundry operations and composes one of the more visible and troublesome particulate air pollutants. Because of its good adsorbent properties, carbon can be a carrier of gaseous and other particulate pollutants. Particulate carbon surfaces may catalyse some heterogeneous atmospheric reactions, including the important conversion of SO_2 to sulphate.

Fly ash

Much of the mineral particulate matter in a polluted atmosphere is in the form of oxides and other compounds produced during the combustion of high-ash fossil fuel. Much of the mineral matter in fossil fuels such as coal or lignite is converted during combustion to a fused, glassy bottom ash which presents no air pollution problems. Smaller particles of fly ash enter furnace flues and are efficiently collected in a properly equipped stack system. However, some fly ash escapes through the stack and enters the atmosphere. Unfortunately, the fly ash thus released tends to consist of smaller particles that do the most damage to human health, plants, and visibility.

The composition of fly ash varies widely, depending upon the source of fuel. The predominant constituents are oxides of aluminium, calcium, iron, and silicon. Other elements that occur in fly ash are magnesium, sulphur, titanium, phosphorus, potassium, and sodium. Elemental carbon (soot, carbon black) is a significant fly ash constituent.

The size of fly ash particles is a very important factor in determining their removal from stack gas and their ability to enter the body through the respiratory tract. Fly ash from coal-fired utility boilers has shown a bimodal (two peak) distribution of size, with a peak at about 0.1 µm as illustrated in Fig. 9.4. Although only about 1–2% of the total fly ash mass is in the smaller size fraction, it includes the vast majority of the total number of particles and particle surface area. Submicrometre particles probably result from a volatilisation-condensation process during combustion as reflected in a higher concentration of more volatile elements, such as As, Sb, Hg, and Zn. Furthermore, the very small particles are the most difficult to remove by electrostatic precipitators and baghouses.

Asbestos

Asbestos is the name given to a group of fibrous silicate minerals, typically those of the serpentine group, for which the approximate formula is $Mg_3P(Si_2O_5)(OH)_4$. The tensile strength, flexibility, and nonflammability of asbestos have led to many uses including structural materials, brake linings, insulation, and pipe manufacture.

Asbestos is of concern as an air pollutant because when inhaled, it may cause asbestosis (a pneumonia condition), mesothelioma (tumor of the mesothelial tissue lining the chest cavity adjacent to the lungs), and bronchogenic carcinoma (cancer originating with the air passages in the lungs).

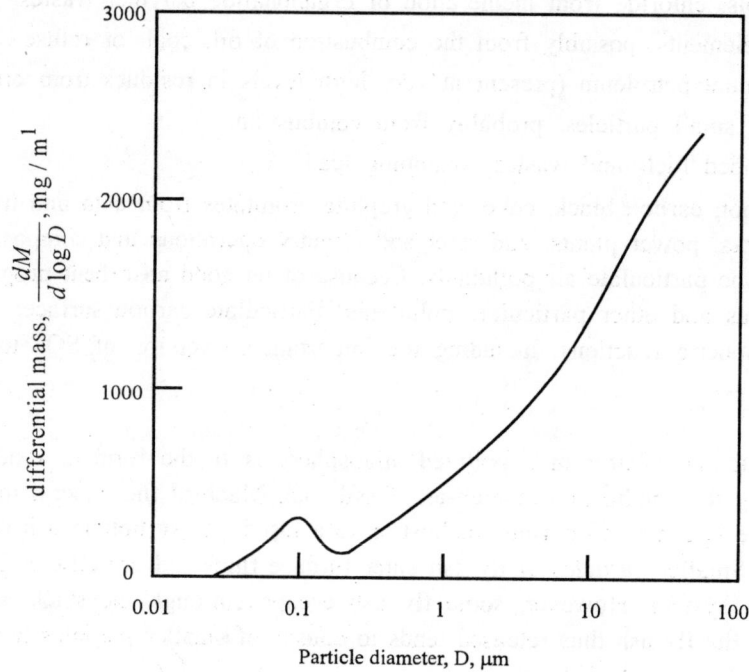

Fig. 9.4. General appearance of particle-size distribution in coal-fired power plant ash.

Toxic Metals

Some of the metals found predominantly as particulate matter in polluted atmospheres are known to be hazardous to human health. All of these except beryllium are so-called "heavy metals". Lead is the toxic metal of greatest concern in the urban atmosphere because it comes closest to being present at a toxic level; mercury ranks second. Others include beryllium, cadmium, chromium, vanadium, nickel, and arsenic (a metalloid).

Atmospheric mercury

Atmospheric mercury is of concern because of its toxicity, volatility, and mobility. Some atmospheric mercury is associated with particulate matter. Much of the mercury entering the atmosphere does so as volatile elemental mercury from coal combustion and volcanoes. Volatile organomercury compounds such as dimethylmercury, $(CH_3)_2Hg$, and monomethylmercury salts, such as CH_3HgBr, are also encountered in the atmosphere.

Atmospheric lead

With the reduction of leaded fuels, atmospheric lead is of less concern than it used to be. However, during the decades that leaded gasoline containing tetraethyllead was the predominant automotive fuel, particulate lead halides were emitted in large quantities. This occurs through the action of dichloroethane and dibromoethane added as halogenated scavengers to prevent the accumulation of lead oxides inside engines. The lead halides formed :

$$Pb(C_2H_5)_4 + O_2 + \text{halogenated scavengers} \rightarrow CO_2 + H_2O +$$
$$PbCl_2 + PbClBr + PbBr_2 \text{ (unbalanced)} \qquad ...(9.8)$$

are volatile enough to exit through the exhaust system but condense in the air to form particles.

Atmospheric beryllium

Beryllium is used for the formulation of specialty alloys used in electrical equipment, electronic instrumentation, space gear, and nuclear reactor components, so that distribution of beryllium is by no means comparable to that of other toxic metals such as lead or mercury. However, because of its "high tech" applications, consumption of beryllium may increase in the future. The toxicity of beryllium and beryllium compounds are widely recognised; it has the lowest allowable limit in the atmosphere of all the elements. One of the main results of the recognition of beryllium toxicity hazards was the elimination of this element from phosphors (coatings which produce visible light from ultraviolet light) in fluorescent lamps.

Radioactive Particles

A significant natural source of radionuclides in the atmosphere is radon, a noble gas product of radium decay. Radon may enter the atmosphere as either of two isotopes, ^{222}Rn (half-life 3.8 days) and ^{220}Rn (half-life 54.5 seconds). Both are alpha emitters in decay chains that terminate with stable isotopes of lead. The initial decay products, ^{218}Po and ^{216}Po are nongaseous and adhere readily to atmospheric particulate matter. Therefore, some of the radioactivity detected in these particles is of natural origin. Furthermore, cosmic rays act on nuclei in the atmosphere to produce other radionuclides, including ^7Be, ^{10}Be, ^{14}C, ^{39}Cl, ^3H, ^{22}Na, ^{32}P, and ^{33}P.

One of the more serious problems in connection with radon is that of radioactivity originating from uranium mine tailings that have been used in some areas as backfill, soil conditioner, and a base for building foundations. Radon produced by the decay of radium exudes from foundations and walls constructed on tailings. Some medical authorities have suggested that the rate of birth defects and infant cancer in areas where uranium mill tailings have been used in residential construction is significantly higher than normal. The combustion of fossil fuels introduces radioactivity into the atmosphere in the form of radionuclides contained in fly ash. Large coal-fired power plants lacking ash-control equipment may introduce up to several hundred millicuries of radionuclides into the atmosphere each year, far more than either an equivalent nuclear or oil-fired power plant.

The radioactive noble gas ^{85}Kr (half-life 10.3 years) is emitted into the atmosphere by the operation of nuclear reactors and the processing of spent reactor fuels. In general, other radionuclides produced by reactor operation are either chemically reactive and can be removed from the reactor effluent, or have such short half-lives that a short time delay prior to emission prevents their leaving the reactor. Widespread use of fission power will inevitably result in an increased level of ^{85}Kr in the atmosphere. Fortunately, biota cannot concentrate this chemically unreactive element.

The above-ground detonation of nuclear weapons can add large amounts of radioactive particulate matter to the atmosphere. Among the radioisotopes that can be detected in rain falling after atmospheric nuclear weapon detonation are ^{91}Y, ^{141}Ce, ^{144}Ce, ^{146}Nd, ^{147}Pm, ^{149}Pm, ^{151}Sm, ^{153}Sm, ^{155}Eu, ^{156}Eu,

^{89}Sr, ^{90}Sr, ^{113m}Cd, ^{129m}Te, ^{131}I, ^{132}Te, and ^{140}Ba. (Note that "m" denotes a metastable state that decays by gamma-ray emission to an isotope of the same element.) The rate of travel of radioactive particles through the atmosphere is a function of particle size. Appreciable fractionation of nuclear debris is observed because of differences in the rates at which various components of nuclear variety move through the atmosphere.

The Composition of Organic Particles

Organic atmospheric particles occur in a wide variety of compounds. For analysis, such particles can be collected onto a filter; extracted with organic solvents; fractionated into neutral, acid, and basic groups; and analysed for specific constituents by chromatography and mass spectrometry. The neutral group contains predominantly hydrocarbons, including aliphatic, aromatic, and oxygenated fractions. The aliphatic fraction of the neutral group contains a high percentage of long-chain hydrocarbons, predominantly those with 16–28 carbon atoms. These relatively unreactive compounds are not particularly toxic and do not participate strongly in atmospheric chemical reactions. The aromatic fraction, however, contains carcinogenic polycyclic aromatic hydrocarbons, which are discussed below. Aldehydes, ketones, epoxides, peroxides, esters, quinones, and lactones are found among the oxygenated neutral components, some of which may be mutagenic or carcinogenic. The acidic group contains long-chain fatty acids and non-volatile phenols. Among the acids recovered from air-pollutant particulate matter are lauric, myristic, palmitic, stearic, behenic, oleic, and linoleic acids. The basic group consists largely of alkaline N-heterocyclic hydrocarbons, such as acridine.

Polycyclic aromatic hydrocarbons

Polycyclic aromatic hydrocarbons (PAH) in atmospheric particles have received a great deal of attention because of the known carcinogenic effects of some of these compounds. Prominent among these compounds are benzo(a)pyrene, chrysene, benzo(e)pyrene, benz(e)acephenanthrylene, benz(a)anthracene, benzo(j)fluoranthene, and indenol.

Elevated levels of PAH compounds of up to about 20 $\mu g/m^3$ are found in the atmosphere. Elevated levels of PAHs are most likely to be encountered in polluted urban atmospheres and in the vicinity of natural fires, such as forest and prairie fires. Coal furnace stack gas may contain over 1000 $\mu g/m^3$ of PAH compounds and cigarette smoke almost 100 $\mu g/m^3$.

Atmospheric polycyclic aromatic hydrocarbons are found almost exclusively in the solid phase, largely sorbed to soot particles. Soot itself is a highly condensed product of PAHs. Soot contains 1–3% hydrogen and 5–10% oxygen, the latter due to partial surface oxidation. Benzo(a)pyrene adsorbed on soot disappears very rapidly in the presence of light yielding oxygenated products; the large surface area of the particle contributes to the high rate of reaction. Oxidation products of benzo(a) pyrene include epoxides, quinones, phenols, aldehydes, and carboxylic acids.

Effects of Particles

Atmospheric particles have numerous effects. The most obvious of these is reduction and distortion of visibility. They provide active surfaces upon which heterogeneous atmospheric chemical reactions can occur and nucleation bodies for the condensation of atmospheric water vapour, thereby exerting a significant influence upon weather and air pollution phenomena.

The most visible effects of aerosol particles upon air quality result from their optical effects. Particles smaller than about 0.1 μm in diameter scatter light much like molecules; that is, Rayleigh scattering. Generally, such particles have an insignificant effect upon visibility in the atmosphere. The light-scattering and intercepting properties of particles larger than 1 μm are approximately proportional to the particle's cross-sectional area. Particles of 0.1 μm to 1 μm cause interference phenomena because they are about the same dimensions as the wavelengths of visible light, so their light-scattering properties are especially significant.

Atmospheric particles inhaled through the respiratory tract may damage health. Relatively large particles are likely to be retained in the nasal cavity and in the pharynx, whereas very small particles are likely to reach the lungs and be retained by them. The respiratory system possesses mechanisms for the expulsion of inhaled particles. In the ciliated region of the respiratory system, particles are carried as far as the entrance to the gastrointestinal tract by a flow of mucus. Macrophages in the nonciliated pulmonary regions carry particles to the ciliated region.

The respiratory system may be damaged directly by particulate matter that enters the blood system or lymph system through the lungs. In addition, the particulate material or soluble components of it may be transported to organs some distance from the lungs and have a detrimental effect on these organs. Particles cleared from the respiratory tract are, to a large extent, swallowed into the gastrointestinal tract.

A strong correlation has been found between increases in the daily mortality rate and acute episodes of air pollution. In such cases, high levels of particulate matter are accompanied by elevated concentrations of SO_2 and other pollutants, so that any conclusions must be drawn with caution.

Water as Particulate Matter

Droplets of water are very widespread in the atmosphere. Although a natural phenomenon, such droplets can have significant and sometimes harmful effects. The most important such consequence is reduction of visibility, with accompanying detrimental effects on driving, flying, and boat navigation. Water droplets in fog act as carriers of pollutants. The most important of these are solutions of corrosive salts, particularly ammonium nitrates and sulphates, and solutions of strong acids. The pH of water in acidic mist droplets has been as low as 1.7, far below that of acidic precipitation. Such acidic mist can be especially damaging to the respiratory tract because it is very penetrating.

Arguably the most significant effect of water droplets in the atmosphere is as aquatic media in which important atmospheric chemical processes occur. The single most significant such process may well be the oxidation of S(IV) species to sulphuric acid and sulphate salts. The S(IV) species so oxidised include $SO_2(aq)$, HSO_3^-, and SO_3^{2-}. Another important oxidation that takes place in atmospheric water droplets is the oxidation of aldehydes to organic carboxylic acids.

The hydroxyl radical, $HO\cdot$, is very important in initiating atmospheric oxidation reactions such as those noted above. Hydroxyl radical as $HO\cdot$ can enter water droplets from the gas-phase atmosphere, it can be produced in water droplets photochemically, or it can be generated from H_2O_2 and $\cdot O_2^-$ radical-ion, which dissolve in water from the gas phase, then produce $HO\cdot$ by solution chemical reaction :

$$H_2O_2 + \cdot O_2^- \rightarrow HO\cdot + O_2 + OH^- \qquad \qquad ...(9.9)$$

Several solutes can react photochemically in aqueous solution (as opposed to the gas phase) to produce hydroxyl radical. One of these is hydrogen peroxide

$$H_2O_2(aq) + hv \rightarrow 2HO\cdot(aq) \qquad\qquad ...(9.10)$$

Nitrite as NO_2^- or HNO_2, nitrate (NO_3^-), and iron (III) as $Fe(OH)^{2+}$ (aq) can also react photochemically in aqueous solution to produce HO·. It has been observed that ultraviolet radiation at 313 nm and simulated sunlight can react to produce HO· radical in authentic samples of water collected from cloud and fog sources. Based on the results of this study and related investigations, it may be concluded that the aqueous-phase formation of hydroxyl radical is an important, and in some cases dominant, means by which this key atmospheric oxidant is introduced into atmospheric water droplets.

Iron is an inorganic solute of particular importance in atmospheric water. This is because of the participation of iron(III) in the atmospheric oxidation of sulphur(IV) to sulphur(VI)—that is, the conversion of $SO_2(aq)$, HSO_3^-; and SO_3^{2-} to sulphates and H_2SO_4.

Control of Particulate Emissions

The removal of particulate matter from gas streams is the most widely practised means of air pollution control. A number of devices have been developed for this purpose, which differ widely in effectiveness, complexity, and cost. The selection of a particle removal system for a gaseous waste stream depends upon the particle loading, nature of particles (size distribution), and type of gas scrubbing system used.

Particle removal by sedimentation and inertia

The simplest means of particulate matter removal is sedimentation, a phenomenon that occurs continuously in nature. Gravitational settling chambers may be employed for the removal of particles from gas streams by simply settling under the influence of gravity. These chambers take up large amounts of space and have low collection efficiencies, particularly for small particles.

Gravitational settling of particles is enhanced by increased particle size, which occurs spontaneously by coagulation. Thus, over time, the size of particles increases and the number of particles decreases in a mass of air that contains particles. Brownian motion of particles less than about 0.1 µm in size is primarily responsible for their contact, enabling coagulation to occur. Particles greater than about 0.3 µm in radius do not diffuse appreciably and serve primarily as receptors of smaller particles.

Inertial mechanisms are effective for particle removal. These depend upon the fact that the radius of the path of a particle in a rapidly moving, curving air stream is larger than the path of the stream as a whole. Therefore, when a gas stream is spun by vanes, a fan, or a tangential gas inlet, the particulate matter may be collected on a separator wall because the particles are forced outward by centrifugal force. Devices utilising this mode of operation are called dry centrifugal collectors (cyclones).

Particle filtration

Fabric filters, as their name implies, consist of fabrics that allow the passage of gas but retain particulate matter. These are used to collect dust in bags contained in structures called *baghouses*. Periodically the fabric composing the filter is shaken to remove the particles and to reduce back-pressure to acceptable levels. Typically the bag is in a tubular configuration as shown in Fig. 9.5. Numerous other configurations are possible. Collected particulate matter is removed from bags by mechanical agitation, blowing air on the fabric, or rapid expansion and contraction of the bags.

Although simple, baghouses are generally effective in removing particles from exhaust gas. Particles as small as 0.01 μm in diameter are removed, and removal efficiency is relatively high for particles down to 0.5 μm in diameter.

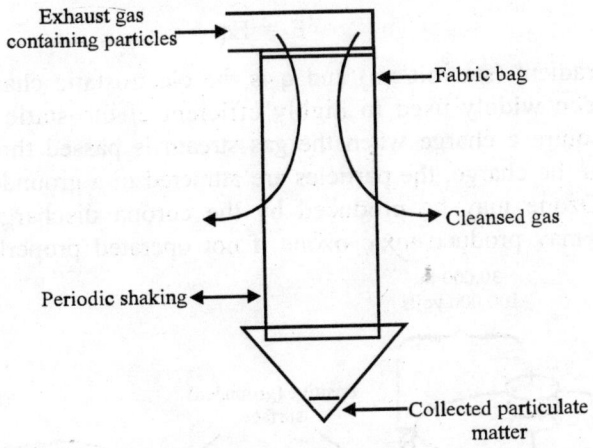

Fig. 9.5. Baghouse collection of particulate emissions.

Scrubbers

A venturi scrubber passes gas through a converging section, throat, and diverging section as shown in Fig. 9.6. Injection of the scrubbing liquid at right angles to incoming gas breaks the liquid into very small droplets, which are ideal for scavenging particles from the gas stream. In the reduced-pressure (expanding) region of the venturi, some condensation can occur, adding to the scrubbing efficiency. In addition to removing particles, venturis may serve as quenchers to cool exhaust gas and as scrubbers for pollutant gases.

Ionising wet scrubbers place an electrical charge on particles upstream from a wet scrubber. Larger particles and some gaseous contaminants are removed by scrubbing action. Smaller particles tend to induce opposite charges in water droplets in the scrubber and in its packing material and are removed by attraction of the opposite charges.

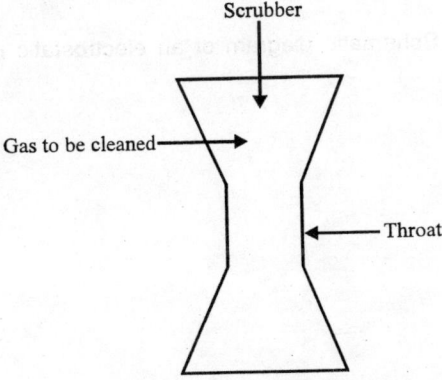

Fig. 9.6. Venturi scrubber.

Electrostatic removal

Aerosol particles may acquire electrical charges. In an electric field, such particles are subjected to a force, F (dynes), given by

$$F = Eq \qquad \qquad ...(9.11)$$

where E is the voltage gradient (statvolt/cm) and q is the electrostatic charge on the particle (in esu). This phenomenon has been widely used in highly efficient electrostatic precipitators, as shown in Fig. 9.7. The particles acquire a charge when the gas stream is passed through a high-voltage, direct-current corona. Because of the charge, the particles are attracted to a grounded surface, from which they may be later removed. Ozone may be produced by the corona discharge. Similar devices used as household dust collectors may produce toxic ozone if not operated properly.

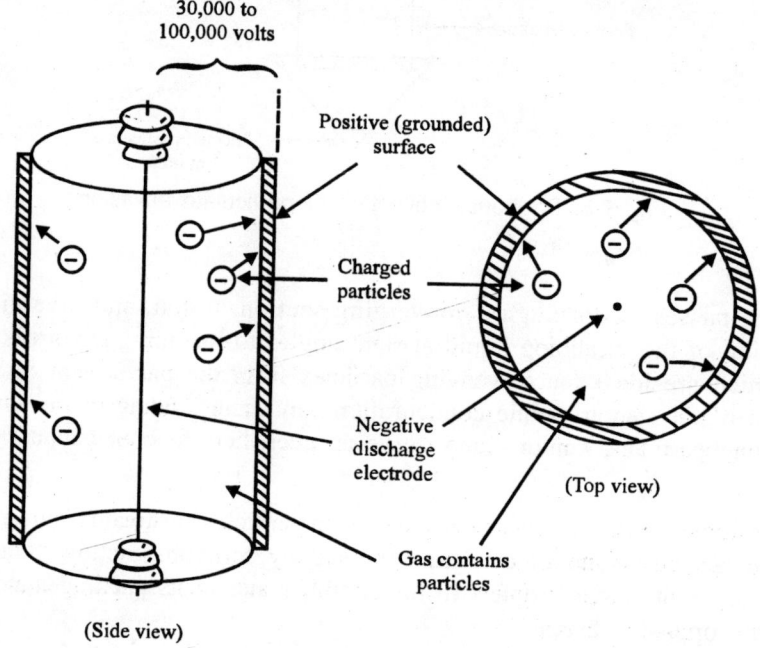

Fig. 9.7. Schematic diagram of an electrostatic precipitator.

CHAPTER 10

Gaseous Inorganic Air Pollution

INTRODUCTION

A number of gaseous inorganic pollutants enter the atmosphere as the result of human activities. Those added in the greatest quantities are CO, SO_2, NO, and NO_2. (These quantities are relatively small, compared to the amount of CO_2 in the atmosphere. Other inorganic pollutant gases include NH_3, N_2O, N_2O_5, H_2S, Cl_2, HCl, and HF. Substantial quantities of some of these gases are added to the atmosphere each year by human activities. Globally, atmospheric emissions of carbon monoxide, sulphur oxides, and nitrogen oxides are of the order of one to several hundred million tonnes per year.

Production and Control of Carbon Monoxide

Carbon monoxide, CO, causes problems in cases of locally high concentrations because of its toxicity. The overall atmospheric concentration of carbon monoxide is about 0.1 ppm corresponding to a burden in the earth's atmosphere of approximately 500 million metric tonnes of CO with an average residence time ranging from 36 to 110 days. Much of this CO is present as an intermediate in the oxidation of methane by hydroxyl radical. It may be seen that the methane content of the atmosphere is about 1.6 ppm, more than 10 times the concentration of CO. Therefore, any oxidation process for methane that produces carbon monoxide as an intermediate is certain to contribute substantially to the overall carbon monoxide burden, probably around two-thirds of the total CO.

Degradation of chlorophyll during the autumn months releases CO, amounting to perhaps as much as 20% of the total annual release. Anthropogenic sources account for about 6% of CO emissions. The remainder of atmospheric CO comes from largely unknown sources. These include some plants and marine organisms known as siphonophores, an order of Hydrozoa. Carbon monoxide is also produced by decay of plant matter other than chlorophyll.

Because of carbon monoxide emissions from internal combustion engines, highest levels of this toxic gas tend to occur in congested urban areas at times when the maximum number of people are exposed, such as during rush hours. At such times, carbon monoxide levels in the atmosphere may become as high as 50–100 ppm. Atmospheric levels of carbon monoxide in urban areas show a positive correlation with the density of vehicular traffic, and a negative correlation with wind speed. Urban atmospheres may show average carbon monoxide levels of the order of several ppm, much higher than those in remote areas.

Control of carbon monoxide emissions

Since the internal combustion engine is the primary source of localised pollutant carbon monoxide emissions, control measures have been concentrated on the automobile. Carbon monoxide emissions may be lowered by employing a leaner air-fuel mixture; that is, one in which the weight ratio of air to fuel is relatively high. At air-fuel (weight:weight) ratios exceeding approximately 16:1, an internal combustion engine emits virtually no carbon monoxide.

Modern automobiles use catalytic exhaust reactors to cut down on carbon monoxide emissions. Excess air is pumped into the exhaust gas, and the mixture is passed through a catalytic converter in the exhaust system, resulting in oxidation of CO to CO_2.

Fate of Atmospheric CO

It is known that the lifetime of carbon monoxide in the atmosphere is not long, perhaps of the order of 4 months. It is generally agreed that carbon monoxide is removed from the atmosphere by reaction with hydroxyl radical, $HO\cdot$:

$$CO + HO\cdot \rightarrow CO_2 + H \qquad \qquad ...(10.1)$$

The reaction produces hydroperoxyl radical as a product:

$$O_2 + H + M \rightarrow HOO\cdot + M \qquad \qquad ...(10.2)$$

$HO\cdot$ is regenerated from $HOO\cdot$ by the following reactions:

$$HOO\cdot + NO \rightarrow HO\cdot + NO_2 \qquad \qquad ...(10.3)$$

$$HOO\cdot + HOO\cdot \rightarrow H_2O_2 + O_2 \qquad \qquad ...(10.4)$$

The latter reaction is followed by photochemical dissociation of H_2O_2 to regenerate $HO\cdot$:

$$H_2O_2 + h\nu \rightarrow 2HO\cdot \qquad \qquad ...(10.5)$$

Methane is also involved through the atmospheric cycle that relates CO, $HO\cdot$, and CH_4.

Soil microorganisms act to remove CO from the atmosphere. Therefore, soil is a sink for carbon monoxide.

Sulphur Dioxide Sources and the Sulphur Cycle

Fig. 10.1. shows the main aspects of the global sulphur cycle. This cycle involves primarily H_2S, SO_2, SO_3, and sulphates. There are many uncertainties regarding the sources, reactions, and fates of these atmospheric sulphur species. On a global basis, sulphur compounds enter the atmosphere to a very large extent through human activities. Approximately 115 million metric tonnes of sulphur per year enters the global atmosphere through anthropogenic activities, primarily as SO_2 from the combustion of coal and residual fuel oil. The greatest uncertainties in the cycle have to do with nonanthropogenic sulphur, which enters the atmosphere largely as H_2S from volcanoes and from the biological decay of organic matter and reduction of sulphate. The quantity added from biological processes may be low. Any H_2S that does get into the atmosphere is converted rapidly to SO_2 by the following overall process :

$$H_2S + 3/2\ O_2 \rightarrow SO_2 + H_2O \qquad \qquad ...(10.6)$$

The initial reaction is hydrogen ion abstraction by hydroxyl radical,

$$H_2S + HO\cdot \rightarrow HS\cdot + H_2O \qquad ...(10.7)$$

followed by the following two reactions to give SO_2:

$$HS\cdot + O_2 \rightarrow HO\cdot + SO \qquad ...(10.8)$$

$$SO + O_2 \rightarrow SO_2 + O \qquad ...(10.9)$$

The primary source of anthropogenic sulphur dioxide is coal, from which sulphur must be removed at great expense to keep sulphur dioxide emissions at acceptable levels. Approximately half of the sulphur in coal is in some form of pyrite, FeS_2, and the other half is organic sulphur. The production of sulphur dioxide by the combustion of pyrite is given by the following reaction :

$$4FeS_2 + 11O_2 \rightarrow 2Fe_2O_3 + 8SO_2 \qquad ...(10.10)$$

Essentially all of the sulphur is converted to SO_2; only 1% or 2% leaves the stack as SO_3.

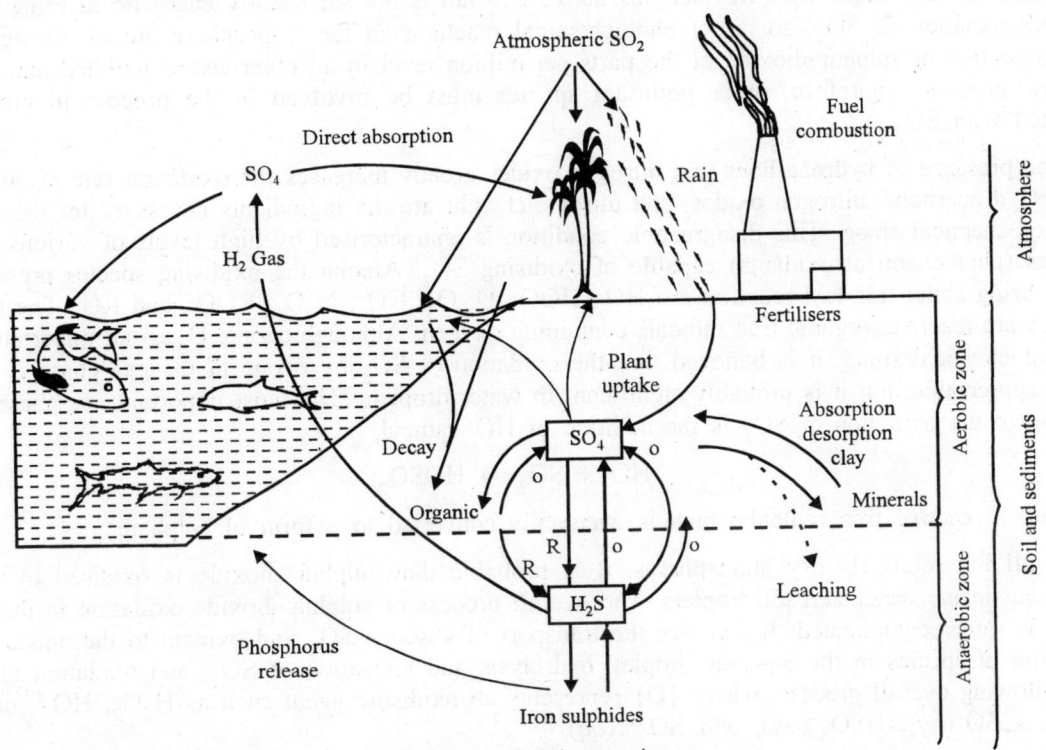

Fig. 10.1. Sulphur cycle.

Sulphur Dioxide Reactions in the Atmosphere

Many factors, including temperature, humidity, light intensity, atmospheric transport,. and surface characteristics of particulate matter, may influence the atmospheric chemical reactions of sulphur dioxide. Like many other gaseous pollutants, sulphur dioxide reacts to form particulate matter, which then settles or is scavenged from the atmospheres by rainfall or other processes. It is known that high levels of air pollution normally are accompanied by a marked increase in aerosol particles and a consequent reduction in visibility. Reaction products of sulphur dioxide are thought to be responsible for some aerosol formation. Whatever the processes involved, much of the sulphur dioxide in the atmosphere ultimately is oxidised to sulphuric acid and sulphate salts, particularly ammonium sulphate and ammonium hydrogen sulphate and it is likely that these sulphates account for the turbid haze. The potential of sulphates to induce climatic change is high and must be taken into account when considering control of sulphur dioxide.

Some of the possible ways in which sulphur dioxide may react in the atmosphere are (i) photochemical reactions; (ii) photochemical and chemical reactions in the presence of nitrogen oxides and/or hydrocarbons, particularly alkenes; (iii) chemical processes in water droplets, particularly those containing metal salts and ammonia; and (iv) reactions on solid particles in the atmosphere. It should be kept in mind that the atmosphere is a highly dynamic system with great variations in temperature, composition, humidity, and intensity of sunlight; therefore, different processes may predominate under various atmospheric conditions.

Photochemical reactions are probably involved in some of the processes resulting in the atmospheric oxidation of SO_2. Light with wavelengths above 218 nm is not sufficiently energetic to bring about the photodissociation of SO_2, so direct photochemical reactions in the troposphere are of no significance. The oxidation of sulphur dioxide at the parts-per-million level in an otherwise unpolluted atmosphere is a slow process. Therefore, other pollutant species must be involved in the process in atmospheres polluted with SO_2.

The presence of hydrocarbons and nitrogen oxides greatly increases the oxidation rate of atmospheric SO_2. Hydrocarbons, nitrogen oxides, and ultraviolet light are the ingredients necessary for the formation of photochemical smog. This disagreeable condition is characterised by high levels of various oxidising species (photochemical oxidants) capable of oxidising SO_2. Among the oxidising species present which could bring about this fast reaction are $HO\cdot$, $HOO\cdot$, O, O_3, NO_3, N_2O_5, $ROO\cdot$, and $RO\cdot$. The latter two species are reactive, organic free radicals containing oxygen. Although ozone, O_3, is an important product of photochemical smog, it is believed that the oxidation of SO_2 by ozone in the gas phase is too slow to be appreciable, but it is probably significant in water droplets. The most important gas-phase reaction leading to the oxidation of SO_2 is the addition of $HO\cdot$ radical,

$$HO\cdot + SO_2 \rightarrow HOSO_2\cdot \qquad ...(10.21)$$

forming a reactive free radical which is eventually converted to a form of sulphate.

In all but relatively dry atmospheres, it is probable that sulphur dioxide is oxidised by reactions occurring inside water aerosol droplets. The overall process of sulphur dioxide oxidation in the aqueous phase is rather complicated. It involves the transport of gaseous SO_2 and oxidant to the aqueous phase, diffusion of species in the aqueous droplet, hydrolysis and ionisation of SO_2, and oxidation of SO_2 by the following overall process, where $\{O\}$ represents an oxidising agent such as H_2O_2, $HO^{\cdot3}$, or O_3 and S(IV) is $SO_2(aq)$, $HSO_3^-(aq)$, and $SO_3^{2-}(aq)$.

$$\{O\}(aq) + S(IV)(aq) \rightarrow 2H^+ + SO_4^{2-} \text{ (unbalanced)} \qquad ...(10.12)$$

In the absence of catalytic species, the reaction with dissolved molecular O_2,

$$1/2\ O_2(aq) + SO_2(aq) + H_2O \rightarrow H_2SO_4(aq) \qquad ...(10.13)$$

is too slow to be significant. Hydrogen peroxide is an important oxidising agent in the atmosphere. It reacts with dissolved sulphur dioxide through the overall reaction,

$$SO_2(aq) + H_2O_2\ (aq) \rightarrow H_2SO_4\ (aq) \qquad ...(10.14)$$

to produce sulphuric acid. The major reaction is thought to be between hydrogen peroxide and HSO_3^-, ion with peroxymonosulphurous acid, $HOOSO_2^-$, as an intermediate.

Ozone, O_3, oxidises sulphur dioxide in water. The fastest reaction is with sulphite ion;

$$SO_3^{2-}\ (aq) + O_3(aq) + H_2O \rightarrow SO_4^{2-}(aq) + O_2 \qquad ...(10.15)$$

reactions are slower with $HSO_3^-(aq)$ and $SO_2(aq)$. The rate of oxidation of aqueous SO_2 species by ozone increases with increasing pH. The oxidation of sulphur dioxide in water droplets is faster in the presence of ammonia, which reacts with sulphur dioxide to produce bisulphite ion and sulphite ion in solution :

$$NH_3 + SO_2 + H_2O \rightarrow NH_4^+ + HSO_3^- \qquad ...(10.16)$$

Some solutes dissolved in water catalyse the oxidation of aqueous SO_2. Both iron(III) and Mn(II) have this effect. The reactions catalysed by these two ions are faster with increasing pH. Dissolved nitrogen species, NO_2, and HNO_2, oxidise aqueous sulphur dioxide in the laboratory. Nitrite dissolved in water droplets may react photochemically to produce $HO \cdot$ radical, and this species in turn could act to oxidise dissolved sulphite.

Heterogeneous reactions on solid particles may also play a role in the removal of sulphur dioxide from the atmosphere. In atmospheric photochemical reactions, such particles may function as nucleation centres. Thus, they act as catalysts and grow in size by accumulating reaction products. The final result would be production of an aerosol with a composition unlike that of the original particle. Soot particles, which consist of elemental carbon contaminated with polynuclear aromatic hydrocarbons produced in the incomplete combustion of carbonaceous fuels, can catalyse the oxidation of sulphur dioxide to sulphate as indicated by the presence of sulphate on the soot particles. Soot particles are very common in polluted atmospheres, so it is very likely that they are strongly involved in catalysing the oxidation of sulphur dioxide.

Oxides of metals such as aluminium, calcium, chromium, iron, lead, or vanadium may also be catalysts for the heterogenous oxidation of sulphur dioxide. These oxides may also adsorb sulphur dioxide. However, the total surface area of oxide particulate matter in the atmosphere is very low so that the fraction of sulphur dioxide oxidised on metal oxide surfaces is relatively small.

Effects of atmospheric sulphur dioxide

Though not terribly toxic to most people, low levels of sulphur dioxide in air do have some health effects. Its primary effect is upon the respiratory tract, producing irritation and increasing airway resistance, especially to people with respiratory weaknesses and sensitised asthmatics. Therefore, exposure to the gas may increase the effort required to breathe. Mucus secretion is also stimulated by exposure to air contaminated by sulphur dioxide. Although SO_2 causes death in humans at 500 ppm, it has not been found to harm laboratory animals at 5 ppm.

Atmospheric sulphur dioxide is harmful to plants. Acute exposure to high levels of the gas kills leaf tissue (leaf necrosis). The edges of the leaves and the areas between the leaf veins are particularly damaged. Chronic exposure of plants to sulphur dioxide causes chlorosis, a bleaching or yellowing of the normally green portions of the leaf. Plant injury increases with increasing relative humidity. Plants incur most injury from sulphur dioxide when their stomata (small openings in plant surface tissue that allow interchange of gases with the atmosphere) are open. For most plants, the stomata are open during the daylight hours, and most damage from sulphur dioxide occurs then. Long-term, low-level exposure to sulphur dioxide can reduce the yields of grain crops, such as wheat or barley. Sulphur dioxide in the atmosphere is converted to sulphuric acid, so that in areas with high levels of sulphur dioxide pollution, plants may be damaged by sulphuric acid aerosols. Such damage appears as small spots where sulphuric acid droplets have impinged on leaves.

One of the more costly effects of sulphur dioxide pollution is its tendency to cause deterioration of building materials. Limestone, marble, and dolomite are calcium and/or magnesium carbonate minerals that are attacked by atmospheric sulphur dioxide. These reactions form products that are either water-soluble or in the form of poorly adherent solid crusts on the rock's surface, adversely affecting the appearance, structural integrity, and life of the building. Although both SO_2 and NO_x attack such stone, chemical analysis of the crusts shows predominantly sulphate salts. Dolomite, a calcium/magnesium carbonate mineral reacts with atmospheric sulphur dioxide as follows :

$$CaCO_3 \cdot MgCO_3 + 2SO_2 + O_2 + 9H_2O \rightarrow CaSO_4 \cdot 2H_2O + MgSO_4 \cdot 7H_2O + 2CO_2 \qquad ...(10.17)$$

Sulphur dioxide removal

A number of processes are being used to remove sulphur and sulphur oxides from fuel before combustion and from stack gas after combustion. Most of these efforts concentrate on coal, since it is the major source of sulphur oxides pollution. Physical separation techniques may be used to remove discrete particles of pyritic sulphur from coal. Chemical methods may also be employed for removal of sulphur from coal. Fluidised bed combustion of coal promises to elimir. $\cdot$ SO_2 emissions at the point of combustion. The process consists of burning granular coal in a bed of finely divided limestone or dolomite maintained in a fluid-like condition by air injection. Heat calcines the limestone,

$$CaCO_3 \rightarrow CaO + CO_2 \qquad ...(10.18)$$

and the lime produced absorbs SO_2 :

$$CaO + SO_2 \rightarrow CaSO_3 \text{ (which may be oxidised to } CaSO_4) \qquad ...(10.19)$$

Many processes have been proposed or studied for the removal of sulphur dioxide from stack gas. Table 10.2 summarises major stack gas scrubbing systems. These include throwaway and recovery systems as well as wet and dry systems. A dry throwaway system used with only limited success involves injection of dry limestone or dolomite into the boiler, followed by recovery of dry lime, sulphites, and sulphates. The overall reaction, shown here for dolomite, is the following :

$$CaCO_3 \cdot MgCO_3 + SO_2 + 1/2O_2 \rightarrow CaSO_4 + MgO + 2CO_2 \qquad ...(10.20)$$

The solid sulphate and oxide products are removed by electrostatic precipitators or cyclone separators. The process has an efficiency of 50% or less for the removal of sulphur oxides. As may be noted from the chemical reactions shown in Table 10.2, all sulphur dioxide removal processes, except for catalytic oxidation, depend upon absorption of SO_2 by an acid-base reaction. The first two processes listed are throwaway processes yielding large quantities of wastes; the others provide for some sort of sulphur product recovery.

Lime or limestone slurry scrubbing for SO_2 removal involves acid-base reactions with SO_2. When sulphur dioxide dissolves in water, equilibrium is established between SO_2 gas and dissolved SO_2 :

$$SO_2(g) \underset{\longleftarrow}{\longrightarrow} SO_2 \ (aq) \qquad ...(10.21)$$

This equilibrium is described by *Henry's law,*

$$[SO_2(aq)] = K \times P_{SO_2} \qquad ...(10.22)$$

where $[SO_2(aq)]$ is the concentration of dissolved molecular sulphur dioxide; K is the Henry's law constant for SO_2; and P_{SO_2} is the partial pressure of sulphur dioxide gas. In the presence of base, Reaction 10.20 is shifted strongly to the right by the following reactions :

$$H_2O + SO_2 \ (aq) \rightleftharpoons H^+ + HSO_3^- \qquad \qquad ...(10.23)$$

$$HSO_3^- \rightleftharpoons H^+ + SO_3^{2-} \qquad \qquad ...(10.24)$$

In the presence of calcium carbonate slurry (as in limestone slurry scrubbing), hydrogen ion is taken up by the reaction

$$CaCO_3 + H^- \rightleftharpoons Ca^{2+} + HCO_3^- \qquad \qquad ...(10.25)$$

Table 10.2. Major stack gas scrubbing systems.

Process	Chemical reactions	Major advantages or disadvantages
Lime slurry scrubbing	$Ca(OH)_2 + SO_2 \rightarrow CaSO_3 + H_2O$	Up to 200 kg of lime are needed per metric tonne of coal, producing huge quantities of waste product.
Limestone slurry scrubbing	$CaCO_3 + SO_2 \rightarrow CaSO_3 + CO_2(g)$	Lower pH than lime slurry, not so efficient.
Magnesium oxide scrubbing	$Mg(OH)_2 \ (slurry) + SO_2 \rightarrow MgSO_3 + H_2O$	The sorbent can be regenerated, off site, if desired.
Sodiumbase scrubbing	$Na_2SO_3 + H_2O + SO_2 \rightarrow 2 \ NaHSO_3$ $2 \ NaHSO_3 + heat \rightarrow Na_2SO_3 + H_2O + SO_2$ (regeneration)	No major technological limitations. Relatively high annual costs.
Double alkali	$2 \ NaOH + SO_2 \rightarrow Na_2SO_3 + H_2O$ $Ca(OH)_2 + Na_2SO_3 \rightarrow CaSO_3(s) + 2 \ NaOH$ (regeneration of NaOH)	Allows for regeneration of expensive sodium alkali solution with inexpensive lime.

The reaction of calcium carbonate with carbon dioxide from stack gas,

$$CaCO_3 + CO_2 + H_2O \rightleftharpoons Ca^{2+} + 2HCO_3^- \qquad \qquad ...(10.26)$$

results in some absorption of CO_2. The reaction of sulphite and calcium ion to form highly insoluble calcium sulphite hemihydrate

$$Ca^{2+} + SO_3^{2-} + 1/2 \ H_2O \rightleftharpoons CaSO_3 \cdot 1/2H_2O(s) \qquad \qquad ...(10.27)$$

also shifts Reactions (10.22) and (10.23) to the right. Gypsum is formed in the scrubbing process by the oxidation of sulphite,

$$SO_3^{2-} + 1/2O_2 \rightarrow SO_4^{2-} \qquad \qquad ...(10.28)$$

followed by reaction of sulphate ion with calcium ion :

$$Ca^{2+} + SO_4^{2-} + 2H_2O \rightleftharpoons CaSO_4 \cdot 2H_2O \ (s) \qquad \qquad ...(10.29)$$

Formation of gypsum in the scrubber is undesirable because it creates scale in the scrubber equipment. However, gypsum is sometimes produced deliberately in the spent scrubber liquid downstream from the scrubber.

When lime, $Ca(OH)_2$, is used in place of limestone (lime slurry scrubbing), a source of hydroxide ions is provided for direct reactions with H^+ :

$$H^+ + OH \rightarrow H_2O \qquad \qquad ...(10.30)$$

The reactions involving sulphur species in a lime slurry scrubber are essentially the same as the those just discussed for limestone slurry scrubbing. The pH of a lime slurry is higher than that of a limestone slurry, so that the former has more of a tendency to react with CO_2, resulting in the absorption of that gas :

$$CO_2 + OH \rightarrow HCO_3^- \qquad \qquad ...(10.31)$$

Current practice with lime and limestone scrubber systems calls for injection of the slurry into the scrubber loop beyond the boilers. A number of power plants are now operating with this kind of system. Experience to date has shown that these scrubbers remove well over 90% of both SO_2 and fly ash when operating properly. (Fly ash is fuel combustion ash normally carried up the stack with flue gas). In addition to corrosion and scaling problems, disposal of lime sludge poses formidable obstacles. The quantity of this sludge may be appreciated by considering that approximately 1 tonne of limestone is required for each 5 tonnes of coal. The sludge is normally disposed of in large ponds, which can present some disposal problems. Water seeping through the sludge beds becomes laden with calcium sulphate and other salts. It is difficult to stabilise this sludge as a structurally stable, nonleachable solid.

Recovery systems in which sulphur dioxide or elemental sulphur are removed from the spent sorbing material, which is recycled, are much more desirable from an environmental viewpoint than are throwaway systems. Many kinds of recovery processes have been investigated, including those that involve scrubbing with magnesium oxide slurry, sodium sulphite solution, ammonia solution, or sodium citrate solution.

Sulphur dioxide trapped in a stack-gas-scrubbing process can be converted to hydrogen sulphide by reaction with synthesis gas (H_2, CO, CH_4),

$$SO_2 + (H_2, CO, CH_4) \rightleftharpoons H_2S + CO_2 \qquad \qquad ...(10.32)$$

The Claus reaction is then employed to produce elemental sulphur :

$$2H_2S + SO_2 \rightleftharpoons 2H_2O + 3S \qquad \qquad ...(10.33)$$

Nitrogen Oxides in the Atmosphere

The three oxides of nitrogen normally encountered in the atmosphere are nitrous oxide (N_2O), nitric oxide (NO), and nitrogen dioxide (NO_2). Nitrous oxide, a commonly used anaesthetic known as "laughing gas," is produced by microbiological processes and is a component of the unpolluted atmosphere at a level of approximately 0.3 ppm. This gas is relatively unreactive and probably does not significantly influence important chemical reactions in the lower atmosphere. Its concentration decreases rapidly with altitude in the stratosphere due to the photochemical reaction

$$N_2O + h\nu \rightarrow N_2 + O \qquad \qquad ...(10.34)$$

and some reaction with singlet atomic oxygen :

$$N_2O + O \rightarrow N_2 + O_2 \qquad \qquad ...(10.35)$$

$$N_2O + O \rightarrow 2NO \qquad \qquad ...(10.36)$$

These reactions are significant in terms of depletion of the ozone layer. Increased global fixation of nitrogen, accompanied by increased microbial production of N_2O, could contribute to ozone layer depletion. Colourless, odourless nitric oxide (NO) and pungent red-brown nitrogen dioxide (NO_2) are very important in polluted air. Collectively designated NO_x, these gases enter the atmosphere from natural sources, such as lightning and biological processes, and from pollutant sources. The latter are much more significant because of regionally high NO_2 concentrations, which can cause severe air quality deterioration. Practically all anthropogenic NO_x enters the atmosphere as a result of the combustion of fossil fuels in both stationary and mobile sources. Globally, somewhat less than 175 million metric tonnes of nitrogen oxides are emitted to the atmosphere from these sources each year, compared to several times that much from widely dispersed natural sources.

Most NO_2 entering the atmosphere from pollution sources does so as NO generated from internal combustion engines. At very high temperatures, the following reaction occurs:

$$N_2 + O_2 \rightarrow 2NO \qquad \qquad ...(10.37)$$

The speed with which this reaction takes place increases steeply with temperature. The equilibrium concentration of NO in a mixture of 3% O_2 and 75% N_2, typical of that which occurs in the combustion chamber of an internal combustion engine, is shown as a function of temperature in Fig. 10.2. At room temperature (27°C) the equilibrium concentration of NO is only 1.1×10^{-10} ppm, whereas at high temperatures it is much higher. Therefore, high temperatures favour both a high equilibrium concentration and a rapid rate of formation of NO. Rapid cooling of the exhaust gas from combustion "freezes" No at a relatively high concentration because equilibrium is not maintained. Thus, by its very nature, the combustion process both in the internal combustion engine and in furnaces produces high levels of NO in the combustion products.

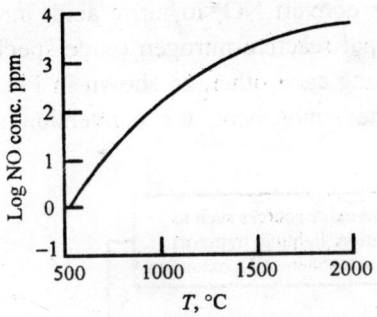

Fig. 10.2. Log of equilibrium NO concentration as a function of temperature in a mixture containing 75% N_2 and 3% O_2.

The mechanism for formation of nitrogen oxides from N_2 and O_2 during combustion is a complicated process. Both oxygen and nitrogen atoms are formed at the very high combustion temperatures by the reactions :

$$O_2 + M \rightarrow O + O + M \qquad \qquad ...(10.38)$$

$$N_2 + M \rightarrow N\cdot + N\cdot + M \qquad \qquad ...(10.39)$$

where M is a highly energetic third body that imparts enough energy to the molecular N_2 and O_2 to break their chemical bonds. The energies required for these reactions are quite high because breakage of the oxygen bond requires 118 kcal/mole and breakage of the nitrogen bond requires 225 kcal/mole. Once formed, O and N atoms participate in the following chain reaction for the formation of nitric oxide from nitrogen and oxygen :

$$N_2 + O \rightarrow NO + N \qquad \qquad ...(10.40)$$

$$N + O_2 \rightarrow NO + O \qquad \qquad ...(10.41)$$

$$N_2 + O_2 \rightarrow 2NO \qquad \qquad ...(10.42)$$

There are, of course, many other species present in the combustion mixture besides those shown. The oxygen atoms are especially reactive toward hydrocarbon fragments by reactions such as the following:

$$RH + O \rightarrow R\cdot + HO\cdot \qquad \qquad ...(10.43)$$

where RH represents a hydrocarbon fragment with an extractable hydrogen atom. These fragments compete with N_2 for oxygen atoms. It is partly for this reason that the formation of NO is appreciably higher at air/fuel ratios exceeding the stoichiometric ratio (lean mixture).

The hydroxyl radical itself can participate in the formation of NO. The reaction is :

$$N + HO\cdot \rightarrow NO + H\cdot \qquad \qquad ...(10.44)$$

Nitric oxide, NO, is a product of the combustion of coal and petroleum containing chemically bound nitrogen. Production of NO by this route occurs at much lower temperatures than those required for "thermal" NO.

Atmospheric reactions of NO$_x$

Atmospheric chemical reactions convert NO_x to nitric acid, inorganic nitrate salts, organic nitrates, and peroxyacetyl nitrate. The principal reactive nitrogen oxide species in the troposphere are NO, NO_2, and HNO_3. These species cycle among each other, as shown in Fig. 10.3. Although NO is the primary form in which NO_x is released to the atmosphere, the conversion of NO to NO_2 is relatively rapid in the troposphere.

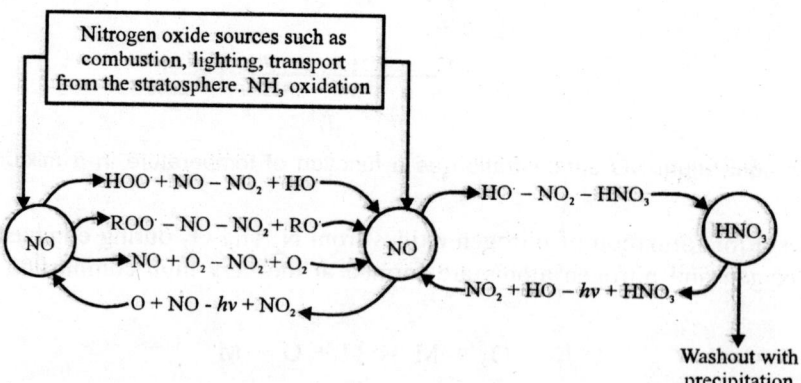

Fig. 10.3. Principle reactions among NO, NO_2, and HNO_3, in the atmosphere. $ROO\cdot$ represents an organic peroxyl radical, such as the methylperoxyl radical, $CH_3OO\cdot$.

Nitrogen dioxide is a very reactive and significant species in the atmosphere. It absorbs light throughout the ultraviolet and visible spectrum penetrating the troposphere. At wavelengths below

398 nm, photodissociation occurs,

$$NO_2 + h\nu \rightarrow NO + O \qquad \qquad ...(10.45)$$

to produce ground state oxygen atoms. Above 430 nm, only excited molecules are formed,

$$NO_2 + h\nu \rightarrow NO_2 \qquad \qquad ...(10.46)$$

whereas at wavelengths between 398 nm and 430 nm, either process may occur. Photodissociation at these wavelengths requires input of rotational energy from rotation of the NO_2 molecule. The tendency of NO_2 to photodissociate is shown clearly by the fact that in direct sunlight the half-life of NO_2 is much shorter than that of any other atmospheric component (only 85 seconds at 40° latitude).

The photodissociation of nitrogen dioxide can give rise to the following significant inorganic reactions, in addition to a host of atmospheric reactions involving organic species:

$$O + O_2 + M(\text{third body}) \rightarrow O_3 + M \qquad \qquad ...(10.47)$$

$$NO + O_3 \rightarrow NO_2 + O_2 \qquad \qquad ...(10.48)$$

$$NO_2 + O_3 \rightarrow NO_3 + O_2 \qquad \qquad ...(10.49)$$

$$O + NO_2 \rightarrow NO + O_2 \qquad \qquad ...(10.50)$$

$$O + NO_2 + M \rightarrow NO_3 + M \qquad \qquad ...(10.51)$$

$$NO_2 + NO_3 \rightarrow N_2O_5 \qquad \qquad ...(10.52)$$

$$NO + NO_3 \rightarrow 2NO_2 \qquad \qquad ...(10.53)$$

$$O + NO + M \rightarrow NO_2 + M \qquad \qquad ...(10.54)$$

Nitrogen dioxide ultimately is removed from the atmosphere as nitric acid, nitrates, or (in atmospheres where photochemical smog is formed) as organic nitrogen. Dinitrogen pentoxide formed in Reaction (10.52) is the anhydride of nitric acid, which it forms by reacting with water :

$$N_2O_5 + H_2O \rightarrow 2HNO_3 \qquad \qquad ...(10.55)$$

In the stratosphere, nitrogen dioxide reacts with hydroxyl radicals to produce nitric acid :

$$HO\cdot + NO_2 \rightarrow HNO_3 \qquad \qquad ...(10.56)$$

In this region, the nitric acid can also be destroyed by hydroxyl radicals,

$$HO\cdot + HNO_3 \rightarrow H_2O + NO_3 \qquad \qquad ...(10.57)$$

or by a photochemical reaction,

$$HNO_3 + h\nu \rightarrow HO\cdot + NO_2 \qquad \qquad ...(10.58)$$

so that HNO_3 serves as a temporary sink for NO_2 in the stratosphere. Nitric acid produced from NO_2 is removed as precipitation, or reacts with bases (ammonia, particulate lime) to produce particulate nitrates.

Harmful effects of nitrogen oxides

Nitric oxide, NO, is biochemically less active and less toxic than NO_2. Like carbon monoxide and nitrite, NO attaches to haemoglobin and reduces oxygen transport efficiency. However, in a polluted atmosphere, the concentration of nitric oxide normally is much lower than that of carbon monoxide so that the effect on haemoglobin is much less.

Acute exposure to NO_2 can be quite harmful to human health. For exposures ranging from several minutes to one hour, a level of 50–100 ppm of NO_2 causes inflammation of lung tissue for a period of 6–8 weeks, after which time the subject normally recovers. Exposure of the subject to 150–200 ppm of NO_2 causes *bronchiolitis fibrosa obliterans*, a condition fatal within 3–5 weeks after exposure. Death generally results within 2–10 days after exposure to 500 ppm or more of NO_2. "Silo-filler's disease", caused by NO_2 generated by the fermentation of ensilage containing nitrate, is a particularly striking example of nitrogen dioxide poisoning. Deaths have resulted from the inhalation of NO_2 containing gases from burning celluloid and nitrocellulose film and from spillage of NO_2 oxidant (used with liquid hydrazine fuel) from missile rocket motors.

Although extensive damage to plants is observed in areas receiving heavy exposure to NO_2, most of this damage probably comes from secondary products of nitrogen oxides, such as PAN formed in smog. Exposure of plants to several parts per million of NO_2 in the laboratory causes leaf spotting and breakdown of plant tissue. Exposure to 10 ppm of NO causes a reversible decrease in the rate of photosynthesis. The effect on plants of long-term exposure to a few tenths of a part per million of NO_2 is less certain.

Nitrogen oxides are known to cause fading of dyes and inks used in some textiles. This has been observed in gas clothes dryers and is due to NO_x formed in the dryer flame. Much of the damage to materials caused by NO_x comes from secondary nitrates and nitric acid. For example, stress-corrosion cracking of springs used in telephone relays occurs far below the yield strength of the nickel-brass spring metal because of the action of particulate nitrates and aerosol nitric acid formed from NO_x.

Concern has been expressed about the possibility that NO_x emitted to the atmosphere by supersonic transport planes could catalyse the partial destruction of the stratospheric ozone layer, which absorbs damaging short-wavelength (240–300 nm) ultraviolet radiation. Detailed consideration of this effect is quite complicated, and only the main features are considered here. In the upper stratosphere and in the mesosphere, molecular oxygen is photo-dissociated by ultraviolet light of less than 242 nm wavelength:

$$O_2 + h\nu \rightarrow O + O \qquad \text{...(10.59)}$$

In the presence of energy-absorbing third bodies, the atomic oxygen reacts with molecular oxygen to produce ozone :

$$O_2 + O + M \rightarrow O_3 + M \qquad \text{...(10.60)}$$

Ozone can be destroyed by reaction with atomic oxygen,

$$O_3 + O \rightarrow O_2 + O_2 \qquad \text{...(10.61)}$$

and its formation can be prevented by recombination of oxygen atoms :

$$O + O + M \rightarrow O_2 + M \qquad \text{...(10.62)}$$

Addition of the reaction of nitric oxide with ozone,

$$NO + O_3 \rightarrow NO_2 + O_2 \qquad \text{...(10.63)}$$

to the reaction of nitrogen dioxide with atomic oxygen,

$$NO_2 + O \rightarrow NO + O_2 \qquad \qquad ...(10.64)$$

results in a net reaction for the destruction of ozone :

$$O + O_3 \rightarrow O_2 + O_2 \qquad \qquad ...(10.65)$$

Along with NO_x, water vapour is also emitted into the atmosphere by aircraft exhausts, which could accelerate ozone depletion by the following two reactions :

$$O + H_2O \rightarrow HO\cdot + HO\cdot \qquad \qquad ...(10.66)$$

$$HO\cdot + O_3 \rightarrow HOO\cdot + O_2 \qquad \qquad ...(10.67)$$

However, there are many natural stratospheric buffering reactions which tend to mitigate the potential ozone destruction from those reactions outlined above. Atomic oxygen capable of regenerating ozone is produced by the photochemical reaction :

$$NO_2 + h\nu \rightarrow NO + O \qquad (\lambda < 420 \text{ nm}) \qquad ...(10.68)$$

A competing reaction removing catalytic NO is :

$$NO + HOO\cdot \rightarrow NO_2 + HO\cdot \qquad \qquad ...(10.69)$$

Current belief is that supersonic aircraft emissions will not cause nearly as much damage to the ozone layer as chlorofluorocarbons.

Control of nitrogen oxides

The level of NO_x emitted from stationary sources such as power plant furnaces generally falls within the range of 50–1000 ppm. NO production is favoured both kinetically and thermodynamically by high temperatures and by high excess oxygen concentrations. These factors must be considered in reducing NO emissions from stationary sources. Reduction of flame temperature to prevent NO formation is accomplished by adding recirculated exhaust gas, cool air, or inert gases. Unfortunately, this decreases the efficiency of energy conversion as calculated by the Carnot Equation.

Low-excess-air firing is effective in reducing NO_x emissions during the combustion of fossil fuels. As the term implies, low-excess-air firing uses the minimum amount of excess air required for oxidation of the fuel, so that less oxygen is available for the reaction

$$N_2 + O_2 \rightarrow 2NO \qquad \qquad ...(10.70)$$

in the high temperature region of the flame. Incomplete fuel burnout, with the emission of hydrocarbons, soot, and CO, is an obvious problem with low-excess-air firing. This may be overcome by a two-stage combustion process consisting of the following steps :

1. A first stage in which the fuel is fired at a relatively high temperature with a substoichiometric amount of air, for example, 90–95% of the stoichiometric requirement. NO formation is limited by the absence of excess oxygen.

2. A second stage in which fuel burnout is completed at a relatively low temperature in excess air. The low temperature prevents formation of NO.

In some power plants fired with gas, the emission of NO has been reduced by as much as 90% by a two-stage combustion process.

Removal of NO_x from stack gas presents some formidable problems. These problems arise largely from the low water solubility of NO, the predominant nitrogen oxide species in stack gas. Possible approaches to NO_x removal are catalytic decomposition of nitrogen oxides, catalytic reduction of nitrogen oxides, and sorption of NO_x by liquids or solids. Uptake of NO_x is facilitated by oxidation of NO to more water-soluble species, including NO_2, N_2O_3, N_2O_4, HNO_2, and HNO_3. A typical catalytic reduction of NO in stack gas involves methane :

$$CH_4 + 4NO \rightarrow 2N_2 + CO_2 + 2H_2O \qquad ...(10.71)$$

Production of undesirable by-products is a major concern in these processes. For example, sulphur dioxide reacts with carbon monoxide used to reduce NO to produce toxic carbonyl sulphide, COS:

$$SO_2 + 3CO \rightarrow 2CO_2 + COS \qquad ...(10.72)$$

Most sorption processes have been aimed at the simultaneous removal of both nitrogen oxides and sulphur oxides. Sulphuric acid solutions or alkaline scrubbing solutions containing $Ca(OH)_2$ or $Mg(OH)_2$ may be used. The species N_2O_3 produced by the reaction

$$NO_2 + NO \rightarrow N_2O_3 \qquad ...(10.73)$$

is most efficiently absorbed. Therefore, since NO is the primary combustion product, the introduction of NO_2 into the flue gas is required to produce the N_2O_3, which is absorbed efficiently.

Acid Rain

As discussed in this chapter, much of the sulphur and nitrogen oxides entering the atmosphere are converted to sulphuric and nitric acids, respectively. When combined with hydrochloric acid arising from hydrogen chloride emissions, these acids cause acidic precipitation (acid rain) that is now a major pollution problem in some areas.

Headwater streams and high-altitude lakes are especially susceptible to the effects of acid rain and may sustain loss of fish and other aquatic life. Other effects include reductions in forest and crop productivity; leaching of nutrient cations and heavy metals from soils, rocks, and the sediments of lakes and streams; dissolution of metals such as lead and copper from water distribution pipes; corrosion of exposed metal; and dissolution of the surfaces of limestone buildings and monuments.

As a result of its widespread distribution and effects, acid rain is an air pollutant that may pose a threat to the global atmosphere.

Ammonia in the Atmosphere

Ammonia is present even in unpolluted air as a result of natural biochemical and chemical processes. Among the various sources of atmospheric ammonia are microorganisms, decay of animal wastes, sewage treatment, coke manufacture, ammonia manufacture, and leakage from ammonia-based refrigeration systems. High concentrations of ammonia gas in the atmosphere are generally indicative of accidental release of the gas.

Ammonia is removed from the atmosphere by its affinity for water and by its action as a base. It is a key species in the formation and neutralisation of nitrate and sulphate aerosols in polluted atmosphere. Ammonia reacts with these acidic aerosols to form ammonium salts :

$$NH_3 + HNO_3 \rightarrow NH_4NO_3 \qquad ...(10.74)$$

$$NH_3 + H_2SO_4 \rightarrow NH_4HSO_4 \qquad \qquad ...(10.75)$$

Ammonium salts are among the more corrosive salts in atmospheric aerosols.

Fluorine, Chlorine, and their Gaseous Compounds

Fluorine, hydrogen fluoride, and other volatile fluorides are produced in the manufacture of aluminium, and hydrogen fluoride is a by-product in the conversion of fluorapatite (rock phosphate) to phosphoric acid, superphosphate fertilisers, and other phosphorus products. The wet process for the production of phosphoric acid involves the reaction of fluorapatite, $Ca_5F(PO_4)_3$, with sulphuric acid :

$$Ca_5F(PO_4)_3 + 5H_2SO_4 + 10H_2O \rightarrow 5CaSO_4 \cdot 2H_2O + HF + 3H_3PO_4 \qquad ...(10.76)$$

It is necessary to recover most of the by-product fluorine from rock phosphate processing to avoid severe pollution problems. Recovery as fluorosilicic acid, H_2SiF_6, is normally practised.

Hydrogen fluoride gas is a dangerous substance that is so corrosive that it even reacts with glass. It is irritating to body tissues, and the respiratory tract is very sensitive to it. Brief exposure to HF vapours at the part-per-thousand level may be fatal. The acute toxicity of F_2 is even higher than that of HF. Chronic exposure to high levels of fluorides causes fluorosis, the symptoms of which include mottled teeth and pathological bone conditions.

Plants are particularly susceptible to the effects of gaseous fluorides. Fluorides from the atmosphere appear to enter the leaf tissue through the stomata. Fluoride is a cumulative poison in plants, and exposure of sensitive plants to even very low levels of fluorides for prolonged periods results in damage. Characteristic symptoms of fluoride poisoning are chlorosis (fading of green colour due to conditions other than the absence of light), edge burn, and tip burn. Conifers (such as pine trees) afflicted with fluoride poisoning may have reddish-brown necrotic needle tips. The sensitivity of some conifers to fluoride poisoning is illustrated by the fact that fluorine produced by aluminum plants in Norway has destroyed forests of *Pinus sylvestris* up to 8 miles distant; trees were damaged at distances as great as 20 miles from the plant.

Silicon tetrafluoride gas, SiF_4, is another gaseous fluoride pollutant produced during some steel and metal smelting operations that employ CaF_2, fluorspar. Fluorspar reacts with silicon dioxide (sand), releasing SiF_4 gas :

$$2CaF_2 + 3SiO_2 \rightarrow 2CaSiO_3 + SiF_4 \qquad \qquad ...(10.77)$$

Another gaseous fluorine compound, sulphur hexafluoride, SF_6, occurs in the atmosphere at levels of about 0.3 parts per trillion. It is extremely unreactive and is used as an atmospheric tracer. It does not absorb ultraviolet light in either the troposphere or stratosphere and is probably destroyed above 50 km by reactions beginning with its capture of free electrons.

Chlorine and hydrogen chloride

Chlorine gas, Cl_2, does not occur as an air pollutant on a large scale, but can be quite damaging on a local scale. It is widely used as a manufacturing chemical, in the plastics industry, for example, as well as for water treatment and as a bleach. Therefore, possibilities for its release exist in a number of locations. Chlorine is quite toxic and is a mucous-membrane irritant. It is very reactive and a powerful oxidising agent. Chlorine dissolves in atmospheric water droplets, yielding hydrochloric acid and hypochlorous acid, an oxidising agent :

$$H_2O + Cl_2 \rightarrow H^+ + Cl^- + HOCl \qquad \qquad ...(10.78)$$

Spills of chlorine gas have caused fatalities among exposed persons.

Hydrogen chloride, HCl, is emitted from a number of sources. Incineration of chlorinated plastics, such as polyvinylchloride, releases HCl as a combustion product.

$$
\begin{array}{ccccccccccc}
\text{Cl} & \text{H} & \text{H} & \text{H} & \text{Cl} & \text{H} & \text{H} & \text{H} & \text{Cl} & \text{H} \\
| & | & | & | & | & | & | & | & | & | \\
\cdots\text{C} & \text{C} & \text{C} & \text{C} & \text{C} & \text{C} & \text{C} & \text{C} & \text{C} & \text{C}\cdots \\
| & | & | & | & | & | & | & | & | & | \\
\text{H} & \text{H} & \text{Cl} & \text{H} & \text{H} & \text{H} & \text{Cl} & \text{H} & \text{H} & \text{H}
\end{array}
\quad \text{polyvinylchloride}
$$

Some compounds released to the atmosphere as air pollutants hydrolyse to form HCl.

Hydrogen Sulphide, Carbonyl Sulphide, and Carbon Disulphide

Hydrogen sulphide is produced by microbial decay of sulphur compounds and microbial reduction of sulphate, from geothermal steam, from wood pulping, and from a number of miscellaneous natural and anthropogenic sources. Most atmospheric hydrogen sulphide is rapidly converted to SO_2 and to sulphates. The organic homologs of hydrogen sulphide, the mercaptans, enter the atmosphere from decaying organic matter and have particularly objectionable odours.

Hydrogen sulphide pollution from artificial sources is not as much of an overall air pollution problem as sulphur dioxide pollution. However, there have been several acute incidents of hydrogen sulphide emissions resulting in damage to human health and even fatalities. The symptoms of poisoning included irritation of the respiratory tract and damage to the central nervous system. Unlike sulphur dioxide, which appears to affect older people and those with respiratory weaknesses, there was little evidence of correlation between the observed hydrogen sulphide poisoning and the age or physical condition of the victim.

Hydrogen sulphide at levels well above ambient concentrations destroys immature plant tissue. This type of plant injury is readily distinguished from that due to other phytotoxins. More sensitive species are killed by continuous exposure to around 3000 ppb H_2S, whereas other species exhibit reduced growth, leaf lesions, and defoliation.

Damage to certain kinds of materials is a very expensive effect of hydrogen sulphide pollution. Paints containing lead pigments, $2PbCO_3 \cdot Pb(OH)_2$ (no longer used), are particularly susceptible to darkening by H_2S. Darkening results from exposure over several hours to as little as 50 ppb H_2S. The lead sulphide originally produced by reaction of the lead pigment with hydrogen sulphide eventually may be converted to white lead sulphate by atmospheric oxygen after removal of the source of H_2S, thus partially reversing the damage.

A black layer of copper sulphide forms on copper metal exposed to H_2S. Eventually, this layer is replaced by a green coating of basic copper sulphate, $CuSO_4 \cdot 3Cu(OH)_2$. The green "patina," as it is called, is very resistant to further corrosion. Such layers of corrosion can seriously impair the function of copper contacts on electrical equipment. Hydrogen sulphide also forms a black sulphide coating on silver.

Carbonyl sulphide, COS, is now recognised as a component of the atmosphere at a tropospheric concentration of approximately 500 parts per trillion by volume, corresponding to a global burden of about 2.4 million tonnes. It is, therefore, a significant sulphur species in the atmosphere. It is possible that the HO· radical-initiated oxidation of COS and carbon disulphide (CS_2) would yield 8–12 million tonnes as S in atmospheric sulphur dioxide per year. Though this is a small yield compared to pollution sources, the HO·-initiated process could account for much of the SO_2 burden in the remote troposphere. Both COS and CS_2 are oxidised in the atmosphere by reactions initiated by the hydroxyl radical. The initial reactions are :

$$HO· + COS \rightarrow CO_2 + HS·$$ ...(10.79)

$$HO· + CS_2 \rightarrow COS + HS·$$ ...(10.80)

The sulphur-containing products undergo further reactions to sulphur dioxide and, eventually, to sulphate species.

CHAPTER 11

Organic Air Pollutants

INTRODUCTION

Organic pollutants may have a strong effect upon atmospheric quality. The effects of organic pollutants in the atmosphere may be divided into two major categories. The first category consists of *direct effects*, such as cancer caused by exposure to vinyl chloride. The second category is the formation of *secondary pollutants*, especially photochemical smog. In the case of pollutant hydrocarbons in the atmosphere, the latter is the more important effect. In some localised situations, particularly the workplace, direct effects of organic air pollutants may be equally important.

Organic contaminants are lost from the atmosphere by a number of routes. These include dissolution in precipitation (rainwater), dry deposition, photochemical reactions, formation of and incorporation into particulate matter, and uptake by plants. Reactions of organic atmospheric contaminants are particularly important in determining their manner and rates of loss from the atmosphere.

Forest trees present a large surface area to the atmosphere and are particularly important in filtering organic contaminants from air. Forest trees and plants contact the atmosphere via plant cuticle layers; that is, the biopolymer "skin" on the leaves and needles of the plants. The cuticle layer is lipophilic, meaning that it has a particular affinity for organic substances, including those in the atmosphere. A study of the uptake of organic substances, including those in the atmosphere. A study of the uptake of organic substances (lindane, triadimenol, bitertanol, 2,4-dichlorophenoxyacetic acid, and pentachlorophenol) by conifer needles has shown that the process consists of two phases: (i) adsorption onto the needle surfaces; and (ii) transport through the cuticle layer into the needle and plant. Uptake increases with increasing lipophilicity of the compounds and with increasing surface area of the leaves. This phenomenon points to the importance of forests in atmospheric purification.

ORGANIC COMPOUNDS FROM NATURAL SOURCES

Natural sources are the most important contributors of organics in the atmosphere, and hydrocarons generated and released by human activities constitute only about 1/7 of the total hydrocarbons in the atmosphere. This ratio is primarily the result of the huge quantities of methane produced by anaerobic bacteria in the decomposition of organic matter in water, sediments, and soil :

$$2\{CH_2O\} \text{ (bacterial action)} \rightarrow CO_2(g) + CH_4(g) \qquad \qquad ...(11.1)$$

Flatulent emissions from domesticated animals, arising from bacterial decomposition of food in their digestive tracts, add about 85 million metric tonnes of methane to the atmosphere and anaerobic conditions in intensively cultivated rice fields produce large amounts of methane. Methane is a natural constituent of the atmosphere.

Methane in the troposphere contributes to the photochemical production of carbon monoxide and ozone. The photochemical oxidation of methane is a major source of water vapour in the stratosphere.

Atmospheric hydrocarbons produced by living sources are called *biogenic hydrocarbons*, and vegetation is the most important natural source of these compounds. A compilation of organic compounds in the atmosphere documented a total of 367 different compounds that are released to the atmosphere from vegetation sources. Other natural sources include microorganisms, forest fires, animal wastes, and volcanoes.

One of the simplest organic compounds given off by plants is ethylene, C_2H_4. This compound is produced by a variety of plants and released to the atmosphere. Because of its double bond, ethylene is highly reactive with hydroxyl radical, HO·, and with oxidising species in the atmosphere. Ethylene from vegetation sources should be considered as an active participant in atmospheric chemical processes.

Most of the hydrocarbons emitted by plants are *terpenes*, which constitute a large class of organic compounds found in essential oils. Essential oils are obtained when parts of some types of plants are subjected to steam distillation. Most of the plants that produce terpenes are conifers (evergreen trees and shrubs, such as pine and cypress), plants of the genus *Myrtus*, and trees and shrubs of the genus *Citrus*. One of the most common terpenes emitted by trees is α-pinene, a principal component of turpentine. The terpene limonene, found in citrus fruit and pine needles, is encountered in the atmosphere around these sources. Isoprene (2-methyl-1,3-butadiene), a hemiterpene, has been identified in the emissions from cottonwood, eucalyptus, oak, sweetgum and white spruce trees. Other terpenes known to be given off by trees include β-pinene, myrcene, ocimene, and α-terpinene.

Terpenes are among the most reactive compounds in the atmosphere. The reaction of terpenes with hydroxyl radical, HO·, is very rapid, and terpenes also react with other oxidising agents in the atmosphere, particularly ozone, O_3. Turpentine, a mixture of terpenes, has been widely used in paint because it reacts with atmospheric oxygen to form a peroxide, then a hard resin. It is likely that compounds such as α-pinene and isoprene undergo similar reactions in the atmosphere to form particulate matter. The resulting Aitken nuclei aerosols probably cause the blue haze in the atmosphere above some heavy growths of vegetation. Smog-chamber experiments have been performed in an effort to determine the fate of atmospheric terpenes.

Although terpenes are highly reactive with hydroxyl radical, it is now believed that much of the atmospheric aerosol formed as the result of reactions of unsaturated biogenic hydrocarbons is the result of processes that start with reactions between them and ozone.

Perhaps the greatest variety of compounds emitted by plants consist of esters. However, they are released in such small quantities that they have little influence upon atmospheric chemistry. Esters are primarily responsible for the fragrances associated with much vegetation.

Pollutant Hydrocarbons

Ethylene and terpenes are hydrocarbons, organic compounds containing only hydrogen and carbon. The major classes of hydrocarbons are alkanes (formerly called paraffins), such as 2,2,3-trimethylbutane;

$$
\begin{array}{ccc}
& H_3C & H \\
& | & | \\
H_3C- & C-C & -CH_3 \\
& | & | \\
& H_3C & CH_3
\end{array}
$$

alkenes (olefins, compounds with double bonds between adjacent carbon atoms), such as ethylene; *alkynes* (compounds with triple bonds), such as acetylene:

$$H-C \equiv C-H$$

and *aromatic* compounds, such as naphthalene :

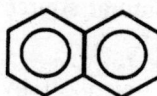

Because of their widespread use in fuels, hydrocarbons predominate among organic atmospheric pollutants. Petroleum products, primarily gasoline, are the source of most of the anthropogenic (originating through human activities) pollutant hydrocarbons found in the atmosphere. Hydrocarbons may enter the atmosphere either directly or as by-products of the partial combustion of other hydrocarbons. The latter are particularly important because they tend to be unsaturated and relatively reactive.

Most hydrocarbon pollutant sources produce about 15% reactive hydrocarbons, whereas those from incomplete combustion of gasoline are about 45% reactive. The hydrocarbons in uncontrolled automobile exhausts are only about 1/3 alkanes, with the remainder divided approximately equally between more reactive alkenes and aromatic hydrocarbons, thus accounting for the relatively high reactivity of automotive exhaust hydrocarbons.

Investigators who study smog formation in smog chambers have developed synthetic mixtures of hydrocarbons that mimic the smog-forming behaviour of hydrocarbons in a polluted atmosphere. The compounds so used provide a simplified idea of the composition of pollutant hydrocarbons likely to lead to smog formation. A typical mixture consists of 0.556 mole fraction of alkanes, including 2-methylbutane, *n*-pentane, 2-methylpentane, 2,4-dimethylpentane, and 2,2,4-trimethylpentane, and 0.444 mole fraction alkenes, including 1-butene, *cis*-2-butene, 2-methyl-1-butene, 2-methyl-2-butene, ethene (ethylene), and propene (propylene).

Alkanes are among the more stable hydrocarbons in the atmosphere. Straightchain alkanes with 1 to more than 30 carbon atoms and branched-chain alkanes with 6 or fewer carbon atoms are commonly present in polluted atmospheres. Because of their high vapour pressures, alkanes with 6 or fewer carbon atoms are normally present as gases, alkanes with 20 or more carbon atoms are present as aerosols or sorbed to atmospheric particles, and alkanes with 6 to 20 carbon atoms per molecule may be present either as vapour or particles, depending upon conditions.

Alkenes enter the atmosphere from a variety of processes, including emissions from internal combustion engines and turbines, foundry operations, and petroleum refining.

Structural formulae of some alkenes is given below :

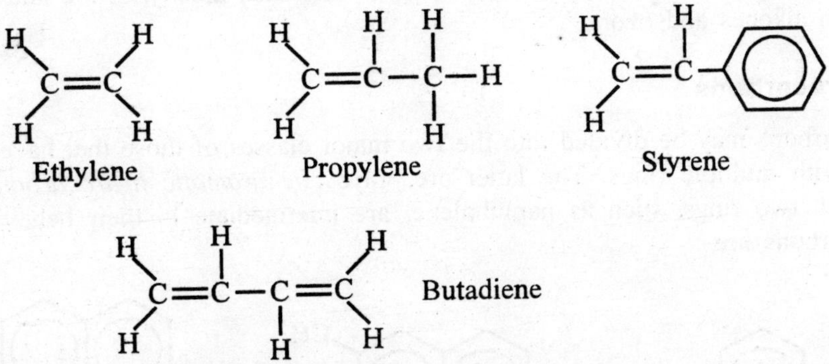

Ethylene   Propylene   Styrene

Butadiene

These compounds are used primarily as monomers, which are polymerised to create polymers for plastics (polyethylene, polypropylene, polystyrene), synthetic rubber (styrenebutadiene, polybutadiene), latex paints (styrenebutadiene), and other applications. All of these compounds, as well as others manufactured in lesser quantities, are released to the atmosphere. In addition to the direct release of alkenes, these hydrocarbons are commonly produced by the partial combustion and "cracking" at high temperatures of alkanes, particularly in the internal combustion engine.

Alkynes occur much less commonly in the atmosphere than do alkenes. Detectable levels are sometimes found in acetylene, used as a fuel for welding torches, and 1-butyne, used in synthetic rubber manufacture :

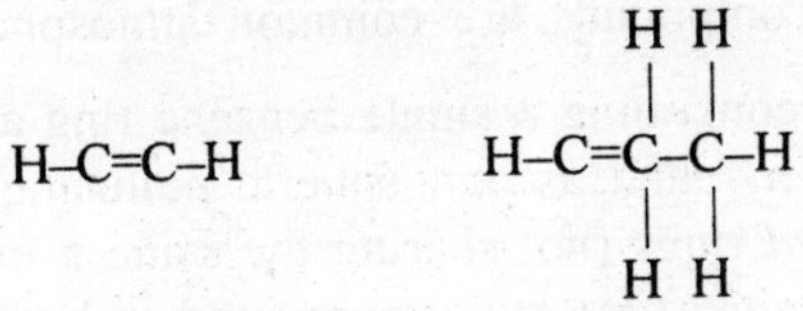

Acetylene   1-Butyne

Unlike alkanes, alkenes are highly reactive in the atmosphere, especially in the presence of NO_x and sunlight. Hydroxyl radical reacts readily with alkenes, either by abstracting a hydrogen atom or by adding to the double bond. If hydroxyl radical adds to the double bond in propylene, for example, the product is :

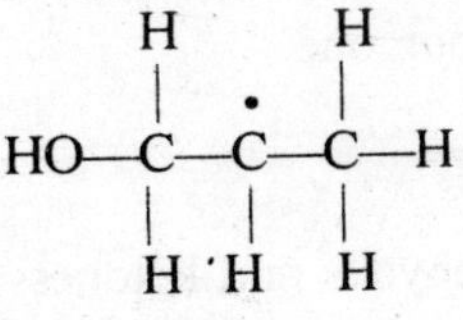

Addition of molecular O_2 to this radical results in the formation of a peroxyl radical :

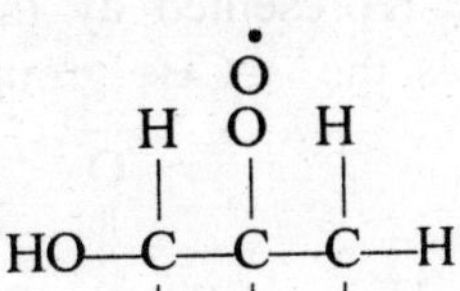

These radicals then participate in reaction chains. Ozone, O_3, adds across double bonds and is rather reactive with alkenes. As shown for the natural alkene limonene, aldehydes are among the products of reactions between alkenes and ozone.

Aromatic Hydrocarbons

Aromatic hydrocarbons may be divided into the two major classes of those that have only one benzene ring and those with multiple rings. The latter are *polycyclic aromatic hydrocarbons*, *PAH*. Aromatic hydrocarbons with two rings, such as naphthalene, are intermediate in their behaviour. Some typical aromatic hydrocarbons are :

Benzene   2,6-Dimethylnaphthalene   Pyrene

Single-ring aromatic compounds are important constituents of lead-free gasoline, which has largely replaced leaded gasoline. Aromatic solvents are widely used in industry. Aromatic hydrocarbons are raw materials for the manufacture of monomers and plasticisers in polymers. Styrene is a monomer used in the manufacture of plastics and synthetic rubber. Cumene is oxidised to produce phenol and acetone, which is a valuable by-product. Because of these applications, plus production of these compounds as combustion by-products, aromatic compounds are common atmospheric pollutants.

Approximately 55 hydrocarbons containing a single benzene ring and approximately 30 hydrocarbon derivatives of naphthalene have been found as atmospheric pollutants. In addition, several compounds containing two or more *unconjugated* rings (not sharing the same π electron cloud between rings) have been detected as atmospheric pollutants. One such compound is biphenyl,

Biphenyl

detected in diesel smoke. It should be pointed out that many of these aromatic hydrocarbons have been detected primarily as ingredients of tobacco smoke and are, therefore, of much greater significance in an indoor environment that in an outdoor one.

Aldehydes and Ketones

Carbonyl compounds, consisting of aldehydes and ketones, are often the first species formed, other than unstable reaction intermediates, in the photochemical oxidation of atmospheric hydrocarbons. The general formulas of aldehydes and ketones are represented by the following, where R and R' represent the hydrocarbon *moieties* (portions), such as the $-CH_3$ group.

$$
\underset{\text{Aldehyde}}{R-\overset{\overset{\displaystyle O}{\|}}{C}-H} \qquad \underset{\text{Ketone}}{R-\overset{\overset{\displaystyle O}{\|}}{C}-R'} \qquad \underset{\text{Carbonyl moiety}}{-\overset{\overset{\displaystyle O}{\|}}{C}-}
$$

Carbonyl compounds are by-products of the generation of hydroxyl radicals from organic peroxyl radicals and the simplest and most widely produced of the carbonyl compounds is formaldehyde,

$$\underset{H \quad H}{\overset{\displaystyle O \atop \displaystyle \parallel \atop \displaystyle C}{}} \quad \text{Formaldehyde}$$

Formaldehydes are used in the manufacture of plastics, resins, lacquers, dyes, and explosives, formaldehyde is uniquely important because of its widespread distribution and toxicity. Humans may be exposed to formaldehyde in the manufacture and use of phenol, urea, and melamine resin plastics and from formaldehyde-containing adhesives in pressed wood products, such as particle board, used in especially large quantities in mobile home construction. However, significantly improved manufacturing processes have greatly reduced formaldehyde emissions from these synthetic building materials. Formaldehyde occurs in the atmosphere primarily in the gas phase.

Acetaldehyde is a widely produced organic chemical used in the manufacture of acetic acid, plastics and raw materials. It is also used as a solvent and for applications in the rubber, leather, and plastics industries. Methylethyl ketone is employed as a low-boiling solvent for coatings and adhesives and for the synthesis of other chemicals.

In addition to their production from hydrocarbons by photochemical oxidation, carbonyl compounds enter the atmosphere from a large number of sources and processes. These include direct emissions from internal combustion engine exhausts, incinerator emissions, spray painting, polymer manufacture, printing, petrochemicals manufacture, and lacquer manufacture. Formaldehyde and acetaldehyde are produced by microorganisms, and acetaldehyde is emitted by some kinds of vegetation.

Aldehydes are second only to NO_2 as atmospheric sources of free radicals produced by the absorption of light. This is because the carbonyl group is a chromophore, a molecular group that readily absorbs light. It absorbs well in the near-ultraviolet region of the spectrum. The activated compound produced when a photon is absorbed by an aldehyde dissociates into a formyl radical and an alkyl radical.

$$\overset{\displaystyle \cdot}{HCO}$$

Formyl radical

and an alkyl radical.

Ketones commonly undergo photochemical dissociation in the atmosphere at one of the bonds joining the carbonyl group to the hydrocarbon moieties :

$$\underset{}{\overset{\displaystyle O \atop \displaystyle \parallel}{R\text{–}C\text{–}R'}} + h\nu \rightarrow \underset{}{\overset{\displaystyle O \atop \displaystyle \parallel}{R\text{–}C\cdot}} + R'\cdot \qquad \qquad ...(11.2)$$

The radicals produced then undergo subsequent reactions with O_2 and other chemical species in the atmosphere.

Miscellaneous Oxygen-Containing Compounds

The oxygen-containing organic compounds consisting of aliphatic alcohols, phenols, ethers, and carboxylic acids are discussed here. These compounds have the general formulas given below, where R and R' represent hydrocarbon moieties and Ar stands specifically for an aromatic moiety, such as the phenyl group (benzene less an H atom) :

$$R—OH \qquad Ar—OH \qquad R—O—R' \qquad R—\overset{\displaystyle O}{\overset{\|}{C}}—OH$$

Aliphatic   Phenols   Ethers   Carboxylic
alcohols   acids

These classes of compounds include many important organic chemicals.

Alcohols

The most common of the many uses of these chemicals is for the manufacture of other chemicals. Methanol is widely used in the manufacture of formaldehyde, as a solvent, and mixed with water as an antifreeze formulation. Ethanol is used as a solvent and as the starting material for the manufacture of acetaldehyde, acetic acid, ethyl ether, ethyl chloride, ethyl bromide, and several important esters. Both methanol and ethanol can be used as motor vehicle fuels, usually in mixtures with gasoline. Ethylene glycol is a common antifreeze compound.

Numerous aliphatic alcohols have been reported in the atmosphere. Because of their volatility, the lower alcohols, especially methanol and ethanol, predominate as atmospheric pollutants. Among the other alcohols released to the atmosphere are 1-propanol, 2-propanol, propylene glycol, 1-butanol, and even octadecanol, $CH_3(CH_2)_{16}CH_2OH$, which is evolved by plants. Alcohols can undergo photochemical reactions, beginning with abstraction of hydrogen by hydroxyl radical. Mechanisms for scavenging alcohols from the atmosphere are relatively efficient because, the lower alcohols are quite water soluble and the higher ones have low vapour pressures.

Some alkenyl alcohols have been found in the atmosphere, largely as by-products of combustion. Typical of these is 2-buten-l-ol,

$$\begin{array}{ccccccc} & H & H & & H & & \\ & | & | & & | & & \\ H{-}C{-}C&=&C{-}C&{-}OH \\ & | & & | & | & \\ & H & & H & H & \end{array}$$

which has been detected in automobile exhausts. Some alkenyl alcohols are emitted by plants. One of these, *cis*-3-hexen-1-ol, $CH_3CH_2CH{=}CHCH_2CH_2OH$, is emitted from grass, trees and crop plants to the extent that it is known as "leaf alcohol". In addition to reacting with HO· radical, alkenyl radicals react strongly with atmospheric ozone, which adds across the double bond.

Phenols

Phenols are aromatic alcohols that have an —OH group bonded to an aromatic ring. They are more noted as water pollutants than as air pollutants.

Some typical phenols that have been reported as atmospheric contaminants are the following :

| Phenol | *o* - cresol | *m* - cresol | *p* - cresol | 1 - napthnol |

Phenols are produced by the pyrolysis of coal and are major byproducts of coking. Thus, in local situations involving coal coking and similar operations, phenols can be troublesome air pollutants.

Ethers

Ethers are relatively uncommon atmospheric pollutants; however, the flammability hazard of diethyl ether vapour in an enclosed work space is well known. In addition to aliphatic ethers, such as dimethyl ether and diethyl ether, several alkenyl ethers, including vinylethyl ether are produced by internal combustion engines. A cyclic ether and important industrial solvent, tetrahydrofuran, occurs as an air contaminant. Methyltertiarybutyl ether, MTBE, has become the octane booster of choice to replace tetraethyllead in gasoline. Because of its widespread distribution, MTBE has the potential to be an air pollutant, although its hazard is limited by its low vapour pressure.

Another possible air contaminant because of its potential uses as an octane booster is diisopropyl ether (DIPE). Its atmospheric chemistry has been studied. As with virtually all organic species in air, it reacts with hydroxyl radical in the troposphere.

Oxides

Ethylene oxide and propylene oxide are the most widely produced industrial chemicals and have a limited potential to enter the atmosphere as pollutants. Ethylene oxide is a moderately to highly toxic sweet-smelling, colourless, flammable, explosive gas used as a chemical intermediate, sterilant, and fumigant. It is a mutagen and a carcinogen to experimental animals. It is classified as hazardous for both its toxicity and ignitability.

Carboxylic acids

Most of the many carboxylic acids found in the atmosphere are probably the result of the photochemical oxidation of other organic compounds through gas-phase reactions or by reactions of other organic compounds dissolved in aqueous aerosols. These acids are often the end products of photochemical oxidation because their low vapour pressures and relatively high water solubilities make them susceptible to scavenging from the atmosphere.

Organohalide Compounds

The organohalides of environmental and toxicological concern exhibit a wide range of physical and chemical properties. The lighter organohalide compounds are the most likely to be found in the atmosphere. On a global basis, the three most abundant organochlorine compounds in the atmosphere are methyl chloride, methyl chloroform, and carbon tetrachloride, which have tropospheric concentrations ranging from a tenth to several tenths of a part per billion. Methyl chloroform is relatively persistent in the atmosphere, with residence times of several years. Therefore, it may pose a threat to the stratospheric ozone layer in the same way as chlorofluorocarbons. Also found are methylene chloride; methyl bromide, CH_3Br; bromoform, $CHBr_3$; assorted chlorofluorocarbons; and halogen-substituted ethylene compounds, such as trichloroethylene, vinyl chloride, perchloroethylene, ($CCl_2 = CCl_2$) and solvent ethylene dibromide ($CHBr = CHBr$).

Chlorofluorocarbons (CFC's)

These compounds are notably stable and non-toxic. They have been widely used in recent decades in the fabrication of flexible and rigid foams and as fluids for refrigeration and air conditioning. Halons are related compounds that contain bromine and are used in fire extinguisher systems. Halons are particularly effective fire extinguishing agents because of the way in which they stop combustion. Halons act by chain reactions that destroy hydrogen atoms which sustain combustion.

Halons are used in automatic fire extinguishing systems, particularly those located in flammable solvent storage areas, and in speciality fire extinguishers, particularly those on aircraft. Because of their potential to destroy stratospheric ozone, the use of halons in fire extinguishers is being curtailed. This has caused concern because of the favourable properties of halons in fire extinguishers, particularly on aircraft. It is possible that hydrogen-containing analogs of halons may be effective as fire extinguishers without posing a threat to ozone.

The nonreactivity of CFC compounds and deliberate or accidental release to the atmosphere, has resulted in CFCs becoming homogeneous components of the global atmosphere. Although quite inert in the lower atmosphere, CFCs undergo photodecomposition by the action of high-energy ultraviolet radiation in the stratosphere, which is energetic enough to break their very strong C–Cl bonds through reactions such as :

$$Cl_2CF_2 + h\nu \rightarrow Cl\cdot + ClCF_2\cdot \qquad ...(11.3)$$

thereby releasing Cl atoms. These atoms react with ozone, destroying it and producing ClO :

$$Cl + O_3 \rightarrow ClO + O_2 \qquad ...(11.4)$$

In the stratosphere, there is an appreciable concentration of atomic oxygen, by virtue of the reaction

$$O_3 + h\nu \rightarrow O_2 + O \qquad ...(11.5)$$

Nitric oxide, NO, is also present. The ClO species may react with either O or NO, regenerating Cl atoms and resulting in chain reactions that cause the net destruction of ozone :

$$ClO + O \rightarrow Cl + O_2 \qquad ...(11.6)$$
$$\underline{Cl + O_3 \rightarrow ClO + O_2} \qquad ...(11.7)$$
$$O + O_3 \rightarrow 2O_2 \qquad ...(11.8)$$

$$ClO + NO \rightarrow Cl + NO_2 \qquad ...(11.9)$$
$$\underline{Cl + O_3 \rightarrow ClO + O_2} \qquad ...(11.10)$$
$$O_3 + NO \rightarrow NO_2 + O_2 \qquad ...(11.11)$$

Both ClO and Cl involved in the above chain reactions have been detected in the 25–45 km altitude region.

The effects of CFCs on the ozone layer may be the single greatest threat to the global atmosphere. Because of the more readily broken H–C bonds that they contain, these compounds are more easily destroyed by atmospheric chemical reactions (particularly with hydroxyl radical) before they reach the stratosphere.

Perfluorocarbons

Perfluorocarbons are completely fluorinated organic compounds, the simplest examples of which are carbon tetrafluoride (CF_4) and hexafluoroethane (C_2F_6). Several hundred metric tonnes of these compounds are produced annually as etching agents in the electronics industry.

Non-toxic perfluorocarbons do not react with hydroxyl radical, ozone, or other reactive substances in the atmosphere, and the only known significant mechanism by which they are destroyed in the atmosphere is photolysis by radiation less than 130 nm in wavelength. Because of their extreme lack of reactivity, they are involved in neither photochemical smog formation nor ozone layer depletion. As a result of this stability, perfluorocarbons are very long-lived in the atmosphere; the lifetime of CF_4 is estimated to be an astoundingly long 50,000 years! The major atmospheric concern with these compounds is their potential to cause greenhouse warming. Taking into account their nonreactivity and ability to absorb infrared radiation, perfluorocarbons have a potential to cause global warming over a very long time span with an aggregate effect per molecule several thousand times that of carbon dioxide.

Chlorinated dibenzo-p-dioxins and dibenzofurans

Polychlorinated dibenzo-*p*-dioxins (PCDDs) and polychlorinated dibenzofurans (PCDFs) are pollutants compounds. These compounds are of considerable concern because of their toxicities. One of the more infamous environmental pollutant chemicals is 2,3,7,8-tetrachlorodibenzo-*p*-dioxin, TCDD, often known simply as "dioxin".

PCDDs and PCDFs enter the air from numerous sources, including automobile engines, waste incinerators, and steel and other metal production. A particularly important source may well be municipal solid waste incinerators. There is evidence to suggest that PCDDs and PCDFs are formed in such incinerators because of the presence of both chlorine (such as from polyvinylchloride plastic in municipal waste) and catalytic metals. It has also been suggested that PCDDs and PCDFs are produced by *de novo* synthesis on carbonaceous fly ash surfaces in the post-combustion region of an incinerator at relatively low temperatures of around 300°C in the presence of oxygen and sources of chlorine and hydrogen.

Atmospheric levels of PCDDs and PCDFs are quite low, in the range of 0.4 to 100 picograms per cubic metre of air. An analysis of pine needle extracts for PCDDs and PCDFs has shown significantly elevated levels of these substances in the vicinity of wood-treating plants that use pentachlorophenol as a wood preservative. Because of their lower volatilities, the more highly chlorinated congeners of these compounds tend to occur in atmospheric particulate matter, in which they are relatively protected from photolysis and reaction with hydroxyl radical, which are the two main mechanisms by which PCDDs and PCDFs are eliminated from the atmosphere. Furthermore, the less highly chlorinated congeners are more reactive because of their higher populations of relatively weaker C–H bonds.

Organosulphur Compounds

Substitution of alkyl or aryl hydrocarbon groups such as phenyl and methyl for H on hydrogen sulphide, H_2S, leads to a number of different organosulphur thiols (mercaptans, R–SH) and sulphides, also called thioethers (R–S–R). Structural formulas of examples of these compounds are shown below:

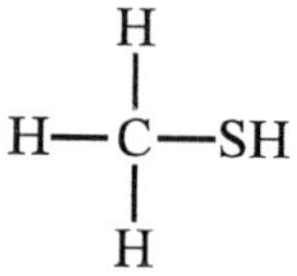

Methanethiol 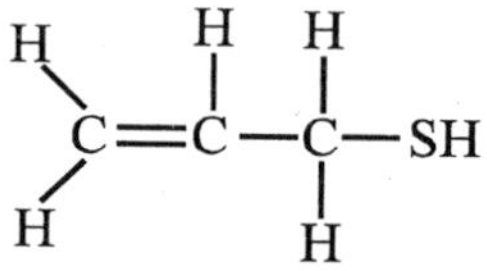   2-Propene-1-thiol 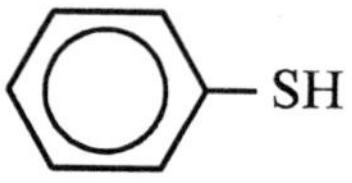   Benzenethiol

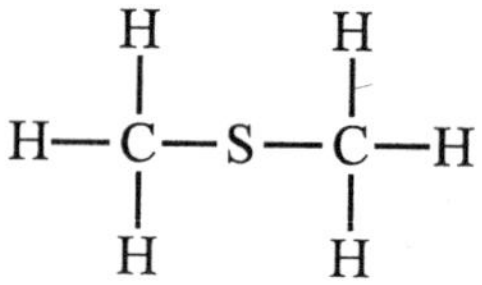

Dimethylsulphide 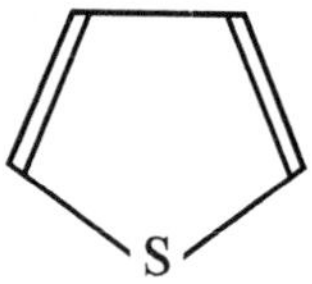   Thiophene 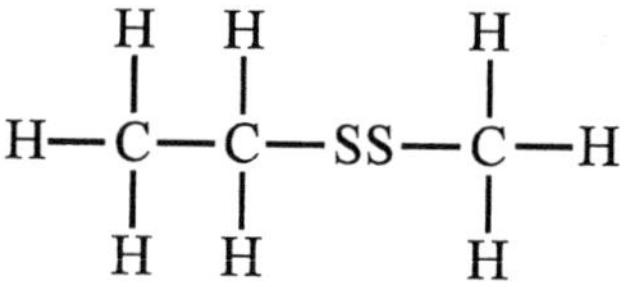   Ethylmethyldisulphide

Methanethiol and other lighter alkyl thiols are fairly common air pollutants that have "ultragarlic" odours; both 1- and 2-butanethiol are associated with skunk odour. Gaseous methanethiol and volatile liquid ethanethiol are used as odourant leak-detecting additives for natural gas, propane, and butane and are also employed as intermediates in pesticide synthesis. Allyl mercaptan (2-propene-1-thiol) is a toxic, irritating volatile liquid with a strong garlic odour. Benzenethiol (phenyl mercaptan) is the simplest of the aryl thiols. It is a toxic liquid with a severely "repulsive" odour.

Although not highly significant as atmospheric contaminants on a large scale, organic sulphur compounds can cause local air pollution problems because of their bad odours. Major sources of organosulphur compounds in the atmosphere include microbial degradation, wood pulping, volatile matter evolved from plants, animal wastes, packing house and rendering plant wastes, starch manufacture, sewage treatment, and petroleum refining.

Although the impact of organosulphur compounds on atmospheric chemistry is minimal in areas such as aerosol formation or production of acid precipitation components, these compounds are the worst of all in producing odour. Therefore, it is important to prevent their release to the atmosphere.

Organonitrogen Compounds

Organic nitrogen compounds that may be found as atmospheric contaminants may be classified as amines, amides, nitriles, nitro compounds, or heterocyclic nitrogen compounds. Structures of common examples of each of these five classes of compounds reported as atmospheric contaminant are :

Methylamine   Dimethyl formamide   Acrylonitrile

Nitrobenzene   Pyridine   Aniline

Amines consist of compounds in which one or more of the hydrogen atoms in NH_3 has been replaced by a hydrocarbon moiety. Lower-molecular-mass amines are volatile. These are prominent among the compounds giving rotten fish their characteristic odour—an obvious reason why air contamination by amines is undesirable. The simplest and most important aromatic amine is aniline, used in the manufacture of dyes, amides, photographic chemicals, and drugs. A number of amines are widely used as industrial chemicals and solvents, so that industrial sources have the potential to contaminate the atmosphere with these chemicals. Decaying organic matter, especially protein wastes, produce amines, so that rendering plants, packing houses, and sewage treatment plants are important sources of these substances.

Aromatic amines are of special concern as atmospheric pollutants, particularly in the workplace, because some are known to cause urethral tract cancer (particularly of the bladder) in exposed individuals. Aromatic amines are widely used as chemical intermediates, antioxidants, and curing agents in the manufacture of polymers (rubber and plastics), drugs, pesticides, dyes, pigments, and inks. In addition to aniline, some aromatic amines of potential concern are the following : Benzidine, 3,3′-Dichlorobenzidine, 1-Naphthylamine, 2-Naphthylamine, and 1-Phenyl-2-naphthylamine.

In the atmosphere, amines can be attacked by hydroxyl radical and undergo further reactions. Amines are bases (electron-pair donors); therefore, their acid-base chemistry in the atmosphere may be important, particularly in the presence of acids in acidic precipitation.

The amide most likely to be encountered as an atmospheric pollutant is dimethylformamide. It is widely used commercially as a solvent for the synthetic polymer, polyacrylonitrile. Most amides have relatively low vapour pressures, which limits their entry into the atmosphere.

Nitriles, which are characterised by the $-C\equiv N$ group, have been reported as air contaminants, particularly from industrial sources. Both acrylonitrile and acetonitrile, CH_3CN, have been reported in the atmosphere as a result of synthetic rubber manufacture. As expected from their volatilities and levels of industrial production, most of the nitriles reported as atmospheric contaminants are low-molecular-mass aliphatic or olefinic nitriles, or aromatic nitriles with only one benzene ring. Acrylonitrile, used to make polyacrylonitrile polymer, is a nitrogen-containing organic chemical produced in very large quantities worldwide.

Among the nitro compounds, RNO_2, reported as air contaminants are nitromethane, nitroethane, and nitrobenzene. These compounds are produced from industrial sources. Highly oxygenated compounds containing the NO_2 group, particularly peroxyacetyl nitrate, PAN, are end products of the photochemical oxidation of hydrocarbons in urban atmospheres.

A large number of heterocyclic nitrogen compounds have been reported in tobacco smoke, and it is inferred that many of these compounds can enter the atmosphere from burning vegetation. Coke ovens are another major source of these compounds. In addition to the derivatives of pyridine, some of the heterocyclic nitrogen compounds are derivatives of pyrrole. Heterocyclic nitrogen compounds occur almost entirely in association with aerosols in the atmosphere. *Nitrosamines*, general formula,

$$\begin{array}{c} R \\ \diagdown \\ N - N = O \\ \diagup \\ R \end{array}$$

deserved special mention as atmospheric contaminants because some are known carcinogens. Both N-nitrosodimethylamine and N-nitrosodiethylamine have been detected in the atmosphere.

CHAPTER 12

Photochemical Smog

INTRODUCTION

This chapter discusses the oxidising smog or photochemical smog that permeates atmospheres. Although smog is the term used to denote a photochemically oxidising atmosphere, the word originally was used to describe the unpleasant combination of smoke and fog laced with sulphur dioxide which was formerly prevalent in London when high-sulphur coal was the primary fuel used in that city. This mixture is characterised by the presence of sulphur dioxide, a reducing compound; therefore, it is a reducing smog or sulphurous smog. In fact, readily oxidised sulphur dioxide has a short lifetime in an atmosphere where oxidising photochemical smog is present.

Characterised by reduced visibility, eye irritation, cracking of rubber, and deterioration of materials, smog became a serious nuisance. It is now recognised as a major air pollution problem in many areas of the world. Smoggy conditions are manifested by moderate to severe eye irritation or visibility below 3 miles when the relative humidity is below 60%. The formation of oxidants in the air, particularly ozone, is indicative of smog formation. Serious levels of photochemical smog may be assumed to be present when the oxidant level exceeds 0.15 ppm for more than one hour. The three ingredients required to generate photochemical smog are ultraviolet light, hydrocarbons, and nitrogen oxides.

Studies made in Log Angeles area in the United States reveal an interesting sequence of concentration levels of various atmospheric pollutants during photo-oxidation reactions that accompany photochemical smog episodes. It is initiated by the photo dissociation of nitrogen dioxide and subsequent oxidation reactions involving unsaturated hydrocarbons, other organic compounds and free radicals to form ozone and organic peroxides. The phenomenon occurs during warm, sunny day with gentle winds and low level inversion. Visibility is reduced markedly. Eye irritation, plant damage, a characteristic odour and accelerated cracking of rubber goods are obvious results.

An important feature of photochemical smog is the production of aerosols that markedly reduce visibility. Aerosols of similar composition are formed by the irradiation of olefins or gasoline in presence of NO_2. It has been experimentally confirmed that the presence of unusually high ozone in the atmosphere is due to solar irradiation. Also ozone is produced during photosensitised reactions, it is destroyed or removed by dark reactions, such as oxidation of NO to NO_2, the formation of PAN and production of ozonides and peroxides of hydrocarbons.

CHEMICAL REACTIONS IN A CONTAMINATED ATMOSPHERE

A variety of chemical reactions occur in the contaminated atmosphere of cities. The products of such reactions have been blamed for many of the unpleasant properties attributed to polluted atmospheres. For example, it has been suggested that the products of the oxidation of hydrocarbons in the presence of nitrogen dioxide and sunlight are responsible for eye irritation, decrease in visibility, plant damage, and cracking of rubber goods.

The substances which can react chemically in the atmosphere fall into two groups: major natural constituents of the atmosphere, which are present in high concentrations, and contaminants, usually present at low concentrations. These groups differ in concentration by a factor of 10^4 to 10^6. This large difference in concentration is convenient for purposes of classification of reactions and calls for careful interpretation of the kinetic and photochemical data available e.g., when nitrogen dioxide is irradiated by near ultraviolent radiation, molecular oxygen and nitric oxide are formed. Some constants for atmospheric reactions are shown in Table 12.1.

Table 12.1. Some constants for atmospheric reaction.

Reactions	Å	E, kcal
$2NO + O_2 \rightarrow 2NO_2$	8.0×10^9	0 or negative
$NO + O_3 \rightarrow NO_2 + O_2$	8.0×10^{11}	2.5
$2NO_2 + O_3 \rightarrow N_2O_6 + O_2$	5.9×10^{12}	7.0
$N_3 + H_3C \rightarrow CH_2 + Products$	3.5×10^5	0.0
$O_3 + H_3C (CH_2)_3 CH \rightarrow Products$	5.6×10^4	0.0
$O_3 + HC = CH \rightarrow Products$	3.0×10^7	4.8
	4.5×10^{13}	—
$CH_3 + CH_3 \rightarrow C_2H_6$	8.0×10^{11}	0.0
$CH_3 + NO \rightarrow Products$	10^{10}	0.0
$CH_3 + O_2 \rightarrow CH_3O_2$	—	0.0

All values have the dimensions cm^3 $mole^1$ s^1 except for the reaction of nitric oxide with oxygen which is third-order and has the dimensions cm^6 $mole^{-2}$ s^{-1}.

Photochemically active radiation in sunlight, near the earth's surface, varies in wavelength from about 2,900 to 7,000 Å. Substances which absorb radiation in this region can serve as primary photochemical reactants for photosensitisers, which function by transferring the absorbed energy to potentially reactant molecules. If a substance is present in small concentrations, it must have a high specific absorption in the 2,900 to 7,000 Å wavelength region if it is to initiate photochemical reactions of any importance. If a substance is present in large concentrations, it can absorb relatively weakly and still initiate reactions of considerable importance to the properties of polluted atmospheres. Many reactions can be postulated which might occur among the constituents of contaminated atmospheres.

If the kinetics of a postulated reaction has been studied, the extent to which the reaction would occur can usually be predicted. If the reaction has not been investigated, considerable information regarding the possibility that the reaction would occur to an appreciable extent can often be obtained by thermodynamic considerations. The free energies and heats of formation of a large number of substances are known or can be estimated. These values can be used to calculate the change in free energy and in heat content accompanying postulated reactions. The free energies and heats of formation of a number of substances which may occur in contaminated atmospheres are shown in Table 12.2.

Table 12.2. Some heat and free energies of formation (values in kcal at 25°C. All substances listed in gaseous state).

Substance	H, kcal	F, kcal
$O(3P_2)$	59.1	55.02
$O(D_2)$	104.3	—
$O_2(L+)$	37.40	—
O_3	34.50	39.40
$H(2S\ 1/2)$	51.9	48.35
HO	−5.93	−4.80
H_2O	−57.8	−34.64
H_2O_2	−33.60	−24.73
S	66.3	56.60
S_8	−5.3	−H
SO_2	−70.90	−71.74
SO_3	−93.90	−88.0
$N(3S\ 1/2)$	85.10	81.00
N_2O	19.65	24.93
NO	21.60	20.66
NO_3	8.00	12.27
N_2O_4	3.06	23.44
N_2O_5	−0.60	26.44
CH_4	−18.24	−12.20
C_2H_4	11.00	—
C_4H_6	−20.96	−10.70
HCHO	−28.70	−24.90
HCO_2H	88.65	−85.1
CO	−26.84	−33.01
CO_2	−04.95	−34.0

A number of reactions which might produce oxygen in the atmosphere are shown in Table 12.3. Oxygen atoms reacting with molecular oxygen produce ozone. Table 12.3 also shows the maximum wavelengths which would supply the energy required by the endothermic reactions. It was assumed in calculating these wavelengths that each reaction as written absorbs one quantum of the radiation.

The thermal reaction of oxygen to form ozone can be used to illustrate the application of such·data. Table 12.2. shows that this reaction would occur with an increase in free energy of 39.4 kcal/mole. The equilibrium constant K for the reaction can be calculated from the equation

$$F = RT \ln K$$

K is found to be about 10^{-29}. Then,

$$(O_3)/(O_2)^{3/2} = 10^{-29}$$

where (O_3) and (O_2) are the concentrations of ozone and oxygen expressed in atmospheres. Substituting 0.2 for (O_2), the equilibrium concentration of ozone from the thermal reaction is found to be about

10^{-30} atm or 10^{-24} ppm. Thus, the thermal reaction would produce a negligible concentration of ozone in the atmosphere.

With the exception of the decomposition of oxygen, all the reactions listed in Table 12.1 from an energetic standpoint alone, proceed towards the right with the light available at the earth's surface. Some could even go in the dark, though this does not mean that they actually do. First, the energy required for the initial intermolecular process (the activation energy) is not taken into account. Second, in the case of those reactions requiring light, the ability of one of the reactants to absorb solar radiation is not indicated. It is for this reason that only a few reactions were included in Table 12.3. The list could be expanded indefinitely.

Table 12.3. Some reactions and their energy deficiencies.

Reaction	H, kcal	Wavelength Å
$O_2 \rightarrow O + O$	+118.2	2,420
$O + O_2 + M \rightarrow O_3 + M$	−24.6	Dark
$SO_2 + O_2 \rightarrow SO_3 + O$	+46.1	6,197
$H_2S + O_2 \rightarrow H_2O + S + O$	+72.4	3,919
$O_2 \rightarrow O_2$	+37.4	7,638
$NO + O_2 \rightarrow NO_2 + O$	+45.5	6,279
$NO_2 + O_2 \rightarrow NO + O_3$	+48.1	5,940
$CH_4 + O_2 \rightarrow CH_3OH + O$	+28.9	9,885
$C_2H_6 + O_2 \rightarrow C_2H_4 + H_2O + O$	+33.3	8,579
$HCHO + O_2 \rightarrow HCO_2H + O$	−0.8	Dark
$CO + O_2 \rightarrow CO_2 + O$	−8.5	Dark

SMOG-FORMING AUTOMOTIVE EMISSIONS

Internal combustion engines used in automobiles and trucks produce reactive hydrocarbons and nitrogen oxides, two of the three key ingredients required for smog to form. At the high temperature and pressure conditions in an internal combustion engine, products of incompletely burned gasoline undergo chemical reactions which produce several hundred different hydrocarbons. Many of these are highly reactive in forming photochemical smog. As shown in Fig. 12.1, the automobile has several potential sources of hydrocarbon emissions other than the exhaust. The first of these to be controlled was the crankcase mist of hydrocarbons composed of lubricating oil and "blowby". The latter consists of exhaust gas and unoxidised carburetted mixture that enters the crankcase from the combustion chambers around the pistons. This mist is destroyed by recirculating it through the engine intake manifold by way of the positive crankcase ventilation (PCV) valve.

A second major source of automotive hydrocarbon emissions is the fuel system, from which hydrocarbons are emitted through fuel tank and carburettor vents. When the engine is shut off and the engine heat warms up the fuel system, gasoline may be evaporated and emitted to the atmosphere. In addition, heating during the daytime and cooling at night causes the fuel tank to breathe and emit gasoline fumes. Such emissions are reduced by fuel formulated to reduce volatility. Automobiles are equipped with canisters of carbon which collect evaporated fuel from the fuel tank and fuel system, to be purged and burned when the engine is operating.

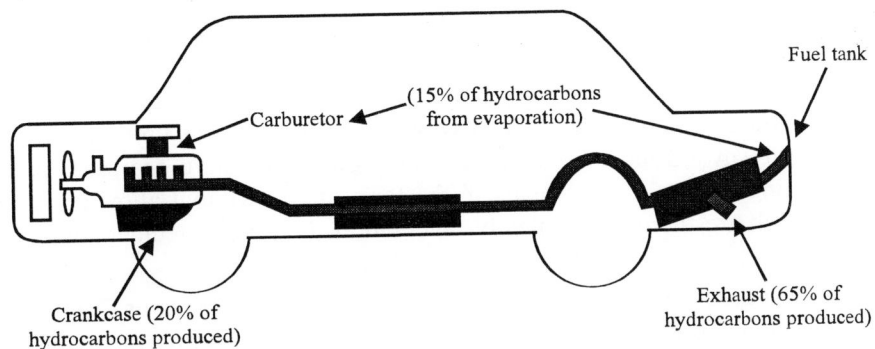

Fig. 12.1. Potential sources of pollutant hydrocarbons from an automobile without pollution control devices.

Control of Exhaust Hydrocarbons

In order to understand the production and control of automotive hydrocarbon exhaust products, it is helpful to understand the basic principles of the internal combustion engine. As shown in Fig. 12.2, the four steps involved in one complete cycle of the four-cycle engine used in most vehicles are the following :

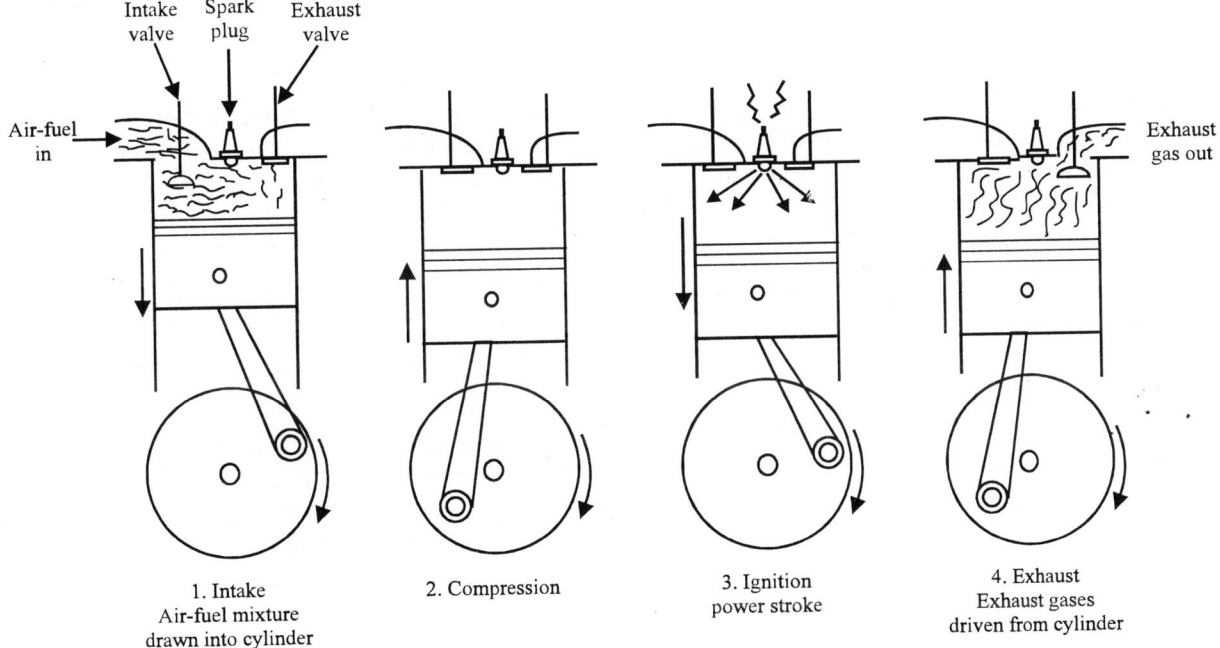

Fig. 12.2. Steps in one complete cycle of a four-cycle internal combustion engine.

1. *Intake:* An air-gasoline mixture produced in the carburettor (or air, followed by fuel injection, in the case of a fuel-injection engine) is drawn into the cylinder through the open intake valve.

2. *Compression:* The combustible mixture is compressed at a ratio of about 7:1. Higher compression ratios favour thermal efficiency and complete combustion of hydrocarbons. However, higher

temperatures, premature combustion ("pinging"), and high production of nitrogen oxides also result from higher combustion ratios.

3. *Ignition and power stroke:* As the fuel-air mixture is ignited by the spark plug near top-dead-centre, a temperature of about 2,500°C is reached very rapidly at pressures up to 40 atm. As the gas volume increases with downward movement of the piston, the temperature decreases in a few milliseconds. This rapid cooling "freezes" nitric oxide in the form of NO without allowing it time to dissociate to N_2 and O_2, which are thermodynamically favoured at the normal temperatures and pressures of the atmosphere.

4. *Exhaust:* Exhaust gases consisting largely of N_2 and CO_2, with traces of CO, NO, hydrocarbons, and O_2, are pushed out through the open exhaust valve, thus completing the cycle.

The primary cause of unburned hydrocarbons in the engine cylinder is wall quench, wherein the relatively cool wall in the combustion chamber of the internal combustion engine causes the flame to be extinguished within several thousandths of a centimetre from the wall. Part of the remaining hydrocarbons may be retained as residual gas in the cylinder, and part may be oxidised in the exhaust system. The remainder is emitted to the atmosphere as pollutant hydrocarbons. Engine misfire due to improper adjustment and deceleration greatly increases the emission of hydrocarbons. Turbine engines are not subject to the wall quench phenomenon because their surfaces are always hot.

Several engine design characteristics favour lower exhaust hydrocarbon emissions. Wall quench, which is mentioned above, is diminished by design that decreases the combustion chamber surface/volume ratio through reduction of compression ratio, more nearly spherical combustion chamber shape, increased displacement per engine cylinder, and increased ratio of stroke relative to bore.

Spark retard also reduces exhaust hydrocarbon emissions. For optimum engine power and economy, the spark should be set to fire appreciably before the piston reaches the top of the compression stroke and begins the power stroke. Retarding the spark to a point closer to top-dead-centre reduces the hydrocarbon emissions markedly. One reason for this reduction is that the effective surface/volume ratio of the combustion chamber is reduced, thus cutting down on wall quench. Second, when the spark is retarded, the combustion products are purged from the cylinders sooner after combustion. Therefore, the exhaust gas is hotter, and reactions consuming hydrocarbons are promoted in the exhaust system.

As shown in Fig. 12.3, the air/fuel ratio in the internal combustion engine has a marked effect upon the emission of hydrocarbons.

As the air/fuel ratio becomes richer in fuel than the stoichiometric ratio, the emission of hydrocarbons increases significantly. There is a moderate decrease in hydrocarbon emissions when the mixture becomes appreciably leaner in fuel than the stoichiometric ratio requires. The lowest level of hydrocarbon emissions occurs at an air/fuel ratio somewhat leaner in fuel than the stoichiometric ratio. This behavior is the result of a combination of factors, including minimum quench layer thickness at an air/fuel ratio somewhat richer in fuel than the stoichiometric ratio, decreasing hydrocarbon concentration in the quench layer with a leaner mixture, increasing oxygen concentration in the exhaust with a leaner mixture, and a peak exhaust temperature at a ratio slightly leaner in fuel than the stoichiometric ratio. Catalytic converters are now used to destroy pollutants in exhaust gases. A reduction catalyst is employed to reduce NO in the exhaust gas and an oxidation catalyst to oxidise hydrocarbons and CO. A dual catalyst system involves running the engine slightly rich in fuel and passing the exhaust gas first over a reduction catalyst to reduce nitrogen oxides. Air is pumped into the exhaust downstream from this device and the exhaust stream passes through an oxidation catalyst where hydrocarbons and carbon monoxide are oxidised.

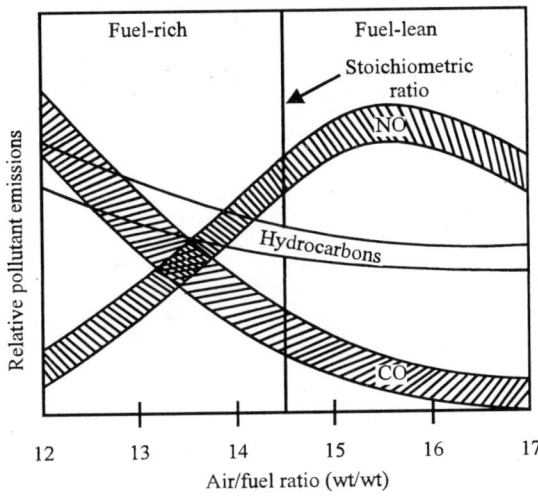

Fig. 12.3. Effects of air/fuel ratio on pollutant emissions from an internal combustion piston engine.

Noble metals (Pd, Pt, Ru) and non-stoichiometric oxides, such as Fe_2O_3 or $CoO·Cr_2O_3$, may be employed as oxidation catalysts. The latter have the property of being able to gain oxygen in their lattice structure, followed by loss of oxygen to the substance being oxidised.

The objective of a reduction catalyst is to reduce NO to harmless N_2. The reduction may be catalysed by noble metals (Pd, Pt, Ru, Rh), base metals (Co, Ni, Cu), or oxides (CuO, $CuCrO_4$). With carbon monoxide as the reducing agent, the two major reactions are :

$$2NO + CO \rightarrow N_2O + CO_2 \qquad ...(12.1)$$

$$N_2O + CO \rightarrow N_2 + CO_2 \qquad ...(12.2)$$

which add up to the following net reaction :

$$2NO + 2CO \rightarrow N_2 + 2CO_2 \qquad ...(12.3)$$

Several undesirable side-reactions, such as the formation of ammonia,

$$2NO + 5H_2 \rightarrow 2H_2O + 2NH_3 \qquad ...(12.4)$$

can occur with a reduction catalyst.

Since lead can poison auto exhaust catalysts, automobiles equipped with catalytic exhaust-control devices require lead-free gasoline, which has become the standard motor fuel.

The Clean Air Act calls for reformulating gasoline by adding more oxygenated compounds to reduce emissions to hydrocarbons and carbon monoxide.

It has been pointed out that much more stringent automotive emission standards for new cars will probably have little benefit in reducing air pollution, and that more effective inspection programmes for vehicles in use should have significant beneficial effects. Some proposed changes in gasoline composition and volatility—reduced vapour pressure and reduced contents of light alkenes and sulphur—will reduce pollution, whereas reduction of aromatic compounds and oxygenated compounds will not have any significant beneficial effects.

SMOG-FORMING REACTIONS OF ORGANIC COMPOUNDS IN THE ATMOSPHERE

Hydrocarbons are eliminated from the atmosphere by a number of chemical and photochemical reactions. These reactions are responsible for the formation of many noxious secondary pollutant products and intermediates from relatively innocuous hydrocarbon precursors. These pollutant products and intermediates make up photochemical smog.

Hydrocarbons and most other organic compounds in the atmosphere are thermodynamically unstable toward oxidation and tend to be oxidised through a series of steps. The oxidation process terminates with formation of CO_2, solid organic particulate matter which settles from the atmosphere, or water-soluble products (for example, acids, aldehydes) which are removed by rain. Inorganic species such as ozone or nitric acid are by-products of these reactions.

Elementary Nature of Photochemical Reactions

A photochemical reaction begins with the act of absorption of radiation. By the Stark-Einstein law of photochemical equivalence, this is a quantum process involving one photon per absorbing molecule. The number of absorbing molecules is, therefore, equal to the number of photons absorbed.

The immediate result of the absorption of a photon is an excited state of the absorbing molecule, which has excess absorbed energy. As most vibrational and rotational spectra lie in the infrared and most electronic spectra in the visible and ultraviolet region, it is the latter region, i.e., the ultraviolet region which is of photochemical importance.

The electronically excited molecules produced by absorption may undergo any of the following changes :

1. They may dissociate.
2. They may react with other molecules on collision.
3. They may internally rearrange or polymerise.
4. They may lose their excitation energy by fluorescence or deactivation and thereby return to their original state.

Any of these possibilities except the last one serve as an initial chemical step or primary process in a photochemical reaction. The three steps involved in an overall photochemical reaction are :

1. Absorption or radiation.
2. Primary reactions.
3. Secondary reactions.

Photochemical Reactions of Methane

Some of the major reactions involved in the oxidation of atmospheric hydrocarbons may be understood by considering the oxidation of methane, the most common and widely dispersed atmospheric hydrocarbon. Like other hydrocarbons, methane reacts with oxygen atoms (generally produced by the photochemical dissociation of NO_2), to generate the all-important hydroxyl radical and an alkyl (methyl) radical

$$CH_4 + O \rightarrow H_3C\cdot + HO\cdot \qquad \qquad ...(12.5)$$

The methyl radical produced reacts rapidly with molecular oxygen to form very reactive peroxyl radicals,

$$H_3C\cdot + O_2 + M \text{ (energy-absorbing third body,}$$

$$\text{usually a molecule of } N_2 \text{ or } O_2) \rightarrow H_3COO\cdot + M \qquad ...(12.6)$$

which with methane as the hydrocarbon is the methoxyl radical, $H_3COO\cdot$. Such radicals participate in a variety of subsequent chain reactions, including those leading to smog formation. The hydroxyl radical reacts rapidly with hydrocarbons to form reactive hydrocarbon radicals,

$$CH_4 + HO\cdot \rightarrow H_3C\cdot + H_2O \qquad ...(12.7)$$

in this case, the methyl radical, $H_3C\cdot$. The following are more reactions involved in the overall oxidation of methane :

$$H_3COO\cdot + NO \rightarrow H_3CO\cdot + NO_2 \qquad ...(12.8)$$

(This is very important kind of reaction in smog formation because the oxidation of NO by peroxyl radicals is the predominant means of regenerating NO_2 in the atmosphere, after it has been photochemically dissociated to NO).

$$H_3CO\cdot + O_3 \rightarrow \text{various products} \qquad ...(12.9)$$

$$H_3CO\cdot + O_2 \rightarrow CH_2O + HOO\cdot \qquad ...(12.10)$$

$$H_3COO\cdot + NO_2 + M \rightarrow CH_3OONO_2 + M \qquad ...(12.11)$$

(The species CH_3OONO_2 is peroxyacetyl nitrate, PAN, a very strong oxidant).

$$H_2CO + h\nu \rightarrow \text{photodissociation products} \qquad ...(12.12)$$

As will be seen throughout this chapter, hydroxyl radical, $HO\cdot$, and hydroperoxyl radical, $HOO\cdot$, are ubiquitous intermediates in photochemical chainreaction processes. These two species are known collectively as odd hydrogen radicals.

Equations such as (12.5) and (12.7) are abstraction reactions involving the removal of an atom, usually hydrogen, by reaction with an active species. Addition reactions of organic compounds are also common. Typically, hydroxyl radical reacts with an alkene such as propylene to form another reactive free radical :

$$...(12.13)$$

Ozone adds to unsaturated compounds to form reactive ozonides :

$$...(12.14)$$

Organic compounds (in the troposphere, almost exclusively carbonyls) can undergo primary photochemical reactions resulting in the direct formation of free radicals. By far the most important of these is the photochemical dissociation of aldehydes :

$$H_3C-\underset{\underset{H}{\|}}{\overset{\overset{O}{\|}}{C}} +h\nu \longrightarrow H_3C^\cdot + H\dot{C}O \qquad ...(12.15)$$

Organic free radicals undergo a number of chemical reactions. Hydroxyl radicals may be generated from organic peroxyl reactions such as,

$$H_3C-\underset{\underset{H}{\|}}{\overset{\overset{\overset{\dot{O}}{O}}{|}}{C}}-CH_3 \longrightarrow H_3C-\overset{\overset{O}{\|}}{C}-CH_3 + HO^\cdot \qquad ...(12.16)$$

leaving an aldehyde or ketone. The hydroxyl radical may react with other organic compounds, maintaining the chain reaction. Gas phase reaction chains commonly have many steps. Furthermore, chain-branching reactions take place in which a free radical reacts with an excited molecule, causing it to produce two new radicals. Chain termination may occur in several ways, including reaction of two free radicals,

$$2HO\cdot \rightarrow H_2O_2 \qquad ...(12.17)$$

adduct formation with nitric oxide or nitrogen dioxide (which, because of their odd numbers of electrons, are themselves stable free radicals),

$$HO\cdot + NO_2 + M \rightarrow HNO_3 + M \qquad ...(12.18)$$

or reaction of the radical with a solid particle surface.

Hydrocarbons may undergo heterogeneous reactions on particles in the atmosphere. Dusts composed of metal oxides or charcoal have a catalytic effect upon the oxidation of organic compounds. Metal oxides may enter into photochemical reactions. For example, zinc oxide photosensitised by exposure to light promotes oxidation of organic compounds.

The kinds of reactions just discussed are involved in the formation of photochemical smog in the atmosphere.

Reaction of Ozone with Unsaturated Hydrocarbons

Fairly rapid reaction occurs between ozone and various olefins at concentrations at which they may occur in contaminated atmospheres. The reaction mechanisms are not precisely known, and a variety of products is obtained. However, the initial rates follow a second order rate law :

$$-d(O_2)/dt = k(O_3) \text{ (Olefin)}$$

The final products include formaldehyde, higher aldehydes, and polymers of unknown composition. The reactions may be tentatively summarised.

$$O_3 + \text{olefin} \rightarrow \text{olefin}-O_3 \text{ complex}$$

Olefin-O_3 complex = decomposition fragments, including free radicals. At high concentrations of reactants, a visible aerosol forms but at concentrations which would probably exist in the smog laden air of cities, no aerosol has been experimentally demonstrated.

The reaction of ozone with acetylene has also been investigated. This reaction also follows a second order rate law, but the velocity constant k is very small. The rates of reaction of ozone with benzene and with paraffin hydrocarbons are so small that they could not be determined by the methods employed in these investigations.

Initial second-order rate constants and half lives for the reaction of ozone with various substances are shown in Table 12.4.

Table 12.4. Rate constant and half-life of initial rate of reaction of ozone with various substances.

Substance reacting with ozone	Half-life if reactant and ozone are 0.2 ppm min.	Half-life if reactant and ozone are 1 ppm min.	Rate constant ppm^l min^l at 25°C
Ethylene	1,000	220	0.0045
1-Hexane	330	66	0.0150
Cyclohexene	37	12	0.0870
Gasoline	250	76	0.0130
Acetylene	500,000	24,000	0.00010
Nitric oxide	0.16	0.03	32.00000
Nitrogen dioxide	65.00	13.00	0.07700

It has been noted that the reaction of nitric oxide with ozone is 2,000 times faster than of 1-hexane and ozone. Nitrogen dioxide reacts with ozone about five times faster than 1-hexane. Thus these last two reactions would proceed at the same rate if, for example, the concentrations of nitrogen dioxide and 1-hexane were 0.2 and 1 ppm, respectively.

Photochemical Reaction of Aldehydes, Ketones and Olefins

Simple aldehydes and ketones absorb radiation which begins between 3,000 and 4,000 Å and extends towards shorter wavelengths, usually terminating between 2,300 and 2,500 Å. The absorption coefficients in the solar region are relatively low. The absorption regions for diketones are displaced toward larger wavelengths to an extent which depends upon the proximity of the groups within the molecule. Thus glyoxal and diacetyl absorb up to about 5,000 Å.

The most important primary photochemical reactions of aldehydes and ketones are to break the molecule into two free radicals. For example, acetaldehyde decomposes into methyl and fromyl radicals.

$$CH_3CHO + h\nu \rightarrow \dot{C}H_3 + HC\dot{O}$$

Acetone decomposes into methyl and acetyl radicals :

$$(CH_3)_2CO + h\nu \rightarrow \dot{C}H_3 + CH_3\dot{C}O$$

The free acetyl radical slowly decomposes into free methyl radicals and carbon monoxide :

$$CH_3\dot{C}O \rightarrow \dot{C}H_3 + CO$$

The free radical HCO is remarkably stable. However, it does very slowly decompose :

$$HC\dot{O} \rightarrow H + CO$$

$$H + O_2 \rightarrow H_2O$$

It might react directly with oxygen :

$$HCO + O_2 \rightarrow HO_2 + CO$$

Large numbers of secondary reactions follow the primary photolysis. Most of the studies of the reaction products have been made in the absence of air. Under these conditions, formaldehyde undergoes a chain reaction giving, among other products, carbon monoxide, carbon dioxide, and water. Among the products of the photochemical decomposition of acetaldehyde in the absence of air are methane, carbon dioxide, glyoxal, diacetyl, formalhyde, and an unidentified material of high molecular weight. Acetone forms carbon monoxide, methane, ethane, diacetyl and material of high molecular weight. In the presence of air, the radicals tend to react rapidly with oxygen. Thus, somewhat different compounds are produced. In the presence of air, acetone appears to give, among other products, acetic acid and dimethyl peroxide. Acetaldehyde apparently gives diacetyl peroxide as a major product. Irradiation with sunlight of mixtures of diacetyl and air has been found to produce ozone, although not enough to account for the ozone concentrations sometimes found in contaminated atmospheres. The reactions involved are not precisely known, but the following presumptions are not unreasonable :

$$CH_3COCOCH_3 \rightarrow 2CH_3CO$$

$$CH_3CO + O_2 \rightarrow CH_3COO_2$$

$$CH_3COO_2 + O_2 \rightarrow CH_3COO + O_3$$

Mono-olefins do not absorb solar radiation in the low atmosphere. However, conjugated diolefins absorb near-ultraviolet fairly strongly to produce electronically excited molecules which react with oxygen.

$$Diolefin + h\nu = excited\ diolefin$$

$$Excited\ diolefin + O_2 = free\ radical\ HO_2$$

Oxygen absorbs weakly the many atmospheric bands in the visible region to produce excited oxygen molecules :

$$O_2 + h\nu \rightarrow O_2$$

By analogy to the reaction, one could write down

$$O_2 + RH \rightarrow R + HO_2$$

OVERVIEW OF SMOG FORMATION

This section addresses the conditions that are characteristic of a smoggy atmosphere and the overall processes involved in smog formation. In atmospheres that receive hydrocarbon and NO pollution accompanied by intense sunlight and stagnant air masses, oxidants tend to form. In air-pollution parlance, gross photochemical oxidant is a substance in the atmosphere capable of oxidising iodide ion to elemental iodine. Sometimes other reducing agents are used to measure oxidants. The primary oxidant in the atmosphere is ozone. Other atmospheric oxidants include H_2O_2, organic peroxides (ROOR'), organic hydroperoxides (ROOH), and peroxyacyl nitrates, such as PAN, mentioned in the preceding section.

Nitrogen dioxide, NO_2, is not regarded as a gross photochemical oxidant. However, it is about 15% as efficient as O_3 in oxidising iodide to iodine(0), and a correction is made in measurements for the positive interference of NO_2. Sulphur dioxide is oxidised by O_3 and produces a negative interference, for which a measurement correction must also be made.

The formation of peroxyacetyl nitrate, PAN, was shown in equation 12.11. PAN and related compounds containing the $-C(O)OONO_2$ moiety, such as peroxybenzoyl nitrate (PBN), a powerful eye irritant and lachrymator, are produced photochemically in atmospheres containing alkenes and NO_x. PAN, especially, is a notorious organic oxidant. In addition to PAN and PBN, some other specific organic oxidants that may be important in polluted atmospheres are peroxypropionyl nitrate (PPN); peracetic acid, $CH_3(CO)OOH$; acetylperoxide, $CH_3(CO)OO(CO)CH_3$; butyl hydroperoxide, $CH_3CH_2CH_2CH_2OOH$; and tert-butylhydroperoxide, $(CH_3)_3COOH$.

As shown in Fig. 12.4, smoggy atmospheres show characteristic variations with time of day in levels of NO, NO_2, hydrocarbons, aldehydes and oxidants.

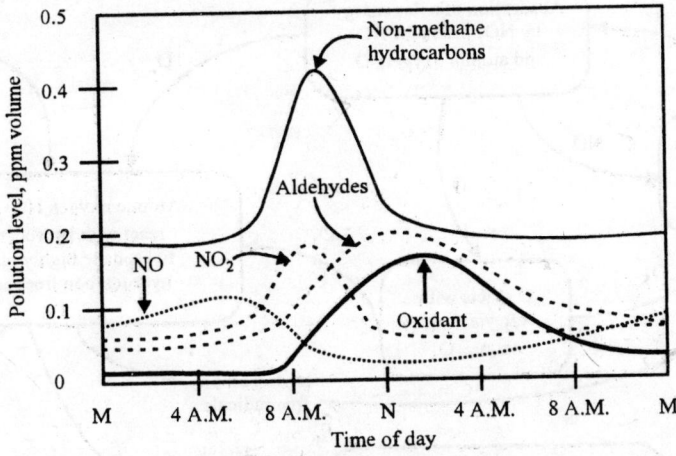

Fig. 12.4. Generalised plot of atmospheric concentrations of species involved in smog formation as a function of time of day.

Examination of the figure shows that, shortly after sunrise, the level of NO in the atmosphere decreases markedly, a decrease that is accompanied by a peak in the concentration of NO_2. During midday (significantly, after the concentration of NO has fallen to a very low level), the levels of aldehydes and oxidants become relatively high. The concentration of total hydrocarbons in the atmosphere peaks sharply in the morning, then decreases during the remaining day-light hours.

An overview of the processes responsible for the behaviour just discussed is summarised in Fig. 12.5. The chemical bases for the processes illustrated in this figure are explained in the following section.

MECHANISMS OF SMOG FORMATION

Some of the primary aspects of photochemical smog formation are discussed here. Since the exact chemistry of photochemical smog formation is very complex, many of the reactions are given as plausible illustrative examples rather than proven mechanisms.

The kind of behaviour summarised in Fig. 12.4 contains several apparent anomalies which puzzled scientists for many years. The first of these was the rapid increase in NO_2 concentration and decrease in

NO concentration under conditions where it was known that photodissociation of NO₂ to O and NO was occurring. Furthermore, it could be shown that the disappearance of olefins and other hydrocarbons was much more rapid than could be explained by their relatively slow reactions with O_3 and O. These anomalies are now explained by chain reactions involving the interconversion of NO and NO₂, the oxidation of hydrocarbons, and the generation of reactive intermediates, particularly hydroxyl radical (HO·).

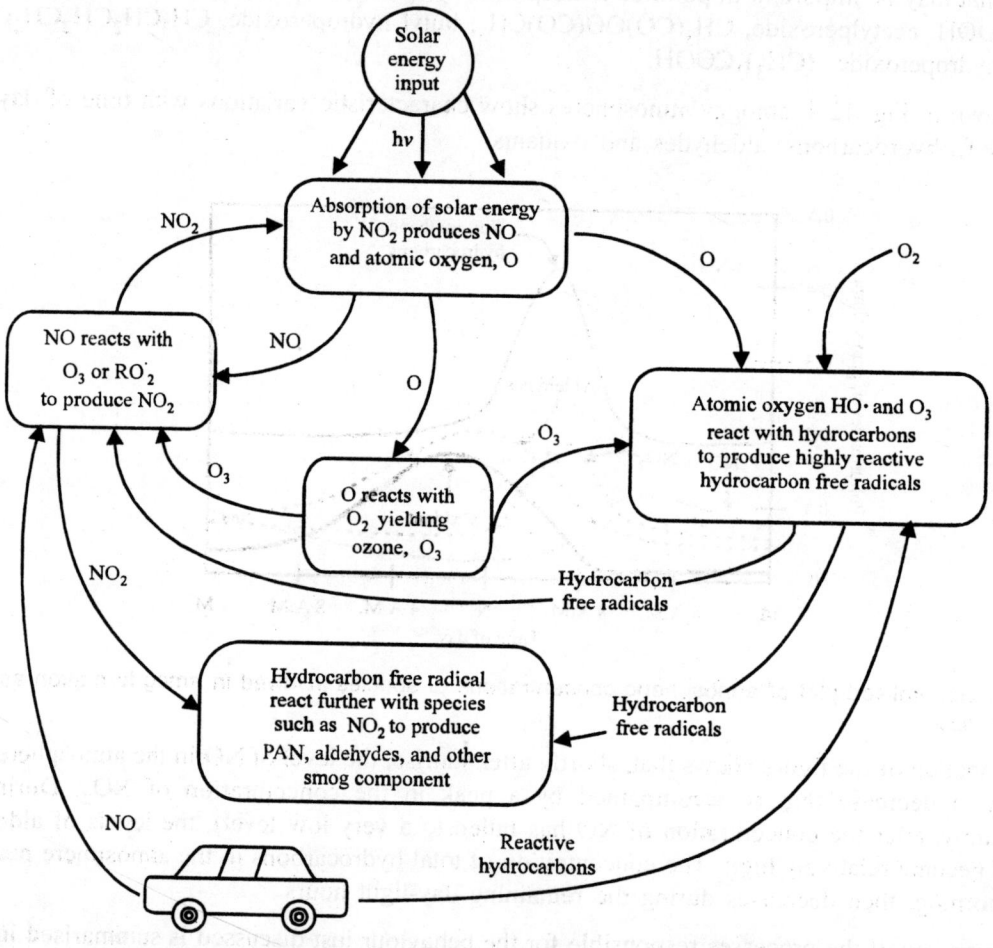

Fig. 12.5. Generalised scheme for the formation of photochemical smog.

Fig. 12.5 shows the overall reaction scheme for smog formation, which is based upon the photochemically initiated reactions that occur in an atmosphere containing nitrogen oxides, reactive hydrocarbons, and oxygen. The time variations in levels of hydrocarbons, ozone, NO, and NO₂ are explained by the following overall reactions :

1. Primary photochemical reaction producing oxygen atoms :

$$NO_2 + h\nu \ (l < 420 \ nm) \rightarrow NO + O$$

...(12.19)

2. Reactions involving oxygen species (M is an energy-absorbing third body) :

$$O_2 + O + M \rightarrow O_3 + M \qquad \qquad ...(12.20)$$

$$O_3 + NO \rightarrow NO_2 + O_2 \qquad \qquad ...(12.21)$$

Because the latter reaction is rapid, the concentration of O_3 remains low until that of NO falls to a low value. Automotive emissions of NO tend to keep O_3 concentrations low along freeways.

3. Production of organic free radicals from hydrocarbons, RH :

$$O + RH \rightarrow R\cdot + \text{other products} \qquad \qquad ...(12.22)$$

$$O_3 + RH \rightarrow R\cdot + \text{and/or other products} \qquad \qquad ...(12.23)$$

(R· is a free radical which may or may not contain oxygen).

4. Chain propagation, braching, and termination by a variety of reactions such as the following :

$$NO + ROO\cdot \rightarrow NO_2 + \text{and/or other products} \qquad \qquad ...(12.24)$$

$$NO_2 + R\cdot \rightarrow \text{products (for example, PAN)} \qquad \qquad ...(12.25)$$

The latter kind of reaction is the most common chain-terminating process in smog because NO_2 is a stable free radical present at high concentrations. Chains may terminate also by reaction of free radicals with NO or by reaction of two R· radicals, although the latter is uncommon because of the relatively low concentrations of radicals compared to molecular species. Chain termination by radical sorption on a particle surface is also possible and may contribute to aerosol particle growth.

A large number of specific reactions are involved in the overall scheme for the formation of photochemical smog. The formation of atomic oxygen by a primary photochemical (equation 12.19) leads to several reactions involving oxygen and nitrogen oxide species :

$$O + O_2 + M \rightarrow O_3 + M \qquad \qquad ...(12.26)$$

$$O + NO + M \rightarrow NO_2 + M \qquad \qquad ...(12.27)$$

$$O + NO_2 \rightarrow NO + O_2 \qquad \qquad ...(12.28)$$

$$O_3 + NO \rightarrow NO_2 + O_2 \qquad \qquad ...(12.29)$$

$$O + NO_2 + M \rightarrow NO_3 + M \qquad \qquad ...(12.30)$$

$$O_3 + NO_2 \rightarrow NO_3 + O_2 \qquad \qquad ...(12.31)$$

There are a number of significant atmospheric reactions involving nitrogen oxides, water, nitrous acid, and nitric acid :

$$NO_3 + NO_2 \rightarrow N_2O_5 \qquad \qquad ...(12.32)$$

$$N_2O_5 \rightarrow NO_3 + NO_2 \qquad \qquad ...(12.33)$$

$$NO_3 + NO \rightarrow 2NO_2 \qquad \qquad ...(12.34)$$

$$N_2O_5 + H_2O \rightarrow 2HNO_3 \qquad \qquad ...(12.35)$$

(This reaction is slow in the gas phase but may be fast on surfaces).

Very reactive HO· radicals can be formed by the reaction of excited atomic oxygen with water,

$$O^* + H_2O \rightarrow 2HO\cdot \qquad \qquad ...(12.36)$$

by photodissociation of hydrogen peroxide,

$$H_2O_2 + h\nu \ (l < 350 \ nm) \rightarrow 2HO\cdot \qquad \qquad ...(12.37)$$

or by the photolysis of nitrous acid,

$$HNO_2 + h\nu \rightarrow HO\cdot + NO \qquad \qquad ...(12.38)$$

Among the inorganic species with which the hydroxyl radical reacts are oxides of nitrogen,

$$HO\cdot + NO_2 \rightarrow HNO_3 \qquad \qquad ...(12.39)$$
$$HO\cdot + NO + M \rightarrow HNO_2 + M \qquad \qquad ...(12.40)$$

and carbon monoxide,

$$CO + HO\cdot + O_2 \rightarrow CO_2 + HOO\cdot \qquad \qquad ...(12.41)$$

The last reaction is significant in that it is responsible for the disappearance of much atmospheric CO and because it produces the hydroperoxyl radical HOO·. One of the major inorganic reactions of the hydroperoxyl radical is the oxidation of NO :

$$HOO\cdot + NO \rightarrow HO\cdot + NO_2 \qquad \qquad ...(12.42)$$

For purely inorganic systems, kinetic calculations and experimental measurements cannot explain the rapid transformation of NO to NO_2 that occurs in an atmosphere undergoing photochemical smog formation and predict that the concentration of NO_2 should remain very low. However, in the presence of reactive hydrocarbons, NO_2 accumulates very rapidly by a reaction process beginning with its photodissociation! It may be concluded, therefore, that the organic compounds form species which react with NO directly, rather than with NO_2.

A number of chain reactions have been shown to result in the general type of species behaviour shown in Fig. 12.4. When aliphatic hydrocarbons, RH, react with O, O_3, or HO· radical,

$$RH + O + O_2 \rightarrow ROO\cdot + HO\cdot \qquad \qquad ...(12.43)$$

$$RH + HO\cdot + O_2 \rightarrow ROO\cdot + H_2O \qquad \qquad ..(12.44)$$

reactive oxygenated organic radicals, ROO· are produced. Alkenes are much more reactive, undergoing reactions with hydroxyl radical,

where R may be one of the number of hydrocarbon moieties or an H atom, with oxygen atoms, or with ozone :

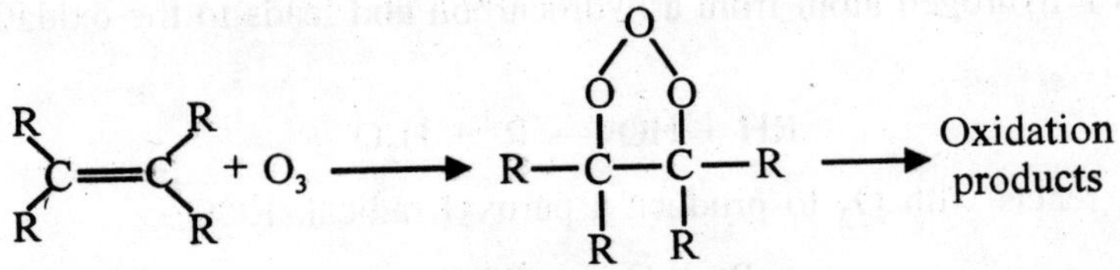

Primary ozonide

...(12.47)

Aromatic hydrocarbons, Ar-H, may also react with O and HO·. Addition reactions of aromatics with HO· are favoured. The product is phenol, as shown by the following reaction sequence :

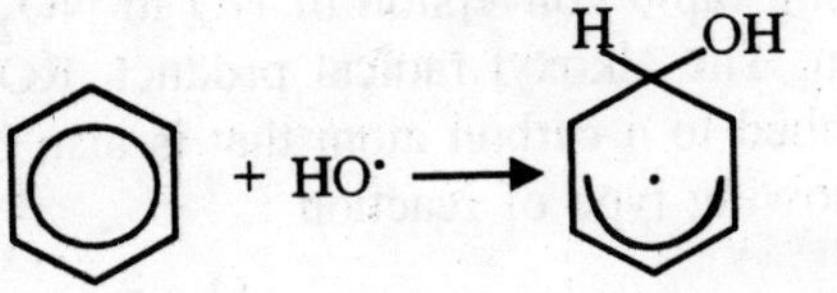

...(12.48)

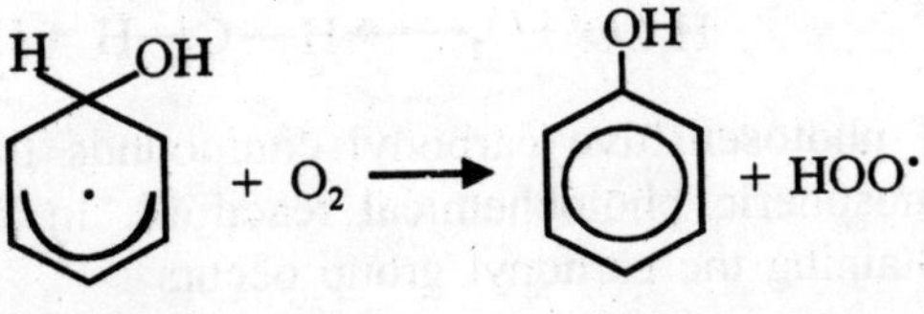

...(12.49)

In the case of alkyl benzenes, such as toluene, the hydroxyl radical attack may occur on the alkyl group, leading to reaction sequences such as those of alkanes.

Aldehydes react with HO·,

$$R-\overset{\overset{\displaystyle O}{\|}}{C}-H +HO^{\cdot}+O_2 \longrightarrow R-\overset{\overset{\displaystyle O}{\|}}{C}-OO^{\cdot} + H_2O \qquad ...(12.50)$$

$$\overset{H}{\underset{H}{>}}C{=}O +HO^{\cdot} +\tfrac{3}{2}O_2 \longrightarrow CO_2 + HOO^{\cdot} + H_2O \qquad ...(12.51)$$

and undergo photochemical reactions :

$$R-\overset{\overset{\displaystyle O}{\|}}{C}-H + h\nu + 2O_2 \longrightarrow ROO^{\cdot} + CO + HOO^{\cdot} \qquad ...(12.52)$$

$$\overset{H}{\underset{H}{>}}C{=}O + h\nu + 2O_2 \longrightarrow CO + 2HOO^{\cdot} \qquad ...(12.53)$$

Hydroxyl radical (HO·), which reacts with some hydrocarbons at rates that are almost diffusion-controlled, is the predominant reactant in early stages of smog formation. Significant contributions are made by hydroperoxyl radical (HOO·) and O_3 after smog formation is well underway.

One of the most important reaction sequences in the smog-formation process begins with the abstraction by HO· of a hydrogen atom from a hydrocarbon and leads to the oxidation of NO to NO_2 as follows :

$$RH + HO· \rightarrow R· + H_2O \qquad \qquad ...(12.54)$$

The alkyl radical, R·, reacts with O_2 to produce a peroxyl radical, ROO· :

$$R· + O_2 \rightarrow ROO· \qquad \qquad ...(12.55)$$

This strongly oxidising species very effectively oxidises NO to NO_2,

$$ROO· + NO \rightarrow RO· + NO_2 \qquad \qquad ...(12.56)$$

thus explaining the once-puzzling rapid conversion of NO to NO_2 in an atmosphere in which the latter is undergoing photodissociation. The alkoxyl radical product, RO·, is not as stable as ROO·. In cases where the oxygen atom is attached to a carbon atom that is also bonded to H, a carbonyl compound is likely to be formed by the following type of reaction :

$$H_3CO· + O_2 \longrightarrow H-\overset{\overset{\textstyle O}{\|}}{C}-H + HOO· \qquad \qquad ...(12.57)$$

The rapid production of photosensitive carbonyl compounds from alkoxyl radicals is an important stimulant for further atmospheric photochemical reactions. In the absence of extractable hydrogen, cleavage of a radical containing the carbonyl group occurs :

$$H_3C-\overset{\overset{\textstyle O}{\|}}{C}-O· \longrightarrow H_3C· + CO_2 \qquad \qquad ...(12.58)$$

Another reaction that can lead to the oxidation of NO is of the following type :

$$R-\overset{\overset{\textstyle O}{\|}}{C}-OO· + NO + O_2 \longrightarrow ROO· + NO_2 + CO_2 \qquad \qquad ...(12.59)$$

Peroxyacyl nitrates (PAN) are highly significant air pollutants formed by an addition reaction with NO_2 ;

$$R-\overset{\overset{\textstyle O}{\|}}{C}-OO· + NO_2 \longrightarrow R-\overset{\overset{\textstyle O}{\|}}{C}-OO-NO_2 \qquad \qquad ...(12.60)$$

When R is the methyl group, the product is peroxyacetyl nitrate. Although it is thermally unstable, peroxyacetyl nitrate does not undergo photolysis rapidly, reacts only slowly with HO· radical, and has a low water solubility. Therefore, the major pathway by which it is lost from the atmosphere is thermal decomposition, the opposite of equation 12.60.

Alkyl nitrates and alkyl nitrites may be formed by the reaction of alkoxyl radicals (RO·) with nitrogen dioxide and nitric oxide, respectively :

$$RO· + NO_2 \rightarrow RONO_2 \qquad \qquad ...(12.61)$$

$$RO· + NO \rightarrow RONO \qquad \qquad ...(12.62)$$

Addition reactions with NO_2 such as these are important in terminating the reaction chains involved in smog formation. Since NO_2 is involved both in the chain-initiation step (equation 12.19) and the chain termination step, only moderate reductions in NO_x emissions alone may not curtail smog formation and in some circumstances may even increase it.

As shown in equation 12.57, the reaction of oxygen with alkoxyl radicals produces hydroperoxyl radical. Peroxyl radicals can react with one another to produce reactive hydrogen peroxide, alkoxyl radicals, and hydroxyl radicals :

$$HOO\cdot + HOO\cdot \rightarrow H_2O_2 + O_2 \qquad ...(12.63)$$
$$HOO\cdot + ROO\cdot \rightarrow RO\cdot + HO\cdot + O_2 \qquad ...(12.64)$$
$$ROO\cdot + ROO\cdot \rightarrow 2RO\cdot + O_2 \qquad ...(12.65)$$

Photolyzable Compounds in the Atmosphere

It may be useful at this time to review the types of compounds capable of undergoing photolysis in the troposphere and thus initiating chain reactions. Under most tropospheric conditions, the most important of these is NO_2:

$$NO_2 + h\nu \ (\lambda < 420 \ nm) \rightarrow NO + O \qquad ...(12.66)$$

In relatively polluted atmospheres, the next most important photodissociation reaction is that of carbonyl compounds, particularly formaldehyde :

$$CH_2O + h\nu \ (\lambda < 335 \ nm) \rightarrow H\cdot + H\dot{C}O \qquad ...(12.67)$$

Hydrogen peroxide photodissociates to produce two hydroxyl radicals :

$$HOOH + h\nu \ (\lambda < 350 \ nm) \rightarrow 2HO\cdot \qquad ...(12.68)$$

Finally, organic peroxides may be formed and subsequently dissociate by the following reactions, starting with a peroxyl radical :

$$H_3COO\cdot + HOO\cdot \rightarrow H_3COOH + O_2 \qquad ...(12.69)$$
$$H_3COOH + h\nu \ (\lambda < 350 \ nm) \rightarrow H_3CO\cdot + HO\cdot \qquad ...(12.70)$$

It should be noted that each of the last three photochemical reactions gives rise to two free radical species per photon absorbed. Ozone undergoes photochemical dissociation to produce excited oxygen atoms at wavelengths less than 315 nm. These atoms may react with H_2O to produce hydroxyl radicals.

Singlet Oxygen

The ground state triplet molecular oxygen, $^3O_2$, can be excited to singlet molecular oxygen, $^1O_2$. Although it was once believed that singlet molecular oxygen was reactive enough with alkenes to play a significant role in photochemical smog formation, current evidence suggests that such is not the case.

REACTIVITY OF HYDROCARBONS

The reactivity of hydrocarbons in the smog formation process is an important consideration in understanding the process and in developing control strategies. It is useful to know which are the most reactive hydrocarbons so that their release can be minimised. Less reactive hydrocarbons, of which propane is a good example, may cause smog formation far downwind from the point of release.

Hydrocarbon reactivity is best based upon the interaction of hydrocarbons with hydroxyl radical. Methane, which is perhaps the least reactive gas-phase hydrocarbon and has an atmospheric half-life exceeding 10 days, is assigned a reactivity of 1.0. (Despite its low reactivity, methane is so abundant in the atmosphere that it accounts for a significant fraction of total hydroxyl radical reactions). In contrast, β-pinene produced by conifer trees and other vegetation, is almost 9,000 times as reactive as methane and d-limonene, produced by orange rind, is almost 19,000 times as reactive. Relative to their rates of reaction with hydroxyl radical, hydrocarbon reactivities may be classified from I through V as shown in Table 12.5.

Table 12.5. Relative reactivities of hydrocarbons and CO with HO· radical.

Reactivity class	Reactivity range	Approximate half-life in the atmosphere	Compounds in increasing order of reactivity
I	< 10	> 10 days	Methane
II	10–100	24 h–10 d	CO, acetylene, ethane
III	100–1000	2.4–24 h	Benzene, propane, n-butane, isopentane, methyl ethyl ketone, 2-methylpentane, toluene, n-propylbenzene, isopropylbenzene, ethene, n-hexane, 3-methylpentane, ethylbenzene
IV	1,000–10,000	15 min–2.4 h	p-xylene, p-ethyltoluene, o-ethyltoluene, o-xylene, methyl isobutyl ketone, m-ethyltoluene, m-xylene, 1,2,3-trimethylbenzene, propene, 1,2,4-trimethylbenzene, 1,3,5-trimethylbezene, cis-2-butene, β-pinene, 1,3-butadiene
V	> 10,000	< 15 min	2-methyl-2-butene, 2,4-dimethyl-2-butene, d-limonene

INORGANIC PRODUCTS FROM SMOG

Two major classes of inorganic products from smog are sulphates and nitrates. Inorganic sulphates and nitrates, along with sulphur and nitrogen oxides, can contribute to acidic precipitation, corrosion, reduced visibility, and adverse health effects.

Although the oxidation of SO_2 to sulphate species is relatively slow in a clean atmosphere, it is much faster under smoggy conditions. During severe photochemical smog conditions, oxidation rates of 5% to 10% per hour may occur, as compared to only a fraction of a per cent per hour under normal atmospheric conditions. Thus, sulphur dioxide exposed to smog can produce very high local concentrations of sulphate, which can aggravate already bad atmospheric conditions.

Several oxidant species in smog can oxidise SO_2. Among the oxidants are O_3, oxidising compounds including NO_3 and N_2O_5, as well as reactive radical species, particularly HO·, HOO·, O, RO·, and ROO·. The two major primary reactions are oxygen transfer,

$$SO_2 + O \text{ (from O, RO·, ROO·)} \rightarrow SO_3 \rightarrow H_2SO_4, \text{ sulphates} \qquad ...(12.71)$$

or addition. As an example of the latter, HO· adds to SO_2 to form a reactive species which can further react with oxygen, nitrogen oxides, or other species to yield sulphates, other sulphur compounds, or compounds of nitrogen :

$$HO· + SO_2 \rightarrow HOSOO· \qquad ...(12.72)$$

The presence of HO· (typically at a level of 3×10^6 radicals/cm^3, but appreciably higher in smoggy atmosphere), makes this a likely route. Addition of SO_2 to RO· or ROO· can yield organic sulphur compounds.

It should be noted that the reaction of H_2S with HO· is quite rapid. As a result, the normal atmospheric half-life of H_2S of about one-half day becomes much shorter in the presence of photochemical smog.

Inorganic nitrates or nitric acid are formed by several reactions in smog. Among the important reactions forming nitric acid are the reaction of N_2O_5 with water (equation 12.35) and the addition of hydroxyl radical to NO_2 (equation 12.39). The oxidation of NO or NO_2 to nitrate species may occur after absorption of gas by an aerosol droplet. Nitric acid formed by these reactions reacts with ammonia in the atmosphere to form ammonium nitrate :

$$NH_3 + HNO_3 \rightarrow NH_4NO_3 \qquad \qquad ...(12.73)$$

Other nitrate salts may also be formed.

Nitric acid and nitrates are among the more damaging end-products of smog. In addition to possible adverse effects on plants and animals, they cause severe corrosion problems. Electrical relay contacts and small springs associated with electrical switches are especially susceptible to damage from nitrate-induced corrosion.

EFFECTS OF SMOG

The harmful effects of smog occur mainly in the areas of (i) human health and comfort; (ii) damage to materials; (iii) effects on the atmosphere; and (iv) toxicity to plants. The exact degree to which exposure to smog affects human health is not known, although substantial adverse effects are suspected. Pungent-smelling, smog-produced ozone is known to be toxic. Ozone at 0.15 ppm causes coughing, wheezing, bronchial constriction, and irritation to the respiratory mucous system in healthy, exercising individuals. Peroxyacyl nitrates and aldehydes found in smog are eye irritants. Materials are adversely affected by some smog components. Rubber has a high affinity for ozone and is cracked and aged by it. Indeed, the cracking of rubber used to be employed as a test for the presence of ozone.

Ozone attacks natural rubber and similar materials by oxidising and breaking double bonds in the polymer according to the following reaction :

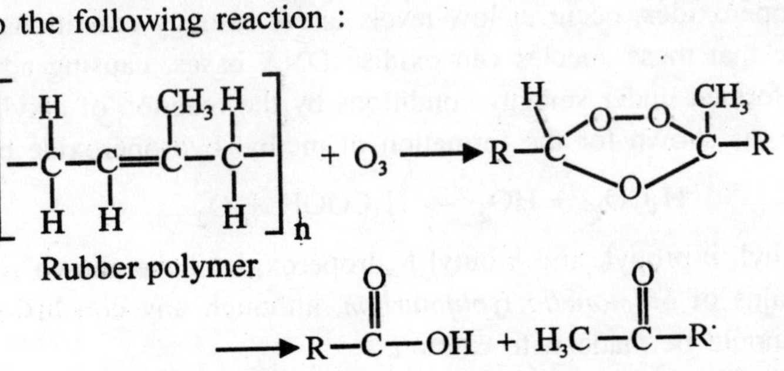

$$...(12.74)$$

This oxidative scission type of reaction causes bonds in the polymer structure to break and results in deterioration of the polymer. Aerosol particles that reduce visibility are formed by the polymerisation of the smaller molecules produced in smog-forming reactions. Since these reactions largely involve the oxidation of hydrocarbons, it is not surprising that oxygen-containing organics make up the bulk of the particulate matter produced from smog. Among the specific kinds of compounds identified in organic smog aerosols are alcohols, aldehydes, ketones, organic acids, esters, and organic nitrates.

Smog aerosols likely form by condensation on existing nuclei rather than by self-nucleation of smog reaction product molecules. In support of this view are electron micrographs of these aerosols showing that smog aerosol particles in the micrometer-size region consist of liquid droplets with an inorganic electron-opaque core (Fig. 12.6). Thus, particulate matter from a source other than smog may have some influence on the formation and properties of smog aerosols.

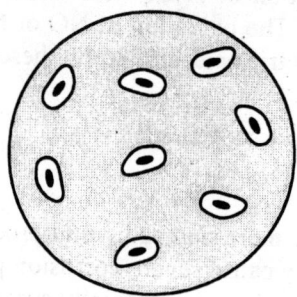

Fig. 12.6. Representation of an electron micrograph of smog aerosol particles collected by a jet inertial impactor, showing electron-opaque nuclei in the centres of the impacted droplets.

In view of worldwide shortages of food, the known harmful effects of smog on plants is of particular concern. These effects are largely due to oxidants in the smoggy atmosphere. The three major oxidants involved are ozone, PAN, and nitrogen oxides. Of these, PAN has the highest toxicity to plants, attacking younger leaves and causing "bronzing" and "glazing" of their surfaces. Exposure for several hours to an atmosphere containing PAN at a level of only 0.02 to 0.05 ppm will damage vegetation. The sulphydryl group of proteins in organisms is susceptible to damage by PAN, which reacts with such groups as both an oxidising agent and an acetylating agent. Fortunately, PAN is usually present at only low levels. Nitrogen oxides occur at relatively high concentrations during smoggy conditions, but their toxicity to plants is relatively low.

Short-chain alkyl hydroperoxides, occur at low levels under smoggy conditions, and even in remote atmospheres. It is possible that these species can oxidise DNA bases, causing adverse genetic effects. Alkyl hydroperoxides are formed under smoggy conditions by the reaction of alkyl peroxy radicals with hydroperoxy radical, $HO_2\cdot$, as shown for the formation of methyl hydroperoxide below :

$$H_3CO_2\cdot + HO_2\cdot \rightarrow H_3COOH + O_2 \qquad ...(12.75)$$

Ames assays of methyl, ethyl, *n*-propyl, and *n*-butyl hydroperoxides have shown some tendency toward mutagenicity on select strains of *Salmonella typhimurium*, although any conclusions drawn from such studies on human health should be made with caution.

The low toxicity of nitrogen oxides and the usually low levels of PAN, hydroperoxides, and other oxidants present in smog leave ozone as the greatest smog-produced threat to plant life. Reduction in plant growth may occur without visible lesions on the plant. Brief exposure to approximately 0.06 ppm of ozone may temporarily cut photosynthesis rates in some plants in half.

EFFECTS OF PHOTOCHEMICAL OXIDANTS ON HUMANS

Both ozone and peroxy acyl nitrate (PAN) cause irritation of the eyes creating lachrymation and affect severely the respiratory tract of human beings. Exposure to 50 ppm of O_3 for several hours will lead to mortality due to pulmonary edema, i.e., accumulation of blood in the lungs.

Primary photochemical pollutant, i.e., NO_2 produces a brownish haze causing nose and eye irritation and pulmonary discomfort. NO_x also cause several chronic diseases of heart, lungs and eyes. Lower concentration of ozone irritates the nose and throat while its higher concentration causes headache, cough, dryness of the throat, chest pain, difficulty in breathing etc. Experimental studies have shown that exposure of ozone upto 0.2 ppm, creates no ill effect but a level of 0.3 ppm appears to be the threshold level at which nose and throat irritation begins. However, exposure to ozone concentrations of 1.0 to 3.0 ppm. for a period of two hours produces extreme fatigue and lack of coordination in central nervous system.

Aromatic hydrocarbons which are conductive to smog formation pose a greater threat than the alicyclic hydrocarbons. Their vapours are much more irritating to the mucous membranes and their inhalation causes systematic injury to the trachea.

Peroxy benzoyl nitrate, PBN, a secondary photochemical pollutant, occurs in polluted atmosphere involving aromatic hydrocarbons, NO_x and ozone. It acts as a powerful eye irritant and lachrymator.

EFFECTS OF PHOTOCHEMICAL OXIDANTS ON PLANTS

PAN and other members of its family (PBN, PPN etc.) are produced photochemically in air having olefins and NO_x. It is very damaging to plants, attacking younger leaves and bringing about "bronzing" and "glazing" of their surfaces. Photochemical smog is characterised by brown hazy fumes which leads to cracking of rubber and extensive damage to plant life. Exposure for several hours to PAN atmosphere at the level 0.02 to 0.05 ppm causes a great loss to vegetation. The sulph hydryl group present in proteins has been susceptible to damage by PAN. It acts both as an oxidising and acetylating agent while reacting with sulph hydryl groups contained in proteins.

Experimental results indicate that ozone is the most toxic photochemical oxidant. Among visible effects of ozone injury to plants are bleached or light flecks or stipples (clusters of dead cells) on the upper surface of leaves. Fully expanded, mature leaves are more susceptible to damage. Leaves tip burn, a disease of white pines is mainly caused by ozone.

Smog which contains O_3, PAN and other photochemical oxidants is regarded to produce early maturity or senescence in plants. A six hour exposure of this smog even at a very low concentration of 0.01 ppm is reported to cause injuries to petunia, lettuce and pinto bean, citrus, forge and salad crops and coniferous trees. PAN causes injury in beets, spinach, celery, pepper, lettuce, alfalfa, aster and primrose etc. It also causes silvering of leaves. NO_x and PAN cause death to forest trees. Actually PAN inhibits "Hill reaction" of photosynthesis and ozone promotes excessive transpiration from the leaves of plants causing dehydration. All these pollutants also destroy the cells of leaves, damage the shoots and interfere with the plant's metabolic processes. Ozone, together with PAN form small drops in air forming smog, thus blanketing the sunshine which inhibits the rate of photosynthesis in plants. Ozone is also harmful for vegetables radish, carrot, tobacco and carnation.

Some sulphates and nitrates which are formed during smog formation due to the oxidation of sulphur containing components (SO_2 and H_2S) and NO_x (N_2O_3, N_2O_5, NO_2), nitric acid and some nitrates are important toxicants of smog. They adversely affect plant growth, damage crops and live stock.

EFFECTS OF PHOTOCHEMICAL OXIDANTS ON MATERIALS

Now scientists believe that degradation of materials commonly attributed to "weathering" is actually the result of attack by photochemical pollutants. The smoke containing fog, dust, mist and soot etc. in the smog reduces the visibility and causes corrosion of metals, stones, building materials, textile, paper, rubber, leather and painted surfaces.

Many organic polymers including rubber, natural and synthetic textiles are subjected to chemical alteration upon exposure to varying quantities of ozone. Ambient air concentrations are frequently high enough to cause these chemical reactions to occur. However, the susceptibility to attack enhances with increasing number of double bonds in olefins. These reactions create carbon chain breaking and carbon-chain cross linking.

Actually long chains of carbon atoms that make up the polymer are broken, and the material becomes fluid like by losing its tensile strength. Even a very low content of ozone results in the formation of new links between parallel carbon-chain, making the material less elastic and more brittle.

Long exposure of ozone to rubber in a relaxed state brings about no characteristic crack. However, exposure to a concentration of 0.01 to 0.02 ppm, ozone causes cracks in rubber if it is under a strain of only 2 to 3%.

So rubber can be protected by applying a coating consisting of rubber and ozone (ozonide) which protects against further penetration of atmospheric ozone. In order to prevent ozone attack and subsequent cracking, anti-ozonant additives are added to the rubber. But they are quite expensive.

Secondary photochemical pollutants also attack cellulose in textile fabrics. The harmful effects of ozone exposure on textiles increase in the order for fabrics made of cotton, nylon, terrycot and polyester. Thus the studies reveal that photochemical-smog problem is by no means unique to Los Angeles. Most metropolitan cities, especially if they have lots of sunshine, are getting periods of smog-alert.

Los Angeles, seems to have run into this problem of civilisation ahead of every body else and more spectacularly. A part of the reason of this critical problem is the urban-sprawl, with its exceedingly enormous number of vehicles, point sources of air pollution. Abundant sunshine and geographical pattern is generally ideal for generating and trapping photochemical pollutants.

Serious out break of smog occurred in Tokyo, New York, Rome and Sydney in 1970, causing spread of diseases as asthma and bronchitis in epidemic form.

Tokyo-Yokohama Asthma, occurred in American soldiers in 1946 living in smoggy atmosphere of Yokohama and Japan. The people also suffered from emphysema, a disease due to structural breakdown of alveoli of lungs. The total surface area available for gaseous exchange is reduced causing chronic breathlessness resulting in death.

CONTROL OF PHOTOCHEMICAL POLLUTANTS

The control of primary precursors, such as NO_x and hydrocarbons will ultimately control ozone and PAN which are secondary pollutants. Today the techniques, like incineration, absorption, adsorption and condensation are devised to control hydrocarbon emissions from stationary sources.

1. *Incineration method:* Hydrocarbon removal efficiencies are greater in flame after burner in which flame is used to complete the oxidation of hydrocarbon into CO_2 and water. The other device makes the use of catalytic after burner in which the catalyst oxidises the hydrocarbon at a lower temperature. The efficiency and fuel cost are lower in this type of incinerator. Catalyst may also cause poisoning during the reaction.

2. *Adsorption method:* In this method, exhaust gases are passed through a bed of granulated adsorber consisting of activated carbon. Hydrocarbon vapours are adsorbed on the surface of activated carbon which remain there until they are periodically removed by passing stream through the system. The hydrocarbons are then condensed to liquids and can be used for further purpose.

3. *Absorption method:* Liquid containing dissolved hydrocarbon is brought intimate contact of exhaust gases. The scrubbed exhaust then passes on, leaving the hydrocarbons trapped in the scrubbing liquid The contact between absorbing liquid and exhaust gases is usually carried in tall scrubbing towers.

4. *Condensation method:* Here the sufficiently low temperature will causes gaseous hydrocarbon i.e. smog forming pollutants to condense to liquids, which are collected and reused.

However, the control of hydrocarbon emissions coming out from automobiles is more complicated. The use of oxidising catalyst would convert CO and hydrocarbons to CO_2 and water while a educing catalyst would convert NO_x to N_2.

Carbon dioxide is the desired carbon-containing end product in the cases given below :

$$\text{Hydrocarbons} \xrightarrow{\text{Combustion}} CO_2 + H_2O \qquad CO \xrightarrow{\text{Combustion}} CO_2$$

New techniques are currently being expanded to convert and utilise smog forming gases and hydrocarbons.

CHAPTER 13

Green House Effect

INTRODUCTION

The term 'green house effect' was first coined by *J. Fourier* in 1827. The effect is also called as 'atmospheric effect', Global warming or 'carbondioxide problem'. Human activities are changing the composition as well as behaviour of the atmosphere at an unprecedented rate. The pollutants from a wide range of human activities are increasing the global atmospheric concentration of certain heat trapping gases, which act like a blanket, trapping heat close to the surface that would otherwise escape through the atmosphere to the outer space. This process is known as green house effect, because it reminds some observers of the heat trapping effect of the glass walls in a horticultural green house.

Thus a green house is that body which allows the short wavelength incoming solar radiation to come in, but does not allow the long wave outgoing terrestrial infra-red radiation to escape.

In a similar way, the earth's atmosphere bottles up the energy of the sun, and is said to act like a 'green house', where CO_2 acts like glass windows. CO_2 and water vapours in the atmosphere transmit short wavelength solar radiation but reflect the longer wavelength heat radiation from warmed surface of the earth. CO_2 molecules are transparent to sunlight but not to the heat radiation. So they trap and re-enforce the solar heat stimulating an effect which is popularly known as green house effect.

The green house effect may therefore be defined as "The progressive warming up of the earth's surface due to blanketing effect of man made CO_2 in the atmosphere". Or, green house effect is "the phenomenon due to which the earth retains heat". Or, green house effect means "the excessive presence of those gases blocked in the infra-red radiation from the earth's surface to the atmosphere leading to an increase in temperature, which in turn would make life difficult on earth".

The four major green house gases, which cause adverse effects are carbon dioxide (CO_2), methane (CH_4), nitrous oxide (N_2O) and chlorofluorocarbons (CFCs). Among these CO_2 is the most common and important green house gas. Here it should also be noted that ozone and SO_2 also act as serious pollutants in causing global warming. The other green house gases such as methane and chlorofluorocarbons contribute about 18% and 14% respectively to the global warming.

All these green house gases are increasing at a rapid rate. Methane, e.g., has approached a level of 1.65 ppm from a pre industrial value of 0.7 ppm and the growth rate now is 1%. N_2O is increasing at the rate of 0.25%, while CFCs at the rate of 5%. It is expected that in a period of about 30 years, i.e., around the year 2020, the non CO_2 effects are going to be one and half times larger than those of CO_2 itself.

HOW THE GREEN HOUSE EFFECT IS PRODUCED

Under normal concentrations of CO_2, the temperature of the earth's surface is maintained by the energy balance of the sun rays that strike the planet and the heat that is radiated back into the outer space. However, when concentration of CO_2 in the atmosphere increases, the thick envelop of this gas prevents the heat from being re-radiated out. The heated earth can re-radiate this absorbed energy as the radiation of longer wavelength.

Thus the thick CO_2 layer acts like the glass panels of a green house or the window glass of a closed car, allowing the sun rays to filter through but preventing, the heat from being escaping in the outer space: thereby warming the troposphere of the atmosphere. This phenomenon can well be understood when one returns to the car after a while whose window glasses are closed and have trapped the heat inside on a warm day. This is what happens in a green house.

Thus green house effect is a phenomenon which is based on the principle of infra-red absorption characteristics of gases. Higher the concentration of CO_2, greater will be the absorption of thermal radiations which meant that more infra-red radiation is trapped and re-emitted back to earth's surface, resulting in a 'heat trap' increasing mean global temperature (Fig. 13.1).

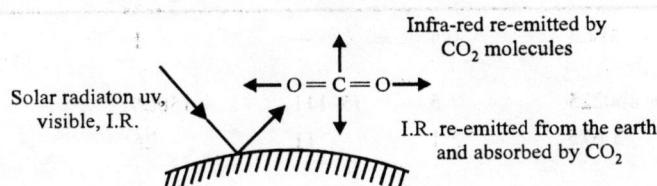

Infra-red re-emitted by CO_2 molecules

Solar radiaton uv, visible, I.R.

$O = C = O$

I.R. re-emitted from the earth and absorbed by CO_2

Fig. 13.1. An illustration of green house effect. I.R. radiation absorbed by CO_2 molecules is re-emitted in all directions deflecting some back to the earth's surface.

Major Sources of Green House Gases

A number of industrial as well as agricultural operations, generate and emit waste gases into the atmosphere. For instance, burning of fossil fuels emit CO_2, growing paddy or live-stock releases methane, the use of aerosols as coolants in refrigerators and air conditioning devices or sprays release chlorofluorocarbons into the atmosphere. These gases create a canopy in the atmosphere and trap the solar radiation reflected back from the earth's surface, leading to atmospheric and climatic changes. In addition to the sources just given, there are many other sources of green house gases. Some of the other sources of green house gases are :

1. A number of factories spread all over the world burn immerse quantity of coal, oil and natural gases and spew huge amount of CO_2, together with another undesirable gases through their chimneys into the atmosphere.
2. Power Stations based on fossil-fuels are significant and wide-spread major sources of man made CO_2.

3. A large fleet of automobiles, railways, aircraft etc. use an immense quantity of diesel and petrol releasing huge amount of CO_2 every year.

4. Burning of fire woods and deforestation are the major sources for the production of CO_2. According to an estimate deforestation has added 90 to 180 billion tonnes of carbon to the atmosphere.

5. An estimate indicates that plants, soils and earth, which are the large storage pools of unoxidised carbon, contain about 2 trillion tonnes of carbon. These trees release carbon as CO_2 after oxidising it.

6. A large scale forest fire either kindled by deliberate or inadvertent actions of man, contributes much to the release of CO_2, favouring green house effect.

7. The reduction in forest cover due to industrial expansion and urbanisation etc. has increased the concentration of CO_2 in the air.

8. Halogenated gases (CFCs etc.) and halons are released to the atmosphere during the operations and maintenance of appliances and equipments using these molecules as coolants and propellants. These gases drastically affect the climatic changes.

Major green house gases and their characteristics are shown in Table 13.1.

Table 13.1. Major green house gases and their characteristics.

Gas	Atmospheric concentration (ppm)	Annual increase (%)	Life span (years)	Relative green house efficiency	Current contribution (%)	Principal sources
Carbon dioxide	351.3	0.4	—	1	57	Coal, oil, natural gas, deforestation
Chlorofluorocarbons	0.000225	5	75–111	15000	25	Foams, coolants
Methane	1.675	1	11	25	12	Wetlands, rice, livestock
Nitrous oxide	0.31	0.2	150	230	6	Fossil fuels deforestation

CORRELATION OF INITIAL RISE IN TEMPERATURE WITH THE INCREASING ATMOSPHERIC CO_2 CONCENTRATION

Among the various emerging environmental issues (like redtides, diesel pollution, nuclear pollution, metal toxicity, new advanced technologies, acid fog and threat to Antarctica etc.) recently UNEP has identified the most vexatious, displeasing and disquieting global warming effect called, "green house effect".

Estimates of rise in temperature with increasing concentration of CO_2 lie in the range from 0.1 to 4.9°C with a mean of about 2–3°C for doubling the CO_2 concentration to 600 ppm. Thus increasing CO_2 levels tend to warm the atmosphere on a global scale.

Since CO_2 content of the atmosphere is increasing day by day at an alarming rate due to various human activities, the global temperature is also increasing gradually. It is expected to increase so drastically that it would reduce the total land area available for human habitation. There are however different opinions, regarding the increasing CO_2 levels and the rise in earth's temperature. According to a study made by computerised models, doubling the CO_2 level will increase the global mean temperature (15°C) by 2°C. According to other views, it will be less than one quarter of degree. UNEP have indicated that a rise of 2°C temperature may disrupt the earth's heat budget and will cause catastrophic consequences. Some researchers believe that changes in the earth's heat budget and will cause catastrophic consequences. Some researchers believe that changes in the earth's mean temperature would be apparent by 2050, when the temperature would rise by 1.5 to 4.5°C.

With the increased CO_2-level, the oceans would be required to absorb and decompose more CO_2 which can raise their normal level of acidity. It would decrease biological productivity of the marine ecosystem, thereby changing the whole oceanic climate.

The impact of green house gases, mainly CO_2, is expected to influence intensely the depletion of ozone layer which would further make the climate hot.

Increasing concentration of CO_2 may increase the total atmospheric pressure. It would broaden the absorption bands and will increase the opacity of the atmosphere to the outgoing longwave terrestrial radiation. It would increase to such an extent that the whole biosphere would come to grinding halt. But this is an extreme situation which is unlikely to occur.

IMPACT OF GREEN HOUSE EFFECT ON GLOBAL CLIMATE

The infra-red absorptive tendency of CO_2 has made possible to treat CO_2 as a key determinant of global temperatures. The CO_2 climatic effects can be studied in two parts :

1. Estimation of the fluctuations in global temperature with the doubling of CO_2 content in the atmosphere.
2. Identification of other climate changes that will accompany global temperature changes.

The green house effect will bring about the following important changes in the climate of the earth :

1. As a result of rise in temperature of the earth due to green house effect the oceans get warm up and sea level would rise flooding low lying regions. A slight increase in sea level could have profound effects on habitation patterns causing many people to move and many of the world's most important cities and ports to come under the threat of floods. In this way, many poor or developing nations may lose large areas of precious coastal land to the rising levels of sea.
2. In temperate regions, the winter will be shorter and warmer and the summer will be longer and hotter. A warmer climate is likely to make some cities extremely hot.
3. There will be enormous increase in rain fall but the problems of desertification, drought and soil erosion will further worsen.
4. The tropics may become wetter and the sub-tropics, which are already dry, are expected to be drier.
5. The rapid increase in industrialisation and urbanisation, coupled with drastic decrease in forest cover, will create a layer of impenetrable gases on the surface of the earth atmosphere converting the planet earth into a hot blast furnace.
6. The plants and animals will also be affected resulting in the disruption of the whole eco-system.
7. The most obvious effect of climate changes will be on agriculture. Because CO_2 is a natural fertiliser, the plants will grow larger and faster with increasing CO_2 in the atmosphere. At first sight, the abnormally fast growth of plants might be expected to be beneficial because the yields of major crops might increase, but with the increase in the yield, the soils may become impoverished or poor more rapidly. The bigger plants with larger yield many cause many complicated problems such as :

(a) Disruption of natural eco-system.

(b) Increase in yield means lower prices to farmers.

(c) Plants will be less rich in nitrogen and hence they are likely to be susceptible to pests.

(d) Soil will become poor or impoverished rapidly. As a result, it will become incapable for yielding good plant growth.

Cause of Fluctuations Occurring in Global Temperature

Actually air pollutants such as CH_4, N_2O, O_3, CFCs, CCl_3F possess intense infra-red absorption which could influence mean global temperature. Aerosols (0.1 to 5.0 μM) and other particulates can absorb and re-emit radiation, the wave length being a function of particle composition. Aerosols could reduce incoming radiation by 10% through back scattering and their increase could possibly lead to the cooling of atmosphere. Dust particles also scatter the out going radiation, although not so effectively. If absorption of sun's rays predominates in the infra-red region corresponding to the earth's emission, the result would be warming and cooling. The rate of increase of atmospheric turbidity has been observed greater than the increasing concentration in CO_2 levels. The most adequate explanation is—that the increase in CO_2 and aerosol content of the air has the potential to fluctuate the world's climate.

Actually dust particles released into the atmosphere become the nuclei for water vapour to condense to form rain. Consequently, the skies become more cloudy and the atmospheric turbidity increases. The more the turbidity, less will be the penetration of solar radiation to the earth. Moreover, the solar radiations are screened directly by dust particles to enter into the atmosphere. It has been reported that about 10% increase in atmospheric turbidity results in the decrease of 0.81% in the total amount of energy absorbed by the atmospheric system which would cause cooling of the climate by about 1°C.

Consequences of green house effect

1. The temperature effect of CO_2 and water vapour combined together have a long range impact on the global climate. With increased level of CO_2, the temperature on the earth's surface rises causing more evaporation of surface water, leading to further temperature increase. It is expected that this combined effect will bring 2–3° rise in surface temperature for doubling of CO_2 concentration around 2080 AD.

2. It has been estimated that if earth's temperature increases from 2.0 to 4.5°C, it will result in receding many glaciers, melting of ice caps in the polar regions found over Greenland and Antarctica and disappearance of deposits of ice on the globe.

3. Because of increased concentration of CO_2 and due to much warmer tropical oceans, there may occur more cyclones and hurricanes and early snow melt in mountains will cause more floods during monsoon.

4. Centre on oceans, within about three decades, rising levels of seas will be able to wash away entire countries and flood cities from Mumbai to Boston. However, little efforts have been done in this direction. If unchecked, it could alter earth's temperature, rainfall and rising levels of the sea.

5. If CO_2 concentration of the air gets doubled, the average temperature of the earth surface would rise to 6.5°F. But it would not happen because of densely accumulated particulate contaminants as smoke, dust and aerosols, which have the ability to reflect some of the sun's thermal radiation into the space.

However, if CO_2 level continues to increase it would accumulate and may inhibit the cooling effect of aerosols and particulate contaminants, consequently temperature of the atmosphere may rise again.

6. Now UNEP has chosen the slogan "Global Warming" : "Global Warning" to alert the public on World Environment Day, June 5, 1989.

7. A slight increase in global temperature can adversely affect the world food production. Biological productivity also decreases due to warming the surface layer. It is, therefore, absolutely necessary that the forest cover should be maintained at least to one third of the total land area of the globe.

8. Now due to population explosion, forest destruction have a serious effect on the levels of atmospheric CO_2. If the present trend of deforestation continues, then the amount of CO_2 in the atmosphere might go up to about 0.0355% by volume.

9. At higher altitudes in the atmosphere, CO_2 undergoes photochemical reactions producing CO, which is drastically dangerous.

$$CO_2 + h\nu \rightarrow CO + O$$
$$(uv)$$

In hot tropical environment, an increase in CO_2 content will influence photosynthetic activities of the plants and consequently plant growth by its direct fertilising effect. However, this adverse fertilising effect should be exploitable by applying modified agricultural practices and technologies and by using superior crop varieties to compensate for the disadvantageous effects of temperature increase.

10. What would be the fate of the future if we do not stop release of CFCs which are responsible for 20% increase in warming. There are enough CFCs which can last upto 120 years.

According to an official estimate the 'green house effect' caused by CFCs would have a far reaching effect or impact on India's 100 crore population living along side the coastline, as it would recede. Small countries like Maldives and other islands might get totally submerged in the sea because of climatic changes. The excessive concentrations of noxious gases like CO_2, SO_2, NO_x, O_3, CH_4 beside CFCs would affect climate stability on the earth.

CONTROL AND REMEDIAL MEASURES OF GREEN HOUSE EFFECT

The green house effect can be controlled by taking the following important measures :

1. Reducing the consumption of fossil fuels such as coal and petroleum. This can be achieved by depending more on non-conventional renewable sources of energy such as wind, solar, nuclear and bio-gas energies.

2. Disposing off the green house gases as they are formed elsewhere than in the atmosphere.

3. Recovering green house gases present already in the atmosphere and disposing off them elsewhere.

4. Learn to adapt and accept the changing climate.

5. International co-operation for attempting the reduction of green house gases.

The following measures may be adopted to reduce the ever increasing green house effect :

1. CO_2-level can be decreased by drastic cut in the consumption of fossil fuels in the highly developed and industrialised countries like UK, USA, Russia, Canada, Japan, France and Germany etc.

   (a) Only the advanced countries should take the responsibilities to decrease the global warming. Not all the countries are advised to implement Toronto Resolutions because then the developing countries will face serious energy-crisis problems.

(b) The developed countries are also cursing India, China and Brazil for global warming due to increase in population density.

2. Advanced researches are necessary to assess and evaluate the potentials of methanol, obtained from methane, which can be efficiently used in transport sector. Methanol may be expected to be a substitute of petroleum.

3. There should be a restriction on the emission of dangerous CO_2 and CFCs, from the factories and automobiles.

4. In tropical and sub-tropical countries, where sunlight shines during most of the period in year, the solar energy may be developed as alternative to the conventional fossil fuels.

5. The developed countries should give sufficient economic aid to the under developed countries to generate solar energy on commercial level.

6. Biogas is the another alternative source of conventional energy for domestic use. The conversion of cow dung from the hearth to the biogas plant will not only provide fuel but will act as a natural fertiliser to the agricultural plants. Thus the developing countries like India, China, Pakistan and Bangladesh can install and operate biogas plants.

7. Enhancing forestation will certainly reduce the CO_2-level, thereby decreasing the 'green house effect'. As trees are the big natural 'sink' of CO_2, they can utilise CO_2 during photosynthesis.

8. Greenary and re-forestation of the landscape is the only immediate solution of every increasing concentration of CO_2 in the atmosphere, so that the green house effect may be decreased.

9. Many researchers have analysed the possibilities of reducing CO_2-emissions at the global level.

10. Flat-plate collector is also able to collect diffused radiation, which makes use of green house effect.

Solar Green Houses and their Future Potential

The solar green house is a classic example of what modern technology should be doing for man; bringing him closer to his environment instead of separating him from it. The solar green house is an encapsulated version of the life processes on this planet, and from this view point man can learn to accept the earth as his solar green house and become more concerned with the quality of life about him.

For a solar green house no more expensive or complicated devices are required. All that needed is a good design and an active and alert technician to regulate the vents, openings and natural energy flow in the dwelling. The exciting net result will hopefully be an acceptable and economical solution to the utilisation of solar energy for space heating, green houses and cold countries residences.

Global Warming: A Serious Threat

By the end of this century, scientists believe, the average global temperature will be higher than ever in the past thousand years. By the end of next century this trend could make the earth 3°–4°C hotter taking the world's climate back to what it was two or three million years ago.

Ironing can reduce global warming

Enriching oceans with iron can help absorb green house gas such as CO_2. The theory is that many parts of the ocean lack iron, which stunts the growth of algae. Enriching enough iron, the algae will bloom luxuriantly soaking the atmospheric CO_2—the chief culprit behind global warming.

Cutting of CO$_2$

Scientists from US have developed a unique strategy to render the carbon dioxide gas (responsible for global warming) emanating from fuel burning plants harmless. It is possible to separate CO$_2$ gas from the smoke of the fossil fuel burning plant and then mix it with certain minerals such as MgO or CaO. In this manner, carbon in the gas CO$_2$ would permanently bind with the minerals keeping it out of the atmosphere. Estimates show that about 5 cents are needed to soak up the CO$_2$ formed in the production of a single kilowatt hour of electricity including every thing from construction of treatment plants to mining and crushing the mixed ores. This is almost equal to 5 cents per kWh to produce the electricity in the first place. However, it is a small price to pay when the polar ice caps or sheets are at stake due to global warming caused by CO$_2$.

Methane on decline

Because of the increase in the atmospheric concentration of methane, a major green house gas is fast levelling off. The annual methane release rates from different sources at global level are shown in Table 13.2.

Table 13.2. Annual methane release rates from different sources at global level.

Source	Methane release rate	
	Amount (g/year)	Per cent
Enteric fermentation	80	14.8
Natural wetlands	115	21.3
Rice paddies	110	20.4
Biomass burning	35	10.2
Termites	40	7.4
Landfills	40	7.4
Oceans	10	1.9
Fresh waters	5	0.9
Methane hydrate desterilisation	5	0.9
Coal mining	35	6.5
Gas drilling	45	8.3
Total	540	100.0

ACID RAINS

Thermal power plants, industries and other sources release thousands of tonnes of oxides of nitrogen and sulphur into the atmosphere everyday. These gases undergo transformation in the atmosphere and form nitrates, sulphates, nitric acid or sulphuric acid droplets. Some of these pollutants, especially the oxides of sulphur can travel 200–300 kms in a day. Thus, the compounds emitted at a place may be carried hundred of kilometres by the wind and deposited on ground or on vegetation directly as 'acid rains'.

$$H_2O + SO_2 \longrightarrow H_2SO_3 + 1/2\ O_2 \longrightarrow H_2SO_4$$

Rain is the purest source of water. 'acid rain' means any precipitation rain, snow or dew, which is more acidic than normal. Acidity of water is measured on a pH-scale, ranging from 0–14 pH. Neutral solutions and freshly prepared distilled water have a pH of 7.0. Acidic solutions have a pH < 7 and alkaline solutions have pH > 7. Lemon juice (citric acid) has a pH of 3. pH scale being logarithmic, an acid of pH 3 is 10 times more acidic than water of pH 5 and 10,000 times more acidic than neutral water. Though rain is the purest water source, it absorbs the atmospheric CO_2 while falling and forms H_2CO_3, the carbonic acid with pH falling to even 5.6. If pH of rain is less than 5.6, it is termed as 'acid rain'. Now-a-days, acid rains with pH < 4.5 are common in many developed countries.

Sources of Acid Rain

The massive emissions of oxides of sulphur and nitrogen—SO_x and NO_x, from thermal power plants and industries react with moisture and other atmospheric constituents to form a cocktail of sulphuric acid and nitric acid. About 70% of the acidity of an acid rain is due to SO_x emissions and 30% due to NO_x emissions. Majority of the fuels, especially coal, contain about 0.5–4% of sulphur and when they are burnt in air and that contains about 80% of nitrogen and nearly 20% of oxygen, huge quantities of SO_x and NO_x are released. The super stack and the mammoth smelters at Sudbury of Ontario release about 2500 tonnes of SO_2 everyday. This largest stack of the world alone contributes about 1% of the total SO_2 released.

SO_2 and NO_x, swept up into the atmosphere can travel thousands of kilometres before being finally deposited as acid rain. For example, SO_2 can remain in the atmosphere upto 40 hours while a sulphate particle for 20 days. It is estimated that 87% of SO_4 in New York and New Jersey and 92% of it in New England has been carried in by long distant transport from the middle west. The origin of acid rains affecting the lakes, rocks and soils of Adirondack in New York is the industrial sector of the Midwest. Here, the rocks are mainly granites and gneisses which have no buffering capacity and hence the acid rains falling on these Adirondack slopes have greater levels of acidity than the rain itself.

Effects of Acid Rain

'Acid rain' is dangerous to man, material and vegetation and can disturb the ecological balance on a global scale. The effects may be summarised as follows :

1. Green algae and many forms of bacteria, which are essential to aquatic life/system are killed due to acidity.

2. High acidity results in the reproductive failure, reduced growth and in the killing of fish. Change in pH prevents hatching of fish eggs and destruction of trout and salmon. Brook trout is the most acid tolerant while rainbow trout the least.

3. At low pH, decomposition of organic matter is less and hence results in the accumulation of organic matter in the water bodies like lakes and streams and hence increases the degree of water pollution. Self purification capacity of water body decreases.

4. Acidity increases the concentrations of heavy metals such as lead, cadmium, copper, zinc, aluminium, chromium and manganese in water. These are highly toxic and hence badly affect the quality of water.

5. Acidity affects germination of seeds. Growth of trees also is adversely affected which results in vanishing of greenery and destruction of forests.

6. Acidity also affects soil by decreasing its fertility. Plant nutrients like potassium are leached out of the soil whereas toxic elements like zinc accumulate. Beneficial micro-organisms are killed or reduced. Earth worms, known as 'farmer's friends' cannot survive in acidic environment.

7. Acid rains corrode buildings, monuments, statues, bridges, fences, railings and art treasures. It is an irreparable loss to mankind.

8. In India, Mumbai and other western parts have experienced acid rains. Taj Mahal, one of the wonders of the world is badly affected by SO_x emissions from Mathura refinery and elsewhere and the damage is named as 'stone cancer' or 'stone leprosy'. Acid rains have already caused severe damage to the environment. About 500 lakes of USA, Canada and Sweden are 'dead' and about 10,000 lakes are fishless. Similarly, 30% of West Germany's forests are dying.

Control of acid rains

The only remedial measure to control acid rains is to control the emissions of the oxides of sulphur and nitrogen from industries and power plants, by using proper control equipments and stingent legislations. Periodic application of lime to neutralise acidity is a solution but is expensive and cannot be applied on a large scale.

Black snow

During the Gulf war of 1991, thousands of tonnes of smoke and dust were released due to the burning of hundreds of oil wells. A portion of these pollutants carried by wind were finally deposited on the shining skin of the Himalayas as a 'black snow'. Hundreds of acres of the Himalayas were covered by a black carpet of smoke and soot. This would not only damage the beauty of the Himalayas but also would lead to higher melting of snow due to the higher absorption of solar radiation by the black surface. The Himalayas are the extremely important components of India and is the main mountain system of Asia, the flora and fauna of which may greatly be affected even when exposed to very low concentrations of air pollutants.

Ozone Holes

Earth has a protective umbrella in the form of ozone layer, of 24 km thickness in the stratosphere about 15 km away from earth's surface. The concentration of ozone in this stratospheric ozone layer is about 10 ppm compared to 0.05 ppm (about 0.0000017%) in the troposphere. Higher ozone concentrations in the troposphere are highly injurious to man and vegetation on earth whereas the stratospheric ozone layer is essential for life to sustain on earth. The ozone layer absorbs the dangerous ultraviolet radiation (especially the uv-B rays with wavelengths from 200 to 280 nm) from the sun and converts it to heat and chemical energy. It is this activity that is responsible for the rise in temperature. This layer is not of uniform thickness. Its profile is shaped like that of earth, being highest at equator and lowest at the poles.

In nature, ozone is continuously formed and destroyed through photochemical interaction and an equilibrium in ozone concentrations is ensured. However, this equilibrium is upset due to the discharge of anthropogenic air pollutants such as CFCs, the chlorofluorocarbons, into the atmosphere. The CFCs release free radicals of chlorine, fluorine or bromine which destroy the stratospheric ozone as a result of which the ozone layer is thinned. The patches of thinned ozone layer are known as "ozone holes". By definition the 'ozone hole' represents only a depletion of ozone concentration but not an empty space in the atmosphere. CFCs are the main pollutants responsible for ozone depletion. Their sources and photochemical activity are mentioned further.

Sources of CFCs

CFCs (chlorofluorocarbons) are a group of synthetic chemicals. These miraculous refrigerants could be traced as the origin of CFCs. The low toxicity, non-inflammability and least chemical reactivity of CFCs made them very popular in being used as refrigerants. CFCs are also used as propellants for dispersing aerosols. The easy handling of CFCs made them popular in almost every field as aero-propellants cleaning solvent, plastic foams, in fast food packaging, in dry cleaning industries, for sterilising surgical instruments, in medicinal and oral inhalation products and for cleaning and degreasing electronic equipments. Thus, the demand of CFCs increased their production at a hefty rate. They are highly corrosive and toxic and hence their use was very much limited.

Photochemistry of ozone depletion

CFCs and halons are highly stable. However, the uv radiation between 1750 and 2200Å present in the stratosphere decomposes them. The chlorine, fluorine or bromine molecules of CFCs and halons are converted into their reactive free radical form by photochemical reactions, as follows :

$$CFCl_3 \xrightarrow{\quad h\nu/sunlight \quad} CFCl_2 + Cl$$

$$CFCl_2 \xrightarrow{\quad h\nu/sunlight \quad} CFCl + Cl$$

$$CF_2Cl_2 \xrightarrow{\quad h\nu/sunlight \quad} CF_2Cl + Cl$$

$$CClF_2 \xrightarrow{\quad h\nu/sunlight \quad} CClF + F$$

These free F or Cl radicals are released during the reaction in the 'ozonosphere'. Chlorine is also ejected into the atmosphere by volcanic eruptions and a fraction of it reaches the ozonosphere. Oxides of nitrogen generally inactivate chlorine but the lowering stratosphere temperature changes NO_x into non-reactive nitric acid. Thus Cl or F are free to react with ozone, disintegrating it into $O_2 + O$. Each atom of chlorine can destroy more than 1,00,000 molecules of ozone catalytically, converting O_3 to O_2.

$$CF_2Cl_2 \xrightarrow{\hspace{3cm}} CF_2Cl + Cl$$

$$Cl + O_3 \xrightarrow{\hspace{3cm}} ClO + O_2$$

$$O_3 \xrightarrow{\quad h\nu \quad} O_2 + O$$

$$ClO + O \xrightarrow{\hspace{3cm}} Cl + O_2$$

This is more efficient catalytically. The NO_x cycle also give the same result.

$$NO + O_3 \xrightarrow{\hspace{3cm}} NO_2 + O_2$$

$$NO + O_2 \xrightarrow{\hspace{3cm}} NO_2 + O$$

$$NO_2 + O_3 \xrightarrow{\hspace{3cm}} NO_3 + O_2$$

$$NO_2 + O \xrightarrow{\hspace{3cm}} NO + O_2$$

The above chain reaction takes about 40 years before the full effect is felt. The tiny ice particles during winter favour the conversion of chlorine into chlorinemonoxide, which behaves as a catalytic compound. The total ozone decreases by about 6.5% during the chain, although natural concentration of ozone is maintained by balancing of O_3-forming and O_3-destroying reactions. But, due to man-made materials, the abundance of chlorine monoxide-rich air in stratosphere continues to rise. This chlorine monoxides reacts with nascent oxygen and free chlorine is formed. Thus the cycle continues destroying the ozone level (Fig. 13.2).

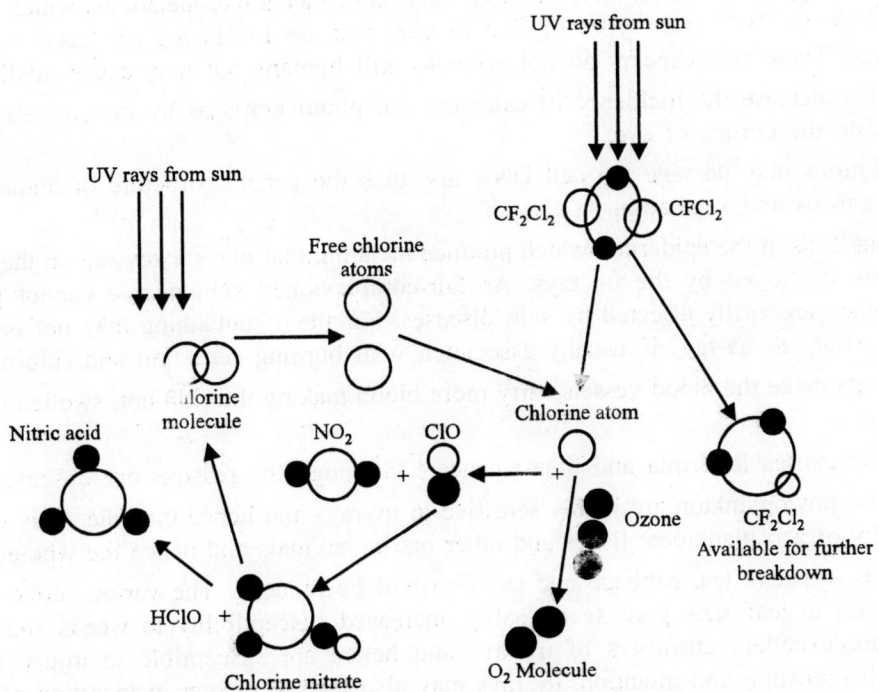

Fig. 13.2. Process of ozone depletion.

Recently it has been realised that unchecked, uncontrolled use of CFCs can spell an ecological disaster for the world in not too distant a future. Unlike other chemicals, CFCs cannot be removed from the atmosphere by the usual scavenging processes like photodissociation, rainfall, oxidation etc. Recently National Aeronautics and Space Administration (NASA) scientists concluded that CFCs and halons are the real culprits in depleting ozone layer.

If this ozone layer is depleted, uv rays can travel through the ozone holes easily. Gaseous pollutants such as oxides of nitrogen and CFCs produced due to human activities are capable of bringing about reactions which cause the decomposition of ozone in the upper atmosphere. Supersonic jets and jumbo jets flying in the troposphere exhaust smoke that floats like thin clouds in the air. The CFCs in such clouds react with sunlight to form smog. This raises the temperature of the atmosphere and also causes the thinning of ozone layer.

Ozone destruction is a function of several parameters. Greatest reductions would occur at distances of above 40 km where the atmosphere is photochemically active. Ozone destructions also depend on the geographical locations. For example, the approximate ozone reductions are 4% in tropics, 9% in the temperate zones and 14% in the polar regions. The appreciable decrease in rainfall level in the British Island and the increasing draughts in the world indicate that ozone depletion and global warming have taken place in a slow but sure way.

Effects of ozone holes

1. The impacts of depleted ozone layer on humans depend mainly on their reaction to uv-β rays. It is expected that every 1% loss in ozone leads to 2% increase in diseases. The most significant effect on human beings would be an increase in various skin cancers like melanoma which is malignant and can cause death. Two other types of skin cancers that are increasing are basal and squamous-cell carcinomas. These two cancers do not normally kill humans but may cause misfigurement.

2. It may also increase the incidence of cataracts and photokeratis as uv rays are easily absorbed by the lens and the cornea of eye.

3. uv-β radiations may damage the cell DNA and thus the genetic structure of humans, animals and other organisms and vegetation.

4. Langerhans cells in the epidermis which produce melanin, that plays a key role in the human immune system, are destroyed by the uv rays. As fair-complexioned skin people cannot produce enough melanin, they are easily affected by skin diseases. For them sunbathing may not be a pleasure any more. Exposure to uv-rays is usually associated with burning sensation and skin aging.

5. uv-radiations make the blood vessels carry more blood making the skin hot, swollen or red and cause sun burns.

6. uv-radiation causes leukemia and breast cancers, although the reasons are obscure.

7. Many micro-phytoplankton are highly sensitive to uv-rays and hence may die. This would affect the productivity of zoo planktons, fishes and other marine animals and hence the whole aquatic system.

8. Crop yields, especial tea, cabbage and soybean will be reduced. The various effects on vegetation are reduction in leaf size, poor seed quality, increased susceptibility to weeds and diseases. Plant proteins are excellent absorbers of uv-rays and hence are susceptible to injury associated with chlorophyll reduction and mutation. uv rays may also lead to greater evaporation of surface waters through the stomata of vegetation and hence may decrease the soil-moisture content.

9. Ozone depletion changes the special composition of solar electromagnetic radiation. The increased solar uv-radiation activates the green house effect affecting the global energy and radiation balance. Formation of hydrogen peroxide, H_2O_2 increases due to which acidity of rain or precipitation increases.

10. It was observed that the uv-β radiations may damage even the inanimate materials.

The ozone layer, if not protected, would enormously affect the productivity and stability of eco-systems and the overall environmental equilibrium.

Control of ozone depletion

1. More than 80% of ozone depletion may be attributed to the large scale release of CFCs into the atmosphere. Hence CFCs must be controlled. There are lot of efforts on earth to decrease the use of CFCs. The 'Montreal Protocol' was a major development in the prevention of the seemingly imminent disaster. There is severe amount of research for alternative technology and substitute chemicals. New techniques have been developed to decrease the leakage of these gases. In Japan success is in the vicinity in finding the alternatives for chlorofluorocarbons. Japanese companies - Mitsubishi Electric and Taiyo Senyo have claimed to have jointly developed an alternative of CFCs.

2. The Satellite Research Institute of Frankfort, Germany has developed a method to use hydrogen as a propellent in aerosol sprays which is environmentally friendly and is a safe alternative to CFCs/butane.

After the Montreal Protocol, 'ozone treats' was held at London in 1990 and it was decided to totally phase out CFCs and halons by 2000 AD. As majority of ozone depletion substances (ODS) are contributed by the developed countries, they should take the responsibility in mitigating this menace. For example, in India, the per capita consumption of CFCs was 8.8 grams in 1990 and at the most may reach 25 g by 2003 AD whereas the Montreal Protocol limit was 300 gram. The developed countries which are the main contributors of ODS should provide funds for research and development of alternatives to CFCs that are safer, cheaper and have zero ODP. However, successful the above research may be, if we want to survive we need to protect the ozone layer by containing ourselves from using CFCs.

CHAPTER 14

Hydrosphere

INTRODUCTION

Hydrosphere covers more than 75 per cent of the earth's surface either as oceans (salt water) or as fresh water. Hydrosphere includes sea, rivers, oceans, lakes, ponds, streams etc. Most of the earth's surface water is in the oceans, which contain about 35 parts per thousand of dissolved salt. Of the remainder, most of fresh water with a salt content of 0.2% is found either in lakes and ponds (still water) or in rivers and streams (running water). Fresh water is also available in the form of rains, snow, dew etc.

Evaporation of water from oceans, cloud formation and precipitation are responsible for worldwide water supply through hydrological cycles. Water is essential to all life. Life was first originated in water. Water is an excellent solvent. It has a high specific heat and this property of high heat capacity of water is functionally important to aquatic organisms.

Water possesses the highest heat of fusion and heat of evaporation, collectively known as latent heat, of all known substances that are liquid at ordinary temperatures. The latent heat of water moderates the temperature of the biosphere. It also plays an important role in the evaporation of water and its condensation as rain and as dew in the hydrological or water cycle.

Water is a poor thermal conductor as compared to metals. It has a high viscosity which allows organisms to swim using relatively simple movements. It also protects the aquatic animals and plants from the mechanical disturbance. The surface tension of water is greatest among all common liquids, except mercury. The very high surface tension is useful in the movement of water through and into the organisms.

No other compound can be compared to water as solvent. It is to be regarded as universal solvent, because more things can be dissolved in water than in any other liquid. The inorganic compounds are mostly soluble in water and also dissociate to form electrically charged particles, called ions. It is probable that all natural elements are soluble in water, at least in trace amounts, and they are all found in natural water at some place or the other on the earth's surface. Many organic compounds are also soluble in water. Thus water is the main medium by which chemical constituents are transported from one part of an ecosystem to the other.

Water is also a buoyant medium. Organisms can exist in it without specialised supportive structures. Organisms living at sea level experience a pressure of about 15 psi (= 1 atmosphere = 760 mm. of Hg).

Pressure increase with increasing depth of water at the rate of one atmosphere for every 10 metres of descent. The solubility, ionic dissociation and surface tension are all influenced by the pressure. Water is slightly compressible with increased pressure.

1. *Salinity of water:* Salinity is the total amount of solid material in gms contained in one kilogram of water, when all the carbonate has been converted into oxide, bromine and iodine replaced by chlorine, and all organic matter completely oxidised.

2. *Natural water:* It contain ions such as Na^+, K^+, Mg^{2+}, Cl^-, SO_4^{2-}, PO_4^{3-}, NO_3^-, CO_3^{2-} and HCO_3^- etc. All these ions are responsible for the salinity of the water. The salinity of fresh water varies from place to place. Salinity of marine water is about 3–3.5%. The salinity of various salt lakes varies from 25% to even 30% or more which greatly restricts life in them.

SOLUBILITY OF GASES IN WATER

Most gases, especially those that are essential for life are soluble in water. Oxygen dissolved in water is used by most living organisms for respiration. Oxygen is a limiting factor for aquatic animals as the saturation concentration of oxygen in water is governed by temperature and salinity. The lower the temperature, the greater the oxygen retaining capacity of water, whether it is fresh water or sea water (saline water).

Respiration, decomposition of organic materials and stream pollution all tend to reduce the amount of available oxygen. The deepest layers of water usually have low oxygen concentration in deeper lakes and ocean areas. This is due to continuous decomposition of organic materials (debris), respiration of organisms and complete absence of photosynthetic activity.

Because oxygen is needed for respiration by all heterotrops, it tends to be taken out of solution continuously by organisms. Oxygen is transferred to deeper water by diffusion or through circulation of water.

Some organisms are independent of oxygen concentration in water, because they may come on the surface and obtain atmospheric oxygen. Examples of such organisms are whales, alligator, crocodiles, seals etc.

Nitrogen is significantly less soluble in water than oxygen. It is chemically inert and does not react with water. Some bacteria, fungi, blue green algae etc. can use nitrogen to satisfy their nitrogen requirements.

Carbon dioxide is produced by the decomposition of organic matter and the respiratory activity of aquatic plants and animals. CO_2 gas is most essential for the photosynthesis of green plants. In the process of photosynthesis, CO_2 combines chemically with water in presence of chlorophyll of green plants and light to form carbohydrates.

CO_2 also combines with water to form carbonic acid (H_2CO_3) which influences the pH or H^+ ion concentration of water. Carbonic acid is an unstable compound and dissociates to produce H^+ ions and HCO_3^- ions. The bicarbonate ions may further dissociate to give more of H^+ ions and carbonate (CO_3^{2-}) ions.

$$CO_2 + H_2O \rightleftharpoons H_2CO_3 \rightleftharpoons H^+ + HCO_3^- \rightleftharpoons 2H^+ + CO_3^{2-}$$

The amount of free CO_2 in water is of great ecological importance, because it governs the precipitation of calcium in the form of $CaCO_3$. The precipitation of $CaCO_3$ takes place when temperature as well as salinity are high and amount of free or uncombined CO_2 is low. This means that more CO_3^{2-} ions are present to combine with the Ca^{2+} ions. These conditions have actually been found to exist in shallow tropical waters, where evaporation is high. It increases the salinity and photosynthetic activity of plants and thus reduces the amount of free CO_2 in water. In deep oceanic water, temperature are low and there are no photosynthetic plants. Thus the CO_2 content of water is high. Deep water fauna, therefore, posses very fragile skeleton because the precipitation of $CaCO_3$ is minimum.

HYDROLOGICAL CYCLES

More than 70% of the human body and approximately 80% of micro-organism are made up of water. Water is thus of great importance, and it is required for all life processes on earth. Water, particularly the movement of water, is also important because it receives much of the pollution that humans generate. In fact, the cyclic movement of water is instrumental to various geological changes on the surface of earth. The above cycle, known as hydrologic cycle, is a responsible and reliable agent for the distribution of nutrients in the environment. But what is more important is that the cycle absorbs huge amount of solar energy and releases them during precipitation—and thus maintains a temperate climate on earth.

The major pathway of the hydrologic cycle is an interchange between the earth's surface and the atmosphere via precipitation and evaporation, the energy for which is derived from the sun. An oversimplified diagram of hydrologic cycle is shown in Fig. 14.1.

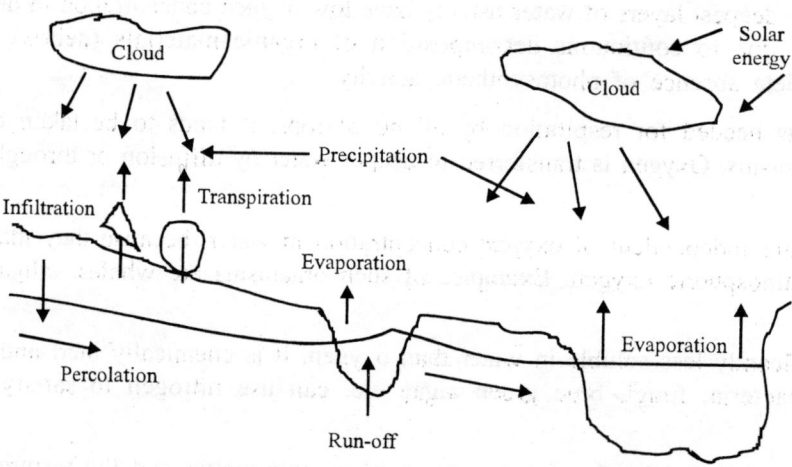

Fig. 14.1. Simplified diagram showing hydrologic cycle.

The total amount of water on earth is estimated to be about 1.5×10^9 km^3, while the total precipitation is about 5.2×10^5 km^3/year. About one quarter of this precipitation occurs over land. Though the total amount of water is quite impressive, it should be appreciated that 97.3% of the above constitutes the water in oceans. Out of the remaining 2.7%, which is considered to be fresh, more than 75% (i.e., 2.1% of the total) is tied up in polar ice caps, glaciers, and as soil and atmospheric moisture. Only 0.62% of the total (i.e., little less than 25% of the fresh water) is available in freshwater lakes, rivers and in ground water supplies. It has been estimated that if the ice-bound water melts completely, it would make a 50 m deep water layer over the entire surface of earth. At present more than 25% of stream run-off are utilised by the humans—the utilisation is likely to go up to about 75% at the end of this century.

The surface run-off is a significant agent in moving nutrients from one place to another. Ground water, which is available below the water table is about 38 times greater in amount than surface water in streams and lakes. The distribution of water on earth, both in respect of time and space, is not uniform and such non-uniform distribution creates inconveniences and disputes.

It has been estimated that about 3×10^{20} kcal of energy per year is absorbed by the hydrologic cycle during evaporation, and are released subsequently for utilisation by biotic and other components that benefit from downstream flow. The amount of energy stored and distributed by the hydrologic cycle is about one fourth of the total solar energy received.

Hydrological Cycle

Water is not locked permanently in the various components of the earth but it is constantly moving through various pathways in the atmosphere, biosphere and lithosphere, thus uniting these components of the ecosphere into a whole. This natural flow of water through various components resulting in the global circulation is called water cycle or hydrological cycle.

Solar energy evaporates the water from oceans, rivers and lakes into the atmosphere where it forms the clouds. The winds transport these clouds to various parts of the earth. The vapours in the clouds condense and precipitate in the form of either dew, rain, snow or hail on the earth. A large part of the precipitation takes place over the oceans themselves, while the remaining precipitates on the land masses. The water falling on the land masses is the potential supply which is determined by the routes by which it is again returned to the atmosphere. Of the water falling on the land surfaces, some is evaporated again and some is returned by surface run-off to drains, streams, rivers and lakes reaching finally to the oceans. The remaining water is infiltered into the soil, and percolates deep into the ground water levels from where a part of it may seep into streams, lakes, or directly to the oceans. Some water on the land is absorbed by plants and consumed by the animals. This water, however, is released again into the atmosphere by respiration and evapotranspiration. The global water cycle operates at a rapid rate with average 10 days residence time of water vapours in the atmosphere. The total global water is regarded to be present in a series of storage tanks interconnected by the transfer processes of evaporation, moisture transport, condensation, precipitation and run-off.

As regard to the freshwater resources on the earth, precipitation on the land masses is critical since, out of total precipitation, about 75% falls directly on the oceans, and only 25% comes to land surface. The distribution of this precipitation on land is highly uneven, therefore, the pattern of natural flow may be an important factor from the resource point of view. In fact, the water management practices are based on the manipulation of hydrological cycle on a local scale.

Human interference with hydrological cycle

Man influences the hydrological cycle in various ways. The impact on the global scale is little because of the inability of man to control the energy distribution on the earth and global climate which are the main forces in governing the hydrological cycle. But man has scored a good success in manipulating the cycle by controlling the processes like run-off, evaporation, precipitation and infiltration of water on a local scale.

The deforestation is sometimes carried out in the catchment or watershed areas of the basins to increase the water yield by augmenting the run-off. But this, unfortunately, decreases the infiltration of water into the soil as there is lack of the obstacles and organic matter which facilitate capturing of the surface moisture. The direct heavy rain on the bare surface may lead to the pulverisation of surface materials which block the soil pores and reduce infiltration.

The agriculture fields where the soil is not covered throughout the year also promote run-off. Deforestation increases the soil water because of the reduction in the consumption of water by the plants. In some areas, where heavy precipitation is associated with steep slopes, floods may occur with the decrease of time lag between precipitation and run-off. The percolation and run-off depends mainly on the soil types. In sandy or permeable soils, though, percolation is more, the upper loose profile of soil is lost with the run-off along with the nutrients. This process is called soil erosion. In the clays or less permeable soils, soil saturation may be caused with development of anaerobic conditions, particularly in the cold and humid regions.

In the areas, where precipitation is low with marked seasonality and variability with higher evaporation rates, the manipulation of water cycle may result in the development of desert conditions. The run-off is increased by the removal of vegetation at the expense of infiltration, the water left in the soil evaporates, and the underground water is tapped beyond the amount of recharge. All this leads to the gradual scarcity of water in the area. As the water table goes down, the wells are further deepened. Ground water becomes non-renewable resource as the withdrawal increases beyond the recharge. Rivers and streams get dried up fast because of the lowering of water table. Continued cropping and grazing increase the risk of wind erosion, and symptoms of drought become pronounced. Eventually, the productive areas turn into deserts.

We are now trying for the change in course of our flowing waters and diverting them to the areas of need, which may cause enormous ecological problems. The creation of dams for storage of flowing waters is also of ecological concern. This increases the natural evaporation of water by increasing the surface area of water. Dams can induce earthquakes as has been reported in several parts of the world. The downstream quality of water is affected and there may be some ecological consequences where the river meets sea. We try to augment the flow of rivers in lean seasons by increased melting of ice covers on the mountains, which may also alter the ecology and micro-climate of the area.

Another important aspect of interference with the hydrological cycle is artificial rain which is carried out by cloud seeding. Our success in this field is limited to only at small scale. If later some effective methods are developed and applied to a large area, it may result in disastrous flooding in one place and serious drought in another.

Water balance on the earth

The hydrological cycle shows that the input of water in any component of the biosphere should be equal to output. On the land surfaces the water balance equation is as follows :

$$r = E + f_w + G$$

where,

r = rainfall (precipitation)

E = evaporation

f_w = surface run-off

G = flow of moisture from the earth's surface to lower layers.

The amount of moisture as G can be divided as the percolation in the soil (f_p) and changes in the moisture content in the upper layers of the lithosphere (soil moisture and aquifers, b). If we combine surface run-off (f_w) and percolation in the soil (f_p) as f, the water balance equation can be written as :

$$r = E + f + b$$

For an average yearly period b is relatively small and can be ignored, hence the water balance equation will become :

$$r = E + f$$

The yearly water balance on various continents and on the whole land is given in Table 14.1.

Table 14.1. Water balance on land masses.

Continent	Precipitation cm/years	Evaporation cm/years	Run-off cm/years
Europe	77	49	28
Asia	63	37	26
Africa	72	58	14
North America	80	47	33
South America	80	94	66
Australia	45	41	4
All land average	160	48.5	31.5

In calculating the water balance on the whole earth, there is no significance of the distribution of water, and the precipitation should essentially be equal to evaporation on the yearly basis.

$$r = E$$

The water balance on the earth is schematically shown in Fig. 14.2.

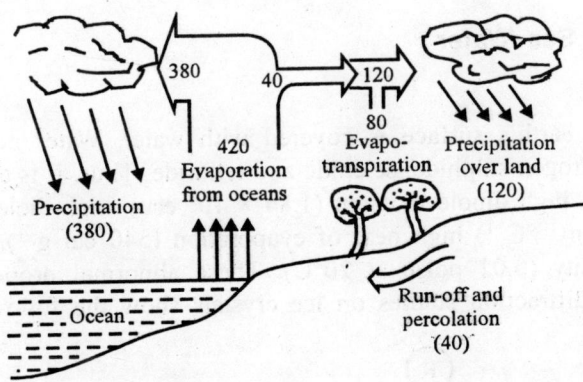

Fig. 14.2. Annual water budget on the earth. (Values are in Tm/Year, Tm = Tera metre cube = 10^{12} m^3).

The total evaporation from the land (80 Tm/year) and oceans (420 Tm/year) is about 500 Tm/year. The total precipitation on the earth is also 500 Tm/year out of which 380 Tm/year falls over the oceans and rest 120 Tm/year, on the land masses. This shows that there is net gain of 40 Tm/year of water on the land masses which comes from the oceans. This extra 40 Tm/year of water is redistributed, and flows to oceans through run-off and percolation.

SEA WATER

The sea contains a great variety of salt and minerals in solution. Of the 35 parts of dissolved salts present in 1000 parts of sea water (i.e. 3.5%), more than 27 parts consist of common salt, the sodium chloride (NaCl) alone. The other main constituents are magnesium chloride ($MgCl_2$), magnesium sulphate ($MgSO_4$), potassium sulphate (K_2SO_4) and calcium carbonate ($CaCO_3$). It has been estimated that the whole sea contains enough salt to cover the continents with a layer 100m thick.

The degree of saltiness of oceans and seas is known as salinity, which is also expressed as the number of parts per thousand of salt dissolved in 1000 parts of water. The salinity varies from place to place and depends upon the following two factors :

1. The rate of evaporation.

2. The amount of fresh water added by rivers and rainfall.

Maximum salinity is found in tropical areas where heat of the sun is greatest and rainfall lowest. At the equator, the temperature is high, but the rate of evaporation is low, because of high humidity and frequent rains. Thus salinity is not found to be the greatest on the equator. In the polar regions, the effects of low temperatures and water obtained from melting of glaciers reduce the salinity. The Arctic ocean has the lowest salinity (about 20%). In Baltic sea the salinity is only about 11 per thousand. This is due to the fact that rate of evaporation is slow and many streams and rivers add fresh water into the sea. The Baltic sea, therefore, freezes rapidly. The average salinity in the Arabian sea is 38%, in Bay of Bengal it is about 30%, while near the mouth of Ganga it is about 20%. In Red sea, the salinity varies from 37% to even 41%. The salinity of the Dead sea is about 20 per thousand, which is probably due to the fact that temperatures in the region are high and hence the rate of evaporation is also high. Moreover, there is very little rainfall and Dead sea receives water from only a few streams. It is therefore, very difficult to swim in Dead sea. You can just float.

Physical Chemistry of Sea Water

Composition

About three-fourth of the earth's surface is covered with water. Water possesses abnormal physical properties compared to hydrogen sulphide, selenide and telluride. Thus, it is a liquid at room temperature with m.p. 0°C, b.p. 100°C, high dipole moment (1.84×10^3 esu), high dielectric constant (80), density (1.0), specific heat (1 cal g^{-1} °C^{-1}) high heat of evaporation (540 cal g^{-1}), surface tension (73 dynes cm^{-1} at 20°C) and viscosity (0.01 poise at 20°C). These abnormal properties of water are due to hydrogen bonding. X-ray diffraction studies on ice crystals show the following structure of water :

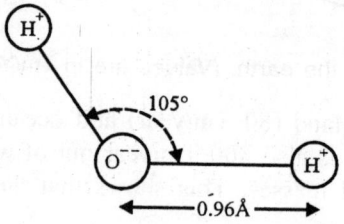

Each water molecule is surrounded by four water molecule neighbours in a tetrahedral arrangement. The seas and oceans are the products of gigantic acid-base titrations during the early stages of formation of the earth. Acids (HCl, H_2SO_4) which leaked out from the interior of the earth, through volcanoes, were titrated with bases liberated by the weathering of primary rock.

The surface water courses, viz. ponds, lakes, streams and rivers, are not linked to one another like the oceans. Thus they retain their separate chemical entities which are not, however, vastly different. While in the former the major anionic constituents are carbonate, sulphate and silicate, chloride is the major species in sea water. Again, in the former the dominating cation is calcium, and sodium in the latter. On the whole, sea water contains about 2000 times more dissolved salts than fresh water.

The average chemical composition (world mean) of river and lake water is expressed in % weight as follows :

CO_3^{2-}	35.2	SO_4^{2-}	12.1	Cl^-	5.7	SiO_2	11.7	NO_3^-	0.9
Ca^{2+}	20.4	Mg^{2+}	3.4	Na^+	5.8	K^+	2.1	$(Fe, Al)_2O_3$	2.7

Chemically speaking, sea water is a solution of 0.5 M NaCl and 0.05 M $MgSO_4$, containing traces of all conceivable matter in the universe. The oceans are the final sink for many substances involved in numerous geochemical processes, as well as the waste dumped as a result of human activities. They receive the run-off from the continents and materials washed from the atmosphere, but are the importance habitat of the bulk of the earth's biosphere. Ocean waters are more or less well mixed with the result that, in contrast to the variable composition of fresh water, the ratios of the major constituents of sea water are almost the same all over the globe although the total salt content (salinity) is variable from place to place. The most abundant elements (except H and O), viz: Na, Cl, Mg, constitute 90% of the matter in sea water, K, Ca and S (SO_4^{2-}), 3%, and the other elements together, 7%. Thus, in the Atlantic ocean, Pacific ocean and Mediterranean sea, Na/Cl = 0.55–0.56; Mg/Cl = 0.06–0.07; K/Cl 0.02, and so on.

The elemental composition of sea water with principal species and residence times is given in order of abundance, in Table 14.2.

Table 14.2. The elemental composition of sea water.

Element	Abundance mg/L (ppm)	Principal species	Residence time (year)
Na	10.5×10^3	Na^+	6×10^3
Mg	1.35×10^3	Mg^{2+}, $MgSO_4$	4.5×10^7
Cs	400	Ca^{2+}, $CaSO_4$	8.0×10^4
K	380	K^+	1.1×10^7
Sr	8	Sr^{2+}, $SrSO_4$	1.1×10^7
Rb	0.12	Rb^+	2.7×10^5
Ba	0.03	Ba^{2+}, $BaSO_4$	8.4×10^4
Fe	0.01	$Fe(OH)_3$ (s)	1.4×10^2
Zn	0.01	Zn^{2+}, $ZnSO_4$	1.8×10^5
Al	0.01	$Al(OH)_3$	1.0×10^2
Mo	0.01	MoO_4^{2-}	5.0×10^5

(Contd...)

Element	Abundance mg/L (ppm)	Principal species	Residence time (year)
U	0.003	$UO_2(CO_3)_3^{4-}$	5.0×10^5
Cu	0.002	Cu^{2+}, $CuSO_4$	5.0×10^4
Ni	0.002	Ni^{2+}, $NiSO_4$	1.8×10^4
Mn	0.002	Mn^{2+}, $MnSO_4$	1.4×10^2
V	0.002	$VO_2(OH)$	1.0×10^4
Ti	0.001	—	1.6×10^2
Co	5×10^4	Co^{2+}, $CoSO_4$	1.8×10^4
Cd	11×10^{-5}	Cd^{2+}, $CdSO_4$	5.8×10^5
Cr	5×10^{-5}	—	3.5×10^2
Ag	4×10^{-3}	$AgCl_2$, $AgCl_3^{2-}$	2.1×10^4
Hg	3×10^{-3}	$Hg_3^-Cl_3$, $HgCl_3^{4-}$	4.2×10^4
Pb	3×10^{-5}	Pb^{2+}, $PbSO_4$	2.0×10^3
Cl	19×10^3	Cl^-	—
S	885	SO_4^{2-}	—
Br	65	Br^-	—
C	28	HCO_3^-, H_2CO_3, CO^{2-}, Organic compounds	—
B	4.6	$B(OH)_3$, $B(OH)_2O^-$	—
F	1.3	F	—
As	0.003	$HAsO_4^{2-}$, H_2AsO_4, H_3AsO_4, H_3AsO_3	—

Sea water equilibrium

The sea water equilibria are extremely complicated and a baffling to our thermodynamic concepts since we have to deal with a system having unconventional parameters, viz. average temperatures 5°C (0°–30°C) and pressure 200 atm (1 atm at the surface to 1000 atm at the bottom). It is estimated that the oceans have undergone about 0.5 million rotations during 500 million years, assuming a rotation period of 1000 years. This means that the liquid phase has been thoroughly mixed. The constant interactions with atmosphere, biosphere and sediment and unique dimensions to the equilibrium processes. It should be noted that

$$pH = 8.1 \pm 0.2$$
$$pE = 12.5 \pm 0.2$$

pH

The constant pH of sea water (8.1 ± 0.2) all over the globe is intriguing for an equilibrium chemist. Three possible explanations are :

1. Buffering action of the H_2CO_3–HCO_3^-–CO_3^{2-} system

$$CO_2 + H_2O \rightleftharpoons H_2CO_3 \rightleftharpoons H^+ + HCO_3^-$$

$$HO_3^- \rightleftharpoons H^+ + CO_3^{2-} \qquad ...(14.1)$$

2. Buffering action of $B(OH)_3$–$B(OH)_4^-$ system.

3. Ion-exchange equilibria of dissolved cations with silicate phase in the marine sediment.

$$3Al_2Si_2O_5(OH)_4 + 4SiO_2 + 2K^+ + 2Ca^{2+} + 9H_2O$$
$$\text{(s)} \qquad \text{(s)}$$
$$\rightleftharpoons 2KCaAi_2Si_5O_{16}(H_2O)_6 + 6H^+ \qquad ...(14.2)$$
$$\text{(s)}$$

The ion-exchange equilibrium is presumably the major buffering factor in the sea/ocean.

pE

pH is defined as the negative log of hydrogen ion activity in aqueous solution: $pH = -\log(a_{H^+})$. Similarly, pE is defined as negative log of the activity of electron in aqueous solution :

$$pE = -\log(a_{e^-}). \qquad ...(14.3)$$

The thermodynamic definition is based on the reaction :

$$2H^+(aq) + 2e^- \rightleftharpoons H_2(g) \qquad ...(14.4)$$

The free energy change for this reaction is exactly zero when all components of the reaction are at unit activity. This reaction is important in the sense that it forms the basis for the fee energies of formation of all ions in aqueous solution and also for defining the free energy changes for redox processes in water.

pE is a little more difficult to visualise than pH. Thus at 25°C in pure water in a medium of zero ionic strength, $a_{H^+} = 10^{-7}$ and $pH = 7.0$. But $a_{e^-} = 1.0$ and $pE = 0.0$ when $H^+(aq)$ at unit activity is in equilibrium with H_2 at 1 atm pressure. If the electron activity is increased ten-fold, i.e., $a_e = 10$, pE will be -1.0.

The relation between pE and electrode potential may now be shown. It must be remembered that pE is not the negative log of the electrode potential, E but the negative log of electron activity. According to the *Nernst* equation for electrode potential :

$$E = E° + \frac{0.0591}{n} \log \frac{[\text{reactants}]}{[\text{products}]} \text{ at } 25°C \qquad ...(14.5)$$

From thermodynamic considerations

$$pE = \frac{E}{0.0591} \text{ at } 25°C \qquad ...(14.6)$$

and the *Nernst* eq. becomes :

$$pE = pE^0 + \frac{1}{n} \log \frac{[\text{reactants}]}{[\text{products}]} \qquad ...(14.7)$$

At 25°C $pE = E/0.0591$ and $pE^0 = E^0/0.0591$ $\qquad ...(14.8)$

Some typical reduction half-reactions with their standard electrode potentials and pE^0 values are cited :

$$Cu^{2+} + 2e^- \rightarrow Cu^0 \qquad E^0 = +0.337 \text{ V}; pE^0 = 5.71 \qquad ...(14.9)$$

$$2H^+ + 2e^- \rightarrow H_2 \qquad E^0 = 0.000 \text{ V}; \qquad pE^0 = 0.00 \qquad ...(14.10)$$

$$Pb^{2+} + 2e^- \rightarrow Pb^0 \qquad E^0 = -0.26 \text{ V}; \qquad pE^0 = -2.13 \qquad ...(14.11)$$

The redox system as above with more + ve E^0 or pE^0 value will proceed more readily than the one with lower value of E^0 or pE^0. Thus Cu^{2+} has a greater tendency to interact with electrons than Pb^{2+} and hence Pb metal gets coated with a layer of Cu when introduced into a solution of Cu^{2+}

$$Cu^{2+} + Pb \rightarrow Cu + Pb^{2+} \qquad ...(14.12)$$

pE values in natural aquatic systems: The limits of pE in water are set by the reactions :

$$2H_2O \rightleftharpoons O_2 + 4H^+ + 4e^- \text{ (oxidation)} \qquad ...(14.13)$$

and

$$2H_2O + 2e^- \rightleftharpoons H_2 + 2OH^- \text{ (reduction)} \qquad ...(14.14)$$

The oxidising limit of the stability of water is defined by the following equation :

$$1/4O_2 + H^+ + e^- \rightarrow 1/2H_2O : pE^0 = +20.75 \qquad ...(14.15)$$

$$pE = pE^0 + \log \ (P_{O_2}^{1/4}[H^+])$$

$$pE = 20.75 - pH \qquad ...(14.16)$$

In neutral water (aerobic condition $P_{O_2} = 0.21$ atm) pE = +13.75.

This may be compared with pE value in anaerobic water in which CH_4 and CO_2 are produced by microorganisms. Assuming pH = 7.0 and $P_{CO_2} = P_{CH_4}$, pE is calculated to be −4.13 and P_{O_2}, 10^{-72} atm.

In sea water pE is determined by the equilibrium with atmospheric oxygen (25°C, zero activity) :

$$1/2O_2(g) + 2H^+ + 2e^- \rightleftharpoons H_2O; \ \log K = 41.55 \qquad ...(14.17)$$

$$\log K = \log \ (H_2O) - 1/2 \log pO_2 - 2 \log \ (H^+) - 2 \log \ (H^+) - 2 \log \ (e^-)$$

$$= -0.01 - 1/2 \log \ (0.21) + 2pH + 2pE$$

$$= -0.01 + 0.34 + 16.4 + 2pE \qquad ...(14.18)$$

Hence pH = 1/2 (25.0) = 12.5 (for sea water)

$$...(14.19)$$

This value of pE of sea water is not very sensitive to the usual small variations of pH and oxygen concentration. Table 14.3. highlights the sinks for water pollutants in the hydrosphere and soil.

OCEAN CURRENTS

The regular movements of water from one part of the ocean to another are called ocean currents. The ocean currents are mainly caused by the difference in density of sea water because of variations in temperature and salinity. The position of the land masses and the shape as well as depth of the ocean basins also influence the ocean currents.

There is a great heat at the equator and less at the poles. Thus sea near the equatorial region is heated more than that near the poles. The unequal heating sets up convection currents in the ocean. The water in the equatorial seas expands and flows down hill towards the poles. The water from the polar regions is cooler. It sinks and flows towards the equator on the sea floor. The convective effect is similar to that which produces currents of air in the atmosphere.

Table 14.3. Sinks for water pollutants in the hydrosphere and soil.

Chemical group	Sources	Major sinks	Examples
Pesticides	Agricultural operations; public health programmes	Photoxidation on surface of soil or water	Dieldrin, 2, 4-D
		Hydrolysis in water ways	DDT
		Oxidation and reduction catalysed by organic and mineral fractions in soils and sediments	Organophosphate pesticides
		Adsorption on particles in soil or suspended in water.	DDT
		Microbially mediated degradation in soils and sediments	Organophosphate and carbamate insecticides
Hydrocarbons	Industrial operations, petroleum spills	Photoxidation on surface of soil or water.	Some petroleum components
		Adsorption on particles in soil or suspended in waterways	Polycyclic aromatic hydrocarbons
		Evaporation	Low boiling point materials
		Microbially mediated degradation	Polycyclic aromatic hydrocarbons, naphthalene
Halogenated hydrocarbons	Industrial operations	Adsorption onto particles in waterways	Polychlorinated biphenyls (PCBs)
Synthetic polymers	Industrial operations	Auto-oxidation on surface of soil or water.	Butyl rubber cellulose nitrate, cellulose acetate, styrene polymers
		Microbially, mediated degradation in sediments and soils	
Fertilisers	Agricultural operations Nitrogen	Chemodenitrification	Nitrate
		Volatilisation from topsoil to air	Ammonium
		Microbially mediated reactions, production of nitrous oxide and nitrogen gas, which may diffuse into atmosphere	Nitrate, ammonium
	Phosphorus	Microbially mediated reduction of phosphate; precipitation from solution in ground and surface waters	Phosphate

The ocean currents, which may be warm or cold, also affect the temperature of the place. Warm currents make the coastal areas warm and raise the temperature, while cold currents make the coastals areas cold and lower the temperature part of coastal Europe are very warm in January as compared with the normal for their latitudes. London on latitude 51°N is quite warm, while New York on latitude 40°N is very cold in spite of the fact that it is nearer the equator. The former is kept warm by the North Atlantic Drift, while the later is affected by the cold labrador current. Similarly, warm ocean currents increase the rain fall in British Columbia in Japan (from the South-East Monsoon) and Queensland (from the North-East Monsoon). The effect of current is more marked when the wind blows from the sea to the land.

CHAPTER 15

Lithosphere

INTRODUCTION

Lithosphere means the mantle of rocks constituting the earth's crust. The earth is a cold, spherical solid planet of the solar system, which spins on its axis and revolves round the sun at a certain constant distance. The solid component of the earth is called *lithosphere*, which is related with edaphic factor and includes mainly soil, earth, rocks, and mountains etc. The lithosphere mainly contains three layers—crust, mantle and outer and inner core. The core is the central fluid or vapourised sphere having diameter of about 2500 km from the centre. Core is probably composed of nickel-iron. The mantle extends about 2900—3000 km above the core and is in the molten state. The outer-most solid zone of the earth is known as crust. It is about 8–40 km above the mantle. The surface of crust is covered with the soil. In so far as environmental chemistry is concerned, the soil is probably the most significant part of the lithosphere.

SOIL

The soil is the upper most part of the earth's crust, and is a mixture of organic as well as weathered rock and materials necessary for the growth of plants. The organic matter and weathered rock materials are formed through the physical, chemical and biological processes occurring slowly and slowly for a long period at earth's surface. Soil is a store house of minerals, a reservoir of water, a conserver of soil fertility, a producer of vegetative crops, a home of wild life and livestock. The food that we eat, the fibres which make our clothes, the material used in making our house or shelter, all are derived from the plants that grow directly in the soil. The soil also supplies nutrients to the plants and is one of the most important ecological factor, called edaphic factor. The science which deals with the study of soil is called soil science, pedology or edaphology.

Besides being the source of nutrients and water to plants, the soil is also the medium for the detritus food chain. A number of soil microbes such as bacteria, algae, fungi, protozoa etc., bound in or on soil particles decompose the nutrients released in detritus, and taken back into the plants through their roots. Soil or mud is also a source of nutrients for all aquatic plants, rooted or submerged or free-floating. Soil is the means of support for all terrestrial organisms. For example, plants are anchored to the soil by their root system. Animals walk on soil and are supported by it. Many animals take shelter under the soil.

Rocks

Soil is formed by different kinds of rocks. Rocks are made of different kinds of minerals. Some important minerals are :

1. *Asbestos:* It is a heat resistant mineral and is used to make fireproof, materials.
2. *Calcite:* This mineral makes up the chalk and limestone.
3. *China clay or kaolin:* It is used in making pottery and fine paper.
4. *Diamond:* It is a form of carbon and is the hardest substance known.
5. *Fluorspar or fluorite:* It is mainly used in the steel industry.
6. *Graphite:* Like diamond, it is another form of carbon and is used in making lead pencils.
7. *Gypsum:* Gypsum is used in the manufacture of chalk and plaster of Paris.
8. *Haematite:* It is deep blood red coloured ore of iron.
9. *Halite:* It is used in home as salt for cooking and in industry to make soda and chlorine.
10. *Quartz:* It is the mineral that makes up sand and gravel. It is used in making electrical instruments.
11. *Talc:* It is also known as soapstone and is the softest mineral. Talcum powder is made from talc.

Kinds of Soils

There are three kinds of soil forming rocks. These are: (i) igneous rocks; (ii) sedimentary rocks; and (iii) metamorphic rocks.

Igneous rocks

Igneous or firelike rocks are formed by cooling of molten magma (lava). Magma is hot. Molten rock from the earth's interior which solidifies as it cools, forming the igneous rock. Lava from volcanoes is nothing but magma that is still hot and flowing. Igneous rocks are also of three types—granite, diorite and basalt. The granite is usually light in colour, coarse to medium grain sized and contains quartz, feldspar, mica, amphibole and iron oxides as the principal minerals. Diorite is grey to dark in colour, coarse to medium grain sized and consists mainly of feldspar, amphibole and iron oxide. Basalt is dark to black in colour, dense to fine grain sized and contains feldspar, pyroxene and iron oxide as the principal minerals.

Sedimentary rocks

Sedimentary or settled rocks are the most common rocks found on the surface of the earth. They are formed by deposition of weathered minerals which are derived from igneous rocks. Tiny particles or sediments are collected by water as it travels through rocks and soil to the sea. Layers of this sediment gradually build up on the sea bed to form rocks such as sandstone and shale. Sedimentary rocks are also of three important types—shales, sandstone and limestone. Shales are light to dark in colour and mostly contain quartz and clay minerals. Sandstone is light to red in colour, granular and porous. They contain quartz, iron oxide, calcium carbonate and clay minerals etc. Limestone is light grey, red brown or black in colour and contains calcite, dolomite, iron oxide and clay minerals etc.

Metamorphic rocks

Metamorphic or changed rocks are formed by change of pre-existing igneous and sedimentary rocks through heat and pressure. As the earth's crust moves, more rocks are changed by heat, pressure or chemical action and become a new kind of rocks called metamorphic rocks. In this way, sedimentary rocks, known as shales become metamorphic rocks, called slate. Similarly, limestone is converted into marble. Metamorphic rocks are usually of five important types—gneiss, schist, slate, quartzite and marble. The *gneiss* is in the form of light and dark bands and contains quartz, feldspar, mica and iron oxide as the principal minerals. *Schist* consists of feldspar, iron oxide, pyroxene etc. *Slate* is grey to black in colour and has compact and uniform texture. It contains quartz and clay minerals. *Quartzite* is light to brown in colour; have compact and uniform texture. It contains quartz, iron oxide, calcium carbonate and clay minerals. *Marble* is light, red, green or black in colour; have compact and fine to coarse texture. It contains calcite, dolomite, iron oxide and clay minerals etc.

It should be noted that :

1. Geologists have classified nearly 3000 minerals.
2. Of the 92 elements (basic chemical substances) that occur naturally in the earth's crust, 22, including gold and silver, are sometimes found in free states. The others occur in combinations with one or more other elements.
3. Sedimentary rocks cover about 75% of the world's land areas.
4. Igneous rocks make up 95% of the top 16 km (10 miles) of the earth's crust.
5. The moon rocks collected by astronauts are igneous in type. There are no sedimentary rocks on the moon, because moon has no atmosphere and so weathering and erosion do not occur.
6. The earth is a cold, spherical, solid planet of the solar system, which spins on its axis and revolves round the sun at a certain constant distance.
7. The solid component of the earth is known as lithosphere.
8. The core is the central fluid or vapourised sphere having diameter of about 2500 km from the centre and it is generally made of nickel-iron.
9. The mantle extends about 2900 km above the core. This is in molten state.
10. The crust is the outermost solid zone of the earth. It is about 8 to 40 km above the mantle.
11. The crust is very complex and its surface is covered with the solid supporting rich and varied biotic communities.
12. The soil is one of the most important ecological factor, called edaphic factor.
13. Soil is the most characteristic feature of terrestrial environment.
14. Soil is loose, friable, unconsolidated top layer of earth's crust.
15. Soil is a mixture of weathered rock materials or minerals and organic detritus, both of which are formed as a result of physical, chemical and biological process taking place slowly and slowly for a long period at the surface of the earth.
16. Soil is not only the source for nutrients and water to the plants, but is also a medium for the detritus food chain. Nutrients released in detritus are decomposed by various microbes such as algae, bacteria, fungi, protozoa etc, bound in or on soil particles.
17. Soil in the form of mud is the main source of nutrients for all aquatic plants (hydrophytes), rooted or submerged or free floating.
18. Soil is also the means of support for all terrestrial organisms. Plants are anchored to the soil by their root systems. Animals walk upon it and are supported by it, as many animals like nematodes, polychaetes, arachnids, insects, rodents etc live under the soil.

PROCESS OF SOIL FORMATION

The important processes from which a matured soil is formed are: (i) physical weathering; (ii) chemical weathering; and (iii) biological weathering. Soil formation is started by weathering or disintegration of parent rocks by some physical, chemical or biological agents. As a result of weathering, soil rocks are broken down in small particles, called regoliths, which under the influence of various other pedogenic processes get converted into mature soil.

Physical Weathering

The physical weathering agents are primarily climatic in character. They exert mechanical effect on the rocks as a result of which fragments are broken down into particles of decreasing particle size (regoliths). Physical weathering does not cause any chemical transformation of rock minerals and commonly occur in deserts, high altitudes, high latitudes and in localities with marked topographic relief and sparse vegetation cover. The temperature, water, ice, gravity, and wind are some of the climatic physical weathering agents. For example, temperature causes break down of those rocks which have heterogeneous structure, because of differential expansion and contraction coefficient of materials composing the rocks. Water causes mechanical weathering of rocks by rain water, torrent water and wave action. In its freezing and ice melting states water causes rock weathering by frost action and glacier formation. Water in the form of frost or ice is an important physical weathering agent of rocks. It seeps into rock crevices, freezes because of sudden decrease in temperature of rocks, expands about 9% of its original volume, exerts a pressure of about 150 tonne/ft^2 and hence cracks the rock into pieces of small particle size.

The landslides and rock slippages caused by earthquakes and faulting are due to gravitation weathering. Here rock is fragmented by abrasion and forces of impact.

The rapid stormy wind carrying suspended particles of sand also causes the abrasion of exposed rock. It serves as a mechanical carrier in moving the particles over the surface of earth as dunes or drifts and in transporting large amounts of fine suspended particles to long distances.

Chemical Weathering

The physical weathering produces a greater surface area of rock exposed to chemical weathering, which occurs simultaneously with physical weathering and continues much beyond that. Chemical weathering consists of chemical decomposition or transformation of parent mineral materials into new mineral materials or secondary mineral products. For example, as a result of chemical weathering, feldspars (primary minerals containing aluminium and silicon) are converted into clay (secondary minerals). It should be noted that chemical weathering is not effective in deserts, because the presence of moisture and air are essential for chemical weathering. Chemical weathering takes place through the following important reactions :

Solution

Water is an important chemical weathering agent in most kinds of rocks through its solvent action. Water soluble minerals like gypsum, limestone etc. of the weathered rock get weathered by the solvent action which increases in the presence of carbon dioxide and organic acids, formed by the decay of organic remains of plants and animals. The solution of these minerals gets absorbed on the surface of negatively charged colloidal particles or removed by leaching.

Hydrolysis

Hydrolysis mostly takes place in combination with other reactions (such as oxidation, reduction or carbonation) and involves the chemical action of water with strong bases producing hydroxides of iron, magnesium, aluminium and calcium etc. For example,

$$K_2Al_2Si_6O_{16} + 2H_2O + CO_2 \rightarrow Al_2O_3.2SiO_2.2H_2O + 4SiO_2 + K_2CO_3$$

Orthoclase (primary mineral)   Kaolin (Secondary clay mineral)

The process of hydrolysis releases Ca, Mg, K, Na and silicates into the soil solution. These materials enhance the growth of plants in the soil.

Oxidation

In this process oxygen reacts with minerals to produce oxides. The latter when dissolved in water weaken the rock and bring about weathering. Oxidation occurs best in well aerated and well drained soil. For example, oxidation of iron in the minerals gives red ferric oxide.

$$4FeO + O_2 \rightarrow 2Fe_2O_3$$

Oxides and sulphides of iron, aluminium and manganese etc. are easily oxidised and cause the chemical weathering of rocks.

Reduction

The red ferric oxide (Fe_2O_3) may also be reduced to grey ferrous oxide (FeO). The reduction mainly occurs in deep zones of earth crust, which are poorly aerated.

$$2Fe_2O_3 \rightarrow 4FeO + O_2$$

Carbonation

The process of carbonation is the combination of CO_2 and H_2O to form carbonic acid (H_2CO_3). The latter combines with hydroxides of Ca, Mg and other minerals of the rock to form carbonates as well as bicarbonates.

$$CO_2 + H_2O \rightleftharpoons H_2CO_3 \qquad\qquad Ca(OH)_2 + CO_2 \rightarrow CaCO_3 + H_2O$$

$$CaCO_3 + H_2O + CO_2 \rightarrow Ca(HCO_3)_2$$

The sparingly soluble carbonates of these minerals either accumulate deeper in the rock material or carried away, depending upon the amount of water passing through.

Hydration

In this process water molecules get attached to the rock material. As a result of hydration, volume of the original material increases and hydrated mineral becomes soft. The soft rock can be readily weathered.

$$2Fe_2O_3 + 3H_2O \rightarrow 2Fe_2O_3.3H_2O$$

Haematite   Limonite

Biological Weathering

A number of micro-organisms such as bacteria, fungi, protozoa, lichens and mosses transform the rock into a dynamic system, storing energy and synthesising organic matter. As a result, physical structure as well as mineral composition of the rock undergo some change. The lichens and mosses extract mineral nutrients such as Mg, Ca, Na, K, Fe, P, Si, Al etc. from the rock. These nutrients are combined with organic matter and thereby return to the developing soil when the vegetation decomposes.

The soil formed by weathering of soil forming rocks is known as primary soil or embryonic. The latter may mature into residual soil or sedimentary soil (mature soil lying immediately over the parent rock), immature soil (partly weathered material without maturation) and secondary or transported soil (weathered parent material has been transported to different places by moraine soil, glacial drift and tilt (agency of glacier), alluvial soil (streams and rivers), colluvial soil (gravitational forces as landslides), aeolian soil (wind), sand dunes (sand storms), and marine soil (oceanic waves) etc.

The soils have also been classified as zonal, intrazonal and azonal soils. Soils are called zonal soils, because climate is the important control over the process of soil formation. The soils are called intrazonal soils, because local geological conditions preclude the normal pattern of soil formation. Alluvial soil is an example of azonal soil. This soil is rich in colloids, and because it is not leached, it may also be rich in nutrient materials. It however, lacks humus, which is finely divided, amorphous, incompletely decomposed black coloured matter added to the mineral matter of soil. Humus includes two kinds of organic matter :

1. Partially decomposed organic matter derived from litter. The litter is composed of dead leaves, twigs, woods, dead roots and various plant products.
2. Excreta of soil animals, which feed on litter or fresh plant material.

The humus is completely decomposed or reduced into simple compounds such as CO_2, H_2O, minerals and salts, by a process known as mineralisation. It should be noted that main sources of litter are forests, grass lands and aquatic plants. The tree litter of forests mainly contains carbohydrates (such as cellulose, lignin and hemicellulose) some fats, resins, waxes and proteins. The inorganic constituents of litter are Ca, Mg, Fe, Mn, Si, Cu, Al, K, P and N etc.

Soils can also be classified as mineral soils (rich in mineral particles), peat and muck (rich in organic matter accumulation in wet areas), mor (low in basic minerals) and mull (rich in basic minerals), depending upon the organic matter content. On this basis, humus has also been classified as mor humus and mull humus. Mor humus is acidic, while mull humus is neutral or slightly alkaline. Moder humus is an intermediate form between the mor and mull humus. Moder humus has a richer and varied fauna.

SOIL PROFILE

At any place where parent material is weathering over a period of time, there develops layers of soil one over the other in progressive state of maturity. Such a vertical section of soil is called soil profile. The soil profiles are characteristic of mature soil and are made up of a succession of horizontal layers (horizons), each of which varies in thickness, colour, texture, structure, acidity, porosity and composition. In general soils have four horizons—an organic or O-horizon and three mineral horizons (A, B, C horizons).

1. *Horizon O:* The uppermost horizon of soil profile is called horizon-O or litter zone. It is usually not present in the soils of deserts, grasslands and cultivated fields, but present in soils of forests.
2. *Horizon A:* Underlying the litter zone is the 'Horizon A' or top soil. It contains undecomposed, partially decomposed and completely decomposed humus from upper to lower sides. This horizon is usually sandy.

3. *Horizon B:* It is also known as sub-soil and is formed with clayey soil. It contains a little humus also.

4. *Horizon C:* It is at the bottom of soil profile and contains weathered rock of parent material. It is light coloured and is virtually lacking in organic material. Below this zone hard rocks are found.

Composition of Soil

The chief components of the soil are: (i) mineral matter (approx. 40%); (ii) organic matter (approx. 10%); (iii) soil water (approx. 25%); and (iv) soil air (approx. 25%). In fact, soil is a mixture of various inorganic and organic chemical compounds. The chief inorganic constituents of soil are the compounds of Ca, Al, Mg, Fe, Si, K and Na. Small amounts of the compounds of Mn, Cu, Zn, Co, B, I and F etc. are also present in the soil. Soil solution also contains complex mixtures of minerals as carbonates, sulphates, chlorides, nitrates and also the organic salts of Ca, Mg, K, Na etc.

Component of soil

The chief organic component of soil is humus, which contains a large number of organic compounds such as amino acids, proteins, aromatic compounds, sugars, alcohols, fats, oils, waxes, resins, tannins, lignin, pigments, purines and many other. As a result, humus is a black coloured, homogeneous complex material. As soil is composed of crystalloids and colloids, it therefore exhibits all the physico-chemical properties related with crystalloids and colloids.

In soil, water is an important solvent and transporting agent. In addition, it also maintains soil texture, arrangement and compactness of soil particles and makes soil suitable for plants and animals. Water comes in the soil mainly through dew, rain, snow, hail and irrigation etc. Water may be chemically combined (e.g., water of crystallisation of mineral grains and water of hydration of clay mineral particles or uncombined (e.g., gravitational water, capillary water, hygroscopic water and water vapour). Uncombined water is held in soil by adhesion (attraction of water molecules to solid particles) and cohesion (attraction of water molecules to each other).

Water which flows down due to force of gravity is known as gravitational water. A certain amount of water is also retained within the intercellular spaces of the soil particles in the form of capillary network. Such water is called capillary water. Water molecules present in the form of thin sheets of water around soil particles form hygroscopic water. Uncombined soil water occurs as moisture or water vapours in the soil atmosphere.

The gases found in the pore spaces of soil profiles form the soil atmosphere. Oxygen, nitrogen and carbon dioxide are the three main gases present in the soil atmosphere. Soil air differs from atmospheric air in having more of moisture and carbon dioxide but less of oxygen. The temperature, pressure, rainfall, wind etc. are the various factors which affect the soil atmosphere.

Soil Texture—Mineral Matter

The shape and size of the soil particles differ from particle to particle. On the basis of their size and shape, soils can be classified into six types as shown in Table 15.1.

Most of the soils are, however, the mixtures of these particles. Soils can also be classified into five texture groups, on the basis of proportions of various particles present in them. These are :

1. Coarse textured soils, which are loose and mainly contain sand and gravel.

2. Moderately coarse soils, which contain sandy loans.

3. Medium textured soils, are the mixture of sand, silt and clay, high enough to hold water as well as plant nutrients.

4. Moderately fine textured soils, which are high in clays.

5. Fine textured soils contain more than 40–45 per cent clay.

Table 15.1. Classification of soil on the basis of size and shape.

Type	Particle diameter
1. Gravel	More than 5000 m.m.
2. Fine gravel	From 5000 to 2000 m.m.
3. Coarse sand	From 2000 to 0.200 m.m.
4. Fine sand	From 0.200 to 0.020 m.m.
5. Silt	From 0.020 to 0.002 m.m.
6. Clay	Less than 0.002 m.m.

The important characteristics of sandy soils are :

1. It is loose soil because pores between particles are large.

2. Roots grow and spread well in such soils.

3. Such soils have more capacity to absorb water.

4. Water flows rapidly through such soils.

5. They have low water holding capacity.

6. They show less water logging capacity and presence of CO_2 but more soil air.

This soil is, however, not rich in nutrients and so it is less fertile. The important characteristics of clay soil are :

1. They have less interspaces between particles.

2. The soil absorbs less water and has more run-off water.

3. This soil is solid, so plant roots penetrate with difficulty.

4. Such soils have more water holding capacity and presence of CO_2 but little of soil air.

The important characteristics of loam soil are :

1. It is a mixture of both sand and clay particles.

2. Best soil for better growth and developments of plants, because it has more soil water, easy water movement, easy penetration of roots and more water holding capacity.

PALAEO-ENVIRONMENTAL EVOLUTION OF THE ATMOSPHERE: INDIAN SETTING

This area of research is somewhat new in India, and in the last few years, more Indian scientists have got interested in this field. However, in this field of palaeontological studies on fossil biota in old rocks, which is related to the evolutionary studies of the atmosphere, some work has been done over the last decade or so. The study of the atmosphere is an interdisciplinary field and the evolutionary sequence is based on sporadic records preserved in the form of chemical and biological fossils. The major part of the atmosphere one observes today is secondary in origin, which came into existence as a result of degassing processes that operated on the earth over aeons of time. By the study of some of these fossil records preserved in the Pre-cambrian rocks, it may be possible to find clues to the type of environments hosting these relict or remnants.

Detailed work has been carried out on phytolitic remnants in the younger *Pre-cambrian rocks* and on the remains of early life forms in some *Archaean* and early *Proterozoic rocks* in Karnataka. Studies related to fossil biota in some Indian rocks have been rather extensive. Stromatolitic and microphytolitic structures have been noticed in several formations such as Bijawar, Cuddapah and Vindhyan and their homotaxial equivalents in the Himalayas. Micro-fossil assemblages are found in some of the oldest rocks of the Sargur complex in Karnataka similar to those found in other parts of the world such as Onwarwacht and Zimbabwe formations. But it should be pointed out that the relation of these developments in the early life forms to the anoxygenic or oxygenic atmosphere were poorly focussed in these studies.

During the early stages of the earth at a time ($\approx$ 4 billion years ago) when it had sufficiently cooled to support the synthesis of organic molecules, anoxygenic conditions seem to have prevailed. These conditions continued for reasonably long period of time, at least some time before 3.6 billion years. By this time, the micro-fossils seem to have developed and subsequently got embedded in the *Archaean rocks* such as the Hosur rocks in Karnataka. To understand the transition from anoxygenic to oxygenic environment in the evolution of atmosphere, it is important to appreciate the physical and chemical environments that affected the *Pre-cambrian litho-biosphere interfaces*. The *Archaean* and early *Proterozoic* eras were characterised by extensive volcanism and the Archaean atmosphere was rich in CO_2, water vapour, Hel/H_2S and ammonia, derived from volcanic emanations. Argon is being used as a tracer to study the evolution of these gas emanations over the last three billion years and preliminary studies are in progress at Physical Research Laboratory, Ahmedabad. The results indicate that the evolution of the atmosphere on earth was not catastrophic, rather it evolved gradually over the last 3.5 billion years. Also, the rate determining parameters for the evolution of terrestrial atmosphere decreased somewhat exponentially from the beginning till today. It was pointed out that in contradistinction to other planets the terrestrial mantle seemed to have separated into the lower and upper mantle. This separation of mantle had profound influence on the isolation of gases i.e. their escape into the atmosphere was inhibited. Therefore except for gas loss in the primary differentiation, the residual primordial gases and the subsequently generated radiogenic gases are efficiently trapped in the earth's interior.

The middle *Archaean* and early *Proterozoic* seem to be the beginning of the appearance of carbonate rocks as a result of precipitation of these rocks from sea waters. With the reduction of CO_2 in the atmosphere; photosynthetic processes were initiated during the early to middle Proterozoic, resulting in the change from oxygen-poor environment to the oxygen-rich conditions. By about 2.5 billion years, the oxygen levels became sufficiently high relative to PAL and the ozone shielding layer got well established. During the late Proterozoic, the oxygen levels seem to be just sufficient for supporting chemical sedimentation.

THERMAL AND RHEOLOGICAL MODELLING OF THE INDIAN CONTINENTAL LITHOSPHERE

Knowledge of the thermal state and rheology of the lithosphere is crucial to the understanding of various geological processes, such as origin and emplacement of igneous bodies, formation of sedimentary basins, and various tectonic activities. The present thermal state can be obtained using the surface heat-flow and heat generation data and also the vertical distribution of magnetisation, electrical conductivity, teleseismic delays, Q-structure, presence of partial melt zones etc. Temperature estimates obtainable from the analysis of metamorphic mineral assemblages can be used to infer palaeothermal states. Some of the work done on the Indian continental lithosphere in the last decade is reviewed.

Crustal Temperatures

Crustal temperatures have been obtained using the heat equation in one dimension with the conductivity model as $K = K_o (1 + CT)$ and the heat source distribution as (i) $A(Z) = A_o$ for $Z\Sigma (0, D)$ and zero elsewhere; (ii) $A(Z) = A_o \exp(-Z/D)$; and (iii) $A(Z) = A_o (1-Z/L)^m$. They used the surface heat flow and heat generation data as given in Table 15.2, and calculated crustal temperatures for all the above models.

Table 15.2. Heat flow, heat generation and other related data.

Region	Heat flow (HFU)	Heat-generation (HGU)	Significant rocks of the region	Age data	
Kolar	1.05 ± .10	3.60 ± .24	Amphibolites of the Schist belt and the gneisses surrounding them.	Kolar Schist belt ~ 3200 million years.	
Singhbhum	1.30 ± 0.12	2.75 ± .22	Singhbhum granite near its northern border and the iron ore group of the Singhbhum thrust belt.	Singhbhum granite ~ 2700 million years. Singhbhum thrust : Two Orogenic cycles close at 2700 million years and 850 million years, respectively.	
Khetri	1.76 ± .10	5.24 ± .42	Phyllites of the Alwar-Ajabgarh groups of the Delhi supergroup.	Close of Delhi orogenic cycle ~ 1600 million years. Delhi supergroup affected by ~ 850 – 580 million years metamorphic/igneous activities.	
Agnigundala	1.80 ± .13	6.25 ± .23	Phyllites and argilites of the Nallamalai group of the Cuddapah supergroup.	Base of Cuddapah supergroup ~ 850–580 million years Dolerite intrusions in Cumbum Slate (Nallamalai)	
Jharia	1.90 ± .14	6.70 ± .50	Sedimentary rocks of the Lower Gondwana, and Pre-cambrian gneisses of the basement.	Lower Gondwana sediments Regional metamorphism of gneiss	– Upper Carbo niferous/ Upper Permian. ~893–1086 million years.
Karadi-kuttam	1.31 ± .09	7.00 ± .22	Gneisses	Pre-cambrian	

1 HFU = 10^{-6} cal cm^{-2} s^{-1} = 41.84 mWm^{-2}

1 HGU = 10^{-18} cal cm^{-8} s^{-1} = .418 Wm^{-3}

The following general conclusions can be drawn :

1. In contrast to other shield regions, higher Moho temperatures are obtained in the Indian shield region.

2. Moho temperatures in the southern part is ~ 550–600°C, increasing to 850–900°C in the northern part.

3. Significant lateral variations in the physical properties are manifest in the lateral variations in the heat-flow data (Table 15.3).

4. Exponential model of the radiogenic heat distribution is preferred as it satisfies the observed linear heat flow/heat generation relationship. This model gives higher Moho temperatures. Other physical manifestations of the thermal structure are :

(a) Pn wave velocity change between southern and northern part is ~14 km/s.

(b) The Curie isotherm depth is about 30–40 km in the southern part whereas in the northern part it is ~20–25 km. This is reflected in the MAGSAT over India.

(c) In the southern part, partial melting does not occur whereas in the northern part of the shield crust, partial melting zones could exist.

Table 15.3. Surface heat flow data in the Indian Peninsula.

Geological unit	No. of observations	Heat flow (mWm^{-2}) Range	Average	Age
Dharwar Schist belts	7	26–40	32	Archaean
Peninsular gneisses	2	46–55	50.5	Archaean
Banded gneissic complex and Aravalli supergroup	3	—	41	Archaean and lower Proterozoic
Delhi supergroup and Bijawar	7	37–74	54.5	Upper Proterozoic
Singhbhum thrust zone	3	59–63	61	Upper Proterozoic
Cuddapah basin	4	27–75	54.5	Upper Proterozoic
Gondwana basins				Sedimentation lasted from Permo-Carboniferous to lower Cretaceous
(a) Damodar valley	5	69–79	74	Dykes and sills intruded (140–140 100 m.y.)
(b) Son-Mahanadi valleys	2	59–107	81	Dykes and sills intruded (140–100 and 70–65 million year)
(c) Satpura basin	2	49–61	55	Dykes and sills intruded (70–65 million year)
(d) Godavari valley	5	63–104	81.6	More or less free from intrusions
Deccean traps	2		43	Upper Cretaceous to early paleocene
Cambay basin				Tertiary
(a) Northern part	6	75–93	83	Tertiary
(b) Southern part	2	—	61	Tertiary
Assam basin	2	—	60.5	Tertiary

Crustal Temperatures

Using the available pressure and temperature estimates from the major elements partitioning amongst coexisting mineral assemblages, palaeothermal state has been estimated. Metamorphic P and T estimates for the South Indian cratonic rocks has been determined. The granulite facies terrain in the charnockite regions show temperature range of 800–850°C and pressure range ~ 6–10 kb. Higher pressures, 7–9 kb are observed in some parts of Tamil Nadu, whereas in Andhra Pradesh and Chennai area of Tamil Nadu the pressures are 7–8 kb. In the transitional zone between charnockitic and non-charnockitic regions, temperature and pressure differences are ~ 20°C and a few hundred bars, respectively.

Taking the pressure, temperature and age of metamorphism of Sargur Schist belt as 6–7 kb, 750–850°C and 3 billion years respectively, the average crustal temperature gradient and Moho temperature have been obtained as 28°C/km and 1118°C respectively. The basal reduced heat flow is $1.08.10^{-6}$ Cal/cm^2/s which is much higher than its present value $.79.10^{-6}$ Cal/cm^2/s. The basal heat flow is related to the mean mantle temperature, which is determined to be 2350°C, based on parameterised mantle convection model. Thus the mantle below the Indian shield is cooling at the rate of 25°C per billion year.

Lithospheric Temperatures

Using the conductivity model of olivine, K(T) = (31 +.21T) for T ≤ 500 k and (31 + .21T) + 5.5 10 (T –500), T ≥ 500 k, it is deduced that the temperature at a depth of 100 km is 1425°C in Khetri belt whereas 1260°C in the Singhbhum region. These estimates are higher than the average shield lithospheres, implying that lithospheric thickness is less than 100 km in the western side of the northern part of the Indian shield and more than 100 km in its eastern side, if the base of the lithosphere is defined as 1300°C isotherm.

Rheological Modelling

Using the flow law of olivines, Σ=B [sinh $(\alpha \sigma)^n$ exp [—(E+PU)/RT] [where B, N and α are constants; E, U and R are activation energy, activation volume and gas constant; p = ρsZ (ρ-density, g-gravity and Z-the depth)] and the temperature T as determined by heat flow data, the variation of shear stress (σ) with depth is obtained for different strain rates (Σ). The characterisation of lithosphere into brittle and ductile zone is indicated using the critical shear stress-depth relationship : σ (kb) = .8 + .6P (kb). The results (Tables 15.4 and 15.5) show the thickness of the elastic zone and basal shear stress at the depth of 100 km, for wide variety of strain rates 10^{-14}—10^{-14}/s (applicable to asthenospheric convection) and 10^{-14}—10^{-11}/s (applicable to vertical uplift). Earthquake occurrences are limited within the elastic zone as borne out by the available focal depth estimates. The thickness of elastic zone varies along the length of Himalayas, 26 km in the western part and 31 km in the eastern part. In the southern part of the shield the thickness of the elastic lithosphere is about 40–50 km.

Proterozoic Sedimentary Basin Evolution

Between Eparchaean interval and Cambrian period (2000 to 600 million years ago), the cresent shaped Cuddapah basin on the South Indian shield was formed with largest development of sedimentary sequences. It is one of the best studied Proterozoic sedimentary basins. Geological, geochronological and geophysical data have been utilised to advance a thermal model for the evolution of this intracontinental, platform basin on the Archaean lithosphere.

Table 15.4. Basal shear stress (bars).

Strain rate (s^{-1})	Khetri		Singhbhum	
	Dry olivine model	Wet olivine model	Dry olivine model	Wet olivine model
10^{-11}	10.0	3.56	25.29	9.66
10^{-12}	5.3	1.73	13.34	4.70
10^{-13}	2.80	.84	7.04	2.29
10^{-14}	1.48	.41	3.71	1.12
10^{-15}	.78	.20	1.96	.54
10^{-16}	.41	.13	1.03	.26

Table 15.5. Thickness of elastic zone (km).

Strain rate (s^{-1})	Khetri		Singhbhum	
	Dry olivine model	Wet olivine model	Dry olivine model	Wet olivine model
10^{-11}	30	29	36	35.5
10^{-12}	29	27.5	34.5	34.0
10^{-13}	27	26	33.0	32.5
10^{-14}	26	25	31.5	31.0
10^{-15}	25	24	30.5	30.0
10^{-16}	24	23	29.5	29.0

A basal heat source leading to formation of dense eclogite rocks at depth within the lithosphere has been shown to produce required subsidence by flexuring to accomodate the observed sedimentary sequences.. Radial stresses due to gravitational loading of the Archaean crust have shown to be sufficient to produce elastic failure at the base of Cuddapah basin, which would lead to normal faulting and opening of deep fractures through which magmatic episodes would follow.

CHAPTER 16

Geosphere and Geochemistry

INTRODUCTION

The geosphere, or solid earth, is that part of the earth upon which humans live and from which they extract most of their food, minerals, and fuels. Once thought to have an almost unlimited buffering capacity against the perturbations of humankind, the geosphere is now known to be rather fragile and subject to harm by human activities. For example, some billions of tonnes of earth material are mined or otherwise disturbed each year in the extraction of minerals and coal. Two atmospheric pollutant phenomena—excess carbon dioxide and acid rain have the potential to cause major changes in the geosphere. Too much carbon dioxide in the atmosphere may cause global heating ("greenhouse effect"), which could significantly alter rainfall patterns and turn currently productive areas of the earth into desert regions. The low pH characteristic of acid rain can bring about drastic changes in the solubilities and oxidation-reduction rates of minerals. Erosion caused by intensive cultivation of land is washing away vast quantities of topsoil from fertile farmlands each year. In some areas of industrialised countries, the geosphere has been the dumping ground for toxic chemicals. Ultimately, the geosphere must provide disposal sites for the nuclear wastes of more than 400 nuclear reactors now operating worldwide. It may be readily seen that the preservation of the geosphere in a form suitable for human habitation is one of the greatest challenges facing humankind.

The interface between the geosphere and the atmosphere at earth's surface is very important to the environment. Human activities on the earth's surface may affect climate, most directly through the change of surface albedo, defined as the percentage of incident solar radiation reflected by a land or water surface. For example, if the sun radiates 100 units of energy per minute to the outer limits of the atmosphere and the earth's surface receives 60 units per minute of the total, then reflects 30 units upward, the albedo is 50 per cent. Some typical albedo values for different areas on the earth's surface are: evergreen forests, 7–15%; dry, plowed fields, 10–15%; deserts, 25–35%; fresh snow, 85–90%; asphalt, 8%. In some heavily developed areas, anthropogenic (human-produced) heat release is comparable to the solar input.

One of the greater impacts of humans upon the geosphere is the creation of desert areas through abuse of land with marginal amounts of rainfall. This process, called desertification is manifested by declining groundwater tables, salinisation of topsoil and water, reduction of surface waters, unnaturally high soil erosion, and desolation of native vegetation. The problem is severe in some parts of the world, particularly Africa's Sahel (southern rim of the Sahara), where the Sahara advanced southward at a particularly rapid rate during the period 1968–1973, contributing to widespread starvation in Africa during the 1980s. Large, arid areas of the western US are experiencing at least some desertification as the result of human activities and a severe drought during the latter 1980s and early 1990s. As the populations of the Western states increase, one of the greatest challenges facing the residents is to prevent additional conversion of land to desert.

The most important part of the geosphere for life on earth is soil. It is the medium upon which plants grow, and virtually all terrestrial organisms depend upon it for their existence. The productivity of soil is strongly affected by environmental conditions and pollutants.

With increasing population and industrialisation, one of the more important aspects of human use of the geosphere has to do with the protection of water sources. Mining, agricultural, chemical, and radioactive wastes all have the potential for contaminating both surface water and groundwater. Sewage sludge spread on land may contaminate water by release of nitrate and heavy metals. Landfills may likewise be sources of contamination. Leachates from unlined pits and lagoons containing hazardous liquids or sludges may pollute drinking water.

It should be noted, however, that many soils have the ability to assimilate and neutralise pollutants. Various chemical and biochemical phenomena in soils operate to reduce the harmful nature of pollutants. These phenomena include oxidation-reduction processes, hydrolysis, acid-base reactions, precipitation, sorption, and biochemical degradation. Some hazardous organic chemicals may be degraded to harmless products on soil, and heavy metals may be sorbed by it. In general, however, extreme care should be exercised in disposing of chemicals, sludges, and other potentially hazardous materials on soil, particularly where the possibility of water contamination exists.

NATURE OF SOLIDS IN THE GEOSPHERE

The earth is divided into layers, including the solid iron-rich inner core, molten outer core, mantle, and crust. Environmental chemistry is most concerned with the lithosphere, which consists of the outer mantle and the crust. The latter is the earth's outer skin that is accessible to humans. It is extremely thin compared to the diameter of the earth, ranging from 5 to 40 km thick.

Most of the solid earth crust consists of rocks. Rocks are composed of minerals, where a mineral is a naturally-occurring inorganic solid with a definite internal crystal structure and chemical composition. A rock is a solid, cohesive mass of pure mineral or an aggregate of two or more minerals.

Structure and Properties of Minerals

The combination of two characteristics is unique to a particular mineral. These characteristics are a defined chemical composition, as expressed by the mineral's chemical formula, and a specific crystal structure. The crystal structure of a mineral refers to the way in which the atoms are arranged relative to each other. It cannot be determined from the appearance of visible crystals of the mineral, but requires structural methods such as X-ray structure determination. Different minerals may have the same chemical composition, or they may have the same crystal structure, but both may not be identical for truly different minerals.

Physical properties of minerals can be used to classify them. The characteristic external appearance of a pure crystalline mineral is its crystal form. Because of space constrictions on the ways that minerals grow, the pure crystal form of a mineral is often not expressed. Colour is an obvious characteristic of minerals, but can vary widely due to the presence of impurities. The appearance of a mineral surface in reflected light describes its lustre. Minerals may have a metallic lustre or appear partially metallic (or sub-metallic), vitreous (like glass), dull or earthy, resinous, or pearly. The colour of a mineral in its powdered form as observed when the mineral is rubbed across an unglazed porcelain plate is known as streak. Hardness is expressed on Mohs scale, which ranges from 1 to 10 and is based upon 10 minerals that vary from talc, hardness 1, to diamond, hardness 10. Cleavage denotes the manner in which minerals break along planes and the angles in which these planes intersect. For example, mica cleaves to form thin sheets. Most minerals fracture irregularly, although some fracture along smooth curved surfaces or into fibres or splinters. Specific gravity, density relative to that of water, is another important physical characteristic of minerals.

Kinds of Minerals

Although over two thousand minerals are known, only about 25 rock-forming minerals make up most of the earth's crust. The nature of these minerals may be better understood with a knowledge of the elemental composition of the crust. Oxygen and silicon make up 49.5% and 25.7% by mass of the earth's crust, respectively. Therefore, most minerals are silicates such as quartz, SiO_2, or orthoclase, $KAlSi_3O_8$. In descending order of abundance, the other elements in the earth's crust are aluminium (7.4%), iron (4.7%), calcium (3.6%), sodium (2.8%), potassium (2.6%), magnesium (2.1%), and other (1.6%). Table 16.1 summarises the major kinds of minerals in the earth's crust.

Secondary minerals are formed by alteration of parent mineral matter. Clays are silicate minerals, usually containing aluminium, that constitute one of the most significant classes of secondary minerals. Olivine, augite, hornblende, and feldspars all form clays.

Table 16.1. Major mineral groups in the earth's crust.

Mineral group	Examples	Formula
Silicates	Quartz	SiO_2
	Olivine	$(Mg, Fe)_2 SiO_4$
	Potassium feldspar	$KAlSi_3O_8$
Oxides	Corundum	Al_2O_3
	Magnetite	Fe_3O_4
Carbonates	Calcite	$CaCO_3$
	Dolomite	$CaCO_3.MgCO_3$
Sulphides	Pyrite	FeS_2
	Galena	PbS
Sulphates	Gypsum	$CaSO_4 \cdot 2H_2O$
Halides	Halite	$NaCl$
	Fluorite	CaF_2
Native elements	Copper	Cu
	Sulphur	S

Evaporites

Evaporites are soluble salts that precipitate from solution under special arid conditions, commonly as the result of the evaporation of seawater. The most common evaporite is halite, NaCl. Other simple evaporite minerals are sylvite (KCl), thenardite (Na_2SO_4), and anhydrite ($CaSO_4$). Many evaporites are hydrates, including bischofite ($MgCl_2 \cdot 6H_2O$), gypsum ($CaSO_4 \cdot 2H_2O$), kieserite ($MgSO_4 \cdot H_2O$), and epsomite ($MgSO_4 \cdot H_2O$). Double salts, such as carnallite ($KMgCl_3 \cdot 6H_2O$), kainite ($KMgClSO_4 \cdot 11/4\ H_2O$), glaserite ($K_3Na(SO_4)_2$), polyhalite ($K_2MgCa_2(SO_4)_4 \cdot 2H_2O$), and loeweite ($Na_{12}Mg_7(SO_4)_{13} \cdot 15H_2O$), are very common in evaporites.

The precipitation of evaporites from marine and brine sources depends upon a number of factors. Prominent among these are the concentrations of the evaporite ions in the water and the solubility products of the evaporite salts. The presence of a common ion decreases solubility; for example, $CaSO_4$ precipitates more readily from a brine that contains Na_2SO_4 than it does from a solution that contains no other source of sulphate. The presence of other salts that do not have a common ion increases solubility because it decreases activity coefficients. Differences in temperature result in significant differences in solubility.

The nitrate deposits that occur in the hot and extraordinarily dry regions of northern Chile are chemically unique because of the stability of highly oxidised nitrate salts. The dominant salt, which has been mined for its nitrate content for use in explosives and fertilisers, is Chile saltpeter, $NaNO_3$. Traces of highly oxidised $CaCrO_4$ and $Ca(ClO_4)_2$ are also encountered in these deposits, and some regions contain enough $Ca(IO_3)_2$ to serve as a commercial source of iodine.

Volcanic Sublimates

A number of mineral substances are gaseous at the magmatic temperatures of volcanoes and are mobilised with volcanic gases. These kinds of substances condense near the mouths of volcanic fumaroles and are called sublimates. Elemental sulphur is a common sublimate. Some oxides, particularly of iron and silicon, are deposited as sublimates. Most other sublimates consist of chloride and sulphate salts. The cations most commonly involved are monovalent cations of ammonium ion, sodium, and potassium; magnesium; calcium; aluminium; and iron. Fluoride and chloride sublimates are sources of gaseous HF and HCl formed by their reactions at high temperatures with water, such as the following :

$$2H_2O + SiF_4 \rightarrow 4HF + SiO_2 \qquad \qquad ...(16.1)$$

Igneous, Sedimentary, and Metamorphic Rock

At elevated temperatures deep beneath Earth's surface, rocks and mineral matter melt to produce a molten substance called magma. Cooling and solidification of magma produces igneous rock. Common igneous rock are granite, basalt, quartz (SiO_2), pyroxene ($(Mg, Fe)SiO_3$), feldspar ($(Ca, Na, K)AlSi_3O_8$), olivine ($(Mg, Fe)_2SiO_4$), and magnetite (Fe_3O_4). Igneous rocks are formed under water-deficient, chemically reducing conditions of high temperature and high pressure. Exposed igneous rocks are under wet, oxidising, low-temperature and low-pressure conditions. Since such conditions are opposite the conditions under which igneous rocks were formed, they are not in chemical equilibrium with their surroundings when they become exposed. As a result, such rocks disintegrate by a process called weathering. Weathering tends to be slow because igneous rocks are often hard, non-porous, and of low reactivity. Erosion from wind, water, or glaciers picks up materials from weathering rocks and deposits it as sediments or soil. A process called lithification describes the conversion of sediments to sedimentary rocks. In contrast to the parent igneous rocks, sediments and sedimentary rocks are porous, soft, and chemically reactive. Heat and pressure convert sedimentary rock to metamorphic rock.

Sedimentary rocks may be detrital rocks consisting of solid particles eroded from igneous rocks as a consequence of weathering; quartz is the most likely to survive weathering and transport from its original location chemically intact. A second kind of sedimentary rocks consists of chemical sedimentary rocks produced by the precipitation or coagulation of dissolved or colloidal weathering products. Organic sedimentary rocks contain residues of plant and animal remains. Carbonate minerals of calcium and magnesium—limestone or dolomite—are especially abundant in sedimentary rocks. Important examples of sedimentary rocks are the following :

1. Sandstone produced from sand-sized particles of minerals such as quartz.
2. Conglomerates made up of relatively larger particles of variable size.
3. Shale formed from very fine particles of silt or clay.
4. Limestone, $CaCO_3$, produced by the chemical or biochemical precipitation of calcium carbonate :

$$Ca^{2+} + CO_3^{2-} \rightarrow CaCO_3(s)$$

$$Ca^{2+} + 2HCO_3^- + h\nu(\text{algal photosynthesis}) \rightarrow \{CH_2O\} \text{ (biomass)}$$
$$+ CaCO_3(s) + O_2(g)$$

5. Chert consisting of micro-crystalline SiO_2.

Rock cycle

The interchanges and conversions among igneous, sedimentary, and metamorphic rocks, as well as the processes involved therein, are described by the rock cycle. A rock of any of these three types may be changed to a rock of any other type. Or a rock of any of these three kinds may be changed to a different rock of the same general type in the rock cycle. The rock cycle is illustrated in Fig. 16.1.

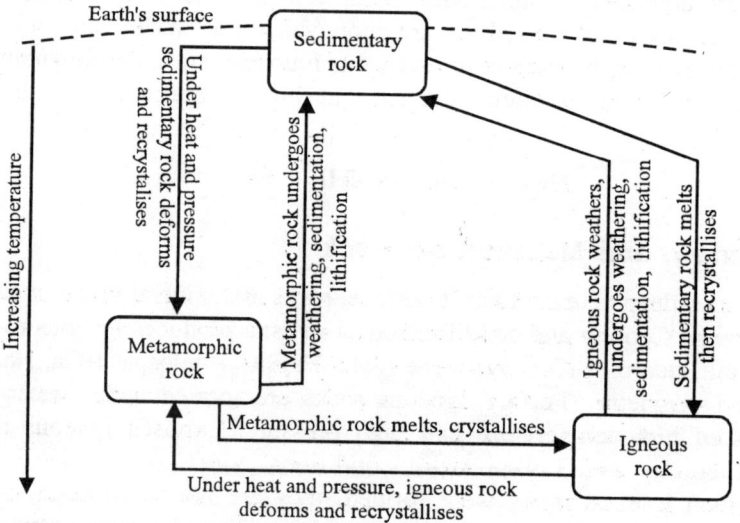

Fig. 16.1. The rock cycle.

Stages of weathering

Weathering can be classified into early, intermediate, and advanced stages. The stage of weathering to which a mineral is exposed depends upon time; chemical conditions, including exposure to air, carbon dioxide, and water; and physical conditions, such as temperature and mixing with water and air.

Reactive and soluble minerals such as carbonates, gypsum, olivine, feldspars, and iron(II)-rich substances can survive only early weathering. This stage is characterised by dry conditions, low leaching, absence of organic matter, reducing conditions, and limited time of exposure. Quartz, vermiculite, and smectites can survive the intermediate stage of weathering manifested by retention of silica, sodium, potassium, magnesium, calcium, and iron(II) not present in iron(II) oxides. These substances are mobilised in advanced-stage weathering, other characteristics of which are intense leaching by fresh water, low pH, oxidising conditions (iron(II) $\rightarrow$ iron(III)), presence of hydroxy polymers of aluminium, and dispersion of silica.

SEDIMENTS

Vast areas of land, as well as lake and stream sediments, are formed from sedimentary rocks. The properties of these masses of material depend strongly upon their origins and transport. Water is the main vehicle of sediment transport, although wind can also be significant. Hundreds of millions of tonnes of sediment are carried by major rivers each year.

The action of flowing water in streams cuts away stream banks and carries sedimentary materials for great distances. Sedimentary materials may be carried by flowing water in streams as the following :

1. Dissolved load from sediment-forming minerals in solution.
2. Suspended load from solid sedimentary materials carried along in suspension.
3. Bed load dragged along the bottom of the stream channel.

The transport of calcium carbonate as dissolved calcium bicarbonate provides a straightforward example of dissolved load. Water with a high dissolved carbon dioxide content (usually present as the result of bacterial action) in contact with calcium carbonate formations contains Ca^{2+} and HCO_3^- ions. Flowing water containing calcium as such temporary hardness may become more basic by loss of CO_2 to the atmosphere, consumption of CO_2 by algal growth, or contact with dissolved base, resulting in the deposition of insoluble $CaCO_3$:

$$Ca^{2+} + 2HCO_3^- \rightarrow CaCO_3(s) + CO_2(g) + H_2O \qquad ...(16.2)$$

Most flowing water that contains dissolved load originates underground, where it dissolves minerals from the rock strata that it flows through.

Most sediments are transported by streams as suspended load, obvious in the observation of "mud" in the flowing water of rivers draining agricultural areas or finely divided rock in Alpine streams fed by melting glaciers. Under normal conditions, finely divided silt, clay, or sand make up most of the suspended load, although larger particles are transported in rapidly flowing water. The degree and rate of movement of suspended sedimentary material in streams are functions of the velocity of water flow and the settling velocity of the particles in suspension.

Bed load is moved along the bottom of a stream by the action of water "pushing" particles along. Particles carried as bed load do not move continuously. The grinding action of such particles is an important factor in stream erosion.

Typically, about 2/3 of the sediment carried by a stream is transported in suspension, about 1/4 in solution, and the remaining relatively small fraction as bed load. The ability of a stream to carry sediment increases with both the overall rate of flow of the water (mass per unit time) and the velocity of the water. Both of these are higher under flood conditions, so floods are particularly important in the transport of sediments.

Streams mobilise sedimentary materials through erosion, transport materials along with stream flow, and release them in a solid form during deposition. Deposits of stream-borne sediments are called alluvium. As conditions such as lowered stream velocity begin to favour deposition, larger, more settleable particles are released first. This results in sorting such that particles of a similar size and type tend to occur together in alluvial deposits. Much sediment is deposited in flood plains where streams overflow their banks.

CLAYS

Clays are extremely common and important in mineralogy. Furthermore, in general, clays predominate in the inorganic components of most soils and are very important in holding water and in plant nutrient cation exchange. All clays contain silicate and most contain aluminium and water. Physically, clays consist of very fine grains having sheet-like structures. For purposes of discussion here, clay is defined as a group of micro-crystalline secondary minerals consisting of hydrous aluminium silicate that have sheet-like structures. Clay minerals are distinguished from each other by general chemical formula, structure, and chemical and physical properties. The three major groups of clay minerals are the following :

1. Montmorillonite, $Al_2(OH)_2Si_4O_{10}$.
2. Illite, $K_{0-2}Al_4(Si_{8-6}Al_{0-2})O_{20}(OH)_4$.
3. Kaolinite, $Al_2Si_2O_5(OH)_4$.

Many clays contain large amounts of sodium, potassium, magnesium, calcium, and iron, as well as trace quantities of other metals. Clays bind cations such as Ca_2^+, Mg^{2+}, K^+, Na^+, and NH_4^+, which protects the cations from leaching by water but keeps them available in soil as plant nutrients. Since many clays are readily suspended in water as colloidal particles, they may be leached from soil or carried to lower soil layers.

Olivine, augite, hornblende, and feldspars are all parent minerals that form clays. An example is the formation of kaolinite $(Al_2Si_2O_5(OH)_4)$ from potassium feldspar rock $(KAlSi_3O_8)$:

$$2KAlSi_3O_8(s) + 2H^+ + 9H_2O \rightarrow Al_2Si_2O_5(OH)_4(s) + 2K^+ \ (aq) + 4H_4SiO_4(aq) \qquad ...(16.3)$$

The layered structures of clays consist of sheets of silicon oxide alternating with sheets of aluminium oxide. The silicon oxide sheets are made up of tetrahedra in which each silicon atom is surrounded by four oxygen atoms. Of the four oxygen atoms in each tetrahedron, three are shared with other silicon atoms that are components of other tetrahedra. This sheet is called the tetrahedral sheet. The aluminium oxide is contained in an octahedral sheet, so named because each aluminium atom is surrounded by six oxygen atoms in an octahedral configuration. The structure is such that some of the oxygen atoms are shared between aluminium atoms and some are shared with the tetrahedral sheet.

Structurally, clays may be classified as either two-layer clays in which oxygen atoms are shared between a tetrahedral sheet and an adjacent octahedral sheet, and three-layer clays, in which an octahedral sheet shares oxygen atoms with tetrahedral sheets on either side. These layers composed of either two or three sheets are called unit layers. A unit layer of a two-layer clay typically is around 0.7 nanometres (nm) thick, whereas that of a three-layer clay exceeds 0.9 nm in thickness. The structure of the two-layer clay kaolinite is represented in Fig. 16.2. Some clays, particularly the montmorillonites, may absorb large quantities of water between unit layers, a process accompanied by swelling of the clay.

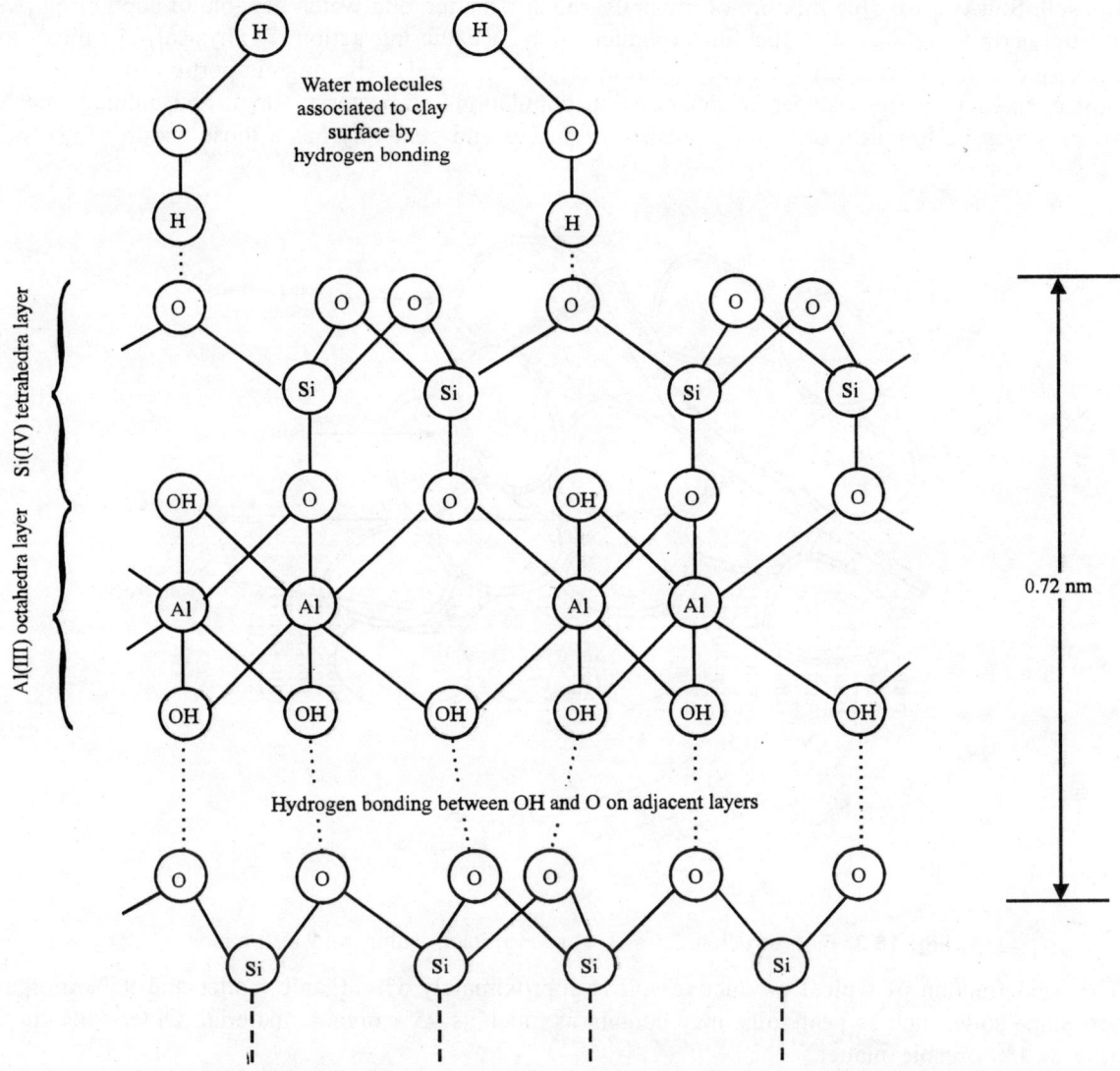

Fig. 16.2. Representation of the structure of kaolinite, a two-layer clay.

The clay minerals may attain a net negative charge by ion replacement, in which Si(IV) and Al(III) ions are replaced by metal ions of similar size but lesser charge. Compensation must be made for this negative charge by association of cations with the clay layer surfaces. Since these cations need not fit specific sites in the crystalline lattice of the clay, they may be relatively large ions, such as K^+, Na^+, or NH_4^+. These cations are called exchangeable cations and are exchangeable for other cations in water. The amount of exchangeable cations, expressed as milliequivalents (millimoles of monovalent cations) per 100 g of dry clay, is called the cation-exchange capacity, CEC, of the clay and is a very important characteristic of colloids and sediments that have cation-exchange capabilities.

SOIL

Insofar as environmental chemistry and life on earth are concerned, the most important part of the earth's crust is soil. Soil is a variable mixture of minerals, organic matter, and water, capable of supporting plant life on the earth's surface. It is the final product of the weathering action of physical, chemical, and biological processes on rocks, which largely produces clay minerals. The organic portion of soil consists of plant biomass in various stages of decay. High populations of bacteria, fungi, and animals such as earthworms may be found in soil. Soil contains air spaces and generally has a loose texture (Fig. 16.3).

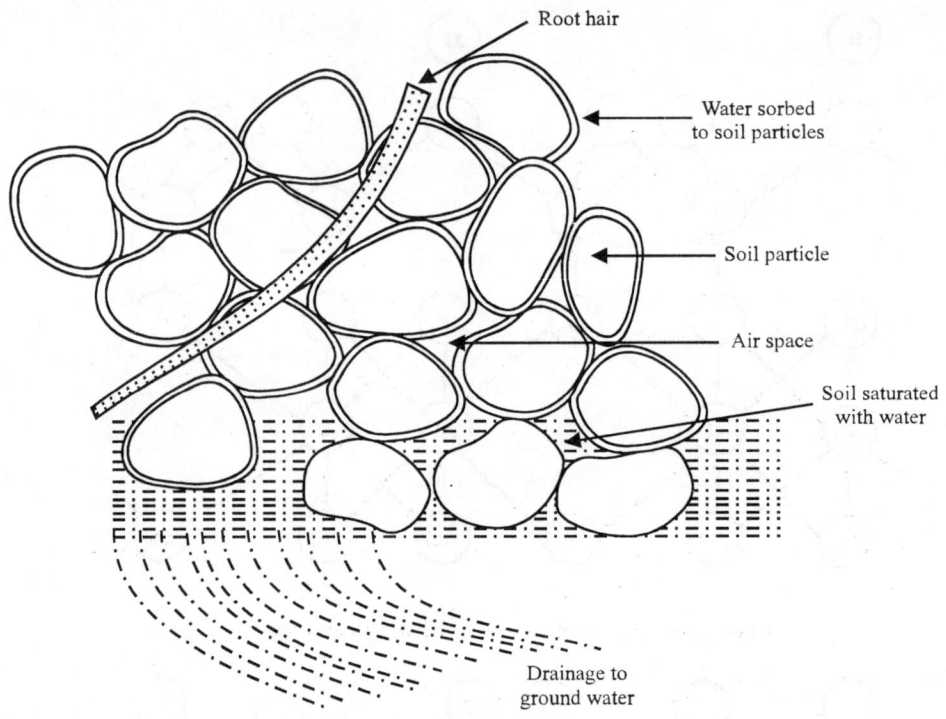

Root hair

Water sorbed
to soil particles

Soil particle

Air space

Soil saturated
with water

Drainage to
ground water

Fig. 16.3. Fine structure of soil, showing solid, water, and air phases.

The solid fraction of typical productive soil is approximately 5% organic matter and 95% inorganic matter. Some soils, such as peat soils, may contain as much as 95% organic material. Other soils contain as little as 1% organic matter.

Typical soils exhibit distinctive layers with increasing depth (Fig. 16.4). These layers are called horizons. Horizons form as the result of complex interactions among processes that occur during weathering. Rainwater percolating through soil carries dissolved and colloidal solids to lower horizons where they are deposited. Biological processes, such as bacterial decay of residual plant biomass, produces slightly acidic CO_2, organic acids, and complexing compounds that are carried by rainwater to lower horizons where they interact with clays and other minerals, altering the properties of the minerals. The top layer of soil, typically several inches in thickness, is known as the 'A' horizon, or topsoil. This is the layer of maximum biological activity in the soil and contains most of the soil organic matter. Metal ions and clay particles in the 'A' horizon are subject to considerable leaching. The next layer is the 'B' horizon, or sub-soil. It receives material such as organic matter, salts, and clay particles leached from the topsoil. The 'C' horizon is composed of weathered parent rocks from which the soil originated.

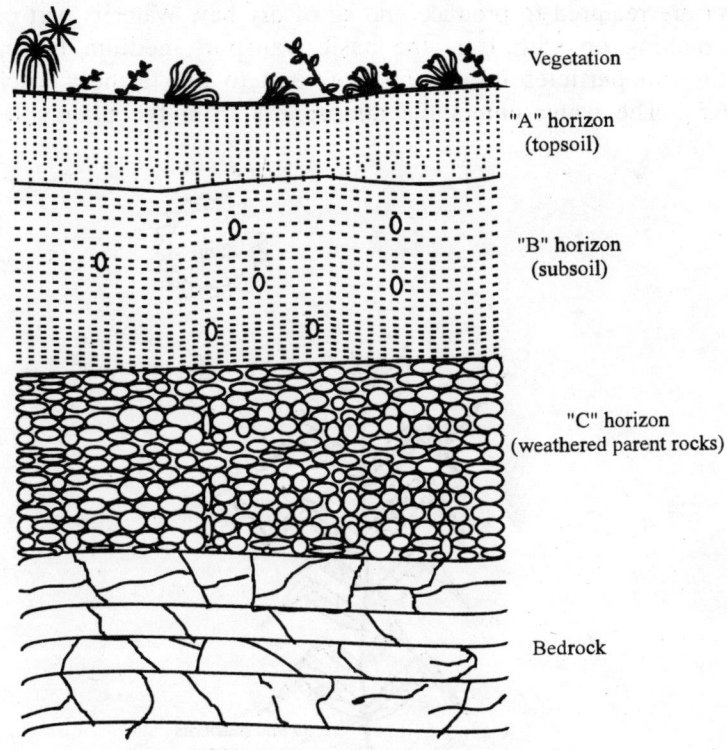

Vegetation

"A" horizon
(topsoil)

"B" horizon
(subsoil)

"C" horizon
(weathered parent rocks)

Bedrock

Fig. 16.4. Soil profile showing soil horizons.

Soils exhibit a large variety of characteristics that are used for their classification for various purposes, including crop production, road construction, and waste disposal. The parent rocks from which soils are formed obviously play a strong role in determining the composition of soils. Other soil characteristics include strength, workability, soil particle size, permeability, and degree of maturity. One of the more important classes of productive soils is the podzol type of soil formed under relatively high rainfall conditions in temperate zones of the world. These generally rich soils tend to be acidic (pH 3.5–4.5) so that alkali and alkaline earth metals and, to a lesser extent aluminium and iron, are leached from their 'A' horizons, leaving kaolinite as the predominant clay mineral. At somewhat higher pH in the 'B' horizons, hydrated iron oxides and clays are re-deposited.

From the engineering standpoint, especially, the mechanical properties of soil are emphasised. These properties, which may have important environmental implications in areas such as waste disposal, are largely determined by particle size. According to the United Classification System (UCS), the four major categories of soil particle sizes are the following: Gravels (2–60 mm) > sands (0.06–2 mm) > silts (0.06–0.006 mm) > clays (less than 0.002 mm). In the UCS classification scheme, clays represent a size fraction rather than a specific class of mineral matter.

Water and Air in Soil

Large quantities of water are required for the production of most plant materials. For example, several hundred kg of water are required to produce one kg of dry hay. Water is part of the three-phase, solid-liquid-gas system making up soil. It is the basic transport medium for carrying essential plant nutrients from solid soil particles into plant roots and to the farthest reaches of the plant's leaf structure (Fig. 16.5). The water enters the atmosphere from the plant's leaves, a process called transpiration.

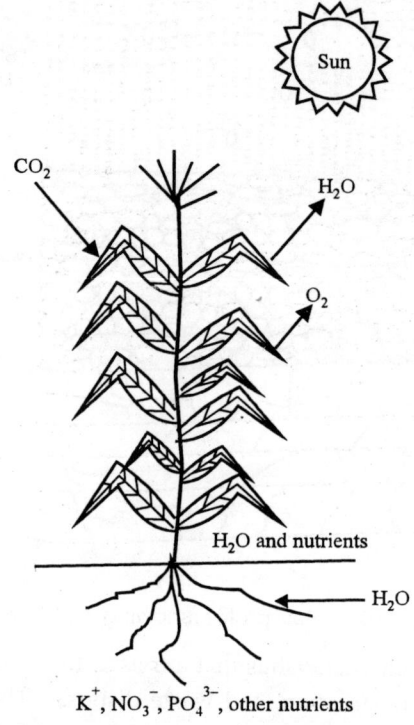

Fig. 16.5. Plants transport water from the soil to the atmosphere by transpiration. Nutrients are also carried from the soil to the plant extremities by this process. Plants remove CO_2 from the atmosphere and add O_2 by photosynthesis. The reverse occurs during plant respiration.

Normally, because of the small size of soil particles and the presence of small capillaries and pores in the soil, the water phase is not totally independent of soil solid matter. The availability of water to plants is governed by gradients arising from capillary and gravitational forces. The availability of nutrient solutes in water depends upon concentration gradients and electrical potential gradients. Water present in larger spaces in soil is relatively more available to plants and readily drains away. Water held in smaller pores, or between the unit layers of clay particles is held much more strongly. Soils high in organic matter may hold appreciably more water than other soils, but it is relatively less available to plants because of physical and chemical sorption of the water by the organic matter.

There is a very strong interaction between clays and water in soil. Water is absorbed on the surfaces of clay particles. Because of the high surface/volume ratio of colloidal clay particles, a great deal of water may be bound in this manner. Water is also held between the unit layers of the expanding clays, such as the montmorillonite clays. As soil becomes waterlogged (water-saturated), it undergoes drastic changes in physical, chemical, and biological properties. Oxygen in such soil is rapidly used up by the respiration of micro-organisms that degrade soil organic matter. In such soils, the bonds holding soil colloidal particles together are broken, which causes disruption of soil structure. Thus, the excess water in such soils is detrimental to plant growth, and the soil does not contain the air required by most plant roots. Most useful crops, with the notable exception of rice, cannot grow on waterlogged soils.

One of the most marked chemical effects of waterlogging is a reduction of pE by the action of organic reducing agents acting through bacterial catalysts. Thus, the redox condition of the soil becomes much more reducing, and the soil pE may drop from that of water in equilibrium with air (+ 13.6 at pH 7) to 1 or less. One of the more significant results of this change is the mobilisation of iron and manganese as soluble iron(II) and manganese(II) through reduction of their insoluble higher oxides :

$$MnO_2 + 4H^+ + 2e^- \rightarrow Mn^{2+} + 2H_2O \qquad ...(16.4)$$

$$Fe_2O_3 + 6H^+ + 2e^- \rightarrow 2Fe^{2+} + 3H_2O \qquad ...(16.5)$$

Although soluble manganese generally is found in soil as Mn^{2+} ion, soluble iron(II) frequently occurs as negatively charged iron-organic chelates. Strong chelation of iron(II) by soil fulvic acids apparently enables reduction of iron(III) oxides at more positive pE values than would otherwise be possible. This causes an upward shift in the $Fe(II)$-$Fe(OH)_3$ boundary.

Some soluble metal ions such as Fe^{2+} and Mn^{2+} are toxic to plants at high levels. Their oxidation to insoluble oxides may cause formation of deposits of Fe_2O_3 and MnO_2, which clog tile drains in fields.

Roughly 35% of the volume of typical soil is composed of air-filled pores. Whereas the normal dry atmosphere at sea level contains 21% O_2 and 0.03% CO_2 by volume, these percentages may be quite different in soil air because of the decay of organic matter :

$$\{CH_2O\} + O_2 \rightarrow CO_2 + H_2O \qquad ...(16.6)$$

This process consumes oxygen and produces CO_2. As a result, the oxygen content of air in soil may be as low as 15%, and the carbon dioxide content may be several per cent. Thus, the decay of organic matter in soil increases the equilibrium level of dissolved CO_2 in groundwater. This lowers the pH and contributes to weathering of carbonate minerals, particularly calcium carbonate. As already discussed CO_2 also shifts the equilibrium of the process by which roots absorb metal ions from soil.

Inorganic Components of Soil

The weathering of parent rocks and minerals to form the inorganic soil components results ultimately in the formation of inorganic colloids. These colloids are repositories of water and plant nutrients, which may be made available to plants as needed. Inorganic soil colloids often absorb toxic substances in soil, thus playing a role in detoxification of substances that otherwise would harm plants. The abundance and nature of inorganic colloidal material in soil are obviously important factors in determining soil productivity.

The uptake of plant nutrients by roots often involves complex interactions with the water and inorganic phases. For example, a nutrient held by inorganic colloidal material has to traverse the mineral/water interface and then the water/root interface. This process is often strongly influenced by the ionic structure of soil inorganic matter.

The most common elements in the earth's crust are oxygen, silicon, aluminium, iron, calcium, sodium, potassium, and magnesium. Therefore, minerals composed of these elements—particularly silicon and oxygen—constitute most of the mineral fraction of the soil. Common soil mineral constituents are finely divided quartz (SiO_2), orthoclase ($KAlSi_3O_8$), albite ($NaAlSi_3O_8$), epidote ($4CaO \cdot 3(AlFe)_2O_3 \cdot 6SiO_2 \cdot H_2O$), geothite ($FeO(OH)$), magnetite ($Fe_3O_4$), calcium and magnesium carbonates ($CaCO_3$, $CaCO_3 \cdot MgCO_3$), and oxides of manganese and titanium.

Organic Matter in Soil

Though typically comprising less than 5% of a productive soil, organic matter largely determines soil productivity. It serves as a source of food for microorganisms; undergoes chemical reactions such as ion exchange; and influences the physical properties of soil. Some organic compounds even contribute to the weathering of mineral matter, the process by which soil is formed. For example, $C_2O_4^{2-}$, oxalate ion, produced as a soil fungi metabolite, occurs in soil as the calcium salts whewellite and weddelite. Oxalate in soil water dissolves minerals, thus speeding the weathering process and increasing the availability of nutrient ion species. This weathering process involves oxalate complexation of iron or aluminium in minerals, represented by the reaction

$$3H^+ + M(OH)_3(s) + 2CaC_2O_4(s) \rightarrow M(C_2O_4)_2^-(aq) + 2Ca^{2+}\ (aq) + 3H_2O \qquad ...(16.7)$$

in which M is Al or Fe. Some soil fungi produce citric acid, and other chelating organic acids, which react with silicate minerals and release potassium and other nutrient metal ions held by these minerals.

The strong chelating agent 2-ketogluconic acid is produced by some soil bacteria. By solubilising metal ions, it may contribute to the weathering of minerals. It may also be involved in the release of phosphate from insoluble phosphate compounds.

Biologically active components of the organic soil fraction include polysaccharides, amino sugars, nucleotides, and organic sulphur and phosphorus compounds. Humus, a water-insoluble material that biodegrades very slowly, makes up the bulk of soil organic matter. The organic compounds in soil are summarised in Table 16.2.

Table 16.2. Major classes of organic compounds in soil.

Compound type	Composition	Significance
Humus	Degradation-resistant residue from plant decay, largely C, H, and O	Most abundant organic component, improves soil physical properties, exchanges nutrients, reservoir of fixed N
Fats, resins, and waxes	Lipids extractable by organic solvents	Generally, only several per cent of soil organic matter, may adversely affect soil physical properties by repelling water, perhaps phytotoxic
Saccharides	Cellulose, starches, hemi-cellulose, gums	Major food source for soil micro-organisms, help stabilise soil aggregates
N-containing organics	Nitrogen bound to humus, amino acids, amino sugars, other compounds	Provide nitrogen for soil fertility
Phosphorus compounds	Phosphate esters, inositol phosphates (phytic acid), phospholipids	Sources of plant phosphate

The accumulation of organic matter in soil is strongly influenced by temperature and by the availability of oxygen. Since the rate of biodegradation decreases with decreasing temperature, organic matter does not degrade rapidly in colder climates and tends to build up in soil. In water and in waterlogged soils, decaying vegetation does not have easy access to oxygen, and organic matter accumulates. The organic content may reach 90% in areas where plants grow and decay in soil saturated with water.

The presence of naturally occurring polycyclic aromatic (PAH) compounds is an interesting feature of soil organic matter. These compounds, some of which are carcinogenic, found in soil include fluoranthene, pyrene, and chrysene. PAH compounds in soil result in part from combustion from both natural sources (grass fires) or pollutant sources. Terpenes also occur in soil organic matter. Extraction of soil with ether and alcohol yields the pigments β-carotene, chlorophyll, and xanthophyll.

Soil Humus

Of the organic components listed in Table 16.2, soil humus is by far the most significant. Humus, composed of a base-soluble fraction called humic and fulvic acids and an insoluble fraction called humin, is the residue 'eft when bacteria and fungi biodegrade plant material. The bulk of plant biomass consists of relatively degradable cellulose and degradation-resistant lignin, which is a polymeric substance with a higher carbon content than cellulose. Among lignin's prominent chemical components are aromatic rings connected by alkyl chains, methoxyl groups, and hydroxyl groups. The process by which humus is formed is called humification. Soil humus is similar to its lignin precursors, but has more carboxylic acid groups. Part of each molecule of humic substance is non-polar and hydrophobic, and part is polar and hydrophilic. Such molecules are called amphiphiles, and they form micelles in which the non-polar parts compose the inside of small colloidal particles and the polar functional groups are on the outside. Amphiphilic humic substances probably also form bilayer surface coatings on mineral grains in soil.

An increase in nitrogen/carbon ratio is a significant feature of the transformation of plant biomass to humus through the humification process. This ratio starts at approximately 1/100 in fresh plant biomass. During humification, micro-organisms convert organic carbon to CO_2 to obtain energy. Simultaneously, the bacterial action incorporates bound nitrogen with the compounds produced by the decay processes. The result is a nitrogen/carbon ratio of about 1/10 upon completion of humification. As a general rule, therefore, humus is relatively rich in organically bound nitrogen.

Humic substances influence soil properties to a degree out of proportions to their small percentage in soil. They strongly bind metals, and serve to hold micronutrient metal ions in soil. Because of their acid-base character, humic substances serve as buffers in soil. The water-holding capacity of soil is significantly increased by humic substances. These materials also stabilise. aggregates of soil particles, and increase the sorption of organic compounds by soil.

Humic materials in soil strongly sorb many solutes in soil water and have a particular affinity for heavy polyvalent cations. Soil humic substances may contain levels of uranium more than 10^4 times that of the water with which they are in equilibrium. Thus, water becomes depleted of its cations (or purified) in passing through humic-rich soils. Humic substances in soils also have a strong affinity for organic compounds with low water-solubility such as DDT or atrazine, a herbicide widely used to kill weeds in corn fields.

$$Atrazine$$

The structure of Atrazine is shown with a central triazine ring bearing a Cl substituent, an ethylamino group ($C_2H_5-N(H)-$) and an isopropylamino group ($-N(H)-C(H)(CH_3)_2$).

In some cases, there is a strong interaction between the organic and inorganic portions of soil. This is especially true of the strong complexes formed between clays and humic (fulvic) acid compounds. In many soils, 50–100% of soil carbon is complexed with clay. These complexes play a role in determining the physical properties of soil, soil fertility, and stabilisation of soil organic matter. One of the mechanisms for the chemical binding between clay colloidal particles and humic organic particles is probably of the flocculation type in which anionic organic molecules with carboxylic acid functional groups serve as bridges in combination with cations to bind clay colloidal particles together as a floc. Support is given to this hypothesis by the known ability of NH_4^+, Al^{3+}, Ca^{2+}, and Fe^{3+} cations to stimulate clay-organic complex formation. The synthesis, chemical reactions, and biodegradation of humic materials are affected by interaction with clays. The lower-molecular-weight fulvic acids may be bound to clay, occupying spaces in layers in the clay.

GEOCHEMISTRY

Geochemistry deals with chemical species, reactions, and processes in the lithosphere and their interactions with the atmosphere and hydrosphere. The branch of geochemistry that explores the complex interactions among the rock/water/air/life systems that determine the chemical characteristics of the surface environment is environmental geochemistry. Obviously, geochemistry and its environmental subdiscipline are very important in environmental chemistry.

Physical Aspects of Weathering

Weathering is a geochemical phenomenon. Rocks tend to weather more rapidly when there are pronounced differences in physical conditions – alternate freezing and thawing and wet periods alternating with severe drying. Other mechanical aspects are swelling and shrinking of minerals with hydration and dehydration as well as growth of roots through cracks in rocks. Temperature is involved in that the rates of chemical weathering (below) increase with increasing temperature.

Chemical Weathering

As a chemical phenomenon, weathering can be viewed as the result of the tendency of the rock/water/mineral system to attain equilibrium. This occurs through the usual chemical mechanisms of dissolution/precipitation, acid-base reactions, complexation, hydrolysis, and oxidation-reduction.

Weathering occurs extremely slowly in dry air. Water increases the rate of weathering by many orders of magnitude for several reasons. Water, itself, is a chemically active substance in the weathering process. Furthermore, water holds weathering agents in solution so that they are transported to chemically active sites on rock minerals and contact the mineral surfaces at the molecular and ionic level.

Prominent among such weathering agents are CO_2, O_2, organic acids (including humic and fulvic acids), sulphur acids [$SO_2(aq)$,H_2SO_4], and nitrogen acids (HNO_3, HNO_2). Water provides the source of H^+ ion needed for acid-forming gases to act as acids, as shown by the following :

$$CO_2 + H_2O \rightarrow H^+ + HCO_3^- \qquad\qquad ...(16.8)$$

$$SO_2 + H_2O \rightarrow H^+ + HSO_3^- \qquad\qquad ...(16.9)$$

Rainwater is essentially free of mineral solutes. It is usually slightly acidic due to the presence of dissolved carbon dioxide or more highly acidic because of acid-rain forming constituents. As a result of its slight acidity and lack of alkalinity and dissolved calcium salts, rainwater is chemically aggressive toward some kinds of mineral matter, which it breaks down by a process called chemical weathering. Because of this process, river water has a higher concentration of dissolved inorganic solids than does rainwater.

The processes involved in chemical weathering may be divided into the following major categories :

1. Hydration/dehydration, for example :

$$CaSO_4(s) + 2H_2O \rightarrow CaSO_4 \cdot 2H_2O(s)$$

$$2Fe(OH)_3 \cdot xH_2O(s) \rightarrow Fe_2O_3(s) + (3 + 2x)H_2O$$

2. Dissolution, for example :

$$CaSO_4 \cdot 2H_2O(s) \text{ (water)} \rightarrow Ca^{2+} (aq) + SO_4^{2-}(aq) + 2H_2O$$

3. Oxidation, such as occurs in the dissolution of pyrite :

$$4FeS_2(s) + 15O_2(g) + (8 + 2x)H_2O \rightarrow 2Fe_2O_3 \cdot xH_2O + 8SO_4^{2-}(aq) + 16H^+ (aq)$$

or in the following example in which dissolution of an iron(II) mineral is followed by oxidation of iron(II) to iron (III) :

$$Fe_2SiO_4(s) + 4CO_2(aq) + 4H_2O \rightarrow 2Fe^{2+} + 4HCO_3^- + H_4SiO_4$$

$$4Fe^{2+} + 8HCO_3^- + O_2(g) \rightarrow 2Fe_2O_3(s) + 8CO_2 + 4H_2O$$

The second of these two reactions may occur at a site some distance from the first, resulting in a net transport of iron from its original location. Iron, manganese, and sulphur are the major elements that undergo oxidation as part of the weathering process.

4. Dissolution with hydrolysis, as occurs with the hydrolysis of carbonate ion when mineral carbonates dissolve :

$$CaCO_3(s) + H_2O \rightarrow Ca^{2+}(aq) + HCO_3^-(aq) + OH^-(aq)$$

Hydrolysis is the major means by which silicates undergo weathering, as shown by the following reaction of foresterite :

$$Mg_2SiO_4(s) + 4CO_2 + 4H_2O \rightarrow 2Mg^{2+} + 4HCO_3^- + H_4SiO_4$$

The weathering of silicates yields soluble silicon as species such as H_4SiO_4, and residual silicon-containing minerals (clay minerals).

5. Acid hydrolysis, which accounts for the dissolution of significant amounts of $CaCO_3$ and $CaCO_3 \cdot MgCO_3$ in the presence CO_2-rich water :

$$CaCO_3(s) + H_2O + CO_2(aq) \rightarrow Ca^{2+}(aq) + 2HCO_3^-(aq)$$

6. Complexation, as exemplified by the reaction of oxalate ion, $C_2O_4^{2-}$ on muscovite, $K_2(Si_6Al_2)Al_4O_{20}(OH)_4$:

$$K_2(Si_6Al_2)Al_4O_{20}(OH)_4(s) + 6C_2O_4^{2-}(aq) + 20H \rightarrow 6AlC_2O_4^+(aq) + 6Si(OH)_4 + 2K^+$$

Reactions such as these largely determine the kinds and concentrations of solutes in surface water and groundwater. Acid hydrolysis, especially, is the predominant process that releases elements such as Na^+, K^+, and Ca^{2+} from silicate minerals.

GROUNDWATER IN THE GEOSPHERE

Groundwater Fig. (16.6) is a vital resource in its own right that plays a crucial role in geochemical processes, such as the formation of secondary minerals. The nature, quality and mobility of groundwater are all strongly dependent upon the rock formations in which the water is held. Physically, an important characteristic of such formations is their porosity, which determines the percentage of rock volume available to contain water. A second important physical characteristic is permeability, which describes the ease of flow of the water through the rock. High permeability is usually associated with high porosity. However, clays tend to have low permeability even when a large percentage of the volume is filled with water.

Most groundwater originates as meteoric water from precipitation in the form of rain or snow. If water from this source is not lost by evaporation, transpiration, or to stream run-off, it may infiltrate into the ground. Initial amounts of water from precipitation onto dry soil are held very tightly as a film on the surfaces and in the micropores of soil particles in a belt of soil moisture. At intermediate levels, the soil particles are covered with films of water, but air is still present in larger voids in the soil. The region in which such water is held is called the unsaturated zone or zone of aeration and the water present in it is *vadose* water. At lower depths in the presence of adequate amounts of water, all voids are filled to produce a zone of saturation, the upper level of which is the water table. Water present in a zone of saturation is called groundwater. Because of its surface tension, water is drawn somewhat above the water table by capillary-sized passages in soil in a region called the capillary fringe.

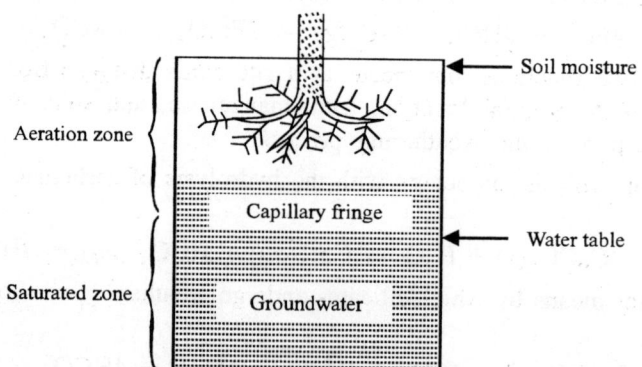

Fig. 16.6. Some major features of the distribution of water underground.

The water table is crucial in explaining and predicting the flow of wells and springs and the levels of streams and lakes. It is also an important factor in determining the extent to which pollutant and hazardous chemicals underground are likely to be transported by water. The water table can be mapped by observing the equilibrium level of water in wells, which is essentially the same as the top of the saturated zone. The water table is usually not level, but tends to follow the general contours of the surface topography. It also varies with differences in permeability and water infiltration. The water table is at surface level in the vicinity of swamps and frequently above the surface where lakes and streams are encountered. The water level in such bodies may be maintained by the water table. Influent streams or reservoirs are located above the water table; they lose water to the underlying aquifer and cause an upward bulge in the water table beneath the surface water.

Groundwater flow is an important consideration in determining the accessibility of the water for use and transport of pollutants from underground waste sites. Various parts of a body of groundwater are in hydraulic contact so that a change in pressure at one point will tend to affect the pressure and level at another point. For example, infiltration from a heavy, localised rainfall may affect the water table at a point remote from the infiltration. Groundwater flow occurs as the result of the natural tendency of the water table to assume even levels by the action of gravity.

Groundwater flow is strongly influenced by rock permeability. Porous or extensively fractured rock is relatively highly pervious, meaning that water can migrate through the holes, fissures, and pores in such rock. Because water can be extracted from such a formation, it is called an aquifer. By contrast, an aquiclude is a rock formation that is too impermeable or unfractured to yield groundwater. Impervious rock in the unsaturated zone may retain water infiltrating from the surface to produce a perched water table that is above the main water table and from which water may be extracted. However, the amounts of water that can be extracted from such a formation are limited and the water is vulnerable to contamination.

Water Wells

Most groundwater is tapped for use by water wells drilled into the saturated zone. The use and misuse of water from this source has a number of environmental implications. In the US, about two-thirds of the groundwater pumped is consumed for irrigation; lesser amounts of groundwater are used for industrial and municipal applications.

As water is withdrawn, the water table in the vicinity of the well is lowered. This drawdown of water creates a zone of depression. In extreme cases the groundwater is severely depleted and surface land levels can even subside (which is one reason that Venice, Italy is now very vulnerable to flooding). Heavy drawdown can result in infiltration of pollutants from sources such as septic tanks, municipal refuse sites, and hazardous waste dumps. When soluble iron or manganese are present in groundwater, exposure to air at the well wall can result in the formation of deposits of insoluble iron and manganese oxides produced by bacterially catalysed processes :

$$4Fe^{2+}(aq) + O_2(aq) + 10H_2O \rightarrow 4Fe(OH)_3(s) + 8H^+ \qquad ...(16.10)$$

$$2Mn^{2-}(aq) + O_2(aq) + (2x + 2)H_2O \rightarrow 2MnO_2 \cdot xH_2O(s) + 4H^+ \qquad ...(16.11)$$

Deposits of iron and manganese that result from the processes outlined above coat the surfaces from which water flows into the well with a coating that is relatively impermeable to water. The deposits fill the spaces that water must traverse to enter the well. As a result, they can seriously impede the flow of water into the well from the water-bearing aquifer. This creates major water source problems for municipalities using groundwater for water supply. As a result of this problem, chemical or mechanical cleaning, drilling of new wells, or even acquisition of new water sources may be required.

CHAPTER 17

Soil Chemistry

INTRODUCTION

Soil and agricultural practices are strongly tied to the environment. Cultivation of land and agricultural practices can influence both the atmosphere and the hydrosphere. Although this chapter deals primarily with soil, the topic of agriculture in general is introduced for perspective.

Agriculture, the production of food by growing crops and livestock, provides for the most basic of human needs. No other industry impacts as much as agriculture does on the environment. Agriculture is absolutely essential to the maintenance of the huge human populations now on earth. The displacement of native plants, destruction of wildlife habitat, erosion, pesticide pollution, and other environmental aspects of agriculture have enormous potential for environmental damage. Survival of humankind on earth demands that agricultural practice become as environmentally friendly as possible. On the other hand, growth of domestic crops removes (at least temporarily) greenhouse gas carbon dioxide from the atmosphere and provides potential sources of renewable resources of energy and fibre that can substitute for petroleum-derived fuels and materials.

Agriculture can be divided into the two main categories of crop farming, in which plant photosynthesis is used to produce grain, fruit, and fibre; and live-stock farming, in which domesticated animals are grown for meat, milk, and other animal products. The major divisions of crop farming include production of cereals, such as wheat, corn, or rice; animal fodder, such as hay; fruit; vegetables; and specialty crops, such as sugarcane, sugar beets, tea, coffee, tobacco, cotton, and cacao. Livestock farming involves raising of cattle, sheep, goats, swine, asses, mules, camels, buffalo, and various kinds of poultry. In addition to meat, livestock produce dairy products, eggs, wool, and hides. Freshwater fish and even crayfish are raised on "fish farms". Beekeeping provides honey.

Agriculture is based on domestic plants engineered by early farmers from their wild plant ancestors. Without perhaps much of an awareness of what they were doing, early farmers selected plants with desired characteristics for the production of food. This selection of plants for domestic use brought about a very rapid evolutionary change, so profound that the products often barely resemble their wild ancestors. Plant breeding based on scientific principles of heredity is a very recent development dating from early in the present century. One of the major objectives of plant breeding has been to increase yield. In some cases the goal is to increase nutritional value, such as in the development of corn high in lysine, an amino acid essential for human nutrition, so that corn becomes a more complete food.

The development of hybrids has vastly increased yields and other desired characteristics of a number of important crops. Basically, hybrids are the offspring of crosses between two different true-breeding strains. Often quite different from either parent strain, hybrids tend to exhibit "hybrid vigour" and to have significantly higher yields. The most success with hybrid crops has been obtained with corn (maize). Corn is one of the easiest plants to hybridise because of the physical separation of the male flowers, which grow as tassels on top of the corn plant, from female flowers, which are attached to incipient ears on the side of the plant. Despite past successes by more conventional means and early disappointments with "genetic engineering", application of recombinant DNA technology may eventually overshadow all the advances in plant breeding made to date.

In addition to plant strains and varieties, numerous other factors are involved in crop production. Weather is an obvious factor, and shortages of water, chronic in many areas of the world, are mitigated by irrigation. Here automated techniques and computer control are beginning to play an important, often more environmentally-friendly, role by minimising the quantities of water required. The application of chemical fertiliser has vastly increased crop yields. The judicious application of pesticides, especially herbicides, but including insecticides and fungicides as well, has increased crop yields and reduced losses greatly. Use of herbicides has had an environmental benefit in reducing the degree of mechanical cultivation of soil required. Indeed, "no-till" and "low-till" agriculture are now widely practiced on some crops.

The crops that provide for most of human caloric food intake, as well as much food for animals, are cereals, which are harvested for their starch-rich seeds. In addition to corn, mentioned above, wheat used for making bread and related foods, and rice, consumed directly, other major cereal crops include barley, oats, rye, sorghum, and millets. As applied to agriculture and food, vegetables are plants or their products that can be eaten directly by humans. A large variety of different parts of plants are consumed as vegetables. These include leaves (lettuce), stems (asparagus), roots (carrots), tubers (potato), bulb (onion), immature flower (broccoli), immature fruit (cucumber), mature fruit (tomato), and seeds (pea). Fruits, which are bodies of plant tissue containing the seed, may be viewed as a sub-classification of vegetables. Common fruits include apple, peach, apricot, citrus (orange, lemon, lime, grapefruit), banana, cherry, and various kinds of berries.

The rearing of domestic animals may have significant environmental effects. The Netherlands' pork industry has been so successful that hog manure and its by-products have caused serious problems. Goats and sheep have destroyed pasture-land in the Near East, Northern Africa, Portugal, and Spain. Of particular concern are the environmental effects of raising cattle. Significant amounts of forest land have been converted to marginal pastureland to raise beef. Production of one pound of beef requires about 4 times as much water and 4 times as much feed as does production of 1 pound of chicken. An interesting aspect of the problem is emission of greenhouse-gas methane by anaerobic bacteria in the digestive systems of cattle and other ruminant animals; cattle rank right behind wetlands and rice paddies as producers of atmospheric methane. However, because of the action of specialised bacteria in their stomachs, cattle and other ruminant animals are capable of converting otherwise unusable cellulose to food.

Pesticides, particularly insecticides and herbicides are an integral part of modern agricultural production.

Soil is the most fundamental requirement for agriculture. To humans and most terrestrial organisms, soil is the most important part of the geosphere. Though only a tissue-thin layer compared to the earth's total diameter, soil is the medium that produces most of the food required by most living things. Good soil—and a climate conducive to its productivity—is the most valuable asset a nation can have.

In addition to being the site of most food production, soil is the receptor of large quantities of pollutants, such as particulate matter from power plant smokestacks. Fertilisers, pesticides, and some other materials applied to soil often contribute to water and air pollution. Therefore, soil is a key component of environmental chemical cycles.

As already discussed, soils are formed by the weathering of parent rocks as the result of interactive geological, hydrological, and biological processes. Soils are porous and are vertically stratified into horizons as the result of downward-percolating water and biological processes, including the production and decay of biomass. Soils are open systems that undergo continual exchange of matter and energy with the atmosphere, hydrosphere, and biosphere.

THE SOIL SOLUTION

The soil solution is the aqueous portion of soil that contains dissolved matter from soil chemical and biochemical processes and from exchange with the hydrosphere and biosphere. This medium transports chemical species to and from soil particles and provides intimate contact between the solutes and the soil particles. In addition to providing water for plant growth, it is an essential pathway for the exchange of plant nutrients between roots and solid soil.

Obtaining a sample of soil solution is often very difficult because the most significant part of it is bound in capillaries and as surface films. The most straightforward means is collection of drainage water. Displacement with a water-immiscible fluid, mechanical separation by centrifugation, or pressure or vacuum treatment may be used to collect soil solution.

Dissolved mineral matter in soil is largely present as ions. Prominent among the cations are H^+, Ca^{2+}, Mg^{2+}, K^+, Na^+, and usually very low levels of Fe^{2+}, Mn^{2+}, and Al^{3+}. The last three cations may be present in partially hydrolysed form, such as $FeOH^+$, or complexed by organic humic substance ligands. Anions that may be present are HCO_3^-, CO_3^{2-}, HSO_4^-, SO_4^{2-}, Cl^-, and F^-. In addition to being bound to H^+ in species such as bicarbonate, anions may be complexed with metal ions, such as in AlF^{2+}. Multivalent cations and anions form ion pairs with each other in soil solutions. Examples of these are $CaSO_4$ and $FeSO_4$.

ACID-BASE AND ION EXCHANGE REACTIONS IN SOILS

One of the more important chemical functions of soils is the exchange of cations. The ability of a sediment or soil to exchange cations is expressed as the cation-exchange capacity (CEC), the number of milli-equivalents (meq) of monovalent cations that can be exchanged per 100 g of soil (on a dry-weight basis). The CEC should be looked upon as a conditional constant, since it may vary with soil conditions such as pE and pH. Both the mineral and organic portions of soils exchange cations. Clay minerals exchange cations because of the presence of negatively charged sites on the mineral, resulting from the substitution of an atom of lower oxidation number for one of higher number; for example, magnesium for aluminium. Organic materials exchange cations because of the presence of the carboxylate group and other basic functional groups. Humus typically has a very high cation-exchange capacity. The cation-exchange capacity of peat may range from 300 to 400 meq/100 g. Values of cation-exchange capacity for soils with more typical levels of organic matter are around 10 to 30 meq/100 g.

Cation exchange in soil is the mechanism by which potassium, calcium, magnesium, and essential trace-level metals are made available to plants. When nutrient metal ions are taken up by plant roots, hydrogen ion is exchanged for the metal ions. This process, plus the leaching of calcium, magnesium, and other metal ions from the soil by water containing carbonic acid, tends to make the soil acidic :

$$\text{Soil\} } Ca^{2+} + 2CO_2 + 2H_2O \rightarrow \text{Soil\} } (H^+)_2 + Ca^{2+} \text{ (root)} + 2HCO_3^- \qquad ...(17.1)$$

Soil acts as a buffer and resists changes in pH. The buffering capacity depends upon the type of soil.

Production of Mineral Acid in Soil

The oxidation of pyrite in soil causes formation of acid-sulphate soils sometimes called "cat clays" :

$$FeS_2 + 7/2 \; O_2 + H_2O \rightarrow Fe^{2+} + 2H^+ + 2SO_4^{2-} \qquad ...(17.2)$$

Cat clay soils may have pH values as low as 3.0. These soils, which are commonly formed when neutral or basic marine sediments containing FeS_2 become acidic upon oxidation of pyrite when exposed to air. For example, soil reclaimed from marshlands and used for citrus groves has developed high acidity detrimental to plant growth. In addition, H_2S released by reaction of FeS_2 with acid is very toxic to citrus roots.

Soils are tested for potential acid-sulphate formation using a peroxide test. This test consists of oxidising FeS_2 in the soil with 30% H_2O_2,

$$FeS_2 + 15/2 \; H_2O_2 \rightarrow Fe^{3+} + H^+ + 2SO_4^{2-} + 7H_2O \qquad ...(17.3)$$

then testing for acidity and sulphate. Appreciable levels of sulphate and a pH below 3.0 indicate potential to form acid-sulphate soils. If the pH is above 3.0, either little FeS_2 is present or sufficient $CaCO_3$ is in the soil to neutralise the sulphuric acid and acidic Fe^{3+}.

Pyrite-containing mine spoils (residue left over from mining) also form soils similar to acid-sulphate soils of marine origin. In addition to high acidity and toxic H_2S, a major chemical species limiting plant growth on such soils is Al. Aluminium ion liberated in acidic soils is very toxic to plants.

Adjustment of Soil Acidity

Most common plants grow best in soil with a pH near neutrality. If the soil becomes too acidic for optimum plant growth, it may be restored to productivity by liming, ordinarily through the addition of calcium carbonate :

$$\text{Soil\} } (H^+)_2 + CaCO_3 \rightarrow \text{Soil\} } Ca^{2+} + CO_2 + H_2O \qquad ...(17.4)$$

In areas of low rainfall, soils may become too basic (alkaline) due to the presence of basic salts such as Na_2CO_3. Alkaline soils may be treated with aluminium or iron sulphate, which release acid on hydrolysis :

$$2Fe^{3+} + 3SO_4^{2-} + 6H_2O \rightarrow 2Fe(OH)_3(s) + 6H^+ + 3SO_4^{2-} \qquad ...(17.5)$$

Sulphur added to soils is oxidised by bacterially mediated reactions to sulphuric acid :

$$S + 3/2 \; O_2 + H_2O \rightarrow 2H^+ + SO_4^{2-} \qquad ...(17.6)$$

and sulphur is used, therefore, to acidify alkaline soils. The huge quantities of sulphur now being removed from fossil fuels to prevent air pollution by sulphur dioxide may make the treatment of alkaline soils by sulphur much more attractive economically.

Ion Exchange Equilibria in Soil

Competition of different cations for cation exchange sites on soil cation exchangers may be described semi-quantitatively by exchange constants. For example, soil reclaimed from an area flooded with seawater will have most of its cation exchange sites occupied by Na^+, and restoration of fertility requires binding of nutrient cations such as K^+ :

$$Soil\} \ Na^+ + K^+ \rightleftharpoons Soil\} \ K^+ + Na^+ \qquad \text{...(17.7)}$$

The exchange constant is K_c,

$$K_c = \frac{N_K[Na^+]}{N_{Na}[K^+]} \qquad \text{...(17.8)}$$

which expresses the relative tendency of soil to retain K^+ and Na^+. In this equation, N_K and N_{Na} are the equivalent ionic fractions of potassium and sodium, respectively, bound to soil, and $[Na^+]$ and $[K^+]$ are the concentrations of these ions in the surrounding soil water. For example, a soil with all cation exchange sites occupied by Na^+ would have a value of 1.00 for N_{Na}; with one-half of the cation exchange sites occupied by Na^+, N_{Na} is 0.5; etc. The exchange of anions by soil is not nearly as clearly defined as is the exchange of cations. In many cases, the exchange of anions does not involve a simple ion-exchange process. This is true of the strong retention of orthophosphate species by soil. At the other end of the scale, nitrate ion is very weakly retained by the soil.

Anion exchange may be visualised as occurring at the surfaces of oxides in the mineral portion of soil. At low pH, the oxide surface may have a net positive charge, enabling it to hold anions such as chloride by electrostatic attraction.

At higher pH values, the metal oxide surface has a net negative charge due to the formation of OH^- ion on the surface, caused by loss of H^+ from the water molecules bound to the surface.

In such cases, it is possible for anions such as HPO_4^{2-} to displace hydroxide ion and bond directly to the oxide surface.

MACRO-NUTRIENTS IN SOIL

One of the most important functions of soil in supporting plant growth is to provide essential plant nutrients—macro-nutrients and micro-nutrients. Macro-nutrients are those elements that occur in substantial levels in plant materials or in fluids in the plant. Micro-nutrients are elements that are essential only at very low levels and generally are required for the functioning of essential enzymes.

The elements generally recognised as essential macro-nutrients for plants are carbon, hydrogen, oxygen, nitrogen, phosphorus, potassium, calcium, magnesium, and sulphur. Carbon, hydrogen, and oxygen are obtained from the atmosphere. The other essential macro-nutrients must be obtained from soil. Of these, nitrogen, phosphorus, and potassium are the most likely to be lacking and are commonly added to soil as fertilisers.

Calcium-deficient soils are relatively uncommon. Liming, a process used to treat acid soils, provides a more than adequate calcium supply for plants. However, calcium uptake by plants and leaching by carbonic acid may produce a calcium deficiency in soil. Acid soils may still contain an appreciable level of calcium which, because of competition by hydrogen ion, is not available to plants. Treatment of acid soil to restore the pH to near-neutrality generally remedies the calcium deficiency. In alkaline soils, the presence of high levels of sodium, magnesium, and potassium sometimes produces calcium deficiency because these ions compete with calcium for availability to plants.

Although magnesium makes up 2.1% of the earth's crust, most of it is rather strongly bound in minerals. Generally, exchangeable magnesium is considered available to plants and is held by ion-exchanging organic matter or clays. The availability of magnesium to plants depends upon the calcium/magnesium ratio. If this ratio is too high, magnesium may not be available to plants, and magnesium deficiency results. Similarly, excessive levels of potassium or sodium may cause magnesium deficiency.

Sulphur is assimilated by plants as the sulphate ion, SO_4^{2-}. In addition, in areas where the atmosphere is contaminated with SO_2, sulphur may be absorbed as sulphur dioxide by plant leaves. Atmospheric sulphur dioxide levels have been high enough to kill vegetation in some areas. However, some experiments designed to show SO_2 toxicity to plants have resulted in increased plant growth where there was an unexpected sulphur deficiency in the soil used for the experiment.

Soils deficient in sulphur do not support plant growth well, largely because sulphur is a component of some essential amino acids and of thiamine and biotin. Sulphate ion is generally present in the soil as immobilised insoluble sulphate minerals or as soluble salts, which are readily leached from the soil and lost as soil water run-off. Unlike the case of nutrient cations such as K^+, little sulphate is adsorbed to the soil (that is, bound by ion exchange binding) where it is resistant to leaching while still available for assimilation by plant roots.

Soil sulphur deficiencies have been found in a number of regions of the world. Whereas most fertilisers formerly contained sulphur, its use in commercial fertilisers is declining. If this trend continues, it is possible that sulphur will become a limiting nutrient in more cases.

As already noted, the reaction of FeS_2 with acid in acid-sulphate soils may release H_2S, which is very toxic to plants and which also kills many beneficial micro-organisms. Toxic hydrogen sulphide can also be produced by reduction of sulphate ion through micro-organism-mediated reactions with organic matter. Production of hydrogen sulphide in flooded soils may be inhibited by treatment with oxidising compounds, one of the most effective of which is KNO_3.

NITROGEN, PHOSPHORUS, AND POTASSIUM IN SOIL

Nitrogen, phosphorus, and potassium are plant nutrients that are obtained from soil. They are so important for crop productivity that they are commonly added to soil as fertilisers. The environmental chemistry of these elements and their production as fertilisers is discussed below.

Nitrogen

Fig. 17.1 summarises the primary sinks and pathways of nitrogen in soil. In most soils, over 90% of the nitrogen content is organic. This organic nitrogen is primarily the product of the biodegradation of dead plants and animals. It is eventually hydrolysed to NH_4^+, which can be oxidised to NO_3^- by the action of bacteria in the soil.

Nitrogen bound to soil humus is especially important in maintaining soil fertility. Unlike potassium or phosphate, nitrogen is not a significant product of mineral weathering. Nitrogen-fixing organisms ordinarily cannot supply sufficient nitrogen to meet peak demand. Inorganic nitrogen from fertilisers and rainwater is often largely lost by leaching. Soil humus, however, serves as a reservoir of nitrogen required by plants. It has the additional advantage that its rate of decay, hence its rate of nitrogen release to plants, roughly parallels plant growth—rapid during the warm growing season, slow during the winter months.

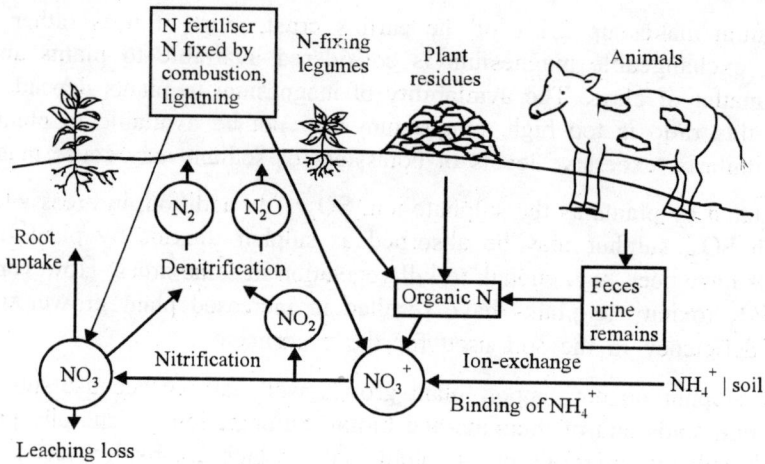

Fig. 17.1. Nitrogen sinks and pathways in soil.

Nitrogen is an essential component of proteins and other constituents of living matter. Plants and cereals grown on nitrogen-rich soils not only provide higher yields, but are often substantially richer in protein and therefore, more nutritious. Nitrogen is most generally available to plants as nitrate ion, NO_3^-. Some plants such as rice may utilise ammonium nitrogen readily; however, other plants have a preference for nitrate nitrogen. When nitrogen is applied to soils in the ammonium form, nitrifying bacteria perform an essential function in converting it to available nitrate ion.

Plants may absorb excessive amounts of nitrate nitrogen from soil. This phenomenon occurs particularly in heavily fertilised soils under drought conditions. Forage crops containing excessive amounts of nitrate can poison ruminant animals such as cattle or sheep. Plants having excessive levels of nitrate can endanger people when used for ensilage, an animal food consisting of finely chopped plant material such as partially matured whole corn plants, fermented in a structure called a silo. Under the reducing conditions of fermentation, nitrate in ensilage may be reduced to toxic NO_2 gas, which can accumulate to high levels in enclosed silos. There have been many cases reported of persons being killed by accumulated NO_2 in silos.

Nitrogen fixation is the process by which atmospheric N_2 is converted to nitrogen compounds available to plants. Human activities are resulting in the fixation of a great deal more nitrogen than would otherwise be the case. Artificial sources now account for 30–40% of all nitrogen fixed. These include chemical fertiliser manufacture; nitrogen fixed during fuel combustion; combustion of nitrogen-containing fuels; and the increased cultivation of nitrogen-fixing legumes. A major concern with this increased fixation of nitrogen is the possible effect upon the atmospheric ozone layer by N_2O released during denitrification of fixed nitrogen.

Prior to the widespread introduction of nitrogen fertilisers, soil nitrogen was provided primarily by legumes. These are plants such as soybeans, alfalfa, and clover, which contain on their root structures bacteria capable of fixing atmospheric nitrogen. Leguminous plants have a symbiotic (mutually advantageous) relationship with the bacteria that provide their nitrogen. Legumes may add significant quantities of nitrogen to soil, up to 10 pounds per acre per year, which is comparable to amounts commonly added as synthetic fertilisers. Soil fertility with respect to nitrogen may be maintained by rotating plantings of nitrogen-consuming plants with plantings of legumes, a fact recognised by agriculturists as far back as the Roman era.

The nitrogen-fixing bacteria in legumes exist in special structures on the roots called root nodules. The rod-shaped bacteria that fix nitrogen are members of a special genus called Rhizobium. These bacteria may exist independently, but cannot fix nitrogen except in symbiotic combination with plants. Although all species of Rhizobium appear to be very similar, they exhibit a great deal of specificity in their choice of host plants. Curiously, legume root nodules also contain a form of haemoglobin, which must somehow be involved in the nitrogen-fixation process.

Nitrate pollution of some surface waters and groundwater has become a major problem in some agricultural areas. Although fertilisers have been implicated in such pollution, there is evidence that feedlots are a major source of nitrate pollution. The growth of livestock populations and the concentration of livestock in feedlots have aggravated the problem. Such concentrations of cattle, coupled with the fact that a steer produces approximately 18 times as much waste material as a human, have resulted in high levels of water pollution in rural areas with small human populations. Streams and reservoirs in such areas frequently are just as polluted as those in densely populated and highly industrialised areas.

Nitrate in farm wells is a common and especially damaging manifestation of nitrogen pollution from feedlots because of the susceptibility of ruminant animals to nitrate poisoning. The stomach contents of ruminant animals such as cattle and sheep constitute a reducing medium (low p^E) and contain bacteria capable of reducing nitrate ion to toxic nitrite ion :

$$NO_3^- + 2H^+ + 2e^- \rightarrow NO_2^- + H_2O \qquad \qquad ...(17.9)$$

The origin of most nitrate produced from feedlot wastes is amino nitrogen present in nitrogen-containing waste products. Approximately one-half of the nitrogen excreted by cattle is contained in the urine. Part of this nitrogen is proteinaceous and the other part is in the form of urea, NH_2CONH_2. As a first step in the degradation process, the amino nitrogen is probably hydrolysed to ammonia, or ammonium ion :

$$RNH_2 + H_2O \rightarrow R–OH + NH_3(NH_4^+) \qquad \qquad ...(17.10)$$

This product is then oxidised through micro-organism-catalysed reactions to nitrate ion :

$$NH_3 + 2O_2 \rightarrow H^+ + NO_3^- + H_2O \qquad \qquad ...(17.11)$$

Under some conditions, an appreciable amount of the nitrogen originating from the degradation of feedlot wastes is present as ammonium ion. Ammonium ion is rather strongly bound to soil (recall that soil is a generally good cation exchanger), and a small fraction is fixed as non-exchangeable ammonium ion in the crystal lattice of clay minerals. Because nitrate ion is not strongly bound to soil, it is readily carried through soil formations by water. Many factors, including soil type, moisture, and level of organic matter, affect the production of ammonia and nitrate ion originating from feedlot wastes, and a marked variation is found in the levels and distributions of these materials in feedlot areas.

Phosphorus

Although the percentage of phosphorus in plant material is relatively low, it is an essential component of plants. Phosphorus, like nitrogen, must be present in a simple inorganic form before it can be taken up by plants. In the case of phosphorus, the utilisable species is some form of orthophosphate ion. In the pH range that is present in most soils, $H_2PO_4^-$ and HPO_4^{2-} are the predominant orthophosphate species.

Orthophosphate is most available to plants at pH values near neutrality. It is believed that in relatively acidic soils, orthophosphate ions are precipitated or sorbed by species of Al(III) and Fe(III). In alkaline soils, orthophosphate may react with calcium carbonate to form relatively insoluble hydroxyapatite :

$$3HPO_4^{2-} + 5CaCO_3(s) + 2H_2O \rightarrow Ca_5(PO_4)_3(OH) \ (s) + 5HCO_3^- + OH^- \qquad ...(17.12)$$

In general, because of these reactions, little phosphorus applied as fertiliser leaches from the soil. This is important from the standpoint of both water pollution and utilisation of phosphate fertilisers. Extensive research has been devoted to explaining the nature of the chemical interactions determining the availability of orthophosphates in soils.

Potassium

Relatively high levels of potassium are utilised by growing plants. Potassium activates some enzymes and plays a role in the water balance in plants. It is also essential for some carbohydrate transformations. Crop yields are generally greatly reduced in potassium-deficient soils. The higher the productivity of the crop, the more potassium is removed from soil. When nitrogen fertilisers are added to soils to increase productivity, removal of potassium is enhanced. Therefore, potassium may become a limiting nutrient in soils heavily fertilised with other nutrients.

Potassium is one of the most abundant elements in the earth's crust, of which it makes up 2.6%; however, much of this potassium is not easily available to plants. For example, some silicate minerals such as leucite, $K_2O \cdot Al_2O_3 \cdot 4SiO_2$, contain strongly bound potassium. Exchangeable potassium held by clay minerals is relatively more available to plants.

MICRO-NUTRIENTS IN SOIL

Boron, chlorine, copper, iron, manganese, molybdenum (for N-fixation), and zinc are considered essential plant micro-nutrients. These elements are needed by plants only at very low levels and frequently are toxic at higher levels. There is some chance that other elements will be added to this list as techniques for growing plants in environments free of specific elements improve. Most of these elements function as components of essential enzymes. Manganese, iron, chlorine, and zinc may be involved in photosynthesis. Though not established for all plants, it is possible that sodium, silicon, and cobalt may also be essential plant nutrients.

Iron and manganese occur in a number of soil minerals. Sodium and chlorine (as chloride) occur naturally in soil and are transported as atmospheric particulate matter from marine sprays. Some of the other micro-nutrients and trace elements are found in primary (unweathered) minerals that occur in soil. Boron is substituted isomorphically for Si in some mica and is present in tourmaline, a mineral with the formula $NaMg_3Al_6B_3Si_6O_{27}(OH,F)_4$. Copper is present, isomorphically substituted for other elements in feldspars, amphiboles, olivines, pyroxenes, and micas; it also occurs as trace levels of copper sulphides in silicate minerals. Molybdenum occurs as molybdenite (MoS_2). Vanadium is isomorphically substituted for Fe or Al in oxides, pyroxenes, amphiboles, and micas. Zinc is present as the result of isomorphic substitution for Mg, Fe, and Mn in oxides, amphiboles, olivines, and pyroxenes and as traces of zinc sulphide in silicates. Other trace level elements that occur as specific minerals, sulphide inclusions, or by isomorphic substitution for other elements in minerals are chromium, cobalt, arsenic, selenium, nickel, lead, and cadmium.

The trace elements listed above may be coprecipitated with secondary minerals that are involved in soil formation. Such secondary minerals include oxides of aluminium, iron, and manganese (precipitation of hydrated oxides of iron and manganese very efficiently removes many trace metal ions from solution); calcium and magnesium carbonates; smectites; vermiculites; and illites.

Some plants accumulate extremely high levels of specific trace metals. Those accumulating more than 1.00 mg/g of dry weight are called hyperaccumulators.

FERTILISERS

Crop fertilisers contain nitrogen, phosphorus, and potassium as major components. Magnesium, sulphate, and micro-nutrients may also be added. Fertilisers are designated by numbers, such as 6–12–8, showing the respective percentages of nitrogen expressed as N (in this case 6%), phosphorus as P_2O_5 (12%), and potassium as K_2O (8%). Farm manure is relatively poor in nutrients, corresponding to an approximately 0.5–0.24–0.5 fertiliser. The organic fertilisers such as manure must undergo biodegradation to release the simple inorganic species (NO_3^-, $H_xPO_4^{x-3}$, K^+) assimilable by plants.

Most modern nitrogen fertilisers are made by the Haber process, in which N_2 and H_2 are combined over a catalyst at temperatures of approximately 500°C and pressures up to 1000 atmosphere :

$$N_2 + 3H_2 \rightarrow 2NH_3 \qquad \text{...(17.13)}$$

The anhydrous ammonia product has a very high nitrogen content of 82%. It may be added directly to the soil, for which it has a strong affinity because of its water solubility and formation of ammonium ion :

$$NH_3(g) \text{ (water)} \rightarrow NH_3(aq) \qquad \text{...(17.14)}$$

$$NH_3(aq) + H_2O \rightarrow NH_4^+ + OH^- \qquad \text{...(17.15)}$$

Special equipment is required, however, because of the toxicity of ammonia gas. Aqua ammonia, a 30% solution of NH_3 in water, may be used with much greater safety. It is sometimes added directly to irrigation water. It should be pointed out that ammonia vapour is toxic and NH_3 is reactive with some substances. Improperly discarded or stored ammonia can be a hazardous waste.

Ammonium nitrate, NH_4NO_3, is a common solid nitrogen fertiliser. It is made by oxidising ammonia over a platinum catalyst, converting the nitric oxide product to nitric acid, and reacting the nitric acid with ammonia. The molten ammonium nitrate product is forced through nozzles at the top of a prilling tower and solidifies to form small pellets while falling through the tower. The particles are coated with a water repellent. Ammonium nitrate contains 33.5% nitrogen. Although convenient to apply to soil, it requires considerable care during manufacture and storage because it is explosive. Ammonium nitrate also poses some hazards. It is mixed with fuel oil to form an explosive that serves as a substitute for dynamite in quarry blasting and construction.

Urea,

$$\underset{\displaystyle H_2N-C-NH_2}{\overset{\displaystyle O}{\overset{\displaystyle \|}{}}}$$

is easier to manufacture and handle than ammonium nitrate. It is now the favoured solid nitrogen-containing fertiliser. The overall reaction for urea synthesis is

$$CO_2 + 2NH_3 \rightarrow CO(NH_2)_2 + H_2O \qquad ...(17.16)$$

involving a rather complicated process in which ammonium carbamate, $NH_2CO_2NH_4$, is an intermediate.

Other compounds used as nitrogen fertilisers include sodium nitrate, calcium nitrate, potassium nitrate, and ammonium phosphates. Ammonium sulphate, a by-product of coke ovens, used to be widely applied as fertiliser. The alkali metal nitrates tend to make soil alkaline, whereas ammonium sulphate leaves an acidic residue.

Phosphate minerals are found in several states. The principal mineral is fluorapatite, $Ca_5(PO_4)_3F$. Because the phosphate from fluorapatite is relatively unavailable to plants, this mineral is frequently treated with phosphoric or sulphuric acid to produce superphosphates :

$$2Ca_5(PO_4)_3F(s) + 14H_3PO_4 + 10H_2O \rightarrow 2HF(g) + 10Ca(H_2PO_4)_2 \cdot H_2O \qquad ...(17.17)$$

$$2Ca_5(PO_4)_3F(s) + 7H_2SO_4 + 3H_2O \rightarrow 2HF(g)$$
$$+ 3Ca(H_2PO_4)_2 \cdot H_2O + 7CaSO_4 \qquad ...(17.18)$$

The superphosphate products are much more soluble than the parent phosphate minerals. The HF produced as a by-product of superphosphate production can create air pollution problems.

Phosphate minerals are rich in trace elements required for plant growth, such as boron, copper, manganese, molybdenum, and zinc. Ironically, these elements are lost in processing phosphate for fertilisers and are sometimes added for fertilisers later.

Ammonium phosphates are excellent, highly soluble phosphate fertilisers. Liquid ammonium polyphosphate fertilisers consisting of ammonium salts of pyrophosphate, triphosphate, and small quantities of higher polymeric phosphate anions in aqueous solution are becoming very popular as phosphate fertilisers. The polyphosphates are believed to have the additional advantages of chelating iron and other micro-nutrient metal ions, thus making the metals more available to plants.

Potassium fertiliser components consist of potassium salts, generally KCl. Such salts are found as deposits in the ground or may be obtained from some brines. These salts are all quite soluble in water. One problem encountered with potassium fertilisers is the luxury uptake of potassium by some crops, which absorb more potassium than is really needed for their maximum growth. In a crop where only the grain is harvested, leaving the rest of the plant in the field, luxury uptake does not create much of a problem because most of the potassium is returned to the soil with the dead plant. However, when hay or forage is harvested, potassium contained in the plant as a consequence of luxury uptake is lost from the soil.

WASTES AND POLLUTANTS IN SOIL

Soil receives large quantities of waste products. Much of the sulphur dioxide emitted in the burning of sulphur-containing fuels ends up on soil as sulphates. Atmospheric nitrogen oxides are converted to nitrates in the atmosphere, and the nitrates eventually are deposited on soil. Soil sorbs NO and NO_2 readily, and these gases are oxidised to nitrate in the soil. Carbon monoxide is converted to CO_2 and possibly to biomass by soil bacteria and fungi. Particulate lead from automobile exhausts is found at elevated levels in soil along heavily travelled highways. Elevated levels of lead from lead mines and smelters are found on soil near such facilities.

Soil is the receptor of many hazardous wastes from landfill leachate, lagoons, and other sources. In some cases, land farming of degradable hazardous organic wastes is practiced as a means of disposal and degradation. The degradable material is worked into the soil, and soil microbial processes bring about its degradation. As already discussed, sewage and fertiliser-rich sewage sludge may be applied to soil.

Volatile organic compounds (VOC) such as benzene, toluene, xylenes, dichloromethane, trichloroethane, and trichloroethylene, are common soil pollutants in industrialised and commercialised areas. One of the more common sources of these contaminants is leaking underground storage tanks. Landfills built before current stringent regulations were enforced and improperly discarded solvents are also significant sources of soil VOCs.

Measurements of levels of polychlorinated biphenyls (PCBs) in soils that have been achieved for several decades provide interesting insight into the contamination of soil by pollutant chemicals and subsequent loss of these substances from soil.

Some pollutant organic compounds are believed to become bound with humus during the humification process that occurs in soil. This largely immobilises and detoxifies the compounds. Binding of pollutant compounds by humus is particularly likely to occur with compounds that have structural similarities to humic substances, such as phenolic and anilinic compounds illustrated by the following two examples :

2,4-Dichlorophenol   4-Chloroaniline

Such compounds can become covalently bonded to humic substance molecules, largely through the action of microbial enzymes. After binding they are known as bound residues and are highly resistant to biological and chemical attack.

Adsorption by soil is a key step in the degradation of a pesticide. The degree of adsorption and the speed and extent of ultimate degradation are influenced by a number of factors. Some of these, including solubility, volatility, charge, polarity, and molecular structure and size, are properties of the medium. Adsorption of a pesticide by soil components may have several effects. Under some circumstances, it retards degradation by separating the pesticide from the microbial enzymes that degrade it, whereas under other circumstances the reverse is true. Purely chemical degradation reactions may be catalysed by adsorption. Loss of the pesticide by volatilisation or leaching is diminished. The toxicity of a herbicide to plants may be strongly affected by soil sorption.

The forces holding a pesticide to soil particles may be of several types. Physical adsorption involves Van der Waals forces arising from dipole-dipole interactions between the pesticide molecule and charged soil particles. Ion exchange is especially effective in holding cationic organic compounds, such as the herbicide paraquat,

$$H_3C-\overset{+}{N}\underset{}{\bigcirc}\underset{}{\bigcirc}\overset{+}{N}-CH_3;\ 2Cl^-$$

to anionic soil particles. Some neutral pesticides become cationic by binding with H^+ and are bound as the protonated positive form. Hydrogen bonding is another mechanism by which some pesticides are held to soil. In some cases, a pesticide may act as a ligand coordinating to metals in soil mineral matter.

The three primary ways in which pesticides are degraded in or on soil are chemical degradation, photochemical reactions, and, most important, biodegradation. Various combinations of these processes may operate in the degradation of a pesticide.

Chemical degradation of pesticides has been observed experimentally in soils and clays sterilised to remove all microbial activity. For example, clays have been shown to catalyse the hydrolysis of o,o-dimethyl-o-2,4,5-trichlorophenyl thiophosphate (also called Trolene, Ronnel, Etrolene, or trichlorometafos), an effect attributed to $-OH$ groups on the mineral surface :

$$(CH_3O)_2\overset{\overset{S}{\|}}{P}-O-\underset{Cl}{\underset{}{\bigcirc}}\overset{Cl}{\underset{Cl}{}}-Cl \xrightarrow[\substack{\text{Mineral}\\\text{surfaces}}]{H_2O} HO-\underset{Cl}{\underset{}{\bigcirc}}\overset{Cl}{}-Cl + \overset{\overset{S}{\|}}{P}(OH)_3 + 2CH_3OH$$

Many other purely chemical hydrolytic reactions of pesticides occur in soil.

A number of pesticides have been shown to undergo photochemical reactions; that is, chemical reactions brought about by the absorption of light. Frequently, isomers of the pesticides are formed as products. Many of the studies reported apply to pesticides in water or on thin films, and the photochemical reactions of pesticides on soil and plant surfaces remain largely a matter of speculation.

Biodegradation and the Rhizosphere

Although insects, earthworms, and plants may play roles in the biodegradation of pesticides and other pollutant organic chemicals, microorganisms have the most important role.

In recent years it has become apparent that the rhizosphere is a particularly important part of soil in respect to biodegradation of wastes. The rhizosphere is the layer of soil in which plant roots are particularly active. It is a zone of increased biomass and is strongly influenced by the plant root system and the micro-organisms associated with plant roots. The rhizosphere may have more than 10 times the microbial biomass per unit volume compared to non-rhizospheric zones of soil. This population varies with soil characteristics, plant and root characteristics, moisture content, and exposure to oxygen. If this zone is exposed to pollutant compounds, micro-organisms adapted to their biodegradation may also be present.

Plants and micro-organisms exhibit a strong synergistic relationship in the rhizosphere, which benefits the plant and enables highly elevated populations of rhizospheric micro-organisms to exist. Epidermal cells sloughed from the root as it grows and carbohydrates, amino acids, and root-growth-lubricant mucigel secreted from the roots all provide nutrients for micro-organism growth. Root hairs provide a hospitable biological surface for colonisation by micro-organisms.

The biodegradation of a number of synthetic organic compounds has been demonstrated in the rhizosphere. Understandably, studies in this area have focused on herbicides and insecticides that are widely used on crops. Among the organic species for which enhanced biodegradation in the rhizosphere has been demonstrated are the following (associated plant or crop shown in parentheses): 2,4-D herbicide (wheat, African clover, sugarcane, flax), parathion (rice, bush bean), carbofuran (rice), atrazine (corn), diazinon (wheat, corn, peas), volatile aromatic alkyl and aryl hydrocarbons and chlorocarbons (reeds), and surfactants (corn, soybean, cattails). It is interesting to note that enhanced biodegradation of polycyclic aromatic hydrocarbons (PAH) was observed in the rhizosperic zones of prairie grasses. This observation is consistent with the fact that in nature such grasses burn regularly and significant quantities of PAH compounds are deposited on soil as a result.

SOIL EROSION AND CONSERVATION

Erosion may be defined as the detachment and transportation of soil. In most of the cases water is the transporting agent, sometimes wind also plays its part. Running water, wind, waves of sea and moving ice etc. cause a certain amount of erosion called geological erosion, which is a natural process, going on very slow through ages and is both constructive and destructive. Man for fulfilment of his ever-increasing needs, started to abuse (misuse) nature and smashed the natural balance between the soil loss and soil manufacturing process, thus giving rise to the serious problem of soil erosion.

Causes of Soil Erosion

1. Heavy destruction of the protective cover like trees and grass by indiscriminate cutting down of trees, forest fires, overgrazing and burning of grasslands.
2. Malpractices in the use of land such as up and successive growing of crops which accelerate soil erosion, removal of organic matter and other valuable nutrients by burning and injudicious cropping practices and faulty methods of irrigation.

Types of Erosion

By water

Erosion by water can be broadly classified as :

1. *Sheet erosion:* This usually occurs on landscape of gentle slopes. The top fertile layer of soil which is most suited for cultivation is ripped of in thin films every year by water.
2. *Rill erosion:* It is the second stage of sheet erosion. In this stage, small finger like rills begin to appear on the landscape. Year after year the rills increase slowly not only in number but also in their shape and size.
3. *Gully erosion:* It is the advanced stage of rill erosion. Within a few year's time the entire landscape is overlaid with a network of gullies, rendering the soil unfertile.
4. *Stream bank erosion:* The bank of the stream or rivers get eroded every year by flowing water. During the time when the streams and rivers get flooded, erosion of this type takes a serious form.

5. *Sea or shore erosion:* The tidal waters of sea cause considerable soil erosion along the coast, particularly so, during the rainy season. Because of this, the low-lying better shore lands are turned into vast strips.

6. *Slip erosion:* This is caused by hydraulic pressure exerted by moisture penetrating into the soil during heavy rains and being unable to percolate further down due to impermeable soil or rock strata, a great mass of overlyng soil on steep lands comes down bodily.

By wind

1. *Erosion by wind:* The high velocity of winds beat over the open lands with increasing force and the soil particles blown from the surface land are taken kilometres and kilometres away. Many times, particles of fine sand are carried away and deposited on vast strachs of neighbouring fertile rich land, thus requiring them useless for any crop growth.

Topography, soil type, vegetation cover are important factors which affect erosion.

Soil Conservation

Soil is one of the exhaustible natural resources. Their fertility can be restored but soil building requires ages. The top soil which is a nation's asset can be wasted away in no time due to faulty land use. It is not the farmer alone who suffers from soil erosion; but the unchecking erosion produces poverty and undermines the strength of nation. Soil conservation is the only way to protect the productive lands of the world.

Main approaches to soil conservation are as follows :

1. Conditioning the soil to make it resistant to detachment and transportation and more absorptive for surface water.

2. Covering the soil to protect it from rain impact and wind.

3. Slowing down run-off and wind.

4. Providing safe ways for the disposal of unavailable surface run-off.

Protecting soil against erosion by water

The different methods employed are as follows :

1. *Contour bunding:* It is the construction of small bunds across the slope of the land on a contour so that the long slope is out into a series of small ones making the water 'walk' rather than run. Different types and alignment of bunds are made according to need.

To attain its full benefits there are certain follow-up practices i.e., improved methods of tillage: (i) ploughing; (ii) harrowing; (iii) sowing; (iv) inter-culturing; (v) manuring; (vi) crop rotation and fallow; and (vii) strip cropping.

2. *Terracing:* It is primarily done on sloping lands. By intercepting the surface runoff before it attains sufficient velocity to erode the soil, terraces reduce soil and water loss, prevents formation of gullies and help in the reclamation of gullied land.

3. *Biological methods:* Control of erosion through crops or vegetation is called biological method.

(a) *Contouring:* It refers to the tillage practices of applying all treatments on the contour in agricultural, pasture or range lands.

(b) *Crop rotation:* It reduces erosion and increases the fertility of the soil, e.g., leguminous crops add nitrogen in soil.

(c) *Strip cropping:* Crop grown in strips or bands at right angles to the slope of the land helps in checking water erosion.

(d) *Erosion resisting crop:* Close growing crops such as legume, pulse or grass acts as erosion resisting crops.

(e) *Cover crops:* These are the green manure crops which protect the soil from erosion and help in absorption of soluble nitrates which would otherwise be lost, e.g., alfa-alf, clovera, kudza etc.

Forestry, woodlands and grassland management can be a powerful weapon for reducing erosion.

Protecting soil against erosion by wind

To check the wind erosion, three principal methods of reducing the surface velocity of wind have been employed.

1. *Vegetation method:* This is the most effective method of controlling erosion. It may be intertilled crops, close growing woody plants such as shrubs and trees. They are grown in the form of wind breaks and shelter belts.

Best species :

(a) *Grass:* Saccharum munja

(b) *Shrub:* Zizyphus jujuba

(c) *Trees:* Acacia arabica Casuarine spp.

2. *Tillage practices:*

(a) *Subble mulch farming:* The advantages claimed for this system are: (i) soil protection from direct on slaught of rain; (ii) better water absorption; (iii) reduced soil evaporation loss; and (iv) increased yield.

(b) *Primary and secondary tillage:* By this method, a rough and cloudy surface is produced on the soil surface. In order to obtain these conditions, cultivation is carried out as soon as possible after a rain.

(c) *Strip cropping or cover crop or grass:* During the period of heavy wind movements strip cropping helps wind erosion hazards.

3. *Mechanical means:* This consists of construction of trenches which should be covered with vegetation to reduce effect of wind erosion. Cultivation at right angles to the direction of wind is also helpful.

Soil and Water Resources

The conservation of soil and the protection of water resources are strongly interrelated. Most fresh water falls initially on soil, and the condition of the soil largely determines the fate of the water and how much is retained in a usable condition. The land area upon which rainwater falls is called a watershed. In addition to collecting the water, the watershed determines the direction and rate of flow and the degree of water infiltration into groundwater aquifers. Excessive rates of water flow prevent infiltration, lead to flash floods, and cause soil erosion. Measures taken to enhance the utility of land as a watershed also fortunately help prevent erosion. Some of these measures involve modification of the contour of the soil, particularly terracing, construction of waterways, and construction of water-retaining ponds. Waterways are planted with grass to prevent erosion, and water-retaining crops and bands of trees can be planted on the contour to achieve much the same goal. Reforestation and control of damaging grazing practices conserve both soil and water.

AGRICULTURE

Genetic Engineering and Agriculture

The nuclei of living cells contain the genetic instruction for cell reproduction. These instructions are in the form of a special material called deoxyribonucleic acid, DNA. In combination with proteins, DNA makes up the cell chromosomes. Such manipulation falls into the category of recombinant DNA technology. Recombinant DNA gets its name from the fact that it contains DNA from two different organisms, recombined together. This technology promises some exciting developments in agriculture.

The "green revolution" of the mid-1960s used conventional plant-breeding techniques of selective breeding, hybridisation, cross-pollination, and back-crossing to develop new strains of rice, wheat, and corn, which, when used with large quantities of chemical fertilisers, yielded spectacularly increased crop yields. For example, India's output of grain increased 50%. By working at the cellular level, however, it is now possible to greatly accelerate the process of plant breeding. Thus, plants may be developed that resist particular diseases, grow in seawater, or have much higher productivity. The possibility exists for developing entirely new kinds of plants.

One exciting possibility with genetic engineering is the development of plants other than legumes which fix their own nitrogen. For example, if nitrogen-fixing corn could be developed, the savings in fertiliser would be enormous. Furthermore, since the nitrogen is fixed in an organic form in plant root structures, there would be no pollutant runoff of chemical fertilisers.

Another promising possibility with genetic engineering is increased efficiency of photosynthesis. Plants utilise only about 1% of the sunlight striking their leaves, so there is appreciable room for improvement in that area.

Cell-culture techniques can be applied in which billions of cells are allowed to grow in a medium and develop mutants which, for example, might be resistant to particular viruses or herbicides or have other desirable qualities. If the cells with the desired qualities can be regenerated into whole plants, results can be obtained that might have taken decades using conventional plant-breeding techniques.

Despite the enormous potential of the "green revolution", genetic engineering, and more intensive cultivation of land to produce food and fibre, these technologies cannot be relied upon to support an uncontrolled increase in world population and may even simply postpone an inevitable day of reckoning with the consequences of population growth. Changes in climate resulting from global warming greenhouse effect, ozone depletion (by chlorofluorocarbons, or natural disasters, such as massive volcanic eruptions or collisions with large meteorites can, and almost certainly will, result in worldwide famine conditions in the future that no agricultural technology will be able to alleviate.

AGRICULTURE AND HEALTH

Some authorities hold that soil has an appreciable effect upon health. An obvious way in which such an effect might be manifested is the incorporation into food of micronutrient elements essential for human health. One such nutrient (which is toxic at overdose levels) is selenium. It is definitely known that the health of animals is adversely affected in selenium-deficient areas, as it is in areas of selenium excess. Human health might be similarly affected.

There are some striking geographic correlations with the occurrence of cancer. Some of these correlations may be due to soil type. A high incidence of stomach cancer has been shown to occur in areas with certain types of soil in the Netherlands, the United States, France, Wales, and Scandinavia. These soils are high in organic matter content, are acidic, and frequently are waterlogged. A "stomach cancer-prone life style" has been described, which includes consumption of home-grown food, consumption of water from one's own well, and reliance on native and uncommon foodstuffs.

One possible reason for the existence of "stomach cancer-producing soils" is the production of cancer-causing secondary metabolites by plants and micro-organisms. Secondary metabolites are biochemical compounds that are of no apparent use to the organism producing them. It is believed that they are formed from the precursors of primary metabolites when the primary metabolites accumulate to excessive levels.

The role of soil in environmental health is not well known, nor has it been extensively studied. The amount of research on the influence of soil in producing foods that are more nutritious and lower in content of naturally occurring toxic substances is quite small compared to research on higher soil productivity. It is to be hoped that the environmental health aspects of soil and its products will receive much greater emphasis in the future.

Chemical Contamination

Sometimes human activities contaminate food grown on soil. Most often this occurs through contamination by pesticides. It has been found that milk from several sources contained very high levels of heptachlor. This pesticide causes cancer and liver disorders in mice; therefore, it is a suspected human carcinogen.

CHAPTER 18

Natural Resources, Energy and Environment

INTRODUCTION

Technology, natural resources, energy, and the environment are intimately related. Perturbations in one usually cause perturbations in the other two. For example, reductions in automotive exhaust pollutant levels with the use of catalytic devices, have resulted in increased demand for platinum metal, a scarce natural resource, and greater gasoline consumption than would be the case if exhaust emissions were not a consideration (a particularly pronounced effect in the earlier years of emissions control). The availability of many metals depends upon the quantity of energy used and the amount of environmental damage tolerated in the extraction of low-grade ores. Many other such examples could be cited. Because of these intimate interrelationships, technology, resources and energy must be considered when environmental chemistry is discussed.

The challenge facing modern technologically-based societies is to achieve and maintain a high standard of living and quality of life while not ruining, and, indeed, while enhancing the Earth support system that sustains the society. Ultimately, this has to be done on a global basis, although much can be done nationally, locally, and individually. In this respect, the education and training of people who will be directing the technology on which society operates is of utmost importance. Traditional narrowly-based education in areas such as chemistry, economics, engineering, and even ecology are not suitable to prepare people for this challenge. The development of a "booming economy" defined in a traditional sense and characterised by high rates of production, consumption, and development can be very bad for the environment, and ultimately unsustainable. Engineering solutions that do not consider environmental protection are similarly undesirable. Some people would advocate the exact opposite, proposing a purely "environmental" solution and a "back-to-the-land" approach. This would result in great hardship, and is unacceptable to the majority of people. What is needed, then, are systems that interweave all the aspects of a modern industrialised society with an environmentally enlightened approach to Earth and its limited resources.

The term sustainable development has been used to describe industrial development that can be sustained without environmental damage and to the benefit of all people. Although the term has become widely used, it has been pointed out that some consider the term to be "an oxymoron without substance". Clearly, if humankind is to survive with a reasonable standard of living, something like "sustainable development" must evolve in which use of non-renewable resources is minimised insofar as possible, and the capability to produce renewable resources (for example, by promoting soil conservation to maintain the capacity to grow biomass) is enhanced. This will require significant behavioural changes, particularly in limiting population growth and curbing humankind's appetite for increasing consumption of goods and energy.

This chapter provides a brief overview of technology and touches upon the major aspects of natural resources and energy resources as they relate to environmental chemistry.

TECHNOLOGY

In studying environmental chemistry it is absolutely essential to consider technology, engineering, industry, and manufacturing because of the enormous influence that they have on the environment. Humans will use technology to provide the food, shelter, and goods that they need for their well-being and survival. The challenge is to interweave technology with considerations of the environment and ecology so that the two are mutually advantageous.

Technology, properly applied, is an enormously positive influence for environmental protection. The most obvious such application is in air and water pollution control. Necessary as "end-of-pipe" measures are for the control of pollution, it is much better to direct technology in manufacturing processes to prevent the formation of pollutants. Technology is being used increasingly to develop highly efficient processes of energy conversion, renewable energy resource utilisation, and conversion of raw materials to finished goods. In the transportation sector, properly applied technology in areas such as high speed train transport can enormously increase the speed, energy efficiency, and safety of the means by which people and goods are moved.

Technology refers to the ways in which humans apply knowledge and make things with materials and energy. In the modern era, technology is to a large extent the product of engineering based on scientific principles. Science deals with the discovery, explanation, and development of theories pertaining to inter-related natural phenomena of energy, matter, time, and space. Based on the fundamental knowledge of science, engineering provides the plans and means to achieve specific practical objectives. Technology uses these plans to carry out the desired objectives.

The 1800s saw an explosion in technology. Among the major advances during that century were widespread use of steam power, steam-powered railroads, the telegraph, telephone, electricity as a power source, textiles, use of iron and steel in building and bridge construction, cement, photography and invention of the internal combustion engine, which revolutionised transportation in the following century. During the last 100 years, advancing technology has been characterised by vastly increased uses of energy; greatly enhanced speed in manufacturing processes, information transfer, computation, transportation, and communication; automated control; an enormous new variety of chemicals; and, more recently, the widespread application of computers to manufacturing, communication, and transportation.

The technological advances of the present century are largely attributable to new and improved materials developed for a variety of new applications. For example, since before World War II, airliners have been constructed of special strong alloys of aluminium; these are being supplanted by even more advanced composites. Synthetic materials that have had a significant impact on modern technology include plastics, fibre reinforced materials, composites, and ceramics.

Until very recently, technological advances were made largely without heed to environmental impacts. Now, however, the greatest technological challenge is to reconcile technology with environmental consequences. The survival of human-kind and of the planet that supports it now requires that the established two-way interaction between science and technology become a three-way relationship including environmental protection.

INDUSTRY AND MANUFACTURING

The things that humans do to earn money with which to buy the goods and services that they need for existence are classified according to industries. An industry is an enterprise that makes a kind of good, or provides a particular service. Various industries have developed because specialisation of human activities makes for the greatest efficiency in providing goods and services. The kinds of industries that a country or region has depends upon the availability of needed attributes, such as raw materials, human resources, and transportation systems. Increasingly, environmental considerations are involved in determining kinds of industries that are required and that thrive.

Classification of Industries

Industries fall into various classes. In an early stage of development, basic-need industries providing essentials of food, clothing, fuel, and shelter are emphasised, but become less important relative to more discretionary industries as wealth increases. The most common example of this is the high percentage of people engaged in agriculture at early stages of development, which dwindles to a very low figure as the industrial and economic base becomes more developed. Somewhat arbitrarily, industries may be divided among the following categories :

1. *Food production:* agriculture and fishing.
2. *Extractive mineral industries:* that are involved with the mining of minerals, such as those used as sources of metals (energy sources are addressed in separate categories here).
3. *Renewable resource industries:* forestry, production of non-food crops, such as cotton.
4. *Renewable energy industry:* a small, but of necessity, growing industry dealing with the utilisation of renewable energy resources, such as solar energy, wind power, and biomass energy.
5. *Extractive energy industry:* consisting of coal mining, uranium ore mining, petroleum, and natural gas.
6. *Manufacturing:* conversion of raw materials or articles to higher-value goods.
7. *Construction:* Building and erection of dwellings, buildings, railroads, highways, and other components of the societal infrastructure.
8. *Utilities:* Electricity distribution systems, natural gas.
9. *Communications:* Telecommunications, media communications.
10. *Transportation:* Rail, highway, air, barge ship.
11. *Wholesale and retail trade:* which provides the interface between the production of goods and their sale for consumption and use.
12. *Finance:* Banks and other entities that provide the financial resources and transactions required for industries and trade.
13. *Services:* Law, medicine, motels, recreation, and many others.
14. *Government:* National, regional (state provincial), city, and local entities that provide needed services and regulation, such as environmental regulation.

15. *Education:* Schools from pre-school through graduate and professional schools providing the intellectual and human resources needed for a smoothly running modern society.

Manufacturing

Once a device or product is designed and developed, it must be made—synthesised or manufactured. This may consist of the synthesis of a chemical from raw materials, casting of metal or plastic parts, assembly of parts into a device or product, or any of the other things that go into producing a product that is needed in the marketplace.

Manufacturing activities have a tremendous influence on the environment. Energy, petroleum to make petrochemicals, and ores to make metals must be dug from, pumped from, or grown on the ground to provide essential raw materials. The potential for environmental pollution from mining, petroleum production, and intensive cultivation of soil is enormous. Huge land-disrupting factories and roads must be built to transport raw materials and manufactured products. The manufacture of goods carries with it the potential to cause significant air and water pollution and production of hazardous wastes. The earlier in the design and development process that environmental considerations are taken into account, the more "environmentally friendly" a manufacturing process will be.

RAW MATERIALS AND MINING

Manufacturing requires a steady flow of raw materials—minerals, fuel, wood, and fibre. These can be provided from either extractive (non-renewable) or renewable sources. A brief overview of these sources is given in this section and they are discussed in greater detail later in the chapter.

The extractive industries are those in which irreplaceable resources are taken from the Earth's crust. This is normally done by mining, but may also include pumping of crude oil and withdrawal of natural gas. The raw materials so obtained may be divided broadly into the categories of inorganic minerals, such as iron ore, clay used for fibrebrick, and gravel, and materials of organic origin, such as coal, lignite, or crude oil.

Geological and geochemical factors are crucial in mining, particularly in locating ore deposits. Deposits of metals often occur in masses of igneous rock that have been extruded in a solid or molten state into the surrounding rock strata; such masses are called batholiths. Other geological factors to consider include age of rock, fault zones, and rock fractures. The crucial step of finding ore deposits falls in the category of mining geology.

Surface mining, which for coal is usually called strip mining, is used to extract minerals that occur near the surface. A common example of surface mining is quarrying of rock. Coal is commonly strip-mined with giant shovels that are employed primarily for removing overburden, leaving the thinner coal seam to be loaded with smaller equipment.

Gravel, sand, and some other minerals, such as gold, often occur in so-called placer deposits to which they have been carried and deposited by running water. Extraction of minerals from placer deposits has obvious environmental implications. Mining of placer deposits can be accomplished by dredging from a boom-equipped barge. Another means that can be used is hydraulic mining with large streams of water. One interesting approach for more coherent deposits is to cut the ore with intense water jets, then suck up the resulting small particles with a pumping system.

For many minerals underground mining is the only practical means of extraction. An underground mine can be very complex and sophisticated. The structure of the mine depends upon the nature of the deposit. It is, of course, necessary to have a shaft that reaches to the ore deposit. Horizontal tunnels extend out into the deposit, and provision must be made for sumps to remove water and for ventilation. Factors that must be considered in designing an underground mine include the depth, shape and orientation of the ore body, as well as the nature and strength of the rock in and around it; thickness of overburden; and depth below the surface.

Usually, significant amounts of processing are required before a mined product is used or even moved from the mine site. Such processing, and the by-products of it, can have significant environmental effects. Even rock to be used for aggregate and for road construction must be crushed and sized. Crushing is also a necessary first step for further processing of ores. Some minerals occur to an extent of a few per cent or even less in the rock taken from the mine and must be concentrated onsite so that the residue does not have to be hauled far. For metals mining, these processes, as well as roasting, extraction, and similar operations are covered under the category of extractive metallurgy.

TRANSPORTATION

Ways of getting around and of moving materials and belongings have always been central to human existence and lifestyle. From prehistoric times humans devised means of transportation, usually powered by human muscles, animals, or wind. For both land and water transport, the development around 1800 of steam engines light enough to fit on self-propelled vehicles provided an enormous impetus to transportation. Steam-powered ships and boats freed water transport from the vagaries of the wind and enabled movement of boats upstream, though in earlier years of steam transport, at a fearful price from boiler explosions and fires. The mating of the steam engine mounted on a locomotive with steel rails enabled the construction of railroads, which totally revolutionised movement of people and freight over land and completely changed human economic and social systems. The next huge advance was enabled by the development of successful internal combustion (gasoline) engines in the late 1800s. These relatively light and compact power plants made the automobile a practical reality and made air transport possible. As the 1900s progressed, rail transport gave way (unfortunately in some respects) to a large extent to airplanes, automobiles, buses, and trucks. Now, at least technically, transport through space is possible.

In more industrialised nations in the modern era, private transport, commuting, and travel over relatively short distances is largely by private automobile. The flexibility, convenience, and independence afforded by private automobiles have made them extremely popular. Technical advances have greatly improved automobile efficiency, comfort, and safety (relative to size). These conveniences have come at a high cost to the environment and, because of automobile accidents, in human lives. Land has been taken over to build highways, and the greater mobility afforded by automobiles has resulted in the conversion of vast areas of agricultural land and wildlife habitat to suburban housing areas. Increased energy demand for private automobiles and pollution from auto exhausts have had marked detrimental effects on the environment.

Movement of people and high-value freight by airplane is now reliable, safe, and relatively inexpensive. Technological advances in aircraft and aircraft engine design, as well as operation and control, made possible by better materials and computers have given a tremendous impetus to air transportation. It is clearly the mode of choice for overseas travel and long-distance travel over land, but definitely needs to be integrated with rail transport for distances of up to a few hundred kilometres.

Advanced technology is being used, and still has a huge unrealised potential, for improvement of transportation systems. Computerised control has enabled automobiles to be much more efficient while emitting far smaller amounts of pollutants. Advanced air traffic control systems enable the operation of more aircraft in smaller spaces and much closer intervals with much greater safety than was previously possible. Mixed modes of transport can be very successful; an example is the movement of truck trailers and their contents over long distances by rail followed by final distribution over short distances by tractor trailer truck. A systems approach that enables tradeoffs to be made among speed, energy consumption, noise, pollution, convenience, and other factors offers much promise in the transportation area.

HIGH TECH

What is high technology, "high tech"? Some have said, "To a caveman, it is the wheel". Certainly, in their day, the telegraph and the steam locomotive were high technology, as was the biplane aircraft of World War I. In present times, however, there is a group of technologies that can be labeled as high tech from the perspective of the modern era. In speaking of high tech a number of terms come to mind, including the following: voice synthesis, bionics, plasma systems, artificial intelligence, computerised language translation, remote sensing, cryogenics, photovoltaics, sonar, computer-controlled prosthetic devices, lasers, composite materials, ceramics, light emitting diodes, fibre optics, genetic engineering, robotics, cryptography, CAT scanners, MRI imaging, digital audio, digital video, as well as computer aided design (CAD), instruction (CAI), graphics (CAG), and manufacturing (CAM. Many of these high tech applications have significant potential to impact environmental quality and, if properly applied, to improve it.

Central to all high tech areas are computers and the microship, integrated circuits, and microprocessors that make them possible. Any other field of modern high technology is dependent upon computers. Computers are essential to modern telecommunications, they make robots possible, they are required for the development of new formulations for exotic materials, and they maintain the exacting conditions under which such materials must be made.

Endeavours in space certainly can be classified as high tech. The realities of high costs and other more pressing priorities have prevented the development of "factories on the moon" or permanently populated orbiting space stations that many predicted when the "space age" became reality in the late 1950s. The exotic space battles visualised in "star wars" concepts have given way to a large extent to the decline of major nuclear powers and to the realities of grubby little ethnic wars fought by more conventional means. The greatest practical success in space so far has been in the launching of telecommunication satelites that have revolutionised global communications. Of particular importance to the environment, space technology is used very successfully in weather forecasting and in studying threats to the global climate, particularly greenhouse warming and atmospheric ozone depletion.

Advanced biotechnology uses biochemical processes to perform tasks and to make products that would otherwise be impossible. Biotechnology directed through bioengineering makes use of enzymes, recombinant DNA, gene splicing, cloning and other biological phenomena. The most exciting area of biotechnology in recent years is the one dealing with gene splicing or recombinant DNA wherein DNA material from one organism is inserted into another to give an organism with desired characteristics, such as the ability to make a specific protein. Several significant products have been produced by gene splicing. One such product is Humulin, a form of insulin identical to human insulin; others are human growth hormone, TPA, and interferon. There is a high level of activity in biotechnology as applied to agriculture. Particularly promising are prospects to develop plants that are resistant to insects or to herbicides applied to competing plants, plants that can be made to fix nitrogen (through symbiotic bacteria growing on their roots), and growth promoters for plants, and perhaps for animals. Properly directed, biotechnology has a high potential to benefit the environment.

METALS

Technological aspects pertinent to environmental chemistry were briefly introduced above. The next part of this chapter deals with resources; that is, the materials potentially available for human use. In discussing minerals and fossil fuels in the remainder of this chapter, two terms related to available quantities are used which should be defined. The first of these is resources, which refers to quantities that are estimated to be ultimately available. The second term is reserves, which refers to well-identified resources that can be profitably utilised with existing technology.

With an adequate supply of all of the important elements and energy, almost any needed material can be manufactured. Most of the elements, including practically all of those likely to be in short supply are metals. Some metals are considered especially crucial because of their importance to industrialised societies, uncertain sources of supply, and price volatility in world markets. One of these is antimony, used in auto batteries, fire-resistant fabrics, and rubber. Chromium, another crucial metal, is used to manufacture stainless steel (especially for parts exposed to high temperatures and corrosive gases), jet aircraft, automobiles, hospital equipment, and mining equipment. The platinum-group metals (platinum, palladium, iridium, rhodium) are used as catalysts in the chemical industry in petroleum refining, and in automobile exhaust antipollution devices.

Mining and processing of metal ores (Table 18.1) involve major environmental concerns, including disturbance of land, air pollution from dust and smelter emissions, and water pollution from disrupted acquifers. This problem is aggravated by the fact that the general trend in mining involves utilisation of less rich ores. Ores as low as 0.1% copper may eventually be processed. Increased demand for a particular metal, coupled with the necessity to utilise lower grade ores, has a vicious multiplying effect upon the amount of ore that must be mined and processed, and accompanying environmental consequences.

NON-METAL MINERAL RESOURCES

A number of minerals other than those used to produce metals are important resources. There are so many of these that it is impossible to discuss them all in this chapter; however, mention will be made of the major ones. As with metals, the environmental aspects of mining many of these minerals are quite important. Typically, even the extraction of ordinary rock and gravel can have important environmental effects.

Various clays are also used for clarifying oils, as catalysts in petroleum processing, as fillers and coatings for paper, and in the manufacture of firebrick, pottery, sewer pipe, and floor tile. The main types of clays that have industrial uses are shown in Table 18.2.

Table 18.1. Important metals their ores, properties, and uses.

Metals	Properties[a]	Major uses	Ores and aspects of resources[a]
Aluminium	mp 660°C, bp 2467°C, sg 2.70, malleable, ductile	Metal products, including autos, aircraft, electrical equipment. Conducts electricity better than copper per unit weight and is used in electrical transmission lines.	From bauxite ore containing 35–55% Al_2O_3.
Chromium	mp 1903°C, bp 2642°C, sg 7.14, hard, silvery colour	Metal plating, stainless steel, wear-resistant and cutting-tool alloys, chromium chemicals, including chromate used as an anticorrosive and cooling-water additive.	From chromite having the general formula [Mg(II), Fe(II)][Cr(III), Al(III), Fe(III)]$_2$O$_4$.
Cobalt	mp 1495°C, bp 2880°C, sg 8.71, bright silvery	Manufacture of hard, heat-resistant alloys such as satellite, permanent magnet alloys, driers, pigments glazes, catalysts, animal-feed additive.	From a variety of minerals, such as linnaeite, Co_3S_4, and as a by-product of other metals.
Copper	mp 1083°C, bp 2582°C, sg 8.96, dense, ductile, malleable	Electrical conductors, alloys, chemicals. Many uses.	Occurs in low percentages as sulphides, oxides, and carbonates in other minerals.
Gold	mp 1063°C, bp 2660°C, sg 19.3	Jewellery, basis of currency, electronics, increasing industrial uses.	In various minerals at a very low 10 ppm for ores.
Iron	mp 1535°C, bp 2885°C, sg 7.86, silvery metal in (rare) pure form	By far the most widely produces metal usually as steel, a high-tensile-strength material containing 0.3–1.7% C. Made into many alloys for special purposes.	Occurs as hematite (Fe_2O_3) geothite ($Fe_2O_4 \cdot H_2O$), and magnetite (Fe_3O_4).
Lead	mp 327°C, bp 1750°C, sg 11.35, silvery colour	Fifth most widely used metal. Storage batteries, gasoline additives, pigments, ammunition.	Major source is galena, PbS.
Manganese	mp 1244°C, bp 2040°C, sg 7.3, hard, brittle, gray-white	Sulphur and oxygen scavenger in steel, manufacture of alloys, dry cells, chemicals.	Found in a variety of minerals, primarily oxides, in manganese nodules on the ocean floor.
Mercury	mp –38°C, bp 357°C, sg 13.6, shiny liquid metal	Instruments, electronic apparatus, electrodes, chemical compounds (such as fungicides and slimicides).	From cinnabar, HgS.
Molybdenum	mp 2620°C, bp 4825°C sg 9.01, ductile, silvery-gray	Alloys, pigments, catalysts, chemicals, and lubricants.	Molybdenite (MoS_2) and wulfenite ($PbMoO_4$) are major ores.
Nickel	mp 1455°C, bp 2835°C, sg 8.90, silvery colour	Alloys, coins, storage batteries, catalysts (e.g., for hydro-genation of vegetable oil).	Found in ores associated with iron.

(Contd...)

Metals	Properties[a]	Major uses	Ores and aspects of resources[a]
Silver	mp 961°C, bp 2193°, sg 10.5, shiny metal	Photographic materials, electronics, sterling ware, jewellery, bearings, dentistry.	Found with sulphide minerals, by-product of copper, lead, and zinc smelting.
Tin	mp 232°C, bp 2687°C, sg 7.31	Coatings, solders, bearing alloys, bronze, chemicals.	Found in many compounds associated with granitic rocks and chrysolites.
Titanium	mp 1677°C, bp 3277°C, sg 4.5, silvery colour	Strong, corrosion-resistant, used in aircraft and their engines, valves, pumps, paint pigments.	Ranks ninth in elemental abundance, commonly as TiO_2.
Tungsten	mp 3380°C, bp 5530°C, sg 19.3 gray	Very strong, high boiling point, used in alloys, drill bits, turbines, nuclear reactors, tungsten carbide.	Found as tungstates, such as scheelite ($CaWO_4$).
Vanadium	mp 1917°C, bp 3375°C, sg 5.87, gray	Used to make strong steel alloys.	Occurs in igneous rocks, primarily as V (III), primarily by-product of other metals.
Zinc	mp 420°C, bp 907°C, sg 7.14, bluish white	Widely used in alloys (brass), galvanised, paint pigments, chemicals. Fourth in metal production worldwide.	Found in many ore minerals, including sulphides, oxides, and silicates.

[a]Mp, melting point; bp, boiling point; sg, specific gravity.

Table 18.2. Major types of clays and their uses.

Type of clay	Percent use	Composition	Uses
Miscellaneous	72	variable	Filler, brick, tile, portland cement, many others
Fireclay	12	variable; can be fired at high temperatures without warping	Refractories, pottery, sewer pipe, tile, brick
Kaolin	8	$Al_2(OH)_4Si_2O_5$; is white and can be fired without losing shape or colour	Paper filler, refractories, pottery, dinnerware, petroleum-cracking catalyst
Bentonite and fuller's earth	7	variable	Drilling muds, petroleum catalyst, carriers for pesticides, sealers, clarifying oils
Ball clay	1	Variable, very plastic	Refractories, tile, whiteware

Fluorine compounds are widely used in industry. Large quantities of fluorspar, CaF_2, are required as a flux in steel manufacture. Synthetic and natural cryolite, Na_3AlF_6, is used as a solvent for aluminium oxide in the electrolytic preparation of aluminium metal. Sodium fluoride is used for water fluoridation.

Micas are complex aluminium silicate minerals which are transparent, tough, flexible, and elastic. Muscovite, $K_2O \cdot 3Al_2O_3 \cdot 6\,SiO_2 \cdot 2\,H_2O$, is a major type of mica. Better grades of mica are cut into sheets and used in electronic apparatus, capacitors, generators, transformers, and motors. Finely divided mica is widely used in roofing, paint, welding rods, and many other applications.

In addition to consumption in fertiliser manufacture, phosphorus is used for supplementation of animal feeds, synthesis of detergent builders, and preparation of chemicals such as pesticides and medicines.

The most common phosphate minerals are fluorapatite, $Ca_5(PO_4)_3F$ and hydroxyapatite, $Ca_5(PO_4)_3(OH)$. Ions of Na, Sr, Th, and U are found substituted for calcium in apatite minerals. Small amounts of PO_4^{3-} may be replaced by AsO_4^{3-}, and the arsenic must be removed for food applications. Approximately 17% of world phosphate production is from igneous minerals, primarily fluorapatites. About three-fourths of world phosphate production is from sedimentary deposits, generally of marine origin. Vast deposits of phosphate, accounting for approximately 5% of world phosphate production, are derived from guano droppings of seabirds and bats.

Pigments and fillers of various kinds are used in large quantities. The only naturally occurring pigments still in wide use are those containing iron. These minerals are coloured by limonite, an amorphous brown-yellow compound with the formula $2Fe_2O_3 \cdot 3H_2O$, and hematite, composed of gray-black Fe_2O_3. Along with varying quantities of clay and manganese oxides, these compounds are found in ocher, sienna, and umber.

Manufactured pigments include carbon black, titanium dioxide, and zinc pigments. Carbon black, manufactured by the partial combustion of natural gas, are used primarily as a reinforcing agent in tyre rubber.

Minerals are used as fillers for paper, rubber, roofing, battery boxes, and many other products. Among the minerals used as fillers are, carbon black, diatomite, barite, fuller's earth, kaolin, mica, limestone, pyrophyllite, and wollastonite $(CaSiO_3)$.

Although sand and gravel are the cheapest of mineral commodities per ton, the average annual dollar value of these materials is greater than all but a few mineral products because of the huge quantities involved. In tonnage, sand and gravel production is exceeded only by that of fossil fuels.

At present, old river channels and glacial deposits are used as sources of sand and gravel. Many valuable deposits of sand and gravel are covered by construction and lost to development. Transportation and distance from source to use are especially crucial for this resource.

Environmental problems involved with defacing land can be severe, although bodies of water used for fishing and other recreational activities frequently are formed by removal of sand and gravel. The biggest single use for sulphur is in the manufacture of sulphuric acid.

However, the element is employed in a wide variety of other industrial and agricultural product.

WOOD—A MAJOR RENEWABLE RESOURCE

Fortunately, one of the major natural resources in the world, wood, is a renewable resource. Wood ranks first worldwide as a raw material for the manufacture of other products, including lumber, plywood, particle board, cellophane, rayon, paper, methanol, plastics and turpentine. Chemically, wood is a complicated substance consisting of long cells having thick walls composed of polysaccharide such as cellulose,

Cellulose polymer

The polysaccharides in cell walls account for approximately three-fourths of solid wood, wood from which extractable materials have been removed by an alcohol-benzene mixture. Wood typically contains a few tenths of a per cent ash (mineral residue left from the combustion of wood).

A wide variety of organic compounds can be extracted from wood by water, alcohol-benzene, ether, and steam distillation. These compounds include tannins, pigments, sugars, starch, cyclitols, gums, mucilages, pectins, galactans, terpenes, hydrocarbons, acids, esters, fats, fatty acids, aldehydes, resins, sterols, and waxes. Substantial amounts of methanol (sometimes called wood alcohol) are obtained from wood, particularly when it is pyrolysed. Methanol, once a major source of liquid fuel, is now being used to a limited extent as an ingredient of some gasoline blends.

A major use of wood is in paper manufacture. The widespread use of paper is a mark of an industrialised society. The manufacture of paper is a highly advanced technology. Paper consists essentially of cellulosic fibres tightly pressed together. The lignin fraction must first be removed from the wood, leaving the cellulosic fraction. Both the sulphuric and alkaline processes for accomplishing this separation have resulted in severe water and air pollution problems, although substantial progress has been made in alleviating these.

Wood fibres and particles can be used for making fibreboard, paper-base laminates (layers of paper held together by a resin and formed into the desired structures at high temperatures and pressures), particle board (consisting of wood particles bonded together by a phenol-formaldehyde or urea-formaldehyde resin) and non-woven textile substitutes consisting of wood fibres held together by adhesives. Chemical processing of wood enables the manufacture of many useful products, including methanol and sugar. Both of these substances are potential major products from the wood wastes produced in India.

THE ENERGY PROBLEM

Since the first "energy crisis" of 1973–1974, much has been said and written, many learned predictions have gone awry and some concrete action has even taken place. Prophecies of catastrophic economic disruption, people "freezing in the dark", and freeways given over to bicycles (though perhaps not a bad idea) have not been fulfilled. Nevertheless, uncertainties over petroleum availability and price, and disruptions such as the 1990 Gulf War have caused energy to be one of the major problems of modern times.

In India, concern over energy supplies and measures taken to ensure alternate supplies reached a peak in the late 1970s. Significant programmes on applied energy research were undertaken in the areas of renewable energy sources, efficiency, and fossil fuels. The financing of these efforts reached a peak around 1980, then dwindled significantly after that date.

The solutions to energy problems are strongly tied to environmental considerations. For example, a massive shift of the energy base to coal in nations that now rely largely on petroleum for energy would involve much more strip mining, potential production of acid mine water, use of scrubbers, and release of greenhouse gases (carbon dioxide from coal combustion and methane from coal mining). Similar examples could be cited for most other energy alternatives.

Dealing with the energy problem requires a heavy reliance on technology, which was discussed in general terms at the beginning of this chapter. Computerised control of transportation and manufacturing processes enables much more efficient utilisation of energy. New and improved materials allow higher peak temperatures and therefore greater extraction of usable energy in thermal energy conversion processes. Innovative manufacturing processes have greatly lowered the costs of photovoltaic cells used to convert sunlight directly to energy.

Clearly, chemists must be involved in developing alternative energy sources. Chemical processes are used in the conversion of coal to gaseous and liquid fuels. New materials developed through the applications of chemistry will be employed to capture solar energy and convert it to electricity. The environmental chemist has a key role to play in making alternative energy sources environmentally acceptable. The energy problem poses both questions and opportunities for students entering a career in chemistry. It is important, therefore, that these students know the basics of energy resources and alternative energy resources as well as their environmental aspects.

WORLD ENERGY RESOURCES

At present, most of the energy consumed by humans is produced from fossil fuels. Estimates of the amounts of fossil fuels available differ. Although world coal resources are enormous and potentially can fill energy needs for a century or two, their utilisation is limited by environmental disruption from mining and emissions of carbon dioxide and sulphur dioxide. These would become intolerable long before coal resources are exhausted. Assuming only uranium-235 as a fission fuel source, total recoverable reserves of nuclear fuel are roughly about the same as fossil fuel reserves. These are many orders of magnitude higher if the use of breeder reactors is assumed. Extraction of only 2% of the deuterium present in the Earth's oceans would yield about a billion times as much energy by controlled nuclear fusion as was originally present in fossil fuels! This prospect is tempered by the lack of success in developing a controlled nuclear fusion reactor. Geothermal power, currently utilised in northern California, Italy and New Zealand, has the potential for providing a high percentage of energy worldwide. The same limited potential is characteristic of several renewable energy resources, including hydroelectric energy, tidal energy, and wind power. All of these will continue to contribute significant, but relatively small, amounts of energy. Renewable, non-polluting solar energy comes as close to being an ideal energy source as any available. It almost certainly has a bright future.

ENERGY CONSERVATION

Any consideration of energy needs and production must take energy conservation into consideration. This does not have to mean cold classrooms with thermostats set at 60°F in mid-winter, nor swelteringly hot homes with no air-conditioning, nor total reliance on the bicycle for transportation, although these, and even more severe, conditions are routine in many countries. The fact remains that India has wasted energy at a deplorable rate. Obviously, a great deal of potential exists for energy conservation that will ease the energy problem.

Transportation is the economic sector with the greatest potential for increased efficiencies. The private auto and airplane are only about one-third as efficient as buses or trains for transportation. Transportation of freight by truck is terribly inefficient compared to rail transport (as well as dangerous, labour-intensive, and environmentally disruptive). Major shifts in current modes of transportation in India will not come without anguish, but energy conservation dictates that they be made.

Household and commercial uses of energy are relatively efficient. Here again, appreciable savings can be made. The all-electric home requires much more energy (considering the percentage wasted in generating electricity) than a home heated with fossil fuels. The sprawling ranch-house style home uses much more energy per person than does an apartment unit or row house. Improved insulation, sealing around the windows, and other measures can conserve a great deal of energy. Electric generating plants centrally located in cities can provide waste heat for commercial and residential heating and cooling and, with proper pollution control equipment, can use refuse for a significant fraction of fuel.

As scientists and engineers undertake the crucial task of developing alternative energy sources to replace dwindling petroleum and natural gas supplies, energy conservation must receive proper emphasis. In fact, zero energy-use growth, at least on a per capita basis, is a worthwhile and achievable goal. Such a policy would go a long way toward solving many environmental problems. With ingenuity, planning, and proper management, it could be achieved while increasing the standard of living and quality of life.

ENERGY CONVERSION PROCESSES

As shown in Fig. 18.1, energy occurs in several forms and must be converted to other forms. The efficiencies of conversion vary over a wide range. Conversion of electrical energy to radiant energy by incandescent light bulbs is very inefficient – less than 5% of the energy is converted to visible light and the remainder is wasted as heat. At the other end of the scale, a large electrical generator is around 80% efficient in producing electrical energy from mechanical energy. The once much-publicised Wankel rotary engine converts chemical to mechanical energy with an efficiency of about 18%, compared to 25% for a gasoline-powered piston engine and about 37% for a diesel engine. A modern coal-fired steam-generating power plant converts chemical energy to electrical energy with an overall efficiency of about 40%.

One of the most significant energy conversion processes is that of thermal energy to mechanical energy in a heat engine such as a steam turbine. The Carnot equation,

$$\text{Per cent efficiency} = \frac{T_1 - T_2}{T_1} \times 100$$

states that the per cent efficiency is given by a fraction involving the inlet temperature (for example, of steam), T_1, and the outlet temperature, T_2. These temperatures are expressed in Kelvin (°C + 273)

Typically, a steam turbine engine operates with approximately 810 K inlet temperature and 330 K outlet temperature. These temperatures substituted into the Carnot equation give a maximum theoretical efficiency of 59%. However, because it is not possible to maintain the incoming steam at the maximum temperature and because mechanical energy losses occur, overall efficiency of conversion of thermal energy to mechanical energy in a modern steam power plant is approximately 47%. Taking into account losses from conversion of chemical to thermal energy in the boiler, the total efficiency is about 40%.

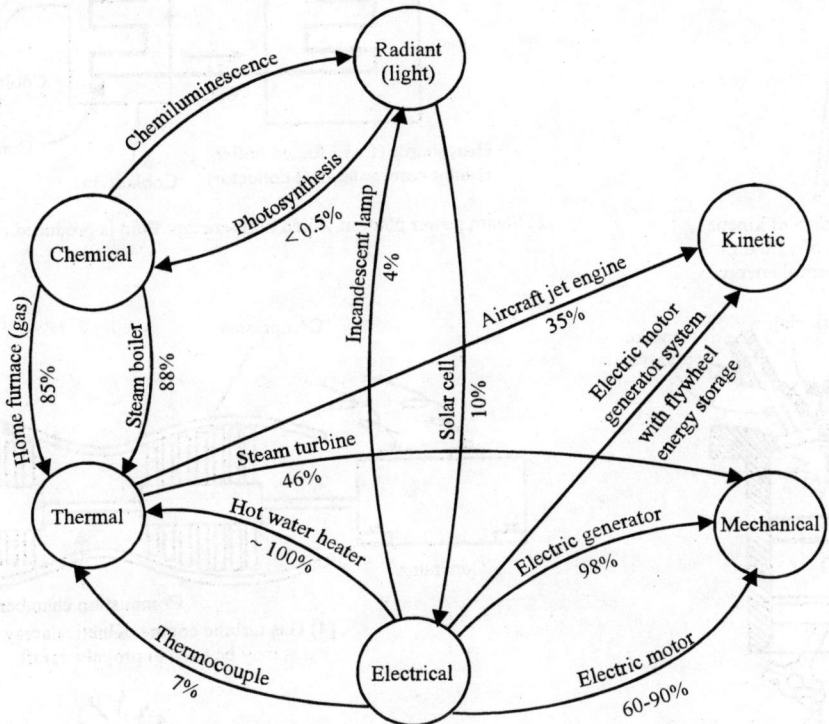

Fig. 18.1. Kinds of energy and examples of conversion between them, with conversion efficiency percentages.

Some of the greatest efficiency advances in the conversion of chemical to mechanical or electrical energy have been made by increasing the peak inlet temperature in heat engines. The use of superheated steam has raised T_1 in a steam power plant from around 550 K in 1900 to about 850 K at present. Improved materials and engineering design, therefore, have resulted in large energy savings.

The efficiency of nuclear power plants is limited by the maximum temperatures attainable. Reactor cores would be damaged by the high temperatures used in fossil-fuel-fired boilers and have a maximum temperature of approximately 620K. Because of limitation, the overall efficiency of conversion of nuclear energy to electricity is about 30%. Most of the 60% of energy from fossil-fuel-fired power plants and 70% of energy from nuclear power plants that is not converted to electricity is dissipated as heat, either to the atmosphere or to bodies of water and streams. The latter is thermal pollution, which may either harm aquatic life or, in some cases, actually increase bioactivity in the water to the benefit of some species. This waste heat is potentially very useful in applications like home heating, water desalination, and aquaculture (growth of plants in water). Some devices for the conversion of energy are shown in Fig. 18.2.

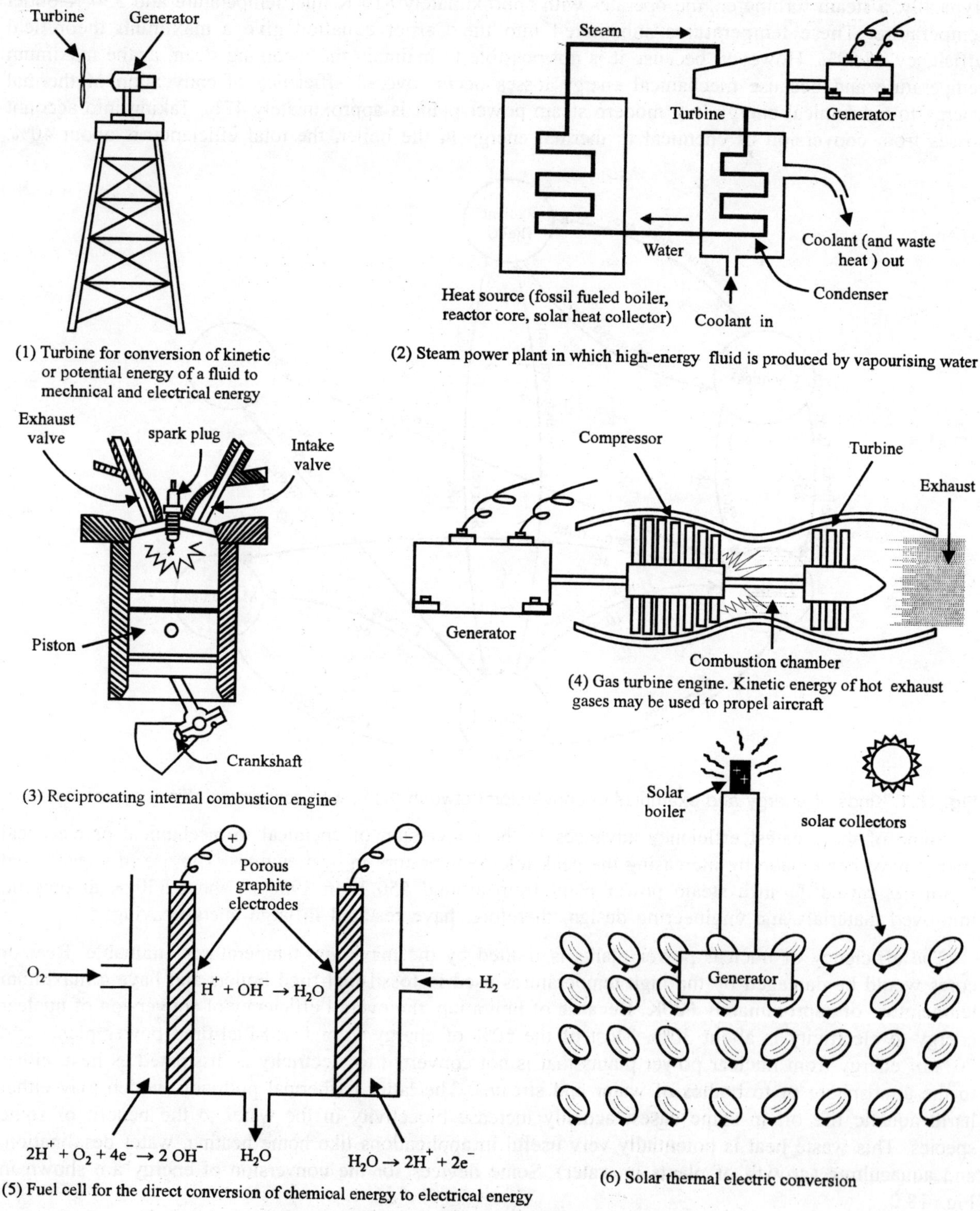

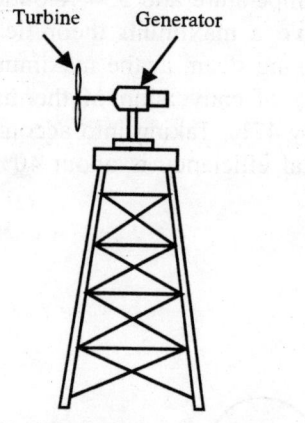

Turbine   Generator

(1) Turbine for conversion of kinetic
or potential energy of a fluid to
mechnical and electrical energy

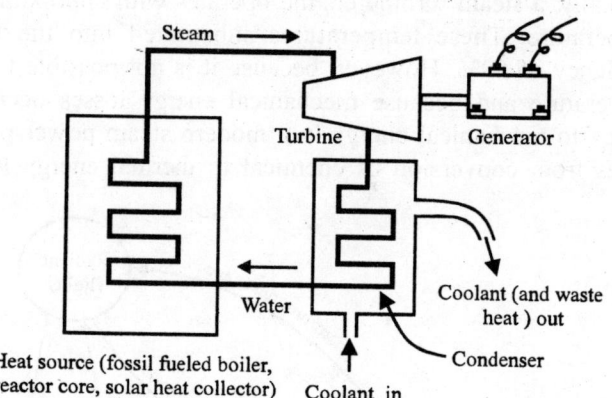

Steam

Turbine   Generator

Coolant (and waste
heat) out

Water   Condenser

Heat source (fossil fueled boiler,
reactor core, solar heat collector)   Coolant in

(2) Steam power plant in which high-energy fluid is produced by vapourising water

Exhaust
valve   spark plug   Intake
valve

Piston

Crankshaft

(3) Reciprocating internal combustion engine

Compressor   Turbine

Exhaust

Generator

Combustion chamber

(4) Gas turbine engine. Kinetic energy of hot exhaust
gases may be used to propel aircraft

Porous
graphite
electrodes

O_2 → $H^+ + OH^- \rightarrow H_2O$ ← H_2

$2H^+ + O_2 + 4e^- \rightarrow 2\,OH^-$ H_2O $H_2O \rightarrow 2H^+ + 2e^-$

(5) Fuel cell for the direct conversion of chemical energy to electrical energy

Solar
boiler

solar collectors

Generator

(6) Solar thermal electric conversion

Fig. 18.2. Some energy conversion devices.

Substantial advances have been made in energy conversion technology over many decades and more can be projected for the future. Through the use of higher temperatures and larger generating units, the overall efficiency of fossil-fueled electrical power generation has increased approximately ten-fold since 1900, from less than 4% to a maximum of around 40%. An approximately four-fold increase in the energy-use efficiency of rail transport occurred during the 1940s and 1950s with the replacement of steam locomotives with diesel locomotives. During the coming decades, increased efficiency can be anticipated from such techniques as combined power cycles in connection with generation of electricity. Magnetohydrodynamics (Fig. 18.2 probably will be developed as a very efficient energy source used in combination with conventional steam generation. Entirely new devices such as thermonuclear reactors for the direct conversion of nuclear fusion energy to electricity will very likely be developed.

PETROLEUM AND NATURAL GAS

The availability of petroleum and natural gas governs the energy growth and status of the country. The industrial revolution was initially fuelled by coal, but subsequently the emphasis was shifted to oil and gas which are cleaner fuels and transported more easily. Petroleum accounts for about 80% of the total energy consumption in the USA, the largest consumer of petroleum in the world.

Petroleum is used in various sectors, viz., agriculture, industry, transportation and communication. These sectors depend upon petroleum in many ways.

Petroleum is an oily, inflammable liquid made up mostly of hydrocarbon-compounds containing only hydrogen and carbon. The hydrocarbon content of petroleum ranges from 50–98 per cent. The rest is made up chiefly of organic compounds containing O_2, N or sulphur.

To meet the growing demand for natural gas, its production is considered to be increased about two–folds during Eighth Five year Plan. To enhance the production of natural gas and to minimise the flaring of gas, necessary infrastructural facilities such as augmentation of capacity of existing pipelines, adequate compensation and evacuation of facilities are considered to be developed. The increase in the quantum of oil import may cause further deficit in foreign exchange of the country. Thus, it is felt that we must restrict/reduce the level of oil imports by adopting following strategies :

1. Improved technology must be opted for efficient use of petroleum products, viz., hydrocarbon vapour recovery plant; isobaric double recycle (IDR) a new urea synthesis process, use of colliery methane as a new energy source; incineration of solid toxic waste for thermal applications; methanisation of sugarcane molasses stillage and cogeneration.

2. Consumer sector must manage their demands to reduce oil consumption.

3. Substitution of petroleum products by coal, natural gas, electricity, non-conventional energy sources, etc., must be encouraged.

Liquid petroleum is found in rock formations ranging in porosity from 10% to 30%. Up to half of the pore space is occupied by water. The oil in these formations must flow over long distances to an approximately 6-inch diameter well from which it is pumped. The rate of flow depends on the permeability of the rock formation, the viscosity of the oil, the driving pressure behind the oil, and other factors. Because of limitations in these factors, primary recovery of oil yields an average of about 30% of the oil in the formation, although it is sometimes as little as 15%. More oil can be obtained using secondary recovery techniques, which involve forcing water under pressure into the oil-bearing formation to drive the oil out. Primary and secondary recovery together typically extract somewhat less than 50% of the oil from a formation. Finally, tertiary recovery can be used to extract even more oil. This normally uses injection of pressurised carbon dioxide, which forms a mobile solution with the oil and allows it to flow more easily to the well. Other chemicals, such as detergents, may be used to aid in tertiary recovery. A recovery efficiency of 60% through secondary or tertiary techniques could double the amount of available petroleum. Much of this would come from fields which have already been abandoned or essentially exhausted using primary recovery techniques.

Shale oil is a possible substitute for liquid petroleum. Shale oil is a pyrolysis product of oil shale, a rock containing organic carbon in a complex structure called kerogen.

Shale oil may be recovered from the parent mineral by retorting the mined shale in a surface retort or by burning the shale underground with an *in situ* process. Both processes present environmental problems. Surface retorting requires the mining of enormous quantities of mineral and disposal of the spent shale, which has a volume greater than the original mineral. *In situ* retorting limits the control available over infiltration of underground water and resulting water pollution. Water passing through spent shale becomes quite saline, so there is major potential for saltwater pollution.

Natural gas, consisting almost entirely of methane, has become more attractive as an energy source. This is because of uncertainties regarding natural gas availability, coupled with the potential for the discovery and development of truly enormous new sources of this premium fuel.

In addition to its use as a fuel, natural gas can be converted to many other hydrocarbon materials. It can be used as a raw material for the *Fischer-Tropsch* synthesis of gasoline. The discovery and development of truly massive sources of natural gas, such as may exist in geopressurised zones, could provide abundant energy reserves for India, though at substantially increased prices.

Oil Shale

Oil is also found in oil shale, a compact sedimentary rock from which petroleum is obtained by the process of destructive distillation. The shale is first of all crushed, then it is heated in a furnace into which air is not admitted. The temperature is kept high enough so that chemical decomposition can take place. The principal products are oil, gases, water, solution of acids and other substances.

Tar Sand

Another form of oil deposited that cannot be tapped by ordinary method is tar sand. Two largest known tar sand deposits in the world are the tar-belt of Eastern Benzuela and Athavasca. Tar sands in the northern part of Canada's Alberta province, the Athavascan region contain one of the largest known deposits of petroleum in the world.

Tar sand is mined and then washed in hot water to remove the tar from sand. The tar is then heated and cracked into simpler molecules which are upgraded and landed to produce synthetic oil.

COAL

It is a complex mixture of compounds of carbon, hydrogen and oxygen and some free carbon. Small amounts of nitrogen and sulphur compounds also occur in coal. It is found in deep coal mines under the surface of earth.

On a worldwide basis, coal is substantially more abundant than oil or gas, the total coal reserve being estimated to be about 7.4×10^{12} MT (metric tonnes) which is equivalent to 4.7×10^{22} calories. This may be compared with the total world energy consumption from all fuels, as on $1975 = 6.0 \times 10^{19}$ calories.

Geologists believe that forests and plant life got buried inside the earth during earthquakes and volcanoes and got covered with sand, clay and water million of years ago. In the absence of air under humid conditions and at high temperature, the buried organic and vegetable matter degraded (got converted) into a black or brown, brittle or soft substance which chiefly contains carbon along with other organic compounds containing hydrogen, oxygen, nitrogen and sulphur. The process of formation of coal known as carbonisation is a very slow process which takes thousands of years.

Types of Coal

The coal found in the earth at different places has different percentage of carbon and so is classified according to the state of mineralisation into following types :

Types of coal	% carbon content
Peat	11
Lignite	38
Bituminous, soft or household coal	65
Anthracite or hard coal	96

Bituminous coal is used as a common household fuel in India. Lignite is the brown variety of coal. Peat has the lowest percentage of carbon whereas anthracite has the highest.

It is estimated that bituminous and anthracite reserves in world are about 4,017,000 million tonnes. The established and estimated reserve of peat coal and brown coal on the earth together amount to some 8.8 million tonnes.

Importance of Coal

Coal is important because it might be used as a source of energy as such, or it can be converted into other forms of energy like coal gas, electricity and oil (synthetic petrol). Coal is also an important source of a large number of organic compounds which are used in the manufacture of dyes, drugs, explosives, synthetic fibres and synthetic detergents.

Use of Coal

1. Coal finds use as a fuel.
2. Coal finds use in the manufacture of fuel gases like coal gas.
3. Coal finds use in the manufacture of synthetic petrol and synthetic natural gas.
4. Coal finds use as a source of organic compounds like benzene, toluene, phenol, aniline, naphthalene and anthracene.
5. Coal finds use to make coke.

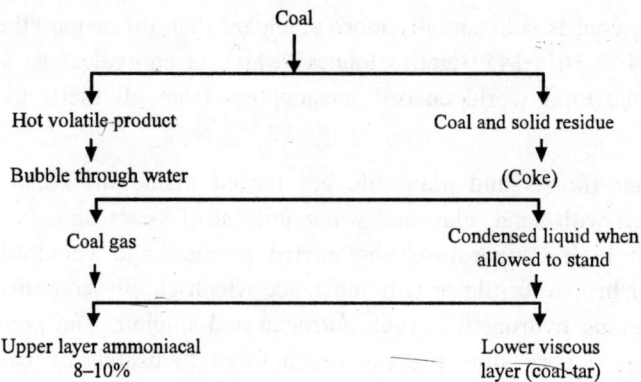

Coal Products and Their Uses

Bituminous coal is an excellent fuel with a high heating value. Unfortunately, most bituminous coals have a high percentage of sulphur (an average of 2–3%), so the use of this fuel presents environmental problems.

Products	Uses
Coal gas	Industrial fuel (mixture of hydrogen, methane, carbon monoxide and carbon dioxide.
Ammoniacal liquor	Manufacture of NH_3 and fertilisers.
Coke	As a reducing agent in metallurgy, for making gaseous fuels like water gas and producer gas, calcium carbide.
Coal tar	
(i) Benzene	As a solvent, motor fuel, manufacture of synthetic fibres, dyes and pigments.
(ii) Toluene	Manufacture of artificial sweatener, saccharing explosive (Trinitro toluene).
(iii) Phenol and cresol	As a disinfectant, manufactured of bake lite, drugs.
(iv) Aniline naphthalene anthracene	Preparation of dyes and pigments.
(v) Pitch	Construction of metalled roads.

The extent to which coal can be used as a fuel depends upon solutions to several problems, including: (i) minimising the environmental impact of coal mining; (ii) removing ash and sulphur from coal prior to combustion; (iii) removing ash and sulphur dioxide from stack gas after combustion; (iv) conversion of coal to liquid and gaseous fuels free of ash and sulphur; and, most important; (v) whether or not the impact of increased carbon dioxide emissions upon global climate can be tolerated. Progress is being made on minimising the environmental impact of mining. As more is learned about the processes by which acid mine water is formed, measures can be taken to minimise the production of this water pollutant. Particularly on flatter lands, strip-mined areas can be reclaimed with relative success. Inevitably, some environmental damage will result from increased coal mining, but the environmental impact can be reduced by various control measures. Washing, floatation, and chemical processes can be used to remove some of the ash and sulphur prior to burning. Approximately half of the sulphur in the average coal occurs as pyrite, FeS_2, and half as organic sulphur. Although little can be done to remove the latter, much of the pyrite can be separated from most coals by physical and chemical processes.

The maintenance of air pollution emission standards requires the removal of sulphur dioxide from stack gas in coal-fired power plants. Stack gas desulphurisation presents some economic and technological problems; the major processes available for it are summarised.

Magnetohydrodynamic power combined with conventional steam generating units has the potential for a major breakthrough in the efficiency of coal utilisation. A schematic diagram of a magnetohydrodynamic (MHD) generator is shown in Fig. 18.3. This device uses a plasma of ionised gas at around 2400°C blasting through a very strong magnetic field of at least 50,000 gauss to generate direct current. The ionisation of the gas is accomplished by injecting a "seed" of cesium or potassium salts. In an MHD generator, the ultra-high-temperature gas issuing through a supersonic nozzle contains ash, sulphur dioxide, and nitrogen oxides which severely erode and corrode the materials used. This hot gas is used to generate steam for a conventional steam power plant, thus increasing the overall efficiency of the process. The seed salts combine with sulphur dioxide and are recovered along with ash in the exhaust. Pollutant emissions are low. The overall efficiency of combined MHD-steam power plants should reach 60%, one and one-half times the maximum of present steam-only plants. Despite some severe technological difficulties, there is a chance that MHD power could become feasible on a large scale.

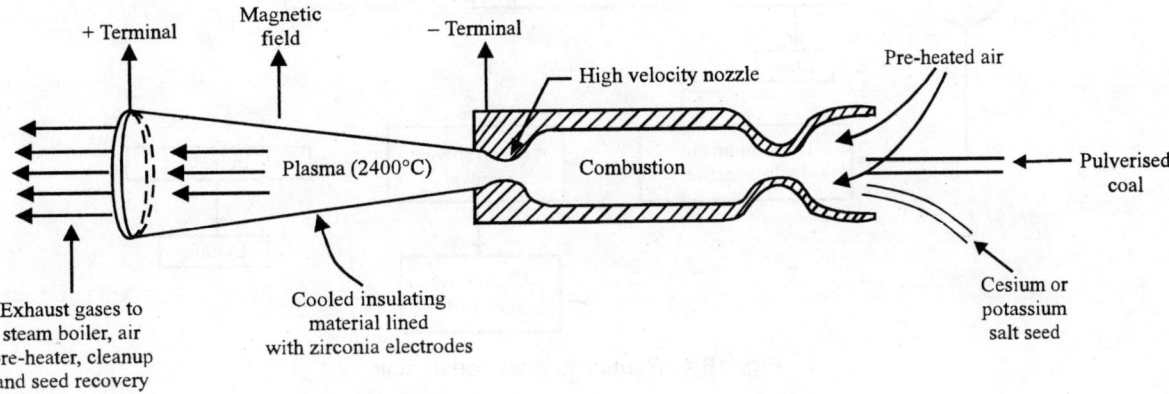

Fig. 18.3. A magnetohydrodynamic power generator.

Coal Conversion

Coal can be converted to gaseous, liquid, or low-sulphur, low-ash solid fuels, such as coal char (coke) or solvent-refined coal (SRC). In their end-use applications, all of these are less polluting than coal, and the gases and liquids made from coal can be used with distribution systems and equipment designed for natural gas or petroleum. A major advantage of coal conversion is that it enables use of high-sulphur coal which otherwise could not be burned without intolerable pollution or expensive stack gas cleanup. The major routes for coal conversion are shown in Fig. 18.4.

Coal conversion is an old idea. The early coal-gas plants used coal pyrolysis (heating in the absence of air) to produce a hydrocarbon-rich product particularly useful for illumination. Later the water-gas process was developed, in which steam was added to hot coal to produce a mixture consisting primarily of H_2 and CO. It was necessary to add volatile hydrocarbons to this "carbureted" water-gas to bring its illuminating power up to that of gas prepared by coal pyrolysis. The water-gas method accounted for 57% of manufactured gas. The gas was made in low-pressure, low-capacity gasifiers which by today's standards would be inefficient and environmentally unacceptable (several locations of these old plants have been designated as hazardous waste sites because of residues of coal tar and other wastes. During World War II, Germany developed a major synthetic petroleum industry based on coal, which reached a peak capacity of 100,000 barrels per day. A plant now operating in Sasol, South Africa, converts several tens of thousands of tonnes of coal per day to synthetic petroleum.

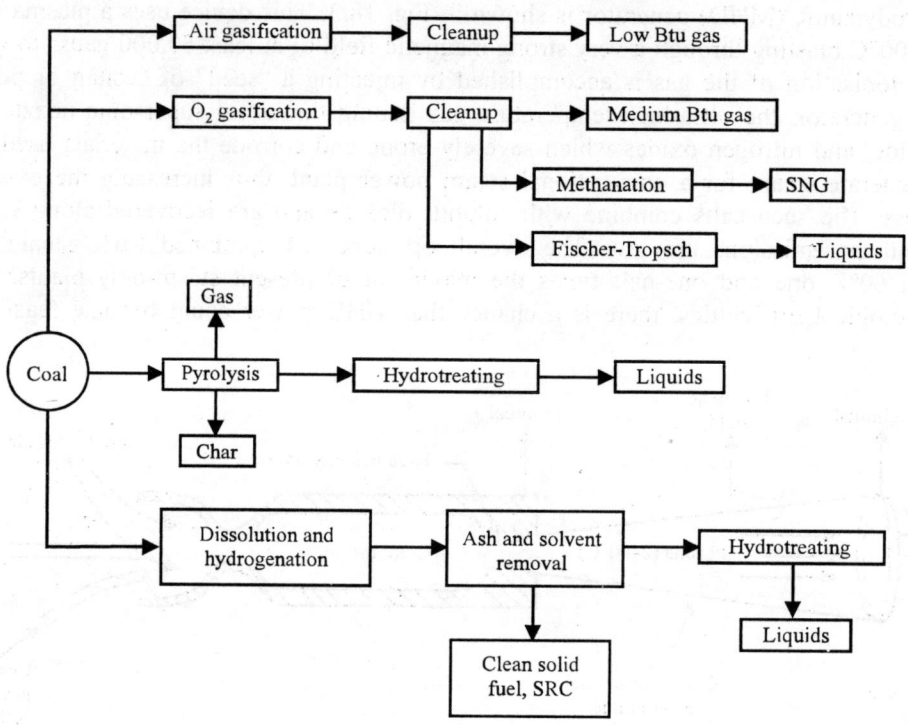

Fig. 18.4. Routes to coal conversion.

Some of the chemistry involved in coal conversion may be understood by considering production of high-heat-content (high-Btu) synthetic natural gas, SNG. The steps in a typical process for SNG production are :

1. Steam and oxygen react with coal in a gasifier, producing a low-Btu gas consisting of carbon monoxide, carbon dioxide, hydrogen, and some methane. The major reactions, simplifying coal as C, are :

$$2C + O_2 \rightarrow 2CO \qquad \Delta H = -26.4 \text{ kcal} \qquad \ldots(18.1)$$
$$\text{(per mole of carbon at 25°C}$$
$$2CO + O_2 \rightarrow 2CO_2 \qquad \ldots(18.2)$$
$$C + H_2O \rightarrow CO + H_2 \qquad \Delta H = +31.4 \text{ kcal} \qquad \ldots(18.3)$$
$$C + 2H_2 \rightarrow CH_4 \qquad \Delta H = -17.9 \text{ kcal} \qquad \ldots(18.4)$$

2. The gas product is freed from tar and dust by scrubbing with water.

3. The shift reaction,

$$H_2O + CO \rightarrow H_2 + CO_2 \qquad \ldots(18.5)$$

over a shift catalyst increases the molar ratio of H_2 to CO to an optimum value of approximately 3.5:1.

4. Acid gases (H_2S, COS, CO_2) are removed in an alkaline scrubber. Complete sulphur removal is necessary to avoid poisoning the methanation catalyst.

5. The catalytic hydrogenation of CO (methanation step) products methane gas :

$$CO + 3H_2 \rightarrow CH_4 + H_2O \qquad \qquad ...(18.6)$$

The two broadest categories of coal conversion are gasification and liquefaction. Arguably the most developed route for coal gasification is the Texaco process, which gasifies a water slurry of coal at temperatures of 1250°C to 1500°C and pressures of 350 to 1200 pounds per square inch. Chemical addition of hydrogen to coal can liquefy it and produce a synthetic petroleum product. This can be done with a hydrogen donor solvent, which is recycled and itself hydrogenated with H_2 during part of the cycle. Such a process forms the basis of the successful Exxon Donor Solvent process, which has been used in a 250 ton/day pilot plant.

A number of environmental implications are involved in the widespread use of coal conversion. These include strip mining, water consumption in arid regions, lower overall energy conversion compared to direct coal combustion, and increased output of atmospheric carbon dioxide. These plus economic factors have prevented coal conversion from being practised on a very large scale.

NUCLEAR ENERGY

The atom bombs dropped on Hiroshima and Nagasaki in August 1945 unleashed before the world the awesome power of the atom and brought the Second World War to halt. The nuclear age dawned and held out promise for the production of abundant energy. U-235 produces per gram, heat energy equivalent to 14 barrels of crude oil or 3 tonnes of coal. But this promise has not materialised to the extent expected and is showing a small growth in percentage of electrical energy.

Nuclear power (fission) reactors are based on the fission of U-235 nuclei by thermal neutrons.

$$^{235}_{92}U + {}^{1}_{0}n \longrightarrow {}^{133}_{51}Sb + {}^{99}_{41}Nb + 2.5\,{}^{1}_{0}n$$

Producing two radioactive fission products and average of 2.5 neutrons, and an average energy of 200 MeV per fission. The energy from these nuclear reaction is used to heat water in the reactor and produce steam to drive a steam turbine.

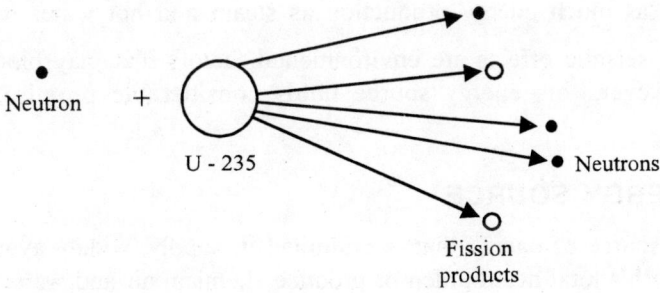

Nuclear reactors operate at about 625 K, compared to 800 K in a fossil fuel power plant. The thermal efficiency of nuclear power generation is low, and the overall efficiency for production of efficiency does not exceed 30%. This means 70% of nuclear fission energy has to be disposed of in the environment.

The major constraint in the widespread use of nuclear fission power is the yield of large quantities of radioactive fission waste products. The latter remains lethal for thousands of years. No perfect disposal method has yet been devised. Past experience shows that human error and negligence are likely to be the atmospheric release of radioactive wastes.

Nuclear fission reactions are based on the deuterium-deuterium reaction :

$$_1^2H + _1^2H \longrightarrow _2^3He + _0n + 1MeV$$

and the denterium–tritium reaction :

$$_1^2H + _1^3H \longrightarrow _2^4He + _0n + 17.6 \text{ MeV}$$

The second reaction is energetically more viable. Tritium is obtained from nuclear reactions of Li-6. Deuterium is however, unlimited in stock. $_2^3He$, products in the reaction above, reacts with neutrons which are abundant in a nuclear fusion reactor, to form H required for the second reaction.

The $_1^2H$–$_1^2H$ reaction promises an endless source of energy without any radioactive wastes, but the technological problems for harnessing fusion energy will take several years to solve.

GEOTHERMAL ENERGY

Underground heat in the form of steam, hot water, or hot rock used to produce steam is already being used as an energy resource. This energy was first harnessed for the generation of electricity at Larderello, Italy, and has since been developed in Japan, Russia, New Zealand, the Philippines, and at the Geysers in northern California.

Underground dry steam is relatively rare but is the most desirable from the standpoint of power generation. More commonly, energy reaches the surface as superheated water and steam. In some cases, the water is so pure that it can be used for irrigation and livestock; in other cases, it is loaded with corrosive, scale-forming salts. Utilisation of the heat from contaminated geothermal water generally requires that the water be reinjected into the hot formation after heat removal to prevent contamination of surface water.

The utilisation of hot rocks for energy requires fracturing of the hot formation, followed by injection of water and withdrawal of steam. This technology is still in the experimental state, but promises approximately ten times as much energy production as steam and hot water sources.

Land subsidence and seismic effects are environmental factors that may hinder the development of geothermal power. However, this energy source holds considerable promise, and its development continues.

SUN: AN IDEAL ENERGY SOURCE

Solar power is an ideal source of energy that is unlimited in supply, widely available, and inexpensive. It does not add to the Earth's total heat burden or produce chemical air and water pollutants. On a global basis, utilisation of only a small fraction of solar energy reaching the earth could provide for all energy needs. In the United States, for example, with conversion efficiencies ranging from 10–30%, it would only require collectors ranging in area from one-tenth down to one-thirtieth that of the state of Arizona to satisfy present U.S. energy needs. (This is still an enormous amount of land, and there are economic and environmental problems related to the use of even a fraction of this amount of land for solar energy collection. Certainly, many residents of Arizona would not be pleased at having so much of the state devoted to solar collectors, and some environmental groups would protest the resultant shading of rattlesnake habitat).

Solar power cells (photovoltaic cells) for the direct conversion of sunlight to electricity have been developed and are widely used for energy in space vehicles. With present technology, however, they remain too expensive for large-scale generation of electricity, although the economic gap is narrowing. Most schemes for the utilisation of solar power depend upon the collection of thermal energy, followed by conversion to electrical energy. The simplest such approach involves focusing sunlight on a steam-generating boiler. Parabolic reflectors can be used to focus sunlight on pipes containing heat-transporting fluids. Selective coatings on these pipes can be used so that only a small percentage of the incident energy is reradiated from the pipes.

The direct conversion of energy in sunlight to electricity is accomplished by special solar voltaic cells. Such devices based on crystalline silicon have operated with a 15% efficiency for experimental cells and 11–12% for commercial units, at a cost of 25–50 cents per kilowatt-hour (kWh), about 5 times the cost of conventionally-generated electricity. Part of the high cost results from the fact that the silicon used in the cells must be cut as small wafers from silicon crystals for mounting on the cell surfaces. Significant advances in costs and technology are being made with thin-film photovoltaics, which use an amorphous silicon alloy. A new approach to the design and construction of amorphous silicon film photovoltaic devices uses three layers of amorphous silicon to absorb, successively, short wavelength ("blue"), intermediate wavelength ("green"), and long wavelength ("red") light as shown in Fig. 18.5. Thin film solar panels constructed with this approach have achieved solar-to-electricity energy conversion efficiencies just over 10%, lower than those using crystalline silicon, but higher than other amorphous film devices. The low cost and relatively high conversion efficiencies of these solar panels should enable production of electricity at only about twice the cost of conventional electrical power, which would be competitive in some situations.

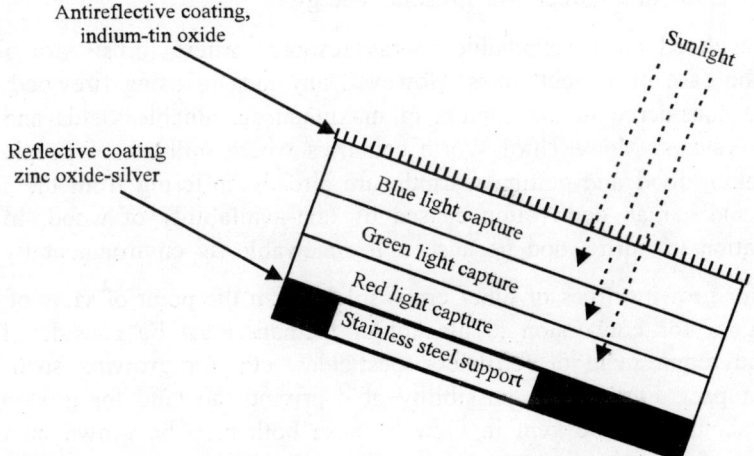

Fig. 18.5. High-efficiency thin film solar photovoltaic cell using amorphous silicon.

A major disadvantage of solar energy is its intermittent nature. However, flexibility inherent in an electric power grid would enable it to accept up to 15% of its total power input from solar energy units without special provision for energy storage. Existing hydroelectric facilities may be used for pumped-water energy storage. In conjunction with solar electricity generation. Heat or cold can be stored in water, in a latent form in water (ice) or eutectic salts, or in beds of rock. Enormous amounts of heat can be stored in water as a supercritical fluid contained at high temperatures and very high pressures deep underground. Mechanical energy can be stored with compressed air or flywheels.

Hydrogen gas, H_2, is an ideal chemical fuel that may serve as a storage medium for solar energy. Electricity generated by solar means can be used to electrolyse water :

$$2H_2O + \text{electrical energy} \rightarrow 2H_2(g) + O_2(g) \qquad \qquad ...(18.7)$$

The hydrogen fuel product, and even oxygen, can be piped some distance and the hydrogen burned without pollution or it may be used in a fuel cell (Fig. 18.2). This may, in fact, make possible a "hydrogen economy". Disadvantages of using hydrogen as a fuel include the fact that it has a heating value per unit volume of about one-third that of natural gas and that it is explosive over a wide range of mixtures with air.

No really insurmountable barriers exist to block the development of solar energy, such as might be the case with fusion power. In fact, the installation of solar space and water heaters became widespread in the late 1970s. With the installation of more heating devices and the probable development of some cheap, direct solar electrical generating capacity, it is likely that during the coming century solar energy will be providing an appreciable percentage of energy needs in areas receiving abundant sunlight.

BIOMASS ENERGY

Biomass energy or Bioconversion refers to the direct burning of wood, waste paper, manure, agriculture or any form of biomass or converting them to a fuel. Certain microorganisms when they digest biomass in the absence of air, produce either alcohol or methane gas, which themselves give energy on combustion. Since biomass is obtained through the process of photosynthesis, biomass energy is considered to be another form of indirect use of solar energy.

Firewood can be considered as a sustainable energy resource where forests are plenty relative to population, which was the case in ancient times. However, any idea of using firewood as a large-scale energy resource must be considered in the context of maximum sustainable yields and preservation of the bio-diversity of eco-systems. Many Third World countries where millions of people still depend on firewood as fuel for cooking food and getting warmth, are already suffering from the severe ecological effects of deforestation and human deprivation caused by non-availability of wood. In this back-drop, any blanket recommendation to burn wood because it is renewable, is environmentally irresponsible.

Similarly, proposals for growing trees or other crops solely from the point of view of biomass energy, either for direct burning or for conversion to alcohol or methane-must be considered in the light of possible soil erosion and requirement of fertilisers, pesticides, etc. for growing such crops and their possible environmental impact. Further, the possibility of depriving the land for growing food crops in favour of energy crops should also be kept in view because both may be grown on the same land.

Biomass energy should be preferred wherever energy can be produced as a byproduct of waste disposal (e.g. Sawmill waste, Sugar refinery waste, Municipal refuse, etc.). However, even while doing this, advantages of recycling, refuse or composting of the wastes should be kept in view.

Use of alcohol as fuel is being vehemently promoted in some grain-growing regions of developed countries such as U.S. Alcohol is produced by fermentation of grains, starches, sugar or similar food products. However, developing and under-developed countries should be very cautious in entertaining such ideas. Since many parts of the world suffer from shortage of grains, production of alcohol from grains may lead to shortage of food for their rapidly growing population, thereby leading to widespread malnutrition. Further, while alcohol, by itself, is clean-burning, we should remember that producing fuel grade alcohol requires distillation for which cheap, dirty-burning fuels such as soft coal are used as fuels. The net energy yield of alcohol is therefore modest, because fuel equivalent of about 0.5 gallon of alcohol is used for every gallon of alcohol produced.

Anaerobic digestion of sewage sludge and animal manure is a biomass utilising method which creates a valuable synergism between recycling and energy production. Such a method yields biogas and a nutrient-rich compost which is a good organic fertiliser which can be recycled back to the land in order to maintain the fertility of the fields growing forage for the animals. Further, we can thus get cheap energy which is sustainable and is free from adverse environmental consequences. In China, millions of small farmers maintain a simple digester in the form of a sealed pit into which they put agricultural wastes. The biogas generated is used as a domestic fuel. This is certainly a better alternative than using wood as a fuel.

BIOGAS

The ultimate product in a biogas plant is a mixture of gases like methane (50% to 60%), carbon dioxide (25% to 30%) hydrogen and traces of other gases. It is only due to methane that biogas burns with a smokeless blue flame. The gas is produced mainly by microbial fermentation of cellulosic waste material. The substrate (agricultural and animal waste) used in biogas plant consists mainly of (i) cellulose; (ii) lignin; and (iii) hemicellulose.

Process of Formation of Biogas

The microbial processes involved in the conversion of substrate into biogas are in the order : (i) aerobic; (ii) micro-aerophilic; (iii) anaerobic; and (iv) strictly anaerobic.

As soon as the oxygen, already present in the system, is consumed, the strictly anaerobic reactions start. On the whole, the process can be divided into two phases, i.e., (i) Non-methanogenic process; and (ii) Methanogenic process.

Non-methanogenic process

1. Due to the activity of micro-organisms, the complex substrates are first hydrolysed to simple sugars like hexoses and pentoses which are further metabolised to lower fatty acids, i.e., formic, acetic, propionic and butyric acids, carbon dioxide and hydrogen.

2. The lignin content of the substrate is least affected as the initial step in the lignin degradation is aerobic and molecular oxygen is absent inside the biogas plant. So, the relative content of lignin in the digested slurry increases considerably.

3. The microorganisms become active during the first phase are *Bacteriodes succinogenes, Butyrivibrio fibrisolvens, Clostridium lochheadic, Cillobacterium cellulosolvens, Ruminococcus flavefaciens, R. albens, Streptococcus, Selenomonas ruminatium, Bacteroids* etc. These organisms are either microaerophilic or anaerobic.

Methanogenic phase

1. The end products of the first phase act as the substrate for the microbes active in the second phase.

2. Unlike the first phase, the second one is strictly anaerobic. Even traces of oxygen can inhibit the methanogenic reaction.

3. The important methanogenic bacteria are *Methanobacterium ruminantium, M. mobilis, M. formicicum, Methanobacillus omelianskii, Methanosarcina bakery, Methanococcus vanniclii* etc.

4. Most of them use carbon dioxide, formate and acetate as the source of carbon.

5. There are some methanogenic bacteria which also utilise propionate, butyrate and alcohols as energy and carbon source.

Biochemistry of Methane Formation

1. The methanogenic bacteria are reported to have two unique components which are absent in all other living organisms, e.g., co-enzyme-M and Factor 420. Co-enzyme-M (CO-H) is a methyl transfer coenzyme and is heat stable. It has been chemically characterised as 2, 2'-dithioethane sulphonic acid and the active form is described as 2-Mercaptod ethane sulphuric acid ($HSCH_2CH_2SO_3$).

2. Factor 420 F_{420} is a low molecular weight fluorescent compound with a maximum absorption peak at 420 mm.

3. The extreme oxygen sensitivity of the methanogenic bacteria is due to the presence of this factor, as in the oxidised form, the enzyme attached to it becomes unstable and it cannot do its function.

4. The oxidation of formate and hydrogen in *Methanobacterium ruminantium* is mediated via F_{420} and coupled to reduction of NADP (Nicotinamide Adenine Dinucleotide Phosphate). The reducing power so generated is used for reduction of carbon dioxide to methane (Fig. 18.6). The terminal reduction step of methane formation is shown below :

$$CH_3-S-COM \xrightarrow[\text{Methyl reductase}]{M_2,\ Mg^{++},\ ATP} CH_4 + HS-CO-M$$

Effect of Temperature

All the reactions occurring in a biogas plant are *mesophilic*, i.e., they occur in the range of 30°C to 40°C. This is one of the reasons why biogas production decreases considerably during winter months. If the temperature of the system is increased, the biogas yield also increases. Some of the methanogenic bacteria like Methanobacterium thermo-autotrophicus are able to grow at 65°C–70°C and the rate of growth is much higher than for the mesosphilic ones. However, the bacteria present in a biogas plant are most active between temperature range of 35°C–40°C, which is close to that in the alimentary tract of mammals, their natural habitat.

GOBAR GAS PLANT

Methane gas, a combustible fermentation product of cattle dung, has been used as a domestic fuel and had been suggested as an energy source in India.

Design of Gobar Gas Plant

The plant has two main parts, viz., (i) a digester; and (ii) a gas holder (Fig. 18.7(a)).

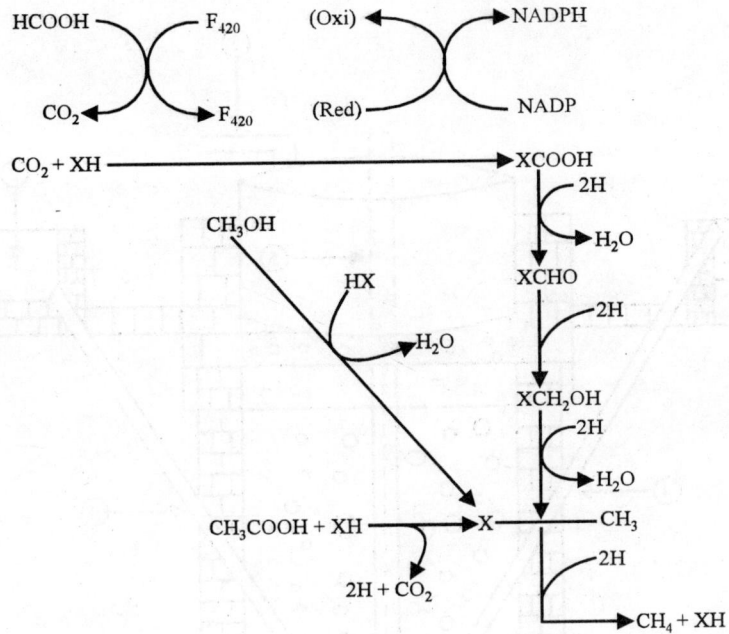

Fig. 18.6. Formation of methane.

The digester

1. The digester in which fermentation of cattle dung takes place is dug below the surface level like a well. It is brick lined and made water proof with cement.

2. The dimensions vary according to the gas generating capacity and the quantity of raw material fed daily.

3. The digester is vertically divided into two semi-cylindrical parts by a brick-wall partition. The raw material is fed in the form of a slurry (dung : water = 4 : 5) from an inlet chamber to the bottom of the digester by means of a cement pipe.

4. Initially the first semi-cylindrical half fills with the slurry and after several days of use it overflows into the second half.

5. The second semi-cylindrical half is connected to the outlet chamber through a cement pipe, which is at a little lower surface level than the inlet chamber.

6. Slurry from the outlet chamber is directed to compost pit.

7. When both semi-cylindrical halves are full of slurry, if more raw material is added from the inlet chamber, an equal amount of exhausted slurry flows out of the outlet chamber and the process continues.

Gas holder

1. The holder, usually a mild steel drum, fits in the upper part of the digester, where there is no vertical partition, like an inverted cup. Because of its own weight gas holder sinks into the slurry and rests on a projected collar made for the purpose.

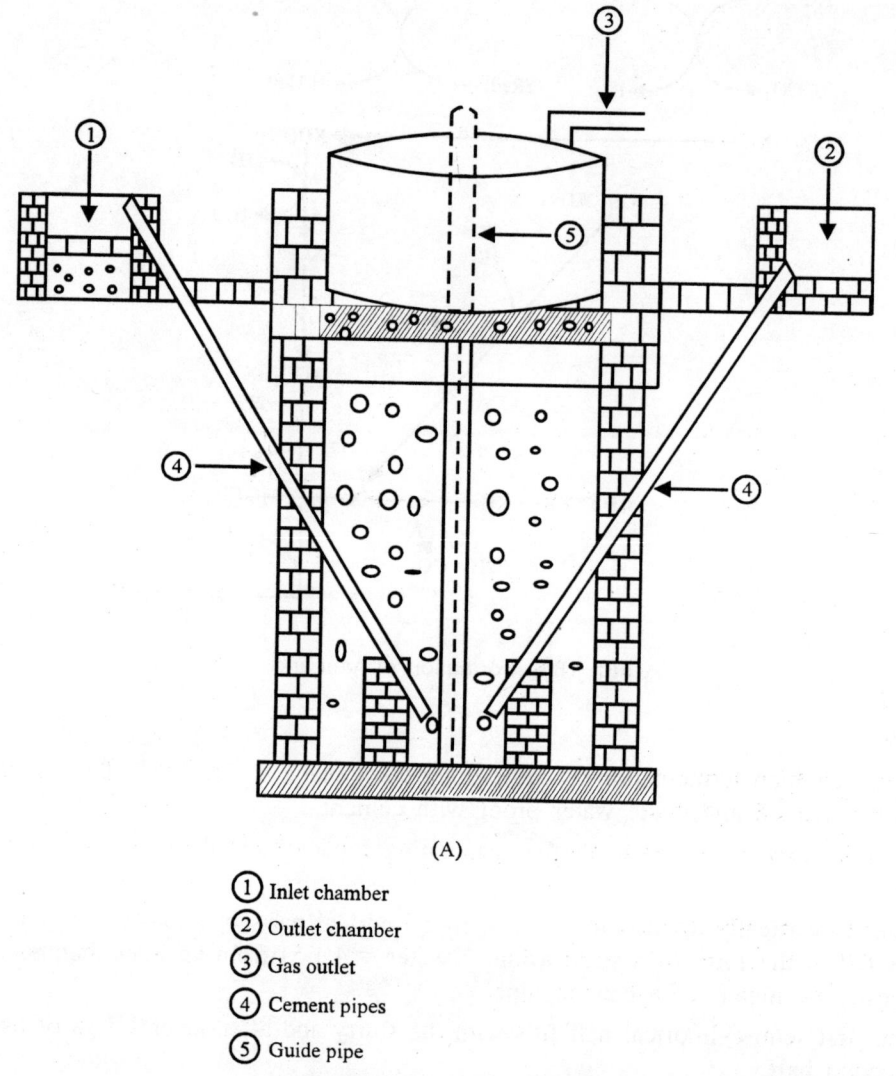

(A)

① Inlet chamber
② Outlet chamber
③ Gas outlet
④ Cement pipes
⑤ Guide pipe

Fig. 18.7(a). The KVIC plant.

2. Upon fermentation, the gas bubbles collect in the gas holder and due to gas pressure the holder rises and floats freely on the surface of slurry.

3. A delivery valve is fixed at the top of gas holder for distribution of gas. The gas holder is held straight with the help of a guide pipe fitted in the centre of the frame and is fixed at the bottom in the masonary work.

4. In the KVIC design the gas holder made from mild steel, suffers the problem of fast corrosion as it comes in direct contact with the slurry.

5. Other material like ferro-cement, reinforced polyester, treated wood, plywood, plastic, polythene with bamboo-basket reinforcement, etc. have been tried for the gas holder, but none has proved to be a suitable replacement for mild steel.

6. For mild steel gas holder, KVIC has suggested both mechanical and chemical treatments to partially overcome the problem of corrosion.

Mechanical treatment

In this treatment, daily cleaning of scum around the holder is suggested as a preventive measure.

Chemical treatment

This treatment involves coating of synthetic enamels. But scratches to the surface increased the rate of corrosion. Besides this, longtime contact with slurry deteriorates the enamel itself. This leads to an increased maintenance cost.

In view of the above problems of treatment of gas holder and in order to cut down the maintenance cost of the plant, the KVIC gobar gas plant has been redesigned. Some investigations on the corrosion of gas hold indicate that the modified design is more economical with an increases life span of the gas holder.

MODIFIED GOBAR GAS PLANT

In the modified gas plant (Fig. 18.7(b)), the digester remains almost the same but tank for the gas holder and the main digester are separated horizontally by a concrete partition, forming a ceiling for the digester well and a floor for the gas holder tank. A leakproof pipe open from both ends is fixed in the centre of concrete floor. This pipe connects the digester and the gas holder and leads the gas formed in the digester into the gas holder above the liquid level. This arrangement enables filling of gas holder tank with pure water, which in turn adds to the life span of gas holder by reducing the corrosive effect of dung slurry.

Moreover, daily cleaning of the gas holder is not required. The rusting of mild steel gas holder may further be reduced by pouring a few litres of burnt mobile oil from automobile engine on the surface of water after installing the gas holder.

The arrangement of dung feeding and exit of exhausted slurry are similar except the surface level of inlet and outlet chambers. The main difference in the construction of the improved plant is that the gas holder tank is constructed above the ground level. Three or four guide pipes are fixed at the periphery instead of one at the centre. Besides, this, openings of inlet and outlet pipes have been extended to the bottom of the digester, thereby increasing the effective fermentation volume.

Further investigations on commercial exploitation of the plant are in progress to work out the economical aspects of the new design.

Gasohol

A major option for converting photosynthetically-produced biochemical energy to forms suitable for internal combustion engines is the production of either methanol or ethanol. Either can be used by itself as fuel in a suitably designed internal combustion engine. More commonly, these alcohols are blended in proportions of up to 20% with gasoline to give gasohol, a fuel that can be used in existing internal combustion engines with little or no adjustment.

Gasohol offers some advantages as a fuel. It boosts octane rating and reduces emissions of carbon monoxide. From a resource viewpoint, because of its photosynthetic origin, alcohol may be considered a renewable resource rather than a depletable fossil fuel. The manufacture of alcohol can be accomplished by the fermentation of sugar obtained from the hydrolysis of cellulose in wood wastes and crop wastes. Fermentation of these waste products offers an excellent opportunity for recycling. Cellulose has significant potential for the production of renewable fuels.

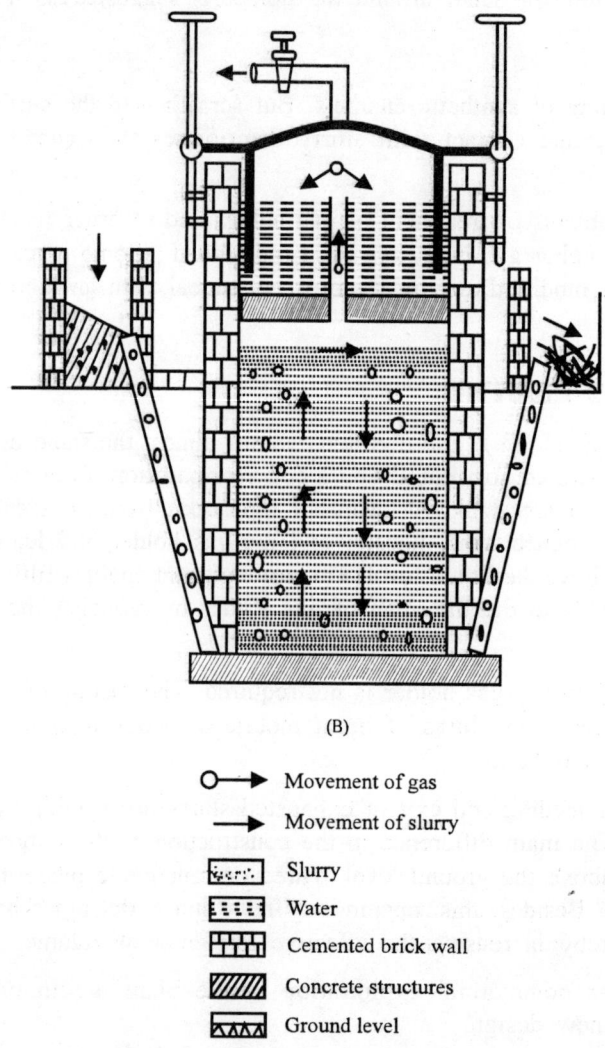

(B)

O⟶	Movement of gas
⟶	Movement of slurry
▨	Slurry
▦	Water
▦	Cemented brick wall
▨	Concrete structures
⋀⋀⋀	Ground level

Fig. 18.7(b). Modified gobar gas plant.

Ethanol is most commonly manufactured by fermentation of carbohydrates. Brazil, a country rich in potential to produce biomass, such as sugarcane, has been a leader in the manufacture of ethanol for fuel uses, with 4 billion litres produced in 1982. At one time Brazil had over 450,000 automobiles that could run on pure alcohol, although many have been converted back to gasoline. Reformulated gasoline blended to reduce air pollution by the addition of oxygenated compounds, including octane-boosting ethanol, is scheduled for use.

Methanol, which can be blended with gasoline, can also be produced from biomass by the destructive distillation of wood or by converting biomass, such as wood, to CO and H_2, and synthesizing methanol from these gases.

FUTURE ENERGY SOURCES

As discussed in this chapter, a number of options are available for the supply of energy in the future. The major possibilities are summarised in Table 18.3.

Table 18.3. Possible future sources of energy.

Source	Principles
Coal conversion	Manufacture of gas, hydrocarbon liquids, alcohol, or solvent-refined coal from coal.
Oil shale	Retorting petroleum-like fuel from oil shale.
Geothermal	Utilisation of underground heat.
Gas-turbine topping cycle	Utilisation of hot combustion gases in a turbine, followed by steam generation.
MHD	Electrical generation by passing a hot gas plasma through a magnetic field.
Thermionics	Electricity generated across a thermal gradient.
Fuel cells	Conversion of chemical to electrical energy.
Solar heating and cooling	Direct use of solar energy for heating and cooling through the application of solar collectors.
Solar cells	Use of silicon semiconductor sheets for the direct generation of electricity from sunlight.
Solar thermal electric	Conversion of solar energy to heat followed by conversion to electricity.
Nuclear fission	Conversion of energy released from fission of heavy nuclei to electricity.
Breeder reactors	Nuclear fission combined with conversion of non-fissionable nuclei to fissionable nuclei.
Nuclear fusion	Conversion of energy released by the fusion of light nuclei to electricity.
Bottoming cycles	Utilisation of waste heat from power generation for various purposes.
Solid waste	Combustion of trash to produce heat and electricity.
Photosynthesis	Use of plants for the conversion of solar energy to other forms by a biomass intermediate.
Hydrogen	Generation of H_2 by thermochemical means for use as an energy-transporting medium.

HYDROPOWER

Water power has been used since ancient times by diverting water from natural streams or rivers over various kinds of paddle wheels or turbines. The power output from waterwheels being low people started building high dams from the last century to obtain a substantial head of hydrostatic pressure. Thus, the water under high pressure, flows through the base of the dam and drives turbo-generators producing hydroelectric power. In U.S., about 300 large dams generate 9.5% of its total electrical power production.

Although hydroelectric power is basically a non-polluting renewable energy source, it is still associated with serious problems :

1. Dams have drowned out beautiful stretch of rivers, wildlife habitat, forests, productive farmlands, and areas of historic, archeological, and geological significance. The construction of big dams have also rendered several farmers and tribals homeless and without any livelihood.

2. The reservoir behind the Aswan High Dam in Egypt has caused the spread of a parasitic worm which caused a debilitating disease. Further, the increase in humidity over a large area because of the reservoir is causing rapid deterioration of ancient monuments and artefacts which were existing over many centuries.

3. Since water flow from the dam is regulated as per the requirement of power, dams play havoc downstream because water levels may change from extremes of near flood levels to virtual dryness and back to flood even in a single day. Other ecological factors are also affected because sediments rich in nutrients settle in the reservoir and only small amounts reach the river's mouth.

4. Devastating earthquakes, observed near Koyana in India, are attributed to the Koyana dam (Maharashtra) by some Scientists.

Many developing countries have great potential for large hydel power projects but due to the above problems, there is lot of opposition from people as well as from Environmental protection organisations.

WIND POWER

Wind mills have been used since ancient times. A large number of different designs were tested, but most practical one seems to be the age-old concept of airplane type propeller blades turning generator geared to the shaft. Modestly sized "wind turbines", comprising of machines with blade diameter of about 17 m which can generate about 100 Kilowatts, have proved to be most practical. "Wind Farms" consisting of arrays of 50 to several thousand such machines, are now producing power in a number of places around the world. California, with 17,000 machines generating 1500 megawatts is the world's largest producer of wind-generated power. This supplants the need for two nuclear power plants. By the turn of this decade, European countries like Britain, Netherlands, Germany, Italy and Denmark will have a combined wind generated power capacity of over 3,000 megawatts. With gradual improvements in design and reliability, wind farms are now generating power at as little as 6 cents per kilowatt-hour, which is certainly competitive with traditional energy sources. Power-generating wind turbines are now installed in 95 countries right from the tropics to the Arctic. However, still there is lot of potential which is untapped. Many regions of the world have areas where winds are constant enough to render wind turbines practical. Wind farms in different locations can be connected to the already existing electrical grid so that it can provide backup for each other since the wind is invariably flowing somewhere or the other. Wind farms can also provide a sustainable complement and back-up to direct solar power facilities.

Wind power is a non-polluting, renewable and hence sustainable source of energy. However, it has the following drawbacks :

1. Location of wind farms on migratory routes could spell hazard to birds and disaster for some avian populations.

2. Their appearance on the landscape and their continual whirring and whistling can be irritating.

OCEAN THERMAL ENERGY CONVERSION (OTEC)

In oceans, a thermal gradient (i.e. temperature difference) of about 20°C exists between surface water heated by the sun and colder deep water. This temperature difference can be harnessed to produce power. This concept is known as Ocean Thermal Energy Conversion (OTEC). An OTEC power plant can be built on a brage (i.e., a sailing vessel) that could travel anywhere in the ocean. It uses the warm surface water to heat and vapourise a low boiling liquid such as ammonia. The increased pressure of the vapoursised liquid would drive turbo-generators. The ammonia vapour leaving the turbines would then be condensed by cold deep water which is about 100 m below the surface and is returned back to start the cycle again. The electrical power so generated could be used to produce hydrogen and shipped to the shore. Alternatively, an energy intensive industry can be located on factory ships that would anchor alongside the OTEC Plant. A few OTEC Plants have been tested.

Owing to the small temperature difference between the surface water and deep water, the conversion efficiency is as low as 2–3%. This low efficiency by itself is immaterial since the primary energy source viz., the temperature difference between the surface and deep waters over most of the ocean, is freely available. However, this low efficiency coupled with other drawbacks such as high capital costs, persistent maintenance problems and fouling of pipes and pumps due to marine organisms, results in meager energy yields thereby rendering OTEC power uneconomical at the present state of the OTEC technology.

CHAPTER 19

Nature and Sources of Hazardous Wastes

INTRODUCTION

Humans have always been exposed to hazardous substances going back to prehistoric times when they inhaled noxious volcanic gases or succumbed to carbon monoxide from inadequately vented fires in cave dwellings sealed too well against Ice-Age cold. Slaves in ancient Greece developed lung disease from weaving mineral asbestos fibres into cloth to make it more degradation resistant. Some archaeological and historical studies have concluded that lead wine containers were a leading cause of lead poisoning in the more affluent ruling class of the Roman Empire. Alchemists who worked during the Middle Ages often suffered debilitating injuries and illnesses resulting from the hazards of their explosive and toxic chemicals. As the production of dyes and other organic chemicals developed from the coal tar industry in Germany during the 1800s, pollution and poisoning from coal tar by-products was observed. By around 1900 the quantity and variety of chemical wastes produced each year was increasing sharply with the addition of wastes such as spent steel and iron pickling liquor, lead battery wastes, chromic wastes, petroleum refinery wastes, radium wastes, and fluoride wastes from aluminium ore refining. As the century progressed into the World War II era, the wastes and hazardous by-products of manufacturing increased markedly from sources such as chlorinated solvents manufacture, pesticides synthesis, polymers manufacture, plastics, paints, and wood preservatives. A hazardous substance is a material that may pose a danger to living organisms, materials, structures, or the environment by explosion or fire hazards, corrosion, toxicity to organisms, or other detrimental effects. A simple definition of a hazardous waste is that it is a hazardous substance that has been discarded, abandoned, neglected, released or designated as a waste material, or one that may interact with other substances to be hazardous.

NATURE AND SOURCES OF HAZARDOUS WASTE

Classification of Hazardous Substances and Wastes

Many specific chemicals in widespread use are hazardous because of their chemical reactivities, fire hazards, toxicities, and other properties. There are numerous kinds of hazardous substances, usually consisting of mixtures of specific chemicals. These include the following :

1. Explosives, such as dynamite, or ammunition.
2. Compressed gases, such as hydrogen and sulphur dioxide.

3. Flammable liquids, such as gasoline and aluminium alkyls.

4. Flammable solids, such as magnesium metal, sodium hydride, and calcium carbide that burn readily, are water-reactive, or spontaneously combustible.

5. Oxidising materials, such as lithium peroxide, that supply oxygen for the combustion of normally non-flammable materials.

6. Corrosive materials, including oleum, sulphuric acid and caustic soda, which may wound exposed flesh or cause disintegration of metal containers.

7. Poisonous materials, such as hydrocyanic acid or aniline.

8. Etiologic agents, including causative agents of anthrax, botulism, or tetanus.

9. Radioactive materials, including plutonium, cobalt-60, and uranium hexafluoride.

Characteristics and listed wastes

The characteristics of listed wastes are :

1. Ignitability, characteristic of substances that are liquids whose vapours are likely to ignite in the presence of ignition sources, non-liquids that may catch fire from friction or contact with water and which burn vigorously or persistently, ignitable compressed gases, and oxidisers.

2. Corrosivity, characteristic of substances that exhibit extremes of acidity or basicity or a tendency to corrode steel.

3. Reactivity, characteristic of substances that have a tendency to undergo violent chemical change (examples are explosives, pyrophoric materials, water-reactive substances, or cyanide or sulphide-bearing wastes).

4. Toxicity, defined in terms of a standard extraction procedure followed by chemical analysis for specific substances.

In addition to classification by characteristics, EPA designates more that 450 listed wastes which are specific substances or classes of substances known to be hazardous. Each such substance is assigned an EPA hazardous waste number in the format of a letter followed by 3 numerals, where a different letter is assigned to substances from each of the following lists :

1. *F-type wastes from non-specific sources:* For example, quenching waste-water treatment sludges from metal heat treating operations where cyanides are used in the process.

2. *K-type wastes from specific sources:* For example, heavy ends from the distillation of ethylene dichloride in ethylene dichloride production.

3. *P-type acute hazardous wastes:* Wastes that have been found to be fatal to humans in low doses, or capable of causing or significantly contributing to an increase in serious irreversible or incapacitating reversible illness. These are mostly specific chemical species such as fluorine or 3-chloropropane nitrile.

4. *U-type miscellaneous hazardous wastes:* These are predominantly specific compounds such as calcium chromate or phthalic anhydride.

Hazardous wastes

Three basic approaches to defining hazardous wastes are: (i) a qualitative description by origin, type, and constituents; (ii) classification by characteristics largely based upon testing procedures; and (iii) by means of concentrations of specific hazardous substances. Wastes may be classified by general type such as "spent halogenated solvents" or by industrial sources such as "pickling liquor from steel manufacturing"

Various countries have different definitions of hazardous waste. Radioactive wastes are a problem for any country with a significant nuclear power or weapons industry. Special problems are posed by mixed waste containing both radioactive and chemical wastes.

Some of the hazardous wastes are those from specific sources produced by industries such as the manufacture of inorganic pigments, organic chemicals, pesticides, explosives, iron and steel, and non-ferrous metals, and from processes such as petroleum refining or wood preservation; some examples are given below :

1. Bottoms sediment sludge from the treatment of waste-waters from wood-preserving processes that use creosote and/or pentachlorophenol.
2. Waste-water treatment sludge from the production of chrome yellow and orange pigments.
3. Heavy ends (residue) from the distillation of vinyl chloride in vinyl chloride monomer production.
4. 2,6-Dichlorophenol waste from the production of 2,4-D.
5. Pink/red water from TNT operations.
6. Slop oil emulsion solids from the petroleum refining industry.
7. Ammonia lime still sludge from coking operations.
8. Electrolytic anode slimes/sludges from primary zinc production.

The second largest category of wastes generated are reactive wastes, followed by corrosive wastes and toxic wastes.

Hazardous wastes and air and water pollution control

Somewhat paradoxically, measures taken to reduce air and water pollution (Fig. 19.1) have had a tendency to increase production of hazardous wastes. Most water treatment processes yield sludges or concentrated liquors that require stabilisation and disposal. Air scrubbing process likewise produce sludges. Baghouses and precipitators used to control air pollution all yield significant quantities of solids, some of which are hazardous.

Origin and Amounts of Wastes

In a non-regulatory sense there is no sharp demarcation between hazardous and non-hazardous wastes. Some wastes, such as soluble toxic heavy metal salt wastes, are obviously hazardous. By comparison, discarded leaves and tree trimmings would be regarded as posing no danger. But, if properly treated and immobilised, the heavy metal wastes are of little danger, whereas discarded tree limbs pose a fire hazard under certain circumstances. Materials that by themselves are non-hazardous may interact with hazardous substances to increase the dangers from the latter. For example, soluble humic substances from the decay of tree leaves may solubilise and transport heavy metal ions.

Staggering amounts of wastes of all kinds are produced by human activities. Such wastes include municipal refuse, sewage sludge, agricultural residues, and toxic, chemically reactive by-products of manufacturing processes. An idea of quantities of solid wastes generated can be obtained by considering mining and milling wastes. The quantities of such wastes are enormous because large quantities of rock must be removed to get to the ore and because the metal or other economically valuable constituent is usually a small percentage of the ore. Therefore, by-products such as overburden and beneficiation wastes accumulate in vast amounts.

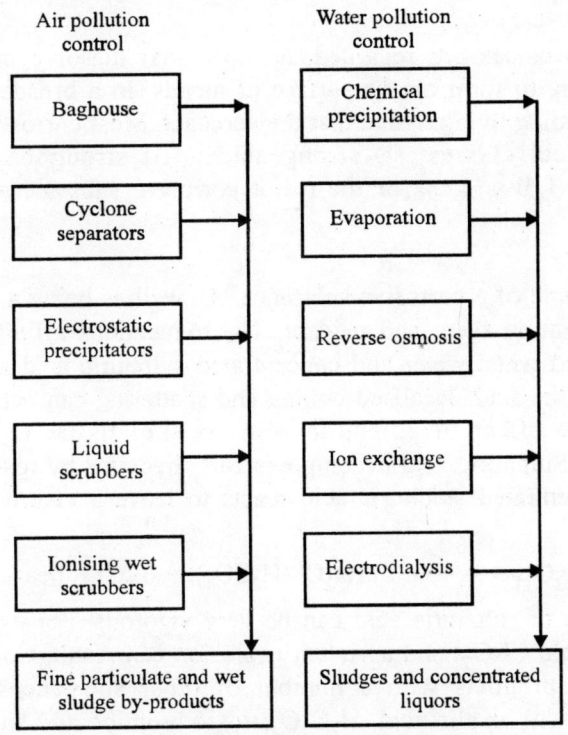

Potentially hazardous waste substances

Fig. 19.1. Potential contributions of air and water pollution control measures to hazardous wastes production.

Non-hazardous solid wastes

It is appropriate to consider "non-hazardous" waste (solid waste, the municipal refuse and garbage produced by human activities) along with hazardous waste because it may not be non-hazardous in all cases and situations, and it may interact with hazardous wastes. Furthermore, the amounts of solid waste produced each year are so enormous that capacity to deal with the problem is under severe strain. Disposal of about 92% of municipal refuse is in landfills. However, as total quantities of solid waste have increased, the landfill capacity to handle waste has decreased.

The potential of incineration for handling municipal refuse is very high because it can reduce waste mass by 75% and volume by 90%. However, environmental concern about organic pollutants (particularly dioxins) in stack emissions and heavy metals in incinerator ash have slowed municipal incinerator development. Recycling can certainly reduce quantities of sold waste, perhaps as much as 50%, but it is not the panacea claimed by its most avid advocates. The overall solution to the solid waste problem must involve several kinds of measures, particularly: (i) reduction of wastes at the source; (ii) recycling as much waste as is practical; (iii) reducing the volume of remaining wastes by measures such as incineration; (iv) treating residual material as much as possible to render it non-leachable and innocuous;

and (v) placing the residual material in landfills, properly protected from leaching or release by other pathways.

Corrosive Substances

Conventionally, corrosive substances are regarded as those that dissolve metals or cause oxidised material, such as rust from iron, to form on the surface of metals. In a broader sense, corrosives cause deterioration of materials, including living tissue, that they contact. Most corrosives belong to at least one of the four following chemical classes: (i) strong acids; (ii) strong bases; (iii) oxidants; (iv) dehydrating agents. Table 19.1 lists some of the major corrosive substances and their effects.

Sulphuric acid

Sulphuric acid is a prime example of a corrosive substance. As well as being a strong acid, concentrated sulphuric acid is also a dehydrating agent and oxidant. The tremendous affinity of H_2SO_4 for water is illustrated by the heat generated when water and concentrated sulphuric acid are mixed. If this is done incorrectly by adding water to the acid, localised boiling and spattering can occur that result in personal injury. The major destructive effect of sulphuric acid on skin tissue is removal of water with accompanying release of heat. Sulphuric acid decomposes carbohydrates by removal of water. In contact with sugar, for example, concentrated sulphuric acid reacts to leave a charred mass. The reaction is

$$C_{12}H_{22}O_{11} \xrightarrow{H_2SO_4} 11H_2O \ (H_2SO_4) + 12C + heat \qquad ...(19.1)$$

Some dehydration reactions of sulphuric acid can be very vigorous. For example, the reaction with perchloric acid produces unstable Cl_2O_7, and a violent explosion can result. Concentrated sulphuric acid produces dangerous or toxic products with a number of other substances, such as toxic carbon monoxide (CO) from reaction with oxalic acid, $H_2C_2O_4$; toxic bromine and sulphur dioxide (Br_2, SO_2) from reaction with sodium bromide, NaBr; and toxic, unstable chlorine dioxide (ClO_2) from reaction with sodium chlorate, $NaClO_3$.

Table 19.1. Examples of some corrosive substances.

Name and formula	Properties and effects
Nitric acid, HNO_3	Strong acid and strong oxidiser, corrodes metal, reacts with protein in tissue to form yellow xanthoproteic acid, lesions are slow to heal.
Hydrochloric acid, HCl	Strong acid, corrodes metals, gives off HCl gas vapour, which can damage respiratory tract tissue.
Hydrofluoric acid, HF	Corrodes metals, dissolves glass, causes particularly bad burns to flesh
Alkali metal hydroxides, NaOH and KOH	Strong bases, corrode zinc, lead, and aluminium, substances that dissolve tissue and cause severe burns.
Hydrogen peroxide, H_2O_2	Oxidiser, all but very dilute solutions cause severe burns.
Interhalogen compounds such as ClF, BrF_3	Powerful corrosive irritants that acidify, oxidise, and dehydrate tissue.
Halogen oxides such as OF_2, Cl_2O, Cl_2O_7	Powerful corrosive irritants that acidify, oxidise, and dehydrate tissue.
Elemental fluorine, chlorine, bromine (F_2, Cl_2, Br_2,)	Very corrosive to mucous membranes and moist tissue, strong irritants.

Contact with sulphuric acid causes severe tissue destruction resulting in severe burns, which may be difficult to heal. Inhalation of sulphuric acid fumes or mists damages tissues in the upper respiratory tract and eyes. Long-term exposure to sulphuric acid fumes or mists has caused erosion of teeth.

Toxic substances

Toxicity is of the utmost concern in dealing with hazardous substances. This includes both long-term chronic effects from continual or periodic exposures to low levels of toxicants, and acute effects from a single large exposure. For regulatory and remediation purposes a standard test is needed to measure the likelihood of toxic substances getting into the environment and causing harm to organisms.

Chemical Classes of Hazardous Substances

Another way of viewing hazardous substances in the context of their chemical properties is to divide them into classes of chemicals. A number of elements are used industrially in their elemental forms, in many cases for chemical synthesis. Some of these elements pose hazards of flammability, corrosivity, reactivity, or toxicity. Elemental hydrogen, H_2, is extremely flammable and forms explosive mixtures with air. Three of the halogens–fluorine, chlorine, and bromine–are widely produced as elemental F_2, Cl_2, and Br_2, respectively. Fluorine is the strongest elemental oxidant and extremely reactive. It is very corrosive to the skin and inhalation of F_2 can cause severe lung damage. Chlorine, one of the most widely produced industrial chemicals, is a reactive oxidant that forms acid in water and is a corrosive poison to tissue, especially in the respiratory tract. Bromine is a volatile brown liquid which is corrosive to skin in both the liquid and vapour form. Elemental white phosphorus is a reactive substance that may catch fire spontaneously in air. It is a systemic poison. Elemental lithium, sodium, and potassium react with a large number of chemicals and burn readily to give off caustic oxide and hydroxide fumes. Elemental mercury vapour is especially toxic by inhalation. Some metals, commonly known as heavy metals, are particularly toxic in their chemically combined forms. These include lead, cadmium, mercury, beryllium, and arsenic.

Many inorganic compounds are hazardous because of reactivity (NH_4ClO_4), corrosivity (HNO_3) and toxicity (KCN). Many organometallic compounds, which have a metal atom or metalloid atom (such as silicon or arsenic) bonded directly to carbon in a hydrocarbon group or in carbon monoxide, CO, are volatile, reactive, and toxic.

Organic compounds

There are millions of known organic compounds, most of which can be hazardous in some way and to some degree. Most organic compounds can be divided among hydrocarbons, oxygen-containing compounds, nitrogen-containing compounds, organohalides, sulphur-containing compounds, phsophorus-containing compounds, or combinations thereof.

Physical Forms and Segregation of Wastes

Three major categories of wastes based upon their physical forms are organic materials, aqueous wastes, and sludges. These forms largely determine the course of action taken in treating and disposing of the wastes. The level of segregation, a concept illustrated in Fig. 19.2. is very important in treating, storing, and disposing of different kinds of wastes. It is relatively easy to deal with wastes that are not mixed with other kinds of wastes; that is, those that are highly segregated. For example, spent hydrocarbon solvents can be used as fuel in boilers. However, if these solvents are mixed with spent organochlorine solvents, the production of contaminant hydrogen chloride during combustion may prevent fuel use and require disposal in special hazardous waste incinerators. Further mixing with inorganic sludges adds mineral matter and water. These impurities complicate the treatment processes required by producing mineral ash in incineration or lowering the heating values of the material incinerated because of the presence of water. Among the most difficult types of wastes to handle and treat are those with the least segregation, of which a "worst case scenario" would be "dilute sludge consisting of mixed organic and inorganic wastes," as shown in Fig. 19.2.

Concentration of wastes is an important factor in their mànagement. A waste that has been concentrated or preferably never diluted is generally much easier and more economical to handle than one that is dispersed in a large quantity of water or soil. Dealing with hazardous wastes is greatly facilitated when the original quantities of wastes are minimised and the wastes remain separated and concentrated insofar as possible.

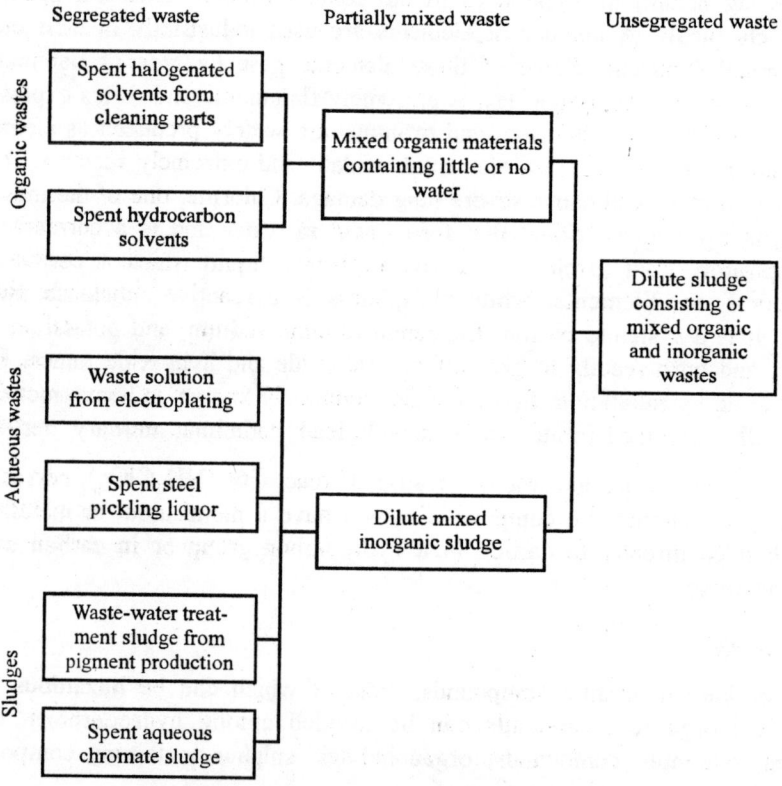

Fig. 19.2. Illustration of waste segregation.

Generation, Treatment and Disposal

Hazardous waste management refers to a carefully organised system in which wastes go through appropriate pathways to their ultimate elimination or disposal in ways that protect human health and the environment. The management of hazards posed by hazardous substances and wastes is a crucial part of the operation of any modern chemical industry. It is a significant and increasing part of the cost of any business dealing with chemical products and processes. Personnel working with such products and processes must have a good understanding of hazardous substances and hazardous wastes. Three main aspects of hazardous waste management involve generation, treatment, and disposal as illustrated in Fig. 19.3. The effectiveness of a hazardous waste system is a measure of how well it reduces the quantities and hazards of wastes, ideally approaching zero for both. In decreasing order of effectiveness, the options for handling hazardous wastes are the following :

1. Measures that prevent generation of wastes.
2. Recovery and recycle of waste constituents.
3. Destruction and treatment, conversion to nonhazardous waste forms.
4. Disposal (storage, landfill).

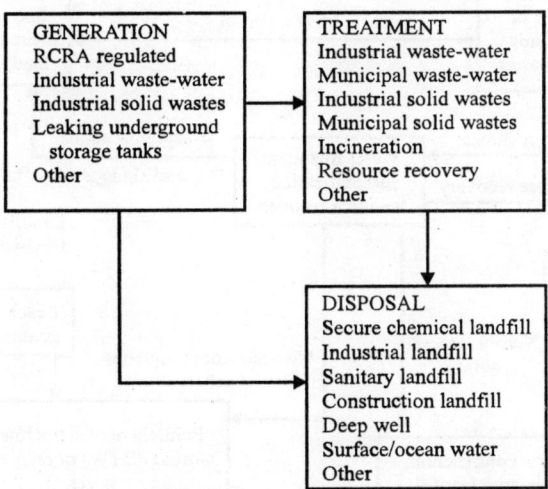

Fig. 19.3. System of generation, treatment, and disposal of hazardous wastes.

Treatment, storage, and disposal facilities

A crucial part of the regulation of hazardous wastes in India pertains to treatment, storage and disposal facilities (TSDF). Treatment alters the physical, chemical or biological character or composition of a waste to make it safer. Storage refers to the holding of hazardous wastes for a temporary period pending treatment or disposal. Disposal refers to the ultimate fate of hazardous substances or their treatment products.

Waste reduction and waste minimisation

Many hazardous waste problems can be avoided at early stages by waste reduction and waste minimisation. As these terms are most commonly used, waste reduction refers to source reduction-less waste-producing materials in, less waste out. Waste minimisation can include treatment processes, such as incineration which reduce the quantities of wastes for which ultimate disposal is required.

Waste treatment

An overall scheme for the treatment of hazardous wastes is shown in Fig. 19.4. Under the category of treatment it is necessary to consider both municipal waste-water and municipal solid wastes along with hazardous wastes. The goal of many industrial waste-water and sludge treatment processes is to produce an effluent that meets standards for release to a municipal waste-water treatment plant and in some cases to produce solids that can be co-disposed with municipal solid wastes. Incineration of municipal solid wastes may produce some solids, particularly fly ash, that have to be treated as hazardous.

The scheme outlined in Fig. 19.4 may serve as a frame of reference for subsequent discussions of waste treatment. The ideal treatment process reduces the quantity of hazardous waste material to a small fraction of the original amount and converts it to a non-hazardous form. However, most treatment processes yield material, such as sludge from waste-water treatment or incinerator ash, which requires disposal and which may be hazardous to some extent.

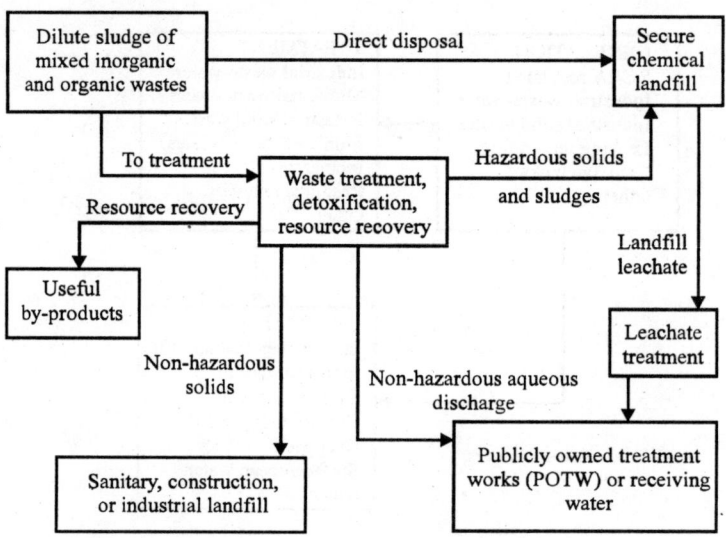

Fig. 19.4. Treatement options for mixed hazardous wastes.

Direct disposal of minimally treated hazardous wastes is becoming more severely limited with new regulations coming from the Hazardous and Solid Waste Amendments. Under its "land-ban" rules, this Act prohibits the land disposal of more than 400 chemicals and waste streams unless they are treated or can be shown not to migrate during the time that they remain hazardous. The ultimate objective of these rules is to reduce the amounts of hazardous wastes generated, although quantities are expected to increase during the next decade. More emphasis in treatment is being placed on recovery of recyclable materials and production of innocuous by-products. There are strong regulatory and economic incentives to generate fewer wastes in manufacturing by modification of processes, product substitution, recycling, and careful control throughout the manufacturing system.

Hazardous Substances and Health

In recent years, the health aspects of hazardous substances have received increased attention by the public and by legislative bodies. A basic question is the linkage between the health of people living near Superfund sites and the chemicals found in the sites.

CHAPTER 20

Environmental Chemistry of Hazardous Wastes

INTRODUCTION

The environmental chemistry of hazardous waste materials in the environment may be considered on the basis of the definition of environmental chemistry according to the following factors :

1. Origin.
2. Transport.
3. Reactions.
4. Effects.
5. Ultimate fate.

In addition, consideration must be given to the distribution of hazardous wastes among the geosphere, hydrosphere, atmosphere, and biosphere, as shown for pollutants in Fig. 20.1

ORIGIN OF HAZARDOUS WASTES

The origin of hazardous wastes refers to their points of entry into the environment. These may consist of the following :

1. Deliberate addition to soil, water, or air by humans.
2. Evaporation or wind erosion form waste dumps into the atmosphere.
3. Leaching from waste dumps into groundwater, streams, and bodies of water.
4. Leakage, such as from underground storage tanks or pipelines.
5. Evolution and subsequent deposition by accidents, such as fire or explosion.
6. Release from improperly operated waste treatment or storage facilities.

TRANSPORT OF HAZARDOUS WASTES

The transport of hazardous wastes is largely a function of their physical properties, the physical properties of their surrounding matrix, the physical conditions to which they are subjected, and chemical factors. Highly volatile wastes are obviously more likely to be transported through the atmosphere and more soluble ones to be carried by water. Wastes will move farther, faster in porous sandy formations than in denser soils. Volatile wastes are more mobile under hot, windy conditions and soluble ones during periods of heavy rainfall. Wastes that are more chemically and biochemically reactive will not move as far as less reactive wastes before breaking down.

The major physical properties of waste that determine their amenability to transport are volatility, solubility, and the degree to which they are sorbed to solids, including solid and sediments.

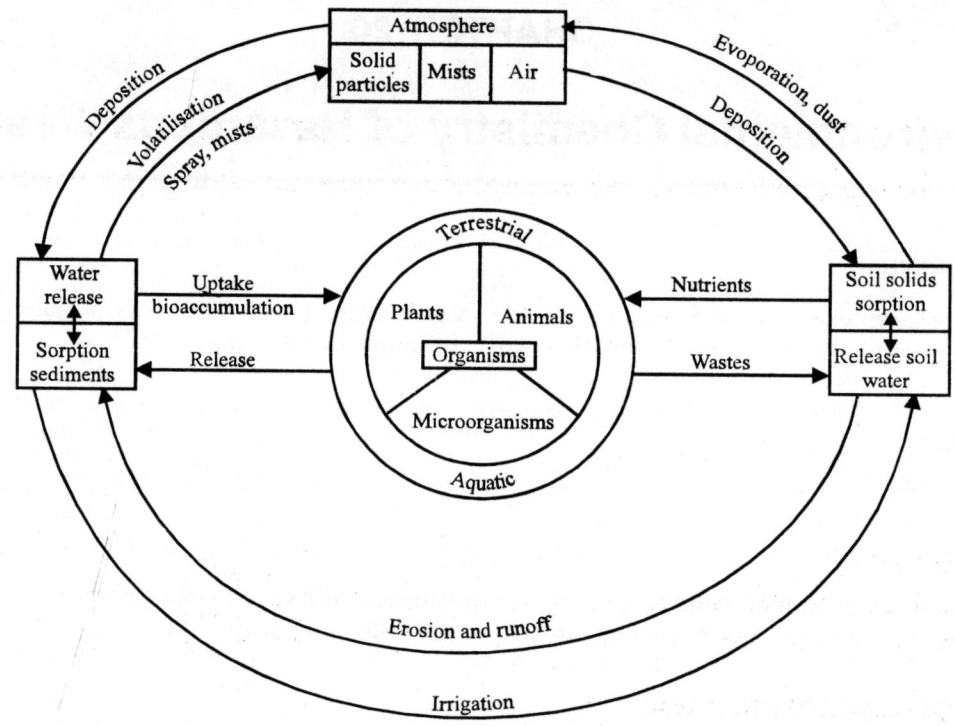

Fig. 20.1. Scheme of interactions of hazardous wastes in the environment.

The distribution of hazardous waste compounds between the atmosphere and the geosphere or hydrosphere is largely a function of compound volatilities. Compound volatilities are usually measured by vapour pressures, which vary over a wide range. A parameter called evaporation rate is used on Material Safety Data Sheets (MSDSs) to express the likelihood of a compound going into the vapour state. Evaporation rate is based upon the vapour pressure at 20°C of butyl acetate, a solvent that is widely used in making lacquers, plastics, and safety glass. The value of the vapour pressure of butyl acetate under these conditions is 10 mm Hg and the evaporation rate of a compound is given as,

$$\text{Evaporation rate} = \frac{\text{Vapour pressure of compound}}{10 \text{ mm Hg}} \qquad ...(20.1)$$

where the vapour pressure of the compound is given in mm Hg. Examples of readily vaporisable hazardous waste compounds are acetone (evaporation rate 22), ethyl ether (evaporation rate 44), and normal pentane (evaporation rate 42.6). By way of contrast, the evaporation rate of the PCB Arochlor 1254 is only 6×10^{-5}.

Usually, in the hydrosphere, and often in soil, hazardous waste compounds are dissolved in water; therefore, the tendency of water to hold the compound is a factor in its mobility. For example, although ethyl alcohol has a higher evaporation rate and lower boiling temperature (4.3 and 77.8°C, respectively) than toluene (2.2 and 110.6°C), vapour of the latter compound is more readily evolved from soil because of its limited solubility in water compared to ethanol, which is totally miscible with water.

Chemical Factors

As an illustration of chemical factors involved in transport of wastes, consider largely cationic inorganic species. Inorganic species can be divided into three groups based upon their retention by clay minerals. Elements that tend to be highly retained by clay include cadmium, mercury, lead, and zinc. Potassium, magnesium, iron, silicon, and NH_4^+ ions are moderately retained by clay, whereas sodium, chloride, calcium, manganese, and boron ions are poorly retained. The retention of the last three elements is probably biased in that they are leached from clay, so that negative retention (elution) is often observed. It should be noted, however, that the retention of iron and manganese is a strong function of oxidation state in that the reduced forms of Mn and Fe are relatively poorly retained, whereas the oxidised forms of $Fe_2O_3.xH_2O$ and MnO_2 are very insoluble and stay on soil as solids.

EFFECTS OF HAZARDOUS WASTES

The effects of hazardous wastes in the environment may be divided among effects on organisms, effects on materials, and effects on the environment. These are addressed briefly here and in greater detail in latter sections.

The ultimate concern with wastes has to do with their toxic effects on animals, plants, and microbes. Virtually all hazardous waste substances are poisonous to a degree, some extremely so. The toxicity of a waste is a function of many factors, including the chemical nature of the waste, the matrix in which it is contained, circumstances of exposure, the species exposed, manner of exposure, degree of exposure, and time of exposure.

As many hazardous wastes are corrosive to materials, usually because of extremes of pH or because of dissolved salt content. Oxidant waste can cause combustible substances to burn uncontrollably. Highly reactive waste can explode, causing damage to materials and structures. Contamination by wastes, such as by toxic pesticides in grain, can result in substances becoming unfit for use.

In addition to their toxic effects in the biosphere, hazardous wastes can damage air, water, and soil. Wastes that get into air can cause deterioration of air quality, either directly, or by the formation of secondary pollutants. Hazardous waste compounds dissolved in, suspended in, or floating as surface films on the surface of water can render it unfit for use and for sustenance of aquatic organisms.

Soil exposed to hazardous wastes can be severely damaged by alteration of its physical and chemical properties and ability to support plants. For example, soil exposed to concentrated brines from petroleum production may become unable to support plant growth, so that the soil becomes extremely susceptible to erosion.

FATES OF HAZARDOUS WASTES

The fates of hazardous waste substances are addressed in more detail in subsequent sections. As with all environmental pollutants, such substances eventually reach a state of physical and chemical stability, although that may take many centuries to occur. In some cases, the fate of a hazardous waste material is a simple function of its physical properties and surroundings.

The fate of a hazardous waste substance in water is a function of the substance's solubility, density, biodegradability, and chemical reactivity. Dense, water-immiscible liquids may simply sink to the bottom of bodies of water or aquifers and accumulate there as "blobs" of liquid. This has happened, for example, with hundreds of tonnes of PCB wastes that have accumulated in sediments in various rivers of India. Biodegradable substances are broken down by bacteria, a process for which the availability of oxygen is an important variable. Substances that readily undergo bioaccumulation are taken up by organisms, exchangeable cationic materials become bound to sediments, and organophilic materials may be sorbed by organic matter in sediments.

The fates of hazardous waste substances in the atmosphere are often determined by photochemical reactions. Ultimately, such substances may be converted to nonvolatile, insoluble matter and precipitate from the atmosphere on to soil or plants.

HAZARDOUS WASTES IN THE GEOSPHERE

The sources, transport, interactions, and fates of contaminant hazardous wastes in the geosphere involve a complex scheme, some aspects of which are illustrated in Fig. 20.2. The primary environmental concern regarding hazardous waste in the geosphere is the possible contamination of groundwater aquifers by waste leachates and leakage form wastes. As the figure shows, there are a number of possible contamination sources. The most obvious one is leachate from landfills containing hazardous wastes. In some cases, liquid hazardous materials are placed in lagoons, which can leak into aquifers. Leaking sewers can also result in contamination, as can the discharge from septic tanks. Hazardous wastes spread on land can result in aquifer contamination by leachate.

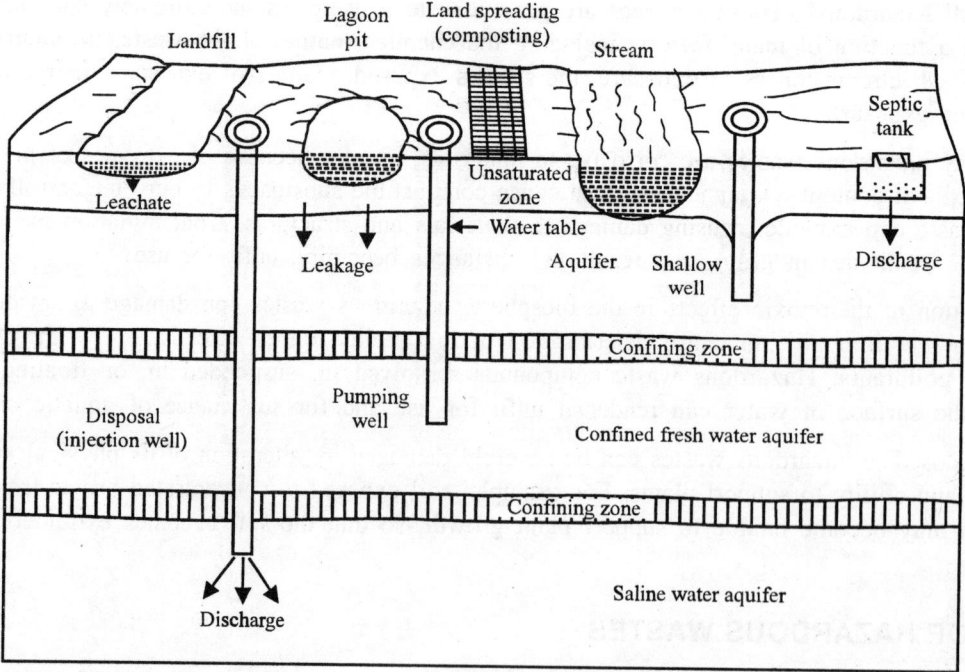

Fig. 20.2. Sources, disposal, and movement of hazardous wastes in the geosphere.

Hazardous chemicals are sometimes deliberately disposed of underground in waste disposal wells. This means of disposal can result in interchange of contaminated water between surface water and groundwater at discharge and recharge points.

The transport of contaminates in the geosphere depends largely upon the hydrologic factors governing the movement of water underground and the interactions of hazardous waste constituents with geological strata, particularly unconsolidated earth material. As shown in Fig. 20.3, groundwater contaminated with hazardous waste tends to flow as a relatively undiluted plug or plume along with the groundwater in an aquifer. The groundwater flow rate depends upon the water gradient and aquifer characteristics, such as permeability and cross-section area. The rate of flow is generally relatively slow; 1 meter per day would be considered fast. As already discussed, contaminated groundwater can result in contamination of a surface water source. This can occur at a discharge area where the groundwater flows into a lake or stream.

As discussed in the preceding section, hazardous waste dissolved in groundwater can be attenuated by soil or rock by means of various sorption mechanisms. Mathematically, the distribution of a solute between groundwater or leachate water and soil is expressed by a distribution coefficient, K_d,

$$K_d = \frac{C_s}{C_w} \qquad \qquad ...(20.2)$$

where C_s is the equilibrium concentration of the species in the solid phase and C_w is its concentration in water. This equation assumes that the relative degree of sorption is independent of C_w: that is, it assumes a linear sorption isotherm. For the more common case of nonlinear sorption isotherms, C_s is expressed as a function of the equilibrium concentration of sorbate in water, C_{eq}, by the *Freundlich equation*,

$$C_s = K_F C_{eq}^{1/n} \qquad \qquad ...(20.3)$$

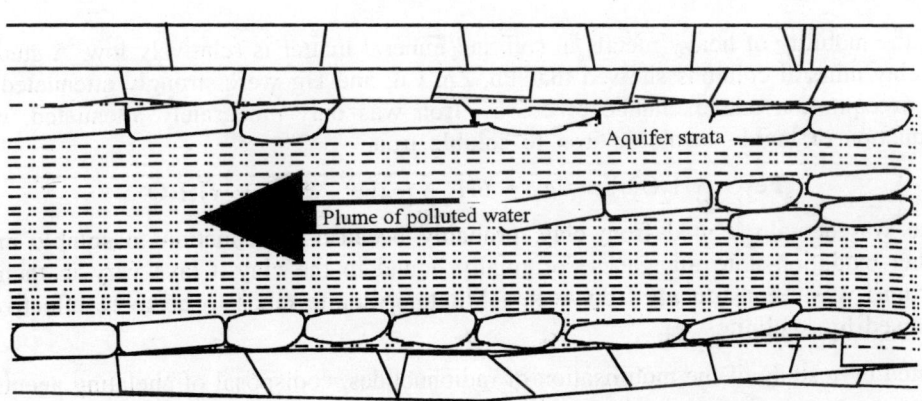

Fig. 20.3. Plug-flow of hazardous wastes in groundwater.

where and K_F and $1/n$ are empirical constants. The degree of attenuation depends upon the surface properties of the solid, particularly its surface area. The chemical nature of the attenuating solid is also important because attenuation is a function of the organic matter (humus) content, presence of hydrous metal oxides, and the content and types of clays present. The chemical characteristics of the leachate also affect attenuation greatly. For example, attenuation of metals is very poor in acidic leachate because precipitation reactions, such as,

$$M^{2+} + 2OH^- \rightarrow M(OH)_2(s) \qquad ...(20.4)$$

are reversed in acid:

$$M(OH)_2\ (s) + 2H^+ \rightarrow M^{2+} + 2H_2O \qquad ...(20.5)$$

Organic solvents is leachates tend to prevent attenuation of organic hazardous waste constituents.

Sorption of nonionic organic matter by soil depends upon the organic content of the soil. In a study of the sorption of trichloromethane, 1,1,1– trichloroethane trichloroethylene, and perchloroethylene on tertiary shale, Jurassic shale, peat, lignite, bituminous coal, and anthracite coal, it has been shown that the type of organic matter in soil is also quite important. According to the study, organic matter with a low oxygen and high hydrogen content, such as that in unweathered shales, is about an order of magnitude more effective in sorbing organic solutes that is more oxidised organic matter, such as that in weathered shales.

The degree of attenuation of a pollutant by soil depends upon the water content of the soil. Above the water table there is an unsaturated zone of soil in which attenuation is more highly favoured. Normally soil has a greater surface areas at liquid-solid interfaces in this zone so that absorption and ion-exchange processes are favoured. Aerobic degradation is possible in the unsaturated zone, enabling more rapid and complete degradation of biodegradable hazardous wastes.

Heavy metals are particularly damaging to groundwater and their movement through the geosphere is of considerable concern. Heavy-metal ions may be sorbed by the soil, held by ion-exchange processes, interact with organic matter in soil, undergo oxidation-reduction processes leading to mobilisation or immobilisation, or even be volatilised as organometallic compounds formed by methylating bacteria. A large number of factors affect heavy-metal mobility and attenuation in soil. These include pH, pE, temperature, cation exchange capacity, nature of soil mineral matter, and kinds of soil organic matter present.

Normally, the mobility of heavy metals in soil and mineral matter is relatively low. A study of relative mobilities in clay mineral columns showed that Pb, Zn, Cd, and Hg were strongly attenuated by the clay, primarily by precipitation and exchange processes. Iron was only moderately attenuated, which has to be due to reduction of highly insoluble iron to soluble iron.

$$Fe_2O_3.xH_2O(s) + 2e^- + 6H^+ \rightarrow 2Fe^{2+} + (3 + x)H_2O \qquad ...(20.6)$$

Manganese was actually eluted from clay, probably because of reduction to soluble manganese of insoluble higher-oxidation-state manganese originally bound to the clay. Clays vary in their abilities to remove hazardous waste constituents from water. Montmorillonite tends to be more effective than illite, which is followed by kaolinite.

As illustrated by a study of the mobilisation of radionuclides, codisposal of chelating agents with heavy metals can have a strong effect upon the mobility of metal ions in soil. The presence of chelating agents resulted from the use of salts of ethylenediaminetetraacetic acid (EDTA) in decontaminating facilities exposed to nuclear wastes. Other chelating agents used for decontamination and later codisposed with radioactive materials include diethylenetriaminepentaacetic acid (DTPA) and nitrilotriacetic acid (NTA). An important consideration regarding chelating agents bound to radionuclides (as heavy metals) in soil, sediments, and groundwater is the biodegradation of the chelating agents. Although this phenomenon has not been studied in detail, indications are that under subsurface conditions, the biodegradability of the chelating agents listed above is in the order NTA > DTPA > EDTA.

Whereas metal cations are readily held by ion exchange processes and precipitation on soil,

$$2Soil\}^-H + Co^{2+} \rightarrow (Soil\}^-)_2Co^{2+} + 2H^+ \qquad \qquad ...(20.7)$$

$$Co^{2+} + 2OH^- \rightarrow CoOH)_2(s) \qquad \qquad ...(20.8)$$

chelated anionic species, such as CoY^{2-} (where Y^{4-} is the chelating EDTA anion), are not strongly retained by the negatively charged functional groups in soil.

Radionuclides have been buried in shallow trenches on the grounds and thus, ample time has elapsed to observe the effects of this means of radioactive waste disposal. The predominant geological formation at these burial sites is *Conasauga shale*. This bedrock material has a very high sorptive capacity for most of the radionuclides produced as by products of nuclear fission, particularly those that are cationic. Despite this, some migration of radionuclides has been observed from sites used to dispose of solid and liquid wastes. Some of this migration has been attributed to the high rainfall in the area, shallow groundwater levels, fractures in the underlying rock that allow for rapid infiltration of dissolved wastes, and other physical factors.

In addition to the factors listed above as contributing to the migration of radionuclides from waste disposal trenches, it has been found that chelating agents used for decontamination, as well as naturally occurring humic substance chelators, are responsible for migration in excess of that expected. Most notably, Co has been found outside the disposal trenches. Levels of radioactive contamination from this isotope adjacent to the disposal trenches have been observed as high as 1×10^5 disintegrations per minute (dpm) per gram (45, 000 picocuries/g) in soil and as high as 1×10^3 dpm/mL in soil water. In addition, traces of various isotopes of the alpha emitters, uranium plutonium radium, thorium, and californium, have been found outside the disposal area. Experiments have been conducted to determine the values of K_d for the distribution of Co between waste sampled from wells at the disposal sites and the shale at the sites. (The distribution coefficient is a measure of the affinity of the solute for a solid phase; the higher its value, the greater the tendency of the solute to be sorbed by the solid.) For well waste samples ranging in pH from 6.0 to 8.5, values of K_d were measure as 7 to 70 with an average of about 35. This is in marked contrast to the value 7.0×10^4 for K_d obtained with a standard Co solution prepared from inorganic cobalt in the absence of a chelating agent, indicative of a tremendous affinity of inorganic cobalt for shale. Similar solution prepared containing 1×10^{-5} M EDTA and cobalt at the same pH gave K_d values of only 2.9. The actual EDTA concentration found in the well samples was 3.4×10^{-7} M, thus explaining why the distribution coefficient in the well water samples was some what higher than that observed in the experimental samples containing 1×10^{-5} M EDTA. Species other than EDTA possess the potential to mobilise radionuclides or heavy metals. Of these, phthalic and palmitic acid

Phthalic acid   Palmitic acid

were found in leachate from the disposal trenches. Other species that might be codisposed with radionuclides which could also increase radionuclide mobilities are citrate, fluoride, oxalate, and gluconate salts. In another study of radioactive waste disposal sites, it was observed that organic chelating agents, particularly EDTA, were present and dramatically increase the migration of radionuclides from the site. According to this study water samples at the disposal site where EDTA containing plutonium wastes were disposed, showed leaves of 300, 000 picocuries/litre, far in excess of those found in the absence of the chelating agent.

The evidence just cited suggests that strong chelating agent would have a tendency to transport heavy metal ions from disposal sites. Co-disposal of EDTA with radionuclides and heavy metals should be avoided.

HAZARDOUS WASTES IN THE HYDROSPHERE

Fig. 20.4, illustrates a typical pathway for the entry of hazardous waste materials into the hydrosphere. Other sources consist of precipitation from the atmosphere with rainfall, deliberate release to streams and bodies of water, runoff from soil, and mobilisation from sediments. Once in an aquatic system, hazardous waste species are subject to a number of chemical and biochemical processes, including acid-base, oxidation-reduction, precipitation-dissolution, and hydrolysis reactions, as well as biodegradation.

The presence of organic matter in waste has a tendency to increase the solubility of hazardous organic substance. Typically, the solubility of hexachlorobenzene is 1.8 μg/L in pure water at 25°C whereas it is 2.3 μg/L in creek water containing organic solutes and 4 to 4.5 μg/L in landfill leachate.

In considering the process that hazardous waste undergo in water, it is important to recall the nature of aquatic systems and the unique properties of water. Water in the environment is far from pure. Just as the atmosphere is a constantly changing mass of bodies of moving air with different temperatures, pressures, and humidities, bodies of waste are highly dynamic systems. Rivers, impoundments, and groundwater aquifers are subject to the input and loss of a variety of materials from both natural and anthropogenic sources.

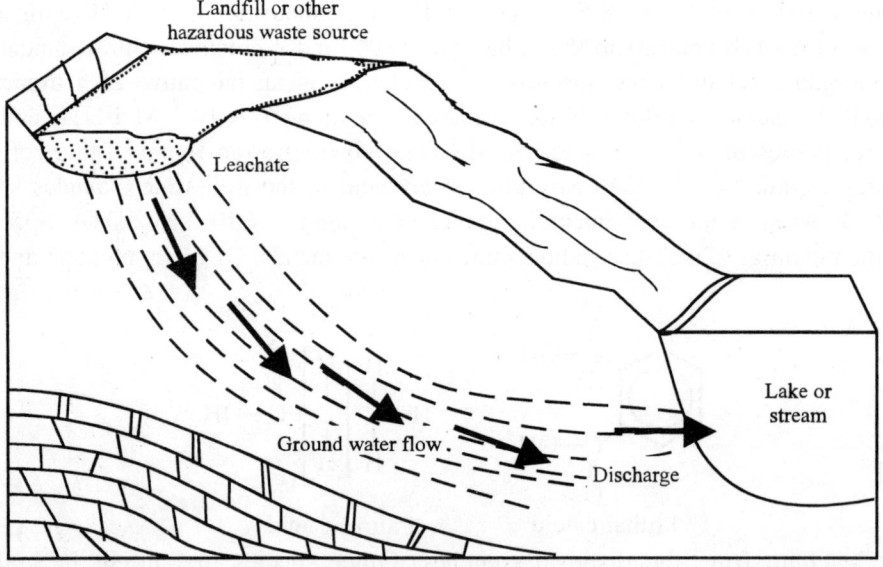

Fig. 20.4. Discharge of groundwater contaminated from hazardous waste landfill into a body of water.

These material may be gases, liquids, or solids. They interact chemically with each other and with living organisms—particularly bacteria–in the water. They are subject to dispersion and transport by stream flow, convection currents, and other physical phenomena. Hazardous substances or their by-products in water may undergo bioaccumulation through food chains involving aquatic organisms.

Several physical, chemical, and biochemical processes are particularly important in determining the transformations and ultimate fates of hazardous chemical species in the hydrosphere. These include hydrolysis reactions, through which a molecule is cleaved with the addition of H_2O; precipitation reactions, generally accompanied by aggregation of colloidal particles suspended in water; oxidation-reduction reactions, generally mediated by microorganisms sorption of hazardous solutes by sediments and by suspended mineral and organic matter; biochemical processes, often involving hydrolysis and organic matter; biochemical processes, often involving hydrolysis and oxidation-reduction reactions; photolysis reaction; and miscellaneous chemical phenomena.

Hydrolysis reactions of two hazardous waste compounds, an acid anhydride and an ester, are illustrated by the following reactions :

$$H-\underset{\underset{H}{|}}{\overset{\overset{H}{|}}{C}}-\underset{}{\overset{\overset{O}{\|}}{C}}-O-\underset{}{\overset{\overset{O}{\|}}{C}}-\underset{\underset{H}{|}}{\overset{\overset{H}{|}}{C}}-H + HOH \longrightarrow 2H-\underset{\underset{H}{|}}{\overset{\overset{H}{|}}{C}}-\underset{}{\overset{\overset{O}{\|}}{C}}-O-H \qquad \text{...(20.9)}$$

      Acetic anhydride   Acetic acid

$$\underset{H}{\overset{H}{>}}C=C-\underset{}{\overset{\overset{O}{\|}}{C}}-O-\underset{}{\overset{\overset{H}{|}}{C}}-H + HOH \longrightarrow \underset{H}{\overset{H}{>}}C=C-\underset{}{\overset{\overset{O}{\|}}{C}}-OH + HO-\underset{\underset{H}{|}}{\overset{\overset{H}{|}}{C}}-H \qquad \text{...(20.10)}$$

      Methyl methacrylate   Methacrylic acid   Methanol

The rates at which compounds hydrolyse in water vary widely. Acetic anhydride hydrolyses very rapidly. In fact, the great affinity of this compound for water (including water in skin) is one of the reasons that it is hazardous. Once in the aquatic environment, though, acetic anhydride is converted very rapidly to essentially harmless acetic acid. Many ethers, esters, and other compounds formed originally by the joining together of two or more molecules with the loss of water hydrolyse very slowly, although the rate may be greatly increased by the action of enzymes in microorganisms (biochemical processes). Hydrolysis of some compounds results in the loss of halogen atoms. For example, bis(chloromethyl) ether hydrolyses rapidly to produce HCl and formaldehyde:

$$Cl-\underset{\underset{H}{|}}{\overset{\overset{H}{|}}{C}}-O-\underset{\underset{H}{|}}{\overset{\overset{H}{|}}{C}}-Cl + H_2O \longrightarrow 2H-\overset{\overset{O}{\|}}{C}-H + 2HCl \qquad \text{...(20.11)}$$

Hydrolysis of a large quantity of this chloroether in the aquatic environment could produce harmful amounts of corrosive HCl and formaldehyde.

As already discussed the formation of precipitates in the form of sludges is one of the most common means of isolating hazardous components from an unsegregated waste. Although solid inorganic ionic compounds are often discussed in terms of very simple formulas, such as $PbCO_3$ for lead carbonate, much more complicated species [for example, $2PbCO_3.Pb(OH)_2$] generally result when precipitates are formed in the aquatic environment. For example, a hazardous heavy metal ion in the hydrosphere may be precipitated as a reactively complicated compound, coprecipitated as a minor constituent of some other compound, or be sorbed by the surface of another solid.

The major anions present in natural waters and waste-waters are OH^-, HCO_3^-, and SO_4^{2-}. Since these anions are all capable of forming precipitates with cationic impurities, such pollutants tend to precipitate as hydroxides, carbonates, and sulphates. Sometimes a distinction can be made between hydroxides and hydrated oxides with similar, or identical, empirical formulas. For example, iron hydroxide, $Fe(OH)_3$, is a relatively uncommon species; iron usually is precipitated from water as hydrated iron oxides, such as β ion oxide monohydrate, $Fe_2O_3H_2O$. Basic salts containing OH^- ion along with some other anion are very common in solids formed by precipitation from water. A typical example is azurite, $2CuCO_3.Cu(OH)_2$. Two or more metal ions may be present in a compound, as is the case with chalcopyrite, $CuFeS_2$.

Two aspects of the precipitation process are particularly important in determining the fate of hazardous ionic solutes in water. If precipitation occurs very rapidly and with a high degree of supersaturation, the solid tends to form as a large number of small colloidal particles that may persist in the colloidal state for a long time. In this form, hazardous substances are much more mobile and accessible to organisms than as precipitates. A second important consideration is that many heavy metals are coprecipitated with hydrated iron oxide ($Fe_2O_3.xH_2O$) or manganese oxide ($MnO_2.xH_2O$).

Sorption processes are particularly common methods for the removal of low level hazardous materials from water. Freshly precipitated $MnO_2.xH_2O$ very effectively scavenges other metal ions, such as Ba^{2+}, from water. As shown by the examples in Table. 20.1, oxidation-reduction reactions are very important means of transformation of hazardous wastes in water. The degradation of most organic wastes proceeds by way of oxidation.

Under many circumstances, biochemical processes largely determine the fates of hazardous chemical species in the hydrosphere. The most important such processes are those mediated by microorganisms. In particular, the oxidation of biodegradable hazardous organic wastes in water generally occurs by means of microorganism-mediated biochemical reactions. Bacteria produce organic acids and chelating agents, such as citrate, which have the effect of solubilising hazardous heavy metal ions. Some mobile methylated forms, such as compounds of methylated arsenic and mercury, are produced by bacterial action.

As already discussed photolysis reaction are those initiated by the absorption of light. The effect of photolytic processes on the destruction of hazardous waste in the hydrosphere is minimal, although some photochemical reactions of hazardous waste compounds can occur when the compounds are present as surface films on water exposed to sunlight.

Groundwater is the part of the hydrosphere most vulnerable to damage from hazardous wastes. Although surface water supplies are subject to contamination, groundwater can become almost irreversibly contaminated by the improper land disposal of hazardous chemicals.

HAZARDOUS WASTES IN THE ATMOSPHERE

Some chemicals founds in hazardous waste sites may enter the atmosphere by evaporation or even as windblown particles. Three major areas of interest in respect to hazardous waste compounds in the atmosphere are their pollution potential, atmospheric fate, and residence time.

Table 20.1. Oxidation reduction of wastes in water.

Reaction	Significance
Oxidation half-reactions	
$SO_2(aq) + 2H_2O \rightarrow 4H^+ + SO_4^{2-} + 2e^-$	Conversion of dissolved SO_2 gas to sulphuric acid
$\underset{\parallel}{\overset{O}{CH_3\ C}}\!-\!H + H_2O \rightarrow \underset{\parallel}{\overset{O}{CH_3 C}}\!-\!OH + 2H^+ + 2e^-$	Conversion of acetaldehyde to acetic acid
$\{CH_2O\} + H_2O \rightarrow CO_2 + 4\ H^+ + 4e^-$	Degradation of biomass
$C_nH_{2n+2} + 2_n\ H_2O \rightarrow n\ CO_2 + (6n + 2)\ H^- + (6n + 2)\ e^-$	Degradation of hydrocarbons
Reduction half-reactions	
$O_2(aq) + 4H^- + 4e^- \rightarrow 2H_2O$	Removal of O_2 from water; O_2 is an electron receptor (source of O_2) for oxidation half-reactions above
$Fe_2O_3 \cdot xH_2O(s) + 6H^+ + 2e^- \rightarrow 2Fe^{2+} + (3 + x)\ H_2O$	Formation of soluble Fe^{2+}
$MnO_2(s) + 4H^+ + 2e^- \rightarrow Mn^{2+} + 2H_2O$	Production of soluble Mn^{2+}

For example, the overall reaction for the oxidation of dissolved SO_2 by O_2 is obtained as follows :

$$2\{SO_2(aq) - 2\ H_2O \rightarrow 4\ H^- + SO_4^{2-} + 2e^-\}$$
$$O_2(aq) + 4\ H^- 4e^- \rightarrow 2\ H_2O$$
$$\overline{2SO_2(aq) - O_2(aq) - 2\ H_2O \rightarrow 4H^- + 2SO_4^{2-}}$$

Air Pollution Potential of Hazardous Waste Compounds

The pollution potential of hazardous wastes in the atmosphere depends upon whether they are primary pollutants that have a direct effect or secondary pollutants that are converted to harmful substances by atmospheric chemical processes. Hazardous waste sites do not usually evolve sufficient quantities of pollutants to give significant amounts of secondary pollutants, so primary air pollutants are of the greater concern. Examples of primary air pollutants include toxic organic vapours (vinyl chloride), corrosive acid gases (HCl), and toxic inorganic gases, such as H_2S released by the accidental mixing of waste acid (HCl from waste steel pickling liquor) and waste metal sulphides:

$$2HCl + FeS \rightarrow FeCl_2 + H_2S(g) \qquad ...(20.12)$$

Primary air pollutants are most dangerous in the immediate vicinity of a site, usually to workers involved in deposal or cleanup or people living adjacent to the site. Quantities are rarely sufficient to pose any kind of regional air pollution hazard.

The two major kinds of secondary air pollutants from hazardous wastes are those that are oxidised in the atmosphere to corrosive substances and organic materials that undergo photochemical oxidation. Plausible examples of the former are sulphur dioxide released from the action of waste strong acids on sulphites and subsequently oxidised in the atmosphere to corrosive sulphuric acid,

$$SO_2 + {}^{1/2}O_2 + H_2O \rightarrow H_2SO_4(aerosol) \qquad ...(20.13)$$

and nitrogen dioxide (itself a toxic primary air pollutant) produced by the reaction of waste nitric acid with reducing agents such as metals and oxidised to corrosive nitric acid or converted to corrosive nitrate salts :

$$4HNO_3 + Cu \rightarrow Cu(NO_3)_2 + 2NO_2(g) + 2H_2O \qquad ...(20.14)$$

$$2NO_2(g) + {}^{1/2}O_2 + H_2O \rightarrow 2HNO_3(aerosol) \qquad ...(20.15)$$

$$HNO_3(aerosol) + NH_3(g) \rightarrow NH_4NO_3(aerosol) \qquad ...(20.16)$$

Organic species that produce secondary air pollutants are those that form photochemical smog. The more reactive of these are unsaturated compounds that react with atomic oxygen or hydroxyl radical in air,

$$R–CH=CH_2 + HO\cdot \, RCH_2CH_2O\cdot \qquad ...(20.17)$$

to yield reactive radicals that participate in chain reactions to eventually yield ozone, organic oxidants, noxious aldehydes, and other products characteristic of photochemical smog.

Fate and Residence Time of Hazardous Waste Compounds in the Atmosphere

An obvious means by which hazardous waste species may be removed from the atmosphere is by dissolution in water in the form of cloud or rain droplets. Inorganic acid, base, and salt compounds, such as H_2SO_4, HNO_3, and NH_4NO_3 mentioned above, are readily removed from the atmosphere by dissolution. For vapours of compounds that are not highly soluble in water, solubility information combined with information about rainfall amounts and mixing in the atmosphere can be used to estimate the atmospheric half-life, $\tau_{1/2}$, of the species. Solubility rate may be used to estimate half-lives for substances that are more miscible in water. For poorly water-soluble compounds, such calculations tend to drastically underestimate lifetimes, which indicates that other removal mechanisms must predominate.

The lifetimes of vapourised hazardous waste species removed from the atmosphere through adsorption by aerosol particles is limited to that of the sorbing aerosol particles (typically about 7 days) plus the time spent in the vapour phase before adsorption. This mechanism appears to be valuable only for highly non-volatile constituents such as benzo[a]pyrene.

Sorptive removal by soil, water, or plants on the earth's surface, called dry deposition, is another means for physical removal of hazardous substances from the atmosphere. Predictions of dry deposition rates vary greatly with type of compound, type of surface, and weather conditions. For highly volatile organic compounds, such as low molecular mass organohalide compounds, predicted rates of dry deposition give atmospheric lifetimes many-fold higher than those actually observed, so for such compounds, dry deposition is probably not a common removal mechanism.

Predicted rates of physical removal of a number of volatile organic compounds that are not very soluble in water are far too slow to account for the loss of such compounds from the atmosphere, so chemical processes must predominate. As already discussed the most important of these processes is reaction with hydroxyl radical, HO·, in the troposphere. Ozone can react with compounds having a double bond. Other oxidant species that might react with hazardous waste compounds in the troposphere and stratosphere are atomic oxygen (O), peroxyl radicals (HOO·), alkylperoxyyl radical (ROO·), and NO_3.

Despite the fact that its concentration in the troposphere is relatively low, HO· is so reactive that it tends to initiate most of the reactions leading to the chemical removal of most refractory organic compounds from the atmosphere. As already discussed, hydroxyl radical undergoes abstraction reactions to remove H atoms from organic compounds containing R–H,

$$R–H + HO\cdot \rightarrow \, + R\cdot \, H_2O \qquad ...(20.18)$$

and may react with those containing unsaturated bonds by addition as illustrated in Reaction 20.17. In both cases, reactive free radicals are formed that undergo further reactions, leading to nonvolatile and/

or water-soluble species, which are scavenged from the atmosphere by physical means. These scavengeable species tend to be aldehydes, ketones, or acids. Halogenated organic compounds may lose halogen atoms in the form of halo-oxy radicals and undergo further reactions to form scavengeable species.

In general, reactions with species other than HO· or O_3 are not considered significant in the removal of hazardous organic waste compounds from the troposphere. Perhaps in some cases such reactions do contribute to a very slow removal of such contaminant compounds.

Photolytic transformations involve direct cleavage (photodissociation) of compounds by reactions with visible and ultraviolet radiation:

$$R–X + h\nu \rightarrow R· + X· \qquad \qquad ...(20.19)$$

The extent of these reactions varies greatly with light intensity, quantum yields (chemical reactions per quantum absorbed) and other factors. In order for photolysis to be an important process for removal of a molecule from the atmosphere, the molecule must have a chromophore (light-absorbing group) that absorbs light in a wavelength region of significant intensity in the impinging electromagnetic radiation spectrum. This requirement limits the importance of photolysis as a removal mechanism to only a few classes of compounds, including conjugated alkenes, carbonyl compounds, some halides, and some nitrogen compounds, particularly nitro compounds. However, these do include a number of the more important hazardous waste compounds.

HAZARDOUS WASTES IN THE BIOSPHERE

One of the most crucial aspects of fate and toxic effects of environmental chemicals is their accumulation by organisms from their surroundings. Biodegradation of wastes is their conversion by biological processes to simple inorganic molecules and, to certain extent, to biological materials. The complete conversion of a substance to inorganic species such as CO_2, NH_3, and phosphate is called mineralisation. Detoxification refers to the biological conversion of a toxic substance to a less toxic species, which may still be relatively complex, or biological conversion to an even more complex material. An example of detoxification is illustrated below for the enzymatic conversion of paraoxon (a highly toxic organophosphate insecticide) to *p*-nitrophenol, which has only about 1/200 the toxicity of the parent compound :

Usually the products of biodegradation are molecular forms that tend to occur in nature. Because the organisms that carry out biodegradation do so as a means of extracting free energy for their metabolic and growth needs they form products that are in greater thermodynamic equilibrium with their surroundings. The definition of biodegradation is illustrated by an example in Fig. 20.5. Biodegradation is usually carried out by the action of microorganisms, particularly bacteria and fungi.

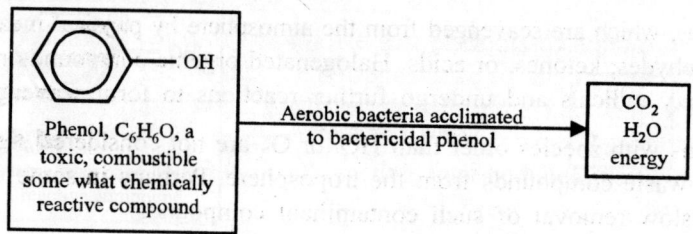

Fig. 20.5. Illustration of biological action on a hazardous waste constituent.

Biodegradation Process

Biotransformation is what happens to any substance that is metabolised and thereby altered by biochemical processes in an organism. Metabolism is divided into the two general categories of catabolism, which is the breaking down of more complex molecules, and anabolism, which is the building up of life molecules from simpler materials. The substances subjected to biotransformation may be naturally occurring or anthropogenic (made by human activities). They may consist of xenobiotic molecules that are foreign to living systems.

An important biochemical process that occurs in the biodegradation of many synthetic and hazardous waste materials is cometabolism. Cometabolism does not serve a useful purpose to an organism in term of providing energy or raw material to build biomass, but occurs concurrently with normal metabolic processes. An example of cometabolism of hazardous wastes is provided by the white rot fungus, *Phanerochaete chrysosporium*, which degrades a number of kinds of organochlorine compounds—including DDT, PCBs, and chlorodioxins—under the appropriate conditions. The enzyme system responsible for this degradation is one that the fungus uses to break down lignin in plant material under normal conditions.

Enzymes in Waste Degradation

Enzymes systems hold the key to biodegradation of hazardous wastes. For most biological treatment process currently in use, enzymes are present in living organisms in contact with the wastes. However, in some cases it is possible to use cell-free extracts of enzymes removed from bacterial or fungal cells to treat hazardous wastes. For this application the enzymes may be present in solution or, more commonly, immobilised in biochemical reactors.

Biodegradation of municipal waste-water and solid wastes in landfills occurs by design. Biodegradation of any kind of waste that can be metabolised takes place whenever the wastes are subjected to conditions conducive to biological process. The most common type of biodegradation is that of organic compounds in the presence of air; that is, aerobic processes. However, in the absence of air, anaerobic biodegradation may also take place. Furthermore, inorganic species are subject to both aerobic and anaerobic biological processes.

Although biological treatment of wastes is normally regarded as degradation to simple inorganic species such as carbon dioxide, water sulphates, and phosphates, the possibility must always be considered of forming more complex or more hazardous chemical species. An example of the latter is the production of volatile, soluble, toxic methylated forms of arsenic and mercury from inorganic species of these elements by bacteria under anaerobic conditions.

For the most part, anthropogenic compounds resist biodegradation much more strongly than do naturally occurring compounds. This is generally due to the absence of enzymes that can bring about an initial attack on the compound. A number of physical and chemical characteristics of a compound are involved in its amenability to biodegradation. Such characteristics include hydrophobicity, solubility, volatility, and affinity for lipids. Some organic structural groups impart particular resistance to biodegradation. These include branched carbon chains, ether linkages, meta-substituted benzene rings, chlorine, amines, methoxy groups, sulphonates, and nitro groups.

Several groups of microorganisms are capable of partial or complete degradation of hazardous organic compounds. Among the aerobic bacteria, those of the *pseudomonas* family are the most widespread and most adaptable to the degradation of synthetic compounds. These bacteria degrade biphenyl, naphthalene, DDT, and many other compounds. Anaerobic bacteria are very fastidious, and they are difficult to study in the laboratory because they require oxygen-free (anoxic) conditions and pE values of less that −3.4 in order to survive. These bacteria catabolise biomass through hydrolytic processes, breaking down proteins, lipids, and saccharides. They are also known to reduce nitro compounds to amines, degrade nitrosamines, promote reductive dechlorination, reduce epoxide groups to alkenes, and break down aromatic structures. *Actinomycetes* are microorganisms that are morphologically similar to both bacteria and fungi. They are involved in the degradation of a variety of organic compounds, including degradation-resistant alkanes, and lignocellulose. Other compounds attacked include pyridines, phenols, nonchlorinated aromatics, and chlorinated aromatics. Fungi are particularly noted for their ability to attack long-chain and complex hydrocarbons and are more successful than bacteria in the initial attack on PCB compounds. Phototrophic microorganisms which include algae, photosynthetic bacteria, and cyanobacteria tend to concentrate organophilic compounds in their lipid stores and induce photochemical degradation of the stored compounds. For example, *Oscillatoria* can initiate the biodegradation of naphthalene by the attachment of −OH groups.

Practically all classes of synthetic organic compounds can be at least partially degraded by various microorganisms. These classes include nonhalogented alkanes, halogenated alkanes (trichloroethane, dichloromethane), nonhalogenated aromatic compounds (hexachlorobenzene, pentachlorophenol), phenols (phenol, cresols), polychlorinated biphenyls, phthalate esters, and pesticides (chlordane, parathion).

Among the most biodegradation-resistant substances are polychlorinated biphenyls, PCBs. Bacteria growing anaerobically in PCB contaminated river sediments exhibited the capacity to partially dechlorinate the more highly chlorinated PCBs. This observation may have some important implication for hazardous waste PCBs in the aquatic and soil environments.

CHAPTER 21

Reduction, Treatment and Disposal of Hazardous Wastes

INTRODUCTION

Many hazardous waste problems can be avoided at early stage by waste reduction (cutting down quantities of wastes from their sources) and waste minimisation (utilisation of treatment processes which reduce the quantities of wastes requiring ultimate disposal).

There are several ways in which quantities of wastes can be reduced, including source reduction, waste separation and concentration, resource recovery, and waste recycling. The most effective approaches to minimising wastes centre around careful control of manufacturing processes, taking into consideration discharges and the potential for waste minimisation at every step of manufacturing. Viewing the process as a whole (as outlined for a generalised chemical manufacturing process in Fig 21.1) often enables crucial identification of the source of a waste, such as a raw material impurity, catalyst, or process solvent. Once a source is identified, it is much easier to take measure to eliminate or reduce the waste.

Modifications of the manufacturing process can yield substantial waste reduction. Some such modifications are of a chemical nature. Changes in chemical reaction conditions can minimise production of by-product hazardous substances. In some cases, potentially hazardous catalysts, such as those formulated from toxic substances, can be replaced by catalysts that are non-hazardous or that can be recycled rather than discarded. Wastes can be minimised by volume reduction, for example, through dewatering and drying sludge.

RECYCLING

Wherever possible, recycling and reuse should be accomplished onsite because it avoids having to move wastes and because a process that produces recyclable materials is often the most likely to have use for them. The four broad areas in which something of value may be obtained from wastes are the following:

1. Direct recycle as raw material to the generator, as with the return to feedstock of raw materials not completely consumed in a synthesis process.

2. Transfer as a raw material to another process; a substance that is a waste product from one process may serve as a raw material for another, sometimes in an entirely different industry.

3. Utilisation for pollution control or waste treatment, such as use of waste alkali to neutralise waste acid.

4. Recovery of energy; for example, from the incineration of combustible hazardous wastes.

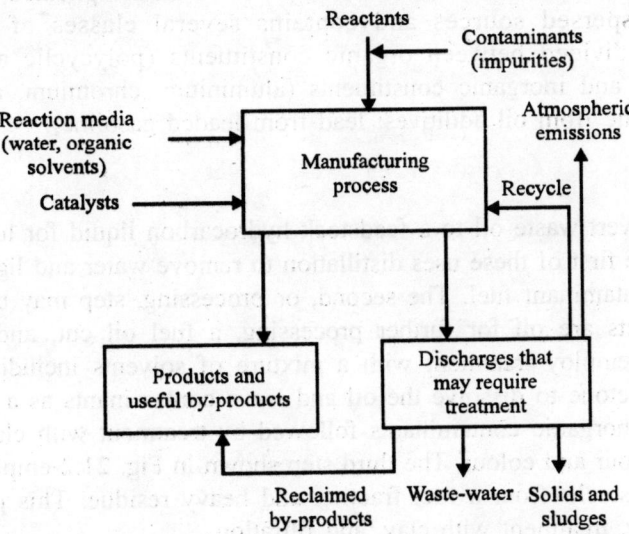

Fig 21.1. Chemical manufacturing process from the viewpoint of discharges and waste minimisation.

Examples of Recycling

Recycling of scrap industrial impurities and products occurs on a large scale with a number of different materials. Most of these materials are not hazardous, but, as with most large-scale industrial operations, their recycle may involve the use or production of hazardous substances. Some of the more important examples are the following :

1. Ferrous metals composed primarily of iron and used largely as feedstock for electric-arc furnaces

2. Nonferrous metals, including aluminium (which ranks next to iron in terms of quantities recycled), copper and copper alloys, zinc, lead, cadmium, tin, silver, and mercury.

3. Metal compounds, such as metal salts.

4. Inorganic substances including alkaline compounds (such as sodium hydroxide used to remove sulphur compounds from petroleum products), acids (steel pickling liquor where impurities permit reuse), and salts (for example, ammonium sulphate form coal coking used as fertiliser).

5. Glass, which makes up about 10% of municipal refuse.

6. Paper, commonly recycled from municipal refuse.

7. Plastic, consisting of a variety of mouldable polymeric materials and composing a major constituent of municipal wastes.

8. Rubber.

9. Organic substances, especially solvents and oils, such as hydraulic and lubricating oils.

10. Catalysts from chemical synthesis or petroleum processing.

11. Materials with agricultural uses, such as waste lime or phosphate containing sludges used to treat and fertilise acidic soils.

Waste oil Utilisation and Recovery

Waste oil generated from lubricants and hydraulic fluids is one of the more commonly recycled materials. The collection, recycling, treatment, and disposal of waste oil are all complicated by the fact that it comes from diverse, widely dispersed sources and contains several classes of potentially hazardous contaminants. These are divided between organic constituents (polycyclic aromatic hydrocarbons, chlorinated hydrocarbons) and inorganic constituents (aluminium, chromium, and iron from wear of metal parts; barium and zinc from oil additives; lead from leaded gasoline).

Recycling waste oil

The processes used to convert waste oil to a feedstock hydrocarbon liquid for lubricant formulation are illustrated in Fig. 21.2. The first of these uses distillation to remove water and light ends that have come from condensation and contaminant fuel. The second, or processing, step may be a vacuum distillation in which the three products are oil for further processing, a fuel oil cut, and a heavy residue. The processing step may also employ treatment with a mixture of solvents including isopropyl and butyl alcohols and methylethyl ketone to dissolve the oil and leave contaminants as a sludge; or contact with sulphuric acid to remove inorganic contaminants followed by treatment with clay to take out acid and contaminants that cause odour and colour. The third step shown in Fig. 21.2 employs vacuum distillation to separate lubricating oil stocks from a fuel fraction and heavy residue. This phase of treatment may also involve hydrofinishing, treatment with clay, and filtration.

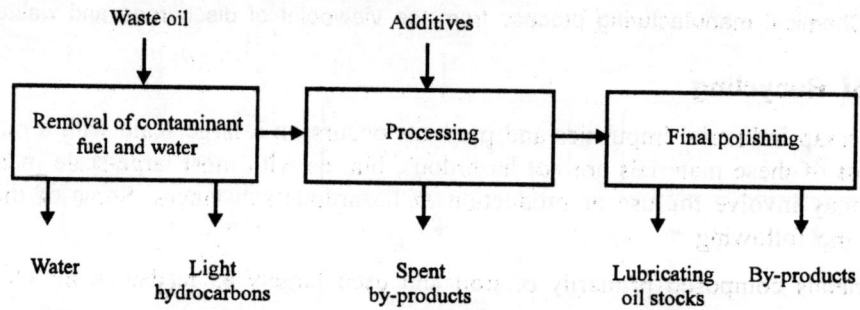

Fig. 21.2. Major steps in reprocessing waste oil

Waste oil fuel

For economic reasons, waste oil that is to be used for fuel is given minimal treatment of a physical nature, including settling, removal of water, and filtration. Metals in waste fuel oil become highly concentrated in its fly ash, which may be hazardous.

Waste Solvent Recovery and Recycle

The recovery and recycling of waste solvents has some similarities to the recycling of waste oil and is also an important enterprise. Among the many solvents listed as hazardous wastes and recoverable from wastes are dichloromethane, tetrachloroethylene, trichloroethylene, 1,1,1-trichloroethane, benzene, liquid alkanes, 2-nitropropane, methylisobutyl ketone, and cyclohexanone. For reasons of both economics and pollution control, many industrial processes that use solvents are equipped for solvent recycle. The basic scheme for solvent reclamation and reuse is shown in Fig. 21.3. A number of operations are used in solvent purification. Entrained solids are removed by settling, filtration, or centrifugation. Drying agents may be used to remove water from solvents and various adsorption techniques and chemical treatment may be required to free the solvent from specific impurities. Fractional distillation, often requiring several distillation steps, is the most important operation in solvent purification and recycle. It is used to separate solvents from impurities, water, and other solvents.

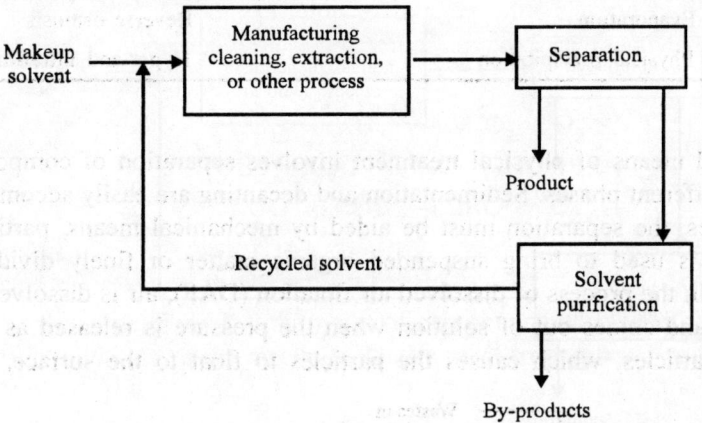

Fig. 21.3. Overall process for recycling solvents

PHYSICAL METHODS OF WASTE TREATMENT

This section addresses predominantly physical methods for waste treatment and the following section addresses methods that utilise chemical processes. It should be kept in mind that most waste treatment measures have both physical and chemical aspects. The appropriate treatment technology for hazardous wastes obviously depends upon the nature of the wastes. These may consist of volatile wastes (gases, volatile solutes in water, gases or volatile liquids held by solids, such as catalysts), liquid wastes (waste-water, organic solvents), dissolved or soluble wastes (water-soluble inorganic species, water-soluble organic species, compounds soluble in organic solvents), semi-solids (sludges, greases), and solids (dry solids, including granular solids with a significant water content, such as dewatered sludges, as well as solids suspended in liquids). The type of physical treatment to be applied to wastes depends strongly upon the physical properties of the material treated, including state of matter, solubility in water and organic solvents, density, volatility, boiling point, and melting point.

As shown in Fig. 21.4, waste treatment may occur at three major levels—primary, secondary, and polishing– somewhat analogous to the treatment of waste-water. Primary treatment is generally regarded as preparation for further treatment, although it can result in the removal of by–products and reduction of the quantity and hazard of the waste. Secondary treatment detoxifies, destroys, and removes hazardous constituents. Polishing usually refers to treatment of water that is removed from wastes so that it may be safely discharged. However, the term can be broadened to apply to the treatment of other products as well so that they may be safely discharged or recycled.

Methods of Physical Treatment

Knowledge of the physical behaviour of wastes has been used to develop various unit operations for waste treatment that are based upon physical properties. These operations include the following :

1.	Phase separation	2.	Phase transfer
	Filtration		Extraction
3.	Phase transition		Sorption
	Distillation	4.	Membrane separations
	Evaporation		Reverse osmosis
	Physical precipitation		Hyper-and ultrafiltration

Phase separations

The most straightforward means of physical treatment involves separation of components of a mixture that are already in two different phases. Sedimentation and decanting are easily accomplished with simple equipment. In many cases, the separation must be aided by mechanical means, particularly filtration or centrifugation. Flotation is used to bring suspended organic matter or finely divided particles to the surface of a suspension. In the process of dissolved air flotation (DAF), air is dissolved in the suspending medium under pressure and comes out of solution when the pressure is released as minute air bubbles attached to suspended particles, which causes the particles to float to the surface.

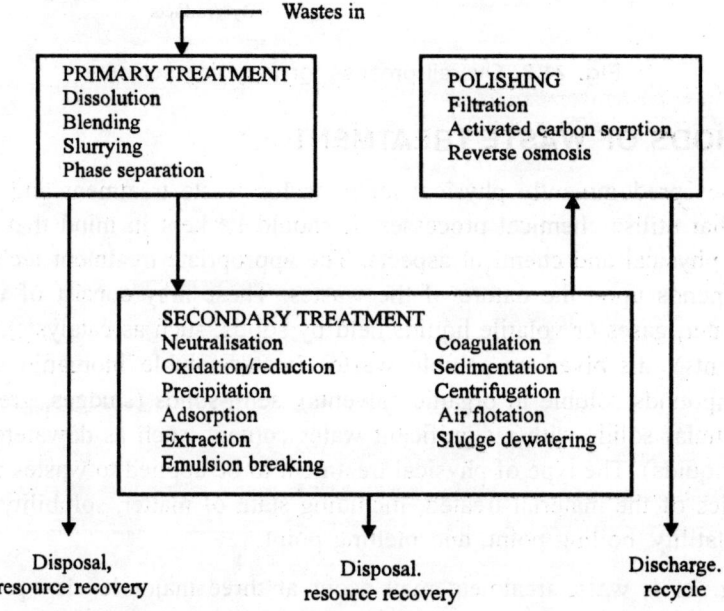

Fig. 21.4. Major phases of waste treatment.

An important and often difficult waste treatment step is emulsion breaking in which colloidal-sized emulsions are caused to aggregate and settle from suspension. Agitation, heat, acid, and the addition of coagulants consisting of organic polyelectrolytes or inorganic substances, such as an aluminium salt, may be used for this purpose. The chemical additive acts as a flocculating agent to cause the particles to stick together and settle out.

Phase transition

A second major class of physical separation is that of phase transition in which a material changes from one physical phase to another. It is best exemplified by distillation, which is used in treating and recycling solvents, waste oil, aqueous phenolic wastes, xylene contaminated with paraffin from histological laboratories, and mixtures of ethylbenzene and styrene. Distillation produces distillation bottoms (still bottoms), which are often hazardous and polluting. These consist of unevaporated solids, semisolid tars, and sludges from distillation. Specific examples are distillation bottoms from the production of acetaldehyde from ethylene.

Evaporation is usually employed to remove water from an aqueous waste to concentrate it. A special case of this technique is thin-film evaporation in which volatile constituents are removed by heating a thin layer of liquid or sludge waste spread on a heated surface.

Drying removal of solvent or water from a solid or semisolid (sludge) or the removal of solvent from a liquid or suspension is a very important operation because water is often the major constituent of waste products, such as sludges obtained from emulsion breaking. In freeze drying, the solvent, usually water, is sublimed from a frozen material. Hazardous waste solids and sludges are dried to reduce the quantity of waste, to remove solvent or water that might interfere with subsequent treatment processes, and to remove hazardous volatile constituents. Dewatering can often be improved with addition of a filter aid, such as diatomaceous earth, during the filtration step.

Stripping is a means of separating volatile components from less volatile ones in a liquid mixture by the partitioning of the more volatile materials to a gas phase of air or steam (steam stripping). The gas phase is introduced into the aqueous solution or suspension containing the waste in a stripping tower that is equipped with trays or packed to provide maximum turbulence and contact between the liquid and gas phases. The two major products are condensed vapour and a stripped bottoms residue. Examples of two volatile components that can be removed from water by air stripping are benzene and dichloromethane. Air stripping can also be used to remove ammonia from water that has been treated with a base to convert ammonium ion to volatile ammonia.

Physical precipitation is used here as a term to describe processes in which a solid forms from a solute in solution as a result of a physical change in the solution, as compared to chemical precipitation in which a chemical reaction in solution produces an insoluble material. The major changes that can cause physical precipitation are cooling the solution, evaporation of solvent, or alteration of solvent composition. The most common type of physical precipitation by alteration of solvent composition occurs when a water-miscible organic solvent is added to an aqueous solution, so that the solubility of a salt is lowered below its concentration in the solution.

Phase transfer

Phase transfer consists of the transfer of a solute in a mixture from one phase to another. An important type of phase transfer process is solvent extraction, a process in which a substance is transferred from solution in one solvent (usually water) to another (usually an organic solvent) without any chemical change taking place. When solvents are used to leach substances from solids or sludges, the process is called leaching. Solvent extraction and the major terms applicable to it are summarised in Fig. 21.5. The same terms and general principles apply to leaching. The major application of solvent extraction to waste treatment has been in the removal of phenol from by-product water produced in coal coking, petroleum refining, and chemical syntheses that involve phenol.

One of the more promising approaches to solvent extraction and leaching of hazardous wastes is the use of supercritical fluids, most commonly CO_2, as extraction solvents. A supercritical fluid is one that has characteristics of both liquid and gas and consists of a substance above its critical temperature and pressure (31.1 °C and 73.8 atm, respectively for CO_2). After a substance has been extracted from a waste in to a supercritical fluid at high pressure, the pressure can be released, resulting in separation of the substance extracted. The fluid can then be compressed again and recirculated through the extraction system. Some possibilities for treatment of hazardous wastes by extraction with supercritical CO_2 include removal of organic contaminants from waste-water, extraction of organohalide pesticides from soil, extraction of oil from emulsions used in aluminium and steel processing, and regeneration of spent activated carbon. Waste oils contaminated with PCBs, metals and water can be purified using supercritical ethane.

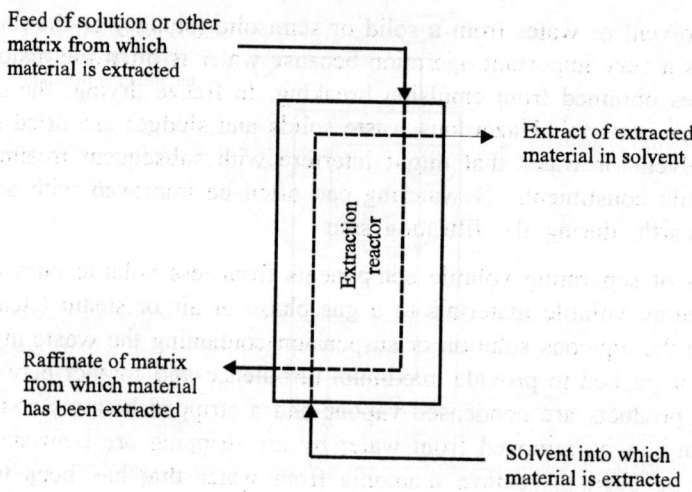

Fig. 21.5. Outline of solvent extraction/leaching process with important terms underlined.

Transfer of a substance from a solution to a solid phase is called sorption. The most important sorbent is activated carbon used for several purposes in waste treatment; in some cases it is adequate for complete treatment. It can also be applied to pretreatment of waste streams going into processes such as reverse osmosis to improve treatment efficiency and reduce fouling. Effluents from other treatment processes, such as biological treatment of degradable organic solutes in water can be polished with activated carbon. Activated carbon sorption is most effective for removing from water those hazardous waste materials that are poorly water-soluble and that have high molar masses, such as xylene, naphthalene, cyclohexane; chlorinated hydrocarbons, phenol, aniline, dyes, and surfactants. Activated carbon does not work well for organic compounds that are highly water-soluble or polar.

Solids other than activated carbon can be used for sorption of contaminants from liquid wastes. These include synthetic resins composed of organic polymers and mineral substance. Of the latter, clay is employed to remove impurities from waste lubricating oils in some oil recycling processes.

Molecular separation

A third major class of physical separation is molecular separation, often based upon membrane processes in which dissolved contaminants or solvent pass through a size-selective membrane under pressure. The products are a relatively pure solvent phase (usually water) and a concentrate enriched in the solute impurities. Hyper filtration allows passage of species with molecular masses of about 100 to 500, whereas ultrafiltration is used for the separation of organic solutes with molar masses of 500 to

1,000,000. With both of these techniques, water and lower molar mass solutes under pressure pass through the membrane as a stream of purified permeate, leaving behind a stream of concentrate containing impurities in solution or suspension. Ultrafiltration and hyperfiltration are especially useful for concentrating suspended oil, grease, and fine solids in water. They also serve to concentrate solutions of large organic molecules and heavy metal ion complexes.

Reverse osmosis is the most widely used of the membrane techniques. Although superficially similar to ultrafiltration and hyperfiltration, it operates on a different principle in that the membrane is selectively permeable to water and excludes ionic solutes. Reverse osmosis uses high pressures to force permeate through the membrane, producing a concentrate containing high levels of dissolved salts.

Electrodialysis, sometimes used to concentrate plating wastes, employs membranes alternately permeable to cations and to anions. The driving force for the separation is provided by electrolysis with a direct current between two electrodes. Alternate layers between the membranes contain concentrate (brine) and purified water.

CHEMICAL TREATMENT

The applicability of chemical treatment to wastes depends upon the chemical properties of the waste constituents, particularly acid-base, oxidation-reduction, precipitation, and complexation behaviour; reactivity; flammability/combustibility; corrosivity; and compatibility with other wastes. The chemical behaviour of wastes translates to various unit operations for waste treatment that are based upon chemical properties and reaction. These include the following :

1. Acid/base neutralisation.
2. Chemical extraction and leaching.
3. Ion exchange.
4. Chemical precipitation.
5. Oxidation.
6. Reduction.

Some of the more sophisticated means available for treatment of wastes have been developed for pesticide disposal.

Acid/Base Neutralisation

Waste acids and bases are treated by neutralisation

$$H^+ + OH^- \rightarrow H_2O \qquad\qquad ...(21.1)$$

Although simple in principle, neutralisation can present some problems in practice. These include evolution of volatile contaminants, mobilisation of soluble substances, excessive heat generated by the neutralisation reaction, and corrosion to apparatus. By adding too much or too little of the neutralising agent, it is possible to get a product that is too acidic or basic.

Lime, $Ca(OH)_2$, is widely used as a base for treating acidic wastes. Because of lime's limited solubility, solutions of excess lime do not reach extremely high pH values. Sulphuric acid, H_2SO_4, is a relatively inexpensive acid for treating alkaline waste. However, addition of too much sulphuric acid can produce highly acidic products; for some applications, acetic acid, CH_3COOH, is preferable. As noted above, acetic acid is a weak acid and an excess of it does little harm. It is also a natural product and biodegradable.

Neutralisation, or pH adjustment, is often required prior to the application of other waste treatment processes. Processes that may require neutralisation include oxidation/reduction, activated carbon sorption, wet air oxidation, stripping, and ion exchange. Micro-organisms usually require a pH in the range of 6–9, so neutralisation may be required prior to biochemical treatment.

Chemical Precipitation

Chemical precipitation is used in hazardous waste treatment primarily for the removal of heavy metal ions from water as shown below for the chemical precipitation of cadmium.

$$Cd^{2+}(aq) + HS^-(aq) \rightarrow CdS(s) + H^+ (aq) \qquad \qquad ...(21.2)$$

Precipitation of metals

The most widely used means of precipitating metal ions is by the formation of hydroxides such as chromium (III) hydroxide.

$$Cr^{3+} + 3OH^- \rightarrow Cr(OH)_3 \qquad \qquad ...(21.3)$$

The source of hydroxide ion is a base (alkali), such as lime $Ca(OH)_2$, sodium hydroxide (NaOH), or sodium carbonate (Na_2CO_3). Most metal ions tend to produce basic salt precipitates, such as basic copper (II) sulphate, $CuSO_4.3Cu(OH)_2$, formed as a solid when hydroxide is added to a solution containing Cu^{2+} and SO_4^{2-} ions. The solubilities of many heavy metal hydroxides reach a minimum value, often at a pH in the range of 9–11, then increase with increasing pH values due to the formation of soluble hydroxo complexes, as illustrated by the following reaction

$$Zn(OH)_2(s) + 2OH^-(aq) \rightarrow Zn(OH)_4^{2-}(aq) \qquad \qquad ...(21.4)$$

The chemical precipitation method that is used most is precipitation of metals as hydroxides and basic salts with lime. Sodium carbonate can be used to precipitate hydroxides ($Fe(OH)_3.xH_2O$), carbonates ($CdCO_3$), or basic carbonate salts ($2PbCO_3.Pb(OH)_2$). The carbonate anion produces hydroxide by virtue of its hydrolysis reaction with water

$$CO_3^{2-} + H_2O \rightarrow HCO_3^- + OH^- \qquad \qquad ...(21.5)$$

Carbonate, alone, does not give as high a pH as do alkali metal hydroxides, which may have to be used to precipitate metals that form hydroxides only at relatively high pH values. The solubilities of some heavy metal sulphides are extremely low, so precipitation by H_2S or other sulphides can be a very effective means of treatment. Hydrogen sulphide is a toxic gas that is itself considered to be a hazardous waste. Iron(II) sulphide (ferrous sulphide) can be used as a safe source of sulphide ion to produce sulphide precipitates with other metals that are less soluble than FeS. However, toxic H_2S can be produced when metal sulphide wastes contact acid :

$$MS + 2H^+ \rightarrow M^{2+} + H_2S \qquad \qquad ...(21.6)$$

Some metals can be precipitated from solution in the elemental metal form by the action of a reducing agent, such as sodium borohydride :

$$4Cu^{2+} + NaBH_4 + 2H_2O \rightarrow 4Cu + NaBO_2 + 8H^+ \qquad ...(21.7)$$

or with more active metals in a process called cementation :

$$Cd^{2+} + Zn \rightarrow Cd + Zn^{2+} \qquad ...(21.8)$$

Coprecipitation of metals

In some cases, advantage may be taken of the phenomenon of coprecipitation to remove metals from wastes. A good example of this application is the coprecipitation of lead from battery industry waste-water with iron hydroxide. Raising the pH of such a waste-water consisting of dilute sulphuric acid and contaminated with Pb^{2+} ion precipitates lead as several species, including $PbSO_4$, $Pb(OH)_2$, and $Pb(OH)_2.2PbCO_3$. In the presence of iron, gelatinous $Fe(OH)_3$ forms, which coprecipitates the lead, resulting in much lower values of lead concentration than would otherwise be possible. Effective removal of lead from battery industry waste-water to below 0.2 ppm has been achieved by first adding an optimum quantity of iron, adjustment of the pH to a range of 9 to 9.5, addition of a polyelectrolyte to aid coagulation, and filtration.

Oxidation/Reduction

Oxidation and reduction can be used for the treatment and removal of a variety of inorganic and organic wastes. Some waste oxidants can be used to treat oxidisable wastes in water and cyanides.

Ozone, O_3, is a strong oxidant that can be generated onsite by an electrical discharge through dry air or oxygen. Ozone employed as an oxidant gas at levels of 1–2 wt % in air and 2–5 wt % in oxygen has been used to treat a large variety of oxidisable contaminants, effluents, and wastes including waste-water and sludges containing oxidisable constituents.

Electrolysis

As shown in Fig. 21.6, electrolysis is a process in which one species in solution (usually a metal ion) is reduced by electrons at the cathode and another gives up electrons to the anode and is oxidised there. In hazardous waste applications, electrolysis is most widely used in the recovery of cadmium, copper, gold, lead, silver, and zinc. Metal recovery by electrolysis is made more difficult by the presence of cyanide ion, which stabilises metals in solution as the cyanide complexes, such as $Ni(CN)_4^{2-}$.

Electrolytic removal of contaminants from solution can be by direct electro-deposition, particularly of reduced metals, and as the result of secondary reactions of electrolytically generated precipitating agents. A specific example of both is the electrolytic removal of both cadmium and nickel from waste-waster contaminated by nickel/cadmium battery manufacture using fibrous carbon electrodes. At the cathode, cadmium is removed directly by reduction to the metal :

$$Cd^{2+} + 2e^- \rightarrow Cd \qquad ...(21.9)$$

At relatively high cathodic potentials, hydroxide is formed by the electrolytic reduction of water :

$$2H_2O + 2e^- \rightarrow 2OH^- + H_2 \qquad ...(21.10)$$

or by the reduction of molecular oxygen, if it is present :

$$2H_2O + O_2 + 4e^- \rightarrow 4OH^- \qquad ...(21.11)$$

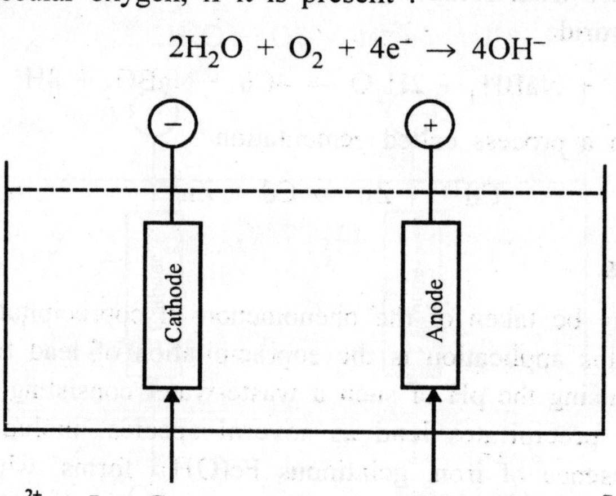

$$Cu^{2+} + 2e^- \rightarrow Cu \qquad\qquad H_2O \rightarrow 1/2\ O_2 + 2e^- + 2H^+$$

Net reaction: $Cu^{2+} + H_2O \rightarrow Cu + 1/2\ O_2 + 2H^+$

Fig. 21.10. Electrolysis of copper solution.

If the localised pH at the cathode surface becomes sufficiently high, cadmium can be precipitated and removed as colloidal $Cd(OH)_2$. The direct electro-deposition of nickel is too slow to be significant, but it is precipitated as solid $Ni(OH)_2$ at pH values above 7.5.

Cyanide, which is often present as an ingredient of electroplating baths with metals such as cadmium and nickel, can be removed by oxidation with electrolytically generated elemental chlorine at the anode. Chlorine is generated by the anodic oxidation of added chloride ion :

$$2Cl^- \rightarrow Cl_2 + 2e^- \qquad ...(21.12)$$

The electrolytically generated chlorine then breaks down cyanide by a series of reactions for which the overall reaction is the following :

$$2CN^- + 5Cl_2 + 8OH^- \rightarrow 10Cl^- + N_2 + 2CO_2 + 4H_2O \qquad ...(21.13)$$

Hydrolysis

One of the ways to dispose of chemicals that are reactive with water is to allow them to react with water under controlled conditions, a process called hydrolysis. Inorganic chemicals that can be treated by hydrolysis include metals that react with water, metals carbides, hydrides, amides, alkoxidesm and halids; and non-metal oxyhalides and sulphides. Examples of the treatment of these classes of inorganic species are given in Table 21.2. Organic chemicals may also be treated by hydrolysis. For example, toxic acetic anhydride is hydrolysed to relatively safe acetic acid:

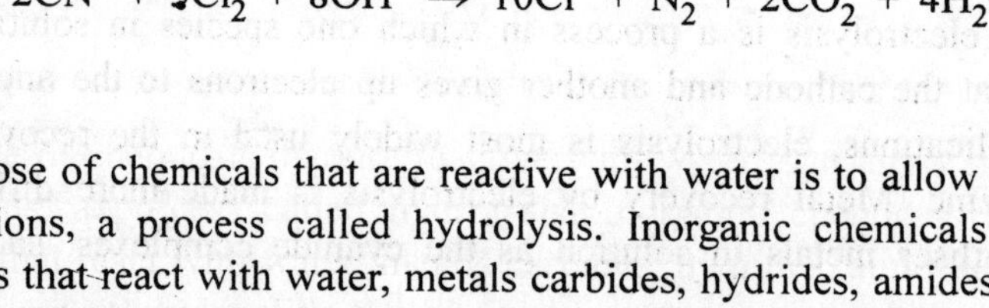

$$\underset{\substack{\text{Acetic anhydride}\\ \text{(an acid anhydride)}}}{H_3C-C(=O)-O-C(=O)-CH_3} + H_2O \rightarrow 2\,H_3C-C(=O)-OH \qquad ...(21.14)$$

Chemical Extraction and Leaching

Chemical extraction or leaching in hazardous waste treatment is the removal of a hazardous constituent by chemical reaction with an extractant in solution. Poorly soluble heavy metal salts can be extracted by reaction of the salt anions with H^+ as illustrated by the following :

$$PbCO_3 + H^+ \rightarrow Pb^{2+} + HCO_3^- \qquad ...(21.15)$$

Acids also dissolve basic organic compounds such as amines and aniline. Extraction with acids should be avoided if cyanides or sulphides are present to prevent formation of toxic hydrogen cyanide or hydrogen sulfide. Non-toxic weak acids are usually the safest to use. These include acetic acid, CH_3COOH, and the acid salt, NaH_2PO_4.

Table 21.2. Inorganic chemicals that may be treated by hydrolysis.

Class of chemical	Reaction with water
Active metals (calcium)	$Ca + 2H_2O \rightarrow H_2 + Ca(OH)_2$
Hydrides (sodium aluminium hydride)	$NaAlH_4 + 4H_2O \rightarrow 4H_2 + NaOH + Al(OH)_3$
Carbides (calcium carbide)	$CaC_2 + 2H_2O \rightarrow Ca(OH)_2 + C_2H_2$
Amides (sodium amide)	$NaNH_2 + H_2O \rightarrow NaOH + NH_3$
Halides (silicon tetrachloride)	$SiCl_4 + 2H_2O \rightarrow SiO_2 + 4HCl$
Alkoxides (sodium ethoxide)	$NaOC_5H_5 + H_2O \rightarrow NaOH + C_2H_5OH$

Chelating agents, such as dissolved ethylene diaminetetra acetate (EDTA, HY^{3-}), dissolve insoluble metal salts by forming soluble species with metal ions :

$$FeS + HY^{3-} \rightarrow FeY^{2-} + HS^- \qquad ...(21.16)$$

Heavy metal ions in soil contaminated by hazardous wastes may be present in a coprecipitated form with insoluble iron(III) and manganese(IV) oxides, Fe_2O_3 and MnO_2, respectively. These oxides can be dissolved from soil by reducing agents, such as solutions of sodium dithuonate/citrate or hydroxylamine. This results in the production of soluble Fe^{2+} and Mn^{2+} and the release of heavy metal ions, such as Cd^{2+} or Ni^{2+}, which are removed with the water.

Ion Exchange

Ion exchange is a means of removing cations or anions from solution onto a solid resin, which can be regenerated by treatment with acids, bases, or salts. The greatest use of ion exchange in hazardous waste treatment is for the removal of low levels of heavy metal ions from waste-water:

$$2H^+ \{CatExchr\} + Cd^{2+} \rightarrow Cd^{2+} \{CatExchr\}_2 + 2H^+ \qquad ...(21.17)$$

Ion exchange is employed in the metal plating industry to purify rinse-water and spent plating bath solutions. Cation exchangers are used to remove cationic metal species, such as Cu^{2+}, from such solutions. Anion exchangers remove anionic cyanide metal complexed [for example $Ni(CN)_4^{2-}$] and chromium (VI) species, such as CrO_4^{2-}. Radionuclides may be removed from radioactive wastes and mixed waste by ion exchange resins.

A special ion exchange resin configuration has been described for the removal of chelatable heavy metals, such as copper, lead, nickel, and zinc, from otherwise innocuous sludge constituents.

PHOTOLYTIC REACTIONS

Photolysis can be used to destroy a number of kinds of hazardous wastes. In such applications, it is most useful in breaking chemical bonds in refractory organic compounds. TCDD, one of the most troublesome and refractory of wastes, can be treated by ultraviolet light in the presence of hydrogen atom donors [H] resulting in reactions such as the following :

$$+ \ h\nu + 2 \ \{H\} \ \rightarrow \qquad\qquad\qquad\qquad \dots(21.18)$$

As photolysis proceeds, the H-C bonds are broken, the C-O bonds are broken and the final product is a harmless organic polymer. An initial photolysis reaction can result in the generation of reactive intermediates that participate in chain reactions that lead to the destruction of a compound. One of the most important reactive intermediates is free radical HO·. In some cases, sensitisers are added to the reaction mixture to absorb radiation and generate reactive species that destroy wastes.

Hazardous waste substances other than TCDD that have been destroyed by photolysis are herbicides (atrazine), 2,4,6-trinitrotoluene (TNT), and polychlorinated biphenyls (PCBs). The addition of a chemical oxidant, such as potassium peroxydisulphate, $K_2S_2O_8$, enhances destruction by oxidising active photolytic products.

THERMAL TREATMENT METHODS

Thermal treatment of hazardous wastes can be used to accomplish most of the common objectives of waste treatment—volume reduction; removal of volatile, combustible, mobile organic matter; and destruction of toxic and pathogenic materials. The most widely applied means of thermal treatment of hazardous wastes is incineration. Incineration utilises high temperatures, an oxidising atmosphere, and often turbulent combustion conditions to destroy wastes. Methods other than incineration that make use of high temperatures to destroy or neutralise hazardous wastes are discussed briefly at the end of this section.

Incineration

Hazardous waste incineration will be defined here as a process that involves exposure of the waste materials to oxidising conditions at a high temperature, usually in excess of 900°C. Normally the heat required for incineration comes from the oxidation of organically bound carbon and hydrogen contained in the waste material or in supplemental fuel

$$C \ (\text{organic}) + O_2 \rightarrow CO_2 + \text{heat}$$

$$4H \ (\text{organic}) + O_2 \rightarrow 2H_2O + \text{heat} \qquad\qquad \dots(21.19)$$

These reactions destroy organic matter and generate heat required for endothermic reactions, such as the breaking of C-Cl bonds in organochlorine compounds.

Incinerable wastes

Ideally, incinerable wastes are predominantly organic materials that will burn with a heating value of at least 5,000 Btu/lb and preferably over 8,000 Btu/lb. Such heating values are readily attained with wastes having high contents of the most commonly incinerated waste organic substances, including methanol, acetonitrile, toluene, ethanol, amyl acetate, acetone, xylene, methyl ethyl ketone, adipic acid, and ethyl acetate. In some cases, however, it is desirable to incinerate wastes that will not burn alone and which require supplemental fuel, such as methane and petroleum liquids. Examples of such wastes are non-flammable organochlorine wastes, some aqueous wastes, or soil in which the elimination of a particularly troublesome contaminant is worth the expense and trouble of incinerating it. Inorganic matter, water, and organic hetero element contents of liquid wastes are important in determining their incinerability.

Hazardous Waste Fuel

Many industrial wastes, including hazardous wastes, are burned as hazardous waste fuel for energy recovery in industrial furnaces and boilers and in incinerators of non-hazardous wastes, such as sewage sludge incinerators. This process is called coincineration, and more combustible wastes are utilised by it than are burned solely for the purpose of waste destruction. In addition to heat recovery from combustible wastes, it is a major advantage to use an existing onsite facility for waste disposal rather than a separate hazardous waste incinerator.

Incineration Systems

The four major components of hazardous waste incineration systems are shown in Fig. 21.7. Waste preparation for liquid wastes may require filtration, settling to remove solid material and water, blending to obtain the optimum incinerable mixture, or heating to decrease viscosity. Solids may require shredding and screening. Atomisation is commonly used to feed liquid wastes. Several mechanical devices, such as rams and augers, are used to introduce solids into the incinerator. The most common kinds of combustion chambers are liquid injection, fixed hearth, rotary kiln, and fluidised bed.

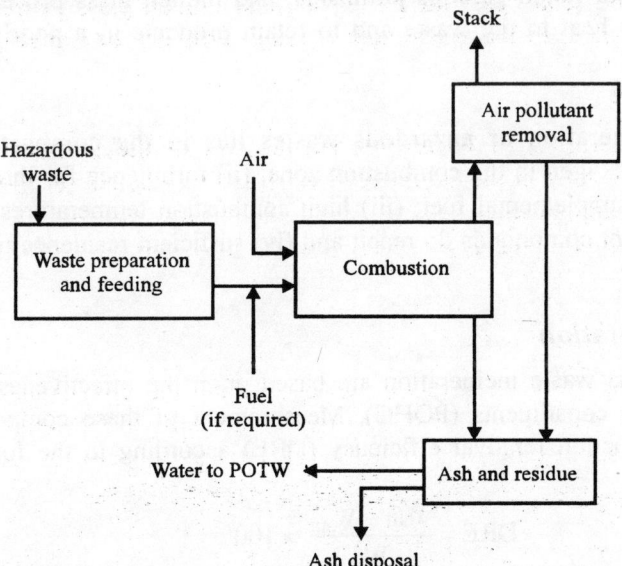

Fig. 21.7. Major components of a hazardous waste incinerator system.

Often the most complex part of a hazardous waste incineration system is the air pollution control system, which involves several operations. The most common operations in air pollution control from hazardous waste incinerators are combustion gas cooling, heat recovery, quenching, particulate matter removal, acid gas removal, and treatment and handling of by-product solids, sludges, and liquids.

Hot ash is often quenched in water. Prior to disposal, it may require dewatering and chemical stabilisation. A major consideration with hazardous waste incinerators and the types of wastes that are incinerated, is the disposal problem posed by the ash, especially in respect to potential leaching of heavy metals.

Types of incinerators

Hazardous waste incinerators may be divided among the following, based upon type of combustion chamber :

1. Rotary kiln in which the primary combustion chamber is a rotating cylinder lined with refractory materials and an after burner downstream from the kiln to complete destruction of the wastes.

2. Liquid injection incinerators that burn pumpable liquid wastes dispersed as small droplets.

3. Fixed-hearth incinerators with single or multiple hearths upon which combustion of liquid or solid wastes occurs.

4. Fluidised-bed incinerators that have a bed of granular solid (such as limestone) maintained in a suspended state by injection of air to remove pollutant acid gas and ash products.

5. Advanced design incinerators including plasma incinerators that make use of an extremely hot plasma of ionised air injected through an electrical arc; electric reactors that use resistance heated incincrator walls at around 2,200°C to heat and pyrolyse wastes by radiative heat transfer; infrared systems, which generate intense infrared radiation by passing electricity through silicon carbide resistance heating elements; molten salt combustion that uses a bed of molten sodium carbonate at about 900°C to destroy the wastes and retain gaseous pollutants; and molten glass processes that use a pool of molten glass to transfer heat to the waste and to retain products in a poorly leachable glass form.

Combustion Conditions

The key to effective incineration of hazardous wastes lies in the combustion conditions. These require : (i) sufficient free oxygen in the combustion zone; (ii) turbulence for thorough mixing of waste, oxidant, and (where used) supplemental fuel; (iii) high combustion temperatures above about 900°C to ensure that thermally resistant compounds do react; and (iv) sufficient residence time (at least 2 seconds) to allow reactions to occur.

Effectiveness of incineration

EPA standards for hazardous waste incineration are based upon the effectiveness of destruction of the principal organic hazardous constituents (POHC). Measurement of these compounds before and after incineration gives the destruction removal efficiency (DRE) according to the formula.

$$DRE = \frac{W_{in} - W_{out}}{W_{in}} \times 100$$

...(21.20)

where W_{in} and W_{out} are the mass flow rates of the POHC input and output (at the stack downstream from emission controls), respectively.

Wet Air Oxidation

Organic compounds and oxidisable inorganic species can be oxidised by oxygen in aqueous solution. The source of oxygen usually is air. Rather extreme conditions of temperature and pressure are required, with a temperature range of 175–327°C and a pressure range of 300–3,000 psig (2070–20,700 kPa). The high pressures allow a high concentration of oxygen to be dissolved in the water and the high temperatures enable the reaction to occur.

Wet air oxidation has been applied to the destruction of cyanides in electroplating waste-waters. The oxidation reaction for sodium cyanide is the following :

$$2Na^+ + 2CN^- + O_2 + 4H_2O \rightarrow 2Na^+ + 2HCO_3^- + 2NH_3 \qquad ...(21.21)$$

A method has been described in which cyanide is oxidised on an aerated carbon bed. The method employs added copper (II) ion and sulphite as catalysts. In addition to destroying CN^-, it also oxidises highly stable complexed cyanide in species such as $Fe(CN)_6^{4-}$.

Organic wastes can be oxidised in supercritical water, taking advantage of the ability of supercritical fluids to dissolve organic compounds. Wastes are contacted with water and the mixture raised to a temperature and pressure required for supercritical conditions for water. Oxygen is then pumped in, sufficient to oxidise the wastes. The process produces only small quantities of CO, and no SO_2 or NO_x. It has reportedly been used to degrade PCBs, dioxins, organochlorine insecticides, benzene, urea, and numerous other materials.

Uv-enhanced Wet Oxidation

Hydrogen peroxide (H_2O_2) can be used as an oxidant in solution assisted by ultraviolet radiation ($h\nu$). For the oxidation of organic species represented in general as $\{CH_2O\}$, the overall reaction is :

$$2H_2O_2 + \{CH_2O\} + h\nu \rightarrow CO_2 + 3H_2O \qquad ...(21.22)$$

The ultraviolet radiation breaks chemical bonds and serves to form reactive oxidant species, such as HO·.

BIODEGRADATION OF WASTES

Biodegradation of wastes is their conversion by biological processes to simple inorganic molecules (mineralisation) and to a certain extent, to biological materials. Usually the products of biodegradation are molecular forms that tend to occur in nature and that are in greater thermodynamic equilibrium with their surroundings than are the starting materials. Detoxification refers to the biological conversion of a toxic substance to a less toxic species. Microbial bacteria and fungi possessing enzyme systems required for biodegradation of wastes are usually best obtained from populations of indigenous micro-organisms at a hazardous waste site where they have developed the ability to degrade particular kinds of molecules. Biological treatment offers a number of significant advantages and has considerable potential for the degradation of hazardous wastes, even *in situ*.

It must be kept in mind, however, that there are many factors that can cause biodegradation to fail as a treatment process. Often physical conditions are such that mixing of wastes, nutrients, and electron acceptor species (such as oxygen) is too slow to permit biodegradation to occur at a useful rate. Low temperatures may make reactions too slow to be useful. Toxicants, such as heavy metals, may inhibit biological activity, and some metabolites produced by the micro-organisms may be toxic to them.

Under the label of bioremediation, the use of microbial processes to destroy hazardous wastes is experiencing a period of very rapid growth. Doubts still exist about claims for its effectiveness in a number of applications. It has been suggested that performance standards are badly needed for judging the effectiveness of bioremediation.

Biodegradability

The biodegradability of a compound is influenced by its physical characteristics, such as solubility in water and vapour pressure, and by its chemical properties, including molar mass, molecular structure, and presence of various kinds of functional groups, some of which provide a "biochemical handle" for the initiation of biodegradation. With the appropriate organisms and under the right conditions, even substances such as phenol that are considered to be biocidal to most micro-organisms can undergo biodegradation.

Recalcitrant or biorefractory substances are those that resist biodegradation and tend to persist and accumulate in the environment. Such materials are not necessarily toxic to oraganisms, but simply resist their metabolic attack. However, even some compounds regarded as biorefractory may be degraded by micro-organisms adapted to their biodegradation; for example, DDT is degraded by properly acclimated *Pseudomonas*. Chemical pretreatment, especially by partial oxidation, can make some kinds of recalcitrant wastes much more biodegradable.

Properties of hazardous wastes and their media can be changed to increase biodegradability. This can be accomplished by adjustment of conditions to optimum temperature, pH (usually in the range of 6–9), stirring, oxygen level, and material load. Biodegradation can be aided by removal of toxic organic and inorganic substances, such as heavy metal ions.

Aerobic Treatment

Aerobic waste treatment processes utilise aerobic bacteria and fungi that require molecular oxygen, O_2. These processes are often favoured by micro-organisms, in part because of the high energy yield obtained when molecular oxygen reacts with organic matter. Aerobic waste treatment is well adapted to the use of an activated sludge process. It can be applied to hazardous wastes such as chemical process wastes and landfill leachates. Some systems used powdered activated carbon as an additive to absorb organic wastes that are not biodegraded by microorganisms in the system.

Contaminated soils can be mixed with water and treated in a bioreactor to eliminate biodegradable contaminants in the soil. It is possible in principle to treat contaminated soils biologically in place by pumping oxygenated, nutrient-enriched water through the soil in a recirculating system.

Anaerobic Treatment

Anaerobic waste treatment in which microorganisms degrade wastes in the absence of oxygen can be practised on a variety of organic hazardous wastes. Compared to the aerated activated sludge process, anaerobic digestion requires less energy; yields less sludge by-product; generates hydrogen sulphide (H_2S), which precipitates toxic heavy metal ions; and produces methane gas, CH_4, which can be used as an energy source.

The overall process for anaerobic digestion is a fermentation process in which organic matter is both oxidised and reduced. The simplified reaction for the anaerobic fermentation of a hypothetical organic substance, "$\{CH_2O\}$", is the following :

$$2\{CH_2O\} \rightarrow CO_2 + CH_4 \qquad \qquad ...(21.23)$$

In practice, the microbial processes involved are quite complex. Most of the wastes for which anaerobic digestion is suitable consist of oxygenated compounds, such as acetaldehyde or methylethyl ketone.

Reductive Dehalogenation

Reductive dehalogenation is a mechanism by which halogen atoms are removed from organohalide compounds by anaerobic bacteria. It is an important means of detoxifying alkyl halides (particularly solvents), aryl halides and organochlorine pesticides, all of which are important hazardous waste compounds. It is the only means by which some of the more highly halogenated waste compounds are biodegraded; such compounds include tetrachloroethene, hexachlorobenzene, pentachlorophenol, and the more highly chlorinated PCB congeners.

The two general processes by which reductive dehalogenation occurs are hydrogenolysis, as shown by the example in equation

$$...(21.24)$$

and vicinal reduction:

$$...(21.25)$$

Vicinal reduction removes two adjacent halogen atoms, and works only on alkyl halides, not aryl halides. Both processes produce innocuous inorganic halide (Cl^-).

LAND TREATMENT AND COMPOSTING

Land Treatment

Soil may be viewed as a natural filter for wastes. Soil has physical, chemical, and biological characteristics that can enable waste detoxification, biodegradation, chemical decomposition, and physical and chemical fixation. Therefore, and treatment of wastes may be accomplished by mixing the wastes with soil under appropriate conditions.

Soil is a natural medium for a number of living organisms that may have an effect upon biodegradation of hazardous wastes. Of these, the most important are bacteria, including those from the general *Agrobacterium, Arhrobacteri, Bacillus, Flavobacterium* and *Pseudomonas. Actinomycetes* and fungi are important organisms in decay of vegetable matter and may be involved in biodegradation of wastes.

Microorganisms useful for land treatment are usually present in sufficient numbers to provide the inoculum required for their growth. The growth of these indigenous micro-organisms may be stimulated by adding nutrients and an electron acceptor to act as an oxidant (for aerobic degradation), accompanied by mixing. The most commonly added nutrients are nitrogen and phosphorus. Oxygen can be added by pumping air underground or by treatment with hydrogen peroxide, H_2O_2. In some cases, such as for treatment of hydrocarbons on or near the soil surface, simple tillage provides both oxygen and the mixing required for optimum microbial growth.

Wastes that are amenable to land treatment are biodegradable organic substances. However, in soil contaminated with hazardous wastes, bacterial cultures may develop that are effective in degrading normally recalcitrant compounds through acclimation over a long period of time. Land treatment is mostly used for petroleum refining wastes and is applicable to the treatment of fuels and wastes from leaking underground storage tanks. It can also be applied to biodegradable organic chemical wastes, including some organohalide compounds. Land treatment is not suitable for the treatment of wastes containing acids, bases, toxic inorganic compounds, salts, heavy metals, and organic compounds that are excessively (soluble, volatile, or flammable.

Composting

Composting of hazardous wastes is the biodegradation of solid or solidified materials in a medium other than soil. Bulking material, such as plant residue, paper, municipal refuse, or sawdust may be added to retain water and enable air to penetrate to the waste material. Successful composing of hazardous waste depends upon a number of factors, including those discussed above under Land Treatment. The first of these is the selection of the appropriate micro-organism or inoculum. Once a successful composting operation is underway, a good inoculum is maintained by recirculating spent compost to each new batch. Other parameters that must be controlled include oxygen supply, moisture content (which should be maintained at a minimum of about 40%), pH (usually around neutral), and temperature. The composting process generates heat, so if the mass of the compost pile is sufficiently high, it can be self-heating under most conditions. Some wastes are deficient in nutrients, such as nitrogen, which must be supplied from commercial sources or from other wastes.

PREPARATION OF WASTES FOR DISPOSAL

Immobilisation, stabilisation, fixation, and solidification are terms that describe techniques whereby hazardous wastes are placed in a form suitable for long-term disposal. These aspects of hazardous waste management are addressed below.

Immobilisation

Immobilisation includes physical and chemical processes that reduce surface areas of wastes to minimise leaching. It isolates the wastes from their environment, especially groundwater, so that they have the least possible tendency to migrate. This is accomplished by physically isolating the waste, reducing its solubility, and decreasing its surface area. Immobilisation usually improves the handling and physical characteristics of wastes.

Stabilisation

Stabilisation means the conversion of a waste from its original form to a physically and chemically more stable material. Stabilisation may include chemical reactions that generate products that are less volatile, soluble, and reactive. Solidification, which is discussed below, is one of the most common means of stabilisation. Stabilisation is required for land disposal of wastes. Fixation is a process that binds a hazardous waste in a less mobile and less toxic form; it means much the same thing as stabilisation.

Solidification

Solidification may involve chemical reaction of the waste with the solidification agent, mechanical isolation in a protective binding matrix, or a combination of chemical and physical processes. It can be accomplished by evaporation of water from aqueous wastes or sludges, sorption onto solid material, reaction with cement, reaction with silicates, encapsulation, or imbedding in polymers or thermoplastic materials.

In many solidification processes, such as reaction with portland cement, water is an important ingredient of the hydrated solid matrix. Therefore, the solid should not be heated excessively or exposed to extremely dry conditions, which could result in diminished structural integrity from loss of water. In some cases, however, heating a solidified waste is an essential part of the overal solidification procedure. For example, an iron hydroxide matrix can be converted to highly insoluble, refractory iron oxide by heating. Organic constituents of solidified wastes may be converted to inert carbon by heating. Heating is an integral part of the process of vitrification.

Sorption to a solid matrix material

Hazardous waste liquids, emulsions, sludges, and free liquids in contact with sludges may be solidified and stabilised by fixing onto solid sorbents, including activated carbon (for organics), fly ash, kiln dust, clays, vermiculite, and various proprietary materials. Sorption may be done to convert liquids and semi-solids to dry solids, improve waste handling, and reduce solubility of waste constituents. Sorption can also be used to improve waste compatibility with substances such as portland cement used for solidification and setting. Specific sorbents may also be used to stabilise pH and pE (a measure of the tendency of a medium to be oxidising or reducing.

The action of sorbents can include simple mechanical retention of wastes, physical sorption, and chemical reactions. It is important to match the sorbent to the waste. A substance with a strong affinity for water should be employed for wastes containing excess waster, and one with a strong affinity for organic materials should be used for wastes with excess organic solvents.

Thermoplastics and organic polymers

Thermoplastics are solids or semi-solids that become liquefied at elevated temperatures. Hazardous waste materials may be mixed with hot thermoplastic liquids and solidified in the cooled thermoplastic matrix, which is rigid but deformable. The thermoplastic material mostly used for this purpose is asphalt bitumen. Other thermoplastics, such as paraffin and polyethylene, have also been used to immobilise hazardous wastes.

Among the wastes that can be immobilised with thermoplastics are those containing heavy metals, such as electroplating wastes. Organic thermoplastics repel water and reduce the tendency toward leaching in contact with groundwater. Compared to cement, thermoplastics add relatively less material to the waste.

A technique similar to that described above uses organic polymers produced in contact with solid wastes to imbed the wastes in a polymer matrix. Three kinds of polymers that have been used for this purpose include polybutadiene, urea-formaldehyde, and vinyl ester-styrene polymers. This procedure is more complicated than is the use of thermoplastics but, in favourable cases, yields a product in which the waste is held more strongly.

Vitrification

Vitrification or glassification consists of imbedding wastes in a glass material. In this application, glass may be regarded as a high-melting-temperature inorganic thermoplastic. Molten glass can be used, or glass can be synthesised in contact with the waste by mixing and heating with glass constituents—silicon dioxide (SiO_2), sodium carbonate (Na_2CO_3), and calcium oxide (CaO). Other constituents may include boron oxide, B_2O_3, which yields a borosilicate glass that is especially resistant to changes in temperature and chemical attack. In some cases, glass is used in conjunction with thermal waste destruction processes, serving to immobilise hazardous waste ash constituents. Some wastes are detrimental to the quality of the glass. Aluminium oxide, for example, may prevent glass from fusing.

Vitrification is relatively complicated and expensive, the latter because of the energy consumed in fusing glass. Despite these disadvantages, it is the best immobilisation technique for some special wastes and has been promoted for solidification of radionuclear wastes because glass is chemically inert and resistant to leaching. However, high levels of radioactivity can cause deterioration of glass and lower its resistance to leaching.

Solidification with cement

Portland cement is widely used for solidification of hazardous wastes. In this application, portland cement provides a solid matrix for isolation of the wastes, chemically binds water from sludge wastes, and may react chemically with wastes (for example, the calcium and base in portland cement react chemically with inorganic arsenic sulphide wastes to reduce their solubilities). However, most wastes are held physically in the rigid portland cement matrix and are subject to leaching.

As a solidification matrix, portland cement is most applicable to inorganic sludges containing heavy metal ions that form insoluble hydroxides and carbonates in the basic carbonate medium provided by the cement. The success of solidification with portland cement strongly depends upon whether or not the waste adversely affects the strength and stability of the concrete product. A number of substances—organic matter such as petroleum or coal; some silts and clays; sodium salts of arsenate, borate, phosphate, iodate, and sulphides; and salts of copper, lead, magnesium, tin, and zinc— are incompatible with portland cement because they interfere with its set and cure and cause deterioration of the cement matrix with time. However, a reasonably good disposal form can be obtained by absorbing organic wastes with a solid material, which in turn is set in portland cement. This approach has been used with hydrocarbon wastes sorbed by an activated coal char matrix.

Solidification with silicate materials

Water-insoluble silicates, (pozzolanic substances) containing oxyanionic silicon such as SiO_3^{2-} are used for waste solidification. These substances include fly ash, flue dust, clay, calcium silicates, and ground-up slag from blast furnaces. Soluble silicates, such as sodium silicate, may also be used. Silicate solidification usually requires a setting agent, which may be portland cement, gypsum (hydrated $CaSO_4$), lime, or compounds of aluminium, magnesium, or iron. The product may vary from a granular material to a concrete-like solid. In some cases, the product is improved by additives, such as emulsifiers, surfactants, activators, calcium chlorides, clays, carbon, zeolites, and various proprietary materials.

Success has been reported for the solidification of both inorganic wastes and organic wastes (including oily sludges) with silicates. The advantages and disadvantages of silicate solidification are similar to those of portland cement discussed above. One consideration that is especially applicable to fly ash is the presence in some silicate materials of leachable hazardous substances, which may include arsenic and selenium.

Encapsulation

As the name implies, encapsulation is used to coat wastes with an impervious material so that they do not contact their surroundings. For example, a water-soluble waste salt encapsulated in asphalt would not dissolve, as long as the asphalt layer remains intact. A common means of encapsulation uses heated, molten thermoplastics, asphalt, and waxes that solidify when cooled. A more sophisticated approach to encapsulation is to form polymeric resins from monomeric substances in the presence of the waste.

Chemical Fixation

Chemical fixation is a process that binds a hazardous waste substance in a less mobile, less toxic form by a chemical reaction that alters the waste chemically. Physical and chemical fixation often occur together. Polymeric inorganic silicates containing some calcium and often some aluminium are the inorganic materials most widely used as a fixation matrix. Many kinds of heavy metals are chemically bound in such a matrix as well as being held physically by it. Similarly, some organic wastes are bound by reactions with matrix constituents. For example, humic acid wastes react with calcium ion in a solidification matrix to produce insoluble calcium humates.

ULTIMATE DISPOSAL OF WASTES

Regardless of the destruction, treatment, and immobilisation techniques used, there will always remain from hazardous wastes some material that has to be put somewhere. This section briefly addresses the ultimate disposal of ash, salts, liquids, solidified liquids, and other residues that must be placed where their potential to do harm is minimised.

Disposal Aboveground

In some important respects disposal aboveground, essentially in a pile designed to prevent erosion and water infiltration, is the best way to store solid wastes. Perhaps its most important advantage is that it avoids infiltration by groundwater that can result in leaching and groundwater contamination common to storage in pits and landfills. In a properly designed aboveground disposal facility any leachate that is produced drains quickly by gravity to the leachate collection system, where it can be detected and treated.

Above-ground disposal can be accomplished with a storage mound deposited on a layer of compacted clay covered with impermeable membrane liners laid somewhat above the original soil surface and shaped to allow leachate flow and collection. The slopes around the edges of the storage mound should be sufficiently great to allow good drainage of precipitation, but gentle enough to deter erosion.

Landfill

Landfill historically has been the most common way of disposing of solid hazardous wastes and some liquids, although it is being severely limited in many nations by new regulations and high land costs. Landfill involves disposal that is at least partially underground in excavated cells, quarries, or natural depressions. Usually fill is continued above ground to utilise space most efficiently and provide a grade for drainage of precipitation.

The greatest environmental concern with landfill of hazardous wastes is the generation of leachate from infiltrating surface water and groundwater with resultant contamination of groundwater supplies. Modern hazardous waste landfills provide elaborate systems to contain, collect, and control such leachate.

There are several components to a modern landfill. A landfill should be placed on a compacted low-permeability medium, preferably clay, which is covered by a flexible-membrane liner consisting of water-tight impermeable material. This liner is covered with granular material in which is installed a secondary drainage system. Next is another flexible-membrane liner above which is installed a primary drainage system for the removal of leachate. This drainage system is covered with a layer of granular filter medium, upon which the wastes are placed. In the landfill, wastes of different kinds are separated by berms, dam-like structures consisting of clay or soil covered with liner material. When the fill is complete, the waste is capped to prevent surface water infiltration and covered with compacted soil. In addition to leachate collection, provision may be made for a system to treat evolved gases, particularly when methane-generating biodegradable materials are disposed of in the landfill.

The flexible-membrane liner made of rubber (including chlorosulphonated polyethylene) or plastic (including chlorinated polyethylene, high-density polyethylene, and polyvinylchloride), is a key component of state-of-art land-fills. It controls seepage out of, and infiltration into the landfill. Obviously, liners have to meet stringent standards to serve their intended purpose. In addition to being impermeable, the liner material must be strongly resistant to biodegradation, chemical attack, and tearing.

Capping is done to cover the wastes, prevent infiltration of excessive amounts of surface water, and prevent release of wastes to overlying soil and the atmosphere. Caps come in a variety of forms and are often multilayered. Some of the problems that may occur with caps are settling, erosion, ponding, damage by rodents, and penetration by plant roots.

Surface Impoundment of Liquids

Many liquid hazardous wastes, slurries, and sludges are placed in surface impoundments, which usually serve for treatment and often are designed to be filled in eventually as a landfill disposal site. Most liquid hazardous wastes and a significant fraction of solids are placed in surface impoundments in some stage of treatment, storage, or disposal.

A surface impoundment may consist of an excavated "pit", a structure formed with dikes, or a combination thereof. The construction is similar to that discussed above for landfills in that the bottom and walls should be impermeable to liquids and provision must be made for leachate collection. The chemical and mechanical challenges to liner materials in surface impoundments are severe, so that proper geological siting and construction with floors and walls composed of low-permeability soil and clay are important in preventing pollution from these installations.

Deep-well Disposal of Liquids

Deep-well disposal of liquids consists of their injection under pressure to underground strata isolated by impermeable rock strata from aquifers. Early experience with this method was gained in the petroleum industry where disposal is required of large quantities of saline waste-water coproduced with crude oil. The method was later extended to the chemical industry for the disposal of brines, acids, heavy metal solutions, organic liquids, and other liquids.

A number of factors must be considered in deep-well disposal. Wastes are injected into a region of elevated temperature and pressure, which may cause chemical reactions to occur involving the waste constituents and the mineral strata. Oils, solids, and gases in the liquid wastes can cause problems such as clogging. Corrosion may be severe. Micro-organisms may have some effects. Most problems from these causes can be mitigated by proper waste pretreatment.

The most serious consideration involving deep-well disposal is the potential contamination of groundwater. Although injection is made into permeable salt-water aquifers presumably isolated from aquifers that contain potable water, contamination may occur. Major routes of contamination include fractures, faults, and other wells. The disposal well itself can act as a route for contamination if it is not properly constructed and cased or if it is damaged.

LEACHATE AND GAS EMISSIONS

Leachate

The production of contaminated leachate is possibility with most disposal sites. Therefore, new hazardous waste landfills require leachate collection/treatment systems and many older sites are required to have such systems retrofitted to them. Modern hazardous waste landfills typically have dual leachate collection systems, one located between the two impermeable liners required for the bottom and sides of the landfill and another just above the top liner of the double-liner system. The upper leachate collection system is called the primary leachate collection system, and the bottom is called the secondary leachate collection system. Leachate is collected in perforated pipes that are imbedded in granular drain material. Chemical and biochemical processes have the potential to cause some problems for leachate collection systems. One such problem is clogging by insoluble manganese and iron hydrated oxides upon exposure to air.

Leachate consists of water that has become contaminated by wastes as it passes through a waste disposal site. It contains waste constituents that are soluble, not retained by soil, and not degraded chemically or biochemically. Some potentially harmful leachate constituents are products of chemical or biochemical transformations of wastes.

The best approach to leachate management is to prevent its production by limiting infiltration of water into the site. Rates of leachate production may be very low when sites are selected, designed, and constructed with minimal production of leachate as a major objective. A well-maintained, low-permeability cap over the landfill is very important for leachate minimisation.

Hazardous Waste Leachate Treatment

The first step in treating leachate is to characterise it fully, particularly with a thorough chemical analysis of possible waste constituents and their chemical and metabolic products. The biodegradability of leachate constituents should also be determined.

The options available for the treatment of hazardous waste leachate are generally those that can be used for industrial waste-waters. These include biological treatment by an activated sludge, or related process, and sorption by activated carbon usually in columns of granular activated carbon. Hazardous waste leachate can be treated by a variety of chemical processes, including acid/base neutralisation, precipitation, and oxidation/reduction. In some cases these treatment steps must precede biological treatment; for example, leachate exhibiting extremes of pH must be neutralised in order for micro-organisms to thrive in it. Cyanide in the leachate may be oxidised with chlorine and organise with ozone, hydrogen peroxide promoted with ultraviolet radiation, or dissolved oxygen at high temperatures and pressures. Heavy metals may be precipitated with base, carbonate, or sulphide.

Leachate can be treated by a variety of physical processes. In some cases, simple density separation and sedimentation can be used to remove water-immiscible liquids and solids. Filtration is frequently required and flotation may be useful. Leachate solutes can be concentrated by evaporation, distillation, and membrane processes, including reverse osmosis, hyperfiltration, and ultrafiltration. Organic constituents can be removed by solvent extraction, air stripping, or steam stripping.

Gas Emissions

In the presence of biodegradable wastes, methane and carbon dioxide gases are produced in landfills by anaerobic degradation (Reaction 21.25). Gases may also be produced by chemical processes with improperly pretreated wastes, as would occur in the hydrolysis of calcium carbide to produce acetylene:

$$CaC_2 + 2H_2O \rightarrow C_2H_2 + Ca(OH)_2 \qquad ...(21.26)$$

Odorous and toxic hydrogen sulphide, H_2S, may be generated by the chemical reaction of sulphides with acids or by the biochemical reduction of sulphate by anaerobic bacteria (*Desulfovibrio*) in the presence of biodegradable organic matter:

$$SO_4^{2-} + 2\{CH_2O\} + 2H^+ \rightarrow H_2S + 2CO_2 + 2H_2O \qquad ...(21.27)$$

Gases such as these may be toxic, combustible, or explosive. Furthermore, gases permenting through landfilled hazardous waste may carry along waste vapours, such as those of volatile aryl compounds and low-molar-mass halogenated hydrocarbons. Of these, the ones of most concern are benzene, 1,2-dibromoethane, 1,2-dichloroethane, carbon tetrachloride, chloroform, dichloromethanem, tetrachloroethane, 1,1,1-trichloroethane, trichloroethlene, and vinyl chloride. Because of the hazards from these and other volatile species, it is important to minimise production of gases and, if significant amounts of gases are produced, they should be vented or treated by activated carbon sorption or flaring.

IN-SITU TREATMENT

In-situ treatment refers to waste treatment processes that can be applied to wastes in a disposal site by direct application of treatment processes and reagents to the wastes. Where possible, *in-situ* treamtent is desirable as a waste site remediation option.

IN-SITU IMMOBILISATION

In-situ immobilisation is used to convert wastes to insoluble forms that will not leach from the disposal site. Heavy metal contaminants including lead, cadmium, zinc, and mercury, can be immobilised by chemical precipitation as the sulphides by treatement with gaseous H_2S or alkaline Na_2S solution. Disadvantages include the high toxicity of H_2S and the contamination potential of soluble sulphide. Although precipitated metal sulphides should remain as solids in the anaerobic conditions of a landfill, unintentional exposure to air can result in oxidation of the sulphide and remobilisation of the metals as soluble sulphate salts.

Oxidation and reduction reactions can be used to immobilise heavy metals *in-situ*. Oxidation of soluble Fe^{2+} and Mn^{2+} to their insoluble hydrous oxides, $Fe_2O_3.xH_2O$ and $MnO_2.xH_2O$, respectively, can precipitate these metal ions and coprecipitate other heavy metal ions. However, sub-surface reducing conditions could later result in reformation of soluble reduced species. Reduction can be used *in-situ* to convert soluble, toxic chromate to insoluble chromium compounds.

Chelation may convert metal ions to less mobile forms, although with most agents chelation has the opposite effect. A chelating agent called Tetran is supposed to form metal chelates that are strongly bound to clay minerals. The humin fraction of soil humic substances likewise immobilises metal ions.

Vapour Extraction

Many important wastes have relatively high vapour pressures and can be removed by vapour extraction. This technique works for wastes in soil above the level of groundwater; that is, in the vadose zone. Simple in concept, vapour extraction involves pumping air into injection wells in soil and withdrawing it, along with volatile components that it has picked up, through extraction wells. The substances vaporised from the soil are removed by activated carbon or by other means. In some cases, the air is used as combustion air in the engines used to run the air pumps, which destroys the organic substances in the air. Vapour extraction is relatively efficient compared to groundwater pumping because of the much higher flow rates of air through soil compared to water. Vapour extraction is most applicable to the removal of volatile organic compounds (VOCs), such as chloromethanes, chloroethanes, chloroethylenes (such as trichloroethylene), benzene, toluene, and xylene.

Solidification In-Situ

In-situ solidification can be used as a remedial measure at hazardous waste sites. One approach is to inject soluble silicates followed by reagents that cause them to solidify. For example, injection of soluble sodium silicate followed by calcium chloride or lime forms solid calcium silicate.

Detoxification In-Situ

When only one, or a limited number of harmful constituents is present in a waste disposal site, it may be practical to consider detoxification *in-situ*. This approach is most practical for organic contaminants including pesticides (organophosphate esters and carbamates), amides, and esters. Among the chemical and biochemical processes that can detoxify such materials are chemical and enzymatic oxidation, reduction, and hydrolysis. Chemical oxidants that have been proposed for this purpose include hydrogen peroxide, ozone, and hypochlorite.

Enzyme extracts collected from microbial cultures and purified have been considered for *in-situ* detoxification. One cell-free enzyme that has been used for detoxification of organophosphate insecticides is parathion hydrolase. The hostile environment of a chemical waste landfill, including the presence of enzyme-inhibiting heavy metal ions, is detrimental to many biochemical approaches to *in-situ* treatment. Furthermore, most sites contain a mixture of hazardous constituents, which might require several different enzymes for their detoxification.

Permeable Bed Treatment

Some ground-water plumes contaminated by dissolved wastes can be treated by a permeable bed of material placed in a trench through which the ground-water much flow. For example, limestone contained in a permeable bed neutralises acid and precipitates some kinds of heavy metals as hydroxides or carbonates. Synthetic ion exchange resins can be used in a permeable bed to retain heavy metals and even some anionic species, although competition with ionic species present naturally in the ground-water can cause some problems with the use of ion exchangers. Activated carbon in a permeable bed will remove some organics, especially less soluble, higher molar mass organic compounds.

Permeable bed treatment requires relatively large quantities of reagent, which argues against the use of activated carbon and ion exchange resins. In such an application it is unlikely that either of these could be reclaimed and regenerated as is done when they are used in columns to treat waste-water. Furthermore, ions taken up by ion exchangers and organic species retained by activated carbon may be released at a later time, causing subsequent problems. Finally, a permeable bed that has been truly effective in collecting waste materials may, itself, be considered a hazardous waste requiring special treatment and disposal.

In-Situ Thermal Processes

Heating of wastes *in-situ* can be used to remove or destroy some kinds of hazardous substances. Both steam injection and radio frequency heating have been proposed for this purpose. Volatile wastes brought to the surface by heating can be collected and held as condensed liquids or by activated carbon.

One approach to immobilising wastes *in-situ* is high temperature vitrification using electrical heating. This process involves pouring conducting graphite on the surface between two electrodes and passing an electrical current between the electrodes. In principle, the graphite becomes very hot and "melts" into the soil leaving a glassy slag in its path. Volatile species evolved are collected and, if the operation is successful, a nonleachable slag is left in place. It is easy to imagine problems that might occur, including difficulties in getting a uniform melt, problems from groundwater infiltration, and very high consumption of electricity.

Soil Washing and Flushing

Extraction with water containing various additives can be used to cleanse soil contaminated with hazardous wastes. When the soil is left in place and the water pumped into and out of it, the process is called flushing; when soil is removed and contacted with liquid, the process is referred to as washing. Here, washing is used as a term applied to both processes.

The composition of the fluid used for soil washing depends upon the contaminants to be removed. The washing medium may consist of pure water, or it may contain acids (to leach out metals or neutralise alkaline soil contaminants), bases (to neutralise contaminant acids), chelating agents (to solubilise heavy metals), surfactants (to enhance the removal of organic contaminants from soil and improve the ability of the water to emulsify insoluble organic species), or reducing agent (to reduce oxidised species). Soil contaminants may dissolve, form emulsions, or react chemically. Inorganic species commonly removed from soil by washing include heavy metals salts; lighter aromatic hydrocarbons, such as toluene and xylenes; lighter organohalides, such as trichloro- or tetrachloroethrylene; and light-to-medium molar mass aldehyde and ketones.

CHAPTER 22

Radioactivity in Environment

INTRODUCTION

Many people are still under the impression that radiation is quite a new element which has entered our present day life. But it will be interesting to note that man has been exposed to radiation since ages. This is an open secret and as suggested by *Barnes and Taylor*, the very existence of man depended on gene mutuations caused by radiation in the animal from which he evolved. Thus the radiation is not new. But it will be correct to say that new sources of radiation have been developed by science.

The types of radiation in general use are at present X-rays, alpha rays, beta rays, gamma rays and special particle beams. The substances which emit these rays are known as the radioactive substances and the property of these substances to emit such rays is referred to as the radioactivity.

Radiation pollution is concerned with radioactive substances, radiation and the environment. There are two rather distinct phases of radiation which require different approaches. On the one hand, we are concerned with the effects of radiation on individuals, population, communities and ecosystems. The other important phase of radio-ecology concerns the fate of radioactive substances released into the environment and the manner by which the ecological community and populations control the distribution of radioactivity.

RADIONUCLIDES

Radiation from human activities has increased since the invention of atom power and introduction of nuclear energy. Different sources produce radiation with different energies and have different biological effects. Ionising radiations occur when uncharged chemicals are changed into charged ion pairs. There are more than 1000 different nuclides in the atmosphere. Some of them are stable in nature while other can split up into parts and give off radiation, which are known as radionuclides. The radionuclides are the products of the natural decay of uranium and are not extracted during milling process. In addition to acid precursors and heavy metals, uranium tilings also contain a variety of radioactive forms of atoms, or radionuclides, including radium-226, thorium-230, lead-210 and polonium-210. Some of these radionuclides have long half lives. Thorium-230 has a half-life of about 80,000 years, which means that radioactivity from thorium will be present in the tillings for hundreds and thousands of years. The pH of the tiling plays a significant role in determining how soluble some of these radionuclides become.

Thorium, particular, becomes more available when pH drops to below 4. Radionuclides with short-half lives, i.e., upto a few days, may be very dangerous when produced, but the danger does not last.

The main exposure pathways for radioactivity from tiling are direct gamma radiation, inhalation of radioactive particulates, and ingestion of radionuclides through the food chain. The environmental contaminants in the tailings, such as metal and acid drainage are of great concern. The primary long-life fission products that enter in the ecosystem in a significant amount, are due to the nuclear weapon tests. Table 22.1 indicates the radionuclides produced and released in the atmosphere during atomic explosions and their periods.

Table 22.1. Some important radionuclides and their half-lives.

Radionuclide	Target issue	Half-life*
Calcium-45	Bone	165 days
Carbon-14	Whole body	5760 years
Cesium-137	Soft tissue, genital organs	27 years
Iodine-129	Thyroid	17 million years
Iodine-131	Thyroid	8 days
Plutonium-239	Bone, liver, spleen	24,400 years
Radium-226	Bone	1620 years
Strontium-90	Bone	28 years
Tritium (^{3}H)	Whole body	12.3 years

* Half-life means the time taken for half the atoms in a given sample to distintegrate.

Other radioactive threats to the environment are the accidents connected with the activities of nuclear-powered vessels and satellites, such as Cosmos 954, which crash-landed in 1978 near the Thelon River in the North-west Territories.

NUCLEAR CONCEPTS AND TERMINOLOGY OF ECOLOGICAL IMPORTANCE

Ionising radiation is any atomic particle or frequency of electromagnetic which is capable of causing ions or free radicals in a material through which it passes. Ionising radiation does this by expelling an electron from its orbit as leaving behind a charged particle or ion; in contrast to light and most solar radiation which does not have this ionising effect. Ions cause most of the physiological damage. Highly reactive ions interact with the molecules of life to change them chemically. If too many ions are produced in a cell or organism, they can alter the enzymes, proteins, mitochondria and genetic materials enough to cause the immediate death of the cell or organism.

When the changes are less severe, the organism or cell may survive but harvour with in it irreversible change that may lead to a delayed death (such as from cancer) or that could be passed to future generations of the cell or organism through genetic materials. The ionising radiations of ecological concern are classified as follows :

1. *Corpuscular radiation:* These consist of streams of atomic or sub-atomic particles which transfer their energy to whatever they strike.

   (a) *α-particles:* These particles are large and travel few centimetres in the air. These cause large amount of ionisation.

(b) *β-particles:* These are small particles with a high speed travelling a number of metres in air. These are capable of entering into tissues for few centimetres.

2. *Electromagnetic radiation:* These include waves of short wavelengths. These could travel long distances and can rapidly penetrate the living tissue causing dispersed ionisation. These include γ-rays. These can penetrate and produce effect even without being taken inside. Therefore, these are known as *external emitters.*

3. *Other types of radiation*

    (a) *Neutrons:* These are long uncharged particles which do not cause radiation by themselves, but they produce radioactivity in non-radioactive materials through which they pass.

    (b) *X-rays:* These are electromagnetic waves very similar to gamma rays but originate from the outer electron shell or radioactive substances, which are not dispersed in nature.

    (c) *Cosmic rays:* These are radiations from the outer space. Those contain α- and β-particles together with γ-rays.

The cosmic rays along with ionising radiation from naturally occurring radioactive substances in soil, water constitute background radiations.

Units of Measurements

1. The basic unit of the quality of a radioactive substance is the Curie (Ci) which is defined as the amount of material in which 3.7×10^{10} atoms disintegrate each second or 2.2×10^{12} disintegrations/minute.

2. The most convenient unit for all types of radiation is the *rad*, which is defined as the absorbed dose of 100 ergs of energy per gram tissues.

3. The roentgen (R) is an older unit which strictly speaking is to be used only for γ and X-rays.

FALL-OUT PROBLEM

The radioactive dust that falls to earth after the atomic explosions is called radioactive fall-out. These materials mix and interact with natural particulate materials in the atmosphere (natural fall-out) and the increased amount of man-made air pollution. This kind of radioactive fallout depends on the type of bomb. First, it may be well to distinguish between the two types of *nuclear weapons* :

1. The fission bomb, in which heavy elements such as uranium and plutonium are split when the release of energy and radioactive fission products, occur.

2. The fusion bomb, on the thermonuclear weapon in which light elements (deuterium) fuse to form heavier elements with the release of energy and neutrons.

According to *Glasstone* (1957), about 10% of the energy of a nuclear weapon is in residual nuclear radiation, some of which becomes widely dispersed in biosphere. The amount of radioactive fallout produced depends not only on the type and size of the weapon but also on the amount of environmental material that gets mixed up in the explosion.

Fall-out from weapons differ from atomic waste materials in that the radionuclides are fused with iron, silica, dust and whatever happens to be in vicinity to form relatively insoluble particles. These particles, which under the microscope often resemble tiny marbles of different colours. The smaller particles adhere tightly to the leaves of plants where they may not only produce variation damage to leaf tissue but may be ingested by grazing animals and dissolved by the digestive juices in the alimentary canal. Thus, this kind of fall-out can enter the food chain directly on the herbivores or primary consumer, trophic levels.

Fall-out from small atomic weapons or nuclear explosions used for peaceful purposes is mostly deposited in a narrow linear pattern downward, but some of the smallest particles may become widely dispersed and come down in rain at long distances. While the total amount of radioactivity decreases with the distance from a nuclear test, it was early discovered that certain biologically significant nuclides especially strontium-90 reached a peak in wild animal populations 50 to 100 miles from the "ground zero" of the explosion.

Studying following the weapons tests on the pacific attolls showed that the kind of radionuclides that enter marine food chains are rather strikingly different from those that enter terrestrial food chains. Elements of radioactive fall-out that form strong complexes with organic matter such as cobalt-60, iron-59, zinc-65 and manganese-54, (which are all nuclides induced by neutron bombardment) and those that are present in particulate or colloidal form transfer in the highest amount of marine organisms. In contrast, it is the soluble fission products such as strontium-90 and cesium-137 that are found in the highest amounts in the land plants and animals.

Fall-out radionuclides (especially 90 Sr and 137 Cs) have been and are being passed on to man via the food chain although concentrations in human tissue are not generally as high as those in ship. Man is somewhat protected by his position in food chain and by food processing and cooking, which remove some of the contamination.

Penetration

Particle radiation

Rays of this type given off by radioactive substances seldom exceed a few MeV and cannot penetrate very deeply into the body. Exposure to such rays are only likely to cause skin burns which are not usually serious except when the doses are very high. However, they are dangerous if they are injected or inhaled, because some have extremely long half life and vital organs can thus become exposed to a cancer hazard.

High frequency quanta or photons (e.g., X or gamma rays)

These rays give up their energy in three main ways depending on the initial energy of the quantsas. Unlike particle radiation X-ray and gamma ray can penetrate deep into the body.

1. *At lowest energies:* The incident photon is totally absorbed in the production of a photo electron (There is also a recoil nucleus).
2. *At moderate energies:* The incident photon collides with an atomic electron and rebounds, when part of the photon energy is transferred to the electron which is thus ejected from the atom. The deflected photon suffers a wavelength shift and is less penetrative.
3. *At high energies (above 1.02 MeV):* A photon passing close to a nucleus is transformed into two particles—an electron and a position (pair production) which fly apart at speed.
4. *Neutrons:* Neutrons being uncharged cannot ionise directly. They only interact with matter by direct collision with a nucleus.
5. *Low energy neutrons:* Can be captured by a nucleus to produce an atomic transmutation.

6. *Fast neutrons:* A proton is ejected from the nucleus with which the neutron collides. Because nuclei only occupy a minute fraction of the volume of an atom a neutron can travel long distances before colliding with a nucleus. The stopping power of a material against fast neutrons depends entirely on the number of nuclei in the path and not on the size of the nuclei or the number of electrons surrounding them. The nucleus of a hydrogen atom has the same chance of stopping a neutron as the nuclei of lead weighing 200 times as much. Since 1 kg of water contains 30 times as many nuclei as 1 kg load the water will absorb more.

Personal Monitoring

Film badge

This consists of a sensitive 'photographic' film sandwich between a base plate and a front surface which consists of different materials of varying thicknesses that act as filters for the incident radiation.

Personnel dosimeter

Two electrodes (one fixed, the other moveable) are charged to the potential. Radiation produces ions which are deposited on the electrode. This reduces the potential hence reduces mutual repulsion. Thus the electrodes move and it is this movement that gives the reading of dose received.

Maximum Permissible Doses

Basic standards have been recommended by the International Commission on Radiological Protection from past experience in the use of x-rays and radium, together with information on radiation injuries to man. They may be expected to produce effects which could be detected only by statistical methods applied to large groups of people. The basic standard is an annual limit on dose, viz. :

1. 5 rems to the whole-body gonads and blood forming organs.
2. 30 rems to bone, skin and thyroid.
3. 15 rems to all other organs or parts of the body.
4. 75 rems to hands, forearms, feet and ankles.

Because of known sensitivity of the foetus, the whole body exposure of women of reproductive age should be restricted so that they do not receive doses at rates in excess of those specified assuming continuous exposures throughout the year. Thus, the dose received by an embryo during the first two months of pregnancy should normally be less than 1 rem, a dose considered acceptable by the Commission. When pregnancy has been diagnosed, the exposure during the remaining seven months should not exceed a further 1 rem.

Derived maximum permissible levels

For radioactivity in the body, the important parameter is the maximum permissible body burden. This is the activity in the whole body which delivers the maximum permissible dose rate appropriate to continuous occupational exposure. The maximum permissible concentrations in air and water, if ingested or inhaled daily will result in the body eventually acquiring the maximum permissible body burden.

For external radiation measurements (internal radiation cannot be easily measured), the annual dose limit of 5 rem year is equivalent to: (i) 5000 millirem for 50 weeks, i.e., 100 millirem/week; and (ii) 100 millirem for 40 hours, i.e., 2.5 millirem/hrs.

The derived level equivalent to 0.3 of the annual dose limit is then equivalent to

$$\frac{3}{10} \times 2.5 \text{ or } 0.75 \text{ millirem / hrs.}$$

EFFECTS OF RADIATION

The knowledge of effects of radiation is mainly based on the following two observations: (i) experiments on animals in laboratories; and (ii) effects of atomic bombs dropped on Hiroshima and Nagasaki.

It is found that the radiations have power of penetrating the tissues of body and in doing so, they leave electrically charged ions in their paths. It is these ions which are responsible for damage done by radiation. The extent of damage depends on various factors such as quantity of radiation received, its duration, age of individual exposed to radiation, the parts of body affected by radiation, etc.

The hazards of radiation on human beings may be divided into the following three categories :

1. Acute damage.
2. Chronic damage.
3. Genetic damage.

Acute Damage

The acute radiation damage occurs from relatively large dose of radiation over a short period of time. Thus the exposure to radiation is for such a short period that it is not possible to give treatment to the damaged tissue. The acute radiation damages include sudden death, death after two or three weeks, loss of hair, widespread ulcers, bleeding from the mouth and gums, inability to deal with even minor infections, etc.

Chronic Damage

The chronic radiation damage occurs from relatively small continuous dose of radiation over a long period of time. The chronic radiation damages include diseases of the blood (leukaemia and anaemia), cancer of the skin and other organs, cataracts (opacities in the lens of the eye), somatic mutations, life-span shortening, etc.

Genetic Damage

The genetic radiation damage represents long term effect of radiation and it indicates changes among future generation. The genetic damage may be caused by acute exposures or probably by chronic exposures. The genetic effect of radiation is very important and very little precise information is available at present for this effect.

It may also be noted that the radiation affects all forms of life, plants and inert materials. In general, the higher organisms are more sensible to the radioactive exposures. The consumption of such animals or plants as food by human beings may lead to serious disasters. The effects of radiation on inert material include catalysing chemical reactions, discolouring crystalline materials, converting pliable materials into brittle materials, etc.

RADIOACTIVE SOURCES

Following are the main sources which are responsible for giving the radioactivity to the water: (i) atomic reactors; (ii) nuclear explosions; (iii) soils and rocks; (iv) use of radioactive substances; and (v) waste of radioactive substances.

Atomic Reactors

The atomic reactors are set up for the production of electric power, for the production of radioactive substances and for research work. The water used in these reactors may be rendered radioactive and if this water is discharged into a river, the water of that river may become radioactive.

Nuclear Explosions

During the last decade or so, the nuclear explosions are carried out for experimental purposes. This has resulted in rain of fission products, commonly known as the radioactive fall-out. The radioactive debris from nuclear tests above ground are readily carried into the atmosphere. As a matter of fact, the atmosphere serves as a major reservoir for nuclear debris. Most of the nuclear explosions are at present carried out underground. But even then, there is danger of pollution of underground water. However, as compared to the atmospheric explosions, the effect of such underground nuclear tests is less.

To illustrate the effects of the atmospheric radioactive fall-out on human life, irrespective of the source, a chart is shown in Fig. 22.1. It indicates the number of ways of transmission of radioactive fall-out.

Soils and Rocks

The soils and rocks contain radioactive substances such as uranium and thorium. If water comes in contact with them, the radioactivity is imparted to the water.

Use of Radioactive Substances

The use of radioactive substances is increasing in medicine, industry and research work. But if the waste from these operations is not disposed off carefully, there are chances of it gaining access to the public water supply.

Waste of Radioactive Substances

Sometimes the waste of radioactive substances is disposed off by filling it in a bore hole in the ground. This may render radioactivity to the underground water. Man-made emissions from various nuclear reactors are given in Table 22.2.

DISPOSAL OF RADIOACTIVE WASTES

The unwanted materials which result from the processes related to the radioactivity are known as the radioactive wastes. Depending upon the intensities of radioactivity, these wastes may be referred to as high level, intermediate level or low level.

Table 22.2. Emissions from various nuclear reactors.

Type of reactor		Emission products	Emission rate Ci/year
Boiling-water reactor	(i)	Noble gases (short-lived + long-lived)	D-18000
	(ii)	Liquid fission products	1–6
	(iii)	Tritium-liquid	5–20
Pressurised-water reactor	(i)	Noble gases (long-lived)	0–2
	(ii)	Liquid fission products	0–11
	(iii)	Tritium-liquid	500–1300
Heavy-water moderated reactor	(i)	Noble gases (short-lived + long-lived)	4000
	(ii)	Fission products and tritium-liquid	5
	(iii)	Tritium-gas	1200
Gas-cooled graphite reactor	(i)	Noble gases (short-lived + long-lived)	100–000
	(ii)	Fission products–liquid	2–100
	(iii)	Tritium-liquid	100–500

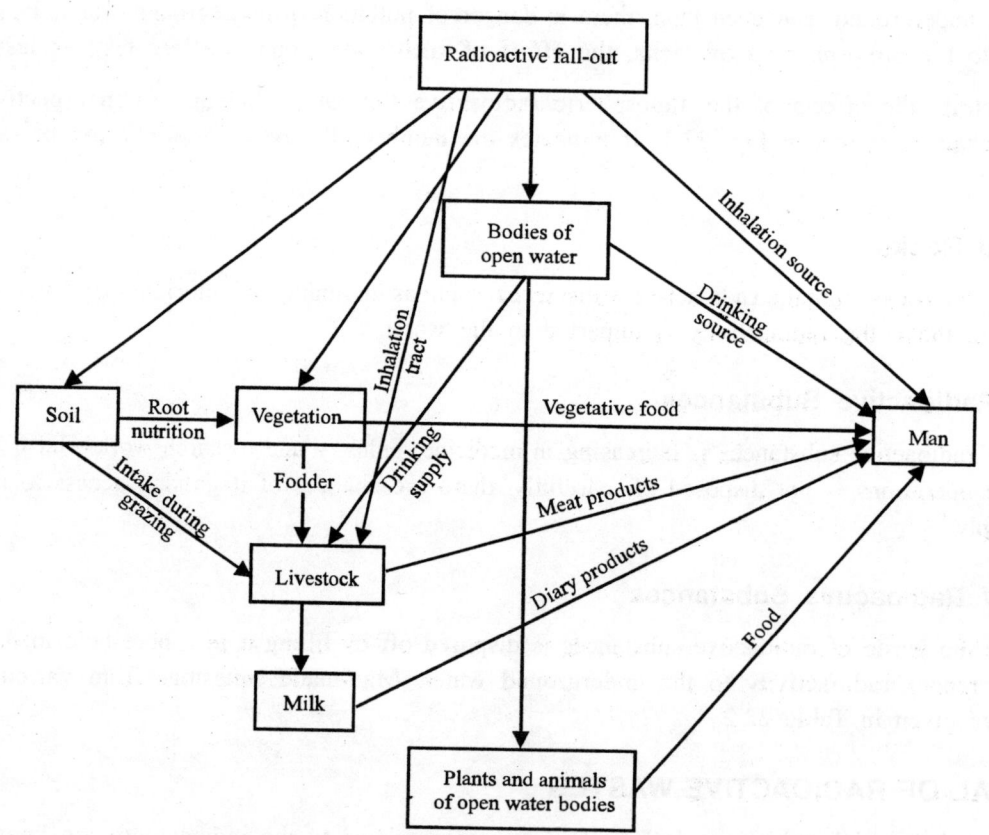

Fig. 22.1. Transmission of radioactive fall-out to man.

The high level wastes are in the form of liquids or solids and they generate large amount of heat. They are too dangerous to be released anywhere in the biosphere. They also cause small earthquakes, if injected into certain types of geological fissures. For intermediate wastes, the arrangement should be made to separate out high level wastes and low level wastes before their disposal. The low level wastes are in the form of liquids, solids and gases and they have very low radioactivity per unit volume.

The high level radioactive wastes cannot be discharged into the environment and they should be properly stored only. The intermediate level radioactive wastes can be discharged into the environment after some dilution. The low level radioactive wastes can be discharged into the environment under controlled conditions and in such a way and at such a rate that the radioactivity equilibrium of biosphere is not disturbed appreciably. Following are the three methods of disposal of the radioactive wastes: (i) dilution; (ii) storage; and (iii) reclamation.

Dilution

The radioactive wastes may be disposed off by dilution with inert materials. The radioactive wastes in gaseous form are diluted to acceptable levels and discharged into the atmosphere through tall stacks. The liquid radioactive wastes are disposed off in rivers and oceans. The design calculations for such disposal are based on the most pessimistic side so as to provide additional safety. The disposal of radioactive wastes by dilution is adopted only for low level and intermediate level radioactive wastes.

Storage

In this method, the radioactive wastes are stored till they decay and become harmless. The storage is done in specially prepared underground tanks of concrete or steel with a non-corrosive material in contact with the wastes. The disposal of radioactive wastes by storage is adopted only for most of the high level wastes.

The time required for decay of high level radioactive wastes ranges from 250 years to 400 years. Hence, if large volumes of such wastes are to be stored, this method of disposal of radioactive wastes becomes impracticable. The alternatives that are considered for tank storage of high level wastes are :

1. Conversion of liquids to inert solids (ceramics) for burial in deep geological strata.
2. Storage of liquids and solids in deep salt mines.

Reclamation

In this method, the radioactive wastes are reclaimed or converted into useful products. This is the most useful method of disposal of radioactive wastes and efforts are being made for finding out more and more beneficial applications of radioactive wastes.

RADIOACTIVITY OF WATER

It is found that the natural waters contain alpha-activity and beta-activity. The former is derived from rocks and minerals while the latter is mainly due to potassium content in water. The most important radio-isotopes are radium from alpha-activity point of view and strontium 90 and cesium 137 from beta-activity point of view. The permissible limits are as follows :

Radium	0.4 µµc/l
Other alpha-emitters	2.4 µµc/l

Radio strontium and radio calcium   20 μμc/l

Other beta-emitters   1000 μμc/l

The unit of measurement of the radioactivity of water is one millionth of a millionth of a curie per litre or micro-microcurie per litre or μμc/l.

Measurement of Radioactivity

The various sensitive instruments and detectors are found out to accurately measure the radioactivity of water. The calculation of radioactivity present in a sample of water is a complicated process as it involves geometrical aspects of the detector, the amount of natural radioactivity, mathematical corrections required, etc.

Recommended Method for Removal of Radioactivity from Water

Following are some of the methods which are studied for the removal of radioactivity from water: (i) phosphate coagulation; (ii) electro-dialysis method; (iii) adding clay materials; (iv) adding metallic dusts; and (v) distillation of water. Each of the above method of removal of radioactivity from water will now be briefly described.

Phosphate coagulation

This method is effective in the removal of strontium, etc. which form insoluble phosphate. However careful control of pH value of water is necessary.

Electro-dialysis method

This method is effective for the removal of soluble constituents. The method requires the removal of colloidal matter at a previous stage.

Adding clay materials

This method is effective at concentrations of 1000 ppm or more, but excessive volume of sludge is to be handled in this method.

Adding metallic dusts

This method is effective in removing radioactive substances in excess of 90 per cent, except cesium 137 and iodine 131.

Distillation of water

This is the most effective method available at present for the removal of radioactivity from water. But it is impracticable on a large scale because of the fact that it will not be feasible to distil large quantities of water.

This new important aspect of radioactivity is changing its character day by day and hence a careful watch on it should be kept by the local authorities supplying water. It should also be remembered that bacteria or poisonous metals in water may give results within one or two weeks. But the use of radioactive water will create effects which last for generations. Any increase in radioactivity, other than natural, should therefore be avoided and where this is not feasible in the interests of national progress, the increased radioactivity level should be kept sufficiently lower than that required for harmless long term effects of radiation.

Treatment of Solid Radioactive Wastes

The aim of radioactive solid waste treatment is to reduce their volume. The main methods of radioactive solid waste treatment are incineration, compacting and fragmentation. Substances before incineration should be carefully sorted to remove the materials producing toxic or corrosive vapours, explosive substances and non-combustible items. The incinerator gases are removed into the atmosphere after proper gas cleaning.

Some kinds of solid wastes can be compressed by use of a press. After compaction the wastes are put into the disposable containers for final disposal. Large-sized or bulky materials can also be cut into pieces before their treatment or disposal. In case of some highly mobile and hazardous wastes like ashes from incinerator, they can be incorporated with concrete or bitumen, which also provide a shield to the radiation, before their transport to burial sites. The solid wastes having a high activity where their storage for a few centuries can bring down the radioactivity to non-hazardous level, can be disposed off by shallow burial or sea dumping taking care that the radioactivity should not be leaked out into the environment. All solid wastes requiring a containment time of more than a few centuries are segregated and can be disposed off away from human environment in relatively deep geologic formation which is capable of ensuring permanent containment of the waste.

Alternatives for disposal of radioactive wastes

Though the final disposal of radioactive high activity wastes in geological formations is a widely favoured practice, suggestions have also been given for alternate methods from time to time for disposal of these wastes. Some important alternatives could be the disposal of these wastes in space, in ice sheets, in a very deep hole of 3–5 km where the rock is still warm, and in deep ocean floor under the sediments. These methods are, however, abandoned or deferred for possible later use, either because they are highly expensive, have greater environmental risk, or because of the present technical knowledge is not enough to evaluate fully their consequence.

Recorded Incidents

1. The reactor fire at Windscale, UK (1957) due to malfunctioning of reactor temperature control system released radioactive gas including I-131 which contaminated grass affecting grazing animals and foodstuff.
2. The accident in Three Mile Island nuclear plants at Harrisburg Pennsylvania USA (1979) due to technical failure in the pressurised water system caused release of radioactive cooling water into the Susque Lanna river and radioactive gases into the atmosphere.
3. Chernobyl disaster in USSR (1986) is of more serious nature. Official accounts are not available, but there were about 40 deaths and radioactive fall-out over wide spread area.

It is well-known that nuclear installation accidents can cause long term effects on the environment. It is, therefore, essential that all nuclear installations observe the safety rules prescribed by the International Commission on Radiological Protection (IRCP) and control their radioactive emissions into environment.

CHAPTER 23

Role of Biotechnology in Environmental Protection

INTRODUCTION

Biotechnology today is regarded as the greatest intellectual enterprise of humankind—a frontline area of research, product and new process development. Biotechnology can be defined as "any technique that uses living organisms (or parts of organisms) to make or modify products, to improve plants or animals, or to develop microorganism for a specific use". Later, the new biotechnology was defined by the US Government in 1991 as: "The industrial use of rDNA, cell fusion and novel bioprocessing techniques" and this definition will go a long way because this may prove to be the most useful industrial tool to the world economy for making money with biology. The European Federation of Biotechnology has described biotechnology as: "The integration of natural resources and engineering in order to achieve the application of microorganisms, as well as parts thereof, and molecular analogues for product and services". Thus in recent terms the biotechnology can be defined as: "The interdisciplinary science encompassing modern techniques and applied to organisms or parts thereof to produce, identify or design substances, or to modify organisms for specific applications".

Undoubtedly biotechnology is a science, which encompasses a very close interface between biology, chemistry, physics, mathematics, information technology, computer science and engineering. Since 1953, after the discovery of the double helix structure of DNA by Watson and Crick, we have advanced our knowledge in the basic understanding of the genetic make up of plant and animal species including man. Major efforts of biotechnologists have now gone towards molecular biology, genome mapping and gene sequencing and analysis. There is now a shift towards structural biology, protein science and protein engineering. With the advent of the recombinant DNA (rDNA) technique and the understanding at the cellular and molecular level of the structure and function of gene, it is now possible to harness the genetic diversity of living orgnisms for the manufacture of useful products from microbes, plants and animals with improved qualities. In the last few years, biotechnology has transformed the chemical industry, agriculture and medicine with remarkable speed. The new biotechnology is already beginning to affect our lives and in future, its influence will profound. The impact of biotechnology on human life and economic progress of various nations worldover has given a major impetus to accelerate research, development and application of this field in relevant socio-economic sectors. The cell fusion techniques, rDNA technology, protein engineering and structural biology have made phenomenal progress as priority research areas.

CURRENT STATUS OF BIOTECHNOLOGY IN ENVIRONMENT PROTECTION

It is important to note that biotechnological applications to environment protection relate to all five options of industrial pollution management mentioned earlier and work for repair as well as prevention. Biodegradation is the ultimate fate of a material that enters the environment. Hence aiming for biodegradability is but obvious. However the current philosophy on the issue of degradation is that :

1. It is not an ideal option
2. It represents waste of material
3. It is valuable only if wastes are hazardous and permanent elimination is sought
4. Products of degradation should come in use if possible (e.g. anaerobic methods, which are getting importance).

The potential of biotechnology in industrial pollution control is diverse and includes :

1. Hazardous waste management with use of specialised cultures or consortia.
2. Bioremediation of polluted land sites, and decontamination or detoxification of spillage.
3. Effluent treatment for variety of industries.
4. Waste gas treatment and deodorisation (removal of phenol, mercaptans, hydrocarbons, hydrogen sulphide etc.).
5. Bioenergy (biogas, ethanol, hydrogen gas) generation from treatment of liquid/solid wastes.
6. Heavy metal recovery from various industrial effluents.
7. Single Cell Protein (SCP)/Biomass/food/mushrooms production from wastes using appropriate biological agents.
8. Modification of processes or new processes/products to prevent pollution (e.g., in tanning/paper/ plastics industries, in desulphurisation of coal and petrol, etc.).
9. Added-value processes involving the conversion of wastes into useful products (e.g. production of catechol from waste streams contaminated with phenols; production of animal feed from wastes of food processing plants, etc.).

ROLE OF BIOTECHNOLOGY IN ENVIRONMENTAL PROTECTION

A symposium held in U.S. way back in 1970 referred to 'Biotechnology' as an 'emerging technology'. Not much confidence was shown in its methods, and it was considered an unfulfilled promise. Important points raised then were :

1. Technical and economic effectiveness is not clear from various studies.
2. Performance data is not published in reported successful cases.
3. Application is mainly in private sector, and literature is therefore not available.
4. Very little quality assurance data is available.
5. Technical data is not easily transformable to other places (having a similar problem).
6. Undue claims or overoptimistic.
7. Uncertainty in prediction of outcome.
8. It is slow and inconsistent.

These misunderstandings seem to have arisen for a number of reasons :

1. Rushing for scale-up (often without complete evaluation of constraints and parameters).
2. Extrapolation of specialised activity to generalised case.
3. Constraints of engineering principles are not realised.
4. Failure to convince about advantages and disadvantages (Biotechnologists are usually consultants and not decision makers).
5. Biotechnology has failed to pay enough attention to complexities of industrial waste-water situations.
6. Right consultancy is not done (a problem in our country).
7. The spread of technology needs demonstration processes and their evaluation, which does not get sufficient attention.
8. Nature of physico-chemical environment affecting biotreatment activity should be an important part of checklist.
9. Undue secrecy.

Environmental biotechnology can offer cheap, compact and effective processes instead of bulky, expensive and space wasting ones. Its philosophy is linked with conservation and by-product recovery, and it is not stimulated by market pressures. While initial costs are high, the treatment may well be less costly overall.

Its full potential is not realised, and laboratory and field successes have not translated into applications. Importantly low-value products, if any, are obtained (e.g. alcohol, methane vs hormones and interferons). Biotechnology can become effective if key technical, legal, economic, business and market issues are successfully tackled. Table 23.1 highlights Eco-friendly Biotechnology.

Table 23.1. Eco-friendly biotechnology.

Other technologies (chemical)	Biotechnology
Raw materials used cause drain on natural resources e.g., crude oil, petroleum	Renewable resources used in most of the cases e.g., starch, cellulose
Processes are hazardous	Processes occur at normal temperature, pressure etc., and are less hazardous
Products are not eco-friendly	Products are eco-friendly
Wastes are more harmful	Less harmful wastes, sometimes even recyclable wastes
Production methods are more polluting	Production methods are less polluting
Pollution control methods only concentrate convert pollutant to another form	Pollution control methods are for total or elimination and not mere conversion
There is drain on energy/resources (petroleum fuels used)	Process consume less energy
Example of options	Examples of options
(a) Fertilisers	(a) Biofertilisers
(b) Pesticides	(b) Biopesticides
(c) Plastics	(c) Bioplastics
(d) Mining	(d) Biomining-bioleaching
(e) Petroleum fuels	(e) Biofuels

APPLICATION OF BIOTECHNOLOGY TO POLLUTION PROBLEMS OF VARIOUS INDUSTRIES

As mentioned earlier, biotechnology can work on five options (three corrective and two preventive) in tackling pollution. Let us see examples of the role played by biotechnology in pollution abatement in some select industries, as an indicator of the direction of efforts.

Food and Allied Industries

Wastes from this industry are characterised by high suspended solids, high BOD/COD, no toxic matter and large volumes (although the quantity and load depends upon size of unit and variety of products). Effluents are generally rich in carbohydrates and may be relatively deficient in nitrogen. The question in the case of these manageable wastes is how economically we treat them. BOD/COD reduction is usually done along with generation of energy of form of biogas and ethanol, or production of SCP/fodder yeast/biomass/aquaculture. Many new efficient biomethanisation reactors are available, and bioconversions of wastes to other chemical products is also possible. Solid wastes is a problem in fruit, vegetable, meat and poultry processing industries. Component separation and recovery of some useful products is common for meat and poultry processing industry which have slaughterhouses, while solid wastes of fruit and vegetable processing/canning industry are suitable for ethanol production or SCP/biomass production.

Paper Industry

Biotechnology has many contributions to offer to the modern pulp and paper industry and microorganisms, enzymes, and newer technologies are being applied at various stages. Some major application areas are : (i) biopulping (fungi used to degrade and reduce lignin contents of cellulose pulp); (ii) mechanical/chemical pulping; (iii) biobleaching (use of enzyme xylanase or fungi producing such enzymes to make pulp brighter) instead of chemical bleaching; (iv) ethanol production from sludge; (v) growing yeasts or fungi on sulphite waste liquors; (vi) decolouration of pulp mill waste liquors with the help of fungal biomass, or degradation of chlorinated lignin derivatives by white rot fungus; (vii) biological deinking of paper (cellulases and hemicellulases to unhook ink from paper) and help its recycling.

Reduced use of chemicals, reduced pollution problems, higher yields, stronger/better quality paper are the advantages with biotechnological applications. Availability of enzymes at low cost has been responsible for increased applications (Table 23.2).

Chemical Industry

These industries will have the widest class of pollutants with each industry having its unique problems. However, xenobiotic ('stranger to life') substances, heavy metals, hazardous chemicals, toxic organics, extreme pH, high salt contents, high level of nitrogen or phosphorus etc. are the common characteristics. Fortunately, biotreatment techniques to tackle many of these problems are available. While an individual category of pollutant can be tackled successfully using these methods, together they pose a problem. Component separation and coordinated efforts using different technologies are required to get the desired results.

Table 23.2. Biotechnological methods for removal of paper effluents.

Process	Earlier method	Biotechnological option
Fibre handling To obtain cleanliness and certain particle size suitable for pulping. The process includes debarking, washing, screening, removal of foreign matter etc.	Mainly it is a physical process.	Pectinase enzymes from Aspergillus can increase efficiency of debarking. It digests pectin from the bark.
Pulping Raw fibre is mechanically processed to get pulp. Wood fibre is broken apart to render more pliable fibre suitable for paper making.	Chemical pulping (kraft pulping using caustic soda and sodium sulphide, or sulphite pulping using sulphurous acid) and mechanical pulping was done earlier. Mechanical pulping gives lower strength pulp, while difficult to degrade wastes is the problem with chemical pulping.	Biopulping improves quality of pulp and environmental credentials of the process. Cellulase less mutants of fungi which degrade lignin but do not attack cellulose are used.
Refining The presence of fine particles decreases the rate of drainage of water. This is likely to affect rate of production at later stage.		Cellulases/hemicellulases from Trichoderama spp. are the enzymes used to improve de-watering and refining of mechanical pulps by removal of secondary fibres a fines.
Bleaching The process is used to get bright, while pulp by removal of residual lignin	Chlorine and chlorine-based, oxidising chemicals are used	Xylanase enzymes or fungi like trichoderm spp or trametous spp are used for bleaching and decrease use of oxidising chemicals by 10–15%.
Deinking Secondary fibre processing involves mainly removal of ink	Normally done by repulping, adding flocculating surfactants and ink solvents	Cellulases can release bound ink from fibre and fines.
Sizing of paper Sizing is done to increase stiffness of paper that is useful as coating or for further paper processing	Starch modification to reduce viscosity, earlier involved use of dilute acids for cooking of starch	An amylase enzyme from bacillus stearothermophilus is used for shortening star polymer and to reduce viscosity.
Pitch removal Sticky (unwanted) lipids from softwood need to be removed as they cause difficulty in processing	Chemical dispersant were used earlier	Lipases from Candida cylindracea hydrolys lipids to constituent fatty acids and substituted polyol
Slirie removal Microbial slime is common problem of a paper mill	Use of every possible chemical biocide is done	For specific type of slime (levan) levan hydrolase enzyme is used.

Petrochemical Industry

Waste-water from the petrochemical industry contains a large number of pollutants, with the composition depending upon the number, type and capacity of individual process plants, feedstock, employed and products manufactured. Water miscible and immiscible liquids and water soluble gases are the major pollutants that need to be removed. Ethylene oxide, isopropanol, acetone, propylene oxide, acrylonitrile, cyclohexane, benzene, xylene, toluene, styrene etc. are the various pollutants normally present. Biotreatment plants are operated on continuous basis and face limitation of carbon source. The usual BOD determination method does not give idea about real degradability since conditions for BOD test and that of biotreatment plant differ. Well-proven bioprocesses for treating pollutants like benzene, toluene, o-xylene, phthalic acid, kerosene, phenol, cresol, catechol, o-dichlorobenzene, resorcinol have been developed by the local industry.

Tanning Industry

Biotechnology can play a significant role in tanning industry, both in preventing generation of wastes and also in effective treatment of wastes. Un-hairing and degreasing can be done with help of enzymes, avoiding chemicals like sulphides, alkylphenol ethoxylates, etc. The use of enzymes can cut down processes like bating and the hide structure will remain least disturbed. The use of fat-digesting enzymes for degreasing is beneficial as it can eliminate use of organic solvents and surfactants. This helps for recovery of proteins and fats from wastes as by-products. Fungi can be used for leaching out chromium from tannery effluents, and to remove toxic tannins present in tannery effluents (Table 23.3).

Table 23.3. Biotechnological methods for removing of toxic tannis from tannery effluents.

Process	Earlier method	Biotechnological option
Soaking		
Washing of dirt, removal of excess fats from skin, the rehydration and swelling of skin layers is achieved.	Sulphides, alkalies, surfactants are used	Alkaline proteases or pancreatic trypsins (also have amylase and lipase) are used to improve water uptake and for removal of fats, gums, dirt.
Dehairing		
Removal of hair and soluble debris to get clean hides and skin	Skins are soaked in 10–15% hydrated lime, inorganic sulphides, inorganic amines	Use of selective proteases allows recovery of hairs by proper removal of hair
Dewooling	Coarse method used is to apply paints (contain lime, sodium chlorite and enzymes) to the flesh side and storing at 25–30°C overnight	Alkaline protease is used
Scouring		
Removal of wool grease and gums to clean the wool and aid in bleaching		Lipases from fungi can be used.
Bating		
The objective is to delime and deswell the collagen of skins and to degrade protein fibres partially to make soft, supple leather	Earlier animal feaces was used (unpleasant and unreliable)	Pancreatic trypsins are used now for bating of leather.

Textile Industry

Textile industry is single largest foreign-exchange earner in India. Biotechnology can help to prevent pollution in this industry in processing of cotton, silk and wool. The industry can be considerably benefited by adopting such ecofriendly processes (Table 23.4).

Table 23.4. Ecofriendly biotechnological process for removing textile effluents.

Process	Earlier method	Biotechnological option
Desizing		
Starch size is used to strengthen the fabric so that it withstands further weaving process. But this size needs to be removed afterwards since sized cloth shows less uptake of dyes, bleaches or texturising chemicals.	Oxidising agents or sodium hydroxide are used but damage the fabric	Superior thermostable and resistant (to chemicals) bacterial amylases are used to degrade starch size.
Scouring & bleaching		
To remove pectins, waxes, colour, residual seed coatings, honeydew sugars and insect secretions which cause stickiness and severe processing problems.		Xylanase enzyme can be used. Commercialisation still not done
Bleaching	Chlorine or hydrogen peroxide is used as bleaching agents. Series of washes are required (between primary bleaching and dyeing) to remove remaining H_2O_2	Enzyme catalyse with wide pH optima is used to degrade excess hydrogen peroxide. Water consumption can be reduced.
Biostoning		
To achieve dye fading, contrast and broken fibres as per fashion	Pumice stone was used. Extreme fading, abrasion more machine wear, more cost of maintenance are the problems	Cellulase enzyme is used for uniform and quality finishing of fabric (100% use by Denim garment processors).
Biopolishing		
To remove fine surface fuzz (protruding fine fibres) and fibrils (which give rough look and stiff touch) from cotton and viscose fabrics		Cellulase enzyme is used to remove fuzz from fabric and to give it a permanent softening effect.
Carbonising		
'Dhoties' are made by weaving polyester and cotton threads. Cotton fibres are then destroyed. This gives light net like effect to the fabric	Earlier 65% sulphuric acid was used	New cellulase enzyme 'softzyme' is used as a replacement to acid treatment.

(Contd...)

Process	Earlier method	Biotechnological option
Silk degumming		
To remove impurities like, wax gums and sericin, a protein substance that covers the silk fibre or fibroin. These impurities made silk coarse and without luster	Conventionally done by use of alkaline soaps	Enzymatic (Degummase) degumming removes sericin (proteolytic action) without harm to fibroin. Better quality silk with luster, treatment at lower temperature, no use of soaps are the advantages.
Jute retting	Conventional retting with microbial attack on jute was uncontrolled process taking days or even weeks	Rettizyme does controlled and reproducible retting in less than 24 hours. Enzyme solution can be recycled.
Wool finishing		
To get increased comfort (reduced prickle, greater softness) and improved surface appearance and pilling performance		Protease enzymes are used.

Pesticide Industry

The tremendous diversity in chemistry of pesticides makes their detoxification process a difficult task. Manufacturing pesticides that are less persistent and more prone to biodegradation, manufacturing and using biopesticides which will have specificity of action and minimal environmental- or bio-hazard, and ultimately aiming for resistance within the crops by use of genetic engineering are all part of clean technology programmes. But till we get success in the prevention of pollution, treatment technologies to efficiently eliminate pollutants needs to be seriously examined. Pesticide industry waste-waters, residual pesticides in fields, and contaminated ground water are required to be decontaminated of pesticides and their intermediates. Microorganisms possessing manipulated genes or enzymes with specific degradative capacity may be used for this purpose. There are various reports of use of enzymes like esterase, phosphatase, alkyl-sulphatase, oxygenase, hydrolase for detoxification of pesticides. Organisms like Pseudomonas putida, Candida tropicalis, Fusarium flocciferum, Aureoasidium pullutants, Aspergillus niger can degrade herbicides of chlorobenzoate class. Though applications are limited today, the potential is proven.

Desulphurisation of Coal and Petroleum Products

Coal has a high sulphur content, compared to petroleum and generates environmentally polluting SO_2 when it is burned. For this reason, the SO_2 must be removed using flue gas desulphurisation equipment after the coal is burned. The expenses involved in flue gas desulphurisation, however, are quite high, so economically developing countries often do not install flue gas desulphurisation equipment for economic reasons. The result is that environmental problems caused by SO_2 emissions are growing more and more severe. The sulphur content contained in coal can be divided into two categories: organic sulphur that is bonded with the carbon atoms and inorganic sulphur that exists in mineral form. Most of the inorganic sulphur contained in coal is pyrite (FeS_2). Desulphurisation using microbes is used for inorganic sulphur can be done. In conventional microbial desulphurisation process (leaching process) iron bacteria known to dissolve pyrite, from trivalent ferrous ions to sulphuric acid, by oxidation. In the leaching process, desulphurisation is thought to proceed by direction oxidation, by which the ferrous oxidation bacteria works directly on the iron pyrite, and by indirect oxidation, which is the chemical action caused by the microorganisms generated by direct oxidation. This leaching process removes the sulphur from coal by using iron bacteria to dissolve the pyrite content, which requires as long as one or two weeks to be completed. For this reason, the leaching method can be applied where large amounts of land are available and the coal can be amply stored. This is not an appropriate process when desulphurisation is needed quickly. New microbial desulphurisation process (microbial floatation separation method) is developed. In this method pulverised coal and liquid fuel called CWD (coal with water) is treated with iron bacteria. Simultaneous separation of coal and removal of pyrites (FeS_2) by attack of bacteria is brought about.

Antibiotics

Pharmaceuticals manufacturing particularly faces the problems of selective conversions, waste generation, and energy usage. Clean technologies can be provided in this sector through biocatalysis. The molecules manufactured are polyfunctional and often unstable. Presence of many functional groups requires protection, deprotection and activation steps.

Although cephalosporins are better antibiotics than the penicillins, their global market is limited, because the processes for producing the intermediates are both complex, costly and environmentally unfriendly. The development of bio-processes for production of cephalosporin intermediates will open the door to environmentally friendly production processes for cephalosporins. Hoechst scientists have developed an enzymatic process for the manufacture of cephalosporin antibiotic intermediate, cephalosporic acid. This has cut down production wastes from 20 tonnes/year to about 2 tonnes/year and the wastes are not toxic.

In the field of antibiotics, enzyme penicillin acylase which today is used to produce semisynthetic penicillin may also be used, for making some other antibiotic production, more safe and environmental friendly.

CHAPTER 24

Environmental Biochemistry

INTRODUCTION

The effects of pollutants and harmful chemicals on living organisms are particularly important in environmental chemistry. These effects are addressed under the topic of *"Toxicological Chemistry"* in Chapter 25 and *"Toxicological Chemistry of Chemical Substances"* in Chapter 26. The current chapter, *"Environmental Biochemistry"*, is designed to provide the fundamental background in biochemistry required to understand toxicological chemistry.

Most people have had the experience of looking through a microscope at a single cell. It may have been an amoeba, alive and oozing about like a blob of jelly on the microscope slide, or a cell of bacteria, stained with a dye to make it show up more clearly. Or, it may have been a beautiful cell of algae with its bright green chlorophyll. Even the simplest of these cells is capable of carrying out a thousand or more chemical reactions. These life processes fall under the heading of *biochemistry*, that branch of chemistry that deals with the chemical properties, composition, and biologically-mediated processes of complex substances in living systems.

Biochemical phenomena that occur in living organisms are extremely sophisticated. In the human body complex metabolic processes break down a variety of food materials to simpler chemicals, yielding energy and the raw materials to build body constituents, such as muscle, blood, and brain tissue. Impressive as this may be, consider a humble microscopic cell of cyanobacteria only about a micrometer in size, which requires only a few simple inorganic chemicals and sunlight for its existence. This cell uses sunlight energy to convert carbon from CO_2, hydrogen and oxygen from H_2O, nitrogen from NO_3^-, sulphur from SO_4^{2-}, and phosphorus from inorganic phosphate into all the proteins, nucleic acids, carbohydrates, and other materials that it requires to exist and reproduce. Such a simple cell accomplishes what could not be done by humans in even a vast chemical factory costing billions of dollars.

Ultimately, most environmental pollutants and hazardous substances are of concern because of their effects upon living organisms. The study of the adverse effects of substances on life processes requires some basic knowledge of biochemistry. Biochemistry is discussed in this chapter, with emphasis upon aspects that are especially pertinent to environmentally hazardous and toxic substances, including cell membranes, DNA, and enzymes.

Biochemical processes not only are profoundly influenced by chemical species in the environment, they largely determine the nature of these species, their degradation, and even their syntheses, particularly in the aquatic and soil environments. The study of such phenomena forms the basis of *environmental biochemistry*.

BIOMOLECULES

The biomolecules that constitute matter in living organisms are often polymers with molecular masses of the order of a million or even larger. As discussed later in this chapter, these biomolecules may be divided into the categories of carbohydrates, proteins, lipids, and nucleic acids. Proteins and nucleic acids consist of macro-molecules, lipids are usually relatively small molecules, carbohydrates range from relatively small sugar molecules to high molar mass macromolecules, such as those in cellulose.

The behaviour of a substance in a biological system depends to a large extent upon whether the substance is hydrophilic ("water-loving") or hydrophobic ("water-hating"). Some important toxic substances are hydrophobic a characteristic that enables them to traverse cell membranes readily. Part of the detoxification process carried on by living organisms is to render such molecules hydrophilic, therefore water-soluble are readily eliminated from the body.

BIOCHEMISTRY AND THE CELL

The focal point of biochemistry and biochemical aspects of toxicants is the cell, the basic building block of living systems where most life processes are carried out. Bacteria, yeasts, and some algae consist of single cells. However, most living things are made up of many cells. In a more complicated organism the cells have different functions. Liver cells, muscle cells, brain cells, and skin cells in the human body are quite different from each other and do different things. Cells are divided into two major categories depending upon whether or not they have a nucleus: eukaryotic cells have a nucleus and *prokaryotic* cells do not. Prokaryotic cells are found predominately in single-celled organisms such as bacteria. Eukaryotic cells occur in multicellular plants and animals—higher life forms.

Major Cell Features

Fig. 24.1 shows the major features of the eukaryotic cell, which is the basic structure in which biochemical processes occur in multicellular organisms. These features are the following :

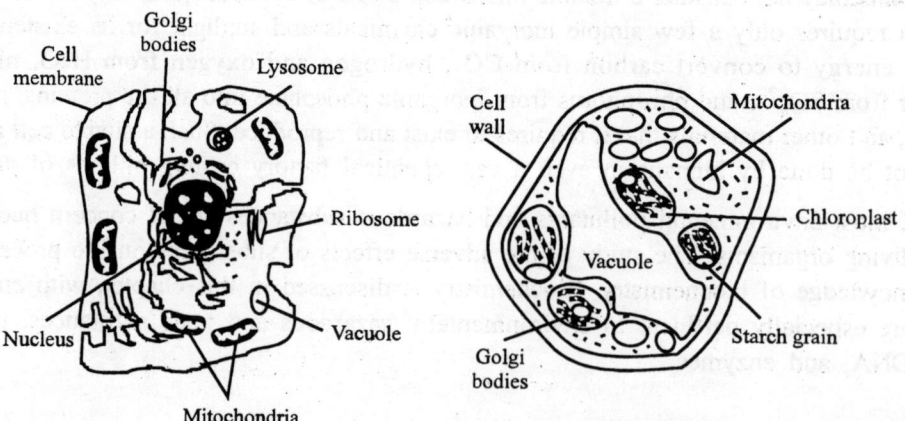

Fig. 24.1. Some major features of the eukaryotic cell in animals (left) and plants (right).

1. *Cell membrane,* which encloses the cell and regulates the passage of ions, nutrients, lipid-soluble ("fat-soluble") substances, metabolic products, toxicants, and toxicant metabolites into and out of the cell interior because of its varying permeability for different substances. The cell membrane protects the contents of the cell from undesirable outside influences. Cell membranes are composed in part of phospholipids that are arranged with their hydrophilic ("water-seeking") heads on the cell membrane surfaces and their hydrophobic ("water-repelling") tails inside the membrane. Cell membranes contain bodies of proteins that are involved in the transport of some substances through the membrane. One reason the cell membrane is very important in toxicology and environmental biochemistry is because it regulates the passage of toxicants and their products into and out of the cell interior. Furthermore, when its membrane is damaged by toxic substances, a cell may not function properly and the organism may be harmed.

2. *Cell nucleus,* which acts as a sort of "control centre" of the cell. It contains the genetic directions the cell needs to reproduce itself. The key substance in the nucleus is deoxyribonucleic acid (DNA). Chromosomes in the cell nucleus are made up of combinations of DNA and proteins. Each chromosome stores a separate quantity of genetic information. Human cells contain 46 chromosomes. When DNA in the nucleus is damaged by foreign substances, various toxic effects, including mutations, cancer, birth defects, and defective immune system function may occur.

3. *Cytoplasm,* which fills the interior of the cell not occupied by the nucleus. Cytoplasm is further divided into a water-soluble proteinaceous filler called cytosol, in which are suspended bodies called cellular organelles, such as mitochondria or, in photosynthetic organisms, chloroplasts.

4. *Mitochondria,* "powerhouses" which mediate energy conversion and utilisation in the cell. Mitochondria are sites in which food materials—carbohydrates, proteins, and fats – are broken down to yield carbon dioxide, water, and energy, which is then used by the cell. The best example of this is the oxidation of the sugar glucose, $C_6H_{12}O_6$:

$$C_6H_{12}O_6 + 6O_2 \rightarrow 6CO_2 + 6H_2O + \text{energy}$$

5. This kind of process is called *cellular respiration.*

6. *Ribosomes,* which participate in protein synthesis.

7. *Endoplasmic reticulum,* which is involved in the metabolism of some toxicants by enzymatic processes.

8. *Lysosome,* a type of organelle that contains potent substances capable of digesting liquid food material. Such material enters the cell through a "dent" in the cell wall, which eventually becomes surrounded by cell material. This surrounded material is called a food vacuole. The vacuole merges with a lysosome, and the substances in the lysosome bring about digestion of the *food material.* The digestion process consists largely of hydrolysis reactions in which large, complicated food molecules are broken down into smaller units by the addition of water.

9. *Golgi bodies,* that occur in some types of cells. These are flattened bodies of material that serve to hold and release substances produced by the cells.

10. *Cells walls* of plant cells. These are strong structures that provide stiffness and strength. Cell walls are composed mostly of cellulose.

11. *Vacuoles* inside plant cells that often contain materials dissolved in water.

12. *Chloroplasts* in plant cells that are involved in photosynthesis (the chemical process which uses energy from sunlight to convert carbon dioxide and water to organic matter). Photosynthesis occurs in these bodies. Food produced by photosynthesis is stored in the chloroplasts in the form of *starch grains.*

PROTEIN

A complex high polymer containing carbon, hydrogen, oxygen, nitrogen, and usually sulphur, and comprised of chains of amino acids connected by peptide linkages (—CO.NH—). Proteins occur in the cells of all living organisms and in biological fluids (blood plasma, protoplasm). They are synthesised by plants largely because of the nitrogen-fixing ability of certain soil bacteria. Their molecular weight may be as high as 40 million (tobacco mosaic virus). They have many important functional forms: enzymes, haemoglobin, hormones, viruses, genes, antibodies, and nucleic acids. They also comprise the basic component of connective tissue (collagen), hair (keratin), nails, feathers, skin, etc. Some have been synthesised in the laboratory.

The sequence of amino acids in the polypeptide chain is of critical importance in genetics. Proteins can be hydrolysed to their constituent amino acids and can be broken down into simpler forms by proteolytic enzymes. They form colloidal solutions, and behave chemically as both acids and bases simultaneously. They are denatured by changes in pH, by heat, ultraviolet radiation, and many organic solvents.

Simple proteins contain only amino acids; conjugated proteins contain amino acids plus nucleic acids, carbohydrates, lipids, etc. On the basis of solubility they can be classified as albumins (water-soluble), globulins (insoluble in water but soluble in aqueous salt solutions), and prolamines (soluble in alcohol-water mixtures but not in alcohol or water alone). A number of proteins have been synthesised, notably the hormone insulin. Proteins are an essential component of the diet, occurring chiefly in meat, eggs, milk and fish. Edible proteins suitable for human food as well as cattle feed can be produced from micro-organisms grown in carbonaceous or nitrogenous media to form yeast-like materials. Paraffinic hydrocarbons (methane) and petroleum-derived ethyl alcohol can be used as growth media for protein biosynthesis.

Industrial applications of proteins include plastics, adhesives and fibres derived from casein and soyabean protein, but these have been declining in recent years. Special forms in which proteins are commercially available include textured proteins for food products, protein hydrolyzate and liquid predigested protein, both for medical use.

The order of amino acids in protein molecules, and the resulting three-dimensional structures that form, provide an enormous variety of possibilities for *protein structure*. This is what makes life so diverse. Proteins have primary, secondary, tertiary, and quaternary structures. The structures of protein molecules determine the behaviour of proteins in crucial areas such as the processes by which the body's immune system recognises substances that are foreign to the body. Proteinaceous enzymes depend upon their structures for the very specific functions of the enzymes.

The order of amino acids in the protein molecule determines its primary structure. *Secondary protein* structures result from the folding of polypeptide protein chains to produce a maximum number of hydrogen bonds between peptide linkages.

Tertiary structures are formed by the twisting of alpha-helics into specific shapes. They are produced and held in place by the interactions of amino side chains on the amino acid residues constituting the protein macro-molecules. Tertiary protein structure is very important in the processes by which enzymes identify specific proteins and other molecules upon which they act. It is also involved with the action of antibodies in blood which recognise foreign proteins by their shape and react to them. This is basically what happens in the case of immunity to a disease where antibodies in blood recognise specific proteins from viruses or bacteria and reject them.

Two or more protein molecules consisting of separate polypeptide chains may be further attracted to each other to produce a *quaternary structure*.

Some proteins are *fibrous proteins*, which occur in skin, hair, wool, feathers, silk, and tendons. The molecules in these proteins are long and threadlike and are laid out parallel in bundles. Fibrous proteins are quite tough and they do not dissolve in water.

Beside from fibrous protein, the other major type of protein form is the globular protein. These proteins are in the shape of balls and oblongs. Globular proteins are relatively soluble in water. A typical globular protein is haemoglobin, the oxygen-carrying protein in red blood cells. Enzymes are generally globular proteins.

Denaturation of Proteins

Secondary, tertiary, and quaternary protein structures are easily changed by a process called *denaturation*. These changes can be quite damaging. Heating, exposure to acids or bases, and even violent physical action can cause denaturation to occur The albumin protein in egg white is denatured by heating so that it forms a semisolid mass. Almost the same thing is accomplished by the violent physical action of an egg beater in the preparation of meringue. Heavy metal poisons such as lead and cadmium change the structures of proteins by binding to functional groups on the protein surface.

CARBOHYDRATES

A compound of carbon, hydrogen and oxygen that contains the saccharose grouping or its first reaction product, and in which the ratio of hydrogen to oxygen is the same as in water. Carbohdyrates are the most abundant class of organic compounds, constituting three-fourths of the dry weight of all vegetation. They are also widely distributed in animals and lower forms of life. They comprise (i) *monosaccharides:* simple sugars, such as fructose (levulose) and its isomer glucose (dextrose), both having the formula $C_6H_{12}O_6$; (ii) *disaccharides:* sucrose ($C_{12}H_{22}O_{11}$), maltose, cellobiose, and lactose; and polysaccharides (high polymeric substances). The last group includes the entire starch and cellulose families, as well as pectin, the seaweed products agar and carrageenan, and natural gums. The simple sugars are crystalline and water-soluble, with a sweet taste; starches are water soluble, tasteless and amorphous; cellulose is insoluble in water and organic solvents, and is only partially crystalline. Galactose, sorbose, xylose, arabinose, and mannose are constituents of more complex sugars. The natural gums are water-soluble plant products composed of monosaccharide units joined by glycosidic bonds (arabic, tragacanth). Carbohydrates are an important natural source of ethyl alcohol, now in extensive use in gasohol and other energy applications.

LIPIDS

Lipids are substances that can be extracted from plant or animal matter by organic solvents, such as chloroform, diethyl ether, or toluene (Fig. 24.2). Whereas carbohydrates and proteins are characterised predominately by the monomers (monosaccharides and amino acids) from which they are composed, lipids are defined essentially by their physical characteristic of organophilicity. The most common lipids are fats and oils composed of triglycerides formed from the alcohol glycerol, $CH_2(OH)CH(OH)CH_2OH$

and a long-chain fatty acid such as stearic acid, $CH_3(CH_2)_{16}COOH$. Numerous other biological materials, including waxes, cholesterol, and some vitamins and hormones, are classified as lipids. Common foods, such as butter and salad oils are lipids. The longer chain fatty acids, such as stearic acid, are also organic-soluble and are classified as lipids.

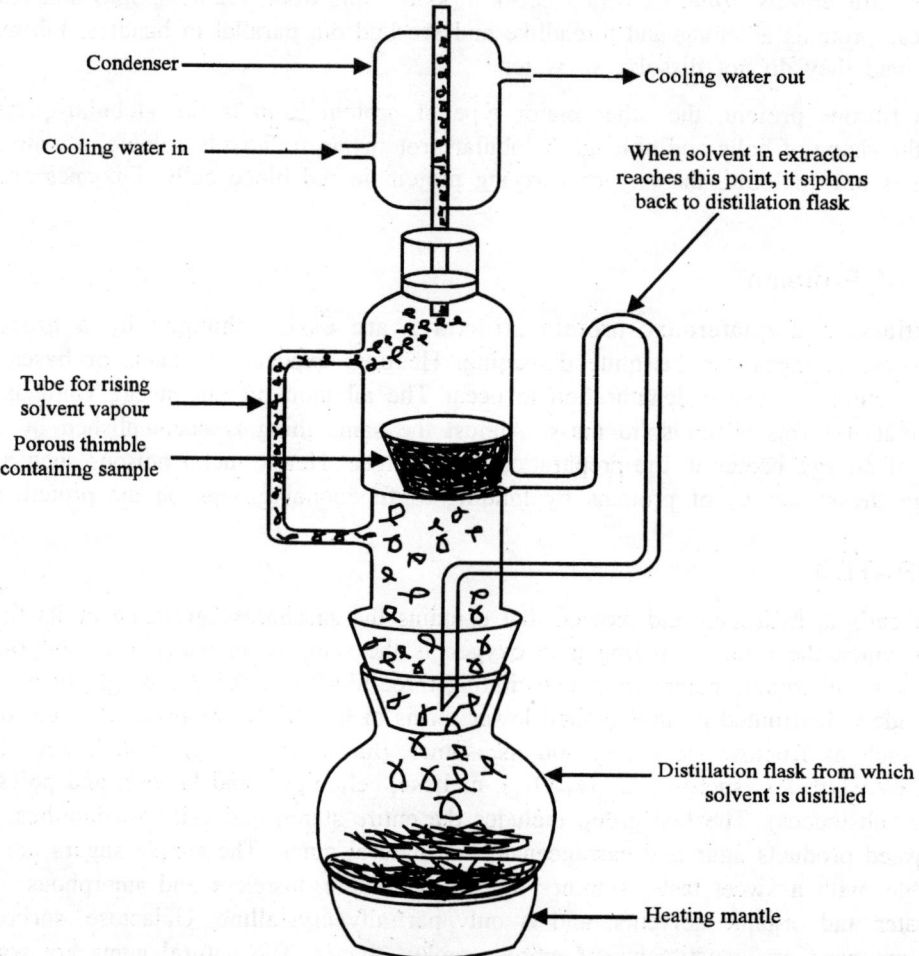

Fig. 24.2. Lipids are extracted from some biological materials with a Soxhlet extractor. The solvent is vapourised in the distillation flask by the heating mantle, rises through one of the exterior tubes to the condenser, and is cooled to form a liquid. The liquid drops onto the porous thimble containing the sample. Siphon action periodically drains the solvent back into the distillation flask. The extracted lipid collects as a solution in the solvent in the flask.

Lipids are toxicologically important for several reasons. Some toxic substances interfere with lipid metabolism, leading to detrimental accumulation of lipids. Many toxic organic compounds are poorly soluble in water, but are lipid-soluble, so that bodies of lipids in organisms serve to dissolve and store toxicants.

An important class of lipids consists of *phosphoglycerides* (glycerophosphatides). These compounds may be regarded as triglycerides in which one of the acids bonded to glycerol is orthophosphoric acid. These lipids are especially important because they are essential constituents of cell membranes. These membranes consist of bilayers in which the hydrophilic phosphate ends of the molecules are on the outside of the membrane and the hydrophobic "tails" of the molecules are on the inside. Waxes are also esters of fatty acids. However, the alcohol in a wax is not glycerol, but is often a very long chain alcohol.

Waxes are produced by both plants and animals, largely as protective coatings. Waxes are found in a number of common products. Lanolin is one of these. It is the "grease" in sheep's wool. When mixed with oils and water, it forms stable colloidal emulsions consisting of extremely small oil droplets suspended in water. This makes lanolin useful for skin creams and pharmaceutical ointments. Carnauba wax occurs as a coating on the leaves of some Brazillian palm trees.

Steroids are lipids found in living systems which all have the ring system for cholesterol. Steroids occur in the bile salts, which are produced by the liver and then secreted into the intestines. Their breakdown products give feces its characteristic colour. Bile salts act upon fats in the intestine. They suspend very tiny fat droplets in the form of colloidal emulsions. This enables the fats to be broken down chemically and digested.

Some steroids are *hormones*. Hormones act as "messengers" from one part of the body to another. As such, they start and stop a number of body functions. Male and female sex hormones are examples of steroid hormones.

ENZYMES

Catalysts are substances that speed up a chemical reaction without themselves being consumed in the reaction. The most sophisticated catalysts of all are those found in living systems. They bring about reactions that could not be performed at all, or only with great difficulty, outside a living organism. These catalysts are called *enzymes*. In addition to speeding up reactions by as much as ten- to a hundred million-fold, enzymes are extremely selective in the reactions which they promote.

Enzymes are proteinaceous substances with highly specific structures that interact with particular substances or classes of substances called *substrates*. Enzymes act as catalysts to enable biochemical reactions to occur, after which they are regenerated intact to take part in additional reactions. The extremely high specificity with which enzymes interact with substrates results from their "lock and key" action based upon the unique shapes of enzymes as illustrated in Fig. 24.3.

This illustration shows that an enzyme "recognises" a particular substrate by its molecular structure and binds to it to produce an *enzyme-substrate complex*. This complex then breaks apart to form one or more products different from the original substrate, regenerating the unchanged enzyme, which is then available to catalyse additional reactions. The basic process for an enzyme reaction is, therefore,

$$\text{enzyme} + \text{substrate} \rightleftharpoons \text{enzyme-substrate complex} \rightleftharpoons \text{enzyme} + \text{product} \qquad ...(24.1)$$

Several important things should be noted about this reaction. As shown in Fig. 24.3, an enzyme acts on a specific substrate to form an enzyme-substrate complex because of the fit between their structures. As a result, something happens to the substrate molecule. For example, it might be split in two at a particular location. Then the enzyme-substrate complex comes apart, yielding the enzyme and products. The enzyme is not changed in the reaction and is now free to react again. Note that the arrows in the formula for enzyme reaction point both ways. This means that the reaction is reversible. An enzyme-substrate complex can simply go back to the enzyme and the substrate. The products of an enzymatic reaction can react with the enzyme to form the enzyme-substrate complex again. It, in turn, may again form the enzyme and the substrate. Therefore, the same enzyme may act to cause a reaction to go either way.

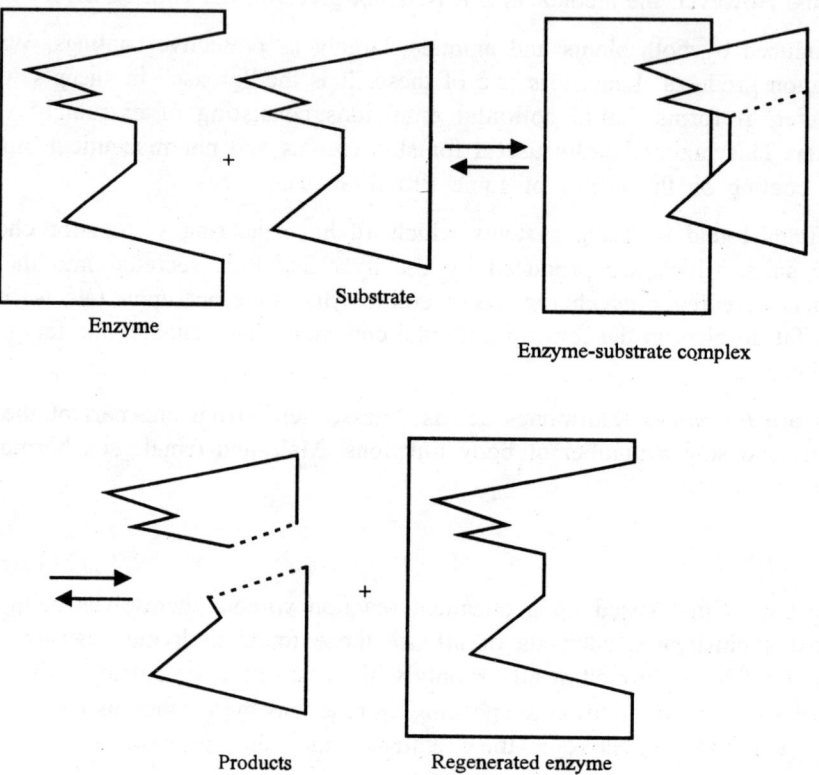

Fig. 24.3. Representation of the "lock-and-key" mode of enzyme action which enables the very high specificity of enzyme-catalysed reactions.

Some enzymes cannot function by themselves. In order to work, they must first be attached to *coenzymes*. Coenzymes normally are not protein materials. Some of the vitamins are important coenzymes.

Enzymes are named for what they do. For example, the enzyme given off by the stomach, which splits proteins as part of the digestion process, is called *gastric proteinase*. The "gastric" part of the name refers to the enzyme's origin in the stomach. The "proteinase" denotes that it splits up protein molecules. The common name for this enzyme is pepsin. Similarly, the enzyme produced by the pancreas that breaks down fats (lipids) is called *pancreatic lipase*. Its common name is steapsin. In general, lipase enzymes cause lipid triglycerides to dissociate and form glycerol and fatty acids.

The enzymes mentioned above are *hydrolysing enzymes*, which bring about the breakdown of high-molecular-weight biological compounds by the addition of water. This is one of the most important types of the reactions involved in digestion. The three main classes of energy-yielding foods that animals eat are carbohydrates, proteins, and fats. Recall that the higher carbohydrates humans eat are largely disaccharides (sucrose, or table sugar) and polysaccharides (starch). These are formed by the joining together of units of simple sugars, $C_6H_{12}O_6$, with the elimination of an H_2O molecule at the linkage where they join. Proteins are formed by the condensation of amino acids, again with the elimination of a water molecule at each linkage. Fats are esters which are produced when glycerol and fatty acids link together. A water molecule is lost for each of these linkages when a protein, fat, or carbohydrate are synthesised. In order for these substances to be used as a food source, the reverse process must occur to break down large, complicated molecules of protein, fat, or carbohydrate to simple, soluble substances which can penetrate a cell membrane and take part in chemical processes in the cell. This reverse process is accomplished by hydrolysing enzymes.

Enzyme action may be affected by many different things. Enzymes require a certain hydrogen ion concentration (pH) to function best. For example, gastric proteinase requires the acid environment of the stomach to work well. When it passes into the much less acidic intestines, it stops working. This prevents damage to the intestine walls, which would occur if the enzyme tried to digest them. Temperature is critical. Not surprisingly, the enzymes in the human body work best at around 98.6°F (37°C), which is the normal body temperature. Heating these enzymes to around 140°F permanently destroys them. Some bacteria that thrive in hot springs have enzymes which work best at relatively high temperatures. Other "cold-seeking" bacteria have enzymes adapted to near the freezing point of water.

One of the greatest concerns regarding the effects of surroundings upon enzymes is the influence of toxic substances. A major mechanism of toxicity is the alteration or destruction of enzymes by toxic agents – cyanide, heavy metals, or organic compounds, such as insecticidal parathion. An enzyme that has been destroyed obviously cannot perform its designated function, whereas one that has been altered may either not function at all or may act improperly. The detrimental effects of toxicants on enzymes are discussed in more detail in *Toxicological Chemistry* (Chapter 25).

NUCLEIC ACIDS

Any of several complex compounds occurring in living cells, usually chemically bound to proteins to form nucleoproteins. Nucleic acids are of high molecular weight and are easily changed by many mild chemical reagents. They contain carbon, hydrogen, oxygen, nitrogen (15–16%), and phosphorus (9–10%).

The fundamental units of nucleic acid are nucleotides; nucleic acids are polynucleotides in which the nucleotides are linked by phosphate bridges. Upon extensive heating in the presence of water (hydrolysis), nucleic acids yield a mixture of purines and pyrimidines, D-ribose or D-deoxyribose, and phosphoric acid. Nucleic acids are subdivided into two types: ribonucleic acid (RNA), containing the sugar D-ribose; and deoxyribonucleic acid (DNA) containing the sugar D-deoxyribose. Good sources of nucleic acids are salmon, thymus, yeasts, and wheat kernel embryo.

The molecule of DNA is sort of like a coded message. This "message", the genetic information contained in, and transmitted by nucleic acids, depends upon the sequence of bases from which they are composed. It is somewhat like the message sent by telegraph, which consists only of dots, dashes, and spaces in between. The key aspect of DNA structure that enables storage and replication of this information is the famed double helix structure of DNA mentioned above.

Nucleic Acids in Protein Synthesis

Whenever a new cell is formed, the DNA in its nucleus must be accurately reproduced from the parent cell. Life processes are absolutely dependent upon accurate protein synthesis as regulated by cell DNA. The DNA in a single cell must be capable of directing the synthesis of up to 3000 or even more different proteins. The directions for the synthesis of a single protein are contained in a segment of DNA called a *gene*. The process of transmitting information from DNA to a newly-synthesised protein involves the following steps :

1. The DNA undergoes *replication*. This process involves separation of a segment of the double helix into separate single strands which then replicate such that guanine is opposite cytosine (and vice versa) and adenine is opposite thymine (and vice versa). This process continues until a complete copy of the DNA molecule has been produced.

2. The newly replicated DNA produces *messenger RNA (m-RNA)*, a complement of the single strand of DNA, by a process called transcription.

3. A new protein is synthesised using *m-RNA* as a template to determine the order of amino acids in a process called *translation*.

Modified DNA

DNA molecules may be modified by the unintentional addition or deletion of nucleotides or by substituting one nucleotide for another. The result is a *mutation* that is transmittable to offspring. Mutations can be induced by chemical substances. This is a major concern from a toxicological viewpoint because of the detrimental effects of many mutations and because substances that cause mutations often cause cancer as well. DNA malfunction may also result in birth defects. The failure to control cell reproduction results in cancer. Radiation from X rays and radioactivity also disrupts DNA and may cause mutation.

RECOMBINANT DNA AND GENETIC ENGINEERING

As noted above, segments of DNA contain information for the specific synthesis of particular proteins. Within the last two decades it has become possible to transfer this information between organisms by means of *recombinant DNA technology*, which has resulted in new industry based on *genetic engineering*. Most often the recipient organisms are bacteria, which can be reproduced (cloned) over many orders of magnitude from a cell that has acquired the desired qualities. Therefore, to synthesise a particular substance, such as human insulin or growth hormone, the required genetic information can be transferred from a human source to bacterial cells, which then produce the substance as part of their metabolic processes.

The first step in recombinant DNA gene manipulation is to lyze, or "open up" a cell that has the genetic material needed and to remove this material from the cell. Through enzyme action the sought-after genes are cut from the donor DNA chain. These are next spliced into small DNA molecules. These molecules, called *cloning vehicles*, are capable of penetrating the host cell and becoming incorporated into its genetic material. The modified host cell is then reproduced many times and carries out the desired biosynthesis.

Early concerns about the potential of genetic engineering to produce "monster organisms" or new and horrible diseases have been largely allayed, although caution is still required with this technology. In the environmental area, genetic engineering offers significant prospects for the production of bacteria engineered to safely destroy troublesome wastes and to produce biological substitutes for environmentally damaging synthetic pesticides.

METABOLIC PROCESSES

Biochemical processes that involve the alteration of biomolecules fall under the category of metabolism. Metabolic processes may be divided into the two major categories of anabolism (synthesis) and catabolism (degradation of substances). An organism may use metabolic processes to yield energy or to modify the constituents of biomolecules.

Energy-Yielding Processes

Organisms can gain energy by the following three processes :

1. *Respiration* in which organic compounds undergo catabolism that requires molecular oxygen (aerobic respiration) or that occurs in the absence of molecular oxygen (anaerobic respiration). Aerobic respiration uses the Krebs cycle to obtain energy from the following reaction :

$$C_6H_{12}O_6 + 6O_2 \rightarrow 6CO_2 + 6H_2O + energy$$

About half of the energy released is converted to short-term stored chemical energy, particularly through the synthesis of adenosine triphosphate (ATP) nucleotide. For longer-term energy storage, glycogen or starch polysaccharides are synthesised, and for still longer-term energy storage lipids (fats) are generated and retained by the organism.

2. *Fermentation*, which differs from respiration is not having an electron transport chain. Yeasts produce ethanol from sugars by fermentation :

$$C_6H_{12}O_6 \rightarrow 2CO_2 + 2C_2H_5OH$$

3. *Photosynthesis* in which light energy captured by plant and algal chloroplasts is used to synthesise sugars from carbon dioxide and water :

$$6CO_2 + 6H_2O + h\nu \rightarrow C_6H_{12}O_6 + 6O_2$$

Plants cannot always get the energy that they need from sunlight. During the dark they must use stored food. Plant cells, like animal cells, contain mitochondria in which stored food is converted to energy by cellular respiration.

Plant cells, which use sunlight as a source of energy and CO_2 as a source of carbon, are said to be *autotrophic*. In contrast animal cells must depend upon organic material manufactured by plants for their food. These are called *heterotrophic cells*. They act as mediators of the chemical reaction between oxygen and food material using the energy from the reaction to carry out their life processes.

METABOLISM OF XENOBIOTIC COMPOUNDS

When toxicants or their metabolic precursors (protoxicants) enter a living organism they may undergo several processes as illustrated in Fig. 24.4. Emphasis is placed on *xenobiotic compounds*, which are those that are normally foreign to living organisms; on chemical aspects; and on processes that lead to products that can be eliminated from the organism. Of particular importance is *intermediary xenobiotic metabolism* which results in the formation of relatively short-lived transient species that are different from both those ingested and the ultimate product that is excreted. These species may have significant toxicological effects. Xenobiotic compounds in general are acted upon by enzymes that function on a material that is in the body naturally—an *endogenous substrate*. For example, flavin-containing mono-oxygenase enzyme acts upon endogenous cysteamine to convert it to cystamine, but also functions to oxidise endogenous nitrogen and sulphur compounds.

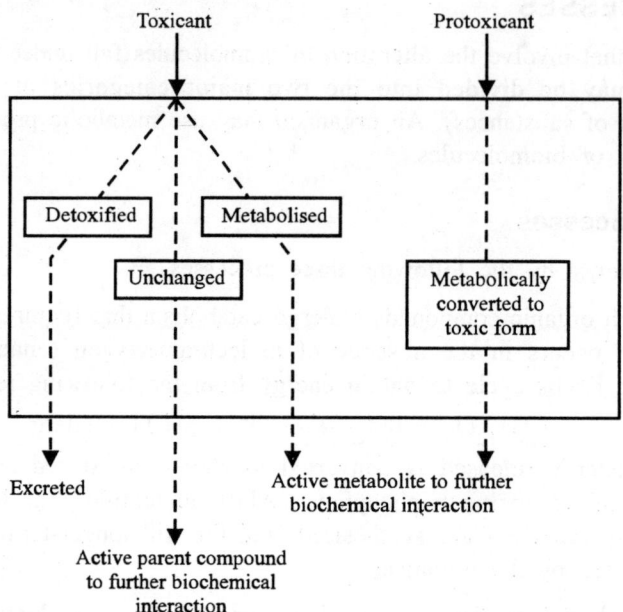

Fig. 24.4. Pathways of xenobiotic species prior to their undergoing any biochemical interactions that could lead to toxic effects.

Bio-transformation refers to changes in xenobiotic compounds as a result of enzyme action. Reaction that are not mediated by enzymes may also be important in some cases. As examples of non-enzymatic transformations, some xenobiotic compounds bond with endogenous biochemical species without an enzyme catalyst, undergo hydrolysis in body fluid media, or undergo oxidation reduction processes. The metabolic Phase I and Phase II reactions of xenobiotics discussed here are enzymatic.

The likelihood that a xenobiotic species will undergo enzymatic metabolism in the body depends upon the chemical nature of the species. Compounds with a high degree of polarity, such as relatively ionisable carboxylic acids, are less likely to enter the body system and, when they do, tend to be quickly excreted. Therefore, such compounds are unavailable, or only available for a short time, for enzymatic metabolism. Volatile compounds, such as dichloromethane or diethyl ether, are expelled so quickly from the lungs that enzymatic metabolism is less likely. This leaves as the most likely candidates for enzymatic metabolic reactions *non-polar lipophilic compounds*, those that are relatively less soluble in aqueous biological fluids and more attracted to lipid species. Of these, the ones that are resistant to enzymatic attack (PCBs, for example) tend to bioaccumulate in lipid tissue.

Xenobiotic species may be metabolised in a wide variety of body tissues and organs. As part of the body's defense against the entry of xenobiotic species, the most prominent sites of xenobiotic metabolism are those associated with entry into the body. The skin is one such organ, as is the lung. The gut wall through which xenobiotic species enter the body from the gastrointestinal tract is also a site of significant xenobiotic compound metabolism. The liver is of particular significance because materials entering systemic circulation from the gastrointestinal tract must first traverse the liver.

Phase I and Phase II Reactions

The processes that most xenobiotics undergo in the body can be divided into the two categories of phase I reactions and phase II reactions. A *phase I reaction* introduces reactive, polar functional groups into lipophilic ("fat-seeking") toxicant molecules. In their unmodified forms, such toxicant molecules tend to pass through lipid-containing cell membranes and may be bound to lipoproteins, in which form they are transported through the body. Because of the functional group present, the product of a phase I reaction is usually more water-soluble than the parent xenobiotic species, and more importantly, possesses a "chemical handle" to which a substrate material in the body may become attached so that the toxicant can be eliminated from the body. The binding of such a substrate is a *phase II reaction*, and it produces a *conjugation product* that is amenable to excretion from the body.

In general, the changes in structure and properties of a compound that result from a Phase I reaction are relatively mild. Phase II processes, however, usually produce species that are much different from the parent compounds. It should be emphasised that not all xenobiotic compounds undergo both phase I and phase II reactions. Such a compound may undergo only a phase I reaction and be excreted directly from the body. Or a compound that already possesses an appropriate functional group capable of conjugation may undergo a phase II reaction without a preceding phase I reaction.

CHAPTER 25

Toxicological Chemistry

INTRODUCTION

Ultimately, most pollutants and hazardous substances are of concern because of their toxic effects. The general aspects of these effects are addressed in this chapter under the heading of toxicological chemistry; the toxicological chemistry of specific classes of chemical substances is addressed in Chapter 26. In order to understand toxicological chemistry, it is essential to have some understanding of biochemistry, the science that deals with chemical processes and materials in living systems.

A poison, or toxicant, is a substance that is harmful to living organisms because of its detrimental effects on tissues, organs, or biological processes. Toxicology is the science of poisons. These definitions are subject to a number of qualifications Whether a substance is poisonous depends upon the type of organism exposed, the amount of the substance, and the route of exposure. In the case of human exposure, the degree of harm done by a poison can depend strongly upon whether the exposure is to the skin, by inhalation, or through ingestion.

Toxicants to which subjects are exposed in the environment or occupationally may be in several different physical forms. This may be illustrated for toxicants that are inhaled. Gases are substances such as carbon monoxide in air that are normally in the gaseous state under ambient conditions of temperature and pressure. Vapours are gas-phase materials that have evaporated or sublimed from liquids or solids. Dusts are respirable solid particles produced by grinding bulk solids, whereas fumes are solid particles from the condensation of vapours, often metals or metal oxides. Mists are liquid droplets.

Often a toxic substance is in solution or mixed with other substances. A substance with which the toxicant is associated (the solvent in which it is dissolved or the solid medium in which it is dispersed) is called the matrix. The matrix may have a strong effect upon the toxicity of the toxicant.

There are numerous variables related to the ways in which organisms are exposed to toxic substances. Another important factor is the toxicant concentration, which may range from the pure substance (100%) down to a very dilute solution of a highly potent poison. Both the duration of exposure per exposure incident and the frequency of exposure are important. The rate of exposure and the total time period over which the organism is exposed are both important situational variables. The exposure site and route also affect toxicity.

It is possible to classify exposures on the basis of acute vs. chronic and local vs. systemic exposure giving four general categories. Acute local exposure occurs at a specific location over a time period of a few seconds to a few hours and may affect the exposure site, particularly the skin, eyes, or mucous membranes. The same parts of the body can be affected by chronic local exposure, for which the time span may be as long as several years. Acute systemic exposure is a brief exposure or exposure to a single dose and occurs which toxicants that can enter the body, such as by inhalation or ingestion, and affect organs such as the liver that are remote from the entry site. Chronic systemic exposure differs in that the exposure occurs over a prolonged time period.

In discussing exposure sites for toxicants it is useful to consider the major routes and sites of exposure, distribution, and elimination of toxicants in the body as shown in Fig. 25.1. The major routes of accidental or intentional exposure to toxicants by humans and other animals are the skin (percutaneous route), the lungs (inhalation, respiration, pulmonary route), and the mouth (oral route); minor routes of exposure are rectal, vaginal, and parenteral (intravenous or intramuscular, a common means for the administration of drugs or toxic substances in test subjects). The way that a toxic substance is introduced into the complex system of an organism is strongly dependent upon the physical and chemical properties of the substance. The pulmonary system is most likely to take in toxic gases or very fine, respirable solid or liquid particles. In other than a respirable form, a solid usually enters the body orally. Absorption through the skin is most likely for liquids, solutes in solution, and semi-solids, such as sludges.

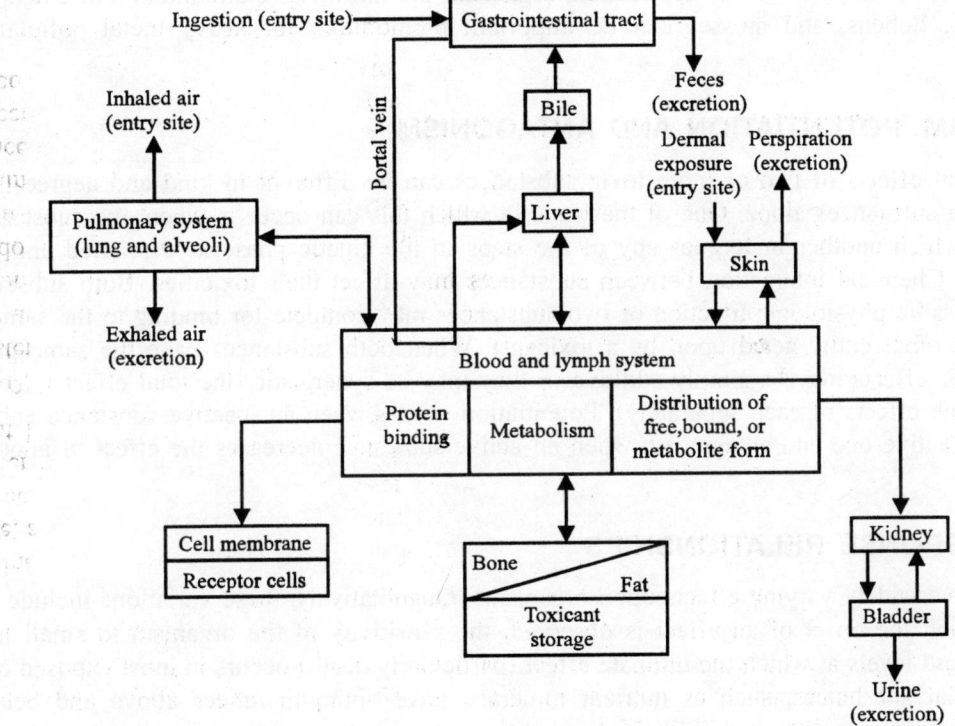

Fig. 25.1. Major sites of exposure, metabolism, and storage, routes of distribution and elimination of toxic substances in the body.

The defensive barriers that a toxicant may encounter vary with the route of exposure. For example, toxic elemental mercury is readily absorbed through the alveoli in the lungs much more readily than through the skin or gastrointestinal tract. Most test exposures to animals are through ingestion or gavage (introduction into the stomach through a tube). Pulmonary exposure is often favoured with subjects that may exhibit refractory behavior when noxious chemicals are administered by means requiring a degree of cooperation from the subject. Intravenous injection may be chosen for deliberate exposure when it is necessary to know the concentration and effect of a xenobiotic substance in the blood. However, pathways used experimentally that are almost certain not to be significant in accidental exposures can give misleading results when they avoid the body's natural defensive mechanisms.

An interesting historical example of the importance of the route of exposure to toxicants is provided by cancer caused by contact of coal tar with skin. The major barrier to dermal absorption of toxicants is the stratum corneum or horny layer. The permeability of skin is inversely proportional to the thickness of this layer, which varies by location on the body in the order soles and palms > abdomen, back, legs, arms > genital (perineal) area. Evidence of the susceptibility of the genital area to absorption of toxic substances is to be found in accounts of the high incidence of cancer of the scrotum among chimney sweeps. The cancer-causing agent is coal tar condensed in chimneys. This material was more readily absorbed through the skin in the genital areas than elsewhere leading to a high incidence of scrotal cancer. (The chimney sweeps' conditions were aggravated by their lack of appreciation of basic hygienic practices, such as bathing and regular changes of underclothing). Organisms can serve as indicators of various kinds of pollutants. In this application, organisms are known as biomonitors. For example, higher plants, fungi, lichens, and mosses can be important biomonitors for heavy metal pollutants in the environment.

SYNERGISM, POTENTIATION AND ANTAGONISM

The biological effects of two or more toxic substances can be different in kind and degree from those of one of the substances alone. One of the ways in which this can occur is when one substance affects the way in which another undergoes any of the steps in the kinetic phase as discussed and illustrated in Fig. 25.9. Chemical interaction between substances may affect their toxicities. Both substances may act upto the same physiologic function or two substances may compete for binding to the same receptor (molecule or other entity acted upon by a toxicant). When both substances have the same physiologic function, their effects may be simply additive or they may be synergistic (the total effect is greater than the sum of the effects of each separately). Potentiation occurs when an inactive substance enhances the action of an active one and antagonism when an active substance decreases the effect of another active one.

DOSE-RESPONSE RELATIONSHIPS

Toxicants have widely varying effects upon organisms. Quantitatively, these variations include minimum levels at which the onset of an effect is observed, the sensitivity of the organism to small increments of toxicant, and levels at which the ultimate effect (particularly death) occurs in most exposed organisms. Some essential substances, such as nutrient minerals, have optimum ranges above and below which detrimental effects are observed (Fig. 25.4).

Factors such as those just outlined are taken into account by the dose-response relationship, which is one of the key concepts of toxicology. Dose is the amount, usually per unit body mass, of a toxicant to which an organism is exposed. Response is the effect upon an organism resulting from exposure to a toxicant. In order to define a dose-response relationship, it is necessary to specify a particular response, such as death of the organism, as well as the conditions under which the response is obtained, such as the length of time from administration of the dose. Consider a specific response for a population of the same kinds of organisms. At relatively low doses, none of the organisms exhibits the response (for example, all live) whereas at higher doses, all of the organisms exhibit the response (for example, all die). In between, there is a range of doses over which some of the organisms respond in the specified manner and others do not, thereby defining a dose-response curve. Dose-response relationships differ among different kinds and strains of organisms, types of tissues, and populations of cells.

Fig. 25.2 shows a generalised dose-response curve. Such a plot may be obtained, for example, by administering different doses of a poison in a uniform manner to a homogeneous population of test animals and plotting the cumulative percentage of deaths as a function of the log of the dose. The dose corresponding to the mid-point (inflection point) of the resulting S-shaped curve is the statistical estimate of the dose that would kill 50 per cent of the subjects and is designatd as LD_{50}. The estimated doses at which 5 per cent (LD_5) and 95 per cent (LD_{95}) of the test subjects die are obtained from the graph by reading the dose levels for 5 per cent and 95 per cent fatalities, respectively. A relatively small difference between LD_5 and LD_{95} is reflected by a steeper S-shaped curve and vice versa. Statistically, 68 per cent of all values on a dose-response curve fall within $\pm$ 1 standard deviation of the mean at LD_{50} and encompass the range from LD_{16} to LD_{84}.

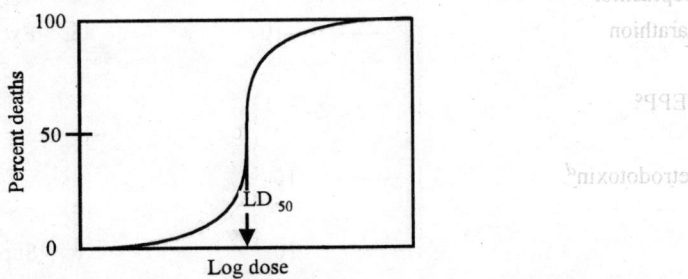

Fig. 25.2. Illustration of a dose-response curve in which the response is the death of the organism. The cumulative percentage of deaths of organisms is plotted on the Y axis.

RELATIVE TOXICITIES

Table 25.1. illustrates standard toxicity ratings that are used to describe estimated toxicities of various substances to humans. In terms of fatal doses to an adult human of average size, a "taste" of a supertoxic substance (just a few drops or less) is fatal. A teaspoonful of a very toxic substance could have the same effect. However, as much as a quart of a slightly toxic substance might be required to kill an adult human.

When there is a substantial difference between LD_{50} values of two different substances, the one with the lower value is said to be the more potent. Such a comparison must assume that the dose-response curves for the two substances being compared have similar slopes.

Non-lethal Effects

So far, toxicities have been described primarily in terms of the ultimate effect, that is, deaths of organisms, or lethality. This is obviously an irreversible consequence of exposure. In many, and perhaps most, cases, sub-lethal and reversible effects are of greater importance. This is obviously true of drugs, where death from exposure to a registered therapeutic agent is rare, but other effects, both detrimental and beneficial, are usually observed. By their very nature, drugs alter biological processes; therefore, the potential for harm is almost always present. The major considerations in establishing drug dose is to find a dose that has an adequate therapeutic effect without undesirable side effects. A dose-response curve can be established for a drug that progresses from non-effective levels through effective, harmful, and even lethal levels. A low slope for this curve indicates a wide range of effective dose and a wide margin of safety (Fig. 25.3). This term applies to other substances, such as pesticides, for which it is desirable to have a large difference between the dose that kills a target species and that which harms a desirable species.

Table 25.1. Toxicity scale with example substances[a].

Substance	Approximate LD_{50}	Toxicity rating
	-10^5	1. Practically non-toxic
DEHP[b] $\longrightarrow$	$-$	$> 1.5 \times 10^4$ mg/kg
Ethanol $\longrightarrow$	-10^4	2. Slightly toxic, 5×10^3
Sodium chloride $\longrightarrow$	$-$	1.5×10^4 mg/kg
Malathion $\longrightarrow$	-10^3	3. Moderately toxic,
Chlordane $\longrightarrow$	$-$	500–5000 mg/kg
Heptachlor $\longrightarrow$	-10^2	4. Very toxic, 50–500 mg/kg
Parathion $\longrightarrow$	-10	5. Extremely toxic,
	$-$	5–50 mg/kg
TEPP[c] $\longrightarrow$	-1	
	$-$	
Tetrodotoxin[d] $\longrightarrow$	10^{-1}	
	$-$	
	10^{-2}	6. Supertoxic, < 5 mg/kg
	$-$	
TCDD[e] $\longrightarrow$	-10^{-3}	
	$-$	
	-10^{-4}	
	$-$	
Botulinus toxin $\longrightarrow$	-10^{-5}	

[a] Doses are in units of mg of toxicant per kg of body mass. Toxicity ratings on the right are given as numbers ranging from 1 (practically non-toxic) through 6 (supertoxic) along with estimated lethal oral doses for humans in mg/kg. Estimated LD_{50} values for substances on the left have been measured in test animals, usually rats, and apply to oral doses.

[b] Bis(2-ethylhexyl)phthalate.

[c] Tetraethylpyrophosphate.

[d] Toxin from pufferfish.

[e] TCDD represents 2,3,7,8-tetrachlorodibenzodioxin, commonly called "dioxin".

REVERSIBILITY AND SENSITIVITY

Sub-lethal doses of most toxic substances are eventually eliminated from an organism's system. If there is no lasting effect from the exposure, it is said to be reversible. However, if the effect is permanent, it is termed irreversible. Irreversible effects of exposure remain after the toxic substance is eliminated from the organism. Fig. 25.3 illustrates these two kinds of effects. For various chemicals and different subjects, toxic effects may range from the totally reversible to the totally irreversible.

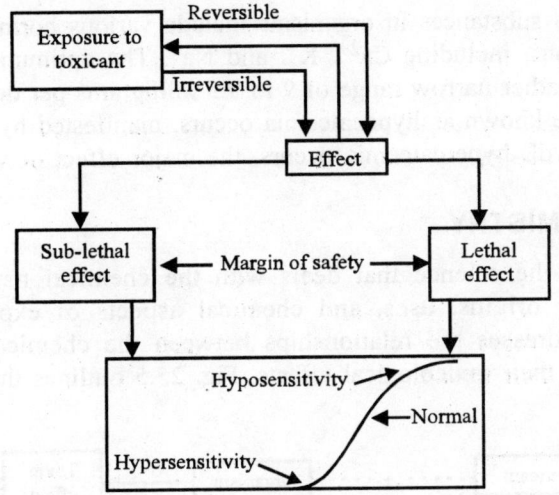

Fig. 25.3. Effects of and responses to toxic substances.

Hypersensitivity and Hyposensitivity

Examination of the dose-response curve shown in Fig. 25.2 reveals that some subjects are very sensitive to a particular poison (for example, those killed at a dose corresponding to LD_5), whereas others are very resistant to the same substance (for example, those surviving a dose corresponding to LD_{95}). These two kinds of responses illustrate hypersensitivity and hyposensitivity, respectively; subjects in the mid-range of the dose-response curve are termed normals. These variations in response tend to complicate toxicology in that there is not a specific dose guaranteed to yield a particular response, even in a homogeneous population.

In some cases, hypersensitivity is induced. After one or more doses of a chemical, a subject may develop an extreme reaction to it. This occurs with penicillin, for example, in cases where people develop such a severe allergic response to the antibiotic that exposure is fatal if countermeasures are not taken.

XENOBIOTIC AND ENDOGENOUS SUBSTANCES

Xenobiotic substances are those that are foreign to a living system, whereas those that occur naturally in a biologic system are termed endogenous. The levels of an endogenous substance must usually fall within a particular concentrating range in order for metabolic processes to occur normally. Levels below a normal range may result in a deficiency response or even death, and the same effects may occur above the normal range. This kind of response is illustrated in Fig. 25.4.

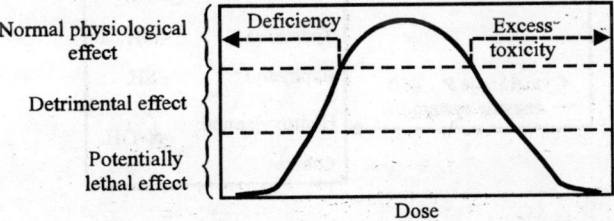

Fig. 25.4. Biological effect of an endogenous substance in an organism showing optimum level, deficiency, and excess.

Examples of endogenous substances in organisms include various hormones, glucose (blood sugar), and some essential metal ions, including Ca^{2+}, K^+, and Na^+. The optimum level of calcium in human blood serum occurs over a rather narrow range of 9 to 9.5 milligrams per decilitre (mg/dL). Below these values a deficiency response known as hypocalcemia occurs, manifested by muscle cramping. At serum levels above about 10.5 mg/dL hypercalcemia occurs, the major effect of which is kidney malfunction.

TOXICOLOGICAL CHEMISTRY

Toxicological chemistry is the science that deals with the chemical nature and reactions of toxic substances, including their origins, uses, and chemical aspects of exposure, fates, and disposal. Toxicological chemistry addresses the relationships between the chemical properties and molecular structures of molecules and their toxicological effects. Fig. 25.5 outlines the terms discussed above the relationships among them.

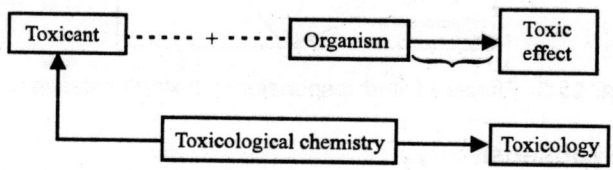

Fig. 25.5. Toxicology is the science of poisons. Toxicological chemistry relates toxicology to the chemical nature of toxicants.

Toxicants in the Body

The processes by which organisms metabolise xenobiotic species are enzyme-catalysed phase I and phase II reactions.

Phase I reactions

Lipophilic xenobiotic species in the body tend to undergo phase I reactions that make them more water-soluble and reactive by the attachment of polar functional groups, such as –OH (Fig. 25.6). Most Phase I processes are "microsomal mixed-function oxidase" reactions catalysed by the cytochrome P–450 enzyme system associated with the endoplasmic reticulum of the cell and occurring most abundantly in the liver of vertebrates.

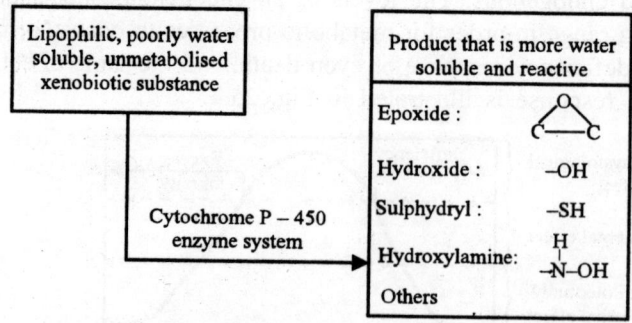

Fig. 25.6. Illustration of Phase I reactions.

Phase II reactions

The polar functional groups attached to a xenobiotic compound in a Phase I reaction provide reaction sites for Phase II reactions. Phase II reactions are conjugation reactions in which enzymes attach conjugating agents to xenobiotics, their Phase I reaction products, and non-xenobiotic compounds (Fig. 25.7). The conjugation product of such a reaction is usually less toxic than the original xenobiotic compound, less lipid-soluble, more water-soluble, and more readily eliminated from the body. The major conjugating agents and the enzymes that catalyse their Phase II reactions are glucuronide (UDP glucuronyltransferase enzymes), glutathione transferase enzyme), sulphate (sulphotransferase enzyme), and acetyl (acetylation by acetyltransferase enzymes). The most abundant conjugation products are glucuronides. A glucuronide conjugate is illustrated in Fig. 25.8, where $-X-R$ represents a xenobiotic species conjugated to glucuronide and R is an organic moiety. For example, if the xenobiotic compound conjugated is phenol, HXR is HOC_6H_5, X is the O atom, and R represents the phenyl group, C_6H_5.

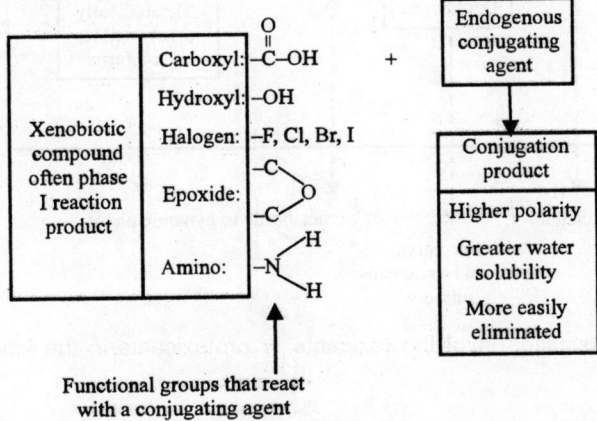

Fig. 25.7. Illustration of phase II reactions.

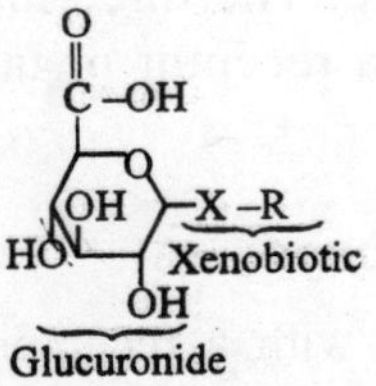

Fig. 25.8. Glucuronide conjugate formed from a xenobiotic, HX–R.

KINETIC PHASE AND DYNAMIC PHASE

Kinetic Phase

The major routes and sites of absorption, metabolism, binding, and excretion of toxic substances in the body are illustrated in Figs. 25.1 and 25.9. Toxicants in the body are metabolised, transported, and excreted; they have adverse biochemical effects; and they cause manifestations of poisoning. It is convenient to divide these processes into two major phases, a kinetic phase and a dynamic phase.

In the kinetic phase, a toxicant or the metabolic precursor of a toxic substance (protoxicant) may undergo absorption, metabolism, temporary storage, distribution, and excretion, as illustrated in Fig. 25.9. A toxicant that is absorbed may be passed through the kinetic phase unchanged as an active parent compound, metabolised to a detoxified metabolite that is excreted, or converted to a toxic active metabolite. These processes occur through phase I and phase II reactions discussed above.

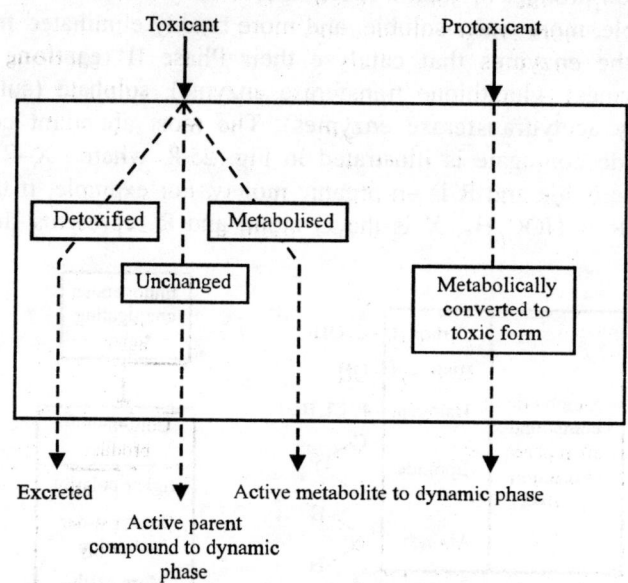

Fig. 25.9. Processes involving toxicants or protoxicants in the kinetic phase.

Dynamic Phase

In the dynamic phase, (Fig. 25.10) a toxicant or toxic metabolite interacts with cells, tissues, or organs in the body to cause some toxic response. The three major sub-divisions of the dynamic phase are the following: (i) primary reaction with a receptor or target organ; (ii) a biochemical response; and (iii) observable effects.

Primary reaction in the dynamic phase

A toxicant or an active metabolite reacts with a receptor. The process leading to a toxic response is initiated when such a reaction occurs; for example, when benzene epoxide, forms an adduct with a nucleic acid unit in DNA (receptor) resulting in alteration of the DNA. This is an example of an irreversible reaction between a toxicant and a receptor. A reversible reaction that can result in a toxic response is illustrated by the binding between carbon monoxide and oxygen-transporting haemoglobin (Hb) in blood :

$$O_2Hb + CO \rightleftharpoons COHb + O_2 \qquad ...(25.1)$$

Biochemical effects in the dynamic phase

The binding of a toxicant to a receptor may result in some kind of biochemical effect. The major biochemical effects are the following.

1. Impairment of enzyme function by binding to the enzyme, coenzymes, metal activators of enzymes, or enzyme subtrates.
2. Alteration of cell membrane or carriers in cell membranes.
3. Interference with carbohydrate metabolism.
4. Interference with lipid metabolism resulting in excess lipid accumulation ("fatty liver").
5. Interference with respiration, the overall process by which electrons are transferred to molecular oxygen in the biological oxidation of energy-yielding substrates.
6. Stopping or interfering with protein biosynthesis by the action of toxicants on DNA.
7. Interference with regulatory processes mediated by hormones or enzymes.

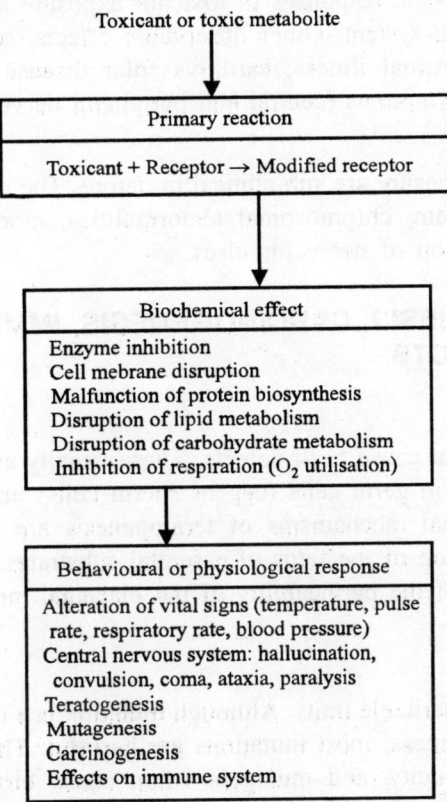

Toxicant or toxic metabolite

Primary reaction

Toxicant + Receptor → Modified receptor

Biochemical effect

Enzyme inhibition
Cell mebrane disruption
Malfunction of protein biosynthesis
Disruption of lipid metabolism
Disruption of carbohydrate metabolism
Inhibition of respiration (O_2 utilisation)

Behavioural or physiological response

Alteration of vital signs (temperature, pulse rate, respiratory rate, blood pressure)
Central nervous system: hallucination, convulsion, coma, ataxia, paralysis
Teratogenesis
Mutagenesis
Carcinogenesis
Effects on immune system

Fig. 25.10. Dynamic phase of toxicant action.

Responses to toxicants

Among the more immediate and readily observed manifestations of poisoning are alterations in the vital signs of temperature, pulse rate, respiratory rate, and blood pressure. Poisoning by some substances may cause an abnormal skin colour (jaundiced yellow skin from CCl_4 poisoning) or excessively moist or dry skin. Toxic levels of some materials or their metabolites cause the body to have unnatural odours, such as the bitter almond odour of HCN in tissues of victims of cyanide poisoning. Symptoms of poisoning manifested in the eye include miosis (excessive or prolonged contraction of the eye pupil), mydriasis (excessive pupil dilation), conjunctivitis (inflammation of the mucus membrane that covers the front part of the eyeball and the inner lining of the eyelids) and nystagmus (involuntary movement of the eyeballs). Some poisons cause a moist condition of the mouth, whereas others cause a dry mouth. Gastrointestinal tract effect inclue . pain, vomiting, or paralytic ileus (stoppage of the normal peristalsis movement of the intestines) occur as a result of poisoning by a number of toxic substances.

Central nervous system poisoning may be manifested by convulsions, paralysis, hallucinations, and ataxia (lack of coordination of voluntary movements of the body), as well as abnormal behaviour, including agitation, hyperactivity, disorientation, and delirium. Severe poisoning by some substances, including organophosphates and carbamates, causes coma, the term used to describes a lowered level of consciousness.

Prominent among the more chronic responses to toxicant exposure are mutations, cancer, and birth defects and effects on the immune system. Other observable effects, some of which may occur soon after exposure, include gastrointestinal illness, cardiovascular disease, hepatic (liver) disease, renal (kidney) malfunction, neurologic symptoms (central and peripheral nervous systems), skin abnormalities (rash, dermatitis).

Often the effects of toxicant exposure are sub-clinical in nature. The most common of these are some kinds of damage to immune system, chromosomal abnormalities, modification of functions of liver enzymes, and slowing of conduction of nerve impulses.

TERATOGENESIS, MUTAGENESIS, CARCINOGENESIS, IMMUNE SYSTEM EFFECTS, AND REPRODUCTIVE EFFECTS

Teratogenesis

Teratogens are chemical species that cause birth defects. These usually arise from damage to embryonic or fatal cells. However, mutations in germ cells (egg or sperm cells) may cause birth defects, such as Down's syndrome. The biochemical mechanisms of teratogenesis are varied. These include enzyme inhibition by xenobiotics; deprivation of the fetus of essential substrates, such as vitamins; interference with energy supply; or alteration of the permeability of the placental membrane.

Mutagenesis

Mutagens alter DNA to produce inheritable traits. Although mutation is a natural process that occurs even in the absence of xenobiotic substances, most mutations are harmful. The mechanisms of mutagenicity are similar to those of carcinogenicity, and mutagens often cause birth defects as well. Therefore, mutagenic hazardous substances are of major toxicological concern.

Biochemistry of mutagenesis

To understand the biochemistry of mutagenesis, it is important to recall from Chapter 24 that DNA contains the nitrogenous bases adenine, guanine, cytosine, and thymine. The order in which these bases occur in DNA determines the nature and structure of newly produced RNA, a substance produced as a step in the synthesis of new proteins and enzymes in cells. Exchange, addition, or deletion of any of the nitrogenous bases in DNA alters the nature of RNA produced and can change vital life processes, such as the synthesis of an important enzyme. This phenomenon, which can be caused by xenobiotic compounds, is a mutation that can be passed on to progeny, usually with detrimental results. There are several ways in which xenobiotic species may cause mutations. For the most part, however, mutations due to xenobiotic substances are the result of chemical alterations of DNA, such as those discussed in the two examples below.

Nitrous acid, HNO_2, is an example of a chemical mutagen that is often used to cause mutations in bacteria. To understand the mutagenic activity of nitrous acid it should be noted that three of the nitrogenous bases—adenine, guanine, and cytosine—contain the amino group, $-NH_2$. The action of nitrous acid is to replace amino groups with a hydroxy group. When this occurs, the DNA may not function in the intended manner, causing a mutation to occur.

Alkylation consisting of the attachment of a small alkyl group, such as $-CH_3$ or $-C_2H_5$, to an N atom on one of the nitrogenous bases in DNA is one of the most common mechanisms leading to mutation. The methylation of "7" nitrogen in guanine in DNA to form N-methylguanine is shown in Fig. 25.11. O-alkylation may also occur by attachment of a methyl or other alkyl group to the oxygen atom in guanine.

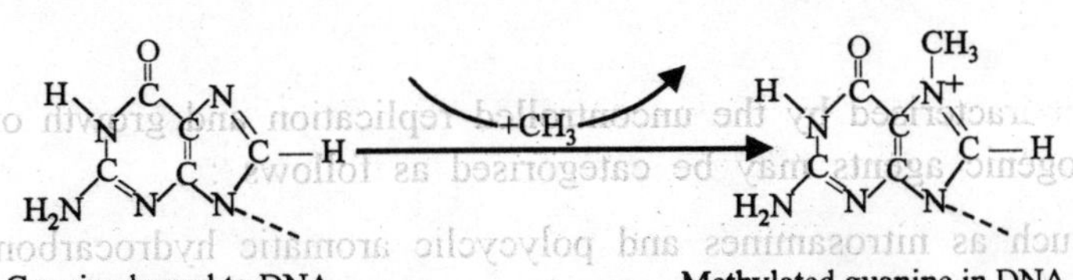

Guanine bound to DNA　　　Methylated guanine in DNA

Fig. 25.11. Alkylation of guanine in DNA.

A number of mutagenic substances act as alkylating agents. Prominent among these are the compounds shown in Fig. 25.12.

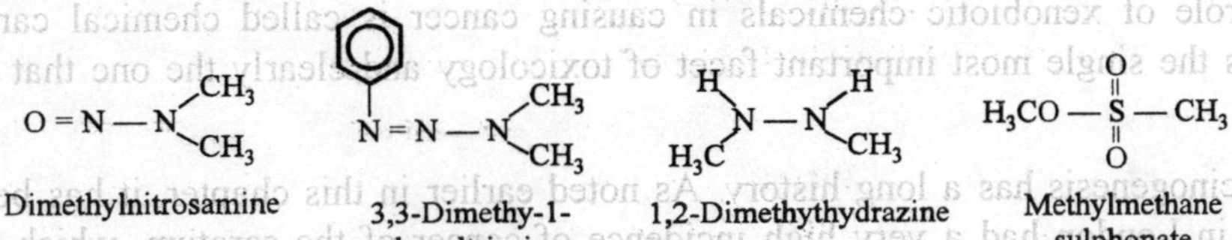

Dimethylnitrosamine　　3,3-Dimethy-1-phenyltriazine　　1,2-Dimethythydrazine　　Methylmethane sulphonate

Fig. 25.12. Examples of simple alkylating agents capable of causing mutations.

Alkylation occurs by way of generation of positively charged electrophilic species that bond to electron-rich nitrogen or oxygen atoms on the nitrogenous bases in DNA. The generation of such species usually occurs by way of biochemical and chemical processes. For example, dimethylnitrosamine (structure in Fig. 25.12) is activated by oxidation through cellular NADPH to produce the following highly reactive intermediate :

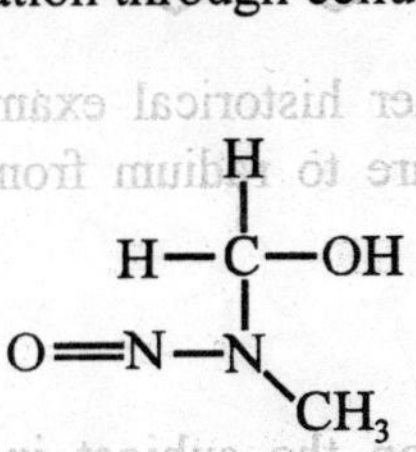

This product undergoes several non-enzymatic transitions, losing formaldehyde and generating a methyl carbocation, $^+CH_3$, that can methylate nitrogenous bases on DNA :

$$\text{...(25.2)}$$

One of the more notable mutagens is tris(2, 3-dibromopropyl)phosphate, commonly called "tris", that was used as a flame retardant in children's sleepwear. Tris was found to be mutagenic in experimental animals and metabolites of it were found in children wearing the treated sleepwear. This strongly suggested that tris is absorbed through the skin and its uses were discontinued.

Carcinogenesis

Cancer is a condition characterised by the uncontrolled replication and growth of the body's own cells (somatic cells). Carcinogenic agents may be categorised as follows :

1. Chemical agents, such as nitrosamines and polycyclic aromatic hydrocarbons.
2. Biological agents, such as hepadnaviruses or retroviruses.
3. Ionising radiation, such as X-rays.
4. Genetic factors, such as selective breeding.

Clearly, in some cases, cancer is the result of the action of synthetic and naturally occurring chemicals. The role of xenobiotic chemicals in causing cancer is called chemical carcinogenesis. It is often regarded as the single most important facet of toxicology and clearly the one that receives the most publicity.

Chemical carcinogenesis has a long history. As noted earlier in this chapter, it has been, observed that chimney sweeps in London had a very high incidence of cancer of the scrotum, which is related to their exposure to soot and tar from the burning of bituminous coal. Also elevated incidences of bladder cancer in dye workers exposed to chemicals extracted from coal tar; 2-naphthylamine,

was shown to be largely responsible. Other historical examples of carcinogenesis include observations of cancer from tobacco juice, oral exposure to radium from painting luminescent watch dials, tobacco smoke, and asbestos.

Biochemistry of carcinogenesis

Large expenditures of time and money on the subject in recent years have yielded a much better understanding of the biochemical bases of chemical carcinogenesis. The overall processes for the induction of cancer may be quite complex, involving numerous steps. However, it is generally recognised that there are two major steps in carcinogenesis: an initiation stage followed by a promotional stage. These steps are further subdivided as shown in Fig. 25.13.

Initiation of carcinogenesis may occur by reaction of a DNA-reactive species with DNA or by the action of an epigenitic carcinogen that does not react with DNA and is carcinogenic by some other mechanism. Most DNA-reactive species are genotoxic carcinogens because they are also mutagens. These substances react irreversibly with DNA. They are either electrophilic or, more commonly, metabolically activated to form electrophilic species, as is the case with electrophilic $^{+}CH_3$ generated from dimethylnitrosamine, discussed under mutagenesis above. Cancer-causing substances that require metabolic activation are called procarcinogens. The metabolic species actually responsible for carcinogenesis is termed an ultimate carcinogen. Some species that are intermediate metabolites between procarcinogens and ultimate carcinogens are called proximate carcinogens. Carcinogens that do not require biochemical activation are categorised as primary or direct-acting carcinogens. Some example procarcinogens and primary carcinogens are shown in Fig. 25.14.

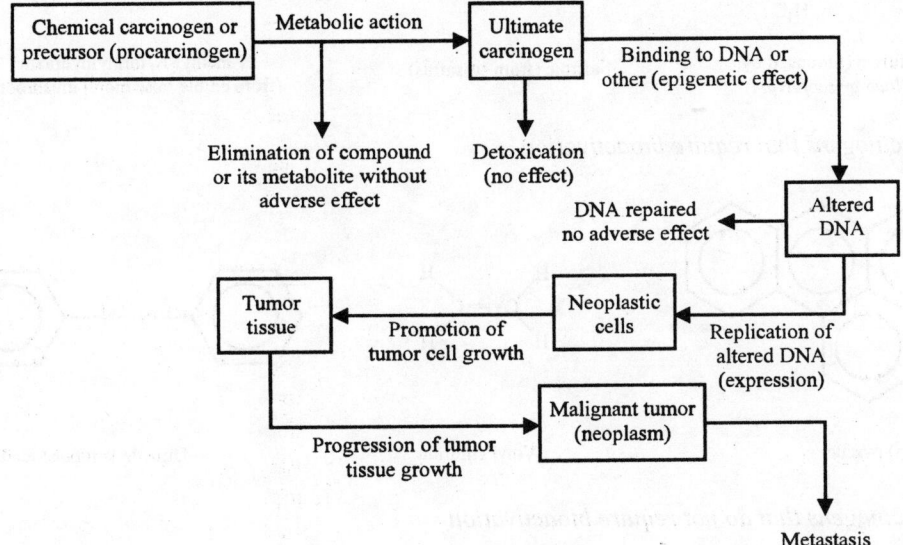

Fig. 25.13. Outline of the process by which a carcinogen or procarcinogen may cause cancer.

Most substances classified as epigenetic carcinogens are promoters that act after initiation. Manifestations of promotion include increased numbers of tumor cells and decreased length of time for tumors to develop (shortened latency period). Promoters do not initiate cancer, are not electrophilic, and do not bind with DNA. The classic example of a promoter is a substance known chemically as decanoyl phorbol acetate or phorbol myristate acetate, a substance extracted from croton oil.

Alkylating agents in carcinogenesis

Chemical carcinogens usually have the ability to form covalent bonds with macromolecular life molecules. Such covalent bonds can form with proteins, peptides, RNA, and DNA. Although most binding is with other kinds of molecules, which are more abundant, the DNA adducts are the significant ones in initiating cancer. Prominent among the species that bond to DNA in carcinogenesis are the alkylating agents which attach alkyl groups–such as methyl (CH_3) or ethyl (C_2H_5)— to DNA. A similar type of compound, arylating agents, act to attach aryl moieties, such as the phenyl group to DNA.

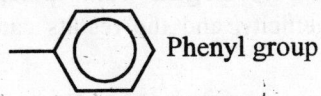 Phenyl group

to DNA. As shown by the examples in Fig. 25.15, the alkyl and aryl groups become attached to N and O atoms in the nitrogenous bases that compose DNA. This alteration in the DNA can initiate the sequence of events that results in the growth and replication of neoplastic (cancerous) cells. The reactive species that donate alkyl groups in alkylation are usually formed by metabolic activation as shown for dimethylnitrosamine in the discussion of mutagenesis earlier in this section.

Naturally occurring carcinogens that require bioactivation

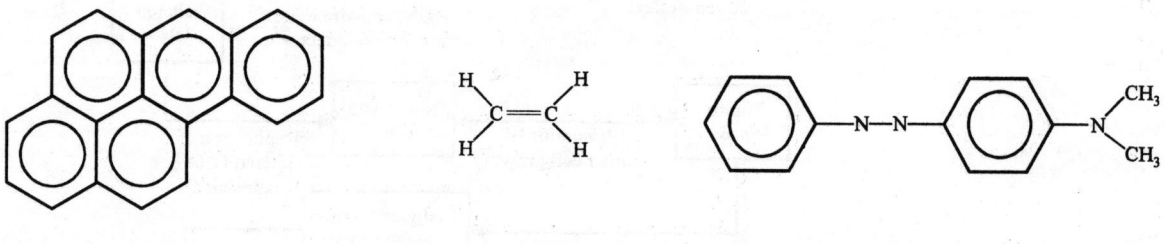

Griseofulvin (produced by *penicillium griseofulvum*)

Saffrole (from sassafras)

N-methyl-N-formylhydrazine (from edible false morel mushroom)

Synthetic carcinogens that require bioactivation

Benzo (a) pyrene

Vinyl chloride

4–Dimethylaminoazobenzene

Primary carcinogens that do not require bioactivation

Bis(chloromethyl)-ether

Dimethyl sulphate

Ethyleneimine

β-Propioacetone

Fig. 25.14. Examples of the major classes of naturally occurring and synthetic carcinogens, some of which require bioactivation, and others of which act directly.

Testing for Carcinogens

Only a few chemicals have definitely been established as a human carcinogens. A well documented example is vinyl chloride, $CH_2=CHCl$, which is known to have caused a rare form of liver cancer (angiosarcoma) in individuals who cleaned autoclaves in the poly(vinyl chloride) fabrication industry. In some cases chemicals are known to be carcinogens from epidemiological studies of exposed humans. Animals are used to test for carcinogenicity, and the results can be extrapolated, although with much uncertainty, to humans.

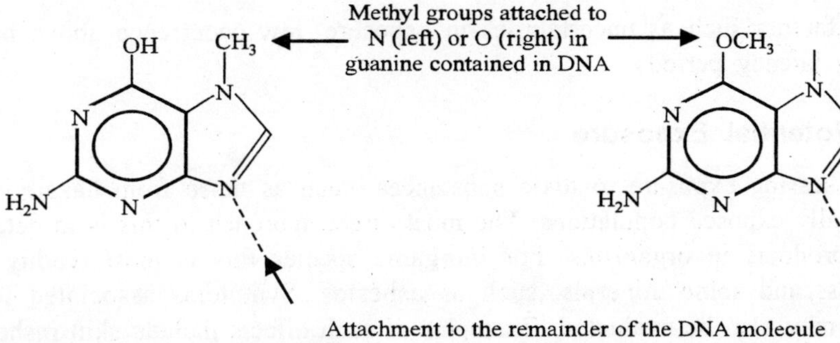

Fig. 25.15. Alkylated (methylated) forms of the nitrogenous base guanine.

Bruce Ames test

Mutagenicity used to infer carcinogenicity is the basis of the *Bruce Ames* test, in which observations are made of the reversion of mutant histidine-requiring *Salmonella* bacteria back to a form that can synthesise its own histidine. The test makes use of enzymes in homogenised liver tissue to convert potential procarcinogens to ultimate carcinogens. Histidine-requiring *Salmonella* bacteria are inoculated onto a medium that does not contain histidine, and those that mutate back to a form that can synthesise histidine establish visible colonies that are assayed to indicate mutagenicity.

According to Bruce Ames, the developer of the test by the same name, animal tests for carcinogens that make use of massive doses of chemicals may give results that cannot be accurately extrapolated to assess cancer risks from smaller doses of chemicals. This is because the huge doses of chemicals used kill large numbers of cells, which the organism's body attempts to replace with new cells. Rapidly dividing cells greatly increase the likelihood of mutations that result in cancer simply as the result of rapid cell proliferation, not genotoxicity.

Immune System Response

The immune system acts as the body's natural defense system to protect it from xenobiotic chemicals; infectious agents, such as viruses or bacteria; and neoplastic cells, which give rise to cancerous tissue. Adverse effects on the body's immune system are being increasingly recognised as important consequences of exposure to hazardous substances. Toxicants can cause immunosuppression, which is the impairment of the body's natural defense mechanisms. Xenobiotics can also cause the immune system to lose its ability to control cell proliferation, resulting in leukemia or lymphoma.

Another major toxic response of the immune system is allergy or hypersensitivity. This kind of condition results when the immune system overreacts to the presence of a foreign agent or its metabolites in a self-destructive manner. Among the xenobiotic materials that can cause such reactions are beryllium, chromium, nickel, formaldehyde, pesticides, resins, and plasticisers.

HEALTH HAZARDS

In recent years attention in toxicology has shifted away from readily recognised, usually severe, acute maladies that developed on a short time scale as a result of brief, intense exposure to toxicants, toward delayed, chronic, often less severe illnesses caused by long-term exposure to low levels of toxicants. Although the total impact of the latter kinds of health effects may be substantial, their assessment is very difficult because of factors such as uncertainties in exposure, low occurrence above background levels of disease, and long latency periods.

Assessment of Potential Exposure

A critical step in assessing exposure to toxic substances, such as those from hazardous waste sites is evaluation of potentially exposed populations. The most direct approach to this is to determine chemicals or their metabolic products in organisms. For inorganic species this is most readily done for heavy metals, radionuclides, and some minerals, such as asbestos. Symptoms associated with exposure to particular chemicals may also be evaluated. Examples of such effects include skin rashes or sub-clinical effects, such as chromosomal damage.

Epidemiological Evidence

Epidemiological studies applied to toxic environmental pollutants, such as those from hazardous wastes, attempt to correlate observations of particular illnesses with probable exposure to such wastes. There are two major approaches to such studies. One approach is to look for diseases known to be caused by particular agents in areas where exposure is likely from such agents in hazardous wastes. A second approach is to look for clusters consisting of an abnormally large number of cases of a particular disease in a limited geographic area, then attempt to locate sources of exposure to hazardous wastes that may be responsible. The most common types of maladies observed in clusters are spontaneous abortions, birth defects, and particular types of cancer.

Epidemiologic studies are complicated by long latency periods from exposure to onset of disease, lack of specificity in the correlation between exposure to a particular waste and the occurrence of a disease, and background levels of a disease in the absence of exposure to a hazardous waste capable of causing the disease.

Estimation of Health Effects Risks

An important part of estimating the risks of adverse health effects from exposure to toxicants involves extrapolation from experimentally observable data. Usually the end result needed is an estimate of a low occurrence of a disease in humans after a long latency period resulting from low-level exposure to a toxicant for a long period of time. The data available are almost always taken from animals exposed to high levels of the substance for a relatively short period of time. Extrapolation is then made using linear or curvilinear projections to estimate the risk to human populations. There are, of course, very substantial uncertainties in this kind of approach.

Risk Assessment

Toxicological considerations are very important in estimating potential dangers of pollutants and hazardous waste chemicals. One of the major ways in which toxicology interfaces with the area of hazardous wastes is in health risk assessment, providing guidance for risk management, cleanup, or regulation needed at a hazardous waste site based upon knowledge about the site and the chemical and toxicological properties of wastes in it. Risk assessment includes the factors of site characteristics; substances present, including indicator species; potential receptors; potential exposure pathways; and uncertainty analysis. It may be divided into the following major components: (i) identification of hazard; (ii) dose-response assessment; (iii) exposure assessment; and (iv) risk characterisation.

CHAPTER 26

Toxicological Chemistry of Chemical Substances

INTRODUCTION

Toxicological chemistry is the science that deals with the chemical nature and reactions of toxic substances, including their origins, uses, and chemical aspects of exposure, fates, and disposal. Toxicological chemistry centres on the relationship between the chemical nature of toxicants and their toxicological effects. This chapter discusses this relationship with regard to some of the major contributors of pollutants and hazardous substances. The first section deals with toxicological aspects of elements (particularly heavy metals) whose presence in a compound frequently means that the compound is toxic. It also discusses the toxicities of some commonly used elemental forms, such as the chemically uncombined elemental halogens. The following section discusses the toxicological chemistry of inorganic compounds, many of which are produced from industrial processes. It also provides a brief discussion of organo-metallic compounds. The last section deals with the toxicology of organic compounds. The toxicological properties of hydrocarbons and oxygen-containing organic compounds are discussed as well as other organic substances containing functional groups, such as alcohols and ketones. This section also discusses the toxicities of organic nitrogen, halide, sulphur, and phosphorus compounds, some of which are used as pesticides or military poisons.

TOXIC ELEMENTS AND ELEMENTAL FORMS

Some of the toxic elements and elemental forms are :

Ozone

Ozone has several toxic effects. Air containing 1 ppm by volume of ozone has a distinct odour. Inhalation of ozone at this level causes severe irritation and headache. Ozone irritates the eyes, upper respiratory system, and lungs. Inhalation of ozone can cause sometimes fatal pulmonary edema. *Pulmonary* refers to lungs and *edema* is an accumulation of fluid in tissue spaces; therefore, pulmonary edema is an abnormal accumulation of fluid in lung tissue. Chromosomal damage has been observed in subjects exposed to ozone.

Ozone generates free radicals in tissue. These reactive species can cause lipid peroxidation, oxidation of sulphydryl (–SH) groups, and other destructive oxidation processes. Compounds that protect organisms from the effects of ozone include radical scavengers, antioxidants, and compounds containing sulphydryl groups.

White Phosphorus

Elemental white phosphorus can enter the body by inhalation, by skin contact, or orally. It is a systemic poison; that is, one that is transported through the body to sites remote from its entry site. White phosphorus causes anaemia, gastrointestinal system dysfunction, bone brittleness, and eye damage. Exposure also causes *phossy jaw*, a condition in which the jawbone deteriorates and becomes fractured.

Elemental Halogens

Fluorine (F_2) is a pale yellow highly reactive gas that is a strong oxidant. It is a toxic irritant and attacks skin and the mucous membranes of the nose and eyes.

Chlorine (Cl_2) gas reacts in water to produce a strongly oxidising solution. This reaction is responsible for some of the damage caused to the moist tissue lining the respiratory tract when the tissue is exposed to chlorine. The respiratory tract is rapidly irritated by exposure to 10–20 ppm of chlorine gas in air, causing acute discomfort that warns of the presence of the toxicant. Even brief exposure to 1,000 ppm of Cl_2 can be fatal.

Bromine (Br_2) is a volatile dark red liquid that is toxic when inhaled or ingested. Like chlorine and fluorine, it is strongly irritating to the mucous tissue of the respiratory tract and eyes and may cause pulmonary edema. The toxicological hazard of bromine is limited somewhat because its irritating odour elicits a withdrawal response.

Iodine (I_2) a solid, is irritating to the lungs much like bromine or chlorine. However, the relatively low vapour pressure of iodine limits exposure to I_2 vapour.

Heavy Metals

Heavy metals are particularly toxic in their chemically combined forms and some, notably mercury, are toxic in the elemental form. The toxic properties of some of the most hazardous heavy metals and metalloids are discussed here.

Although not truly a heavy metal, *beryllium* (atomic mass 9.01) is one of the more hazardous toxic elements. Its most serious toxic effect is berylliosis, a condition manifested by lung fibrosis and pneumonitis, which may develop after a latency period of 5–20 years. Beryllium exposure also causes skin granulomas and ulcerated skin. Beryllium is also a hypersensitising agent.

Cadmium adversely affects several important enzymes; it can also cause painful osteomalacia (bone disease) and kidney damage. Inhalation of cadmium oxide dusts and fumes results in cadmium pneumonitis characterised by edema and pulmonary epithelium necrosis (death of tissue lining lungs).

Lead, widely distributed as metallic lead, inorganic compounds, and organo-metallic compounds, has a number of toxic effects, including inhibition of the synthesis of haemoglobin. It also adversely affects the central and peripheral nervous systems and the kidneys.

Arsenic is a metalloid which forms a number of toxic compounds. The toxic $+3$ oxide, As_2O_3, is absorbed through the lungs and intestines. Biochemically, arsenic acts to coagulate proteins, forms complexes with coenzymes, and inhibits the production of adenosine triphosphate (ATP) in essential metabolic processes.

Mercury vapour can enter the body through inhalation and be carried by the bloodstream to the brain where it penetrates the blood-brain barrier. It disrupts metabolic processes in the brain causing tremor and psychopathological symptoms such as shyness, insomnia, depression, and irritability. Divalent ionic mercury, Hg^{2+}, damages the kidney. Organo-metallic mercury compounds such as dimethylmercury, $Hg(CH_3)_2$, are also very toxic.

TOXIC INORGANIC COMPOUNDS

Cyanide

Both hydrogen cyanide (HCN) and cyanide salts (which contain CN^- ion) are rapidly acting poisons; a dose of only 60–90 mg is sufficient to kill a human. Metabolically, cyanide bonds to iron(III) in iron-containing ferricytochrome oxidase enzyme, preventing its reduction to iron(II) in the oxidative phosphorylation process by which the body utilises O_2. This prevents utilisation of oxygen in cells, so that metabolic processes cease.

Carbon Monoxide

Carbon monoxide, CO, is a common cause of accidental poisonings. At CO levels in air of 10 parts per million (ppm), impairment of judgement and visual perception occur; exposure to 100 ppm causes dizziness, headache, and weariness; loss of consciousness occurs at 250 ppm; and inhalation of 1,000 ppm results in rapid death. Chronic long-term exposures to low levels of carbon monoxide are suspected of causing disorders of the respiratory system and the heart.

After entering the blood stream through the lungs, carbon monoxide reacts with haemoglobin (Hb) to convert oxyhaemoglobin (O_2Hb) to carboxyhaemoglobin (COHb) :

$$O_2Hb + CO \rightarrow COHb + O_2 \qquad ...(26.1)$$

In this case, haemoglobin is the receptor acted on by the carbon monoxide toxicant. Carboxyhaemoglobin is much more stable than oxyhaemoglobin so that its formation prevents haemoglobin from carrying oxygen to body tissues.

Nitrogen Oxides

The two most common toxic oxides of nitrogen are NO and NO_2, of which the latter is regarded as the more toxic. Nitrogen dioxide causes severe irritation of the innermost parts of the lungs resulting in pulmonary edema. In cases of severe exposures, fatal bronchiolitis fibrosa obliterans may develop approximately three weeks after exposure to NO_2. Fatalities may result from even brief periods of inhalation of air containing 200–700 ppm of NO_2. Biochemically, NO_2 disrupts lactic dehydrogenase and some other enzyme systems, possibly acting much like ozone, a stronger oxidant. Free radicals, particularly HO·, are likely formed in the body by the action of nitrogen dioxide and the compound probably causes lipid peroxidation in which the C=C double bonds in unsaturated body lipids are attacked by free radicals and undergo chain reactions in the presence of O_2, resulting in their oxidative destruction.

Nitrous oxide, N_2O is used as an oxidant gas and in dental surgery as a general anaesthetic. This gas is known as "laughing gas". Nitrous oxide is a central nervous system depressant and can act as an asphyxiant.

Hydrogen Halides

Hydrogen halides (general formula HX, where X is F, Cl, Br, or I) are relatively toxic gases. The most widely used of these gases are HF and HCl; their toxicities are discussed here.

Hydrogen fluoride

Hydrogen fluoride, (HF, mp −83.1°C, bp 19.5°C) is used as a clear, colourless liquid or gas or as a 30–60% aqueous solution of hydrofluoric acid, both referred to here as HF. Both are extreme irritants to any part of the body that they contact, causing ulcers in affected areas of the upper respiratory tract. Lesions caused by contact with HF heal poorly, and tend to develop gangrene.

Fluoride ion, F^-, is toxic in soluble fluoride salts, such as NaF, causing fluorosis, a condition characterised by bone abnormalities and mottled, soft teeth. Livestock are especially susceptible to poisoning from fluoride fallout on grazing land; severely afflicted animals become lame and even die. Industrial pollution has been a common source of toxic levels of fluoride. However, about 1 ppm of fluoride used in some drinking water supplies prevents tooth decay.

Hydrogen chloride

Gaseous hydrogen chloride and its aqueous solution, called hydrochloric acid, both denoted as HCl are much less toxic than HF. Hydrochloric acid is a natural physiological fluid present as a dilute solution in the stomachs of humans and other animals. However, inhalation of HCl vapour can cause spasms of the larynx as well as pulmonary edema and even death at high levels. The high affinity of hydrogen chloride vapour for water tends to dehydrate eye and respiratory tract tissue.

Interhalogen Compounds and Halogen Oxides

Interhalogen compounds, including ClF, BrCl, and BrF_3, are extremely reactive and are potent oxidants. They react with water to produce hydrohalic acid solutions (HF, HCl) and nascent oxygen {O}. Too reactive to enter biological systems in their original chemical state, interhalogen compounds tend to be powerful corrosive irritants that acidify, oxidise, and dehydrate tissue, much like those of the elemental forms of the elements from which they are composed. Because of these effects skin, eyes, and mucous membranes of the mouth, throat, and pulmonary systems are especially susceptible to attack.

Major halogen oxides, including fluorine monoxide (OF_2), chlorine monoxide (Cl_2O), chlorine dioxide (ClO_2), chlorine heptoxide (Cl_2O_7), and bromine monoxide (Br_2O), tend to be unstable, highly reactive, and toxic compounds that pose hazards similar to those of the interhalogen compounds discussed above. Chlorine dioxide, the most commonly used halogen oxide is employed for odour control and bleaching wood pulp. As a substitute for chlorine in water disinfection, it produces fewer undesirable chemical by-products, particularly trihalomethanes.

The most important of the oxyacids and their salts formed by halogens are hypochlorous acid, HOCl, and hypochlorites, such as NaOCl, used for bleaching and disinfection. The hypochlorites irritate eye, skin, and mucous membrane tissue because they react to produce active (nascent) oxygen ({O}) and acid as shown by the reaction below :

$$HClO \rightarrow H^+ + Cl^- + \{O\} \qquad \qquad ...(26.2)$$

Inorganic Compounds of Silicon

Silica (SiO_2, quartz) occurs in a variety of types of rocks, such as sand, sandstone, and diatomaceous earth. Silicosis resulting from human exposure to silica dust from construction materials, sand blasting, and other sources has been a common occupational disease. A type of pulmonary fibrosis that causes lung nodules and makes victims more susceptible to pneumonia and other lung diseases, silicosis is one of the most common disabling conditions resulting from industrial exposure to hazardous substances. It can cause death from insufficient oxygen or from heart failure in severe cases.

Silane, SiH_4, and *disilane*, H_3SiSiH_3, are examples of inorganic silanes, which have H-Si bonds. Numerous organic ("organo-metallic") silanes exist in which alkyl moieties are substituted for H. Little information is available regarding the toxicities of silanes.

Silicon tetrachloride, $SiCl_4$, is the only industrially significant compound of the silicon tetrahalides, a group of compounds with the general formula SiX_4, where X is a halogen. The two commercially produced silicon halohydrides, general formula $H_{4-x}SiX_x$, are dichlorosilane (SiH_2Cl_2) and trichlorosilane, ($SiHCl_3$). These compounds are used as intermediates in the synthesis of organo-silicon compounds and in the production of high-purity silicon for semi-conductors. Silicon tetrachloride and trichlorosilane, fuming liquids which react with water to give off HCl vapour, have suffocating odours and are irritants to eye, nasal, and lung tissue.

Asbestos

Asbestos is the name given to a group of fibrous silicate minerals, typically those of the serpentine group, for which the approximate formula is $Mg_3(Si_2O_5)(OH)_4$. Asbestos has been widely used in structural materials, brake linings, insulation, and pipe manufacture. Inhalation of asbestos may cause asbestosis (a pneumonia condition), mesothelioma (tumor of the mesothelial tissue lining the chest cavity adjacent to the lungs), and bronchogenic carcinoma (cancer originating with the air passages in the lungs) so that uses of asbestos have been severely curtailed and widespread programmes have been undertaken to remove the material from buildings.

Inorganic Phosphorus Compounds

Phosphine (PH_3), a colourless gas that undergoes autoignition at 100°C, is a potential hazard in industrial processes and in the laboratory. Symptoms of poisoning from potentially fatal phosphine gas include pulmonary tract irritation, central nervous system depression, fatigue, vomiting, and painful breathing.

Tetraphosphorus decoxide, P_4O_{10}, (often called phosphorus pentoxide, P_2O_5) is produced as a fluffy white powder from the combustion of elemental phosphorus and reacts with water from air to form syrupy orthophosphoric acid. Because of the formation of acid by this reaction and its dehydrating action, P_4O_{10} is a corrosive irritant to skin, eyes and mucous membranes.

The most important of the phosphorus halides, general formulas PX_3 and PX_5, is phosphorus pentachloride used as a catalyst in organic synthesis, as a chlorinating agent and as a raw material to make phosphorus oxychloride ($POCl_3$). Because they react violently with water to produce the corresponding hydrogen halides and oxo phosphorus acids,

$$PCl_5 + 4H_2O \rightarrow H_3PO_4 + 5HCl \qquad \qquad ...(26.3)$$

the phosphorus halides are strong irritants to eyes, skin, and mucous membranes.

The major phosphorus oxyhalide in commercial use is *phosphorus oxychloride* ($POCl_3$), a faintly yellow fuming liquid. Reacting with water to form toxic vapours of hydrochloric acid and phosphonic acid (H_3PO_3), phosphorus oxyhalides is a strong irritant to the eyes, skin, and mucous membranes.

Inorganic Compounds of Sulphur

A colourless gas with a foul rotten-egg odour, *hydrogen sulphide* is very toxic. In some cases inhalation of H_2S kills faster than even hydrogen cyanide; rapid death ensues from exposure to air containing more than about 1000 ppm H_2S due to asphyxiation from respiratory system paralysis. Lower doses cause symptoms that include headache, dizziness, and excitement due to damage to the central nervous system. General debility is one of the numerous effects of chronic H_2S poisoning.

Sulphur dioxide, SO_2, dissolves in water, to produce sulphurous acid, H_2SO_3; hydrogen sulphite ion, HSO_3^- and sulphite ion SO_3^{2-}. Because of its water solubility, sulphur dioxide is largely removed in the upper respiratory tract. It is an irritant to the eyes, skin, mucous membranes and respiratory tract. It is an irritant to the eyes, skin, mucous membranes and respiratory tract. Some individuals are hypersensitive to sodium sulphite (Na_2SO_3), which has been used as a chemical food preservative.

Number one in synthetic chemical production, *sulphuric acid* (H_2SO_4) is a severely corrosive poison and dehydrating agent in the concentrated liquid form; it readily penetrates skin to reach sub-cutaneous tissue causing tissue necrosis with effects resembling those of severe thermal burns. Sulphuric acid fumes and mists irritate eye and respiratory tract tissue and industrial exposure has even caused tooth erosion in workers.

The more important halides, oxides, and oxyhalides of sulphur are listed in Table 26.1. The major toxic effects of these compounds are also given in this Table.

Table 26.1. Inorganic sulphur compounds.

Name of compound	Formula	Properties
Sulphur		
Monofluoride	S_2F_2	Colourless gas, mp 104°C, bp 99°C, toxicity similar to HF
Tetrafluoride	SF_4	Gas, bp −40°C, mp −124°C, powerful irritant
Hexafluoride	SF_6	Colourless gas mp −51°C, surprisingly non-toxic when pure, but often contaminated with toxic lower fluorides
Monochloride	S_2Cl_2	Oily, fuming orange liquid, mp −80°C, bp 138°C, strong irritant to eyes, skin, and lungs
Tetrachloride	SCl_4	Brownish/yellow liquid/gas, mp −30°C, decomposes below 0°C, irritant
Trioxide	SO_3	Solid anhydride of sulphuric acid reacts with moisture or steam to produce sulphuric acid
Sulphuryl chloride	SO_2Cl_2	Colourless liquid, mp −54°C, bp 69°C, used for organic synthesis, corrosive toxic irritant
Thionyl chloride	$SOCl_2$	Colourless-to-orange fuming liquid, mp −105°C, bp 79°C, toxic corrosive irritant
Carbon oxysulphide	COS	Volatile liquid by-product of natural gas or petroleum refining, toxic narcotic
Carbon disulphide	CS_2	Colourless liquid, industrial chemical, narcotic and central nervous system anaesthetic

Organo-metallic Compounds

The toxicological properties of some organo-metallic compounds—pharmaceutical organo-arsenicals, organo-mercury fungicides, and tetraethyllead antiknock gasoline additives—that have been used for many years are well known. However, toxicological experience is lacking for many relatively new organo-metallic compounds that are now being used in semi-conductors, as catalysts, and for chemical synthesis, so they should be treated with great caution until proved safe.

Organo-metallic compounds often behave in the body in ways totally unlike the inorganic forms of the metals that they contain. This is due in large part to the fact that, compared to inorganic forms, organo-metallic compounds have an organic nature and higher lipid solubility.

Organo-lead compounds

Perhaps the most notable toxic organo-metallic compound is tetraethyllead, $Pb(C_2H_5)_4$, a colourless, oily liquid that was widely used as an octane-bosting gasoline additive. Tetraethyllead has a strong affinity for lipids and can enter the body by inhalation, ingestion, and absorption through the skin. Acting differently from inorganic compounds in the body, it affects the central nervous system with symptoms such as fatigue, weakness, restlessness, ataxia, psychosis, and convulsions. Recovery from severe lead poisoning tends to be slow. In cases of fatal tetraethyllead poisoning, death has occurred as soon as one or two days after exposure.

Organo-tin Compounds

The greatest number of organo-metallic compounds in commercial use are those of tin—tributyltin chloride and related tributyltin (TBT) compounds. These compounds have bactericidal, fungicidal, and insecticidal properties. They have particular environmental significance because of their increasing applications as industrial biocides. Organo-tin compounds are readily absorbed through the skin, sometimes causing a skin rash. They probably bind with sulphur groups on proteins and appear to interfere with mitochondrial function.

Carbonyls

Metal carbonyls regarded as extremely hazardous because of their toxicities include nickel tetracarbonyl Ni(CO)$_4$], cobalt carbonyl, and iron pentacarbonyl. Some of the hazardous carbonyls are volatile and readily taken into the body through the respiratory tract or through the skin. The carbonyls affect tissue directly and they break down to toxic carbon monoxide and products of the metal, which have additional toxic effects.

Reaction products of organo-metallic compounds

An example of the production of a toxic substance from the burning of an organo-metallic compound is provided by the oxidation of diethylzinc :

$$Zn(C_2H_5)_2 + 7O_2 \rightarrow ZnO(s) + 5H_2O(g) + 4CO_2(g) \qquad ...(26.4)$$

Zinc oxide is used as a healing agent and food additive. However, inhalation of zinc oxide fume particles produced by the combustion of zinc organo-metallic compounds causes zinc metal fume fever. This is an uncomfortable condition characterised by elevated temperature and "chills".

TOXICOLOGY OF ORGANIC COMPOUNDS

Alkane Hydrocarbons

Gaseous methane, ethane, propane, n-butane, and isobutane (both C_4H_{10}) are regarded as simple asphyxiants that form mixtures with air containing sufficient oxygen to support respiration. The most common toxicological occupational problem associated with the use of hydrocarbon liquids in the workplace is dermatitis caused by dissolution of the fat portions of the skin and characterised by inflamed, dry, scaly skin. Inhalation of volatile liquid 5–8 carbon n-alkanes and branched-chain alkanes may cause central nervous system depression manifested by dizziness and loss of co-ordination. Exposure to n-hexane results in loss of myelin (a fatty substance constituting a sheath around certain nerve fibres) and degeneration of axons (part of a nerve cell through which nerve impulses are transferred out of the cell). This has resulted in multiple disorders of the nervous system (polyneuropathy) including muscle weakness and impaired sensory function of the hands and feet.

Alkene and Alkyne Hydrocarbons

Ethylene, a widely used colourless gas with a somewhat sweet odour, acts as a simple asphyxiant and anaesthetic to animals and is phytotoxic (toxic to plants). The toxicological properties of propylene (C_3H_6) are very similar to those of ethylene. Colourless, odourless, gaseous 1,3-butadiene is an irritant to eyes and respiratory system mucous membranes; at higher levels it can cause unconsciousness and even death. Acetylene, H-C≡C-H, is a colourless gas with a garlic odour. It acts as an asphyxiant and narcotic, causing headache, dizziness, and gastric disturbances. Some of these effects may be due to the presence of impurities in the commercial product.

Benzene and Aromatic Hydrocarbons

Inhaled benzene is readily absorbed by blood, from which it is strongly taken up by fatty tissues. For the non-metabolised compound, the process is reversible and benzene is excreted through the lungs. As shown in Fig. 26.1, benzene is converted to phenol by a Phase I oxidation reaction in the liver. The benzene epoxide intermediate in this reaction is probably responsible for the unique toxicity of benzene, which involves damage to bone marrow.

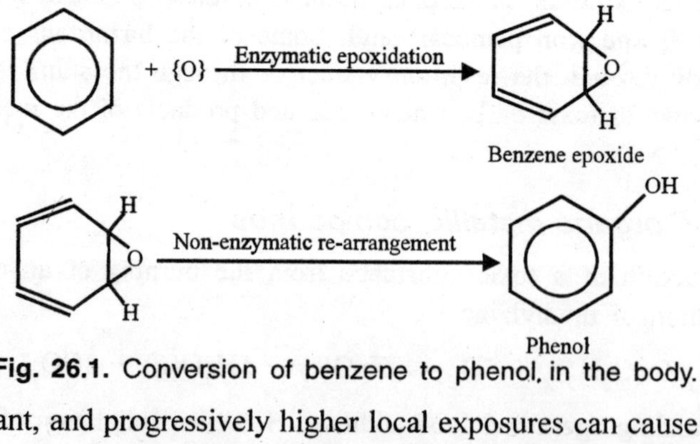

Fig. 26.1. Conversion of benzene to phenol, in the body.

Benzene is a skin irritant, and progressively higher local exposures can cause skin redness (erythema), burning sensations, fluid accumulation (edema) and blistering. Inhalation of air containing about 7 g/m^3 of benzene causes acute poisoning within an hour, because of a narcotic effect upon the central nervous system manifested progressively by excitation, depression, respiratory system failure, and death. Inhalation of air containing more than about 60 g/m^3 of benzene can be fatal within a few minutes.

Long-term exposures to lower levels of benzene cause non-specific symptoms, including fatigue, headache, and appetite loss. Chronic benzene poisoning causes blood abnormalities, including a lowered white cell count, an abnormal increase in blood lymphocytes (colourless corpuscles introduced to the blood from the lymph glands), anaemia, a decrease in the number of blood platelets required for clotting (thrombocytopenia), and damage to bone marrow. It is thought that preleukemia, leukemia, or cancer may result.

Toluene

Toluene, a colourless liquid boiling at 101.4°C, is classified as moderately toxic through inhalation or ingestion; it has a low toxicity by dermal exposure. Toluene can be tolerated without noticeable ill effects in ambient air up to 200 ppm. Exposure to 500 ppm may cause headache, nausea, lassitude, and impaired co-ordination without detectable physiological effects. Massive exposure to toluene has a narcotic effect which can lead to coma. Because it possesses an aliphatic side chain that can be oxidised enzymatically to products that are readily excreted from the body (see the metabolic reaction scheme in Fig. 26.2) toluene is much less toxic than benzene.

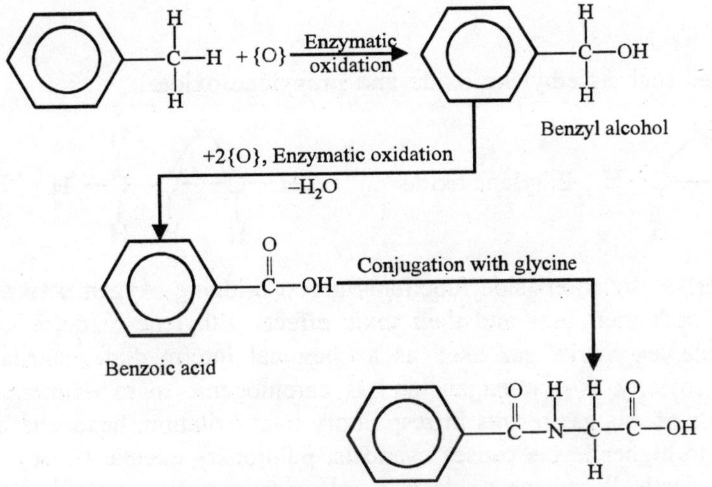

Fig. 26.2. Metabolic oxidation of toluene with conjugation to hippuric acid, which is excreted with urine.

Naphthalene

As is the case with benzene, naphthalene undergoes a Phase I oxidation reaction that places an epoxide group on the aromatic ring. This process is followed by Phase II conjugation reactions in red cell count haemoglobin, and hematocrit in genetically susceptible individuals. Naphthalene causes skin irritation or severe dermatitis in sensitised individuals. Headaches, confusion, and vomiting may result from inhalation or ingestion of naphthalene. Death from kidney failure occurs in severe instances of poisoning.

Polycyclic aromatic hydrocarbons

Benzo[a]pyrene is the most studied of the polycyclic aromatic hydrocarbons (PAHs). Some metabolites of PAH compounds, particularly the 7,8-diol-9,10-epoxide of benzo[a]pyrene shown in Fig. 26.3 are known to cause cancer. There are two stereoisomers of this metabolite, both of which are known to be potent mutagens and presumably can cause cancer.

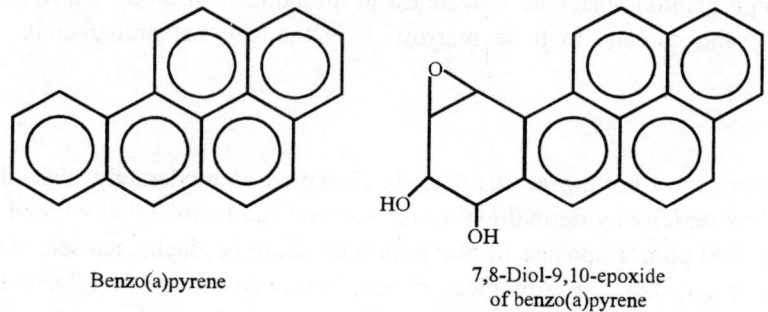

Benzo(a)pyrene

7,8-Diol-9,10-epoxide
of benzo(a)pyrene

Fig. 26.3. Benzo[a]pyrene and its carcinogenic metabolic product.

Oxygen-Containing Organic Compounds

Oxides

Hydrocarbon oxides such as ethylene oxide and propylene oxide,

which are characterised by an eposide functional group bridging oxygen between two adjacent C atoms, are significant for both their uses and their toxic effects. Ethylene oxide, a gaseous colourless, sweet-smelling, flammable, explosive gas used as a chemical intermediate, sterilant, and fumigant, has a moderate to high toxicity, is a mutagen, and is carcinogenic to experimental animals. Inhalation of relatively low levels of this gas results in respiratory tract irritation, headache, drowsiness, and dyspnea, whereas exposure to higher levels causes cyanosis, pulmonary edema, kidney damage, peripheral nerve damage, and even death. Propylene oxide is a colourless, reactive, volatile liquid (bp 34°C) with uses similar to those of ethylene oxide and similar, though less severe, toxic effects. The toxicity of 1,2,3,4-butadiene epoxide, the oxidation product of 1,3-butadiene, is notable in that it is a direct-acting (primary) carcinogen.

Alcohols

Human exposure to the three light alcohols shown in Fig. 26.4 is common because they are widely used industrially and in consumer products.

Methanol, which has caused many fatalities when ingested accidentally or consumed as a substitute for beverage ethanol, is metabolically oxidised to formaldehyde and formic acid. In addition to causing acidosis, these products affect the central nervous system and the optic nerve. Acute exposure to lethal doses causes an initially mild inebriation, followed in about 10–20 hours by unconsciousness, cardiac depression, and death. Sublethal exposures can cause blindness from deterioration of the optic nerve and retinal ganglion cells. Inhalation of methanol fumes may result in chronic, low level exposure.

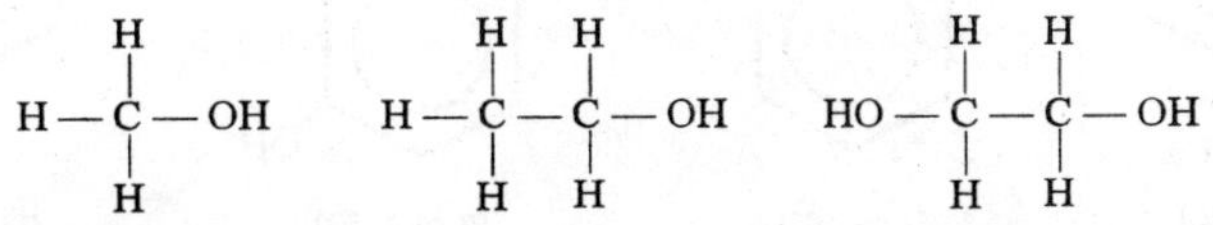

Fig. 26.4. Alcohols such as these three compounds are oxygenated compounds in which the hydroxyl functional group is attached to an aliphatic or olefinic hydrocarbon skeleton.

Ethanol is usually ingested through the gastrointestinal tract, but can be absorbed as vapour by the alveoli of the lungs. Ethanol is oxidised metabolically more rapidly than methanol, first to acetaldehyde (discussed later in this section), then to CO_2. Ethanol has numerous acute effects resulting from central nervous system depression. These range from decreased inhibitions and slowed reaction times at 0.05% blood ethanol, through intoxication, stupor and – at more than 0.5% blood ethanol – death.

Despite its widespread use in automobile cooling systems, exposure to ethylene glycol is limited by its low vapour pressure. However, inhalation of droplets of ethylene glycol can be very dangerous. In the body, ethylene glycol initially stimulates the central nervous system, then depresses it. Glycolic acid, $HOCH_2CO_2H$, formed as an intermediate metabolite in the metabolism of ethylene glycol, may cause acidemia and oxalic acid produced by further oxidation may precipitate in the kidneys as solid calcium oxalate, CaC_2O_4, causing clogging.

Of the higher alcohols, 1-butanol is an irritant, but its toxicity is limited by its low vapour pressure. Unsaturated (alkenyl) allyl alcohol, $CH_2=CHCH_2OH$, has a pungent odour and is strongly irritating to eyes, mouth, and lungs.

Phenols

Fig. 26.5 shows some of the more important phenolic compounds, aryl analogs of alcohols which have properties much different from those of the aliphatic and olefinic alcohols. Nitro groups $(-NO_2)$ and halogen atoms (particularly Cl) bonded to the aromatic rings strongly affect the chemical and toxicological behavior of phenolic compounds.

Although the first antiseptic used on wounds and in surgery, phenol is a protoplasmic poison that damages all kinds of cells and is alleged to have caused "an astonishing number of poisonings" since it came into general use. The acute toxicological effects of phenol are predominantly upon the central nervous system and death can occur as soon as one-half hour after exposure. Acute poisoning by phenol can cause severe gastrointestinal disturbances, kidney malfunction, circulatory system failure, lung edema, and convulsions. Fatal doses of phenol may be absorbed through the skin. Key organs damaged by chronic exposure to phenol include the spleen, pancreas, and kidneys. The toxic effects of other phenols resemble those of phenol.

Aldehydes and ketones

Aldehydes and ketones are compounds that contain the carbonyl (C=O) group, as shown by the examples in Fig. 26.6.

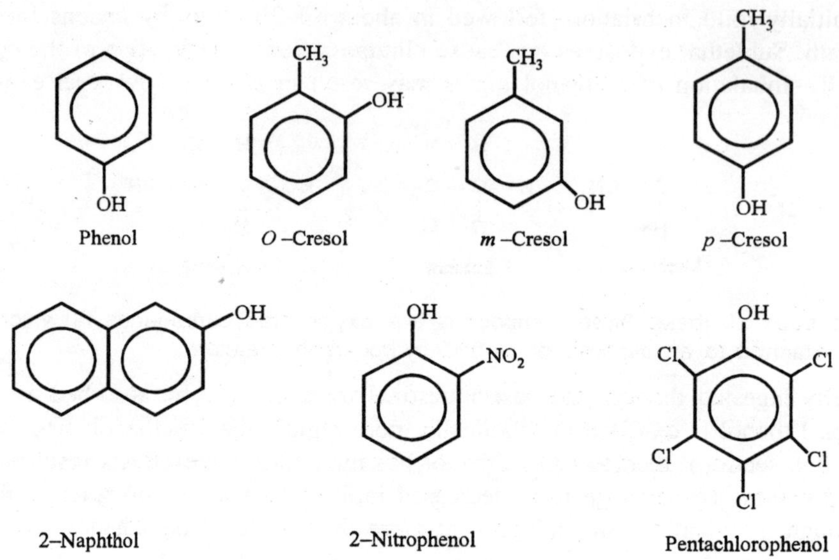

Fig. 26.5. Some phenols and phenolic compounds.

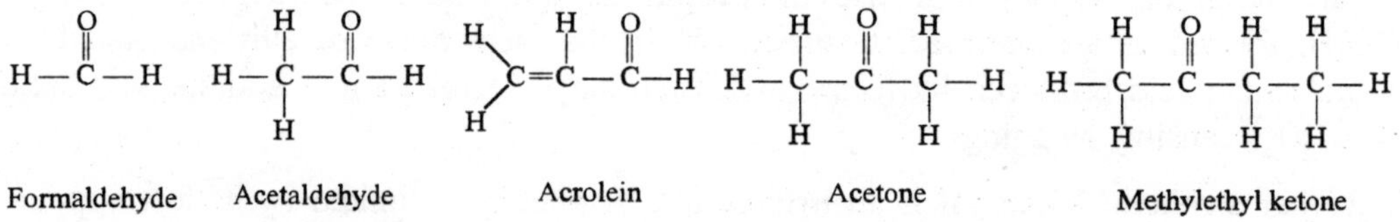

Fig. 26.6. Commercially and toxicologically significant aldehydes and ketones.

Formaldehyde is uniquely important because of its widespread use and toxicity. In the pure form formaldehyde is a colourless gas with a pungent, suffocating odour. It is commonly encountered as formalin, a 37–50% aqueous solution of formaldehyde containing some methanol. Exposure to inhaled formaldehyde via the respiratory tract is usually due to molecular formaldehyde vapour, whereas exposure by other routes is usually due to formalin. Prolonged, continuous exposure to formaldehyde can cause hypersensitivity. A severe irritant to the mucous membrane linings of both the respiratory and alimentary tracts, formaldehyde reacts strongly with functional groups in molecules. Formaldehyde has been shown to be a lung carcinogen in experimental animals. The toxicity of formaldehyde is largely due to its metabolic oxidation product, formic acid.

The lower aldehydes are relatively water-soluble and intensely irritating. These compounds attack exposed moist tissue, particularly the eyes and mucous membranes of the upper respiratory tract. (Some of the irritating properties of photochemical smog are due to the presence of aldehydes). However, aldehydes that are relatively less soluble can penetrate further into the respiratory tract and affect the lungs. Colourless, liquid acetaldehyde is relatively less toxic than acrolein and acts as an irritant and systemically as a narcotic to the central nervous system. Extremely irritating, lachrimating acrolein vapour has a choking odour and inhalation of it can cause severe damage to respiratory tract membranes. Tissue exposed to acrolein may undergo severe necrosis, and direct contact with the eyes can be especially hazardous.

The ketones shown in Fig. 26.6 are relatively less toxic than the aldehydes. Pleasant smelling acetone can act as a narcotic; it causes dermatitis by dissolving fats from skin. Not many toxic effects have been attributed to methyl ethyl ketone exposure. It is suspected of having caused neuropathic disorders in shoe factory workers.

Carboxylic acids

Formic acid, HCO_2H, is a relatively strong acid that is corrosive to tissue. In Europe, decalcifier formulations for removing mineral scale that contain about 75% formic acid are sold; and children ingesting these solutions have suffered corrosive lesions to mouth and esophageal tissue. Although acetic acid as a 4–6% solution in vinegar is an ingredient of many foods, pure acetic acid (glacial acetic acid) is extremely corrosive to tissue that it contacts. Ingestion of, or skin contact with acrylic acid can cause severe damage to tissues.

Ethers

The common ethers have relatively low toxicities because of the low reactivity of the C–O–C functional group which has very strong carbon-oxygen bonds. Exposure to volatile diethyl ether is usually by inhalation and about 80% of this compound that gets into the body is eliminated unmetabolised as the vapour through the lungs. Diethyl ether depresses the central nervous system and is a depressant widely used as an anaesthetic for surgery. Low doses of diethyl ether cause drowsiness, intoxication, and stupor, whereas higher exposures cause unconsciousness and even death.

Acid anhydrides

Strong smelling, intensely lachrimating acetic anhydride,

$$
\begin{array}{ccccccc}
& H & O & & O & H & \\
& | & \| & & \| & | & \\
H- & C- & C- & O- & C- & C- & H \quad \text{Acetic anhydride} \\
& | & & & & | & \\
& H & & & & H &
\end{array}
$$

is a systemic poison. It is especially corrosive to the skin, eyes, and upper respiratory tract, causing blisters and burns that heal only slowly. Levels in the air should not exceed 0.04 mg/m^3 and adverse effects to the eyes have been observed at about 0.4 mg/m^3.

Esters

Many esters (Fig. 26.7) have relatively high volatilities so that the pulmonary system is a major route of exposure. Because of their generally good solvent properties, esters penetrate tissues and tend to dissolve body lipids. For example, vinyl acetate acts as a skin defatting agent. Because they hydrolyse in water, ester toxicities tend to be the same as the toxicities of the acids and alcohols from which they were formed. Many volatile esters exhibit asphyxiant and narcotic action. Whereas many of the naturally occurring esters have insignificant toxicities at low doses, allyl acetate and some of the other synthetic esters are relatively toxic.

Organo-nitrogen Compounds

Organo-nitrogen compounds constitute a large group of compounds with diverse toxicities. Examples of several of the kinds of organo-nitrogen compounds discussed here are given in Fig. 26.8.

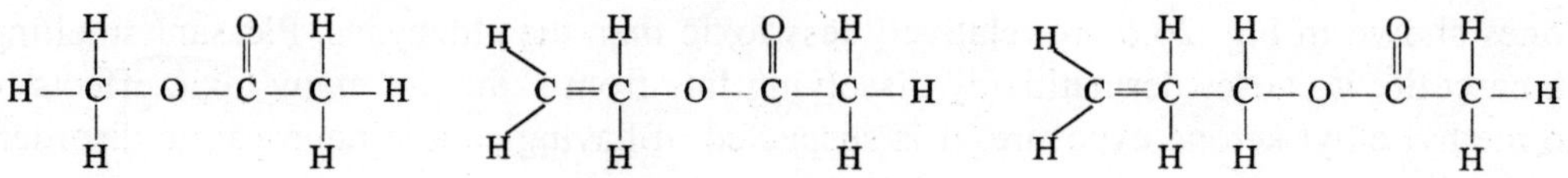

Methyl acetate   Vinyl acetate   Allyl acetate

Fig. 26.7. Examples of esters.

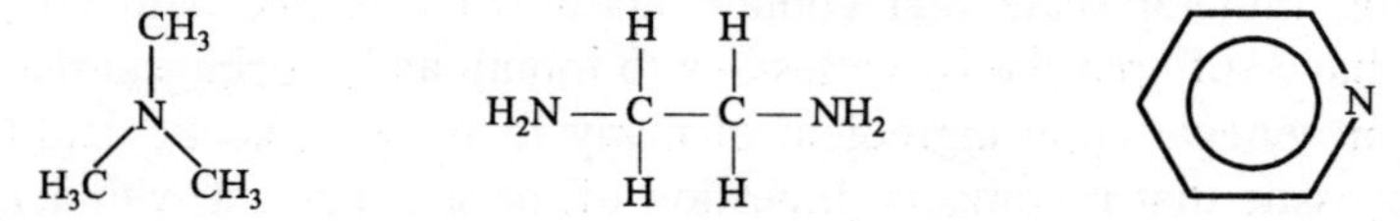

Trimethylamine   Ethylenediamine   Pyridine

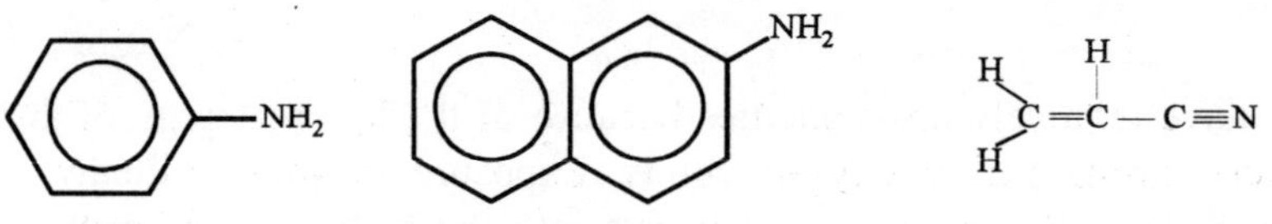

Aniline   2–Naphthylamine   Acrylonitrile

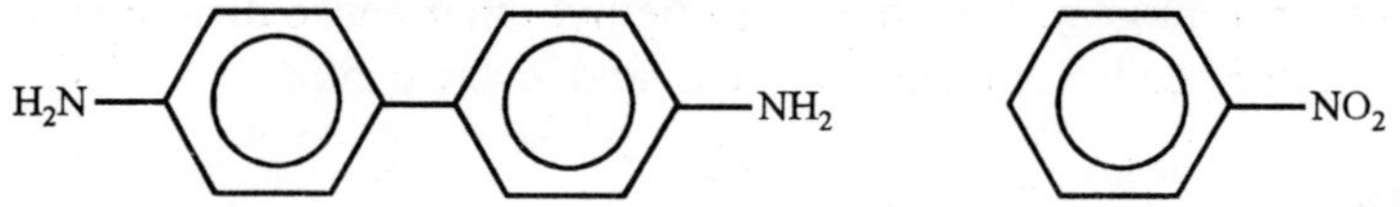

Benzidine   Nitrobenzene

Fig. 26.8. Some toxicologically significant organo-nitrogen compounds.

Aliphatic amines

The lower amines, such as the methylamines, are rapidly and easily taken into the body by all common exposure routes. They are basic and react with water in tissue,

$$R_3N + H_2O \rightarrow R_3NH^+ + OH^- \qquad \qquad ...(26.5)$$

raising the pH of the tissue to harmful levels, acting as corrosive poisons (especially to sensitive eye tissue), and causing tissue necrosis at the point of contact. Among the systemic effects of amines are necrosis of the liver and kidneys, lung haemorrhage and edema, and sensitisation of the immune system. The lower amines are among the more toxic substances in routine, large-scale use.

Ethylenediamine is the most common of the alkyl polyamines, compounds in which two or more amino groups are bonded to alkane moieties. Its toxicity rating is only 3, but it is a strong skin sensitiser and can damage eye tissue.

Carbocyclic aromatic amines

Aniline is a widely used industrial chemical and is the simplest of the carbocyclic aromatic amines, a class of compounds in which at least one substituent group is an aromatic hydrocarbon ring bonded directly to the amino group. There are numerous compounds with many industrial uses in this class of amines. Some of the carbocyclic aromatic amines have been shown to cause cancer in the human bladder, ureter, and pelvis, and are suspected of being lung, liver, and prostate carcinogens. A very toxic colourless liquid with an oily consistency and distinct odour, aniline readily enters the body by inhalation, ingestion and through the skin. Metabolically, aniline converts iron(II) in haemoglobin to iron(III). This causes a condition called methaemoglobinemia, characterised by cyanosis and a brown-black colour of the blood, in which the haemoglobin can no longer transport oxygen in the body. This condition is not reversed by oxygen therapy.

Both 1-naphthylamine (α-naphthylamine) and 2-naphthylamine (β-naphthylamine) are proven human bladder carcinogens. In addition to being a proven human carcinogen, benzidine, 6,6'-diaminobiphenyl, is highly toxic and has systemic effects that include blood hemolysis, bone marrow depression, and kidney and liver damage. It can be taken into the body orally, by inhalation, and by skin sorption.

Pyridine

Pyridine, a colourless liquid with a sharp, penetrating, "terrible" odour, is an aromatic amine in which an N atom is part of a 6-membered ring. This widely used industrial chemical is only moderately toxic with a toxicity rating of 3. Symptoms of pyridine poisoning include anorexia, nausea, fatigue, and, in cases of chronic poisoning, mental depression. In a few rare cases pyridine poisoning has been fatal.

Nitriles

Nitriles contain the $-C \equiv N$ functional group. Colourless, liquid acetonitrile, CH_3CN, is widely used in the chemical industry. With a toxicity rating of 3–4, acetonitrile is considered relatively safe, although it has caused human deaths, perhaps by metabolic release of cyanide. Acrylonitrile, a colourless liquid with a peach-seed (cyanide) odour, is highly reactive because it contains both nitrile and C=C groups. Ingested, absorbed through the skin, or inhaled as vapour, acrylonitrile metabolises to release deadly HCN, which it resembles toxicologically.

Nitro compounds

The simplest of the nitro compounds, nitromethane H_3CNO_2, is an oily liquid that causes anorexia, diarrhea, nausea, and vomiting and damages the kidneys and liver. Nitrobenzene, a pale yellow oily liquid with an odour of bitter almonds or shoe polish, can enter the body by all routes. It has a toxic action much like that of aniline, converting haemoglobin to methaemoglobin, which cannot carry oxygen to body tissue. Nitrobenzene poisoning is manifested by cyanosis.

Nitrosamines

N-nitroso compounds (nitrosamines), characterised by the $>N-N=O$ functional group, have been found in a variety of materials to which humans may be exposed, including beer, whiskey, and cutting oils used in machining. Cancer may result from exposure to a single large dose or from chronic exposure to relatively small doses of some nitrosamines. Nitrosomines was once widely used as an industrial solvent and known to cause liver damage and jaundice in exposed workers.

Isocyanates and methyl isocyanate

Compounds with the general formula R–N=C=O, isocyanates are widely used as industrial chemicals noted for the high chemical and metabolic reactivity of their characteristic functional group. Methyl isocyanate, $H_3C-N=C=O$, was the toxic agent involved in the catastrophic industrial poisoning in Bhopal, India on December 2, 1984, the worst industrial accident in history. In this incident, several tonnes of methyl isocyanate were released, killing 2,000 people and affecting about 100,000. The lungs of victims were attacked; survivors suffered long-term shortness of breath and weakness from lung damage as well as numerous other toxic effects, including nausea and bodily pain.

Organonitrogen pesticides

Pesticidal carbamates are characterised by the structural skeleton of carbamic acid outlined by the dashed box in the structural formula of carbaryl in Fig. 26.9. Widely used on lawns and gardens, insecticidal carbaryl has a low toxicity to mammals. Highly water-soluble carbofuran is taken up by the roots and leaves of plants and poisons insects that feed on the leaves. The toxic effects to animals of carbamates are due to the fact that they inhibit acetylcholinesterase directly without the need to first undergo biotransformation. This effect is relatively reversible because of metabolic hydrolysis of the carbamate ester.

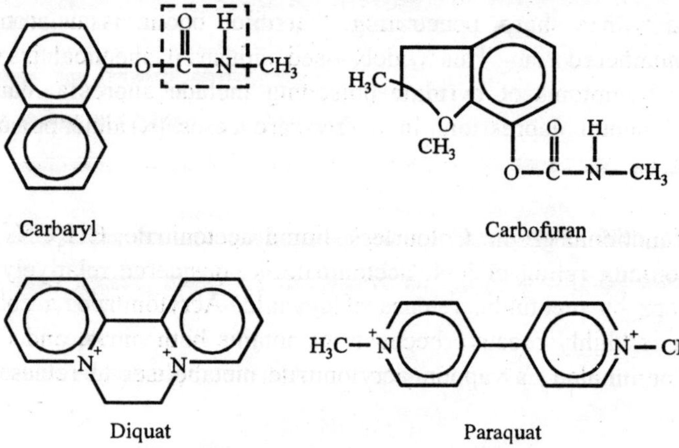

Carbaryl   Carbofuran

Diquat   Paraquat

Fig. 26.9. Examples of organo-nitrogen pesticides.

Reputed to have "been responsible for hundreds of human deaths", herbicidal paraquat has a toxicity rating of 5. Dangerous or even fatal acute exposures can occur by all pathways, including inhalation of spray, skin contact, and ingestion. Paraquat is a systemic poison that affects enzyme activity and is devastating to a number of organs. Pulmonary fibrosis results in animals that have inhaled paraquat aerosols, and the lungs are also adversely affected by non-pulmonary exposure. Acute exposure may cause variations in the levels of catecholamine, glucose, and insulin. The most prominent initial symptom of poisoning is vomiting, followed within a few days by dyspnea, cyanosis, and evidence of impairment of the kidneys, liver, and heart. Pulmonary fibrosis, often accompanied by pulmonary edema and hemorrhaging, is observed in fatal cases.

Organo-halide Compounds

Alkyl halides

The toxicities of alkyl halides, such as carbon tetrachloride, CCl_4, vary a great deal with the compound. Most of these compounds cause depression of the central nervous system, and individual compounds exhibit specific toxic effects. It is a systemic poison that affects the nervous system when inhaled, and the gastrointestinal tract, liver, and kidneys when ingested. The biochemical mechanism of carbon tetrachloride toxicity involves reactive radical species, including

$$Cl-\overset{\displaystyle Cl}{\underset{\displaystyle Cl}{C}}\cdot \quad \cdot OO-\overset{\displaystyle Cl}{\underset{\displaystyle Cl}{C}}-Cl$$

Unpaired
electrons

that react with biomolecules, such as proteins and DNA. The most damaging such reaction occurs in the liver as lipid peroxidation, consisting of the attack of free radicals on unsaturated lipid molecules, followed by oxidation of the lipids through a free radical mechanism.

Alkenyl halides

The most significant alkenyl or olefinic organo-halides are the lighter chlorinated compounds, such as vinyl chloride and tetrachloroethylene :

$$\underset{H}{\overset{H}{>}}C=C\underset{Cl}{\overset{H}{<}} \qquad \underset{Cl}{\overset{Cl}{>}}C=C\underset{Cl}{\overset{Cl}{<}}$$

Vinyl chloride   Tetrachloroethylene

Because of their widespread use and disposal in the environment, the numerous acute and chronic toxic effects of the alkenyl halides are of considerable concern.

The central nervous system, respiratory system, liver, and blood and lymph systems are all affected by vinyl chloride exposure, which has been widespread because of this compound's use in poly(vinyl chloride) manufacture. Most notably, vinyl chloride is carcinogenic, causing a rare angiosarcoma of the liver. This deadly form of cancer has been observed in workers chronically exposed to vinyl chloride while cleaning autoclaves in the poly(vinyl chloride) fabrication industry. The alkenyl organo-halide, 1,1-dichloroethylene, is a suspect human carcinogen based upon animal studies and its structural similarity to vinyl chloride. The toxicities of both 1,2-dichloroethylene isomers are relatively low. These compounds act in different ways in that the *cis* isomers is an irritant and narcotic, whereas the *trans* isomer affects both the central nervous system and the gastrointestinal tract, causing weakness, tremors, cramps, and nausea. A suspect human carcinogen, trichloroethylene has caused liver carcinoma in experimental animals and is known to affect numerous body organs. Like other organo-halide solvents, trichloroethylene causes skin dermatitis from dissolution of skin lipids and it can affect the central nervous and respiratory systems, liver, kidneys, and heart. Symptoms of exposure include disturbed vision, headaches, nausea, cardiac arrhythmias, and burning/tingling sensations in the nerves (paresthesia). Tetrachloroethylene damages the liver, kidneys, and central nervous system. It is a suspect human carcinogen.

Aryl halides

Individual exposed to irritant monochlorobenzene by inhalation or skin contact suffer symptoms to the respiratory system, liver, skin, and eyes. Ingestion of this compound causes effects similar to those of toxic aniline, including incoordination, pallor, cyanosis, and eventual collapse.

The dichlorobenzenes are irritants that affect the same organs as monochlorobenzene; the 1,4-isomer has been known to cause profuse rhinitis (running nose), nausea, jaundice, liver cirrhosis, and weight loss associated with anorexia. *Para*-dichlorobenzene (1,4-dichlorobenzene), a chemical used in air fresheners and mothballs, has become the centre of a controversy regarding the evaluation of carcinogenicity. Some animal tests have suggested that 1,2-dichlorobenzene is a potential cancer-causing substance.

Because of their once widespread use in electrical equipment, as hydraulic fluids, and in many other applications, polychlorinated biphenyls (PCBs) became widespread, extremely persistent environmental pollutants. PCBs have a strong tendency to undergo bioaccumulation in lipid tissue. Polybrominated biphenyl analogs (PBBs) were much less widely used and distributed. However, PBBs were involved in one major incident that resulted in catastrophic agricultural losses when livestock feed contaminated with PBB flame retardant caused massive livestock poisoning in Michigan in 1973.

Organo-halide Pesticides

Exhibiting a wide range of kind and degree of toxic effects, many organo-halide insecticides affect the central nervous system, causing symptoms such as tremor, irregular jerking of the eyes, changes in personality, and loss of memory. Such symptoms are characteristic of acute DDT poisoning. However, the acute toxicity of DDT to humans is very low; it was used for the control of typhus and malaria in World War II, and was applied directly to people. The chlorinated cyclodiene insecticides—aldrin, dieldrin, endrin, chlordane, heptachlor, endosulphan, and isodrin—act on the brain, releasing betaine esters and causing headaches, dizziness, nausea, vomiting, jerking muscles, and convulsions. Dieldrin, chlordane, and heptachlor have caused liver cancer in test animals and some chlorinated cyclodiene insecticides are teratogenic or fetotoxic. Because of these effects, aldrin, dieldrin, heptachlor, and — more recently—chlordane have been prohibited from use in the US.

The major chlorophenoxy herbicides are 2,4-dichlorophenoxyacetic acid (2,4-D), 2,4,5-trichlorophenoxyacetic acid (2,4,5-T or Agent Orange), and Silvex. Large doses of 2,4-dichlorophenoxyacetic acid have been shown to cause nerve damage, (peripheral neutropathy), convulsions, and brain damage. Farmers who had handled 2,4-D extensively have suffered 6 to 8 times the incidence of non-Hodgkins lymphoma as comparable unexposed populations. With a toxicity somewhat less than that of 2,4-D, Silvex is largely excreted unchanged in the urine. The toxic effects of 2,4,5-T (used as a herbicidal warfare chemical called "Agent Orange") have resulted from the presence of 2,3,7,8-tetrachloro-*p*-dioxin (TCDD, commonly known as "dioxin", discussed below), a manufacturing by-product. Autopsied carcasses of sheep poisoned by this herbicide have exhibited nephritis, hepatitis, and enteritis.

TCDD (2,3,7,8-Tetrachlorodibenzo-p-dioxin)

Polychlorinated dibenzodioxins have the same basic structure as that of TCDD (2,3,7,8-tetrachlorodibenzo-*p*-dioxin),

TCDD (2,3,7,8 – tetrachlorodibenzo–*p*–dioxin)

but different numbers and locations of chlorine atoms on the ring structure. Extremely toxic to some animals, the toxicity of TCDD to humans is rather uncertain; it is known to cause a skin condition called chloracne. TCDD has been a manufacturing by-product of some commercial products, a contaminant identified in some municipal incineration emissions, and a widespread environmental pollutant from improper waste disposal. This compound has been released in a number of industrial accidents, the most massive of which exposed several tens of thousands of people to a cloud of chemical emissions spread over an approximately 3-square-mile area at the Givaudan-La Roche Icmesa manufacturing plant near Seveso, Italy, in 1976.

Chlorinated phenols

The chlorinated phenols used in largest quantities have been penta-chlorophenol and the trichlorophenol isomers used as wood preservatives. Although exposure to these compounds has been correlated with liver malfunction and dermatitis, contaminant polychlorinated dibenzodioxins may have caused some of the observed effects.

Organo-sulphur Compounds

Despite the high toxicity of H_2S, not all organo sulphur compounds are particularly toxic. Their hazards are often reduced by their strong, offensive odours that warn of their presence.

Inhalation of even very low concentrations of the alkyl thiols, such as methanethiol, H_3CSH, can cause nausea and headaches; higher levels can cause increased pulse rate, cold hands and feet, and cyanosis. In extreme cases, unconsciousness, coma, and death occur. Like H_2S, the alkyl thiols are precursors to cytochrome oxidase poisons.

An oily water-soluble liquid, methylsulphuric acid is a strong irritant to skin, eyes, and mucous tissue. Colourless, odourless dimethyl sulphate is highly toxic and is a

$H_3C-O-\overset{\displaystyle O}{\underset{\displaystyle O}{\overset{\|}{\underset{\|}{S}}}}-OH$ $H_3C-O-\overset{\displaystyle O}{\underset{\displaystyle O}{\overset{\|}{\underset{\|}{S}}}}-O-CH_3$

Methylsulphuric   Dimethyl
acid   sulphate

primary carcinogen that does not require bioactivation to cause cancer. Skin or mucous membranes exposed to dimethyl sulphate develop conjunctivitis and inflammation of nasal tissue and respiratory tract mucous membranes following an initial latent period during which few symptoms are observed. Damage to the liver and kidney, pulmonary edema, cloudiness of the cornea, and death within 3–4 days can result from heavier exposures.

Sulphur mustards

A typical example of deadly sulphur mustards, compounds used as military poisons, or "poison gases", is mustard oil (bis(2-chloroethyl)sulphide) :

H   H   H   H
| | | |
Cl—C —C—S—C —C—Cl
| | | |
H   H   H   H

Mustard oil

An experimental mutagen and primary carcinogen, mustard oil produces vapours that penetrate deep within tissue, resulting in destruction and damage at some depth from the point of contact; penetration is very rapid, so that efforts to remove the toxic agent from the exposed area are ineffective after 30 minutes. This military "blistering gas" poison causes tissue to become severely inflamed with lesions in the lung can cause death.

Organo-phosphorus Compounds

Organo-phosphorus compounds have varying degrees of toxicity. Some of these compounds, such as the "nerve gases" produced as industrial poisons, are deadly in minute quantities. The toxicities of major classes of organo-phosphate compounds are discussed in this section.

Organo-phosphate esters

Some organo-phosphate esters are shown in Fig. 26.10. Trimethyl phosphate is probably moderately toxic when ingested or absorbed through the skin, whereas moderately toxic triethyl phosphate, $(C_2H_5O)_3PO$, damages nerves and inhibits acetylcholinesterase. Notoriously toxic tri-o-cresyl phosphate, TOCP, apparently is metabolised to products that inhibit acetylcholinesterase. Exposure to TOCP causes degeneration of the neurons in the body's central and peripheral nervous systems with early symptoms of nausea, vomiting, and diarrhea accompanied by severe abdominal pain. About 1–3 weeks after these symptoms have subsided, peripheral paralysis develops manifested by "wrist drop" and "foot drop", followed by slow recovery, which may be complete or leave a permanent partial paralysis.

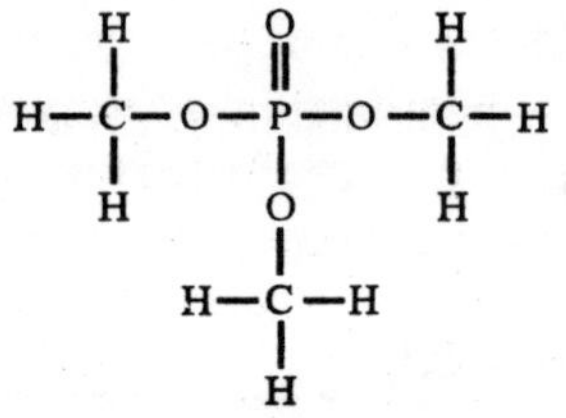

Trimethylphosphate

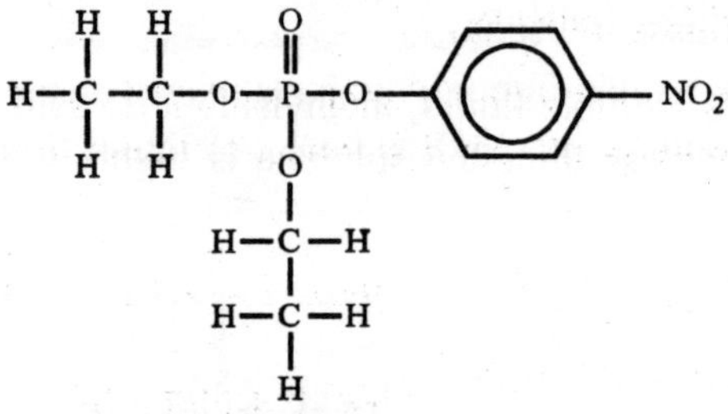

Paraoxon

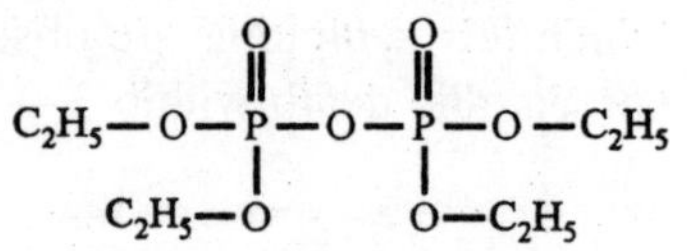

Tetraethylpyrophosphate

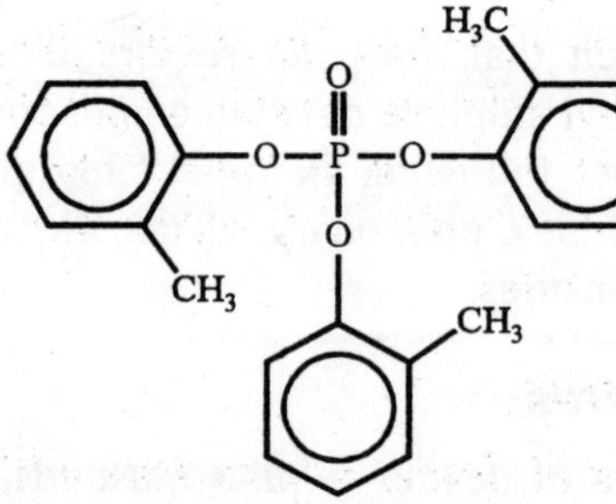

Tri–O–cresylphosphate TOCP

Fig. 26.10. Some organo-phosphate esters.

Briefly used in Germany as a substitute for insecticidal nicotine, tetraethyl pyrophosphate, TEPP, is a very potent acetylcholinesterase inhibitor. With a toxicity rating of 6 (supertoxic), TEPP is deadly to humans and other mammals.

Because esters containing the P=S (thiono) group are resistant to non-enzymatic hydrolysis and are not as effective as P=O compounds in inhibiting acetylcholinesterase, they exhibit higher insect:mammal toxicity ratios than their non-sulphur analogs. Therefore, phosphorothionate and phosphorodithioate esters (Fig. 26.11) are widely used as insecticides. The insecticidal activity of these compounds requires metabolic conversion of P=S to P=O (oxidative desulphuration). Environmentally, organo-phosphate insecticides are superior to many of the organochlorine insecticides because the organo-phosphates readily undergo biodegradation and do not bioaccumulate.

The first commercially successful phosphorothionate/phosphorodithioate ester insecticide was parathion, *O,O*-diehtyl-*O*-*p*-nitrophenylphosphorothionate. This insecticide has a toxicity rating of 6 (supertoxic). Since its use began, several hundred people have been killed by parathion, including 17 of 79 people exposed to contaminated flour in Jamaica in 1976. As little as 120 mg of parathion has been known to kill an adult human and a dose of 2 mg has been fatal to a child. Most accidental poisonings have occurred by absorption through the skin. Methylparathion (a closely related compound with methyl groups instead of ethyl groups) is regarded as extremely toxic.

In order for parathion to have a toxic effect, it must be converted metabolically to paraoxon (Fig. 26.11), which is a potent inhibitor of acetylcholinesterase. Because of the time required for this conversion, symptoms develop several hours after exposure, whereas the toxic effects of TEPP or paraoxon develop much more rapidly. Humans poisoned by parathion exhibit skin twitching and respiratory distress. In fatal cases, respiratory failure occurs due to central nervous system paralysis.

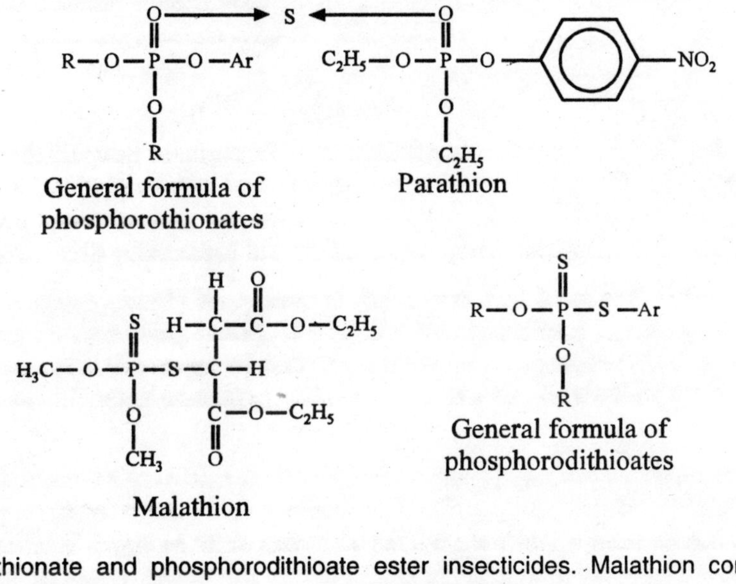

Fig. 26.11. Phosphorothionate and phosphorodithioate ester insecticides. Malathion contains hydrolyzable carboxyester linkages.

Malathion is the best known of the phosphorodithioate insecticides. It has a relatively high insect:mammal toxicity ratio because of its two carboxyester linkage which are hydrolysable by carboxylase enzymes (possessed by mammals, but not insects) to relatively non-toxic products. For example, although malathion is a very effective insecticide, its LD_{50} for adult male rats is about 100 times that of parathion.

Organo-phosphorus military poisons

Powerful inhibitors of acetylcholinesterase enzyme, organo-phosphorus "nerve gas" military poisons include Sarin and VX, for which structural formula are shown below. (The possibility that military poisons such as these might be used in war was a major concern during the 1991 mid-East conflict, which, fortunately, ended without their being employed). A systemic poison to the central nervous system that is readily absorbed as a liquid through the skin, Sarin may be lethal at doses as low as about 0.01 mg/kg; a single drop can kill a human.

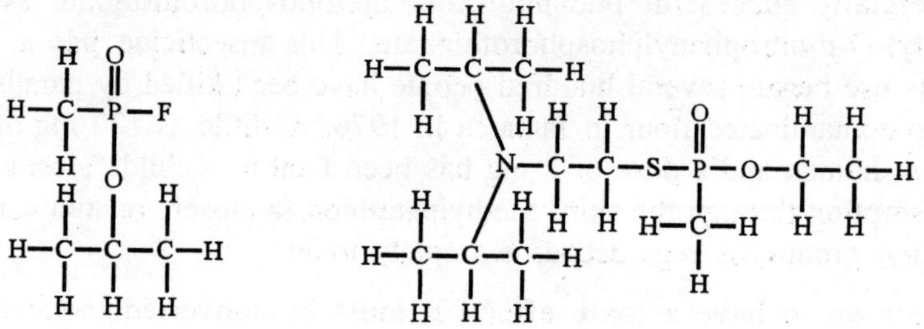

Sarin   VX

CHAPTER 27

Noise Pollution

INTRODUCTION

Noise, commonly defined as unwanted sound, is an environmental phenomenon to which we are exposed before birth and throughout life. Noise can also be considered an environmental pollutant, a waste product generated in conjunction with various anthropogenic activities. Under the latter definition, noise is any sound 'independent of loudness' that can produce an undesired physiological or psychological effect in an individual, and that may interfere with the social ends of an individual or group. These social ends include all of our activities—communication, work, rest, recreation, and sleep.

As waste products of our way of life, we produce two general types of pollutants. The general public has become well aware of the first type—the mass residuals associated with air and water pollution–that remain in the environment for extended periods of time. However, only recently has attention been focused on the second general type of pollution, the energy residuals such as the waste heat from manufacturing processes that creates thermal pollution of our streams. Energy in the form of sound waves constitutes yet another kind of energy residual, but fortunately, one that does not remain in the environment for extended periods of time. The total amount of energy dissipated as sound throughout the earth is not large when compared to other forms of energy; it is only the extraordinary sensitivity of the ear that permits such a relatively small amount of energy to adversely affect us and other biological species.

It has long been known that noise of sufficient intensity and duration can induce temporary or permanent hearing loss, ranging from slight impairment to nearly total deafness. In general, a pattern of exposure to any source of sound that produces high enough levels can result in temporary hearing loss. If the exposure persists over a period of time, this could lead to permanent hearing impairment. Short-term, but frequently serious, effects include interference with speech communication and the perception of other auditory signals, disturbance of sleep and relaxation, annoyance, interference with an individual's ability to perform complicated tasks, and general diminution of the quality of life.

Where once noise levels sufficient to induce some degree of hearing loss were confined to factories and occupational situations, noise levels approaching such intensity and duration are today being recorded on city streets and, in some cases, in and around the home. There are valid reasons why widespread recognition of noise as a significant environmental pollutant and potential hazard or, as a minimum, a detractor from the quality of life, has been slow in coming. In the first place, noise, if defined as unwanted sound, is a subjective experience. What is considered noise by one listener may be considered desirable by another.

Secondly, noise has a short decay time and thus does not remain in the environment for extended periods, as do air and water pollution. By the time the average individual is spurred to action to abate, control, or, at least, complain about sporadic environmental noise, the noise may no longer exist.

Thirdly, the physiological and psychological effects of noise on us are often subtle and insidious appearing so gradually that it becomes difficult to associate cause with effect. Indeed, to those persons whose hearing may already have been affected by noise, it may not be considered a problem at all.

Further, the typical citizen is proud of this nation's technological progress and is generally happy with the things that technology delivers, such as rapid transportation, labour-saving devices, and new recreational devices. Unfortunately, many technological advances have been associated with increased environmental noise, and large segments of the population have tended to accept the additional noise as part of the price of progress.

The engineering and scientific community has already accumulated considerable knowledge concerning noise, its effects, and its abatement and control. In that regard, noise differs from most other environmental pollutants. Generally, the technology exists to control most indoor and outdoor noise. As a matter of fact, this is one instance in which knowledge of control techniques exceeds the knowledge of biological and physical effects of the pollutant.

NOISE SOURCES

Noise is found almost everywhere, not just in factories. Thunder is perhaps the loudest natural sound we hear; it sometimes reaches the threshold of discomfort. Jet aircraft takeoffs are often louder to the listener. Some industrial locations have even louder continuous noise. Community noise is largely produced by transportation sources—most often airplanes and highway vehicles. Noise sources are also in public buildings and residences.

Typical Range of Noise Levels

Variation in noise levels is wide. In rural areas, ambient noise can be as low as 30 dB; even in residential areas in or near cities, this low level is seldom achieved. In urban areas, the noise level can be 70 dB or higher for eighteen hours of each day. Near freeways, 90 to 100 dB levels are not unusual. Many industries have high noise levels. Heavy industries such as iron and steel production and fabricating and mining display high levels; so do refineries and chemical plants, though in the latter, few people are exposed to the highest levels of noise. Automobile assembly plants, saw-mills and planing mills, furniture factories, textile mills, plastic factories, and the like often employ many people in buildings with high noise levels throughout. Hearing impairment of such employees is probable unless corrective measures are taken.

The construction industry often exposes its employees to hazardous noise levels, and at the same time, adds greatly to community noise. Community noise may not be high enough to damage hearing (within buildings) and yet have an unfavourable effect on general health.

Transportation contributes largely to community noise. The public may suffer more than the employees—the crew and passengers of a jetliner do not receive the high noise level found along the takeoff and approach paths. The drivers of passenger cars often are less bothered by their own noise than are their fellow drivers, and they are less annoyed than residents nearby for psychological reasons. Noise levels high enough to be harmful in their immediate area are produced by many tools, toys, and other devices. The dentist's drill, the powder-powered stud-setting tool used in building, home workshop tools, and even hi-fi stereo headphones can damage the hearing of their users. They are often overlooked because their noise is localised. Some typical noise sources are listed in Table 27.1 and are classified by origin.

Table 27.1. Noise levels from various areas.

Noise sources[a]	Noise levels dB or 0.0002 μ bar
Industrial	
Near large gas-regulator, as high as	150
Foundry shake-out floor, as high as	128
Automobile assembly line, as high as	125
Large cooling tower (600')	120–130
Construction and mining	
Bulldozer (10')	90–105
Oxygen jet drill in quarry (20')	128
Rock drill (jumbo)	122
Transportation	
Jet takeoff (100')	130–140
Diesel truck (200')	85–110
Passenger car (25')	70–80
Subway (in car or on platform), as high as	110
Community	
Heavy traffic, business area, as high as	110
Pneumatic pavement-breaker (25')	92–98
Power lawn mower (5'), as high as	95
Barking dog (250'), as high as	65
Household	
Hi-fi in living room, as high as	125
Kitchen blender	90–95
Electric shaver, in use	75–90

[a] Figures in parentheses indicate listening distance. Where a range is given, it describes the difference to be expected between makes or types.

Characteristics of Industrial Noise

Industrial noise varies in loudness, frequency components, and uniformity. It can be almost uniform in frequency response (white noise) and constant in level; large rotating machines and places such as textile mills with many machines in simultaneous operation are often like this. An automobile assembly line usually shows this steady noise with many momentary or impact noises superimposed on it. Other industries show continuous background noise at relatively low levels with intermittently occurring periods of higher noise levels.

Such non-uniform noises are likely to be more annoying and more fatiguing than steady noise, and they are more difficult to evaluate. The terms used to describe them are sometimes ambiguous. Usually the term *intermittent* refers to a noise which is on for several seconds or longer 'perhaps for a several hours' then off for a comparable time.

Industrial noises also vary in their frequency characteristics. Large, slow-moving machines generally produce low-frequency noises; high-speed machines usually produce noise of higher frequency. A machine such as a large motor-generator produces noise over the entire audible frequency range; the rotational frequency is the lowest (1800 RPM produces 30 Hz) but higher frequencies from bearing noise (perhaps brush noise too), slot or tooth noise, wind noise, and the like are also present.

The radiating area of a source affects the amount of sound emitted; not only does the total amount of acoustic energy radiated increase roughly in proportion to the area in vibration, but a pipe or duct passing through a wall emits sound on both sides of the wall. The vibration amplitude can be only a few microinches, yet produce loud sounds. If the natural frequency of an elastic member is near the frequency of the vibration, its amplitude can become large, unless the member is damped or the driving force isolated.

Industrial Noise Sources

In rotating and reciprocating machines, noise is produced through vibration caused by imperfectly balanced parts; bearing noise, wind noise, and other noises also exist. The amplitude of such noises varies with operating speed, usually increasing exponentially with speed. Noise frequencies cover a wide range since normally several harmonics of each fundamental are produced.

Electrical machines produce noise from magnetic as well as mechanical forces. Alternating current machines convert electrical to mechanical energy by cyclically changing magnetic forces which also cause vibration of the machine parts. These magnetic forces change in magnitude and direction as the machine rotates and air gaps and their magnetic reluctance change. The noise frequencies thus produced are related both to line frequency and its harmonics and to rotational speed. The entire pattern is quite complicated. In non-rotating machines (transformers, magnetic relays, and switches), the noise frequencies are the line frequency and its harmonics and the frequencies of vibration of small parts which are driven into vibration when their resonant frequencies are near some driving frequency.

In many machines, more noise is produced by the material being handled than by the machine. In metal-cutting or grinding operations, much noise is produced by the cutting or abrading process and is radiated from both work-piece and machine.

Belt and screw conveyors are sometimes serious noise sources; they are large-area sources; their own parts vibrate and cause noise in operation, and the material they handle produces noise when it is stirred, dropped, or scraped along its path of motion. Vibration from conveyors is conducted into supports and building structure as well. Feeding devices, as for automatic screw machines, often rattle loudly.

Jiggers, shakers, screens, and other vibrating devices produce little audible noise in themselves (partly because their operating frequency is so low), but the material they handle produces much higher frequency noise. Ball mills, tumblers, and the like produce noise from the many impacts of shaken or lifted-and-dropped pieces; their noise frequencies are often low, and much mechanical vibration is around them. Industry uses many pneumatic tools. Some air motors are quite noisy, others less so. Exhausting air is a major noise maker, and the manner in which it is handled has much to do with the noise produced.

Exhausting or venting any gas (in fact, any process which involves high velocity and pressure changes) usually produces turbulence and noise. In liquids, turbulent flow is noisy because of cavitation. Turbulence noise in gas is usually predominantly high frequency; cavitation noise in liquids is normally midrange to low frequency. Both types of noise can span several octaves in frequency range.

Gas and steam turbines produce high-frequency exhaust noise; steam turbines (for improved efficiency) usually exhaust their steam into a condenser; gas turbines sometimes feed their exhausts to mufflers. If such turbines are not enclosed, they can be extremely noisy; turbojet airplane engines are an example.

Impact noises in industry are produced by many processes; materials handling, metal piercing, metal forming, and metal fabrication are perhaps most important. Such noises vary widely because of machine design and location, energy involved in the operation, and particularly because of the rate of exchange of energy.

Not all industrial noises are within buildings; cooling towers, large fans or blowers, transformer substations, external ducts and conveyor housings, materials handling and loading, and the like are outside sources of noise. They often involve a large area and contribute to community noise. Bucket unloaders, discharge chutes, and carshakers, such as those used for unloading ore, coal, and gravel, produce noise which is more annoying because of its lack of uniformity. Fig. 27.1 summarises a range of industrial plant noise levels at the operator's position. Table 27.2. gives some industrial equipment noise sources.

Table 27.2. Industrial equipment noise sources.

System	Source
Heaters	Combustion at burners
	Inspiration of premix air at burners
	Draft fans
	Ducts
Motors	Cooling air fan
	Cooling system
	Mechanical and electrical parts
Air fan coolers	Fan
	Speed alternator
	Fan shroud
Centrifugal compressors	Discharge piping and expansion joints
	Antisurge bypass system
	Intake piping and suction drum
	Air intake and air discharge
Screw compressors (axial)	Intake and discharge piping compressor and gear casings
Speed changers	Gear meshing
Engines	Exhaust
	Air intake
	Cooling fan
Condensing tubing	Expansion joint on steam discharge
Atmospheric vents, exhaust and intake	Discharge jet upstream valves
	Compressors
Piping	Eductors

(Contd...)

System	Source
Excess velocities	Valves
Pumps	Cavitation of fluid
	Loose joints
	Piping vibration
	Sizing
Fans	Turbulent air-flow interaction with the blades and exchanger surfaces
	Vortex shredding of the blades

Mining and construction noise

Both mining and construction employ noisy machines, but construction noise is more troublesome to the general public because of its proximity to urban and residential areas.

Motor trucks, diesel engines, and excavating equipment are used in both kinds of work. Welders and rivetters are widely used in construction, especially in steel-framed buildings and ship building. Pneumatic hammers, portable air compressors, loaders, and conveyors are used in both mining and construction.

	Noise levels, dB(A)								
	80	85	90	95	100	105	110	115	120
1. Pneumatic power tools (grinders, chippers, etc.)			●———————————————●						
2. Moulding machines (I.S., blow moulding, etc.)						●——●			
3. Air blow-down devices (painting, cleaning, etc.)				●————●					
4. Blowers (forced, induced, fan, etc.)	●—————————●								
5. Air compressors (reciprocating, centrifugal				●——●					
6. Metal forming (punch, shearing, etc.)		●————●							
7. Combustion (furnaces, flare stacks) 20 ft		●————————●							
8. Turbogenerators (steam) 6 ft		●—●							
9. Pumps (water, hydraulic, etc.)	●————●								
10. Industrial trucks (LP gas)		●—●							
11. Transformers	●—●								

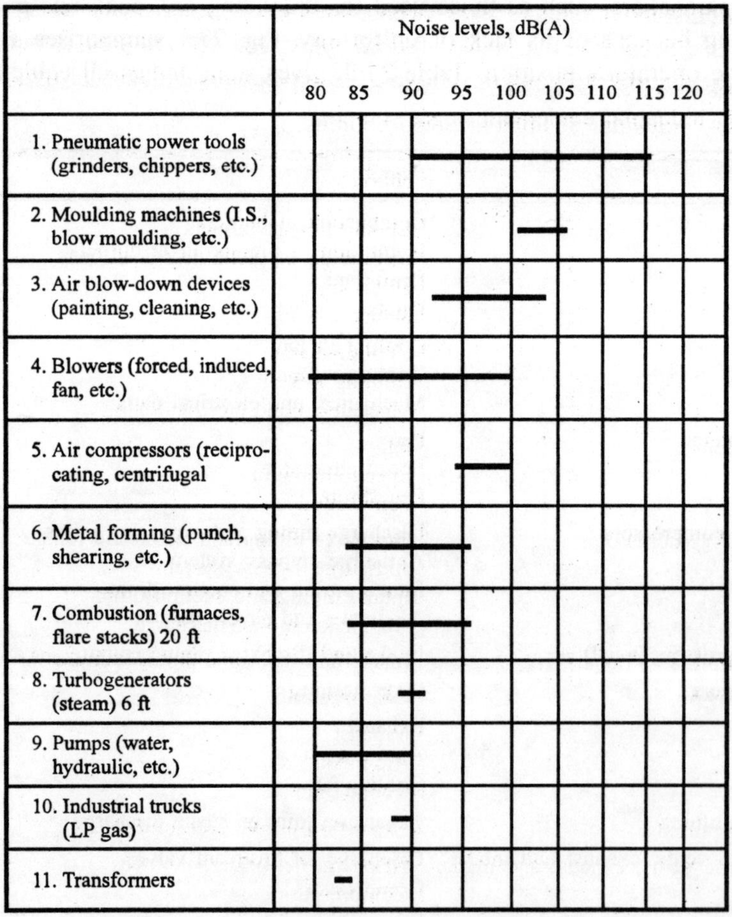

Fig. 27.1. Range of industrial plant noise levels at operator's position.

Crushing and pulverising machines are widely used in mining and in mineral processing. A Portland cement plant has all of these plus ball or tube mills, rotary kilns, and other noise makers as well. Highway and bridge construction use noisy earth-moving equipment; asphalt processing plants produce offensive fumes as well as burner noises; concrete mixing plants produce both dust and noise.

The actual noise-producing mechanisms include turbulence from air discharge; impact shock and vibration from drills, hammers, and crushers; continuous vibrations from shakers, screens, and conveyors; explosive noise; and exhaust turbulence from internal combustion engines.

Transportation noise

Motor vehicles and aircraft are estimated to cause more urban and community noise than all other sources combined, and 60 to 70% of the Indian population lives in locations where such transportation noise is a problem. The number of workers exposed to hazardous noise in their daily work is estimated at between 30 to 40% of the population; most of them are also exposed to the annoying, sleep-destroying general urban noise.

Of all sources, aircraft noise probably causes the most annoyance to the greatest number of people. Airports are located near population centres, and approach and takeoff paths lie above residential areas. Residential buildings are especially vulnerable to aircraft noise since it comes from above, striking roofs and windows, which are usually vulnerable to noise penetration. The individual resident feels that he is pursued by tormenting noise against which he has no protection and no useful channel for protest.

Railway equipment has a high noise output, but causes less annoyance than either highway and street traffic or air traffic. Railway noise is confined to areas adjacent to right-of-way, usually comes from extended sources, and is predictable. It is basically low frequency, thus less annoying than aircraft. Since railway equipment stays on its established routes, protection to residential areas is easily provided. Subway trains can be extremely annoying to their passengers; their noise levels are high; tunnel and station surfaces are highly reflective; and many passengers are present. Newer subway construction is less noisy than in the past (when 100 to 110 dBA was common). Subways, trolleys, and city buses all contribute considerably to urban noise and vibration.

Pumphouses and pipeline distributing terminals compare to other industrial locations, but the pipelines themselves present no noise problem.

Urban noise

The distribution patterns for urban noise are quite complex and differ from city to city; yet, in general, common factors describe them. A noise base exists twenty-four hours per day, consisting of household noises, heating and ventilating noises, ordinary atmospheric noises, and the like; this noise base is usually of low level, from 30 to 35 dB. Here and there are somewhat louder sources of noise; electrical substations, powerplants, shopping centres with roof-mounted equipment, hotels, and other buildings which do not change with the night hours.

During the day and evening hours, this base level increases because of increased residential activity, and also because of general widespread city traffic. A new pattern appears; in busy downtown areas, traffic is heavy, on throughways and main streets, extremely dense traffic occurs during rush periods with heavy traffic continually, some factories are at work, etc. Noise levels in the streets can rise to 85 to 95 dB locally. An intermittent pattern is added from emergency vehicles, aircraft, and the like. The general noise level for the entire city can increase by 10 to 20 dB. The highest noise levels remain local;

after a few blocks, the noise is attenuated through scattering and reflections among buildings, and the many sources blend into the general noise pattern.

The intermittent noise pattern is usually more disturbing than the steady pattern, especially at night. Measurements near main highways and freeways show general traffic noise levels at a distance of 30 metres to be in the 65 to 80 dB range with frequent excursions up to 100 dB or even higher almost always cause by trucks but sometimes by motorcycles. Only at the edges of urban areas does the noise level drop appreciably; and even there, main highways, airports, and such other can prevent a reduction. In most cities, no place is further than a few blocks from some part of the grid of principal streets carrying heavy traffic.

Important contributions are made by entertainment installations. These noise sources include music on streets and in shopping centres, amusement parks and racetracks, paging and public address systems, schools, athletic fields, and even discotheques where performances indoors are often audible several blocks away. Other offenders include sound trucks, advertising devices, and kennels or animal shelters. Because noise sources are distributed over an urban area, the sound-power output of a source can be more informative than the noise level produced at a specific distance.

Specific Noise Sources

Some noise sources are so intense, so widespread, or so unavoidable that they must be characterised as specific cases. Pile driving and building demolition involve violent impacts and large forces and are often done in congested urban areas. Some piles can be sunk with less noisy methods, but sometimes the noise and vibration must simply be tolerated; however, these effects can be minimised and the working hours adjusted to cause the least disturbance. In construction work, explosive-actuated devices can produce high noise levels at the operator's ear. Chain saws and other portable gasoline-powered tools are used close to the operator and, thus, their noise readily reaches the ear.

The dentist's drill, used several hours per day close to the ear, is a hearing hazard. Even musicians (especially in military bands and in amplified rock-type music groups) have shown hearing losses after some time. Usually, of course, music is not continuous, and the intervals of rest for the ear are helpful. Some special industrial processes, such as explosion-forming, shot-peening, and flame-coating, are so noisy that they must be performed in remote locations or behind walls. Many mining, ore-dressing, and other mineral-processing operations are performed in remote locations; but those employees who must be present must be protected.

THE EFFECTS OF NOISE

Human response to noise displays a systematic qualitative pattern, but quantitative responses vary from one individual to another because of age, health, temperament, and the like. Even with the same individual, they vary from time to time because of change in health, fatigue, and other factors. Variation is greatest at low to moderately high sound levels; at high levels, almost everyone feels discomfort. A detailed investigation of the physiological damage to human ears is difficult, but controlled tests on animals indicate the probable type of physiological damage produced by excessively high noise levels.

Reactions to Noise

Specific physiological reactions begin at sound levels of 70 to 75 dB for a 1000 Hz pure tone. At the threshold of such response, the observable reaction is slow but definite after a few minutes. These reactions are produced by other types of stimulation, so they can be considered as reactions to general physiological stress. Most people find that under noisy conditions, more effort is required to maintain attention and that the onset of fatigue is quicker.

Auditory Effects

Within 0.02 to 0.05 seconds after exposure to sound above the 80 dB level, the middle-ear muscles act to control the response of the ear. After about fifteen minutes of exposure, some relaxation of these muscles usually occurs. This involuntary response of the ear—the auditory reflex—provides limited protection against high noise levels. It can not protect against unanticipated impulsive sounds; it is effective only against frequencies below about 2000 Hz. And, in any case, it provides only limited control over the entrance of noise. These muscles relax a few seconds after the noise ceases.

Acoustic Trauma

The outer and middle ear are rarely damaged by intense noise. However, explosive sounds can rupture the tympanic membrane or dislocate the ossicular chain. The permanent hearing loss that results from brief exposure to a very loud noise is termed *acoustic trauma*. Damage to the outer and middle ear may or may not accompany acoustic trauma.

Damage-Risk Criteria

A damage-risk criterion specifies the maximum allowable exposure to which a person can be exposed if risk of hearing impairment is to be avoided.

Psychological Effects of Noise Pollution

Speech Interference

Noise can interfere with our ability to communicate.

Annoyance

Annoyance by noise is a response to auditory experience. Annoyance has its base in the unpleasant nature of some sounds, in the activities that are disturbed or disrupted by noise, in the physiological reactions to noise, and in the responses to the meaning of the messages carried by the noise.

Sleep interference

Sleep interference is a category of annoyance that has received much attention and study. Everyone has been wakened or kept from falling to sleep by loud, strange, frightening, or annoying sounds. Being wakened by an alarm clock or clock radio is common. However, one can get used to sounds and sleep through them. Possibly, environmental sounds only disturb sleep when they are unfamiliar. If so, sleep disturbance depends only on the frequency of unusual or novel sounds. Everyday experience also suggests that sound can induce sleep and, perhaps, maintain it. The soothing lullaby, the steady hum of a fan, or the rhythmic sound of the surf can induce relaxation. Certain steady sounds serve as an acoustical shade and mask disturbing transient sounds.

Common anecdotes about sleep disturbance suggest an even greater complexity. A rural person may have difficulty sleeping in a noisy urban area. An urban person may be disturbed by the quiet when sleeping in a rural area. And how is it that a parent wakes to a slight stirring of his or her child, yet sleeps through a thunderstorm? These observations all suggest that the relations between exposure to sound and the quality of a night's sleep are complicated.

Effects on performance

Noise does not seem to influence the overall rate of work, but high levels of noise can increase the variability of the rate of work. Noise pauses followed by compensating increases in the work rate can occur. Noise is more likely to reduce the accuracy of work than to reduce the total quantity of work. Complex tasks are more likely to be adversely influenced by noise than are simple tasks.

Acoustic privacy

Without acoustical privacy, sound, like a faulty telephone exchange, reaches the wrong number. The result disturbs both the sender and the receiver.

NOISE MEASUREMENTS

Noise measurements are usually conducted for one of three purposes :

1. To understand the mechanisms of noise generation so that engineering methods can be applied to control the noise.
2. To rate the sound field at various locations on a scale related to the physiological or psychological effects of noise on human beings.
3. To rate the sound power output of a source, usually for future engineering calculations, that can estimate the sound pressure it produces at a given location.

This section describes a few frequently used terms and units proposed for the study of sound and noise; most are quite specialised. It also describes techniques and instruments to measure noise.

Frequency Analysis of Noise

The frequency characteristics of sound are important; they describe its annoyance factor as well as its potential for hearing damage. They indicate to the noise control analyst probable sources of noise and suggest means for confining them. To the mechanical or design engineer, frequency analysis can show the source of machine vibrations which produce noise and contribute to damage. Noise is often a symptom of malfunctioning, and the frequency analysis can sometimes describe the malfunction. Spectral characteristics are also useful in describing the transmission of sound through a wall or its absorption by some material.

Speech Interference and Noise Criteria (NC) Curves

Interference with the intelligibility of speech is a serious problem caused by noise; it impairs comfort, efficiency, and safety. The amount of such interference depends on both frequency and level of sound; and a family of curves has been developed to describe various noise environments. These are called *NC curves*.

Vibration and vibration measurement

Noise is usually accompanied by vibration; noise is caused by vibration; noise causes vibration. However, exposure to vibration often contributes to fatigue and, thus, to loss in efficiency and to accidents. Vibration study is an important part of noise control, especially in analysing problems.

Measuring Noise

Background corrections

Sound-measuring instruments do not distinguish between the noise of principal interest and any background noise present. Background noise corrections are needed to determine the contribution of a specific source or a group of sources.

The simplified procedure is as follows: the noise level is measured with the unknown source(s) in operation; then with the other conditions unchanged, the unknown source is stopped and the background level measured. (If several different sources are being measured, they can be turned off in different combinations). Analysts must evaluate the difference between the readings on an energy basis, not simply by taking numerical differences.

Instruments for measuring noise

The basic instrument for measuring noise levels is the sound level meter, sensitive to RMS sound pressures between about 20 and 20,000 Hz. It is equipped with weighting networks, fast and slow response, an attenuator with 10–dB steps, and an indicating meter which spans 16 dBs, from –6 to +10 dB. It operates over a total range of about 30 to 140 dB sound pressure level. Minimum specifications for general-purpose sound level meters are in the International Electrotechnical Commission (IEC).

Most sound level meters have output terminals so that accessories can be attached; these accessories include impact-noise meters, octave-band and 1/3 octave-band filter sets, graphic recorders, and the like. Self-contained analysers are also available, with all components housed in a single unit; these often have variable width settings.

Vibration meters are like sound level meters in that they contain a sensing element, amplifier, attenuator and output meter. They do not have weighting circuits, but do have integrating circuits, so that with a single pickup acceleration, velocity and displacement can be measured. (Sometimes a vibration-measuring accessory operates with a sound level meter; it consists only of a vibration pickup, integrating circuits, and a table of conversions; the electronic circuitry in the sound level meter operates with it.) Vibration meters are usually calibrated in acceleration (cm/sec^2), velocity (cm/sec), or displacement (cm) and display either peak or average values.

Fig. 27.2 shows simplified block diagrams of typical sound level meters, vibration meters, audiometers, and noise monitors; Fig. 27.3 shows a block diagram of the elements of a sound level meter.

Primary calibration of microphones or vibration pickups is normally done in the best equipped laboratories; most manufacturers of the instruments maintain such laboratories. Their standards are traceable to those maintained by government standards agencies. Such calibrations are made with piston phones, shake-tables, electrostatic actuators, and the use of reciprocity methods. Until recent years, the Rayleigh disk was widely used. Reciprocity calibrators are now easily available, and small electronic generators are widely used for field checks.

Audiometers are basically audio-frequency oscillators, adjustable in frequency and in output level, with headsets for the subject. The subject determines the existence of the threshold. Some audiometers are more complex, having recording devices and masking noise facilities. Since loudness is a subjective magnitude, it cannot be measured with a simple instrument. Arbitrary systems have been developed for determining loudness by instrumental measurements, using rather complex empirical criteria and techniques. A self-contained instrument is available, which analyses the sound as it is received, breaks it up into *critical bands* "actually 1/3 octave bands" performs a number of calculations on these data, and produces a single-figure output. This equipment has been incorporated into a system for continuously monitoring impulse and continuous noise and for printing out loudness levels and occurrence times.

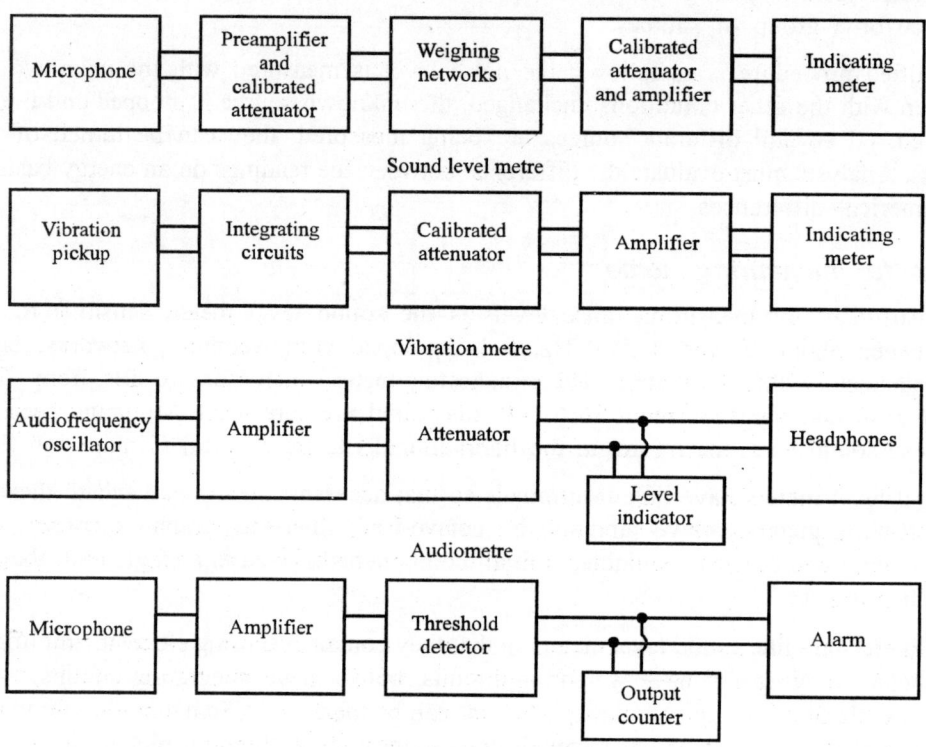

Fig. 27.2. Simplified block diagram of noise-measuring instruments.

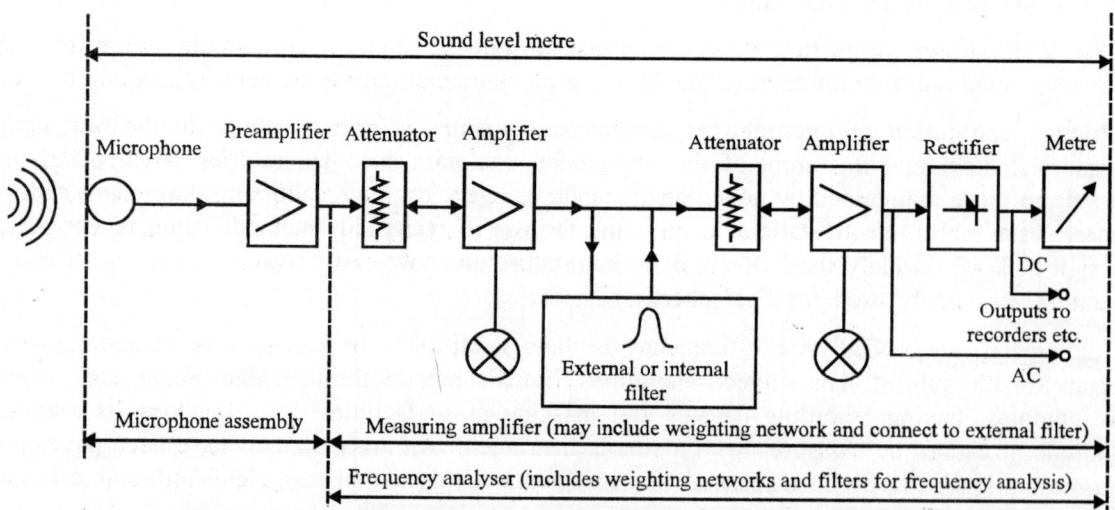

Fig. 27.3. Block diagram of the elements of a sound level metre.

Impact and impulse magnitudes

The brief noise pulses (lasting only a fraction of a second) caused usually by impacts or explosions require special means of measurement. Even the fast setting of a standard sound level meter usually requires about 0.2 second to reach its deflection, and an overshoot also occurs.

Analysts can ideally measure true impulse or impact magnitudes by recording the wave form on an oscillograph for detailed analysis, and then determining frequencies and total energy as well as peak magnitudes. Most impulsive noise meters are somewhat simpler.

A typical impact-noise measuring instrument can determine the peak pressure (by a sample-and-hold circuit or by a peak-reading voltmeter) and the time-average (the average voltage over a chosen period of time). The result is indicated on a pointer-type instrument, where the reading is retained long enough to permit accurate readings. The time-average mode of operation can average over any selected period of time from 2 milliseconds to 1 seconds; this average provides a simple means of approximating the pulse length. The instrument should respond to rise times as short as 100 microseconds.

Analysts have devised simplified techniques for approximating peak values of impact noise based on the dynamic characteristics of the sound level meter. One method for checking at the 140 dB peak level sets the sound level meter on its C or flat weighting, fast response and the 130 dB range. If the pointer swings no higher than a 125 dB indication for impulsive noise, the peaks are probably no higher than 140 dB. This method assumes that the sound pulse is perhaps 25 to 50 milliseconds long; shorter pulses read too low; longer pulses read too high since the amount of energy in the pulse as well as the peak value affects this type of indication. The length of impact and explosive noise pulses is often increased by reflected sound; indoors, it is increased by reverberation.

Monitoring devices (noise dosimeters)

Since noise exposure involves both duration and noise level, continuous observation is required to evaluate exposure if durations and levels vary.

Noise monitors or noise exposure meters respond only when a preset level is reached; some include several circuits with different preset values. They are available in small portable packages to be worn by a worker or in larger (and more accurate) models which are not restricted to the within-limits or exceeding-limit indication. Fig. 27.4 shows a block diagram of a typical dosimeter that complies with the Occupational Safety and Health Administration (OSHA) criteria.

Where variations in level are wide and not easily predictable, integrating devices are used. The Occupational Safety and Health Act (and others) halves the permissible exposure for each 5 dB increase in level above 90 dB; this computation can be made by a simple circuit.

Monitoring devices may simply record the duration of exposure above a preset level for daily examination; they may signal hazardous levels with a flashing light; for airport use, they record times of occurrence of all levels above a preset threshold and sound some alarm when another (and higher) pre-set limit is reached.

Field measurements

Field measurements are measurements where little or no control over operating conditions is possible. Laboratory measurements are made in circumstances where some control is possible and conditions can at least be predicted. Usually less sophisticated equipment is used for field work; consequently, the experience and judgment of the analyst become important.

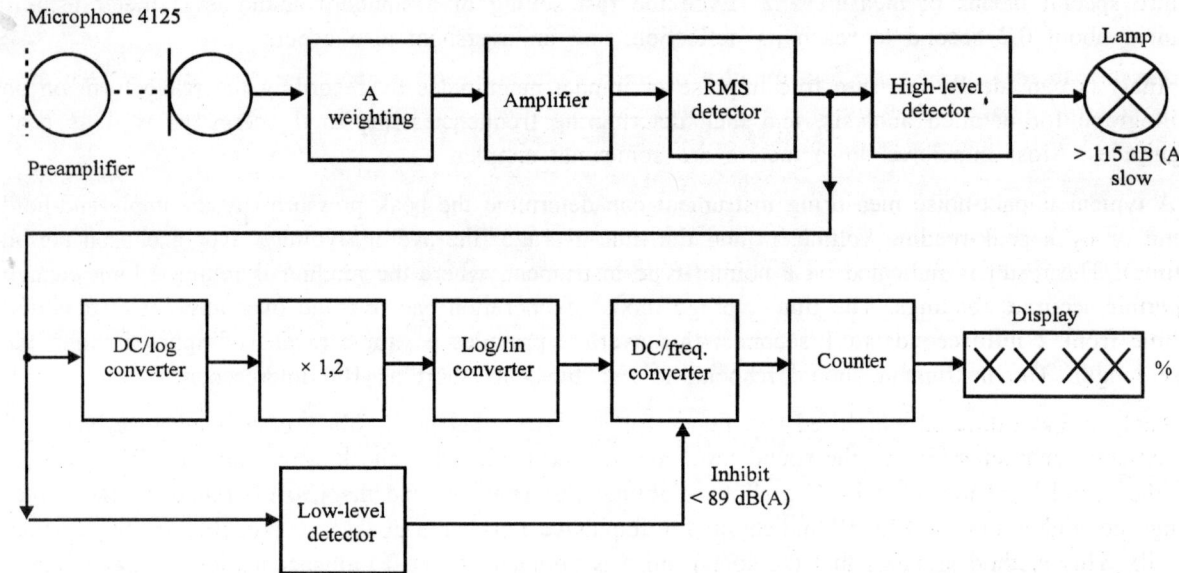

Fig. 27.4. Simplified block diagram of a personal noise dose meter, complying with OSHA criteria.

Background noise is normally present, and often it is not possible to maintain conditions while a machine under test is started and stopped. The effect of background noise should be minimised. Stationary noise sources, if small, can be approached to within a metre or so to raise the level of noise being measured relative to the background. (Questions of directivity, size of source, reflections and interferences, and other experimental problems always appear; such problems must be dealt with individually at the time they appear). For large sources, such handling of experimental problems is seldom possible; where the effect of an entire building or a large cooling tower on a neighborhood, for example, is being studied, the ingenuity and experience of the analyst is taxed to the utmost.

Transient noises—auto horns, vehicle noises, passers by, and the like—must be avoided as far as possible. Measurements within buildings are easier in that traffic is controllable and atmospheric noises do not interfere, but the proximity of the walls introduces danger of reflections in addition to those from the analyst and his equipment. Both the abilities and limitations of the measuring equipment should be known. Its condition should be checked just before use and again just afterward, including a field calibration check.

Practical problems

Whether indoors or out, in good conditions or bad, some problems are common to almost all noise measurements. The convenient measurable range of sound levels extends from about 10 dB above the ambient level (which can be as low as 20 dB in good test rooms) to about 140 dB or so. Whenever measurements are made with background noise within 10 dB of the test noise level, errors tend to increase. By whatever means is convenient, the difference between the test noise level and background should be increased. Work outside these limits is possible, but not easy.

The analyst can increase the sensitivity of the measuring instrument at low levels by using a pre-amplifier (but with danger of introducing electrical noise) or reduce it at high levels with attenuators (but the mechanical vibration in high-noise fields can affect the operation of the instrument). Microphones vary in their characteristics. Condenser types have wide frequency response, good sensitivity, and withstand high temperatures and high-sound levels. They are often deranged by high humidity, they require some auxiliary equipment; and their cost is rather high.

Piezoelectric microphones (made with synthetic ceramic) are excellent for general purpose use. Their stability and sensitivity are good, but their frequency response is less wide than that of condenser microphones. Their cost is relatively low. They are not usable at extremely low or extremely high temperatures.

Dynamic microphones can be used at both lower and higher temperatures than the piezoelectric and at higher humidities than the condenser types. They sometimes lose calibration over long periods of time, however, and their frequency response can vary from temperature change. Their construction employs acoustic resonating chambers, which must remain clean and dust free. None of the three types is completely non-directional.

Frequently the acoustic power of a source is needed but is not supplied by the maker. Ideally this kind of data should be determined in a quiet room or in a free-field situation with low background noise, but practical values can often be approximated by field measurements. For outside sources, if reflecting surfaces (except the ground or pavement) are not close by and the device is the principal noise source, the following equation gives the power level :

$$PWL = SPL + 20 \log_{10} r + 0.5 - Q \text{ dB}$$

Here r is the distance in feet from the source to the sound pressure meter, and Q is a directivity factor.

The directivity factor varies from about 2 for a hard pavement to about 0.5 for absorbing grass; the sound radiated by the machine is usually somewhat directional in itself, too. Several SPL measurements should be taken to strike an average. Values obtained in this way are seldom highly accurate; usually field measurements give low values for PWL. A simpler procedure (for rough estimates) is to measure the SPL at a distance of 1 metre; the PWL value is about 15 units higher. (This procedure can be used on small sources only.) Neither of these procedures works well on extended sources such as cooling towers, large trucks, or the like, nor for distances greater than about 60 metres.

The term here called *directivity factor* is sometimes called *directional gain* or DG. It is the factor by which the power of the source should be multiplied, if its sound radiation were non-directional, to give the measured level in the direction of measurement. The directivity factor must be determined for the location and direction in question for each measurement or machine. Field measurements can be disturbed by wind noise. Analysts can reduce this problem by using a windscreen. Windscreens have various forms, but a common method is to surround the microphone with a skeletal spherical frame about 15 to 20 cm in diameter and stretch fine meshed cloth over it.

At extreme distances, sound behaviour becomes unpredictable; but noise control work seldom requires measurements beyond a mile or so. At high altitudes, low barometric pressure, or high humidity, the density change in the atmosphere affects calculations. For most work, these effects can be ignored.

Multiple sources are usually present, as well as multiple transmission paths; and as in other measurements, the presence of the measuring device can affect the local conditions. Interpretation in both raw and tabulated data can be a problem. Judgement, based on experience, is of the greatest importance in noise-control work.

NOISE ASSESSMENT AND EVALUATION

Workplace Noise

A hearing conservation programme begins with determining a worker's noise exposure. It is the first step in identifying those employees who must be included in the total hearing conservation programme. Several criteria suggest the need for noise assessment :

1. If any area has a past record of excessive noise.
2. If employees complain of discomfort or temporary hearing loss.
3. If employees are unable to converse easily, without shouting, at a distance of 2 feet.

A dosimeter is the most important instrument in noise assessment. It determines the noise level to which employees are exposed by measuring sound over time and analysing the information to produce a noise dose, expressed in a percentage. A noise dose, D, is defined as :

$$D = \frac{C}{T} \qquad \qquad ...(27.1)$$

where D is the noise dose; C, actual duration of exposure in hours; T, noise exposure limit in hours. The allowable exposure time is listed in Table 27.3. The employee exposure exceeds the OSHA limits if the noise dose, D, exceeds unity, or 100%.

Where the daily exposure is due to more than one noise level, the ratios for each level are added to compute the total noise dose as follows :

$$D = \frac{C_1}{T_1} + \frac{C_2}{T_2} + ... \frac{C_n}{T_n} \qquad \qquad ...(27.2)$$

Table 27.3. OSHA hearing conservation table.

A-weighted sound level	Duration (hours)
80 dB *	32
85†	16
90‡	8
95	4
100	2
105	1
110	0.5
115	0.25
120	0.125
125	0.063
130 #	0.031

* Measuring threshold

† Hearing conservation begins—50% dose

‡ Eight-hour criteria level

\# Minimum upper range

Noise dosimeters

The OSHA pushed the development of noise instrumentation to higher capability levels. Noise dosimeters are programmable, data-logging, multithreshold instruments. They communicate directly with printers and personal computers. They provide information such as multithreshold doses, average sound level or LAvg, peak levels, histograms, statistical distributions, and projections.

A dosimeter is a small unit which can be attached to a belt or placed in a shirt pocket. The microphone is placed at the ear level, about 3 to 5 inches from the head, to avoid shadow error. Dosimeters must integrate all noise—continuous, variable, and impulse or impact—in the range between 80 to 130 dBs, using a 5-dB doubling or exchange rate; and they must be A weighted. The following definitions explain these dosimeter terms and keys :

1. *Thresholds:* That sound level below which the instrument assumes no noise.

2. *LTL:* Low threshold level, when set at 80 dB, measures only 80 dB and above; assumes 0s below that level. Used for OSHA hearing conservation compliance.

3. *HTL:* High threshold level, when set at 90 dB, measures only 90 dB and above; assumes 0s below that level. Used for OSHA engineering control compliance.

4. *Criterion Level:* 90 dB, when measured for eight hours, reads 100% dose.

5. *5dB Exchange Rate:* Used by OSHA; every 5 dB increase or decrease either doubles or halves the dose.

6. *A weighting:* Used by OSHA (and others); reads the way the human ear hears sound. Basically the low frequencies and some high are not read as loud as they acoustically are.

7. *LAVB:* Average sound level for the actual time measured based on other than a 3 dB exchange rate, i.e 5 dB-OSHA, 4 dB-DOD, or a 3 dB exchange rate with a threshold. Useful for short-term samples or for making projections.

8. *LEQ:* Average sound level for the actual time measured based on a 3 dB exchange rate and no threshold.

9. *TWA:* An eight-hour dB average regardless of the sample time length. For example, in a four-hour sample measurement, TWA assumes the last four hours are 0s and averages them in the overall reading, making the average dB level lower than it should be appropriate to use if someone works other than an eight-hour day, ie. twelve-hour shifts. TWA takes twelve hours of exposure and condenses it into eight giving the appropriate TWA for eight hours. Not appropriate to use for short samples. (Note: LAvg is always greater than TWA in samples less than eight hours. Samples of exactly eight hours result in equal LAvg and TWA, and samples longer than eight hours produce TWAs that exceed LAvg).

10. *DOSE:* The accumulated exposure obtained, expressed in percent allowable over eight hours.

11. *SEL:* A one-second average of a noise occurrence that is any duration in length. (Used to compare several noise events with different time durations).

The capability for multithreshold dose measurements is helpful. One threshold level is 90 dBA set by OSHA engineering regulations. A concern arises when an employee works in an environment which is always below 90 dBA but where the average sound level is in the high 80s. The dosimeter indicates this area as having no hazard (0% dosage) while the worker is within a few decibels of maximum allowable exposure (100% exposure). Having a second available threshold set at 80 dBA allows the analyst to detect marginal environments while assuring compliance with both the engineering regulation and the hearing conservation amendment. In addition, having these data available provides clues on how to attack noise problems through engineering efforts.

An analyst can determine the effect of using hearing protection devices in the workplace environment by subtracting the NRR value of a device from the LAvg. Histograms, or sound-level–time history, in the printed format are quite useful. The benefits of histograms include identification of noise patterns— and significant break in patterns—during the workday. Patterns help identify sound sources which require corrective action to lower total exposure to the worker. The histogram identifies the exact time when a significant break in these patterns occurs. This information is the basis of documentation presented at compensation hearings.

Sound level metres

This instrument is used to spot-check for excessive noise. When coupled with an octave-band filter, it can also be used to assess the frequency content of the noise. This information is useful for noise control.

Integrating sound level meters

Integrating sound level meters measure average dB levels and are useful for evaluating rapidly changing noise environments. They present a direct A-weighted dB (dBA) reading.

Community Noise

The three major sources of community noise are aircraft, highway traffic and construction activities. Construction noise must be controlled by local ordinances. Control usually involves muffling of air compressors, jack hammers, hand compactors, etc. Since mufflers cost money, contractors will not take it upon themselves to control noise, and outside pressure must be exerted.

The effort should be to divert flight paths away from populated areas and, whenever necessary, to have pilots use less than maximum power when the takeoff carries them over a noise-sensitive area. Often this approach is not enough to prevent significant noise induced damage or annoyance, and aircraft noise remains a real problem in urban areas.

Supersonic aircraft present a special problem. Not only their engines are noisy, but the sonic boom can create property damage and mental anguish. The magnitude of this problem will become known only when (and if) supersonic airlines begin regular service over land. The third major source of community noise is from highway. The car or truck create noise by a number of means : (i) exhaust noise; (ii) tyre noise; (iii) engine intake noise; (iv) gears and transmission; and (v) aerodyanamic (wind) noise.

A modern passenger car is so well muffled that its most important contribution, at a moderate and high speed, is tyre noise. Other cars and motorcycles, on the other hand, contribute exhaust, intake and gear noise. The worst offender on the roads is the heavy truck. Truck noise is generally from all the above sources. In most cases, the total noise generated by vehicles can be correlated directly to the truck volume.

Alternatives for reducing traffic noise

A number of alternatives are available for reducing traffic noise. First, the source can be controlled by making quieter vehicles; second, highway could be routed away from populated areas; and third, noise could be baffled with walls or other types of barriers. Other methods include lowering speed limits, designing for nonstop operation, and reducing all highways to less than 8% grade.

Vegetation, surprisingly, makes a very poor noise screen. The most effective buffers have been to raise or lower the highway, or to build physical barriers beside the road and thus screen the noise. All of these have limitations. For example, noise will bounce off the walls and create little or no noise shadow. In addition, walls hinder highway ventilation, thus contributing to the buildup of dangerous air pollutants. The design noise level for various uses, are shown in Table 27.4.

Table 27.4. Design noise levels.

Land category	Design noise level	Description of land use
A	60 dB Exterior	Activities requiring special qualities of serenity and quiet, such as amphitheatres.
B	70 dB Exterior	Residence, motels, hospitals, schools, parks, libraries.
	55 dB Interior	Residences, motels, hospitals, schools, parks, libraries.
C	75 dB Exterior	Developed land not covered in categories A and B.
D	No limit	Undeveloped land.

Noise in the home

Private dwellings get noisier because of internally produced noise as well as external community noise. The list of gadgets in a modern home are the noise-makers. Some examples of domestic noise-makes are listed in Table 27.5.

Table 27.5. Some domestic noise-makers.

Item	Sound level dB(A)
Vacuum cleaner (3M)	75
Inside quiet car (80 KMPH)	65
Inside sports car (80 KMPH)	80
Flushing toilet	85
Garbage disposal (1M)	80
Window air conditioner (3M)	55
Ringing alarm clock (0.6M)	80
Lawn mower (operator's position)	105
Rock band (3M)	115
Snowmobile (driver's position)	120

Similar products of different brands often will vary significantly in noise levels. When shopping for an appliance, it is just as important therefore to ask "How noisy is it?" as it is to ask him "How much does it cost?" And if he looks at you as if you had two heads, explain to him that he should know the dB(A) at the operator's position for all of his wares. He may actually bother to find out.

Instrumentation

Generally one of the following types of instruments is indicated for community noise.

Noise logging dosimeters/analyser

Used for longer measurements, possibly twenty-four hours or more. These instruments should have the capability to measure exceedance levels and Ldn levels and to store all data for later printout or downloading to a personal computer.

Integrating L_{eq} sound level meter

Used for measuring noise for shorter periods of time where the noise is variable. Additional features are the abilities to add an octave-band filter set, to output for downloading information to a printer or personal computer, and to measure exceedance levels and SEL.

Simple sound level meter

A low cost alternative when a quick on-the-spot level check is all that is required. The meter should have a Max-hold feature allowing for accurate measurements of short-duration noises such as moving vehicles or impact noises.

Traffic noise prediction

Computer programmes are available to predict noise levels generated by a proposed project. They can be used to determine the noise impact on surrounding areas. These programmes (hand calculations are possible for simple cases) can determine the noise levels at distances away from the project and for various atmospheric effects, barriers, and topographical features. Equations for these prediction methods are well documented.

Plant Noise Survey

The following guidelines outline how to conduct a plant noise survey :

1. Review the working area situation thoroughly; the type of sound fields, the number of people affected, and their locations.
2. Determine which machine generates the most sound and find its true sound level.
3. Run the survey under time variations as well as normal plant operations. Note that a change in humidity or outside interference can alter results.
4. Select sound measuring devices carefully, giving particular attention to the types of microphones necessary.
5. Be familiar with the sound measuring equipment before testing. Make sure it is correctly calibrated.
6. Set up the measuring devices properly and have no interferences to testing conditions, if possible. Note that the microphone must be mounted on a tripod at the same height as the worker's ear.
7. Make sure that all equipment aiding in measuring—the meter, recorder, and correcting apparatus— is outside the testing area.

NOISE CONTROL

Source-Path-Receiver Concept

If you have a noise problem and want to solve it, you have to find out something about what the noise is doing, where it comes from, how it travels, and what can be done about it. A straightforward approach is to examine the problem in terms of its three basic elements: that is, sound arises from a source, travels over a path, and affects a receiver or listener.

The source may be one or any number of mechanical devices that radiate noise or vibratory energy. Such a situation occurs when several appliances or machines are in operation at a given time in a home or office.

The most obvious transmission path by which noise travels is simply a direct line-of-sight air path between the source and the listener. For example, aircraft flyover noise reaches an observer on the ground by the direct line-of-sight air path. Noise also travels along structural paths. Noise can travel from one point to another via any one path or a combination of several paths. Noise from a washing machine operating in one apartment may be transmitted to another apartment along air passages such as open windows, doorways, corridors, or duct work. Direct physical contact of the washing machine with the floor or walls sets these building components into vibration. This vibration is transmitted structurally throughout the building, causing walls in other areas to vibrate and to radiate noise.

The receiver may be, for example, a single person, a classroom of students, or a suburban community. Solution of a given noise problem might require alteration or modification of any or all of these three basic elements :

1. Modifying the source to reduce its noise output.

2. Altering or controlling the transmission path and the environment to reduce noise level reaching the listener.

3. Providing the receiver with personal protective equipment.

Control of Noise Source by Design

Reduce impact forces

Many machines and items of equipment are designed with parts that strike forcefully against other parts, producing noise. Often, this striking action or impact is essential to the machine's function. A familiar example is the typewriter—its keys must strike the ribbon and paper in order to leave an inked impression. But the force of the key also produces noise as the impact falls on the ribbon, paper, and platen.

Several steps can be taken to reduce noise from impact forces. The particular remedy to be applied will be determined by the nature of the machine in question. Not all of the steps listed below are practical for every machine and for every impact-produced noise. But application of even one suggested measure can often reduce the noise appreciably. Some of the more obvious design modifications are as follows:

1. Reduce the weight, size, or height of fall of the impacting mass.

2. Cushion the impact by inserting a layer of shock-absorbing material between the impacting surfaces. (For example, insert several sheets of paper in the typewriter behind the top·sheet to absorb some of the noise-producing impact of the keys). In some situations, you could insert a layer of shock-absorbing material behind each of the impacting heads or objects to reduce the transmission of impact energy to other parts of the machine.

3. Whenever practical, one of the impact heads or surfaces should be made of non-metallic material to reduce resonance (ringing) of the heads.

4. Substitute the application of a small impact force over a long time period for a large force over a short period to achieve the same result.

5. Smooth out acceleration of moving parts by applying accelerating forces gradually. Avoid high, jerky acceleration or jerky motion.

6. Minimise overshoot, backlash and loose play in cams, followers, gears, linkages, and others parts. This can be achieved by reducing the operational speed of the machine, better adjustment, or by using spring-loaded restraints or guides. Machines that are well made, with parts machined to close tolerances, generally produce a minimum of such impact noise.

Reduce speeds and pressures

Reducing the speed of rotating and moving parts in machines and mechanical systems results in smoother operation and lower noise output. Likewise, reducing pressure and flow velocities in air, gas, and liquid circulation systems lessens turbulence, resulting in decreased noise radiation. Some specific suggestions that may be incorporated in design are the following :

1. Fans, impellers, rotors, turbines, and blowers should be operated at the lowest blade tip speeds that will still meet job needs. Use large-diameter, low-speed fans rather than small-diameter, high-speed units for quiet operation. In short, maximise diameter and minimise tip speed.

2. All other factors being equal, centrifugal squirrel-cage type fans are less noisy than vane axial or propeller type fans.

3. In air ventilation systems, a 50 percent reduction in the speed of the air flow may lower the noise output by 10 to 20 dB, or roughly one-quarter to one-half of the original loudness. Air speeds less than 3 m/s measured at a supply or return grille produce a level of noise that usually is unnoticeable in residential or office areas. In a given system, reduction of air speed can be achieved by operating at lower motor or blower speeds, installing a greater number of ventilating grilles, or increasing the cross-sectional area of the existing grilles.

Reduce frictional resistance

Reducing friction between rotating, sliding, or moving parts in mechanical systems frequently results in smoother operation and lower noise output. Similarly, reducing flow resistance in fluid distribution systems results in less noise radiation. Some of the more important factors that should be checked to reduce frictional resistance in moving parts are the following :

1. *Alignment:* Proper alignment of all rotating, moving, or contacting parts results in less noise output. Good axial and directional alignment in pulley systems, gear trains, shaft coupling, power transmission systems, and bearing and axle alignment are fundamental requirements for low noise output.

2. *Polish:* Highly polished and smooth surfaces between sliding, meshing, or contacting parts are required for quiet operation, particularly where bearings, gears, cams, rails, and guides are concerned.

3. *Balance:* Static and dynamic balancing of rotating parts reduces frictional resistance and vibration, resulting in lower noise output.

4. *Eccentricity (out-of-roundness):* Off-centering of rotating parts such as pulleys, gears, rotors, and shaft/bearing alignment causes vibration and noise. Likewise, out-of-roundness of wheels, rollers, and gears causes uneven wear, resulting in flat spots that generate vibration and noise.

The key to effective noise control in fluid systems is *streamline flow.* This holds true regardless of whether one is concerned with air flow in ducts or vacuum cleaners, or with water flow in plumbing systems. Streamline flow is simply smooth, nonturbulent, low-friction flow.

The two most important factors that determine whether flow will be stream-line or turbulent are the speed of the fluid and the cross-sectional area of the flow path, that is, the pipe or duct diameter. The rule of thumb for quiet operation is to use a low-speed, large-diameter system to meet a specified flow capacity requirement. However, even such a system can inadvertently generate noise if certain aerodynamic design features are overlooked or ignored. A system designed for quiet operation will employ the following features :

1. *Low fluid speed:* Low fluid speeds avoid turbulence, which is one of the main causes of noise.
2. *Smooth boundary surfaces:* Duct or pipe systems with smooth interior walls, edges, and joints generate less turbulence and noise than systems with rough or jagged walls or joints.
3. *Simple layout:* A well-designed duct or pipe system with a minimum of branches, turns, fittings, and connectors is substantially less noisy than a complicated layout.
4. *Long-radius turns:* Changes in flow direction should be made gradually and smoothly. It has been suggested that turns should be made with a curve radius equal to about five times the pipe diameter or major-cross-sectional dimension of the duct.
5. *Flared sections:* Flaring of intake and exhaust openings, particularly in a duct system, tends to reduce flow speeds at these locations, often with substantial reductions in noise output.
6. *Streamline transition in flow path:* Changes in flow path dimensions or cross-sectional areas should be made gradually and smoothly with tapered or flared transition sections to avoid turbulence. A good rule of thumb is to keep the cross-sectional area of the flow path as large and as uniform as possible throughout the system.
7. Remove unnecessary obstacles: The greater the number of obstacels in the flow path, the more tortuous, turbulent, and hence noisier, the flow. All other required and functional devices in the path, such as structural supports, deflectors, and control dampers, should be made as small and as streamlined as possible to smooth out the flow patterns.

Reduce radiating area

Generally speaking, the larger the vibrating part or surface, the greater the noise output. The rule of thumb for quiet machine design is to minimise the effective radiating surface areas of the parts without impairing their operation or structural strength. This can be done by making parts smaller, removing excess material, or by cutting openings, slots, or perforations in the parts. For example, replacing a large, vibrating sheet-metal safety guard on a machine with a guard made of wire mesh or metal webbing might result in a substantial reduction in noise because of the drastic reduction in surface area of the part.

Reduce noise leakage

In many cases, machine cabinets can be made into rather effective soundproof enclosures through simple design changes and the application of some sound-absorbing treatment. Substantial reductions in noise output may be achieved by adopting some of the following recommendations :

1. All unnecessary holes or cracks, particularly at joints, should be caulked.
2. All electrical or plumbing penetrations of the housing or cabinet should be sealed with rubber gaskets or a suitable non-setting caulk.
3. If practical, all other functional or required openings or ports that radiate noise should be covered with lids or shields edged with soft rubber gaskets to effect an airtight seal.

4. Other openings required for exhaust, cooling, or ventilation purposes should be equipped with mufflers or acoustically lined ducts.

5. Openings should be directed away from the operator and other people.

Isolate and damper vibrating elements

In all but the simplest machines, the vibrational energy from a specific moving part is transmitted through the machine structure, forcing other component parts and surfaces to vibrate and radiate sound—often with greater intensity than that generated by the originating source itself.

Generally, vibration problems can be considered in two parts. First, we must prevent energy transmission between the source and surfaces that radiate the energy. Second, we must dissipate or attenuate the energy somewhere in the structure. The first part of the problem is solved by *isolation*. The second part is solved by *damping*.

The most effective method of vibration isolation involves the resilient mounting of the vibrating component on the most massive and structurally rigid part of the machine. All attachments or connections to the vibrating part, in the form of pipes, conduits, and shaft couplers, must be made with flexible or resilient connectors or couplers. For example, pipe connections to a pump that is resiliently mounted on the structural frame of a machine should be made of resilient tubing and be mounted as close to the pump as possible. Resilient pipe supports or hangers may also be required to avoid by-passing the isolated system (Fig. 27.5).

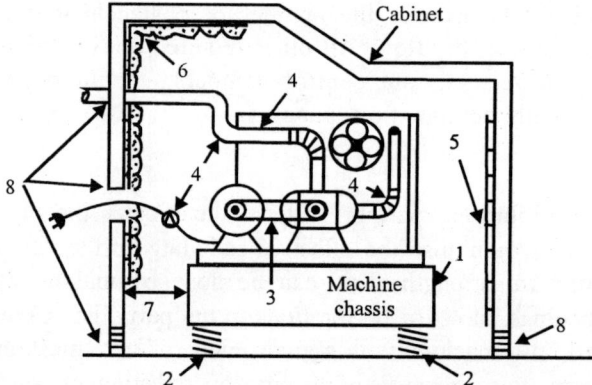

Fig. 27.5. Examples of vibration isolation; 1. Motors, pumps, and fans installed on most massive part of the machine; 2. Resilient mounts or vibration isolators used for the installation; 3. Belt-drive or roller-drive systems used in place of gear trains; 4. Flexible hoses and wiring used instead of rigid piping and stiff wiring; 5. Vibration-damping materials applied to surfaces undergoing most vibration; 6. Acoustical lining installed to reduce noise buildup inside machine; 7. Mechanical contact minimised between the cabinet and the machine chassis; 8. Openings at the base and other parts of the cabinet scaled to prevent noise leakage.

Damping material or structures are those that have some viscous properties. They tend to bend or distort slightly, thus consuming part of the noise energy in molecular motion. The use of spring mounts on motors and laminated galvanised steel and plastic in air-conditioning ducts are two examples.

When the vibrating noise source is not amenable to isolation, as, for example, in ventilation ducts, cabinet panels, and covers, then damping materials can be used to reduce the noise.

The type of material best suited for a particular vibration problem depends on a number of factors such as size, mass, vibrational frequency, and operational function of the vibrating structure. Generally speaking, the following guildelines should be observed in the selection and use of such materials to maximise vibration damping efficiency.

1. Damping materials should be applied to those sections of a vibrating surface where the most flexing, bending, or motion occurs. These usually are the thinnest sections.

2. For a single layer of damping material, the stiffness and mass of the material should be comparable to that of the vibrating surface to which it is applied. This means that single-layer damping materials should be about two or three times as thick as the vibrating surface to which they are applied.

3. Sandwich materials (*laminates*) made up of metal sheets bonded to mastic (sheet metal viscoelastic composites) are much more effective vibration dampers than single-layer materials; the thickness of the sheet-metal constraining layer and the viscoelastic layer should each be about one-third the thickness of the vibrating surface to which they are applied. Ducts and panels can be purchased already fabricated as laminates.

Provide mufflers/silencers

There is no real distinction between mufflers and silencers. They are often used interchangeably. They are in effect acoustical filters and are used when fluid flow noise is to be reduced. The devices can be classified into two fundamental groups: *absorptive mufflers* and *reactive mufflers*. An absorptive muffler is one whose noise reduction is determined mainly by the presence of fibrous or porous materials, which absorb the sound. A reactive muffler is one whose noise reduction is determined mainly by geometry. It is shaped to reflect or expand the sound waves with resultant self-destruction.

Although there are several terms used to describe the performance of mufflers, the most frequently used appears to be *insertion loss* (IL). Insertion loss is the difference between two sound pressure levels that are measured at the same point in space before and after a muffler has been inserted. Each muffler IL is highly dependent on the manufacturer's selection of materials and configuration.

Noise Control in the Transmission Path

After you have tried all possible ways of controlling the noise at the source, your next line of defence is to set up devices in the transmission path to block or reduce the flow of sound energy before it reaches your ears. This can be done in several ways: (i) absorb the sound along the path; (ii) deflect the sound in some other direction by placing a reflecting barrier in its path; or (iii) contain the sound by placing the source inside a sound-insulating box or enclosure.

Selection of the most effective technique will depend upon various factors, such as the size and type of source, intensity and frequency range of the noise, and the nature and type of environment.

Separation

We can make use of the absorptive capacity of the atmosphere, as well as divergence, as a simple, economical method of reducing the noise level. Air absorbs high-frequency sounds more effectively than it absorbs low-frequency sounds. However, if enough distance is available, even low-frequency sounds will be absorbed appreciably.

If you can double your distance from a point source, you will have succeeded in lowering the sound pressure level by 6 dB. It takes about a 10 dB drop to halve the loudness. If you have to contend with a line source such as a railroad train, the noise level drops by only 3 dB for each doubling of distance from the source. The main reason for this lower rate of attenuation is that line sources radiate sound waves that are cylindrical in shape. The surface area of such waves only increases two-fold for each doubling of distance from the source. However, when the distance from the train becomes comparable to its length, the noise level will begin to drop at a rate of 6 dB for each subsequent doubling of distance.

Indoors, the noise level generally drops only from 3 to 5 dB for each doubling of distance in the near vicinity of the source. However, further from the source, reductions of only 1 or 2 dB occur for each doubling of distance due to the reflections of sound off hard walls and ceiling surfaces.

Absorbing materials

Noise, like light, will bounce from one hard surface to another. In noise control work, this is called *reverberation*. If a soft, spongy material is placed on the walls, floors, and ceiling, the reflected sound will be diffused and soaked up (absorbed). Sound-absorbing materials are rated either by their *Sabin absorption coefficients* (α_{SAB}) at 125, 500, 1,000, 2,000, and 4,000 Hz or by a single number rating called the *noise reduction coefficient* (NRC). Area of open window is assumed to transmit all and reflect none of the acoustical energy that reaches it, it is assumed to be 100 percent absorbent. This unit area of totally absorbent surface is called a "sabin". The absorptive properties of acoustical materials are then compared with this standard. The performance is expressed as a fraction or percentage of the sabin (α_{SAB}). The NRC is the average of the α_{SAB}s at 250, 500, 1,000, and 2,000 Hz rounded to the nearest multiple of 0.05. The NRC has no physical meaning. It is a useful means of comparing similar materials.

Sound-absorbing materials such as acoustical tile, carpets, and drapes placed on ceiling, floor, or wall surfaces can reduce the noise level in most rooms by about 5 to 10 dB for high-frequency sounds, but only by 2 or 3 dB for low-frequency sounds. Unfortunately, such treatment provides no protection to an operator of a noisy machine who is in the midst of the direct noise field. For greatest effectiveness, sound-absorbing materials should be installed as close to the noise source as possible.

Acoustical lining

Noise transmitted through ducts, pipe chases, or electrical channels can be reduced effectively by lining the inside surfaces of such passageways with sound-absorbing materials. In typical duct installations, noise reductions on the order of 10 dB/m for an acoustical lining 2.5 cm thick are well within reason for high-frequency noise. A comparable degree of noise reduction for the lower frequency sounds is considerably more difficult to achieve because it usually requires at least a doubling of the thickness and/or length of acoustical treatment.

Barriers and panels

Placing barriers, screens, or deflectors in the noise path can be an effective way of reducing noise transmission, provided that the barriers are large enough in size, and depending upon whether the noise is high frequency or low-frequency. High-frequency noise is reduced more effectively than low-frequency noise. The effectiveness of a barrier depends on its location, its height, and its length. Referring to Fig. 27.6, we can see that the noise can follow various different paths.

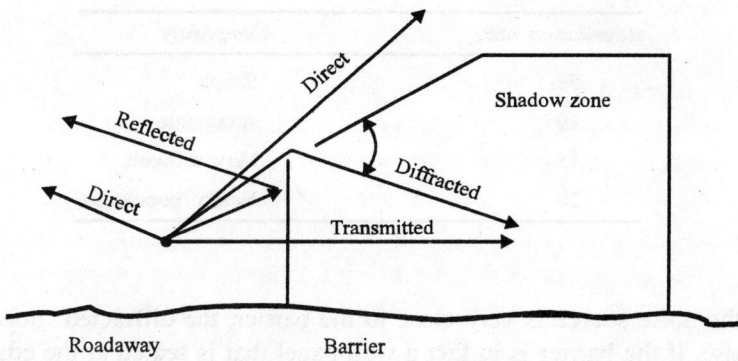

Fig. 27.6. Noise paths from a source to a receiver.

First, the noise follows a direct path to receivers who can see the source well over the top of the barrier. Second, the noise follows a diffracted path to receivers in the shadow zone of the barrier. Third, in the shadow zone, the noise transmitted directly through the barrier may be significant in some cases. The fourth path shown in Fig. 27.6 is the reflected path. After reflection, the noise is of concern only to a receiver on the opposite side of the source. For this reason, acoustical absorption on the face of the barrier may sometimes be considered to reduce this reflected noise; however, this treatment will not benefit any receivers in the shadow zone.

Of these four paths, the noise diffracted over the barrier into the shadow zone represents the most important parameter from the barrier design point of view. Generally, the determination of barrier attenuation or barrier noise reduction involves only calculation of the amount of energy diffracted into the shadow zone. The procedures presented in the barrier nomograph used to predict highway noise are based on this concept.

Another general principle of barrier noise reduction that is worth reviewing at this point is the relation between noise attenuation expressed in (i) decibels; (ii) energy terms; and (iii) subjective loudness. Table 27.6. gives these relationships for line sources.

Table 27.6. Relation between sound level reduction, energy, and loudness for line sources.

To reduce A-level by dB	Remove portion of energy (%)	Divide loudness by
3	50	1.2
6	75	1.5
10	90	2
20	90	4
30	99.9	8
40	99.99	16

As indicated in the loudness column, a barrier attenuation of 3 dB will be barely discerned by the receiver. However, to attain this reduction, 50 percent of the acoustical energy must be removed. To cut the loudness of the source in half, a reduction of 10 dB is necessary. That is equivalent to eliminating 90 percent of the energy initially directed toward the receiver. As indicated previously, this drastic reduction in energy requires very long and high barriers. In summary, when designing barriers, you can expect the complexity of the design to be about as follows :

Attenuation (dB)	Complexity
5	Simple
10	Attainable
15	Very difficult
20	Nearly impossible

Transmission loss

When the position of the noise source is very close to the barrier, the diffracted noise is less important than the transmitted noise. If the barrier is in fact a wall panel that is sealed at the edges, the transmitted noise is the only one of concern. The ratio of the sound energy incident on one surface of a panel to the energy radiated from the opposite surface is called the *sound transmission loss* (TL). The actual energy loss is partially reflected and partially absorbed. Since TL is frequency-dependent, only a complete octave or one-third octave band curve provides a full description of the performance of the barrier.

Enclosures

Sometimes it is much more practical and economical to enclose a noisy machine in a separate room or box than to quiet it by altering its design, operation, or component parts. The walls of the enclosure should be massive and airtight to contain the sound. Absorbent lining on the interior surfaces of the enclosure will reduce the reverberant buildup of noise within it. Structural contact between the noise source and the enclosure must be avoided, or else the source vibration will be transmitted to the enclosure walls and thus short-circuit the isolation. For maximum effective noise control, all of the techniques illustrated in Fig. 27.7 must be employed.

Control of Noise Source by Redress

The best way to solve noise problems is to design them out of the source. However, we are frequently faced with an existing source that, either because of age, abuse, or poor design, is a noise problem. The result is that we must redress, or correct, the problem as it currently exists. The following sections identify some measures that might apply if you are allowed to tinker with the source.

Balance rotating parts

One of the main sources of machinery noise is structural vibration caused by the rotation of poorly balanced parts, such as fans, fly wheels, pulleys, cams, shafts, and so on. Measures used to correct this condition involve the addition of counterweights to the rotating unit or the removal of some weight from the unit. You are probably familiar with noise caused by imbalance in the high-speed spin cycle of washing machines. The imbalance results from clothes not being distributed evenly in the tub. By redistributing the clothes, balance is achieved and the noise ceases. This same principle of balance can be applied to furnace fans and other common sources of such noise.

Reduce frictional resistance

A well-designed machine that has been poorly maintained can become a serious source of noise. General cleaning and lubrication of all rotating, sliding, or meshing parts at contact points should go a long way toward fixing the problem.

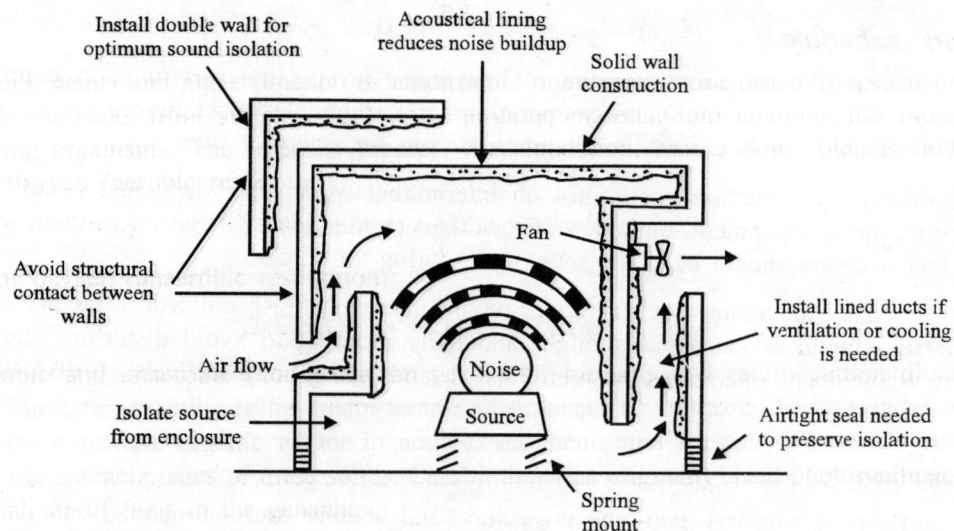

Fig. 27.7. Enclosures for controlling noise.

Apply damping materials

Since a vibrating body or surface radiates noise, the application of any material that reduces or restrains the vibrational motion of that body will decrease its noise output. Three basic types of redress vibration damping materials are available :

1. Liquid mastics, which are applied with a spray gun and harden into relatively solid materials, the most common being automobile "undercoating".

2. Pads of rubber, felt, plastic foam, leaded vinyls, adhesive tapes, or fibrous blankets, which are glued to the vibrating surface.

3. Sheet metal viscoelastic laminates or composites, which are bonded to the vibrating surface.

Seal noise leaks

Small holes in an otherwise noise-tight structure can reduce the effectiveness of the noise control measures. If the designed transmission loss of an acoustical enclosure is 40 dB, an opening that comprises only 0.1 percent of the surface area will reduce the effectiveness of the enclosure by 10 dB.

Perform routine maintenance

We all recognise the noise of a worn muffler. Likewise, studies of automobile tyre noise in relation to pavement roughness show that maintenance of the pavement surface is essential to keep noise at minimum levels. Normal road wear can yield noise increases of the order of 6 dBA.

Protect the Receiver

When all else fails

When exposure to intense noise fields is required and none of the measures discussed so far is practical, as, for example, for the operator of a chain saw or pavement breaker, then measures must be taken to protect the receiver. The following two techniques are commonly employed.

Alter work schedule

Limit the amount of continuous exposure to high noise levels. In terms of hearing protection, it i. preferable to schedule an intensely noisy operation for a short interval of time each day over a perioc of several days rather than a continuous eight-hour run for a day or two.

In industrial or construction operations, an intermittent work schedule would benefit not only the operator of the noisy equipment, but also other workers in the vicinity. If an intermittent schedule is no possible, then workers should be given relief time during the day.

Inherently noisy operations, such as street repair, municipal trash collection, factory operation, anc aircraft traffic, should be curtailed at night and early morning to avoid disturbing the sleep of the community. Remember : operations between 10 P.M. and 7 P.M. are effectively 10 dBA higher than the measured value.

Ear protection

Moulded and pliable earplugs, cup-type protectors, and helmets are commercially available as hearin protectors. Such devices may provide noise reductions ranging from 15 to 35 dB (Fig. 27.8). Earplug are effective only if they are properly fitted by medical personnel. As shown in Fig. 27.8, maximun protection can be obtained when both plugs and muffs are employed. Only muffs that have a certificatior stipulating the attenuation should be used.

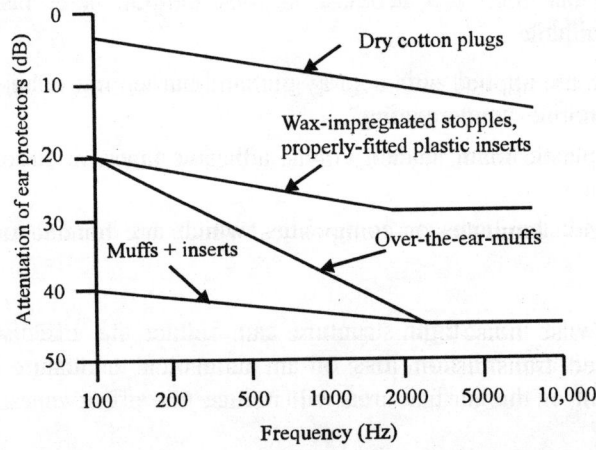

Fig. 27.8. Attenuation of ear protectors at various frequencies.

These devices should be used only as a last resort, after all other methods have failed to lower th oise level to acceptable limits. Ear protection devices should be used while operating lawn mower hippers, and while firing weapons at target ranges. It should be noted that protective ear devices d nterfere with speech communication and can be a hazard in some situations where warning calls ma e a routine part of the operation. A modern ear-destructive device is a portable mini-radio/recorder th ses earphones. In this "reverse" muff, high noise levels are directed at the ear without attenuation. ou can hear someone else's radio/recorder, that person is subjecting him- or herself to noise levels xcess of 90–95 dBA.

CHAPTER 28

Environmental Methods for Chemical Analysis

INTRODUCTION

The air and water in our environment contain a wide assortment of toxic organic and inorganic pollutants. They enter the environment as emissions to the atmosphere or as discharges to water bodies. These may be either in concentrated point sources such as from factory smoke, stacks and sewage discharges, or in diffused forms such as from automobiles' exhausts and runoff from agricultural land. These pollutants eventually endanger the life of both humans and animals. These pollutants can range from parts per billion (ppb) or below in rural areas to hundreds of parts per million (ppm) or higher in large industrial conurbations and urbanised areas.

In recent years, many chemicals previously considered only moderately toxic have been identified very toxic, e.g., potential carcinogens thus have been assigned lower threshold limit values (TLVS). In addition, the number of newly identified toxic substances are increasing every day. Hence, there is an increasing need for rapid screening and monitoring of toxic substances in air and water to meet the requirements of Government Pollution Control Authorities.

In order to control the levels of these pollutants in the environment, it is necessary to know their chemical or bio-chemical route of formation and degradation, the extent of their occurrence in the environment and their eco-toxicity. Analytical Chemistry plays the most vital role in determining the extent of their occurrence in the environment while the degree of their eco-toxicity determines the priority of pollutants and the overall sensitivity required for the analytical method to be employed for their measurement. As the TLVs are lowered, the strain on analytical technology increases, since improved sensitivities and increased specificities are essential in measuring lower concentration of pollutants. Many of the conventional methods are time consuming and do not have adequate sensitivities to measure the lower concentration. Additionally, due to lack of specificities they often lead to unreliable results.

In recent past a great deal of advances have been made in analytical methodology and instrumentation. With the ever increasing advancement of microprocessor technology these analytical instruments have become more versatile. These new generation computer aided instruments (so called 'intelligent instruments') offer highly improved detection limits (down to parts per billion (ppb) to parts per trillion (ppt) level), better precision, accuracy and increased specificity.

TECHNIQUES FOR ANALYSIS IN THE LABORATORY

Gas Chromatography (GC)

In GC, nanogram to microgram quantities of volatile compounds in mixtures are separated by the passage of a vaporised sample in a carrier gas stream through a chromatographic column. Migration occurs through the column at different rates depending on relative boiling points, solubilities or adsorption. This results in difference in retention behaviour of individual components in the mixture. The components eluted at different times are detected by a suitable detector. The resulting signals are recorded and computed for detection and/or determination of various components (Fig. 28.1).

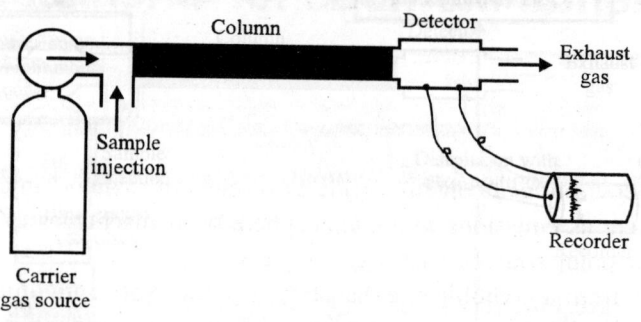

Fig. 28.1. Schematic diagram of gas chromatography.

In gas-solid chromatography (GSC), the columns contain a stationary phase, which is a finely divided surface active solid, capable of selective absorption of sample components. In gas-liquid chromatography (GLC), a non-volatile liquid stationary phase is supported on a fine inert solid or on the walls in case of modern capillary columns.

GC is a fast and sensitive technique. The technique is versatile and can be easily automated. By optimisation of column length and packing in conjunction with careful selection of a selective detector, GC is extensively used in environmental pollution monitoring.

A variety of detectors are used in GC technique. They are Flame ionisation detector (FID) for general detection of organic compounds, photo ionisation detector (PID) for improved detection of organic and for selective detection of phosphorus and sulphur compounds, Electron capture detector (ECD) and Hall electrolytic detector for selective detection of halogenated organic compounds, thermal conductivity detector (TCD) for permanent and natural gases, etc.

GC technique has been widely used as standard method for analysis of water and waste-water. Compounds which are determined by GC include hydro-carbons, benzidines, pesticides, polycycticaromatic hydrocarbons (PAHs), phenols, polychlorinated benzenes (PCBs), etc. Non-purgeables such as phenols, PCBs, PAH, etc. are solvent extracted, concentrated on a Kuderna-Danish concentrator and injected on to GC. Purgeables such as hydro-carbons, and aromatics are adsorbed and concentrated on tenax by purging inert gas in water. The compounds are desorbed by heating and passed on to GC for analysis, the technique known as purge and trap injection (PTI) techniques.

Air samples are also analysed by GC by direct injection, adsorption on tenax and thermal desorption or concentration in a suitable volatile solvent. Compounds determined by GC include, sulphur dioxide, carbon dioxide, phosgene, vinyl chloride, nitrous oxide and hydrocarbons, etc.

Although GC is a suitable technique for volatile compounds, non-volatile compounds are analysed through volatile derivatives. Recently, GC technique has been used for trace elements specification to study the environmental and toxicological impact of trace elements.

High Performance Liquid Chromatography (HPLC)

In principle HPLC is analogous to conventional column chromatography. The technique differs, however, in degree of separation, and sensitivity attainable due to developments in column technology, detector design and use of pressurised solvent delivery system.

An aqueous or non-aqueous mobile phase is pumped using a high pressure pump (5000–6000 psi) at a rate of 1 to 2 ml per minute through a stainless steel small dia (3–5 mm) column packed with silica or bonded silica of 60–70 Å pore size and 3–5 microns particle size. The sample is injected in the mobile phase stream just before it enters the column using a specially designed valve injector. The components in the mixture are separated in the column depending on degree of retention on the column material and solubility in the mobile phase. The elute passes through a detector and signals are recorded and computed.

A large variety of columns and detectors are commercially available. By optimising the chromatographic conditions and choosing a suitable column and a detector, a large number of pollutants can be determined in air as well as in water and waste-water. The most commonly used detectors are variable wave length UV VIS and fluorescence detector. In case of HPCL, it is also possible to determine non-volatile and thermally unstable compounds. Water sample can be analysed directly or by enrichment of compound of interest using solid phase extraction, on-line column switching, etc. Pollutants from air can be extracted by a suitable technique and analysed by HPLC (Fig. 28.2).

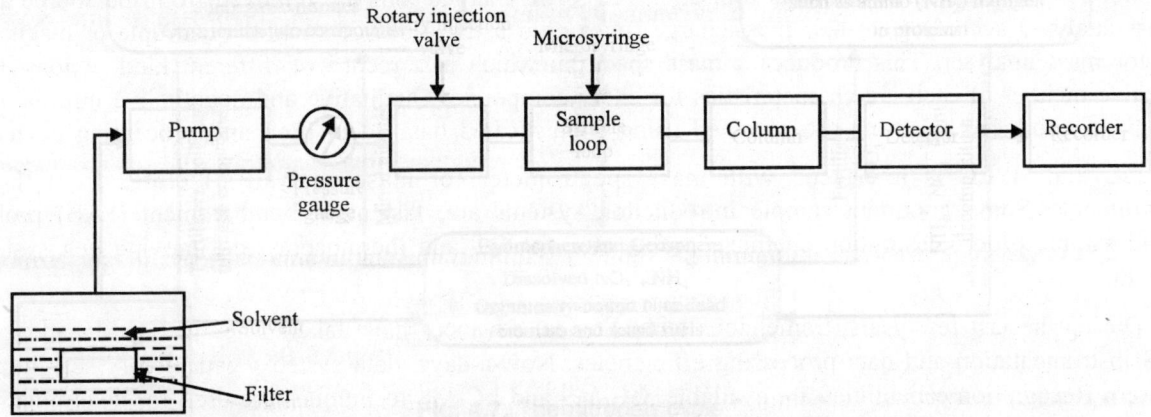

Fig. 28.2. Schematic diagram of a high pressure liquid chromatograph.

HPLC has been widely used for determination of organic compounds, pesticides, PAH etc. Environmental Pollution Control Act (EPCA) has included HPLC as standard method for determination of PAH in water and waste-water.

An interesting application of HPLC is Ion Chromatography which provides an excellent means of simultaneous determination of a variety of anions and cations with detection limits up to ppb levels. The technique has ability of on line analysis. It uses an ion exchange column and a conductivity detector. UV-detector can also be used with a buffer as a mobile phase having low UV absorbance which provides indirect measurement of anions. Ion chromatography has wide utility in monitoring of inorganic ions in air and water.

Gas Chromatography-Mass Spectrometry (GC-MS) and Liquid Chromatography-Mass Spectrometry (LC-MS)

In recent years, GC-MS and LC-MS have been widely accepted as an ideal tools for identification and quantification of pollutants in water and air. These techniques have excellent sensitivity and capability to detect each of several hundred compounds present in complex forms in the environment down to sub ppb levels (Fig. 28.3).

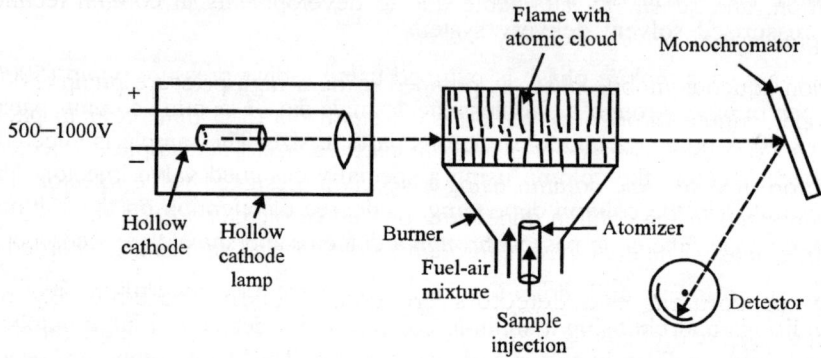

Fig. 28.3. Atomic absorption spectrophotometer.

These techniques utilise a mass spectrometer where sample molecules are ionised into charged ions consisting of parent ions and ionic fragments of the original molecules. The ionisation is brought about by electron impact (EI) or chemical ionisation (CI). The charged ions are removed from the source and mass analysed according to their mass/charge (m/e) ratio using a device such as quadruple or magnetic sector mass analyser. This produces a mass spectrum which is a record of different kind of ions, the relative number of each are characteristics for every compound. Qualitative and quantitative information from the produced spectrum are obtained using sophisticated data acquisition and processing devices.

GC and HPLC is interfaced with mass spectrometer for mass analysis of elutes from these instruments. Some important sample introduction systems are, fast atom bombardment (FAB) probe, solid sample probe, desorption chemical probe, etc., for QC and thermospray and moving belt system for LC.

During the last few years, tremendous technological advances have taken place in GC-MS and LC-MS instrumentation and data processing efficiencies. Now-a-days, data systems with library search and Pattern Recognition capabilities are available. GC-MS and LC-MS techniques are used for determination of organohalides, PAH, pesticides, PCBs and other organic contaminants in water and air.

Atomic Absorption Spectrophotometry (AAS)

Atomic absorption spectrophotometry is based on the specific absorption of light by the unexcited atoms. The source of specific energy for free atom production in this technique is heat, most common in the form of an air-acetylene or nitrous oxide-acetylene flame or a heated graphite furnace atomiser. Such sources are capable to excite only a fraction of all atoms to higher energy, about 99 per cent remain unexcited. In this method, a beam of monochromatic light, the wavelength of which is matched to that of a resonance line of the enexcited atom, is passed through the flame or graphite furnace (Fig. 28.3).

The sample is introduced in as an aerosol by a nebuliser or as liquid or solid into the graphite furnace. As a result of absorption of light, an electronic transition occurs from the ground state to the appropriate excited level. The unabsorbed radiation transmitted by the flame or furnace is passed through a monochromator and its intensity is measured by means of a suitable photometric system. Measurements are made in a manner similar to that in conventional spectrophotometry.

AAS technique is very widely used for trace metal analysis in water and air particulate matters. In last few years, many improvements are made in AAS instruments. It has been made more versatile by incorporation of advanced microprocessors. Graphite furnace AAS offers detection limit down to sub-ppb level.

Other AAS techniques used in environmental monitoring are cold vapour technique capable of analysing mercury down to 0.1 and below, and gaseous hydride method providing determination of arsenic and selenium down to 1 ppb level.

Inductively Coupled Plasma-Atomic Emission Spectroscopy (ICP-AES)

Inductively coupled plasma-atomic emission spectroscopy (ICP-AES) provides accurate and cost effective analysis of minor and trace elements in water waste-water, air particulates and other environmental samples.

The basis of the technique is the measurement of atomic emission by the optical spectroscopic technique. Samples are nebulised and the aerosol that is produced is transported into a radio frequency inductively coupled argon plasma having high temperature (8000–1000 degree K). The excitation occurs and characteristic atomic line emission spectra are produced. The spectra are dispersed by a grating spectrometer and the intensities of the lines are monitored by photo multiplier tube (s). The photocurrent (s) from the photomultiplier tube (s) are processed and controlled by a computer system (Fig. 28.4).

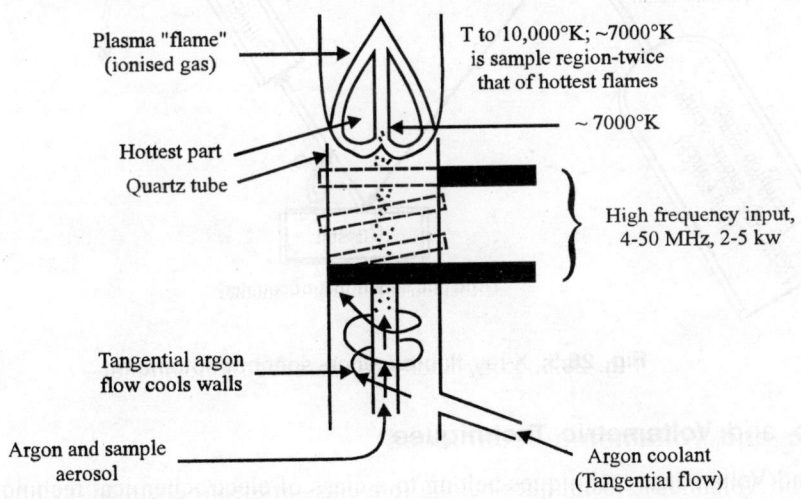

Fig. 28.4. Inductively coupled plasma or optical emission spectroscopy.

A background correction technique is required to compensate for variable background contribution to the determination of trace elements. The technique provides very rapid analysis, typically 15 elements per minute can be determined. It offers very wide dynamic concentration range. It is free from chemical interferences. It suffers from problems of spectral interferences but can be corrected by careful use of computer controlled spectrometer. ICP-AES offers detection limits similar to that of flame AAS, but it is particularly useful for determination of elements such as phosphorus, boron, aluminium, silicon and rare earth elements. ICP-AES is used as a standard method for the determination of trace metals in water and waste-water.

X-Ray Emission Spectroscopy

X-Ray Spectroscopy has also been used for the elemental analysis of environmental samples. In X-ray spectroscopy, the sample is excited by X-ray. This causes emission of secondary (fluorescent) X-rays radiation which is characteristic of individual elements. The measurement of X-ray fluorescence radiation can be accomplished by using either a wavelength dispersive or by energy dispersive X-ray spectrometer (Fig. 28.5).

Both water and air samples have been analysed by X-ray spectroscopy to monitor the trace metal concentration. In both cases, pre-concentration of sample is required to achieve sensitivity. Contaminants from air and water are collected on filter paper or membrane by direct filtration or filtration after precipitation, adsorption or absorption on suitable material, etc. The thin deposit on filter is subjected to X-ray analysis. The technique is generally only amenable to elements of atomic number greater than 12. Detection limits upto 1/10th of TLVs for many elements can be achieved by careful optimisation of the analytical procedure.

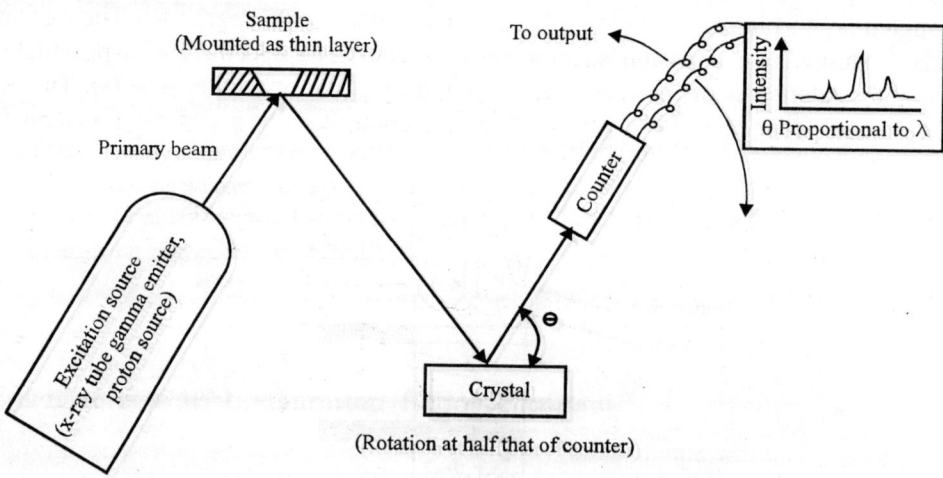

Fig. 28.5. X-ray fluorescence spectrophotometer.

Polarographic and Voltametric Techniques

Polarographic and Voltametric techniques belong to a class of electrochemical techniques that can be used to study solution composition through current potential relationships. Inorganic cations and anions, organometallic species and organic molecules of environmental importance have all been studied. Several analytical methods have been developed using these techniques for trace analysis of inorganic and organic substances in water and air. Polarography is used to investigate solution composition by the reduction or oxidation of ions or molecules, at dropping mercury electrode (DME), whereas other voltametric techniques such as anodic and cathodic stripping voltammetry employ stationary mercury drop indicator electrodes or indicator electrodes of a solid variety such as glassy carbon or gold. The technique is widely used for the determination of copper, lead, cadmium, nickel, zinc, bismuth, tin, and many organic species in water and air particulate matter. It offers extremely low detection limits down to sub ppb level. Recent advances in polarographic instruments have made this technique more versatile and provide improved results.

Ion Selective Electrodes

Ion selective electrodes enable the determination of activity of a given ion in solution potentiometrically. This consists of a selective membrane, an internal reference electrolyte (filling solution) and an internal reference electrode which forms a half cell (Fig. 28.6). An external reference electrodes completes the circuit (Fig. 28.6). When the ion selective electrode comes in contact with a solution of ions, a potential difference is generated between the internal filling solution and the sample solution across the membrane. The activity of the ion is then determined utilising the Nernst Equation.

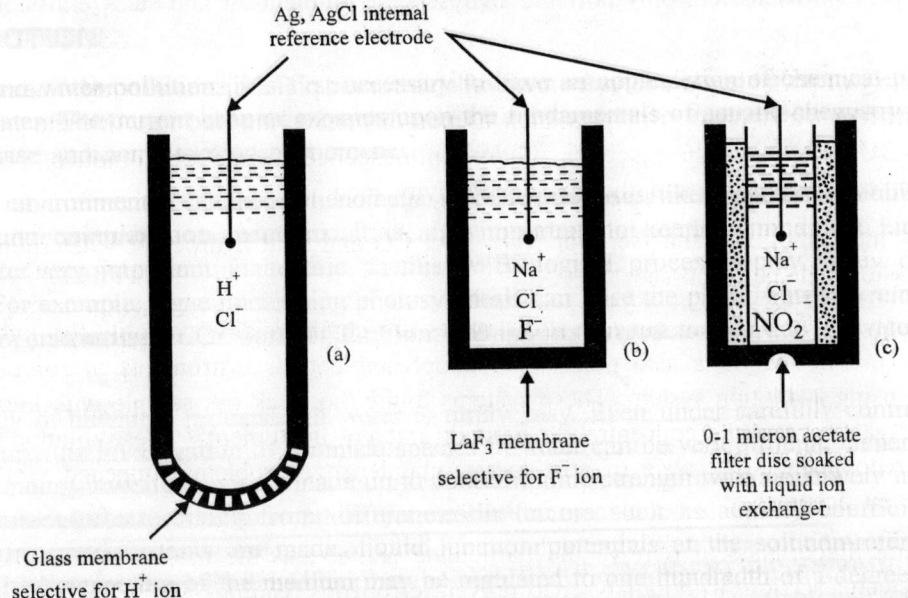

Fig. 28.6. Ion selective electrodes. (a) glass electrode selective for hydrogen ion; (b) lanthanum fluoride solid membrane electrode selective for fluoride ion; (c) liquid ion-exchanger electrode selective for nitrate ion.

This technique is most widely used for rapid determination of a large ions in water samples. It can also be used to monitor various ions in air by collecting the particulate matter and bringing them in solution form.

Ultraviolet-Visible Spectrophotometry (UV-VIS)

This technique is based on measurement of absorption of radiation by molecules or atoms in solution. The method is well known and widely used in determination of a large number of anions and cations for environmental pollution monitoring. With recent advancement in microprocessor technology, many sophisticated instruments, with several added features such as derivative spectrophotometry and computer controlled measurements have been produced.

Infra-red Spectrophotometry (IR)

IR spectrophotometry has very limited application for environmental samples brought to the laboratory which include detection of oil and grease in water and gas analysis. Nevertheless, this technique is widely used in field monitoring.

PORTABLE INSTRUMENTS

Often economics, time consideration and need for flexibility in field studies dictate that many analytical studies be performed in the field itself rather than on samples transported to the laboratory. A large number of portable instruments are widely used in monitoring of water and air. These instruments are based on a number of analytical techniques.

Many of the portable analysers based on ion selective electrode, pH and conductivity measurements, spectrophotometry, etc., are well known and widely used for on-site analysis of water and wastes. Hence, in the present discussion, only portable analysers for ambient air and stack emission monitoring will be included.

A large variety of portable gas analysers are available for air pollution monitoring. They are based on a number of different analytical principles such as non-dispersive infrared (NDIR) and non-dispersive ultraviolet (NDUV), pulsed fluorescence, chemiluminescence, flame photometric, polarographic, electrocatalytic, conductometric and thermal conductivity methods.

Now, a few most important analysers for NO_x, SO_2, CO and total hydrocarbon determinations in air and stack emission will be discussed.

Separate analysers in different concentration ranges are commercially available for ambient air and stack emission monitoring. It is also possible to adopt ambient air instruments to source monitoring application by using a dilution system. These systems dilute the stack gas sample with known volumes of inert carrier gas to meet the concentration limit of ambient instruments. These dilution systems are, however, not recommended because of cost factor, maintenance problems, impeded system response time and reduced accuracy.

It is of prime importance to remove the particulate matter and the moisture from the gas before entering the detector section of the analyser to prevent erroneous readings and/or malfunctioning of the detector. A variety of sample conditioning systems are available incorporating different types of filtering and drying methods. It is critical to select a pre-conditioning system that is compatible with the detection device to be used.

Non-Dispersive Infrared Analysers (NDIR)

Non-dispersive Infrared (NDIR) analysers have been developed to monitor SO_2, NO_x, CO, CO_2 and other gases that absorb in infrared, including hydrocarbons. In a typical NDIR analyser, infrared light from a lamp or glower passes through two gas cells: (i) a reference cell; and (ii) a sample cell. The reference cell generally contains dry nitrogen gas which does not absorb light at a wavelength used in the instrument. As the light passes through the sample cell, pollutant molecules will absorb some of the infrared light. As a result, when the light emerges from the end of the sample cell, it will have less energy than when it entered. It also will have less energy than the light emerging from the reference cell. The energy difference is than sensed by some type of detector, such as a thermistor, thermocouple, or microphone arrangement (Fig. 28.7).

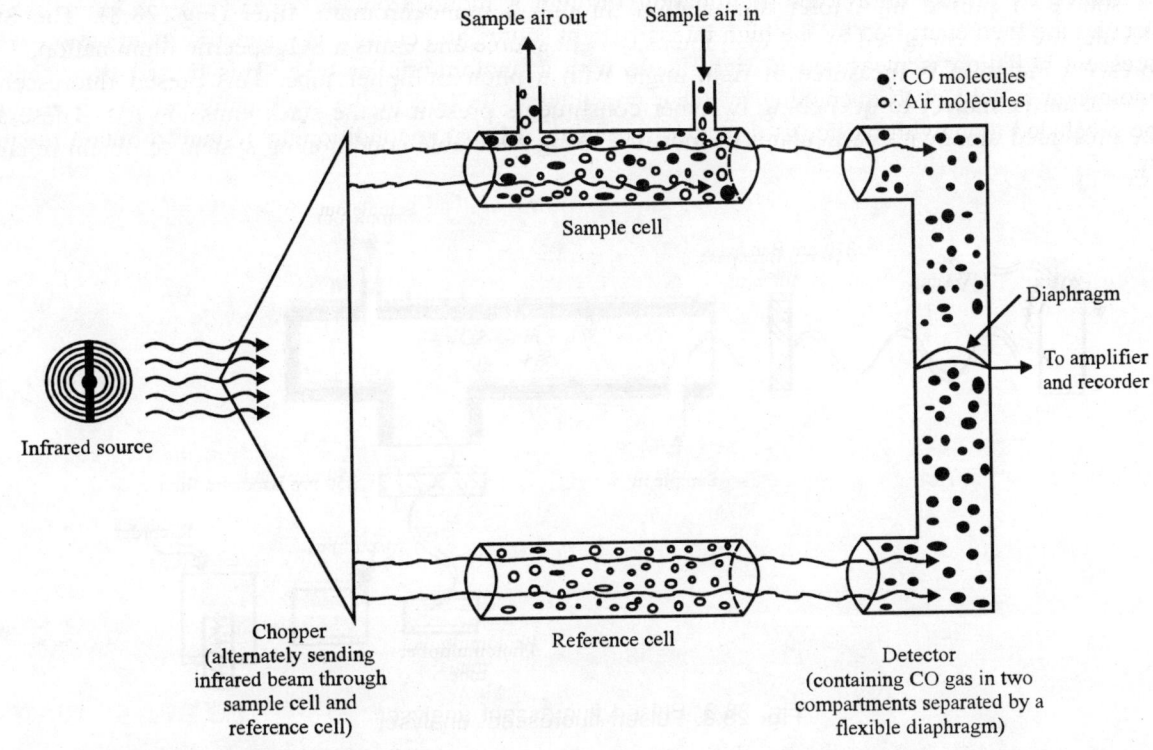

Fig. 28.7. Non-dispersive infrared analyser for monitoring carbon monoxide.

Non-dispersive Ultraviolet Analysers (NDUV)

NDUV instrument uses a reference wave length at 578 nm instead of using a reference cell (as in NDIR systems). Light from a mercury discharge lamp passes through the sample cell to a beam splitter. The beam splitter, actually a semi-transparent mirror, directs the light to two separate photomultiplier tubes. Narrow bandpass filter allows light of only the specific wave lengths to reach each of the photomultipliers. The reflected beam passes through a 578 nm filter and is used to generate the reference signal in the detector. The transmitted beam, however, passes through a 280-nm filter for an NO_2 monitor. Since SO_2 will absorb light at 280 nm (NO_2 at 436 nm) the amount of light or energy reaching the phototube will be less than that reaching the reference phototube. The resultant photomultiplier signals are amplified and processed to give a reading for the pollutant concentrations. Nitric Oxide (NO) does not absorb in the spectral region covered by the instrument and hence it must first be quantitatively converted to NO_2 for subsequent analysis. This is done sequentially by stopping the flow in the NO_2 sample cell, pressurising it with O_2 and waiting approximately 5 minutes for the NO to be converted to NO_2 by the excess oxygen. The NO is then determined from the difference in the readings before and after the reaction with oxygen.

Pulsed Fluorescence Analysers

This analyser is used to monitor SO_2 gas in the ambient air. For stack emission, a dilution probe is required. In Pulsed fluorescence analysers, the sample is introduced in the reaction chamber and exposed to a source of pulsed ultraviolet illumination through a monochromatic filter (Fig. 28.8). The SO_2 molecules are then energised by the high intensity light source and emits a SO_2 specific illumination. This fluorescent radiation is measured at right angle with a photomultiplier tube. This pulsed fluorescence phenomenon is subject to quenching by other constituents present in the stack emission gas. These are to be precluded alongwith particulates or moisture using a suitable conditioning system to obtain reliable data.

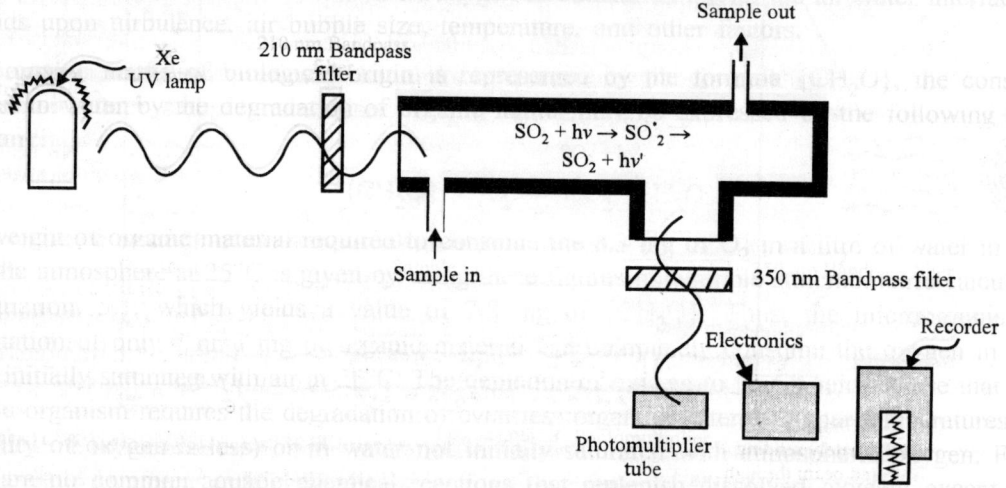

Fig. 28.8. Pulsed fluorescent analyser.

Chemiluminescent Analyser

This type of analyser provides reliable measurement of NO_x in ambient and stack emission monitoring (Fig. 28.9). The measurement is based upon chemiluminescent reaction between nitric oxide and ozone. Light emission results when the electronically excited NO_2 molecules revert to their ground state and this light emission is measured via a photomultiplier tube. This obviously provides measurement of NO but not of NO_x. To provide a NO_x measurement, it is necessary to run the sample stream through a catalytic converter to reduce NO_2 to NO. The NO produced is then reacted with ozone and chemiluminescence is measured to give a total $NO + NO_2$ (NO_x) reading. This is an approved EPA method now a days. The method requires a suitable sample conditioner.

Polarographic Analysers

Polarographic analysers have been called voltammetric analysers or electrochemical transducers. With the proper choice of electrodes and electolytes, instruments have been developed utilising the principles of polarography to monitor SO_2, NO_2, CO, O_2, H_2S, and other gases.

The transducer in these instruments is generally a suitably contained electrochemical cell in which a chemical reaction takes place involving the pollutant molecule. Two basic techniques used in the transducer are: (i) the utilisation of a selective semi-permeable membrane that allows the pollutant molecule to diffuse to an electrolytic solution; and (ii) the measurement of the current change produced at an electrode by the oxidation or reduction of the dissolved gas at the electrode. With proper use, polarographic analyser can be a valuable tool to an air pollution agency's inspection programme or to a source operator wishing to check pollutant levels at various plant locations. Complete systems also are available for continuous monitoring but should be designed carefully so as to give accurate emission data.

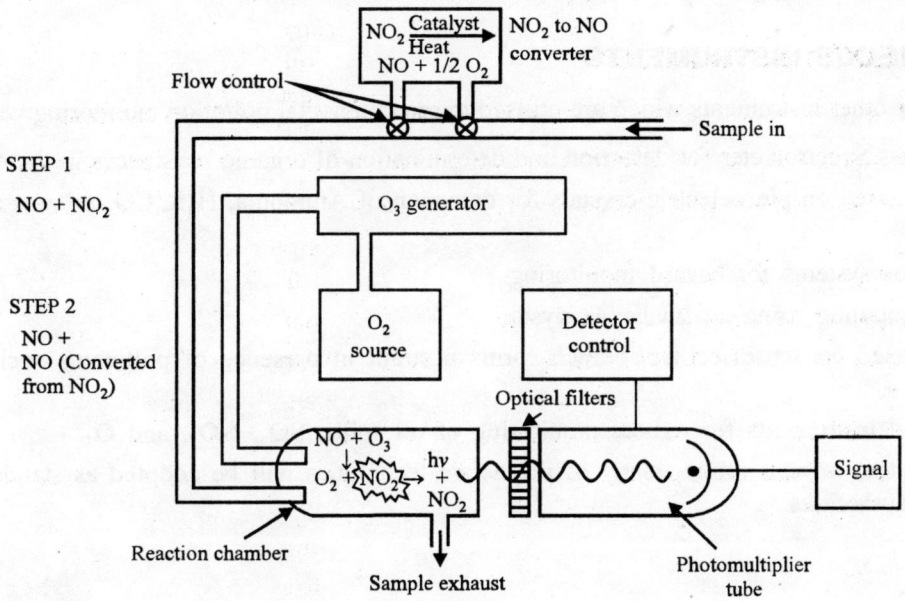

Fig. 28.9. Chemiluminescence analyser.

Hydrocarbon Analysers

Many hydrocarbon analysers are commercially available for simultaneous determination of total hydrocarbon, methane, non-methane hydrocarbon and carbon monoxide in ambient air and stack emission.

In such analysers, air sample is introduced directly into a flame ionisation detector (FID) for total hydrocarbon determination. An aliquot of the sample is introduced into pre-chromatographic or stripper column which removes hydrocarbons other than methane and CO. Methane and CO are passed quantitatively through the analytical column where they are separated. The methane is eluted first and passed unchanged through the catalytic reduction tube into the FID. The CO is eluted next and passed through the catalytic reduction tube where it is reduced to methane before reaching into the FID.

The electrometer output is measured using a recorder or integrator. Non-methane hydrocarbons are determined by substracting methane from total hydrocarbons.

Portable Gas Chromatographs

A large variety of portable gas chromatographs with interchangeable detectors are available for on-site measurement of organic pollutants in concentrations ranging from 0.1 to 2000 ppm.

CONTINUOUS MONITORING

With advances of microprocessor technology, a vast array of sophisticated automatic on-line analysers are commercially available for continuous monitoring of various pollutants both in water and air. Continuous monitors for water are generally based on ion-selective electrodes or spectrophotometry. Air monitors generally utilise gas chromatography in multi-detection mode.

MISCELLANEOUS INSTRUMENTS

There are many other instruments which are utilised in environmental pollution monitoring which include:

1. Mobile Mass Spectrometer for detection and determination of organic substances in the environment.

2. Analysers based on piezoelectric crystals for detection of Ammonia, H_2S, CO, toluene and benzene in the air.

3. Multi-sensor systems for hazard monitoring.

4. Asbestos counting using an Image Analyser.

5. Monitor based on sensitised tape which forms a strain in presence of pollutants such as toluene diisocyanate.

6. Laser based instruments for remote monitoring of oil spills, SO_2, NO_2, and O_3.

It is hoped that in near future, many of the above techniques will be adopted as standard methods by regulatory authorities.

CHAPTER 29

Environmental Impact Assessment

INTRODUCTION

India is going through a process of metamorphosis since independence. During the last 50 years, it has gone through various facets of development and corresponding environmental degradation in the form of depleting forests and diminishing natural resources. For the optimal utilisation of finite natural resources through the use of better technologies and incorporation of suitable remedial measures to check environmental degradation at the project formulation stage itself, environmental clearance of 29 categories of major development projects has been made mandatory since 1994.

Early practices in project assessment were limited and often based on technical feasibility studies and cost benefit analysis. Till recently, the environmental factors were not taken into consideration while examining viability of a project. The growing number of cases of adverse environmental consequences as a result of interactions between socio-economic and industrial development activities have led to a serious rethinking that environmental consequences have to be explicitly considered in the decision making process. Free air and free water, once free gift of nature to human beings, need to be conserved' on sustainable basis. The growing awareness and concern about the need for environmental protection has resulted gradually in the introduction of related legislation and institutional arrangements. India being a developing nation, the socio-economic development is essential for providing satisfactory quality of life to the people. At the same time, environmental degradation need to be checked so that the natural resources are available on sustainable basis. To meet these requirements, in 1992, a Policy statement for Abatement and Control of Pollution as well as National Conservation strategy for conservation of natural resources were enunciated.

In this context, Environment Impact Assessment (EIA) has come to be recognised as an important technique for incorporating the objectives of environmental concerns with the requirements of economic growth and social development. EIA helps in examining the options for choosing an environmentally acceptable course of action including site selection, choice of technology, resource conservation, etc. The International Financing Institutions have also introduced the procedure/development of EIA as a pre-requisite for project funding. Various laws have been enacted in different countries of the world, to make EIA essential for the approval of all the new development projects.

AIMS AND OBJECTIVES OF EIA

Environment Impact Assessment (EIA) is a procedure designed to identify the potential impacts (positive/adverse) of a development project on the surrounding environment.

Environmental clearance of these activities is carried out by the Central/or the State Governments, with the following objectives :

1. Optimal utilisation of finite natural resources through use of better technologies and management packages.

2. Incorporating suitable remedial measures at the project formulation stage to ensure minimum harm to the surrounding environment.

Earlier, obtaining environmental clearance from the Central Ministry was only an administrative requirement intended for mega projects undertaken by the Government or Public Sector undertakings. However, the new Notification issued by the Ministry in January 1994 (as amended in May 1994), makes Environment Impact Assessment (EIA) statutory, for 29 different identified activities (Annexure–I). This EIA notification also includes details of procedures for obtaining environmental clearance and for public involvement, besides setting time schedules for decision making. Format of application form is given in Annexure–II.

The objective of an EIA is to ensure that environmental problems are foreseen and properly addressed. To achieve this aim, decision makers must fully understand the EIA's conclusions. Most of decision-makers overlook information, unless it is presented in terms of its immediate use in a concise form. Therefore, the format of EIA may include :

1. 'Hard' factors and prediction about impacts, comment on the reliability of this information, and summarise consequences of each of the proposed options.

2. Terminology and vocabulary used by the decision-makers should communicate the right message to community affected by the project.

3. Essential findings should be in a concise document, supported by separate background materials wherever it is necessary.

4. The document should be made easy to use, providing visuals wherever possible.

IMPORTANT PRINCIPLES IN MANAGING AN EIA

Principle 1

Focus on the main issues

The scope of the EIA should be limited only to the most likely and most serious of the possible environmental impacts. Some EIA's have resulted in large and complex reports running to several thousand pages. Such extensive work is unnecessary and can be counter-productive, because the EIA's findings must be readily accessible and immediately useful to decision makers and project planners. While suggesting the mitigative measures, it is important to focus on the study only on workable, acceptable solutions to the problems rather than wasting time for considering measures that are impractical or totally unacceptable to the developer or to the government. It is desirable to provide a summary of information relevant to each group of the decision making.

Principle 2

Involve the appropriate persons and groups

It is important, to be selective while involving various groups/people in EIA approaches. Following persons may be associated with this process :

1. Those appointed to manage and undertake the EIA process.
2. Those who can contribute facts, ideas or concerns to the study, including scientists, economists, engineers, policy makers and representatives of interested or affected groups/people.
3. Those who have direct authority to permit, control or alter the project—that is, the decision makers—including the developers, aid agencies or investors, competent authorities, regulators and politicians.

Principle 3

Link information to decisions about the project

An EIA should be organised so that it directly supports the decisions that need to be taken about the proposed project. It should start early enough to provide information to improve the designs, and should progress through the several stages of project planning by providing the relevant information necessary for arriving at a particular decision for the project.

Principle 4

Present clear options for the mitigation of impacts and for sound environment management

To help decision-makers, the EIA, must be designed so as to put forward clear choices on the planning and implementation of the project. It should clearly mention the positive as well as adverse impacts of the project on environment. To mitigate adverse impacts, the EIA could propose the following measures :

1. Pollution control technology or design features.
2. The reduction, treatment, recycling or disposal of wastes.
3. Optimal utilisation of natural resources.
4. Compensation or concessions to affected groups/people.

To enhance environment compatibility, the EIA should suggest

1. Several alternative sites.
2. Changes to the project design and operation.
3. Limitations to its initial size or growth.
4. Separate programmes which contribute in a positive way to local resources or to the quality of the environment.

And to ensure that the implementation of an approved project is environmentally sound, the EIA may prescribe :

1. Monitoring programmes or periodic impact reviews.
2. Contingency plans for regulatory action.
3. The involvement of the local community before taking final decisions.

Principle .5

Details about the consequences of considered alternatives along with reasons to select the particular alternative should be provided to the decision makers.

ENVIRONMENTAL IMPACT ASSESSMENT PROCESS

Screening

Screening is the first and simplest stage of project evaluation. Screening helps to clear the type of projects which in past experience are not likely to cause any serious environmental problems/degradation.

Preliminary Assessment

If screening does not automatically clear a project, the developer may be asked to undertake a preliminary assessment. This involves sufficient research expert advice to :

1. Identify the project's key impacts on the local environment.
2. Generally describe and predict the extent of the impacts.
3. To evaluate their importance to decision-makers.

The preliminary assessment can be used to assist early project planning – for instance, to narrow down the discussion of possible sites; to serve as an early warning that the project may have serious environmental problems, etc. It is in the developer's interest to do a preliminary assessment, as this step can clear some of the projects which do not require a full EIA.

EIA Study

When it is established that EIA is necessary for clearance of the project after preliminary assessment, the developer should go to full-fledged study. The various steps involved in an EIA study include the following :

1. To commission the project to an independent project co-ordinator with an expert team to carry out the detailed study for collection of base line data.
2. Identification of the key decision makers who will plan, finance, permit and control the proposed project.
3. To understand existing laws and regulations that will affect the decisions about the project.
4. Making contact with each of the various decision makers.
5. Determining how and when the EIA's findings will be communicated.

To carry out the detailed study, the study team can follow the steps described below :

1. The EIA study team should ensure that all the issues of importance are taken into account.
2. The team should initiate discussions with the project developers, decision-makers, regulatory agencies, local community leaders on all the possible issues and concerns, raised by the various groups.

3. To collect required information along with supportive baseline data to formulate the project profile.

4. To prepare EIA study report taking into account the identification, predictions of various facets of development project, evaluating and suggesting mitigative measures to save environmental degradation for the successful implementation of the proposed project.

The last step in the EIA process, is documentation of the process and the conclusions. Many technically first-rate EIA studies fail to exert their importance and usefulness because of poor documentation. The EIA can achieve its purpose only if its findings are well communicated to decision-makers.

To summarise, an EIA report typically contains :

1. An executive summary of the EIA findings.

2. A description of the proposed development project.

3. The major environmental and natural resource issues.

4. The project's impacts on the environment (in comparison with a baseline environment as it would be without the project), and how these impacts were identified and predicted.

5. A discussion of options for mitigating adverse impacts and for shaping the project to suit its proposed environment, and an analysis of the trade-offs involved in choosing between alternative actions.

6. An overview of gaps or uncertainties in the information.

7. A summary of the EIA for the general public.

All of this should be contained in a very precise, easy-to-read document, with cross references to background documentation, if any, which may be provided in an appendix.

LIMITATIONS

Even though environmental impact assessment is potentially one of the most valuable, intra-disciplinary, objective decision taken, yet it is unable to quantify the values of various parameters to indicate their effect on the quality of environment. In its present form, EIA suffers from several limitations at conceptual, methodological and procedural levels. Some of these limitations are described below.

Presently EIA is conceived merely as a project level tool and does not address to developmental programmes at the policy and planning levels. The project level decisions are many times constrained by existing policies and plans and the range of possible alternatives in project EIA are often small or non existent. Also there is no objective screening criteria to decide of the type or scale of projects should be subjected to an EIA.

EIA should ideally be undertaken at the policy and planning levels as the environmental consequences of the project often arise due to higher level decisions. But the policy EIA is extremely complex mainly due to the fact that potential range of alternatives to achieve a desired total would be unlimited. This problem can be solved to some extent by adopting an hierarchical approach. Only the number of alternatives are reduced by defining the problem in terms of a series of choices.

The most appropriate stage for implementing EIA is at the level of district planning, since at this stage a reasonable number of alternatives are available to the developer.

Another conceptual limitation of EIA is that it does not incorporate the strategies of preventive environmental intervention. The issues of resource conservation, waste minimisation and of project recovery, improvement in efficiency of equipment etc. should be pursued as explicit goals in EIA.

Experience over the years had shown that EIAs are always conducted with severe limitations of time, manpower, financial resources and data. India has unique problems resulting in widely diversent life-style, varying nature of terrain, flora and fauna. No reliable environmental information base exists to cater to the needs of a comprehensive EIA study. Therefore, data for EIA could be merely reduced by focusing the study on a limited number of relevant issues rather than an elaborate listing of values for all environmental parameters.

A major limitation of the present environmental process is the lack of objective criteria to decide whether a project requires comprehensive EIA or not. Screening is a method of selection which allows elimination from the review process of all those projects that do not require detailed EIA, thus avoiding unnecessary expense and delays in project clearance. Incorporation of screening in the environmental review process requires formulation of project and site related screening criteria for various types of developmental activities.

Most ecological problems are the cumulative result of environmental and social impacts of human activity in the region. Planning for sustainable development in the context of ecosystem, carrying capacity thus requires systematic identification, quantification and management of cumulative trends in significant environmental variables on a regional basis. Functional planning regions need to be identified based on ecological criteria such as climatic and vegetation patterns, soil classification and watershed boundaries rather than political justifications. Within the context of sustainable development, regional EIA could provide the means for estimation of developmental limits imposed by regional carrying capacity.

Legal Framework

Ministry of Environment and Forests has issued a notification in January, 1994 and amended in May, 1994 which makes environmental impact assessment statutory for development of 29 projects. It also enclosed the details of procedures for obtaining environmental clearance and public involvement besides time schedule for decision making.

In addition to the application form as per schedule to the Notification, project proponents were required to furnish the following information for environmental appraisal :

1. EIA/EMP Report.
2. Risk Analysis Report – However, such reports are not required for every category of projects.
3. NOC from the State Pollution Control Board.
4. Commitment regarding availability of water and electricity from the competent authority.
5. Summary of project report/feasibility report.
6. Filled in questionnaire for environmental appraisal of the project.
7. Comprehensive rehabilitation plan, if more than 1000 people are likely to be displaced, otherwise a summary plan would be adequate.

The reports submitted with the application shall be evaluated and assessed by the Ministry. The Ministry has appointed a Committee of Reports in different sectors to assess and make recommendations on the proposed application of the project proponents. The composition of the committee as specified in Schedule-III of the Notification is given in Annexure–III and Annexure-IV.

Public Hearing

Recently, a new dimension has been added in the EIA process known as "public hearing" of the project. The project proponent, when he applies for environmental clearance of project, would submit to the concerned State Pollution Control Board (SPCB) details containing an executive summary of the project, details about effluent and emission discharged during the process/operation in the industry and other details as felt necessary by the SPCB.

The SPCB shall give a notice for environmental public hearing mentioning the date, time and venue, which shall be published in at least two newspapers widely circulated in the region around the project. Suggestions, views, comments and objections of public on the project shall be invited within 30 days from the date of publication of the notice.

All persons including bonafide residents, environmental groups and others located at project site/sites or displacement sites likely to be affected can participate in the public hearing. They can also make oral/ written suggestions to the SPCB for consideration during environmental clearance of the project. Public hearings would be called for in case of projects involving large displacement or having severe environmental impacts.

POST PROJECT MONITORING

Environmental clearance to a project is given by a competent authority along with certain conditions, safeguards, etc. These safeguards are to be implemented by the project authorities while executing the project. The project authorities are required to submit bi-annually progress report on implementation of the safeguards as well as compliance with the prescribed environmental norms. The State Pollution Control Boards/other local authorities periodically monitor the progress of implementation of the project. The reports are also sent to apprise the competent authorities, who have approved the aforesaid project.

ANNEXURE–I

List of Projects Requiring Environmental Clearance from the Central Government

1. Nuclear Power and related projects such as Heavy Water Plants, nuclear fuel complex, rare earths.
2. River valley projects including hydel power, major irrigation and their combination including flood control.
3. Ports, Harbours, Airports (except minor ports and harbours).
4. Petroleum Refineries including crude and product pipelines.
5. Chemical Fertilisers (Nitrogenous and phosphatic other than single superphosphate).
6. Pesticides (Technical).
7. Petrochemical complexes (both Olefinic and Aromatic) and petro-chemical intermediates such as Dimethyl Phthalate (DMT), Caprolactam, Linear Alkyl Benzene (LAB) etc. and production of basic plastics such as Low Density Polyethylene (LDPE), High Density Polyethylene (HDPE), Polypropylene (PP) and Polyvinyl Chloride (PVC).
8. Bulk drugs and pharmaceuticals.
9. Exploration for oil and gas and their production, transportation and storage.
10. Synthetic rubber.

11. Asbestos and asbestos products.

12. Hydrocyanic acid and its derivatives.

13. (i) Primary metallurgical industries (such as production of iron and steel, aluminium, copper, zinc, lead and ferro alloys; and (ii) Electronic arc furnaces (Mini steel plants).

14. Chlor-alkali industry.

15. Integrated paint complex including manufacture of resins and basic raw materials required in the manufacture of paints.

16. Viscose staple fibre and filament yarn.

17. Storage batteries integrated with manufacture of oxides of lead and lead antimony alloy.

18. All tourism projects between 200 m – 500 m of High Tide Line or at locations with an elevation of more than 1000 metres with investment of more than Rs. 5 crores.

19. Thermal Power Plants.

20. Mining projects (major minerals) with leases more than 5 hectares.

21. Highway projects.

22. Tarred Roads in Himalayas and/or Forest areas.

23. Distilleries

24. Raw skins and Hides.

25. Pulp, Paper and newsprint.

26. Dyes.

27. Cement.

28. Foundries (individual).

29. Electroplating.

ANNEXURE–II

Application Form

1. (a) Name and address of the project proposed :

(b) Location of the project

Name of the place :

District, Tehsil :

Latitude/Longitude :

Nearest Airport/Railway Station :

(c) Alternative sites examined and the reasons for selecting the proposed site :

(d) Does the site conform to stipulated land use as per local land use plan :

2. Objective of the Project

3. (a) Land Requirement :

Agriculture land :

Forest land and Density of Vegetation :

Others (Specify) :

(b) (I) Land use in the catchment/within 10 km. radius of the proposed site

(II) Topography of the area indicating gradient, aspects and altitude

(III) Erodability classification of the proposed land

(c) Pollution sources existing in 10 km. Radius and their impact on quality of air, water and land:

(d) Distance of the nearest National Park/Sanctuary, Biosphere reserve/Monuments/Heritage site/ Reserve Forest :

(e) Rehabilitation Plan for quarries/borrow areas :

(f) Green belt plan

4. Climate and Air Quality

(a) Windrose at site

(b) Max./Min./Mean annual temperature.

(c) Frequency of inversion

(d) Frequency of cyclones/tornadoes/cloud burst :

(e) Ambient air quality data

(f) Nature and Concentration of emission of SPM, Gas (CO, CO_2, NO_2, CH_n, etc.) from the project.

5. Water Balance

(a) Water balance at site

(b) Lean season water availability

(c) Source to be tapped with competing users (Rivers, Lake, Ground, Public Supply)

(d) Water quality

(e) Changes observed in quality and quantity of ground water in the last 15 years and present charging and extraction details

(f) (i) Quantum of waste water to be released with treatment details

(ii) Quantum and quality of water in the receiving body before and after disposal of solid waste

(iii) Quantum of waste water to be released on land and type of land

(g) (i) Details of reservoir water quality with necessary catchment treatment plan

(ii) Command area development plan

6. Solid Wastes

(a) Nature and quantity of solid wastes generated

(b) Solid waste disposal method

7. Noise and Vibrations

(c) Sources of noise and vibrations

(d) Ambient noise level

(e) Noise and vibration control measures proposed

(f) Subsidence problem if any with control measures

8. Power requirement indicating source of supply: complete environmental details to be furnished separately, if captive power unit proposed :

9. Peak labour force to be deployed giving details of :
   (a) Endemic health problems in the area due to waste water/air/soil borne diseases
   (b) Health care system existing and proposed
10. Number of villages and population to be displaced :
   (a) Rehabilitation Master Plan :
11. Risk Assessment Report and Disaster Management Plan :
12. (a) Environmental Impact Assessment   Report prepared as
   (b) Environmental Management Plan   Per guidelines of
   (c) Detailed Feasibility Report   MOEF issued from
   (d) Duly filled-in questionnaire   time to time
13. Details of Environmental Management Cell :

I hereby give an undertaking that the data and information given above are true to the best of my knowledge and belief and I am aware that if any part of the data/information submitted is found to be false or misleading at any stage, the project may be rejected and the clearance given, if any, to the project is likely to be revoked at our risk and cost.

Date :

Place :

Signature of the Applicant

with name and full address

Given under the seal of Organisation on

behalf of whom the applicant is signing

In respect of item for which data are not required or is not available as per the declaration of project proponent, the project would be considered on that basis.

ANNEXURE–III

Composition of the Expert Committees for Environmental Impact Assessment

1. The Committees will consist of experts in the following disciplines :
   (a) Eco-system Management
   (b) Air/Water Pollution Control
   (c) Water Resources Management
   (d) Flora/Fauna Conservation and Management
   (e) Land Use Planning
   (f) Social Sciences/Rehabilitation
   (g) Project Appraisal
   (h) Ecology
   (i) Environmental Health

(j) Subject Area Specialists

(k) Representatives of NGOs/persons concerned with Environmental issues.

2. The Chairman will be an outstanding and experienced ecologist or environmentalist or technical professional with wide managerial experience.

3. The representatives of IAA will act as Member-Secretary.

4. Chairman and Members will serve in their individual capacities except those specifically nominated as representatives.

5. The membership of a committee shall not exceed 15.

ANNEXURE–IV

Schedule–IV

(See sub-para 1 of para 2)

Procedure for Public Hearing

1. Process of Public Hearing : Whoever apply for environmental clearance of projects, shall submit to the concerned state Pollution Control Board twenty sets of the following documents, namely :

    (a) An executive summary containing the salient features of the project, both in English as well as local language.

    (b) Form XII prescribed under Water (Prevention and Control of Pollution) Rules, 1975 where discharge of sewage, trade effluents, treatment of water in any form, is required.

    (c) Form I prescribed under Air (Prevention and Control of Pollution) Union Territory Rules, 1983 where discharge of emissions are involved in any process, operation or industry.

    (d) Any other information or document which is necessary in the opinion of the Board for their final disposal of the application.

2. Notice of Public Hearing :

    (a) The State Pollution Control Board shall cause a notice for environmental public hearing which shall be published in at least two newspapers widely circulated in the region around the project, one of which shall be in the vernacular language of locality concerned. The State Pollution Control Board shall mention the date, time and place of public hearing. Suggestions, views, comments and objections of the public shall be invited within thirty days from the date of publication of the notification.

    (b) All persons including bonafide residents, environmental groups and others located at the project site/sites or displacement sites likely to be affected can participate in the public hearing. They can also make oral/written suggestions to the State Pollution Control Board.

3. Explanations :

For the purpose of the paragraph, person means :

    (a) Any person who is likely to be affected by the grant of environmental clearance

    (b) Any person who owns or has control over the project with respect to which an application has been submitted for environmental clearance

    (c) Any association of persons, whether incorporated or not, likely to be affected by the project and or functioning in the field of environment

(d) Any local authority, any part of whose local limits is within the neighbourhood, wherein the project is proposed to be located

1. Composition of the public hearing panel : The composition of the Public Hearing Panel may consist of the following, namely :

(a) Representative of the State Pollution Control Board

(b) The District collector or his nominee

(c) Representative of the State Government dealing with the subject

(d) Representative of the Department of the State Government dealing with environment

(e) Not more than three representatives of the local bodies such as municipalities or panchayats

(f) Not more than three senior citizens of the area nominated by the District collector

2. Access to the Executive Summary: The concerned persons shall be provided access to the Executive Summary of the project at the following places namely :

(a) The District Collectors office

(b) The District Industry centre

(c) In the office of the Chief Executive officers of Zilla Parishad or Commissioner of the Municipal Corporation/Local body as the case may be

(d) In the head office of the concerned State Pollution Control Board and its concerned Regional Office

(e) In the concerned Department of the State Government dealing with the subject of environment.

CHAPTER 30

Environmental Auditing

INTRODUCTION

Environmental Audit (EA) is a structured and comprehensive mechanism for ensuring that the industrial activities do not adversely affect the environmental quality and the economy of the industrial sector improves as a consequence of improved process and energy. It is emphasised that the successful Environmental Audit (EA) programme in a systems approach concept investigates all possibilities of energy saving, raw material saving and all water budgeting through conservation included to demonstrate how the status of production and environmental management systems are investigated, vis-à-vis the Indian regulatory requirements, and how material, energy, water, emissions health and safety audits have been conducted to identify avenues for savings in the cost of production. Process industries in India are diversifying into progressively more capital intensive and energy intensive areas, which are degrading the environmental quality. Considering the future environmental and energy scenarios in Indian context, Environmental Audit (EA) deserves to be adopted as a pre-requisite for sustainable development and friendly environmental management of process industries.

A paradox of modern technological society in India is that efforts towards economic prosperity and increased standard of living could be detrimental to the overall quality of life, due to the encroachment of nature beyond its sustenance level or the injection of pollutants into the environment that exceed its assimilative capability. In India, efforts for environmental protection have often relied on strict regulatory measures with little regard to productivity. In order to guard the environment without seriously blocking technological development, it is a must to develop and adopt environmentally benign strategies and technologies. In this context, Environmental Audit provides an effective management tool considering both environmental management systems and process systems at the same time, thus accounting for productivity as focussed here. In the context of safeguarding the environment, probably no other issue has drawn so much of attention as the environmental audit.

EVOLUTION OF ENVIRONMENTAL AUDIT IN INDIA

The term audit in Indian context is understood as an official examination of accounts with respect to services and vouchers. An important aspect of audit is to verify compliance to company law and other statute wherever made obligatory to auditors. Though it commenced as a system of checking accounts (which normally mean financial aspect), the system of audit has been extended over the years to cover other areas also like, Environment.

Environmental Audit, the process of determining whether all or selected levels of an industry are in compliance with regulatory requirements internal policies and standards has proved to be a successful component of Environmental Management Programme in Western countries. However, in India, the concept of Environmental Audit appears to have emerged since the beginning of nineties. Each practitioner has developed his or her own special blend of procedures reflecting programme objectives and company cultures.

Environmental Audit has assumed more importance in India, as it is statutorily compulsory to submit environmental statement to the concerned state pollution control board. Since last two years, the concept and practice of Environmental audit has developed rapidly within Industries. Most approaches to environmental audit include provisions for systematic examination of performance to ensure compliance with prescribed regulatory requirements.

It may be concluded that even with the introduction of environmental audit concept in India, the option for environmental management in Indian industry is not focussing on introducing anticipative and preventive strategies such as adopting cleaner technologies of industrial production. If the option of more intensive pollution control is to be avoided in future, then it is a must to adopt a strategy of pollution prevention based on technologies that conserve resources, minimise pollution and reuse wastes as secondary resources to the maximum extent possible. *Hence it can be concluded that the main drawback of the prevailing practice of environmental audit is that it does not include the assessment of technology and manufacturing process as an integral part of it. If the same is included, then environmental audit may be regarded as an important instrument for sustainable development through introduction of cleaner–technologies in industrial sectors.*

Therefore, with the realisation of limitations of pollution control approach and felt need of resource conservation, the policy of environmental management should focus on anticipation and prevention.

ENVIRONMENTAL AUDIT IN INDIAN CONTEXT

Environmental Audit in Indian context should take cognisance of the following :

1. Environmental Audit undertaken in developed countries comprises two components, viz., assessment and verification of environmental systems.

2. Endeavour on environmental protection thus far in India has relied on strict regulatory measures with little regard to economic productivity.

3. Preventive and reactive approaches do not complement each other in the current practices of environmental management in India as reflected in legislative, administration and policy formulation.

Thus, the Environmental Audit in the context of free market economy in India is to be defined as "a pragmatic management tool comprising systematic, documented, periodic and objective evaluation of production and environment management systems to ensure resource conserving modes of manufacturing, and cost-effective environmental protection as a consequence of improved material, water and energy effectively for enhanced economic productivity and acceptable environmental quality and health of employees and general public".

The above definition recognises the potential for resource conservation in manufacturing processes with concomitant reduction in the cost of production and pollution control, and comprises comparative analysis of the process technology in use vis-à-vis the state-of-the-art technology available in that sector with concomitant implications on raw materials and energy use effectively, as also the quantum and characteristics of residues.

The definition, interalia, considers shifts towards cleaner production technologies through comparative analysis of various competing alternatives in industrial sectors based on technological, economic, societal and environmental considerations as a part of Environmental Audit.

Components of Environmental Audit

The components considered while formulating a broad based framework for Environmental Audit are :

1. Production of technology/Process of manufacturing.
2. Material usage.
3. Energy usage.
4. Water usage.
5. Air, Waste-water and Noise emissions.
6. Solid and hazardous waste emissions.
7. Health and safety.
8. Legal and social obligations.

Basic Steps in Environmental Audit

Environmental Audit may be conducted in three basic steps :

1. Pre-audit phase.
2. On-site audit phase.
3. Post audit phase.

Various activities of Environmental Audit Methodology have been recast as follows :

1. In Pre-audit phase, an audit plan will be developed and the industry will be visited to collect basic information regarding process, materials flow, energy consumption pattern, water consumption pattern, waste-water sources and emissions compliance status regarding fugitive emissions, stack emissions, noise-levels, waste-water, solid/hazardous wastes, health and safety.
2. During on-site Audit phase, main thrust will be on experiments to evaluate performance, sampling and analysis, material, energy and water flow measurements.
3. Post audit phase involves developing material, water and energy balance diagrams, technology assessment, working out suggestion and recommendations with cost benefit analysis, preparation of an action plan and implementation schedule.

Environmental Audit

Sustainable development with its pre-conditions of economic efficiency, environmental harmony, equity and social justice warrants an environmental audit considering a systems approach comprising two inter-related subsystems, a production system and environmental system for objective evaluation. Assessment of manufacturing processes should be an integral part of environmental audit system, so as to achieve the best possible use of natural and human resources for gaining maximum benefits to all concerned.

Environmental audit should typically be a combination of Process Audit, Material Audit, Energy Audit, Water Audit, Air, Wastewater and Noise emissions Audit, Compliance Audit and Health/Safety Audit. Environmental Audit should aim at assessing the technology, manufacturing processes and environmental management systems. Operational framework for environmental audit is depicted in Fig. 30.1.

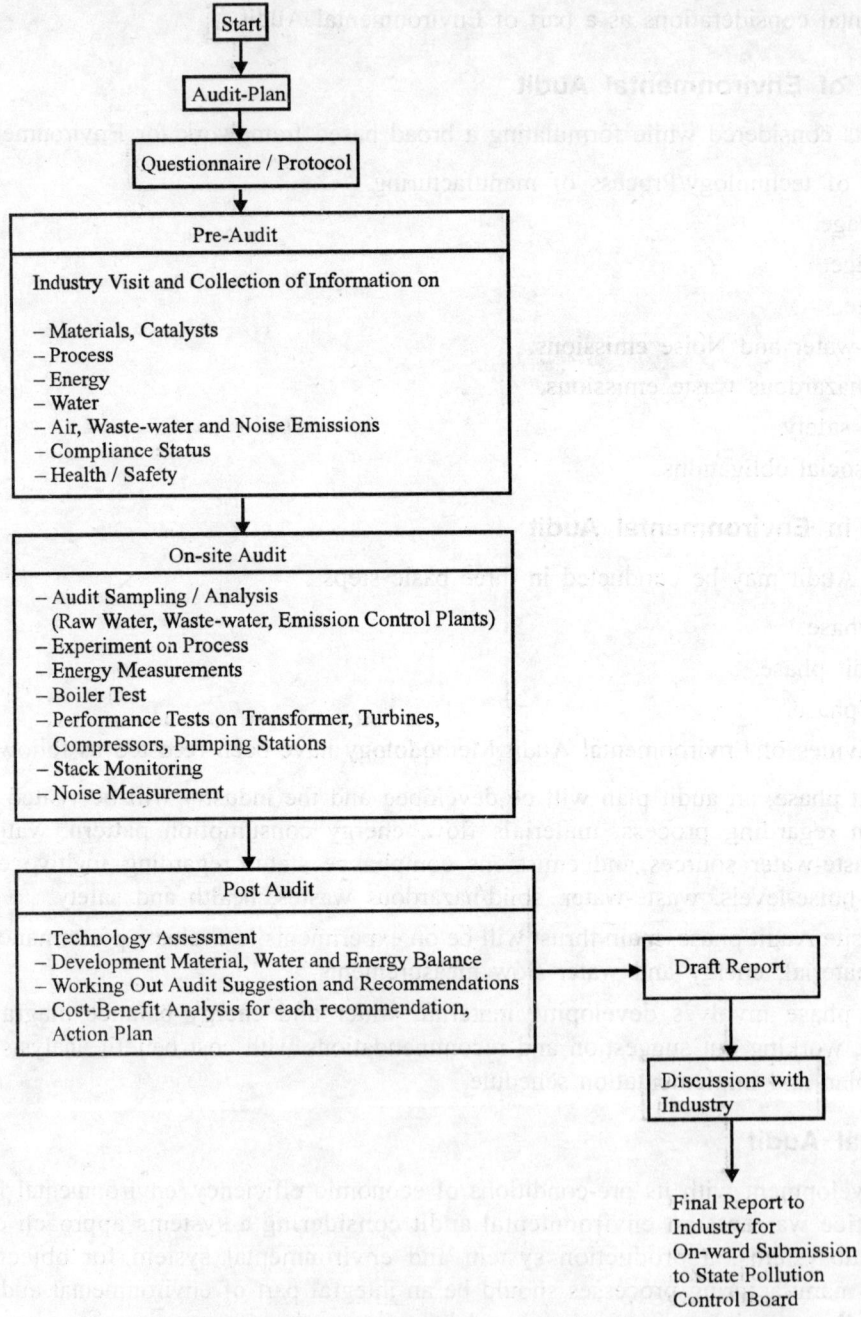

Fig. 30.1. Operational framework for environmental audit.

CASE STUDY

The application of the methodology for Environmental Audit developed has been validated through a case study of fertiliser industry. The selected fertiliser industry is a single streamline plant and comprises 900 MTPD ammonia plant, 1500 MTPD urea plant, and other allied offsite facilities like 2×1000 MTPH high pressure steam boilers, 2×7.5 MW gas turbines, raw water pretreatment and demineralisation plants, cooling towers, inert gas generator, storage tanks, urea silo, bagging plant and emission control plant. Process flow diagrams for ammonia and urea production are shown in Fig. 30.2. and 30.3.

The industry is a gas-based fertiliser complex with Natural Gas requirement of about 1.3 million SM^3/day supplied by the Gas Authority of India Ltd. (GAIL) through a 90 km long pipeline. The raw water requirement of 14.55 MLD is drawn from the reservoir which is 13 km away from the plant. The plant generates about 3000 M^3/day of liquid effluent, which is treated at the effluent treatment plant and then pumped for irrigation to the greenbelt.

The ammonia plant, based on Haldor Topsoe's Steam reformation technology, with a design energy consumption of 7.8 MM K Cal/MT ammonia, has a conventional front-end with desulphurisation, primary reforming, secondary reforming, high and low temperature shift conversion, and carbon dioxide removal by Giammarco-Vetrocoke technology. The urea plant is based on Snamprogetti's total recycle stripping process using ammonia as the self stripping agent.

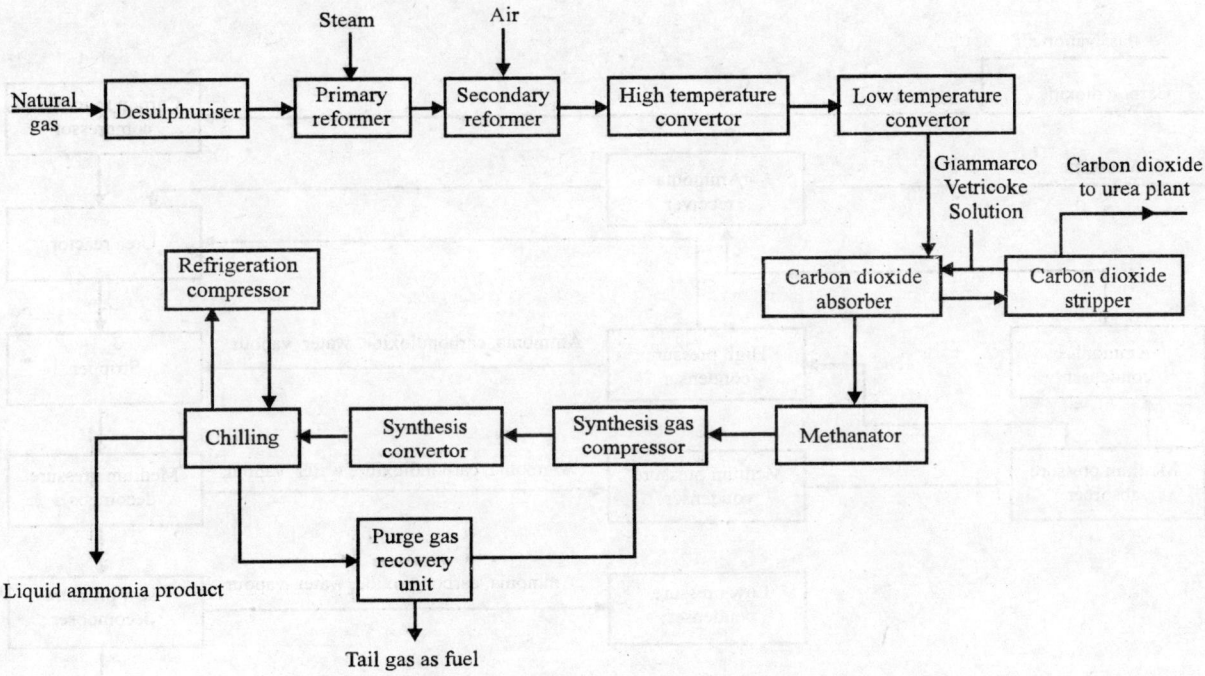

Fig. 30.2. Simplified process flow sheet of ammonia plant.

The salient features of the Ammonia and Urea Plant in the selected fertiliser industry are :

1. Ammonia plant is the first of its kind to use 25/35 Cr-Ni-Nb alloy for the primary reformer tubes at the grass root level, which is capable of operating at lower steam: carbon ratio (2.5 – 3.0) without carbon formation in tubes and results in lower skin temperature and higher overall furnace efficiency.

2. Low heat Giammarco-Vetrocoke (GV) process is adopted for carbon dioxide removal.

3. The use of S-200 converter results in higher conversion per pass and lower pressure drop leading to considerable energy savings. Haldor Topsoe's low pressure synthesis loop of 140 kg/cm² has been used for the first time in India compared to 220-230 kg/cm² in other conventional plants.

4. Installation of Pure Gas Recovery Unit (PGRU) at the grass root level has helped in low energy consumption in primary reformer. Ammonia absorber in PGRU operates at synthesis loop pressure resulting in low ammonia slip of 150 ppm and higher ammonia production.

5. Ammonia preheater in urea plant is installed for the first time in India at design state itself to pre-heat the ammonia entering the urea reactor to 780°C utilising the heat available from Low Pressure decomposer vapours.

6. A vacuum pre-concentrator has been added before the first vacuum concentrator to concentrate urea solution. The pre-concentrator concentrates urea solution from 70 to 88.5% by utilising waste heat from medium pressure decomposer off gases, which results in energy savings to the tune of 0.102 MM K Cal/MT of urea.

7. Excess Low Pressure Steam generated in the urea plant is injected into carbon dioxide compressor turbine to reduce High Pressure steam consumption. The overall requirement of steam is 0.86 MT/MT urea compared to 1.05 MT/MT urea in other conventional plants in India.

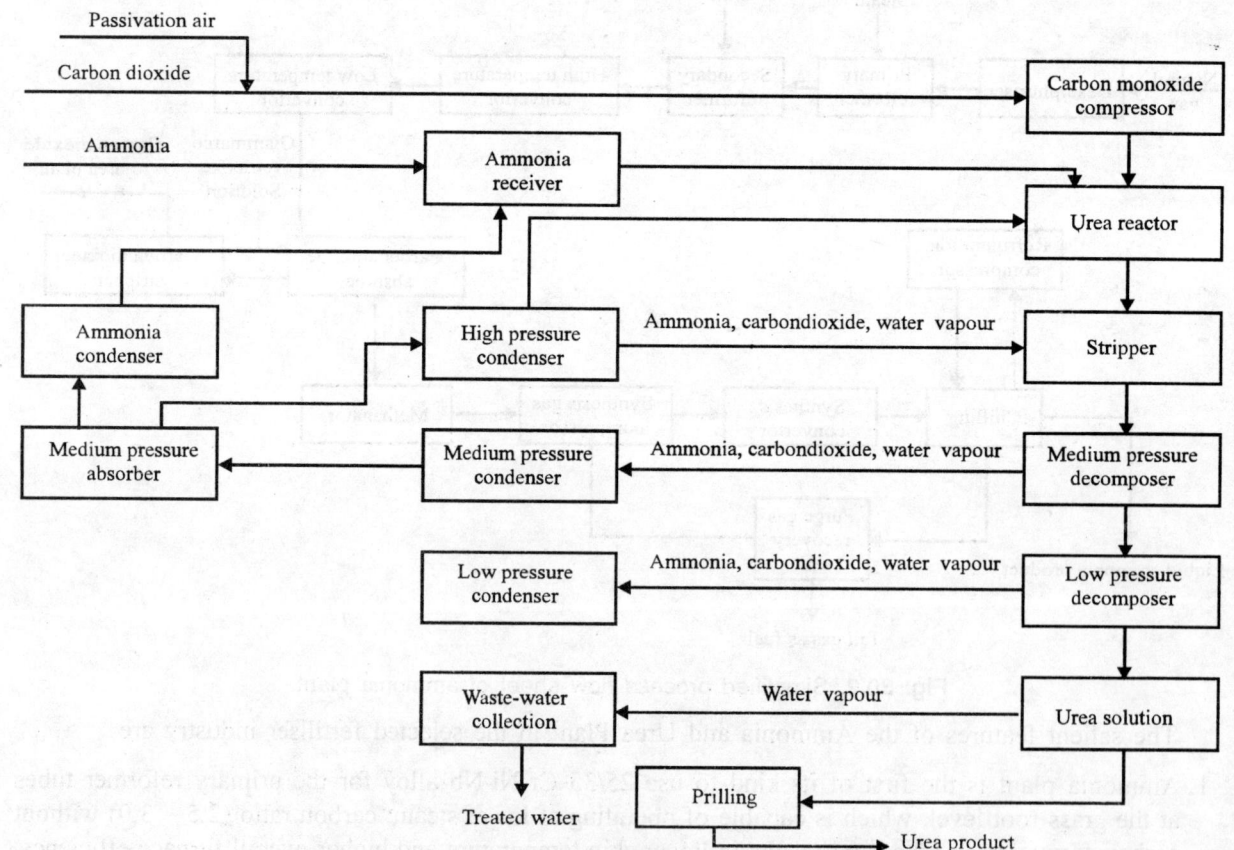

Fig. 30.3. Simplified process flow sheet of urea plant.

The case study has been conducted in three phases: Phase I : Pre-Audit activities; Phase II : On-site activities; and Phase III : Post-audit activities. The pre-audit activities commenced with the development of an audit plan, which included delineation of scope, priority areas for examination, and explanation of the procedure. Questionnaires and protocols were developed for collection of background information.

The on-site audit activities commenced with a meeting with the concerned personnel of the Industry. Material and energy measurements were taken during this phase. Identification of sampling locations for Water Audit, Wastewater and Air Emission Audit, and Compliance Audit was an important step in this phase followed by the sampling and analysis exercises. The post-audit activities included technology assessment; development of material, energy and water balance diagrams; and exploration of conservation potential in the industry.

Technology assessment for fertiliser industry was undertaken as a part of the process audit. A combination of Delphi and analytical Hierarchy Process (AHP) was adopted for ranking of alternative technologies. The parameters selected for ranking of technology options for ammonia and urea production are fuel, total energy and water used per tonne of product manufactured along with stack emissions per tonne of product manufactured. The assessment reveals that the most appropriate technologies for ammonia and urea production are M.W. Kellogg's Technology, and Snamprogetti's ammonia stripping process.

ENERGY AUDIT

Basic data regarding the operational features and working of various process units, their overall energy consumption, and their cost and production figures for the past years were collected. When sufficient data was assembled, existing records of consumption were reviewed and measurements were taken wherever necessary using portable instruments. These figures gave a trend of energy consumption and its cost per unit production over the year. A pie diagram of energy consumption was prepared to indicate the share of various forms of energy in the total energy consumption of the plant. Energy conservation potentials drawn on the basis of this environmental audit for electrical and thermal energy are shown in Table 30.1.

Table 30.1. Energy conservation potential.

| | | Savings from implementation of the suggested measures | | | | Cost of Implemen- tation of the suggested measures (Rs. in lakhs) | Simple Payback period |
| | | Thermal Energy | | Electrical Energy | | | |
S. No.	Item	Ng NM^3	Lakh Rs/Yr	Lakh Kwh	Lakh Rs/Yr		(Years)
1.	Steam distribution and utilisation	66.616	1.99	—	—	3.0	1.50
2.	Electrical systems	—	—	3.25	3.03	10.0	3.30
3.	Electric Drives	—	—	0.52	0.37	3.28	8.86
4.	Lighting	—	—	2.63	1.86	8.18	4.39
	Total	66.616	1.99	6.4	5.26	24.46	—

WATER AUDIT

Water Audit studies aimed to evaluate raw water intake facilitates, the performance of existing water treatment plant, the water consumption in different processes, and the development of a water balance scenario highlighting water conservation measures. Fig. 30.4. depicts the raw water consumption pattern and Fig. 30.5 depicts the suggested water balance diagram to the industry to ensure water conservation.

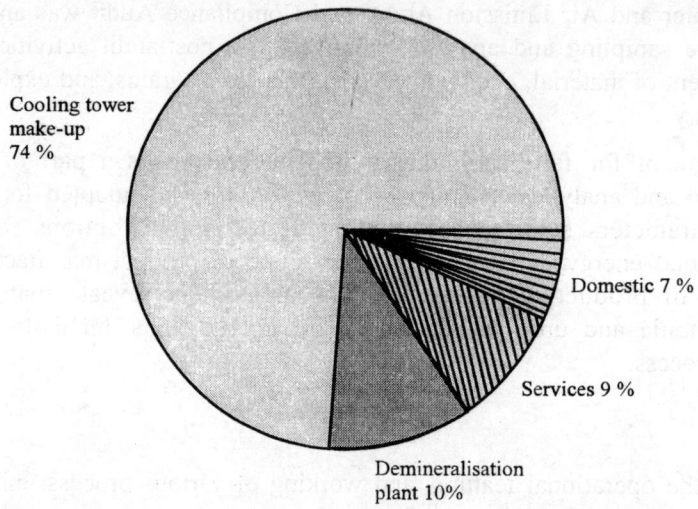

Fig. 30.4. Raw water consumption pattern.

COMPLIANCE AUDIT

Compliance audit studies aimed at investigating the status of the environment management systems and equipment vis-à-vis the regulatory requirements. It was suggested that the proper operation and maintenance practices helps this industry reducing emissions to arrest environmental quality deterioration. Work-zone monitoring was carried out to assess the status of existing air quality within the industrial complex. Stack monitoring was carried out at primary reformer, prill tower, steam generation plant and feed stock preheater stack in ammonia plant.

Monitoring of segregated and combined waste-water discharge were carried out and a performance evaluation of the effluent treatment plant and management system was undertaken. Existing facilities for handling and disposal of solid/hazardous wastes were critically examined. Table 30.2. depicts the outcome of the performance evaluation exercise.

HEALTH AND SAFETY AUDIT

Preliminary information (gained through questionnaire and protocols) on health and safety aspects was collected. Occupational health and accident scenario were gathered through audit exercises, pre audit meetings, and inspections. As a part of the health and safety audit, damage distances were evaluated using EFFECTS software, and suggestions were drawn for improvement. Table 30.3. depicts the outcome of these studies.

Table 30.2. Performance of prilling tower.

Flow rate : 68,300 NM³/hr Temperature : 60°C

Location	Without Dedusting Systems		With Dedusting Systems		Remarks
	Urea dust (Mg/NM³)	*Ammonia (mg/NM³)*	*Urea dust (mg/NH³)*	*Ammonia (mg/NM³)*	
Up wind side	0.56	—	0.21	—	At the time of monitoring the following observations were made :
Down wind side					
Site I	6.58	68.2	2.77	35.9	
				22.9	1. Ammonia releasing vent superimposed once in a while on monitoring site
Site II	7.89	20.2	2.77	87.5	
				127.9	
Average Conc.	7.24	46.3	2.77	68.6	
Pollutant load in Kg/hr	4.93	31.6	1.89	46.8	2. Due to low velocity of wind, waste gas plume (urea dust) is overlapping the monitoring site

Observation: Sampling ports were not available, therefore monitoring was carried out by Roof monitoring method, which is the only possible method to assess the total emissions from prilling tower.

Note :   – Threshold limit value is 35 mg/Nm³ (50 ppm) for Ammonia and 30 mg/NM³ for urea dust.

– The dust levels in exit air from the dedusting system is around 20–25 mg/NM³ air (one month average figure) against statutory norm of 50 mg/NM³

– Dust extraction system

Parameter	Design		Actual		Efficiency
	In (g/m³)	*Out (g/m³)*	*In (g/m³)*	*Out (g/m³)*	
Urea dust	10	0.05	5.54	0.0019	99.96%

Table 30.3. Health/safety audit findings.

Description	1992-93*	1993-94	1994-95**
Manhours worked	12, 95, 424	15, 34, 036	15, 02, 456
Accidents	Nil	1	3
Mandays lost	Nil	23	211

* Manhours are taken from August 1992 which is the month of commercial production

** Manhours figures are upto March 15, 1995 only

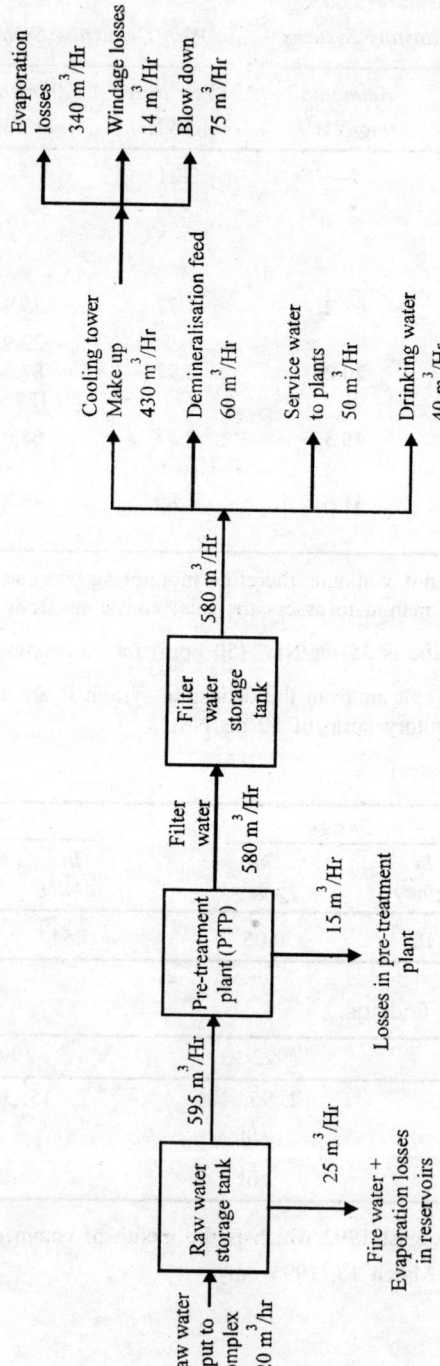

Fig. 30.5. Suggested water balance diagram.

Productivity indicators

Safety	*1994-95*	*1995-96*	*1996-97*
@ Frequency rate	–	0.652	1.996
@@ Severity rate	–	0.014	0.140

$$@ \text{ Frequency} = \frac{\text{Number of accidents} \times 1000000}{\text{Total manhours worked in the year}}$$

$$@@ \text{ Severity rate} = \frac{\text{Total number of days lost in a year due to accidents} \times 1000}{\text{Total number of manhours worked in the year}}$$

Thus, process, material and water audit studies have indicated that it is possible to conserve the basic raw materials. Major conservation opportunities exist in the field of energy usage, and it is possible to save Rs. 7,25.000/annum by adopting suggested environmental audit measures. The suggested measures identified during this environmental audit studies are being implemented by the industry in a phased manner. The industry is already recording savings from the implemented suggestions.

CHAPTER 31

Pollution Control Laws and Acts

INTRODUCTION

In modern industrial society, the control and management of pollution is underpinned by the law. Legal principles can, for a start, provide remedies for those whose interests are damaged or threatened by pollution. Such legal actions originally formed the mainstay of Indian Pollution Law, and continue today to have some significance. Legal control over pollution are dominated by forms of public law which provides various mechanisms of control over sources of pollution.

The Constitution of India contains a direct commitment to environmental protection. Article 48-A of our Constitution stipulates that the State shall endeavour to protect and improve the environment and to safeguard the forests and wildlife of the country. Under Article 51-A(g), citizens are requested to protect and improve the natural environment, including forests, lakes, rivers and wildlife. The Directive Principles under Article 49 and 51-A(f) also recognise the importance of protecting the sites of cultural heritage as part of the total environment. Thus, our Constitution provides the necessary bedrock for framing and enforcement of environmental legislations.

Schedule VII of the Constitution classifies the various legislative subjects into three categories, namely, Union list, State list, and Concurrent list. The legislations in the Union list are enacted by Parliament, while in the case of the State list, the state legislatures are empowered to enact the necessary legislation. The third category, the Concurrent list, specifies the subjects that are to be looked after jointly by the central and state governments. Though the subject environmental pollution control as such does not figure anywhere in these lists various individual sectors are specifically provided for in one or the other of these lists. For instance, while water supplies, irrigation and canal drainage, etc. are State subjects, the regulation and development of inter-state rivers and river valleys are Central subjects. Subjects like public health and sanitation, agriculture, protection against pests and prevention of plant diseases find place in the State list. On the other hand, prevention of the extension from one state to another of infectious or contagious diseases is included in the Concurrent list. Forests and protection of wild animals and birds are also some of the subjects figuring in the Concurrent list. In the national interest and in certain special circumstances, the Constitution enables the Parliament to legislate with respect to matters in the State list (Article 249 and 252). For instance, Parliament is empowered to legislate on all subjects of International Agreements to which our country is a party (Article 253).

The various Acts relating to pollution control and environmental protection can be broadly classified into two categories, namely, central and state enactments. Some of the important Acts are listed in *Annexture – I*.

POLLUTION LAWS AND ACTS

Air Pollution Laws and Acts

The main objective of enacting any pollution law is to control pollutant sources so that ambient pollutant concentrations are reduced to levels considered safe. The Central Air (Prevention and Control of Pollution) Act, 1981 is a significant development in this direction.

This act provides an integrated approach to tackling of problems related to pollution. It empowers the Central Board for the Prevention and Control of Water Pollution, constituted under the Water (Prevention and Control of Pollution) Act, 1974, to exercise the powers and perform the functions of the Central Board for the Prevention and Control of Air Pollution also. It also empowers the state governments to declare air pollution areas and to prohibit the use of any fuel which is likely to cause air pollution. The state governments are also empowered to prohibit the burning of any material which is likely to cause air pollution in any air Pollution control area. The officers of the State Board are empowered to obtain information relating to air pollution and to inspect the concerned premises and to take samples of emissions from sources for analysis.

The Central Air Act will certainly help to maintain performance standards of anti-pollution control measures. In addition, many state governments have their "Environmental Protection Advisory Committees" to take effective measures for pollution control, and these bodies also caution the government about the adverse consequences if such measures are not taken in time.

In many countries, the legal standards for ambient air quality are prescribed by the governments. The purpose of these standards is to reduce the pollutants to a certain level which would avoid undesirable effects. The prescribed standards may vary slightly from country to country depending upon their meteorological and geographical features, and the population density.

WATER POLLUTION LAWS AND ACTS

In India, specific laws have been passed by both State and Central governments to control water pollution. All these laws are based on "standards", a set of parameters used to define "water quality". These standards are usually based on the water quality in the streams, and specify the quality of effluents discharged. The Indian Standards Institution (ISI) has the principal role in specifying the norms for various effluents so that the ambient water quality standards are maintained.

A Water Pollution Act was passed by the Parliament in 1974. The Act makes provision for the constitution of Central and State Boards. The Central Board advises the Central government on matters pertaining to the prevention and control of water pollution and is bound by the directives of the Central government in discharging its duties. The State Board, on the other hand, is bound by the directives of both the Central and State governments. Maharashtra is perhaps the only state in India where a law for the control of water pollution was enacted as early as 1969. Usually the water pollution control comes under the State list, and most of the States follow the guidelines set by the Central Water Pollution Control Board. The Maharashtra Water Pollution Control Board has two types of pollution monitoring. In the first type, surprise checks are conducted once in three months to see whether permissible limits are exceeded; then the offending party is served with a warning notice to explain as to why action should not be taken against it. The case is then taken to the Board for discussion and, in the case of unsatisfactory explanation, to court for prosecution. In the second type of monitoring, various streams and rivers are checked once in a month at various points to see if water quality standards are maintained.

There are, however, several enforcement problems. Since the Central Act is rather silent on funds for the Boards, the State Boards were unable to enforce the Act effectively. Hence, the Water (Prevention and Control of Pollution) Cess Act was passed in 1977 to meet the expenses of the Central and State Boards. The Water Pollution Control Board has no power to take direct action against the erroring party. The court procedures are time-consuming and the delays often prevent quick and preventive action thereby defeating the sole purpose of the Act. Because of the problems inherent in the implementation of the Act, amendments were proposed for strengthening the working of the State Boards.

Inspite of the legislative measures, the pollution of our waterways continues unabated. This is due to the lack of civic sense among people and due to the lack of necessary infrastructure for enforcing implementation of the laws efficiently.

ANNEXURE – I

Environmental Laws in India

General enactments

Water pollution

1. The River Boards Act, 1956.
2. The Merchant Shipping (Amendment) Act, 1987.
3. The Water (Prevention and Control of Pollution) Act, 1974, amended in 1988.
4. The Water (Prevention and Control of Pollution) Cess Act, 1977.
5. The North India Canal and Drainage Act, 1873.
6. The Indian Fisheries Act, 1897.
7. The Damodar Valley-Corporation (Prevention of Pollution of Water) Regulation Act, 1948.
8. The Environment (Protection) Act, 1986.

Air pollution

1. The Air (Prevention and Control of Pollution) Act, 1981, amended in 1987.
2. The Indian Boilers Act, 1923.
3. The Factories Act, 1948, amended in 1987.
4. The Industries (Development and Regulation) Act, 1951.
5. The Mines and Minerals (Regulation and Development) Act, 1947.
6. The Oriental Gas Company Act, 1857.
7. The Indian Explosives Act, 1884.
8. The Explosive Substances Act, 1908.

9. The Motor Vehicles Act, 1938, amended in 1988 and Rules, 1989.
10. The Inflammable Substances Act, 1952.
11. The Petroleum Act, 1934 and Rules, 1979.
12. The Environmental (Protection) Act, 1986.

Marine pollution

1. The Share Nuisance (Mumbai and Colaba) Act, 1953.
2. The Obstruction in Fairways Act, 1891.
3. The Indian Fisheries Act, 1897.
4. The Indian Ports Act, 1908.
5. The Major Port Trusts Act, 1963.
6. The Merchant Shipping (Amendment) Act, 1987.
7. The Territorial Waters, Continental Shelf, Exclusive Economic Zone, and other Maritime Zones Act, 1976.
8. The Coastguards Act, 1978.

Noise pollution

1. The Environmental (Protection) Act, 1986.

Hazardous substances

1. The Poison Act, 1919.
2. The Dangerous Drugs Act, 1930.
3. The Drugs and Cosmetics Act, 1940.
4. The Factories Act, 1948, amended in 1987.
5. The Prevention of Food Adulteration Act, 1954.
6. The Industries (Development and Regulation) Act, 1951.
7. The Insecticides Act, 1968.
8. The Environment (Protection) Act, 1986.
9. The Consumer (Protection) Act, 1986.

Radiation

1. The Atomic Energy Act, 1962.
2. Radiation Protection Rules, 1971.

Pesticides

1. The Insecticides Act, 1968.
2. The Factories Act, 1948.
3. The Poison Act, 1919.

Forest and wildlife conservation

1. The Indian Arms Act, 1978.
2. The Wildlife (Protection) Act, 1972.

3. The Indian Forests Act, 1927.

4. The Forest (Conservation) Act, 1980, amended in 1988.

Others

1. The Urban Land (Ceiling and Regulation) Act, 1976.

2. The Prevention of Food Adulteration Act, 1954.

3. The Ancient Monuments and Archaeological Sites and Remains Act, 1958.

4. The Slum Areas (Improvement and Clearance) Act, 1956.

State enactments

Water pollution

1. The Orissa River Pollution Prevention Act, 1953.

2. The Maharashtra Prevention of Water Pollution Act, 1969.

Smoke control

1. The Bengal Smoke Nuisance Act, 1905.

2. The Gujarat Smoke Nuisance Act, 1963.

3. The Mumbai Smoke Nuisance Act, 1912.

Pest control

1. The Andhra Pradesh Agricultural Pests and Diseases Act, 1919.

2. The Assam Agricultural Pests and Diseases Act, 1954.

3. The Uttar Pradesh Agricultural Diseases and Pests Act, 1954.

4. The Mysore Destructive Insects and Pests Act, 1917.

5. The Kerala Agricultural Pests and Diseases Act, 1958.

Land utilisation and land improvement

1. The Andhra Pradesh Improvement Scheme Act, 1949.

2. The Acquisition of Land for Flood Control and Prevention of Erosion Act, 1955.

3. The Bihar Waste lands (Reclamation) Cultivation and Improvement Act, 1946.

4. The Delhi Restriction of Use of Land Act, 1964.

5. The Madhya Pradesh Nagar tatha Gram Nivesh Adhiniyam, 1973.

6. The Madhya Pradesh Gandhi Basti Kshetra (Sudhar tatha Nirmulan) Adhiniyam, 1976.

7. The Madhya Pradesh Town (Periphery) Control Act, 1960.

8. The Madhya Pradesh Regulation of Uses of Land Act, 1948.

Forest and wildlife conservation

1. The Madras Elephants Preservation Acts, 1873 and 1879.

2. The Nilgiris Game and Fish Preservation Act, 1879.

3. The Indian Arms Act, 1978.

4. The Wild Birds and Game Protection Act, 1887.

5. Notification in 1902 under the Sea Custom Act, 1878.

6. The Wild Birds and Animals Protection Act, 1912.

7. The Bengal Rhinoceros Protection Act, 1932

8. The Punjab Wild Birds and Wild Animals Protection Act, 1933.

9. The Andhra Pradesh Forests Act, 1967.

Note: Further information on any of these can be obtained from the Central Pollution Control Board.

In spite of all these Acts and implementing authorities, serious lacunae have been found in dealing with aspects relating to hazardous chemicals and siting of industries. However, the catastrophic leakage of methyl isocyanate gas from the Union Carbide Factory in Bhopal brought out the need for providing *teeth* to the Acts for deterrent action, where warranted. The necessity to have an umbrella legislation was also felt for taking appropriate measures on various aspects of environmental concern including those that are not covered under the existing laws. Thus, the Environment (Protection) Act, 1986, promulgated under Article 253 of the Constitution, was brought into force with effect from 19 November 1986.

Glossary

Abatement. A measure taken for reducing air, water, land or noise pollution which may involve legislative proceedings and technological applications.

Abiotic. Non-living ecological components of an ecosystem.

Abyssal. Very deep. refers to ocean depth.

Acacia. Refers to the genus of leguminous trees of tropical and sub-tropical origin which may form dominant vegetation on arid areas.

Aggregation. It is coming together of organisms into a group as in locusts.

Autocology. Ecological study of the individual organism in relation to environmental condition.

Autotrophic. Organisms capable of producing organic materials from inorganic chemicals by means of energy conservation e.g., green plants.

Adaptation. A characteristic of an organism that improves its chances of surviving in its environment.

Bacteria. A large group of unicellular or filamentous microscopic organisms, lacking chlorophyll and multiplying rapidly by simple fissure.

Bacteriocide. Something that kills bacteria.

Benthic. Refers to the bottom layer of any body of water, and the organism therein.

Bioassay. The determination of the character and strength of a potentiality toxic compound by shielding its effects on standard test organisms under laboratory condition.

Biocode. Any agent that is able to kill living orgnisms. Sometimes used as a synonym for pesticide.

Biocoen. All the living components of the environment.

Biodegradable. Capable of being broken down by bacteria into basic element.

Biogenesis. The origin of living organisms from other living organisms.

Biogeochemical cycles. The circulation of elements within ecosystems.

Biological control. The control of pests by the use of living organisms.

Biological indicator of pollution. The presence of frequency of a plant, animal, microbe or other form of life can give an idea of the level of pollution in its environment.

Biomass. The total weight of all living matter in particular habitat or area. Biomass is often expressed as grams of organic matter per square metre.

Biosphere. The part of the earth and its atmosphere in which organisms live.

Biota. The animal and plant life found within an environment of geographical region.

Black smoke. Smoke usually containing unburnt carbon.

Bloom. A population explosion of micro-organisms caused by sudden availability of an essential substance.

Biotic factor. Influence on the environment which are the result of the activities of living organisms.

Biotope. A region of relatively uniform environmental condition occupied by a given plant community and its associated animal community.

Boom. A floating device used to contain oil on a body of water.

Climax. The mature community capable or perpetuation under the prevailing climatic and edaphic conditions.

Commensalism. A relation existing between members of different species in which one organism definitely benefits from the association but the other individual does not benefit or adversely affected under normal condition.

Conservation. The rational use of the environment to improve quality of living for man-kind.

Cultural eutrophication. Increasing the rate at which water bodies "die" by pollution from human activities.

Compost. Rotted leaves, bark, twigs, etc., a material of low bulk density that improves poor solid, whether clays or sands.

Decomposer. An organism such as carrion beetle or a fungus that feeds upon and breaks down dead matter.

Decomposition. The separation of organic material into simpler compounds.

Deforestation. The removal of forest and undergrowth to increase the surface of arable land or to use the timber for construction or industrial purposes.

Degradation. The mass movement of water over any type of substrate, wearing away of the parent material.

Denudation. The combined effects of weathering and erosion in wearing away the surface of the land.

Desalination. The removal of salt, especially from sea-water.

De-silting. Removal of silt.

Diversity. A community is said to have a high degree of diversity it contains many species of fairly equal abundance.

Detritus. Dead organic material in solid or particulate form.

Ditch. Soil drainage.

Ecology. The study of the interrelationships between and amongst organisms and environment.

Ecosystem. Formed by the interaction of co-acting organisms and their environment.

Edaphic. Pertaining to soil, especially with regard to its influence on plants and animals.

Effluvium. Effluent what flows out, usually applied only to offensive smells.

Endotoxin. A toxin released by the degeneration or death (lysis) of bacterial cells.

Energy flow. The passage of energy through the tropic levels of a food chain.

Enteric disease. Any disease caused by microbes living in the intestine.

Epilimnion. The upper stratum of water in a lake which usually has the highest oxygen concentration ·and is characterised by temperature gradient of less than one degree °C per metre depth.

Euphotic zone. Layer of water to the depth of light penetration at which photosynthesis just equals respiration in a lake or pond etc.

Euryhaline. Tolerant to wide range of osmotic conditions.

Eutrophic. Refers to lakes which are highly productive in terms of organic matter formed and are well supplied with nutrients.

Eurytopic. Applied to organisms with a wide spread distribution.

Facultative parasite. An organism capable of living either as a parasite on another living organism or of growing on dead organic matter.

Floc. A clump of solids formed in sewage by biological or chemical action.

Flue. A passage of conducting combustion gases in an incinerator installation, also used synonymously with chimney.

Fluorosis. Disease in ruminants caused by over-consumption of fluorine compounds often as a result of air-pollution which deposits fluorine compounds on vegetation that is then consumed.

Fly-ash. Very small, often powder-like particles of glassy materials carried away from a fire, commonly a coal fire.

Fog. Collection of tiny water droplets that float on the air. It is similar to clouds except that clouds do not touch the earth's surface as fog does.

Fumigant. Any rapidly evaporating chemical compound used as a pesticide or a disinfectant.

Fungicide. One of a class of chemicals applied to plants or to their seeds and bulbs to inhibit or prevent the growth of scabs, blotches, moulds, rusts and other fungus disease.

Flotsam. Debris from natural sources or human activity floating on water surfaces.

Gall. Any abnormal growth of plant tissue produced in response to mechanical injury or to the invasion of insects, mites, bacteria or viruses.

Genotype. The assemblage of genes in an organism.

Geochemical anomaly. The local enrichment or depletion of element in soil or rock.

Geothermal. Pertaining to the heat of the interior of the earth.

Germide. Any compound that kills disease carrying micro-organisms.

Glacier. A body of ice originating on land by the compaction of re-crystallisation of snow. Glacier occurs where winter snowfalls exceeds summer melting.

Grit. Heavy inorganic matter, pebbles or sand grains which are removed during pre-treatment process in a grit chamber.

Halomorphic soil. Soils containing excess salt or alkali.

Halophile. An organism that lives in salt water or salty soil.

Haze. A state of atmospheric obscurity due to the presence of fine dust particles in suspension.

Heavy water. Water in which the molecule consists of a combination of heavy hydrogen and oxygen.

Heliographic. Referring to positions on the sum measured in latitude from the sun's equator and in longitude from a reference meridian.

Herb. A non-woody vascular plant having no parts which persists above the ground.

Helionics. Discolouration on leaves caused by high intensities of sun light.

Histic. Applied to soil surface layers that are high in organic carbon and seasonally saturated with water.

Holding pond. A pond or reservoir usually made of earth built to store polluted run-off.

Holophyte. An organism that obtains food by photosynthesis.

Hydrocyclone. A wet cyclone.

Hypertonics. Having a higher osmotic pressure than the solution with plant nutrients.

Hertz (H_2). A unit of frequency equal to one cycle per second.

Humus. Decomposing organic matter in the soil.

Iceberg. Huge masses of ice that break off the lower end of glacier and fall into the sea.

Idiobiology. The study of individual organisms.

Impacter. Refers to dust collectors.

Indicator species. Those species in a habitat which are most sensitive to slight changes in environmental factors.

Industrial effluent. The water or air-born wastes of industry.

Incineration. Burning the sludge to remove water and reduce remaining residues to ash.

Isotherm. A line on maps to connect places that have the same temperature.

Isohyets. A line on a map connecting points receiving equal rainfall during a stated period.

Isthmus. A narrow strip of land which connects larger bodies of land.

Katabatic. Wind caused by cold air flowing downhill, occurs most commonly when a hillside cools at night.

Keratin. A sulphur containing protein constituting epidermal production such as nails, hooves, claws, hair, beaks.

Kerogen. Insoluble organic material found in sedimentary rocks.

Lagoon. A pond where dirty water is allowed to settle and clarity before discharge to a river.

Land reclamation. The treatment of any unusable land.

Laterite. Occurring as a layer or as scattered nodules in tropical soils.

Lentical. A pore, through which gases can diffuse, in the corky outside layer of a woody stem.

Lithosols. Surface deposits with no soil horizons developed.

Limnetic zone. This is the depth at which photosynthesis balances respiration, also known as compensation depth.

Limnic. Refers to sediments deposited in fresh water lakes.

Lithosere. The stages in plant succession beginning on an exposed rock surface.

Littoral. Region of shallow water near sea-shore lying between high and low tide levels.

Loan. Soil having about 30–50 per cent sand, 30–40 per cent silt and 10–20 per cent clay content.

Leaching. Removal of soluble constituents from soil or sludge deposits by percolating water, also the disposal of liquid through porous strata.

Lysozyme. An enzyme which kills bacteria by destroying their cell walls.

Micronutrient. An essential element required by living organisms in relatively minute amounts.

Monitoring. The instrumentation and procedures for continuous measurement of air-pollutants.

Mycology. The study of fungi.

Mixotroph. An organism that is both heterotrophic and autographic e.g., insectivorous plant.

Nekton. Tiny organism that swim strongly in pelagic zone of a lake or sea.

Nitrification. The conversion of oxygen-demanding ammoniacal nitrogen to nitrate nitrogen in waste-water by biological or chemical oxidation.

Obligate anaerobe. An organism obliged to live in the absence of oxygen.

Osmoconformer. An animal whose body fluids conform closely in osmotic pressure to that of the environment.

Osmoregulator. An animal whose body fluid osmotic pressure can remain constant over a wide range of external conditions.

Outfall. That site or place where treated waste-water or effluent is discharged.

Oxidation pond. Artificial lake or pond in which wastes are degraded and consumed by bacteria.

Oxidant. An oxidising agent.

Oncogenic. A substance that causes tumors.

ORP. Oxidation Reduction Potential.

Oligotrophic lake. A deep, clear, poor in nutrients lake. It contains little organic matter and high dissolved oxygen level.

Paleoecology. The study of past environments, their flora and fauna and the interrelationship among them.

Peat. Organic soil, often many feet deep.

Pedogenesis. The origin and development of soils.

Percolation. Movement of water through the soil, usually downward.

Periglacial. The area surrounding the limit of glaciation and subject to intense frost action.

Periphyton. Plants or animals that cling to rooted water plants above the bottom mud.

Phonology. The study of timing of recurring natural phenomenon with particular reference to climatological observations.

Phagotrophic. Ingesting solid food particles.

Potable water. Water of a quality acceptable for human consumption.

Polyclimax. A number of climaxes occurring within a climax region probably owing to the sub-ordination of climatic factors by edaphic ones.

Psammon. The organisms inhabiting the water lying between grains of sand.

Putrefaction. Anaerobic degradation of proteins mediated by microbial enzymatic activity.

Quadrat. A sampling area (often one metre square) used in studying the composition of an area of vegetation.

Qualitative analysis. An analysis which seeks to determine what components are present in significant amounts in a substance or mixture of substances.

Quantitative analysis. An analysis which seeks to determine relative amount of significant compounds present in a substance or mixture of substance.

Reclamation. Treatment of waste-water so that it could be recycled for human use or activity such as recreation, irrigation etc.

Recycling. To return water after some types of treatment for further use, generally implying a closed system.

Reverse osmosis. A membrane separation process in which osmotic pressure is exceeded by an external force which reverses the normal flow of solvents.

River basin. The land area drained by river and its tributaries.

Rock. Any naturally formed aggregate or mass of mineral matter.

Savannah. Grassland with occasional trees.

Scatology. Branch of science dealing with the study of faeces, excreta etc.

Sediment. Settlement of any solid phase from out of a liquid phase.

Sewers. A system of pipes that collect and deliver waste-water to treatment plants or receiving water.

Sierozem. A kind of grayish infertile desert soil.

Sludge. The solid matter that settles to the bottom, floats or becomes suspended in the sedimentation tanks and must be disposed off by filtration and incineration or by transport to appropriate disposal site.

Standing crop. The amount of biotic and abiotic materials present in an ecosystem at a given time.

Stenohaline. Confined to a narrow range of osmotic concentration.

Subnatant. Liquid which remains below the surface of floating solids after floatation.

Sullage. Run-off a sewage waste-water.

Synecology. Study of groups of organisms that are associated together as a unit.

Senescence. The aging process. It can refer to lakes in advance stages of eutriphication.

Scum. Any material that floats to the top of still water, especially sedimentation tank.

Saprobe. An organism which feeds on dead or decaying organic matter.

Salvage. The utilisation of waste material.

Squal. A strong wind that begins suddenly lasts for several minutes that dies away slowly.

Solar house. A house, where solar collectors bring heat down from the root and impart it to a storage tank through a heat exchanger.

Tertiary effluent. Liquid portion leaving tertiary treatment.

Transect. A sampling unit in ecological studies.

Threshold dose. The minimum dose of a given substance necessary to produce an effect.

Toxicant. A chemical that controls pest by killing rather than repelling them.

Trophobiosis. A type of symbiosis in which two organisms (trophobionts) of different species feed one another.

Turbid. Water that is not transparent.

Ulva, sea lettuce. A green seaweed that grows in sewage pollution nutrient rich estuaries.

Uv-light. Ultra-violet radiation.

Up-welling. Vertical movement of water currents, usually near coasts and driven by onshore winds; the process results in transference of nutrients from the depths of oceans to surface layers.

Vector. An organism, often an insect, that carries disease.

Waste. Any solid, liquid or gaseous emission as a result of human activity.

Waste treatment. Physical, chemical and biological processes employed to remove dissolved and suspended solids from waste-water.

Weathering. Mechanical or chemical breakdown of surface rocks into small particles by atmospheric action.

Wetland. A submerged or water saturated lands, both natural or artificial, permanent or temporary with water.

Woodland. Vegetation dominated by trees which form a distinct but open canopy.

Xeromorphic. A plant displaying the ability to restrict water loss during adverse conditions.

Xerad. A plant which live in dry habitat and can endure prolonged drought.

Yeasts. Many species of unicellular fungi and reproduce by budding. This belongs to Ascomycetes and is used in brewing, wine making and baking. Yeasts are used as a source of protein and of vitamins of the B group.

Zoobiotic. Applied to an organism which lives parasitically on animal.

Zooplankton. Plankton without photosynthesis.

Zymase. An enzyme produced by yeast which breakdown sugars into alcohol and carbon dioxide.

Zwitter ion. An ion which has either positive or negative charge depending on pH.

References

Allaby, M., *Natural Environment and the Biological Cycles, Lewis Publishers.*

Alabaster, J.S. and Lloyd, R., *Water Quality Criteria for Freshwater Fish, Butterworths, London.*

Arceivala, S.J., *Waste-water Treatment and Disposal, Marcel Dekker Inc., New York.*

Averett, R.C., *Environmental Ethics, Chapman and Hall.*

Baker, J.M., *Ecological Effects of Oil Pollution, Applied Science, London.*

Besselievre, E.B. and Schwartz, M., *The treatment of Industrial Wastes, McGraw Hill Kogakusha Ltd., Tokyo.*

Brown, R.L., *Pesticides in Clinical Practice, Charles C. Thomas, Publisher, Spring field.*

Bryan, G.W., *Heavy metal contamination in the Sea. Marine Pollution, Academic Press, London.*

Bryan, G.W., *Effects of Pollutants on Aquatic Organisms, Cambridge University Press, Cambridge.*

Budyko. M.I., *Global Ecology, Progress Publishers, Moscow.*

Cattabeni, S., Cavallaro, A. and Galli, G., *Dioxin-Toxicological and Chemical Aspects , S.P. Medical and Scientific Books, New York.*

Coolingwood, R.W., *Biological Indicators of Water Quality; John Wiley and Sons, New York.*

Connell, D.W. and Miller, G.J., *Chemistry and Ecotoxicology of Pollution, John Wiley and Sons, Inc., New York.*

CPCB., *Pollution Control Acts. Rules and Notifications Issued There under, Central Pollution Control Board, New Delhi.*

Curds, C.R., and Hawkes, H.A. (eds.) *Ecological Aspects of Used Water Treatment, The Organisms and their Ecology, Academic Press.*

Dix, H.M., *Environmental Pollution, John Wiley and Sons.*

Dugan, P.R., *Biochemical Ecology of Water Pollution, Plenum Publishing Corporation, London.*

Fruh, G.E., *E.F. Gloyna and W.W. Eckenfelder, Advances in Water Quality Improvement, Univ. of Texas Press, Austin.*

Goldberg, E.D., *A Guide to Marine Pollution, Gordon and Breach, Science Publishers, Inc., New York.*

Holmes, J.W., and Talsma, T., *Land and Stream Salinity. Elsevier Scientific Publishing Co., Amsterdam.*

James, A. and Evison L., *Biological Indicators of Water Quality, John Wiley and Sons, New York.*

Kathern, R.L., *Radioactivity in the Environment: Sources, Distribution and Surveillance, Harwood Academic Publishers, New York.*

Lehr, J.H., Tyler, E.G., Wayne, A.P. and Jack, D., *Domestic Water Treatment, McGraw Hill Book Company. New York.*

Lenihan, J. and Fletcher W.W.(Eds.) *Environment and Man (Vol.5) The Marine Environment, Blackie, Glasgow and London.*

Liptak, Bela., G., *Water Pollution Environmental Engineers Handbook, Chilton Book Company, Radnor, Pennsylvania.*

Lowman, F.G., *Radioactivity in Marine Environment, National Academy of Sciences, Washington DC.*

Masters, G.M., *Introduction to Environmental Science and Technology. John Wiley and Sons, New York.*

McCaull, J. and Crossland, J., *Water Pollution, Harcourt Brace Jovanovich. Inc., New York.*

Neff, J.M., *Polycyclic Aromatic Hydrocarbons in the Aquatic Environment, Applied Science Publishers, London.*

Nemerow, N.L., *Industrial Water Pollution: Origins, Characteristics and Treatment, Addison-Wesley Publishing Company Inc. Philippines.*

Odum, E.P., *Fundamentals of Ecology, Saunders, Philadelphia.*

Palmer, C.M., *Algae and Water Pollution, Castle Housing Publications Ltd.*

Phillips, D.J.H., *Quantitative Aquatic Biological Indicators. Applied Science Publishers, London.*

Rao, K.L., *India's Water Wealth, Orient Longmans, New Delhi, 47-48: 228.*

Reid, G.K., *Ecology of Inland Waters and Estuaries Reinhold Publ. Crop. New York.*

Rudd, R.L., *Pesticides and the Living Landscape, University of Wisconsin Press, Madison, Wisconsin.*

Rudd, R.L. and Herman, S.G., *Environmental Toxicology of Pesticides, Academy Press, Inc. New York.*

Sawyer, C.N. and McCarty P.L., *Chemistry for Environmental Engineering. McGraw-Hill Book Company, Tokyo.*

Sax, N.I., *Industrial Pollution, Van Nostrand Reinhold Ltd.*

Stumm. W. and Morgan, J.J., *Aquatic Chemistry, John Wiley and Sons, New York.*

Stumm, W. and Stumm-Zollineger, E., *Water Pollution Microbiology, Wiley Interscience, New York.*

Teal. J.M., *Effects of Petroleum Hydrocarbons in Marine Organisms and Ecosystems, Pergamon Press, New York.*

Tebbutt, T.H.Y., *Principles of Water Quality Control, Pergamon press, Oxford.*

Vinogradov, A.P., *The Geochemistry of Rare and Dispersed Chemical Elements in Soils, Consultants Bureau, Inc. New York.*

Vishwanathan, R., *Indian J.Med. Res. 45, Sppl. No. 1*

WHO, *Health Hazards of the Human Environment, World Health Organisation. Geneva.*

Allen, S.E., Grimshaw, H.M., Parkinson, J.A. and Christopher, *Chemical analysis of ecological materials. Blackwell Scientific Publications. Oxford.*

Smith, K.M., *Methods for air sampling and analysis, Second ed. American Public Health Association, U.S.A.*

Black, C.A., *Methods of Soil Analysis, John Wiley and Sons, New York.*

Bordne, R.B. and Winter, J., *Microbiological methods for monitoring the environment-Water Cincinnati, Ohio, U.S.A.*

Chaphekar, S.B., Boralkar, D.B. and Shetye, R.P., *Effects of industrial pollutants on plants. Final report of the U.G.C. sponsored research project, Inst. of Science, Bombay.*

Chapman, S.B., *Air Pollution Sources Blackwell Scientific Publications.*

Chapman, H.D. and Pratt, P.F., *Methods of analysis of soils, plants and water, Univ. of California.*

Cox, C.R., *Operation of control of water treatment process. No. 49, WHO, Geneva.*

Curtis, J.T., *Pollution Prevention Univ. of wisconsin Press. Madison.*

Golterman, H.L., *A manual of methods for physical and chemical analysis of water, IBP Hand book No.8, Blackwell Scientific Publications.*

Gopal, B. and Bharadwaj, N., *Elements of Ecology, Vikas Publishing House, New Delhi.*

Hesese, P.R., *A text book of soil Chemical Analysis, Prentice Hall, London*

Jackson, M.L., *Soil chemical analysis. Prentice Hall, London.*

Ledbetter, J.O., *Air Pollution: Analysis. Marcel Dekker Inc., New York.*

Liptak, B.G., *Environmental Engineering Handbook Vol. 2. Air Pollution, Chilton Book Co., U.S.A.*

Manivaskam, N., *Physico-chemical examination of water, sewage and industrial effluent. Pragati Prakashan, Meerut.*

Mansfield, T.A., *Effects of air Pollutants on Plants. Cambridge University Press. Combridge.*

Mason, C.F., *Biology of freshwater Pollution. Longman group Ltd.*

Metham, R.A., *Atmospheric Pollution its history, origin and Prevention Pergamon Press.*

Michael, P., *Ecological Methods of Field and Laboratory Investigations. Tata McGraw Hill Publishing Co. Ltd., New Delhi.*

Munn, R.F., *Environment Impact Assessment. John Wiley and Sons.*

NEERI, *Air Quality monitoring. A course Manual, NEERI, Nagpur.*

Odum, E.P., *Fundamentals of Ecology. W.B. Saunders and Co., New York.*

Painter, D.E., *Air Pollution Technology. Reston Publishing Co. Inc. Reston, Virginia.*

Raw, J.G. and Wooten, D.C., *Environment impact analysis handbook. McGraw Hill and Co.*

Schrowebel, J., *Chemistry of Air Pollution pergamon Press.*

Southwood, T.R.E., *Ecological Methods, with special reference to the study of insect population. London Methuen.*

Southwood, T.R.E., *Ecological Methods. Chapman and Hall.*

Snell, I.D. and snell, C.T., *D.Van Nostrand, New York.*

Trivedy, R.K., *Role of Algae in Biomonitoring of water pollution. Asian Environment.*

Trivedy, R.K., and Goel, P. K., *Chemical and Biological Methods for Water Pollution Studies Environmental Publications Karad.*

Trivedy, R.K. and Goel, P.K., *Air Pollution; Sources, Monitoring, effects and control. Environmental Publications. Post, Box. 60. Maharashtra. India.*

Vollenweider, R.A., *A manual of methods for measuring Primary production in aquatic environment. IBP handbook No. 12 (ed.II) Blackwell Scientific Publications.*

Index